INSTRUMENTAL SURFACE ANALYSIS
of
GEOLOGIC MATERIALS

INSTRUMENTAL SURFACE ANALYSIS *of* GEOLOGIC MATERIALS

Dale L. Perry

Dale L. Perry
Lawrence Berkeley Laboratory
1 Cyclotron Road, MS 70A-115
Berkeley, CA 94720

Library of Congress Cataloging-in-Publication Data

Instrumental surface analysis of geologic materials/Dale L. Perry, editor.
     p. cm.
   Includes bibliographical references.
   ISBN 0-89573-758-2
   1. Mineralogical chemistry. 2. Surface chemistry. I. Perry Dale L.
QE438.S87 1990
552'.06--dc20
90-12310
CIP

British Library Cataloguing in Publication Data
Instrumental surface analysis of geologic materials.
   1. Surface phenomena. Spectroscopy
   I. Perry, Dale L.
   541.33

   ISBN 3-527-27849-4

ISBN 0-89573-758-2 VCH Publishers
ISBN 3-527-27849-4 VCH Verlagsgesellschaft

Distributed in North America by: Distributed Worldwide by:

VCH Publishers Inc.
220 East 23rd Street, Suite 909
New York, New York 10010

VCH Verlagsgesellschaft mbH
P.O. Box 1260/1280
D-6940 Weinheim
Federal Republic of Germany

CONTRIBUTORS

Frank J. Berry
Department of Chemistry
University of Birmingham
P.O. Box 363
Birmingham, B15 2TT U.K.

David Blake
Planetary Biology Branch
NASA/Ames Research Center
MS 239-4 Moffet Field, CA 94035

Harry G. Brittain
Squibb Institute for Medical Research
P.O. Box 191
New Brunswick, NJ 08903

Raymond Browning
Center for Integrated Systems
Stanford University
Stanford, CA 94305

Cecil R. Dybowski
Department of Chemistry
University of Delaware
Newark, DE 19716

Michael F. Hochella, Jr.
Department of Geology
Stanford University
Stanford, CA 94035

Clifford T. Johnston
Department of Soil Science
University of Florida
Gainesville, FL 32611

Murray B. McBride
Department of Agronomy
Cornell University
Ithaca, NY 14853

James B. Metson
Department of Chemistry
University of Aukland
Private Bag
Auckland, New Zealand

Dale L. Perry
Lawrence Berkeley Laboratory
University of California
MS 70A/1115
Berkeley, CA 94720

J. Ashley Taylor
AT&T Bell Laboratories
555 Union Blvd.
Allentown, PA 18103

Charles D. Wagner
Surfex Company
29 Starview Dr.
Oakland, CA 94618

PREFACE

Over the last two decades, the use of sophisticated surface techniques has reemphasized surface chemistry as being an extremely important area of research. Surface chemistry can generally be construed as the chemistry of a material that occurs or exists within the outer few layers of material. The actual depth indicated by the term "surface" can vary, depending on the interest of the investigator coupled with the instrumental surface technique being employed. This depth, however, ranges from a few angstroms to a micron or more for the common surface instruments that are currently in use. If one is studying a pristine, or "clean," surface, the surface study is also essentially one of the bulk. However, one of the great advantages of this type of experimental approach is that the investigator can look at species that have reacted with the outermost layer of a material, thus allowing observation of the interfacial reaction in a stepwise, mechanistic fashion. Thus, reactions with both gases and liquids (and anything contained in them) can be studied from the development of initial, thin films to the development of relatively thick precipitates on the parent material. Chemical reactions and products, along with their concomitant morphological surface changes, can be easily followed.

Basic and applied studies have been performed in such areas as catalysis, electrochemistry, and corrosion; also, large efforts have been made in materials science that have resulted in a greater knowledge and understanding of the properties of various materials. While many advances have been made in these scientific disciplines, progress in research using surface oriented techniques in the geologic sciences has not kept pace. There are several major reasons for this, including the cost and availability of the instrumentation. Another reason is the lack of spectroscopy and other modern instrumentation-related courses in the traditional geology curriculum. These problems, along with a decided lack of the chemistry and physics training in these curricula that is inherently necessary for making the most effective use of the various types of spectroscopies, have thus hampered the geologic research community from gaining the maximum benefits from surface chemistry on a widespread basis.

This is most unfortunate, since the techniques have an extremely wide range of potential applications to geologic problems. Indeed, many of these applications were discussed in the first Symposium on Applications of Modern Surface Techniques in Geochemistry that this author organized and chaired at the 184th National Meeting of the American Chemical Society in Kansas City, Missouri, in 1982. The different types of spectroscopies display varying degrees of sensitivity to surface species and complement one another in terms of the kinds of surface information they yield. Combined, the surface techniques allow one to study bonding of both the geologic substrates themselves and the reaction chemistry (i.e., sorption, oxidation/reduction, etc.) that they can undergo. These studies have obvious relevance to a number of critically important geologic reaction systems such as acid rain and toxic waste; without rigorous studies involving the chemical mechanisms of the reaction/transport processes (many of which involve liquid and gas/solid interfaces) of these two phenomena, any real understanding will be impossible.

The purpose of this book is to present a variety of important surface techniques that are currently available for doing research involving geologic substrates. At the present time, there is *no one treatise available* that summarizes both the various approaches that can be used to study surface-related problems of geologic materials and some of their applications that have been reported to date. While there are other branches of spectroscopy other than those in this book that have been used for surface studies, this treatise has been restricted to those techniques that have been used for the study of *geologic* materials and reactions. Also, techniques that are useful for obtaining both qualitative and quantitative analytical data and techniques that yield chemical bonding and topographical data have been included. It is hoped that by compiling these techniques into one single volume, researchers interested in the general field of surface chemistry of geologic materials will have at their fingertips an overview of several of the more important experimental approaches. This should make for an easier and more efficient means of both designing surface experiments involving geologic materials and for determining which techniques will yield the most information for a particular surface chemical system.

Finally, there are several people that should be acknowledged for their supportive (both moral and fiscal) roles regarding geologic surface work here at Lawrence Berkeley Laboratory. The Office of Basic Energy Sciences, U.S. Department of Energy, has generously provided funding for the research on a continuing basis. Garrett Hyde of the U.S. Bureau of Mines, Department of the Interior, in addition to providing financial support, has helped "spread the word" about the importance of using modern instrumentation approaches to study surface/interface reactions involving geologic materials. Others that deserve thanks for their interest and support are Harold Helgeson, Anthony Lasaga, John Morse, Raymond Siever, and Dimitri Sverjensky. I also especially wish to thank Eleanor Riemer for her role in encouraging me to produce this book.

Dale L. Perry
Berkeley, California

CONTENTS

1. INTRODUCTION

Dale L. Perry
Lawrence Berkeley Laboratory
University of California
Berkeley, CA 94720

1. General

1.1. Modern Advances in Instrumentation

Over the last two decades, the field of surface chemistry has enjoyed a renaissance unparalleled by any other in science. The reason for this is the advent of electronics and high vacuum technologies that have made possible the development of spectroscopic techniques that can be used to directly observe chemical and morphological conditions of a solid surface. In addition to the introduction of ultrahigh vacuum (UHV) techniques such as x-ray photoelectron and Auger spectroscopy, tremendous strides have been made in non-UHV experimental approaches. The utility of infrared spectroscopy, for example, which has been widely used to examine surface reactions since the 1950s, has been greatly enhanced by the Fourier transform approach to the technique. Similarly, Mossbauer spectroscopy, which was initially used to study surface phenomena in the transmission mode only, has demonstrated even greater utility in these studies with the development of conversion electron and emission Mossbauer spectroscopy.

1.2. Potential for Future Advances

While all of these advances in instrumentation have made it possible to make great strides in the understanding of surface chemical phenomena, they only serve as being indicative of things to come; this is especially true with respect to applications for the study of geologic materials. Scanning tunneling microscopy (STM), for example, is just starting to be applied to mineral surfaces to give detailed information about surfaces at the atomic level. Other very new approaches, such as optical second-harmonic generation (SHG), should prove quite effective for probing at submonolayer sensitivities of interfaces that are accessible to light, including solid/liquid interfaces. Rutherford backscattering (RBS) also holds promise for studying mineral reaction systems.

1.3. Purpose of Book

The purpose of the present book is to provide a good introductory overview of a variety of techniques that can be used to study geologic surface/interface reactions. A thorough knowledge of surface reactions is extremely important in several areas of geochemistry, including mineral dissolution and precipitation (including weathering), elemental partitioning at the mineral/aqueous solution interface, and natural heterogeneous catalytic reactions. All of these fundamental geochemical processes impact heavily on the applied fields of mineral extraction, ore deposit formation, and toxic waste transport. Toxic waste transport, for example, is a process in which the sorption of heavy metal ions and organic molecules on geologic substrates plays a very important role. So, a book addressing surface chemical phenomena of geologic materials comes at a most critical time with respect to rigorously studying this social/technological issue.

2. Types of Surface Studies Involving Geologic Materials

2.1. Structure

Experimental approaches to surface studies involving geologic materials can take several forms, but all of them can be placed into four general categories: structural, analytical, electronic, and reaction chemistry studies. When one discusses structural studies of a bulk geologic material using surface-sensitive techniques, it is assumed that the clean surface of the material also represents the bulk structure. The same is true for all the types of spectroscopy described in this book; they can be used for both surface and bulk studies, depending on the experimental circumstances of how a sample is prepared (or reacted) before it is spectroscopically or microscopically examined. Under the heading of structural studies, for example, one can cite the correlation of x-ray photoelectron spectral data with the bulk structures of various aluminosilicates.

With respect to structures of species *on the surface*, the different experimental approaches are very effective. Morphological phenomena on surfaces can be studied using scanning electron microscopy, while Raman and infrared spectroscopy can provide valuable structural information on anion and organic molecule bonding modes to central heavy metal ions. Other techniques such as electron spin resonance and nuclear magnetic resonance also can provide structure-related data.

2.2. Analyses

Analytical chemical studies are focused on both the aspects of what species (qualitative) and how much of them (quantitative) are present on a geologic surface, with the most common area of interest being on elemental analysis. With modern surface instrumentation, however, the analytical approach can be taken still further. A determination can be made as to whether the element is present as a simple elemental cation or anion or as one or more complex, polyatomic cations or anions. Quantitative information can be derived by either internal calibration (making use of relative spectral intensities) or by the use of rigorous analytical standards. In turn, these same analytical approaches can be used to monitor analytical changes on a surface as it undergoes further reactions.

2.3. Electronic States

Most surface instrumental techniques readily lend themselves to the analysis of electronic states of elements that are involved in surface reactions. This is especially true if multiple techniques are used in conjunction with one another to study a particular metal ion system. Surface techniques can be used to study diamagnetic vs. paramagnetic species of metallic ions, i.e. copper(I) vs. copper(II) by x-ray photoelectron spectroscopy. This technique can also differentiate between high- and low-spin electronic configurations of the same ion. A good example is the cobalt(III) ion, which can be either diamagnetic or paramagnetic, depending on the ligands attached to it. Electron spin resonance can give detailed information about paramagnetic chemical species such as transition metal ions and adsorbed radicals.

Additionally, the electronic state can be studied in conjunction with the elemental stoichiometry and crystallographic structure of a mineral to determine the chemical bonding; the most notable examples are the transition metal sulfide minerals.

2.4. Reaction Chemistry

The importance of surface techniques for studying interfacial chemical reactions involving geologic surfaces cannot be overstated. For that reason, a large portion of this book is directed at this application. Chemical reactions on geologic surfaces naturally encompass a wide range of reacting chemical species and geologic substrates. For this reason, such studies can impact on a large number of chemical reaction-dependent geologic areas. Perhaps the largest class of this type of reaction is that of adsorption, a surface-related process that dictates several highly important basic and applied chemical reactions involving geologic materials. Two really good examples of these are weathering of rocks and minerals (during which process surface oxidation plays an important role) and froth flotation processing of ores; still another good example of a basic area involving detailed studies of chemisorption is that of ore body formation. Many other geologic material/chemical interaction studies are highlighted in this book.

3. Geologic Surface/Interface Reactions–Overview

3.1. Surface/Gas and Surface/Liquid Interactions

Geologic surface/interface reactions are ones which typically involve reactions between a geologic surface and a gas or liquid. Schematically, simple and complex (multiple) surface/gas reactions are shown in Fig. 1. In naturally occurring interface reactions in the environment, both classes

Fig. 1. Geologic surface reactions.

of this reaction occur. First, there are the surface/gas reactions, with the most common interaction involving air, with the process resulting in one simple product. Other gases also react with surfaces, depending on the chemical atmospheric conditions. A second type of reaction, that of a surface/liquid reaction, is possible in theory; however, for it to be an exclusive *liquid* reaction with a geologic surface, the liquid would have to have never contained a gas or would have to be depleted of the gas prior to the reaction. Since such a scenario occurring naturally would be difficult to envision, it must be considered a process that can best be effected in the laboratory. Third, there are the surface/gas/liquid interactions. These reactions are ones involving a liquid containing a dissolved gas (or gases) reacting with a surface.

Case I represents the relatively "simple" scenario of a gas reacting with a surface under very controlled conditions, i.e., in an ultrahigh vacuum chamber where the materials are exposed to varying amounts of a gas. The gas could be oxygen, for example, reacting with a sulfide mineral in order to study the oxidation process. Under these carefully controlled conditions, the geologic material also may be annealed and ion sputtered preparatory to the reaction sequence. This approach will yield a single (hopefully, but not always) reaction product under each set of conditions and yields much useful information about stepwise reaction products during a reaction sequence and the mechanism by which a final product is achieved. This experiment is thus more controlled than a geologic surface oxidation reaction effected in the open air. In that situation, the surface oxidation process would not be a chemical process exclusively dependent on oxygen but rather on a mixture of oxygen, carbon dioxide, water vapor, and any other reactive trace gases present. If a reaction consists of a pure, degassed liquid interacting with a clean, fresh surface, and it results in one product on the surface, then Case I is also schematically appropriate.

Case II, however, represents a more complex surface/gas interaction, that of several gases forming different product layers as they react with the previous layer formed. Regions 1, 2, and 3 represent the different products. Region 4 represents the pristine, unreacted mineral. In its contact with air, the surface is altered by atmospheric oxygen, carbon dioxide, and water vapor; this process thus creates an initial surface product, represented by Region 3. As time progresses, this same reaction proceeds or, alternatively, the new mineral surface product reacts with still yet another reactant (or, not unrealistically, several reactants) to form *another* reacted product designated by Region 2. Logical reaction candidates could be sulfur and/or nitrogen oxides. One can thus see that the "surface" of a mineral is not one simple product but more a composite of the oxidation reaction process. A very good example of this is galena, PbS; naturally oxidized galena shows a range of oxidation products on its surface, ranging from elemental sulfur to the sulfate anion. Lead oxides, carbonates, and mixed hydroxide/carbonates have also been observed, depending on the oxidation timescale and chemical conditions.

3.2. Surface/Gas/Liquid Interactions

Case II can also illustrate the underlying reasons for a potentially far more complex situation, that of a heterogeneous surface/gas/liquid reaction, in which chemical interaction of the surface products occurs. This interaction results in a "mixing", thus causing heterogeneous surface conditions. This scenario is more truly illustrative of most natural geologic surface/interface reactions, reactions in which multiple reactants and products are involved. A very good example of Case II becoming important in a role in a surface/gas/liquid interaction with mixing is that of the natural weathering of minerals. Once the various layers of surface products are formed, rainwater can also react with them, altering the newest surface again through dissolution of the top surface product and subsequent reaction of deeper products with the newly formed solution.

Another good example of such heterogeneous reactions is that of ore processing, or froth flotation processing. Metal ions, organic compounds, and other reagents interact with ore surfaces, making the surface-dependent processing extremely complex. Again, concurrent processes such as oxidation-reduction and organic/inorganic complexation are operable, resulting in heterogeneous surface conditions. This situation also occurs when minerals undergo electrochemical surface alterations, depending on what reaction solution is used.

3.3. Caveats Concerning Surface Reaction Complexity

Both the geologic surface/gas and surface/gas/liquid scenarios discussed above serve to illustrate the inherent complexity of surface reactions involving natural geologic materials and their reactions under natural conditions. In light of this, several important warnings (or caveats) should be made concerning both the bases of such complexity and a warning that the complexity cannot be overstated. First, with respect to solid-gas interactions, the composition of the reacting gas will have a dramatic effect on both the intermediate and final reaction products; this is also true of the sequence of different gaseous reactions with a given surface. A reaction involving a mixture of gases A, B, and C in the environment will produce different reaction products than gases A, B, and D. This is especially true if gases C and D display decidedly different reaction characteristics. In the case of C being a nitrogen oxide, for example, that reaction mixture will yield different products on an oxide mineral surface than a set containing nitrogen, water, or oxygen as gas D. Similarly, if carbon dioxide is first reacted with a hydrated metal oxide, a quite stable mixed hydroxide/carbonate mineral phase may form. If water vapor (or some other hydroxide-forming reactant) is reacted *after* the carbon dioxide, such a phase might not result. Of course, various factors, such as surface-reactant gas bond strengths, thermodynamic considerations, and lattice stability energies, come into play in deciding the final surface composition.

Another caveat concerning gas phase interactions with geologic surfaces is that--under natural conditions especially--gaseous surface products can be easily altered. Products chemisorbed on surfaces during cool, dark periods (such as at night) might be desorbed or undergo further reactions in intense, hot sunlight. This is especially true if the gas adsorbed is susceptible to photolytic effects. Depending on the components of a multiple gas layer on the surface, the gases themselves could react with one another to form new compounds.

In the case of geologic surface/liquid interactions, the complexities of the reaction system may be much more severe. This is especially true if the liquid is not a single, pure compound but rather a mixture of different liquids or a solution. Because of multiple, competing reactions at the surface/solution interface, the final surface chemical state is a reflection of the *total, composite reaction.* In solutions involving metal ions, for example, the pH will dictate the species in solution that actually will be reacting with the surface. This can be still further complicated if the reaction system contains metal ions that can undergo oxidation/reduction or if complexing organic ligands are present. Two other experimental parameters--pressure and temperature--can make reactions at the surface even more complicated, often increasing the number of reaction products formed and thus the overall surface heterogeneity. Also, with increased temperature and pressure, the mixing of surface reaction product layers increases.

4. Philosophy and Composition of Book

When a book is first conceived, several important issues must be addressed and resolved; the present book is no exception. First, the choice of a *theme* must be one of the decisions made. In this case, *Instrumental Surface Analysis of Geologic Materials* is indeed not only the title of the book but also the theme. This is meant to encompass that concept in its entirety, i.e. the analytical surface chemistry of geologic materials in terms of *the morphology of the surfaces, the chemical state of the original surfaces, and their chemical compositions after undergoing reactions.* This includes microscopic and spectroscopic investigations of thin films and precipitate layers by various (but complementary) techniques. The techniques include ultrahigh vacuum approaches such as x-ray photoelectron and Auger spectroscopy to study monolayer and thin film surface products as well as techniques such as infrared and electron spin resonance spectroscopy which have been traditionally used to study thicker reaction product layers. Several can even be used for both classes of surface conditions; Mossbauer, for example, can be used in the conventional transmission mode to study thicker precipitates while also being pressed into service as emission and conversion electron Mossbauer spectroscopy to yield data more reflective of the first few surface layers. Taken as a whole, all of the techniques described in this book can give a researcher a relatively complete picture of either a virgin or reacted geologic surface.

A brief explanation should also be given concerning other criteria for selection of the techniques in the book. First, one of the operable words concerning the techniques included in this book is *complementary*. The reader can survey Table 1 and see that all types of chemical species on surfaces can be studied by one or more of the included approaches. Diamagnetic and paramagnetic species, metal ions, organic species,

Table 1-1. A Comparison of Several Types of Surface-Applied Spectroscopy

Spectroscopy	Typical Transition Energy (eV)	Type of Transition	Surface Species Observable
Infrared	$(50 - 2.5) \times 10^{-2}$[a]	Vibrational	Most di- and poly-atomic molecules
Raman	$(50 - 0.6) \times 10^{-2}$[a,b]	Vibrational	Most di- and poly-atomic molecules
Mössbauer	$10^4 - 10^5$[a]	Nuclear	Species containing Mössbauer nuclides
Electron spin resonance	$(14 - 3.8) \times 10^{-5}$[a]	Electron spin	Paramagnetic species
Nuclear magnetic resonance	$(2.5 - 8.3) \times 10^{-7}$[a]	Nuclear spin	Diamagnetic species
X-ray photoelectron	$0.1 - 1500$[a]	Bound electron to the continuum	All elements except hydrogen and helium
Laser emission	$1.3 - 4$	Electronic	All elements that can undergo laser emission
Auger	$0.1 - 2200$	Bound electron to the continuun	All elements except hydrogen and helium

[a]Source: Delgass, W. N.; Haller, G. L.; Kellerman, R.; Lunsford, J. H., "Spectroscopy in Heterogeneous Catalysis," Academic Press: New York, 1979.

[b]The energy range of the incident and scattered radiation is ~ 2 eV.

polyatomic cations and anions, light and heavy atoms in different chemical states--all can be studied using the various types of spectroscopy. In organizing the book and deciding what techniques to include in each chapter, one of the deciding factors was that the chapters complement one another in their coverage of both different surface chemical species and the types of geologic substrates on which the chemistry occurs.

Second, all of the techniques described herein rely on "stand alone" instrumentation; all of it is commercially available and can be placed in a research investigator's laboratory. Thus, techniques which rely on a synchrotron source, for example, have been excluded, since this would necessitate the use of a central facility. Such a facility has the normally attendant problem of a queue waiting to obtain experimental time. Also, in may cases, travel is necessary for an investigator to make use of it. Third, the techniques discussed in this book are ones which are, for the most part, mature enough to have been experimentally developed and generated a substantial amount of published geologic surface literature. Scanning tunneling microscopy, as mentioned above, promises to yield a wealth of information concerning mineral and other geologic surfaces. Unfortunately, only recently have researchers begun to obtain data on geological materials (principally, conducting minerals such as sulfides) using this technique. This is true also for a large number of other experimental surface approaches.

In summary, experimental approaches to studying surface phenomena have contributed greatly to the understanding of many processes and reactions involving geologic materials. Further advances in both instrumentation and their geologic applications continue to be made, being limited only by an investigator's imagination. Using these techniques, research into the understanding of both natural and laboratory-induced geologic processes has expanded rapidly, and, just like all other areas of scientific research, the future promises a continuation in its growth.

Acknowledgments

The author wishes to acknowledge support of his research by the U.S. Department of Energy (OBES and OCRWM) under Contract No. DE-AC03-76SF00098. Additional support has been provided by the Bureau of Mines, U.S. Department of Interior, and the U.S. Nuclear Regulatory Commission.

2. SCANNING ELECTRON MICROSCOPY

David F. Blake
Planetary Biology Branch
NASA / Ames Research Center
Moffett Field, CA 94035

1. Introduction

The scanning electron microscope (SEM) was developed as a practical laboratory instrument by Oatley in 1948 and first introduced commercially by Cambridge Scientific Instruments in 1965. Since that time, the SEM has become one of the most commonplace, powerful, and sophisticated tools available for the examination of the surface and near-surface characteristics of samples. The images produced by SEMs are (or appear to be) compellingly easy to interpret, and sample preparation is in general straightforward. Modern instruments are easier to operate than their predecessors, but proper evaluation of the data they provide demands a knowledge of the principles of operation of the SEM and of electron beam/specimen interactions. Furthermore, the role of the SEM is being both expanded and refined to the point that some capabilities of imaging and analysis are encroaching on what has traditionally been the realm of surface analysis instrumentation.

In this chapter, the general principles of SEM operation will be outlined and some of the more recent advances in instrumentation (with geologic applications) will be described. Techniques that have application to geologic materials and to surfaces and surface-related processes will be emphasized. For a more comprehensive treatment of SEM imaging and microanalysis, the reader is referred to [1 - 4].

2. SEM: How It Works

2.1. The Instrument

2.1.1 Overview. The scanning electron microscope is an instrument that provides relatively high spatial resolution surface and near-surface information from bulk samples. In brief, electrons are ejected from a filament and accelerated and focused into a small probe that is scanned over the surface of a sample. As the electron beam is rastered sequentially over the sample surface, the electron beam inside a CRT display (the viewing screen) is rastered in identical fashion. One of a number of signals received by detectors in the sample chamber is used to modulate the brightness of the electron gun of the viewing CRT. The magnification of the image is controlled by changing the dimensions of the scanned area on the sample while keeping the scanned area of the viewing CRT constant.

The most common signal used for creating images in the SEM is the secondary electron (SE) signal, made up of low energy electrons ejected from the sample surface by the interaction of the primary beam. Because the intensity of the SE signal is (all else being equal) primarily a function of sample surface morphology, SE images appear to be similar to what one would see when viewing a macroscopic surface with the naked eye. This paradigm is commonly quite correct; however, as we shall see, gross misinterpretations can be made.

2.1.2. Electron Sources. There are two types of electron guns that are used as electron sources in SEMs: thermionic guns and field emission guns. In thermionic emission, a flux of electrons is produced by heating a filament to a high enough temperature (1800 - 2800 K) to produce electrons with kinetic energies that exceed the work function of the filament material. The filament is kept at a large negative voltage (-1 to -40 kV) relative to ground (i.e., the anode plate), which accelerates the electrons down the electron optical column. A grid cap or Wehnelt cylinder [5] surrounds the filament and is kept at a negative voltage (minus 2500 to 0 V) relative to it. The combined effect of the heated filament, Wehnelt cylinder, and anode plate is to produce a beam of electrons that converges to a crossover, d_0, just below the aperture of the Wehnelt cylinder (Fig. 2-1). It is the diameter, brightness, and angular divergence of this crossover spot that ultimately determines the performance of the microscope. Two filament materials are presently in use in thermionic guns: tungsten and lanthanum hexaboride (LaB_6). Lanthanum hexaboride has a lower work function than tungsten, and can therefore achieve higher brightness at the same temperature, or equal brightness and longer filament life at lower temperatures. The LaB_6 filaments can be fashioned with smaller emitter areas, resulting in smaller source (i.e., crossover spot) sizes as well.

In a field emission electron gun, the filament consists of a tungsten rod that is chemically etched down to an extremely small tip, typically <10 nm in diameter. The filament is kept at a negative voltage (approximately -3 kV) relative to an "extraction anode" placed very near the filament tip. In this geometry, an electric field of $>10^7$ V/cm is produced that can extract electrons from the filament without the necessity of exceeding the work function of the material [6]. Tungsten is commonly used as a filament material because of its high strength, since weaker materials would be ripped apart by the high electric field. Field emission sources are extremely bright and produce a flux of electrons with a very low inherent energy spread. A low energy spread contributes to low chromatic aberration in the overall probe-forming system and results in higher attainable resolution.

Each of the three filament types has its own set of advantages and disadvantages. Table 2-1 lists operating parameters and significant characteristics. In the case of the bent wire tungsten thermionic filament, the advantages are ease of use, low cost, and a tolerance of poor vacuum

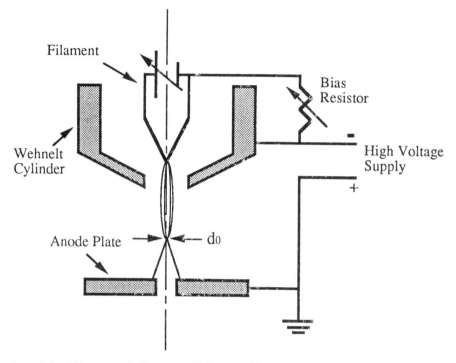

Fig. 2-1. Diagram of filament, Wehnelt cylinder, and anode plate (Redrawn after [1, 5]).

conditions. Disadvantages of the tungsten hairpin filament include low brightness, poor performance at small probe sizes (< 10 nm), and low brightness at low accelerating voltages. For the lanthanum hexaboride (LaB_6) filament, the advantages are improved brightness over the tungsten hairpin filament, higher performance at small probe sizes and low accelerating voltages, and longer filament life. Disadvantages include a requirement for higher vacuum levels (at least $5 * 10^{-7}$ Torr), higher cost than tungsten, and less tolerance to thermal shocks (during filament heating) or vacuum transients.

It is likely that in the future the tungsten bent hairpin filament will be replaced entirely by the lanthanum hexaboride filament since the advantages of LaB_6 far outweigh the disadvantages in almost any application. For the cold field emission gun, the advantages are extreme brightness, high performance at small probe sizes, and acceptable electron emission even at low accelerating voltages. The disadvantages are a requirement for ultrahigh vacuum (~10^{-9} Torr), high cost, and low total beam intensity when the probe diameter at the sample surface exceeds 50 - 100 nm.

Table 2-1. *Characteristics of Electron Gun Types Commonly Used in SEM*

	Thermal Emission Gun		Cold Field Emission Gun
	Tungsten Filament	LaB_6 Filament	Tungsten Filament
Brightness $(A/cm^2 \cdot sr)$	10^6	10^7	10^9
Electron source diameter (nm)	$>10^4$	10^4	<10
Electron energy spread (eV)	~ 2	~ 1	~ 0.2
Operating life (h)	50	1000	>1000
Required vacuum (Torr)	10^{-5}	10^{-7}	10^{-9}

2.1.3. Probe Forming Optics. If the diameter of the first crossover d_0 were small enough to produce the desired resolution in the final scanned image, the electron/optical column could consist of nothing more than a set of scanning coils to raster the beam over the sample. However, typically one wants to demagnify the crossover (electron source) by as much as 3 - 4 orders of magnitude to achieve an ultimate probe size <5 nm (and, in the case of the new field emission SEMs, <1 nm). This is accomplished by using one or two electromagnetic condenser lenses in addition to a probe-forming objective lens. A thorough discussion of the electron/ optical system of a SEM is beyond the scope of this chapter, and the interested reader is referred to [1] and references therein. For the purposes of this discussion, only an brief explanation of the factors that limit resolution will be made. If perfect lenses could be designed, the source size could be demagnified by any desired amount by the action of the lenses. However, real lenses have aberrations and the demagnifying power of the condenser system is limited primarily by these and by the diffraction limit imposed by the wave nature of the electrons themselves. At conventional accelerating voltages, the primary limitations in resolution are imposed by the spherical aberration coefficient (C_s) of the objective lens, by the wavelength (λ) of the electrons, and by the total current (i_{max}) at minimum beam spot size (d_{min}) on the sample. An improvement in resolution is achieved by increasing the accelerating voltage, which decreases λ and increases i_{max} by increasing the source brightness. Likewise, improvements in resolution can be achieved by using higher brightness electron guns and filaments (see Table 2-1). For conventional objective lens/specimen geometries, ultimate achievable resolutions are: tungsten thermionic filament, ~10 nm; LaB_6 thermionic filament, ~5 nm; tungsten field emission source, ~2 nm. By placing the specimen inside the objective lens, as is the case in modern transmission electron microscope/scanning transmission electron microscope (TEM/STEM) instruments, the C_s of the objective lens can be reduced by nearly an order of magnitude. Special purpose Field Emission SEMs (FESEMs) are commercially available that utilize this geometry and that have ultimate resolutions of less than 1 nm. Low-voltage operation of an SEM (which has a number of applications, detailed in Section 2.2.3) places severe limitations on d_{min}, due primarily to increased chromatic aberration (C_c) and decreased beam brightness in the electron source. For high-resolution, low-voltage SEM operation, a field emission source is preferrable, if not mandatory. Modern FESEMs with in-the-lens specimen geometry are capable of less than 5-nm resolution at 1-keV accelerating voltage.

2.1.4. Vacuum System. In traditional surface analysis applications, ultrahigh vacuums (10^{-9} to 10^{-11} Torr) are necessary in order to prevent the buildup of environmental contaminants on the surface being analyzed. In conventional SEMs, ultrahigh vacuums are not necessary and have not in general been available. Diffusion-pumped systems with an ultimate vacuum in the 10^{-5} to 10^{-6} Torr range are sufficient to allow for routine observation of samples without substantial hydrocarbon contamination on the sample surface.

In modern SEMs, vacuum systems have in general been improved for a number of reasons: (1) The use of LaB_6 electron sources requires a vacuum in the mid-10^{-7} Torr range to achieve reasonable filament life. This is usually accomplished by attaching an ion pump to the gun chamber and placing a differentially pumped aperture between the gun and the sample chamber. (2) Windowless light element energy-dispersive x-ray (EDX) detection systems are open to the microscope vacuum during operation. Clean, moderately high vacuums are needed in the sample chamber in order to reduce the contamination of the detector crystal, which is kept at cryogenic temperature. This is usually accomplished by eliminating the oil diffusion pump altogether and evacuating the sample chamber with a low-vibration turbomolecular pump. (3) For both high-resolution surface imaging and light element x-ray microanalysis, it is important that surface contamination in the area of the probe be minimized. This can be achieved through the use of a turbomolecular pump, or through the use of cryogenic traps in the vicinity of the sample. (4) In field emission SEMs, an ultrahigh vacuum is necessary in the gun chamber to minimize contamination of the filament. In systems of this kind, the sample chamber is usually kept at moderate vacuum (10^{-6} Torr) and a series of two or three differentially pumped apertures and ion pumps is used to maintain the vacuum differential along the electron column.

With the increasing availability of low voltage SEMs the trend will undoubtably be toward cleaner and higher vacuum in the sample chamber. Likewise, the common use of light element detectors will lead to cleaner systems in which hydrocarbon polymerization in the vicinity of the electron beam is reduced.

2.1.5. Sample Preparation. Because the SEM is a multipurpose instrument, specimen preparation techniques depend critically upon the kind of information sought from the sample. SEM applications can be divided into two categories: those which seek to image or analyze the as-received surface or near-surface region of a sample, and those which seek to image or analyze phase relationships beneath the surface or within the bulk of the sample. For analyzing the surface of as-received samples, minimal or no sample preparation may be necessary. The size and shape of samples is limited only by the size and geometry of the sample chamber. Indeed, one of the most valuable features of SEM is the tremendous depth of field attained relative to optical microscopy. If one has access to a SEM capable of low-voltage imaging (see Section 2.2.3), it may be possible to view even nonconductive samples without any surface preparation whatsoever. At conventional accelerating voltages however, nonconducting samples will require a thin coating (0.5 - 20 nm) of a conductor of some kind. If a quantitative elemental analysis is required, it is usually necessary to prepare a flat polished surface in order to minimize analytical errors due to x-ray absorption path lengths, etc. However, if this is not feasible, there are techniques available that will allow limited quantitation despite the absence of a flat surface [7 - 9].

For cross-sectional analysis, quantitative elemental analysis or analysis of features of the sample bulk such as secondary phases, etc., sample preparation is more substantial. It is usually necessary to embed the sample in resin of some kind (e.g., fiberglass boat resin) and grind the surface flat with successively smaller grit sizes. Typically, one starts with 250 grit carborundum powder and finishes with 600 grit powder or wet sandpaper. Polishing is then performed on a lap with 0.05-μm alumina or

0.25-μm diamond paste. Diamond paste is preferable because the sample surface is evenly polished despite hardness differences between minerals. Lloyd [3] recommends a chemical/mechanical polishing process that utilizes an alkaline silica sol slurry. This process is probably not necessary for ordinary polishing for x-ray microanalysis or backscattered electron imaging but is essential for more advanced techniques such as backscattered electron channeling contrast. The chemical/mechanical technique removes surface damage and local strain due to mechanical grinding. One way to maximize the information available from a particular sample is to prepare an uncovered (i.e., no glass coverslip) 25-μm petrologic thin section. Optical petrology can then be performed on the same areas that are imaged and analyzed in the SEM. For x-ray microanalysis and backscattered imaging, a perfectly flat surface is highly desirable. However, for secondary electron (SE) imaging, a flat sample appears nearly featureless in the SEM, particularly if it is Au or C coated. A very subtle and sophisticated way to enhance phase discrimination in the SE image is to etch the polished surface in a solution or stain to enhance one phase over another. Weak acids are useful, as are chelating agents such as ethylene diamine tetraacetic acid (EDTA). Useful sources of etchant recipes are semiconductor industry publications or materials science handbooks [e.g.,10].

There are three main reasons for coating a sample prior to observation in a SEM: to eliminate charge buildup on the sample surface, to conduct away heat in order to minimize sample damage, and to enhance surface contrast. For low- to moderate-resolution imaging, it is generally acceptable to sputter a thin (10 - 20 nm) coating of gold onto the surface or to evaporate a thin film of carbon (if microanalyis is anticipated). However, at the high magnifications easily achieved by modern SEMs, the grain structure of sputtered Au films is resolved. Newer sputter coaters utilize Au-Pd alloy or Pt targets and refrigerate the sample so that smaller grain sizes are achieved (see [1], Chapter 10 or [2], Chapter 7 for a thorough discussion of specimen coating techniques). Echlin and others [11 - 13] report results from low-voltage sputter coating experiments using a variety of materials and voltages. These authors found that very low voltages (250 - 400 V) produced finer grain sizes, as did higher melting point metals, such as niobium, tungsten, or tantalum. A second technique, ion-beam sputter coating, can produce even finer grained films, as detailed in [14]. Recently, a chromium coating technique has been developed that allows more or less uniform coating of samples with <1.0 nm of conductive material [15] . This appears to offer the ultimate surface resolution of any of the coating techniques yet developed because the method produces continuous films even in very thin coatings. If it is necessary to perform x-ray microanalysis on a non-conductive specimen, it is preferable to coat the sample with a material that does not contribute to the x-ray spectrum and that only minimally absorbs x-rays generated within the bulk of the sample. The usual technique consists of evaporating a carbon rod in an electric arc to produce a thin (<20 nm) carbon coat. Carbon evaporation onto a substrate has the potential to cause damage by radiative heat transfer, although in most geologic materials this does not pose a problem. Should specimen heating be a problem, procedures outlined in [16] for flash evaporation of carbon fibers may reduce thermal effects to acceptable levels. With the widespread availability of light element detectors, analysis for carbon is becoming commonplace. For carbon analyses, a thin film of carbon on the sample surface may mimic several percent of carbon in the bulk, and so alternatives to carbon coating must be found. It is possible to coat with very thin layers (~0.5 nm) of metals such as Cr, whose x-ray spectral peaks are unlikely to interfere with other elements present in the sample. Procedures are described in [12] for removing Cr artifact peaks from spectra.

2.2. Signals and Detectors

2.2.1. Depth and Lateral Distribution of Primary Beam Electrons.
The lateral and depth resolution of information obtained in an SEM is a function of the distribution of primary beam electrons within the sample. The depth of penetration and lateral spread of electrons into the sample are directly proportional to the primary beam energy, E_0, and inversely proportional to sample density, average atomic number of the sample, and the angle of incidence of the primary beam. Figure 2-2 shows Monte Carlo simulations of electron trajectories in SiO_2 for accelerating voltages of 2, 5, and 10 keV. The depth of penetration of the electrons, and thus the maximum depth of generation of a variety of signals useful in SEM, is decreased significantly at the lowest accelerating voltage.

It is clear from Fig. 2-2 (and also intuitively appealing) that at lower accelerating voltages the spatial resolution of any signal will be higher. However, only recently have instrument designers been able to solve many of the problems attendant to low-voltage operation. Additionally, some types of analysis require higher accelerating voltages, such as x-ray microanalysis (XM), using higher energy x-ray lines.

2.2.2. Backscattered Electrons.
Backscattered electrons (BSEs) are primary beam electrons that have been multiply elastically scattered within the sample and that exit the sample surface with a significant proportion of their original energy. The backscatter coefficient, η, is defined as the ratio of backscattered electrons to primary beam electrons. It is a function of accelerating voltage as well as a variety of sample-dependent parameters, including:

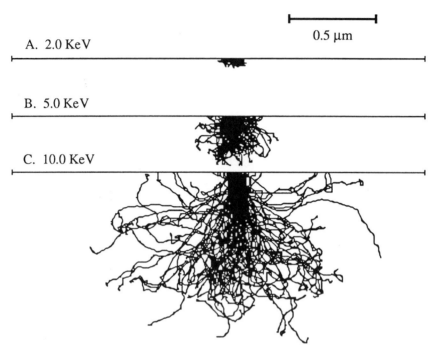

Fig. 2-2. Monte Carlo simulation of primary beam electron trajectories in silicon dioxide. One hundred electron trajectories are figured in each plot. Accelerating voltages: a). 2.0 keV. b). 5.0 keV. c). 10.0 keV.

1. Average atomic number of the sample
2. Mass density of the sample
3. Crystallographic orientation of the sample (e.g., alignment of low index planes with the beam direction)
4. Angle of incidence of the primary beam to the sample surface

Perhaps the most useful characteristic of the backscattered electron (BSE) signal is its dependence on average atomic number. With modern solid-state BSE detectors, it is possible to discriminate differences of as little as 0.1 average atomic number between adjacent sample regions. Figure 2-3 shows a plot, after [17], of the BSE coefficient vs. average atomic number for a variety of geologically significant phases.

2.2.3. Secondary Electrons. In SEM parlance, secondary electrons (SEs) are low-energy valence and conduction band electrons from the sample which are freed from their parent atoms by the action of the primary elec-

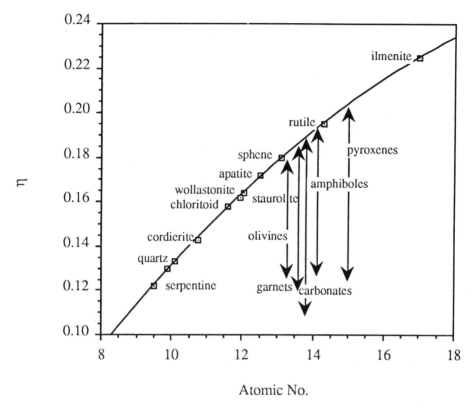

Fig. 2-3. Backscattered electron coefficient vs. atomic number for selected minerals (after [17], with permission). Vertical arrows indicate the range of values for the BSE coefficient within particlular solid solution series.

tron beam. In this usage, the term "secondary electron" is a misnomer, because a number of types of sample-generated electrons are produced (e.g., Auger electrons) that could rightfully be called secondary electrons. SEs are defined as those electrons having an energy less than 50 eV. Most SEs, however, have energies in the range 1 - 10 eV, depending on the nature of the sample (conducting or insulating, metallic, etc.). Because of the low energy of SEs, a single primary beam electron can produce a multitude of SEs before exiting or coming to rest in the sample. The great majority of these SEs are produced in the bulk of the sample and are reabsorbed close to the point of generation. Within a few nanometers of the sample surface, however, SEs have a high probability of escaping the sample surface and can be collected by a detector to produce a SE image. SE generation is a function of a variety of sample parameters, the most important of which are composition (elements present, bonding type, conductivity, etc.) and the surface area of the sample intersected by the beam (surfaces that intersect the electron beam at an angle produce more SEs than surfaces that are normal to the beam). The latter relationship is the reason SE images appear similar to images of macroscopic surfaces viewed by eye: Since the electron beam intersects more of the sample surface when the surface is non normal relative to the beam, sloped features appear brighter. The resulting image shows hills and valleys lighted from above and to the side. Additionally, when the beam intersects the sample surface at an oblique angle, more of the surface is within the escape depth of the secondary electrons generated in the near-surface regions by the beam.

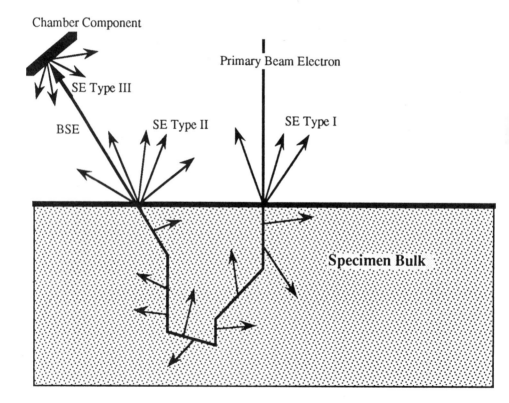

Fig. 2-4. Secondary electron generation from an idealized sample surface. In this simplified diagram, a single primary beam electron is backscattered from the sample. Production of high-resolution SE Type I and low-resolution SE Type II and Type III electrons is illustrated.

This simplistic explanation of the generation of SEs is complicated by the fact that high-energy electrons of all kinds (including BSEs) can produce secondary electrons. Indeed; high-energy primary beam electrons are relatively inefficient at producing SEs relative to lower energy electrons, such as BSEs. The result is that there is a variety of SE signals originating from the sample and elsewhere which may carry conflicting information about the sample surface. Figure 2-4 shows the origin of the most important of these signals. The "SE Type I" signal, generated by the primary electron beam as it enters the sample, carries the desirable high-resolution surface information about the sample. The "SE Type II" signal, generated by BSEs as they exit the sample surface at remote locations from the beam, contains low-resolution information which may be modulated by subsurface structures or compositional heterogeneities. At conventional SEM voltages, the SE Type II signal may exceed the SE Type I signal by a factor of 3 or more [1]. The "SE Type III" signal, produced by BSEs that hit the objective lens polepiece and other areas of the sample chamber, also generates low-resolution information, which can complicate the SE image. For this reason, the interpretation of SE images is not always straightforward.

SE imaging at low accelerating voltages (0.5 - 3.0 keV) has a number of advantages relative to conventional SE imaging in the 10 - 30 keV range. These include: (1) higher lateral spatial resolution on the sample surface, (2) greater "surface" contribution to the SE signal, (3) decreased electron charging artifact when viewing nonconductive surfaces, and (4) the potential to view "material contrast" between differing materials on the sample surface. However, instrumental design considerations limit the performance of conventional SEMs when they are operated at low accelerating voltages. Performance limitations include low beam brightness (beam brightness is a direct function of accelerating voltage), increased effect of chromatic aberration, "space charge" effects at beam crossover points in the electron column, and the increased effects of stray magnetic fields on the electron beam both in the column and the sample chamber. Table 2-1 lists the operating characteristics of various types of electron emitters presently in use in SEMs. As can be seen from the table, a cold field emission electron source is far superior to tungsten and lanthanum hexaboride thermal emitters with regard to brightness, "space charge" effects, and chromatic aberration. For other problems, such as stray magnetic fields, additional design improvements have been incorporated in some of the modern machines, such as through-the-lens SE detectors, lens designs in which the specimen is placed directly within the pole piece gap, etc.

Higher lateral spatial resolution is achieved at low voltage through the generation of an increased number of SE Type I electrons and through the decreased source size of SE Type II electrons. The increased topographic content of the SE signal is the result of an overall reduction in the contribution of SE Type II electrons to the image. The primary source of topographic contrast in the SE signal is the greater number of SEs within escape depth when the sample surface is non normal to the electron beam. Joy [18] points out that at very low accelerating voltages, it is possible for the entire electron beam to be stopped within the escape depth of SEs, theoretically yielding a constant SE signal regardless of sample orienta-

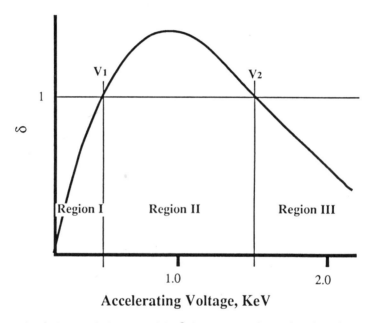

Fig. 2-5. Total electron yield δ from the surface of an idealized sample surface, plotted as a function of accelerating voltage. V_1 and V_2 are crossover points where $\delta = 1$ (total number of electrons entering and leaving the sample surface is equal)

tion. However, there should always be some topographic component to the SE signal due to the increased area intersected by the beam at non normal incidence to the surface.

The reduction or elimination of charging in electrically nonconductive samples is theoretically possible at low accelerating voltages but is difficult to achieve in practice. As was discussed earlier, a single primary beam electron may generate a large number of SEs before coming to a halt in the sample. At higher accelerating voltages (e.g., Region III in Fig. 2-5), the depth of penetration of the primary beam is so much greater than the SE escape depth that the SEs are mostly reabsorbed by the sample and the sample becomes negatively charged due to the trapping of primary beam electrons below the surface. As the accelerating voltage is decreased, the total electron yield from the sample (the ratio of electrons leaving the sample surface to those entering it) gradually increases because the sample volume in which SEs are generated is shifted toward the sample surface. At some point (the upper bound of Region II in Fig. 2-5), total electron yield exceeds unity. Decreasing the accelerating voltage beyond this point results in a positive charge forming on the sample surface because more electrons are leaving the sample than are entering it. A reduction of accelerating voltage beyond the lower bound of Region II results in a resumption of negative charging since the primary beam lacks sufficient energy to generate the necessary number of escaping SEs to produce charge neutrality. In the region just below the upper cross-over of Region II, a slight positive charge is produced on the sample surface. This positive charge is sufficient to cause the reacquisition of low-energy SEs by the sample surface resulting in a dynamic equilibrium in which electron charging is effectively neutralized. In practice, one can find this voltage by observing the sample image while increasing and decreasing the magnification slightly and observing the contrast of the square rastered by the beam at the higher magnification. If the square is dark, the area is charging positively and the accelerating voltage must be increased slightly. If the square is bright, the area is charging negatively and the accelerating voltage must be decreased. The charge equilibrium can also be manipulated by tilting the sample with respect to the beam thus changing the electron emission coefficient.

Using a model sample, Fig. 2-6 shows the influence of accelerating voltage on depth of penetration of the electron beam, the "surface component" of the SE signal, and true "topographic contrast." The sample is a 3-mm electron microscope grid onto which has been placed a ~20 nm thick sputtered Au film. Figure 2-6 (a-d) were recorded at 25 keV, 10 keV, 5 keV, and 1 keV, respectively, on an area of the grid that included areas of self-supporting Au film (Au), Au film over Cu grid bars (Au/Cu), bare Cu grid bar (Cu), and holes (H). In (a), the Au film appears transparent and the surface structure of the grid bar is evident even through the Au film (this despite the fact that SEs can escape from only the upper 1 - 2 nm of the sample). The manifestation of subsurface structure in a "surface" signal occurs because at conventional SEM voltages, the majority of the SE signal is produced by BSEs as they leave the sample surface. The BSE signal from the Cu grid bar below the film is modulated by the (sub)surface bumpiness of the grid bar, and these BSEs as they exit the sample create cascades of SE Type II electrons which mimic the BSE subsurface information. At 10 and 5 keV accelerating voltages, this effect is mitigated somewhat, but true surface structure is still not apparent. At

Fig. 2-6. SE Images of a 200 mesh Cu electron microscope grid covered
by a 20 nm thick foil of Au. (a). 25 keV. Note the transparency of Au
foil (Au) and the appearance of subsurface grid bar morphology through
the foil (Au/Cu). Cu = bare Cu grid bar, H = hole in film. (b). 10 keV.
(c). 5 keV. (d). 1 keV. Note opaque appearance of Au film and the
presence of a square surface coantamination mark (C).

1.0 keV, not only is the Au film opaque, but small surface particles are evident, as is a square contamination mark (C) which shows a few nanometers of hydrocarbon contamination polymerized on the sample surface during focusing at higher magnification. In addition, surface topography (warps in the film, etc.) is readily apparent. It appears from Fig. 2-6 that true surface imaging can be assured only at low voltage.

2.2.4. X-Ray Microanalysis. As primary beam electrons enter the sample bulk, they undergo elastic and inelastic collisions with atoms of the sample. There are two types of inelastic (i.e., energy absorbing) interactions that have the capacity to generate x-ray photons (for a thorough discussion of the physics of x-ray generation, see [1, 19]). "Continuum" or "bremsstrahlung" x-rays are produced when primary beam electrons are decelerated by the coulombic field of sample atoms. These x-rays are not directly useful for elemental analysis since there is a continuous distribution of x-ray energy from zero to the voltage of the primary beam,, irrespective of sample composition. Bremsstrahlung x-rays, however, are valuable for some quantitation procedures since they can be used to normalize for matrix effects such as surface roughness or differential x-ray absorption (see, for example, [7]). A second way in which primary beam electrons can lose energy in the sample is through the ionization of an inner shell electron of a sample atom. The resulting inner shell vacancy is filled when an outer shell electron drops into the vacancy. This electron transition between shells of different energies can result in either the ejection of an Auger electron or the production of an x-ray photon of a characteristic energy. X-Rays produced in this fashion are called "characteristic" x-rays because their energies are characteristic of a particular electron transition in a particular element. X-Ray photons can be detected and measured according to either their energies or their wavelengths, and assigned to particular elements. The types and amounts of sample atoms can be determined either by using standard materials to generate working curves of x-ray intensity vs. element concentration, or by using theoretically derived K factors and applying matrix corrections. In modern SEMs equipped with x-ray detectors and associated computers, it is possible to perform quantitative analyses for all major elements above boron in the periodic table, accurate to within ±5% of their abundance. X-Ray detectors are in general sensitive to the presence of all elements above the 1.0 % level, and lateral spatial resolutions on the order of 1.0 μm can be achieved. There are two types of X-ray detectors commonly in use, the energy-dispersive x-ray (EDX) detector and the wavelength-dispersive x-ray (WDX) detector. EDX detectors are more commonly found on SEM instruments, while WDX detectors are more commonly found on dedicated microanalysis instruments such as the electron microprobe. This brief discussion will concern itself exclusively with EDX detection.

Qualitative Analysis. As the primary electron beam enters the sample bulk, characteristic x-rays are generated in a roughly onion-shaped or bulb-shaped volume whose absolute dimensions are a function of the accelerating voltage, x-ray energy, sample density, and average atomic number of the sample. X-Ray spatial resolution can be determined in a variety of ways including Monte Carlo simulation, which is now possible on most personal computers (e.g., see Fig. 2-2, this chapter, or [2], Chapter 1). A less precise but simpler method is to consult charts of x-ray generation vs. depth, prepared using the extended electron range equation (e.g., [20]). In planning even qualitative analyses, it is critical to ensure that the volume of x-ray production does not exceed the dimensions of the features of interest in the sample. This can be a potential problem because SE and BSE image resolution is typically much higher than can be achieved in x-ray analysis.

In order to produce x-rays from a given element in the sample, it is only necessary to choose an accelerating voltage high enough to produce inner shell ionization events in the element of interest. However, to generate x-rays efficiently, it is a good rule of thumb to exceed the energy of the x-ray line of interest by a factor of about 2.5 or 3.0. For example, if one wants to excite Fe Kα x-rays (6.40 keV) efficiently, an "overvoltage" of 2.5 would yield a recommended accelerating voltage of 16.0 keV.

The above considerations are all that are necessary to produce a flux of x-rays that is characteristic of the sample composition in a qualitative sense. There are, however, a variety of detector artifacts that may affect even the qualitative accuracy of an EDX analysis. These include:

1. *Absorption of low-energy x-rays in the detector window.* Low atomic number elements may be present in the sample and go undetected due to their near-total absorption in the detector window. Because the detector crystal is kept at liquid nitrogen temperature, x-ray detectors require their own sealed or self-pumped vacuum systems in which a thin window maintains vacuum in the cryostat while keeping out contaminants present in the microscope vacuum system. In conventional beryllium window detectors, x-rays lower in energy than Na Kα (1.04 keV) are almost totally absorbed by the window material. Even after qualitative x-ray analysis the true identity of some samples (carbonates, oxides, nitrides, carbides) may be unknown. Windowless detectors or atmospheric pressure thin window detectors are now quite common on SEMs. These "light element" detectors are sensitive to all elements above beryllium in the periodic table. It is becoming less acceptable to report analyses of only the elements heavier than Na in x-ray microanalysis, particularly in the case of geologic materials where light elements such as O and C constitute a large percentage of the total.

2. *Peak broadening due to the statistical nature of the x-ray detection process.* Minor elements may be overlooked because their x-ray spectral peaks are totally occluded by a major element present in the sample. As x-ray photons enter the active area of the detector crystal, their energy is dissipated in the form of electron/hole pairs which create a current pulse within the detector. Because the production of electron/hole pairs is a statistical process, x-rays of identical energy may generate slightly different voltage pulses. The result is that a population of x-ray photons from the same element which have an inherent energy distribution of 1 - 2 eV, produces an x-ray spectral peak which has an energy distribution of 50 - 150 eV. This broadening creates the potential for peak overlaps, which could obscure the presence of minor elements. One can minimize problems of this type by inspecting other regions of the x-ray spectrum for peaks. For example, in classic spectral overlaps such as Pb Mα / Mo Lα / S Kα, one simply needs to inspect the spectrum at higher energies for the Mo Kα or Pb Lα peaks.

3. *Generation of silicon escape peaks.* Due to the physics of the detection process in solid state detectors, artifact peaks may be generated that have the potential to be incorrectly identified as additional elements present in the sample. As sample x-rays enter the detector (which is typically a high-purity, lithium-doped silicon crystal), high-energy photoelectrons are produced that are reabsorbed by the crystal to produce electron/hole pairs. Occasionally, a photoelectron fluoresces a Si Kα x-ray within the detector much as x-rays are fluoresced by primary beam electrons in the sample material. The total energy deposited in the crystal in this case is equal to the original energy of the sample x-ray minus the energy of a Si Kα x-ray (1.74 keV). The result is that small peaks occur about 1.74 keV below major element peaks in the x-ray spectrum. This process is unavoidable; however, most EDX software packages contain routines to remove the counts present in these "silicon escape peaks" and add them back to their parent peaks.

4. *Production of sum peaks.* Due to the finite time associated with x-ray detection and processing, two x-rays entering the detector at nearly the same time may be counted as one x-ray with an energy equal to the sum of the two x-rays. Sum peaks occur at an energy equal to the sum of two major elements (or 2 * a single major element) present in the spectrum. As a simple example, in a sample of pure Fe the Fe Kα peak occurs at 6.40 keV. The sum peak, if it occurs, will be present at 12.80 keV. To avoid sum peaks it is only necessary to reduce the x-ray count rate.

X-Ray mapping is another powerful technique that utilizes qualitative EDX data. In this technique, whenever an x-ray photon is detected within a specified energy range, a bright dot is placed on an scanned "image" of the sample. The result is a two-dimensional representation of element distribution in the near-surface region of the sample. There are a number of caveats associated with the interpretation of x-ray maps. The first is that surface roughness on the sample can cause the differential absorption of x-rays, yielding the false impression of element heterogeneity on the sample surface. One way to control for this is to record a second x-ray map using an energy range near the element peak of interest but containing only background. The x-ray background map should show the effect of surface roughness on the x-ray signal. A second caveat concerns the nature of x-ray background. In EDS x-ray mapping, there is no way to tell whether an x-ray photon within a specified energy range was generated as bremsstrahlung or as a characteristic x-ray. Because the peak to background ratio is in general lower in EDX spectra than in WDX spectra, there is an increased probability that minor amounts of an element will appear to be background or that background will appear to be minor amounts of an element.

Quantitative Analysis. In quantitative analysis, background-subtracted x-ray intensities from sample materials are compared to x-ray intensities acquired from standard materials of known composition. X-Ray data are corrected for a variety of second-order "matrix" effects. Matrix effects have traditionally been placed in three groups ("ZAF" factors), the atomic number effect (Z), the absorption effect (A), and the fluorescence effect (F). The "ZAF" factors compensate for matrix differences between mineral standards and unknowns. The atomic number effect, Z, results from differences in backscattered electron yield and "stopping power" between standard and unknown. A higher average atomic number matrix has a higher backscattered electron yield, resulting in fewer electrons remaining within the sample to generate x-rays. "Stopping power" refers to the energy lost by the electron per unit distance traveled in the specimen. The absorption effect, A, refers to the degree to which the sample matrix absorbs x-rays after they have been generated within the sample by the electron beam. Absorption is a function of the composition of the matrix, the energy of the x-ray of interest, and the absorption path length in the sample. The fluorescence effect, F, refers to the extent to which characteristic and continuum x-rays, once generated, fluoresce characteristic x-rays of matrix elements.

A variety of correction schemes have been devised to produce quantitative results from raw x-ray spectral peaks. The most popular are the ZAF methods ([1] and references therein). There are a number of empirical correction schemes, including the Bence-Albee technique [21], the peak to background technique [7 - 9], and particle shape correction schemes [22, 23].

A description of the theory and procedural details of quantitative x-ray microanalysis is beyond the scope of this chapter. For in-depth descriptions of quantitative techniques the reader is referred to [1, 2, 20]. A few comments are appropriate for the beginning analyst or for those who wish to evaluate x-ray data. To begin with, many of the fitting parameters for the ZAF correction scheme are extrapolations from higher energies or from different systems. For accurate results, therefore, it is most important to use standards that are as nearly identical to the unknowns as is practicable. If one is using a program that prints out the values of the ZAF correction factors, they should be inspected. The individual values of Z, A, and F should not deviate from unity by more than about 10%. The preparation of the sample surface is also extremely important. The surface should be polished and flat, and carbon coating (for conductivity) of both standards and unknowns should be performed at the same time (or at least under the same conditions). If the analyst must work with rough or unpolished surfaces, it is more appropriate to use the peak to background method [7], which does not assume sample flatness. If the sample/detector geometry can be altered, it is desirable to use the highest takeoff angle (the angle between the sample surface and the detector axis) possible in order to reduce the x-ray absorption path length and minimize the effect of errors in the absorption correction. The analysis should be performed at the lowest accelerating voltage possible which still adequately excites the x-rays of interest.

Prior to performing the analysis, a number of short analyses should be performed to evaluate element loss and/or beam damage effects. In planning the analysis, an acquisition time should be chosen such that the desired precision is achieved. Relative precision is generally quoted at the the 95% confidence level for elements of interest. Since the (peak/background) intensity of any x-ray peak can be described using Poisson statistics, the standard deviation of the number of counts in the peak (N) is equal to the square root of the total number of counts ($\sqrt{N}$). The 95% confidence interval is equal to $N \pm 2\sqrt{N}$. Usings this relationship, in order to quote an elemental value (at the 95% level of confidence) to a precision of $\pm 1\%$, the net peak value should be at least 40,000 counts; for $\pm 2\%$ precision, 10,000 counts; and for $\pm 5\%$, 2000 counts.

3. Applications in Geology

3.1. Surface Imaging of Particles

3.1.1. Interplanetary Dust Particles. Interplanetary Dust Particles (IDPs) are 1- to 75-μm grains or aggregates of extraterrestrial material that are collected by high-flying aircraft in the upper stratosphere [24]. While they are called particles, individual IDPs in fact consist of hundreds or thousands of individual grains ranging from subnanometer to microns in size. IDPs are texturally and mineralogically distinct from meteoritic material and are believed to be (in some cases) debris spalled off of cometary bodies [25, 26]. These particles enter the earth's atmosphere at ballistic velocities (5 - 20 km/s) but some are sufficiently small to be slowed down without extensive atmospheric heating. IDPs appear to be fragments of the most primitive and pristine objects in the solar system. They offer ground-based scientists the opportunity to study material that apparently condensed from the solar nebula >4.6 billion years ago and has remained relatively unaltered since that time [26, 27]. Indeed, anomalously high deuterium/hydrogen ratios measured by secondary ion mass spectrometry (SIMS) may indicate that presolar interstellar material is commonly incorporated into many of these particles [28, 29]. While IDPs have been imaged and spectroscopically analyzed utilizing a variety of specialized techniques, SEM/XM remains the primary means of screening and classification [30] and is also one of the most commonly utilized techniques for mineralogic characterization [31, 32]. Figure 2-7 shows conventional SEM and low voltage SEM (LVSEM) micrographs of an IDP, recorded at 30 keV, 5 keV, and 2.0 keV accelerating voltage. The particle has not been coated with a conductive medium of any kind. The conventional voltage image shows little of the high-resolution surface structure and three-dimensional topography evident in the LVSEM images. This particle is of the chondritic filled variety, consisting of hydrated Fe-Mg silicates, glasses, metal oxides and sulfides, and other minerals whose phase relationships and petrology can be used to discern conditions in the early solar system.

3.2. X-Ray Microanalysis of a Particle

One of the most common analyses performed on geologic samples using a SEM/XM system is qualitative EDX analysis. In less than a minute, a complete analysis can be performed for all elements above boron that are present in concentrations greater than 1.0 wt.%. Figure 2-8 shows an EDX spectrum recorded from an IDP similar to that illustrated in Fig. 2-7. This type of information is extremely valuable for initial screening and later complete characterization (indeed, an elemental abundance pattern that is within a factor of 2 of solar or chondritic abundance is sufficient to warrant an assumption of extraterrestrial origin for an IDP). It was emphasized earlier (Section 2.2.4) that in order to perform a quantitative analysis of EDX data, it is necessary to have a flat sample in order to calculate x-ray path lengths for absorption corrections. There are in fact some data reduction strategies that eliminate this restriction and that can be used for quantitative analysis of irregular surfaces or

Fig. 2-7. Comparison of conventional SEM and low voltage SEM SE images of an interplanetary dust particle. (a). Conventional SE image recorded at 30 keV on an uncoated, pristine IDP. Note electrostatic charging, lack of surface detail, and "flatness" of topography. (b). Same as (a), but image recorded at 5.0 keV. (c). Same as (a), but image recorded at 2.0 keV. (d). Higher magnification image from central portion of (c), 2.0 keV. There is an increased resolution of surface generated detail, as well as a reduction of charging and improvement of 3-dimensional appearance of topography.

complexly shaped particles [7, 9, 22]. However, it should be made clear that these techniques are not appropriate for (and were not intended for) quantitation of EDX data from particles in which there is a heterogeneous distribution of elements. For example, if one is analyzing for carbon using

Fig. 2-8. Energy-dispersive x-ray (EDX) spectrum of an interplanetary dust particle. The spectrum consists of the major elements Fe, Si, Mg, and O as well as minor elements Ca and Al, and traces of Cr and Ni (not labeled). In general, it is not possible to perform quantitative analyses of particles of this kind because (1) irregular shape and surface roughness cause differential absorption of the x-ray signal within the sample and (2) the non uniform distribution of elements within the bulk of the particle make proper atomic number and absorption corrections impossible.

the peak to background method [7], it is assumed that carbon x-rays (~0.28 keV) are absorbed to the same extent in the sample bulk as continuum X-rays of equivalent energy. If the carbon is concentrated in a thin surface layer, the peak to background method would greatly overestimate the total amount of carbon in the bulk of the sample. In cases such as this, it is more appropriate simply to show the raw spectra and comment on the limitations of the data.

Fig. 2- 9. LVSEM and BSE images of a carbonaceous interplanetary dust particle. (a) Low-magnification SE image of a rare IDP that contains primarily C and traces of Ti and O. Accelerating voltage, 1.0 keV. (b) Low-magnification BSE image of same particle. Bright spots in the image were found to be a titanium-rich phase. (c) Higher magnification SE image of area circled in (a) above. (d) BSE image of the same area as (c) showing Ti-rich particles.

3.3. Surface and Near-Surface Information from a Particle

Through a combined use of LVSEM and BSE images and qualitative EDX data, it is possible to learn a great deal about a particle, polished section, or fracture surface. Figure 2-9, for example shows a rare variety of IDP that consists almost totally of carbon with minor titanium and oxygen [33]. The LVSEM SE images of the surface are shown in (a) and (c), while (b) and (d) show BSE images at the same magnifications. The bright particles shown in the BSE images are high atomic number phases present within the low atomic number matrix. By placing the electron beam on a number of the bright particles during a qualitative EDX analysis, it was found that they are enriched in Ti. Because of matrix interference, however, it cannot be determined whether the Ti is present as an oxide, carbide, or metal without recourse to other methods such as analytical electron microscopy.

3.4. Second-Phase Precipitates

3.4.1. Pyroxene Exsolution Textures in an Achondritic Meteorite. Figure 2-10 (a - f) show BSE and elemental x-ray images of a clinopyroxene/orthopyroxene exsolution texture in a crystal of pyroxene contained within an achondritic meteorite. Figure 2-10a is a digital image of the BSE signal. The bright phase is a higher average atomic number material (Ca-poor orthopyroxene), and the dark phase is lower average atomic number material (Ca-rich clinopyroxene). Quantitative elemental analyses were performed on the bright and dark lamellae, yielding a composition of $Fe_{.58}Mg_{.38}Ca_{.02}Al_{.01}SiO_3$ for the orthopyroxene (bright phase), and $Fe_{.25}Mg_{.28}Ca_{.43}Al_{.03}Si_{.97}O_3$ for the clinopyroxene (dark phase). The average atomic number for the bright phase is 11.66 and for the dark phase, 11.37. It is not out of the ordinary to generate contrast between phases with as little as 0.1 atomic number difference. Figures 2-10 (b - f) are digital x-ray maps for major elements present within the material. One curious result is seen in the oxygen map shown in Fig. 10e. There should be essentially no difference in oxygen content between the two exsolved phases, yet it appears from the map that more O is contained within the Fe-rich phase. A likely explanation is that in the higher average atomic number phase, which has greater "stopping power" (ability to decelerate primary beam electrons more readily once they enter the sample bulk), oxygen x-rays are generated nearer the surface and have less likelihood of being absorbed. Other explanations could be offered, but this does serve to point out the hazards associated with the interpretation of X-ray maps.

3.4.2. Direct Analysis of Fuid Inclusions in Minerals. A variety of techniques has been employed in recent years for the determination of salinity,

Fig. 2-10. BSE and x-ray dot maps of a pyroxene grain within an achondritic meteorite. (a) BSE image showing atomic number contrast between exsolution lamellae within a portion of the grain. Inset histogram shows digitized grayscale information from image. (b) Ca x-ray map of area shown in (a). (c) Fe X-ray map. (d) Si x-ray map. (e) O x-ray map showing apparent heterogeneity in O content. In fact, the apparent concentration differences are due to matrix effects between the Fe-rich and Fe-poor phases. (f) Mg x-ray map showing relatively little heterogeneity for this element.

temperature of fluid entrapment, and chemical composition of fluid inclusions in minerals. This information can be used to determine the conditions of formation or emplacement of ore bodies, or the composition of ancient hydrothermal mineralizing fluids. Conventional freezing and heating stage measurements using optical microscopy can provide filling temperatures and salinities of individual inclusions [34, 35], but numerous assumptions must be made as to the chemical makeup of the inclusions themselves.

Kelly and Burgio [36] describe a SEM/EDX analysis technique in which inclusions exposed on fracture surfaces are analyzed in the frozen hydrated state, thus providing the opportunity to analyze individual fluid inclusions with regard to both total salinity and cation ratios of dissolved salts. In this technique, EDX spectra are collected from frozen/hydrated fluid inclusions 10 - 20 μm in size exposed on fracture surfaces of minerals quick-frozen to minus 200°C. It is possible to quantitate such

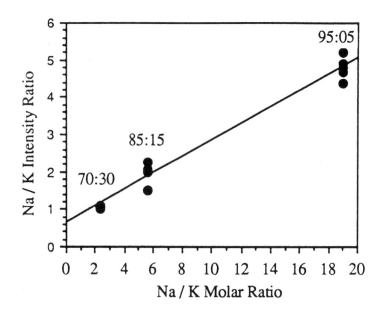

Fig. 2-11. Plot of Na/K x-ray intensity ratio of frozen/hydrated standard droplets vs. molar Na/K ratio. Least-squares best fit line through the replicate analyses is used to quantify Na/K peak to background values from fluid inclusions in mineral crystals.

spectra with working curves generated from frozen standard solutions using the "peak to background" method discussed in Section 2.2.4 [7 - 9]. Peak to background (p/b) ratios for individual elements are relatively insensitive to detector efficiency, beam current fluctuations, and live time correction inaccuracies. In addition, when the background is chosen to be in the same energy range as the characteristic peak of interest, irregularities in the surface of the samples are largely factored out. This is because

Fig. 2-12. EDX elemental spectrum from a frozen/hydrated fluid inclusion in quartz. The high background indicates that the fluid inclusion is still hydrated. Silicon originates from the edges of the inclusion, which were within the volume of excitation of the electron beam. Copper originates as a spectral contaminant from the sample holder.

the background and peak x-rays suffer the same absorption and fluorescence effects in the sample. By using frozen standard solutions of approximately the same average atomic number as the unknowns, the atomic number, or "Z" effects are largely cancelled out.

Cation content is determined by ratioing the p/b values of the elements of interest. Lines are fitted to the standard data by least-squares regression, and elemental ratios are quantified using the resulting regression equations. Figure 2-11 shows (Na p/b)/(K p/b) ratios for a set of frozen standards, vs. Na/K molar ratio. Once these are obtained (during the same analysis session and under the same conditions as the real fluid inclusions), spectra can be recorded from frozen fluid inclusions and cation p/b values can be compared and quantified using the working curves. Figure 12 shows an EDX spectrum recorded from a frozen hydrated fluid inclusion in quartz.

The SEM/EDX cryogenic procedure is attractive in that it may provide for quantitative analysis of individual inclusions in both optically clear and optically opaque minerals. Additionally, by using geothermometers such as those described in [37], an independent determination may be made of filling temperatures of individual fluid inclusions in minerals.

4. Conclusions and Future Directions

This brief introduction to the methods of SEM is intended not so much as a primer but rather as a sampler of useful techniques for the geologist. A variety of books have been written on the subject (included in the reference section) which should be consulted by those interested in particular applications in their specialty areas.

A number of recent innovations in SEM/XM have created dramatic improvements in the techniques and in the breadth of applications in the geological sciences. Some of these are listed below:

1. The introduction of automated field emission SEMs which have made high-resolution low voltage imaging practical. The ability to view surfaces without conductive coatings is especially significant in the geologic sciences, as is the ability to image "true" surface structure rather than a mixture of surface and bulk information.

2. The recent introduction of atmospheric pressure thin window EDX detectors. These light element detectors are retrofittable onto older SEMs that have relatively poor vacuums. In most minerals, carbon or oxygen together comprise at least half of the total. It is now feasible to determine total composition directly by measuring the light elements rather than by appealing to stoichiometry.

3. Dramatic improvements in computer performance have allowed for moderately priced digital control of instrumentation and digital data acquisition and analysis. The result is that complex analyses can now be performed more or less routinely and routine analyses can be performed in unattended mode.

The SEM has not yet reached the theoretical limits of its development. It is to be expected that further improvements in electron optics, detector design, digital signal processing and digital instrument control will make SEMs even more powerful than at present. The ultimate resolution of modern field emission SEMs is already approaching that of moderate-resolution transmission electron microscopes (TEMs), and most samples can be viewed in the FESEM with minimal sample preparation. Backscattered electron detectors have recently been introduced that will operate at low accelerating voltages (~1 keV) producing images with the lateral resolution of secondary electron images, the depth resolution of a surface analysis instrument, and the atomic number contrast typical of a BSE image. In addition to the more or less "typical" signals expected in SEM, future machines will probably provide for the integration of several signals, such as the Auger electron signal, signals derived from ion beam sources contained within the SEM and the like. At least one machine is being developed which will have a scanning tunneling microscope (STM) contained within the sample chamber of the SEM so that correlative signals can be recorded.

5. Acknowledgments

The author would like to thank the following individuals for assistance in preparing images or data for this manuscript: Mr. Terrence Reilly and Mr. Hideo Naito of Nissei Sangyo America, Mountain View, CA for assistance and instrument time on Hitachi S-900 and S-800 FESEMs; Mr. Tom Stark and Mr. Andrew Isaacs of Kevex Corp., San Carlos, CA for assistance and instrument time on an SEM equipped with a quantum light element EDX detector and digital imaging system; Mr. Josef Erlichmann and Dr. Theodore Bunch of NASA/Ames Research Center, Moffett Field, CA for analyses obtained using a Cameca Electron Microprobe; and Dr. Charles Fiori of the National Institutes of Health, Washington, DC for providing the computer program used to generate Monte Carlo electron trajectory simulations. Support of the author through the National Research Council Associateship program is gratefully acknowledged.

6. References

[1] Goldstein, J.I.; Newbury, D.E.; Echlin, P.; Joy D.C.; Fiori, C.; Lifshin, E., "Scanning Electron Microscopy and X-ray Microanalysis." Plenum Press, New York and London, 1981.
[2] Newbury, D.E.; Joy, D.C.; Echlin, P.; Fiori, C.E.; Goldstein, J.I., "Advanced Scanning Electron Microscopy and X-ray Microanalysis." Plenum Press, New York and London, 1985.
[3] Lloyd, G.E., In "Short Course in Applications of Electron Micros-

copy in the Earth Sciences", J.C. White, ed., Mineralogical Association of Canada, Toronto, Canada, 1985, Chapter 3.

[4] Gabriel, B.L., "SEM: A User's Manual for Materials Science." American Society for Metals, Metals Park, OH, 1985.

[5] Hall, C.E., " Introduction to Electron Microscopy." McGraw Hill, New York, 1953.

[6] Crewe, A.V. SEM, Inc., *1969*, I, 31.

[7] Statham, P.J.; Pawley, J.B., SEM, Inc., *1978*, I, 469.

[8] Small, J.A.; Heinrich, K.F.J.; Fiori, C.E.; Myklebust, R.L.; Newbury, D.E.; Dilmore, M.F., SEM, Inc, *1978*, I, 445.

[9] Small, J.A.; Heinrich, K.F.J.; Newbury, D.E.; Myklebust, R.L., SEM, Inc. *1979*, II, 807.

[10] Glang, R.; Gregor, L.V., In "Handbook of Thin Film Technology," McGraw Hill, Inc., New York, 1983, Chapter 7.

[11] Echlin, P., SEM, Inc., *1981*, I, 109.

[12] Echlin, P.; Taylor, S.E., J. Microsc., *1986*, 141(2), 329.

[13] Echlin, P.; Chapman, B.; Stoter, L.; Gee, W.; Burgess, A., SEM, Inc., *1982*, I, 29.

[14] Kemmenoe, B.H.; Bullock, G.R., J. Microsc., *1983*, 132(2), 153.

[15] Peters, K.R., SEM, Inc., *1985*, IV, 1519.

[16] Peters, K.R., J. Microsc., *1984*, 133, 17.

[17] Hall, M.G.; Lloyd, G.E., Am. Mineral., *1981*, 66, 362.

[18] Joy, D.C., Proc. Ann. Mtg., EMSA, *1988*, 46, 200.

[19] Bertin, E.P., "Introduction to X-ray Spectrometric Analysis," Plenum Press, New York, 1978.

[20] Heinrich, K.F.J., "Electron Probe Microanalysis," Van Nostrand, New York, 1981.

[21] Bence, A.E.; Albee, A., J. Geol., *1968*, 76, 382.

[22] Armstrong, J.T.; Buseck, P.R., Theoret. Anal. Chem., *1975*, 47, 2178.

[23] Armstrong, J.T., SEM, Inc., 1978, I, 455.

[24] Brownlee, D.E.; Tomandl, D.A; Hodge, P.W.; Hörz, F., Nature, *1974*, 252, 667.

[25] Brownlee, D.E., Rev. Geophys. Space Phys, *1979*, 17, 1735.

[26] Mackinnon, I.D.R.; Rietmeijer, F.J.M., Rev. Geophys., *1987*, 25, 1527.

[27] Fraundorf, P., Geochim. Cosmochim. Acta, *1981*, 45, 915.

[28] Zinner, E.; McKeegan, K.D.; Walker, R.M., Nature, *1983*, 305, 119.

[29] McKeegan, K.D.; Walker, R.M.; Zinner, E., Geochim. Cosmochim. Acta, *1985*, 49, 1971.

[30] Mackinnon, I.D.R.; McKay, D.S.; Nace, G.A.; Isaacs, A.M., J. Geophys. Res., Suppl., *1982*, 87, A413.

[31] Mackinnon, I.D.R.; Rietmeijer, F.J.M.; McKay, D.S.; Zolensky,

M.E., Microbeam Anal. - 1985, *1985*, 20, 291.

[32] Blandford, G.E.; Thomas, K.L.; McKay, D.S., Meteoritics, *1988*, 23, 113.

[33] Blake, D.F.; Fleming, R.H.; Bunch, T.E., Lun. Planet. Sci., XX, *1989*, 84.

[34] Roeder, E., Econ. Geol., *1962*, 57, 1045.

[35] Roeder, E., Econ. Geol., *1963*, 58, 167.

[36] Kelly, W.C.; Burgio, P.A., Econ. Geol., *1983*, 78, 1262.

[37] Fournier, R.O.; Truesdell, A.H., Geochem. Cosmochim. Acta, *1975*, 37, 1255.

3. X-RAY-INDUCED PHOTOELECTRON AND AUGER SPECTROSCOPY

Dale L. Perry
Lawrence Berkeley Laboratory
University of California
Berkeley, CA 94720

J. Ashley Taylor
AT&T Bell Laboratories
Allentown, PA 18103

Charles D. Wagner
Surfex Company
Oakland, CA 94618

1. Introduction

X-ray photoelectron spectroscopy [XPS, also known by the acronym ESCA (electron spectroscopy for chemical analysis)] is perhaps the most widely known of the modern ultrahigh-vacuum-related surface techniques. Stemming from work reported by Einstein [1] in 1905, the technique takes advantage of the concept that electrons emitted from surfaces being irradiated with soft x-rays can have different kinetic energies as a function of the chemical state of the atom from which they are emitted. The researcher can thus address a wide variety of chemical topics concerning the surface of a geologic material that has undergone a chemical reaction. Information related to surface species and reactions, i.e., oxidation/reduction reactions, chemisorption, hydrolysis, and surface dissolution, can be obtained in considerable detail. The paramount advantage of x-ray photoelectron spectroscopy (along with the other experimental techniques in this book) is the ability to gather such surface information as it relates to geologic material interface reactions with gases, liquids, and the products that are formed.

Auger electron spectroscopy (AES) also has its theoretical and experimental roots in the first part of the twentieth century. In 1923, Pierre Auger, while conducting experimental research on the photoelectric effect, reported the existence of the secondary, or Auger, electron in conjunction with the photoelectron while irradiating noble gases with x-rays [2]. The first published account of electron-induced Auger spectroscopy was that of Lander in 1953 [3], and it is by this experimental approach that the bulk of Auger spectra using modern instruments has been obtained.

Several decades elapsed between the experimental proof of concept and the development of readily available, commercial instrumentation for performing research in x-ray photoelectron and Auger spectroscopy. Siegbahn and his co-workers [4] initiated research in the 1950s that addressed the development of high-resolution spectrometers that made possible the high-precision measurements of kinetic energies of electrons emitted from surfaces. The early research by this group focused on bonding in solids, making use of the rather wide range of binding energies of the core electrons. Although the original instrumentation effectively sampled a rather large area of the surface of a material (several square millimeters), newer, more recent instruments have the capability of performing "small spot" (approaching several square microns) surface analyses.

While the original Auger research was performed using x-rays to obtain spectra, the technique did not come into its own as a powerful surface analytical technique until the 1960s. The use of electron-induced

Auger spectroscopy for surface studies involving deposited cathode-evaporated carbon was reported by Harris [5]. In 1968 Weber and Peria [6] introduced the use of low-energy electron diffraction instruments as an approach for recording Auger spectra. Finally, Palmberg et al. [7] introduced the use of the cylindrical mirror analyzer to obtain the rapid acquisition of Auger spectra coupled with high sensitivity. More recent studies of x-ray-induced Auger spectra (observed during the recording of x-ray photoelectron spectra of samples) have shown that much information can be obtained from surfaces using this combined approach. Advances continue to be made in instrumentation in both x-ray photoelectron and Auger spectroscopy, while the growth in applications of the techniques appears to continue unabated. This is especially true in the areas of analytical and surface chemistry of geologic materials.

The present work is written to provide an introduction to combined x-ray photoelectron and Auger spectroscopy as it relates to geologic materials. In addition to the basic principles of the two techniques, numerous applications involving surface chemical reactions and surface morphology changes are discussed. Interface reactions involving such geologic species as silicates, aluminosilicates, and sulfides are subjects of several previous studies; these materials have been studied from both basic natural alteration (or weathering) processes of the materials under geologic conditions and the addition of chemical species through gaseous or liquid phases. The result of this treatise should be a better understanding by the reader of existing applications of the techniques and an awareness of new applications of which they have not been previously aware. However, it is important that the reader consult more detailed and extensive treatises on x-ray photoelectron [8–15] and Auger [12–16] spectroscopy. He should also refer to various annual reviews that compile the most recent advances in both fields; one of the best ongoing series of annual reviews is that found in *Analytical Chemistry* [17–31].

2. X-Ray Photoelectron and Auger Spectroscopy

2.1. Basic Principles

The principles of conducting the x-ray photoelectron experiment are quite straightforward. A source of soft x-rays (usually either Mg Kα or Al Kα at 1253.6 and 1486.6 eV, respectively) is directed onto the surface of a sample housed in a vacuum chamber of an x-ray photoelectron spectrometer, with the pressure in the chamber being typically in the range of 10^{-8} to 10^{-11} Torr. The ensuing ejected photoelectron (Fig 3-1) is then

Fig. 3-1. Schematic representation for the x-ray photoelectron and Auger electron formation processes.

collected by an analyzer which measures the kinetic energy of that electron. The kinetic energy is translated into a binding energy for the specific atomic orbital of that electron according to Eq. (3-1):

$$E_k = h\upsilon - E_b - \phi_s \qquad (3-1)$$

where E_k is the kinetic energy of the electron, $h\upsilon$ is photon energy supplied by the source anode, E_b is the binding energy of the corresponding elemental orbital, and ϕ_s is the work function [32] of the spectrometer. The resulting spectrum mirrors the top 10–50 Å of the surface (the escape depth of the Auger and photoelectrons which comprise the spectrum) and appears as a collection of peaks as shown in Fig. 3-2 [33]; the 0–1000 eV range covered is referred to as a ''survey'' spectrum and includes the x-ray photoelectron lines associated with the chemical elements present on the sample surface. The sensitivity for detecting elements lies in the range of 0.1–0.5% atomic concentration for both x-ray photoelectron and Auger spectroscopy. In addition to the *qualitative* chemical information obtained about the chemical surface states, *quantitative* information about the surface can be derived also. This can be achieved, however, only with the use of rigorous, precise standards for comparison, and the best absolute value obtained still has an error of ± 1–5%. Because of sample roughness and differences in thickness of contaminant layers, absolute quantitative measurements (such as those typically made with the electron microprobe) are rarely used. Relative atomic concentrations are made extensively by measuring the line intensities for different atoms and dividing them by generalized sensitivity factors or by a set which the investigator has empirically determined for himself [34, 35]. The set of corrected intensities is then normalized to 100%. This gives essentially the atomic

Fig. 3-2. The x-ray photoelectron survey spectrum of galena, PbS, reacted with aqueous sodium dichromate, $Na_2Cr_2O_7$.

ratio of any two elements in the sample. By doing so, the investigator can get a true picture of elemental concentration changes from one reacted surface to another involving the same elements.

In addition to the ejection of the photoelectron shown in Fig. 3-1, there is the emission of the Auger electron. Formed by the relaxation of the energetic ion left after the emission of the photoelectron, the process involves an outer electron falling into the "hole" created by the lost photoelectron. A secondary, or Auger, electron is then emitted which exhibits a kinetic energy that is the difference between the initial ion and the doubly charged final ion. In general, the Auger lines observed in a combined XPS/Auger spectrum are more electronically (and thus geometrically) complex than the photoelectron lines.

There are four principal series of Auger lines that appear along with x-ray photoelectron spectra, those being the KLL, LMM, MNN, and NOO lines. The first letter designates the hole level of the ejected photoelectron, while the last two letters designate the level of the double vacancies created by the Auger process. Thus, the Auger process in Fig. 3-1 represents that of the KLL series of the oxygen atom. Table 3-1 lists a group of both x-ray photoelectron and Auger lines that are available for studies of the major elements in a variety of mineral systems. While not intended to be exhaustive, the list does contain a rather wide array of possible lines for study concerning those minerals.

2.2. Sample Considerations

In addition to the instrumental requirements for performing x-ray photoelectron and Auger experiments, there are also practical considerations as to what samples can be studied. The solid samples must be vacuum amenable and not be susceptible to decomposition in the 10^{-8} to

Table 3-1. Selected Suitable Lines for Combined X-Ray Photoelectron/Auger Studies of Selected Mineral Systems

Element	Photoelectron Line	Auger Line	Mineral System
Aluminum[a]	1s	$KL_{23}L_{23}$	Aluminates, aluminosilicates
Silicon[a]	1s	$KL_{23}L_{23}$	Silicates, aluminosilicates
Oxygen[b]	1s	KVV	Oxides
Sulfur[c]	1s	KVV	Sulfides
Iron[b]	$2p_{3/2}$	L_3VV	Iron oxides, hydroxides, sulfides
Selenium[b]	$3d_{5/2}$	$L_3M_{45}M_{45}$	Selenium oxides and selenium oxyanions
Phosphorus[c]	1s	$KL_{23}L_{23}$	Phosphates

[a]Au Mα (2122.9 eV) source.

[b]Al Kα (1486.6 eV) source.

[c]Ag Lα (2984.3 eV) source.

10^{-11} Torr range. The samples can be in the form of both powders and continuous solids such as wafers, chips, and thin films. In the case where an unreacted sample is being used as a standard, it should be as fresh a specimen as possible and free from contamination. Reacted samples that are to be studied should be introduced into the spectrometer without disturbing or altering the reacted surface, since this would produce a spectrum which would not be representative of the true, chemical state in the spectrometer chamber, while powders have to be mounted on ultrahigh vacuum-amenable tape, pressed between indium foil, or pressed into pellets. All of these techniques have well-documented advantages and disadvantages, and the reader is encouraged to consult standard references that discuss the preparation and manipulation of samples in detail [33].

While this precludes studying samples which are actually wet at room temperature, it does not dictate a *pro forma* exclusion of samples that have been reacted with solvents such as water. Alterations of a surface that has been in the presence of (and thus contains molecules of) liquid reagents can indeed occur; however, if they are chemically bonded to the surface, their chemical state will be reflected in the spectrum. Indeed, thousands of studies involving normally liquid molecules reacting with solid materials exist in the surface chemistry literature. The sample can also be studied at a cooled temperature to minimize change in the sample surface. Conversely, however, the investigator must take the same precautions in studying these solvent/reaction systems as he does in studying any other ones. The most effective way of monitoring any surface changes in the course of an experiment is to conduct the experiment *as a function of time*. If the spectra remain unchanged and the samples remain visibly constant (i.e., no discoloration occurs) during the course of experiments on a sample, the sample can be assumed to be stable. This is the same caveat that must be observed for studying all chemical systems by any type of spectroscopy.

There are a number of other well-documented problems involving surface alterations during the x-ray photoelectron and Auger experiment in addition to dehydration that can confront the investigator, one of the chief ones being that of charging. This problem can be a pronounced one, especially in cases where the sample being studied is an insulator. Recalling that the basic, initial process involved in the x-ray photoelectron experiment is the ejection of a negatively charged electron, one realizes that this leaves a positively charged sample surface. If the sample is not able to achieve electrical equilibrium by an electron flow from ground or by electron acquisition from the vacuum space, a positive charge accumulates on the surface which cannot be dispersed. This effect manifests itself in both subtle and gross ways. In cases where the charging is rather weak or intermediate, increases in the binding energy and photoelectron linewidth are observed. Severe levels of charging, however, may result in the appearance of multiple photoelectron lines, and many such cases have been reported in the literature. This occurs especially with heterogeneous samples such as conductors embedded in insulators. These effects can be minimized by the use of a flood gun which provides a source of compensating electrons to the surface [33, 36].

There are several methods for treating or correcting peak shifts due to charging. The most common one is the use of the so-called "adventitious" carbon 1s line which is introduced to the sample surface both by atmospheric handling and by the spectrometer system itself. This line has been experimentally determined to occur at 284.8 eV, so its shift in a spectrum can be assumed to result from charging. This shift is then used to adjust the binding energy values for the other lines in the spectrum. Other often used techniques include the use of an internal standard element (such as the silicon 2p line in aluminosilicates) for appropriate samples and the use of gold decoration. In the latter technique, a very thin film (or trace) of gold is evaporated onto the sample after it has been studied; the ensuing spectrum is then calibrated against the binding energies of the intense gold $4f_{7/2,5/2}$ lines.

One potentially serious experimental problem for studying some chemical surface systems is that of metal ion reduction. Metal ions which are susceptible to x-ray beam reduction during the course of XPS/Auger studies are ones that are in a rather high oxidation state initially and may form stable, lower oxidation states. Hexavalent uranium, for example, can undergo photoreduction to the uranium(IV) species, which in turn can be reduced still further to the metal itself. One of the most thoroughly studied metal-ion-reduced systems is that of the copper(II) ion. This reaction has been reviewed [37], and it seems to be the most severe for copper(II) oxide and copper(II) halides. Several mechanisms for the reduction of the former compound have been published [38–40].

Another problem sometimes encountered is that of sputter-induced metal ion reduction. In an attempt to clean or depth profile a surface, an investigator may use argon ion sputtering, or bombardment, for this purpose. The result may be the reduction of a high oxidation state metal ion to a lower oxidation state. Again, the sputter reduction of copper(II) has been studied by Yamada et al. [41]. Brundle and his co-workers [42] have demonstrated that both Fe_3O_4 and Fe_2O_3 can be reduced by sputtering.

One way of circumventing the problem is to vary the angle between the plane of the sample surface and the analyzer entrance angle [33]. Thus the investigator can obtain a limited depth profile of a material without altering the surface.

By inspection of the survey spectrum shown in Fig. 3-2, one can see that differences exist among the various x-ray photoelectron lines with respect to their intensity and shape. Additionally, the Auger lines are found along with the photoelectron lines, and it is this ability to generate both x-ray photoelectron and Auger data in the same spectrum that makes the Auger parameter discussed below such a useful tool. A detailed, high-resolution study of these lines, along with their associated satellites and fine structure, yields a wealth of information about the chemistry and bonding of geologic surfaces. The following discussion of several of the aspects of the spectral parameters will give a clearer picture of their application in understanding surface properties and surface phenomena.

2.3. Spectral Parameters

2.3.1. Binding Energy and Auger Shifts. The numerical positions of the x-ray photoelectron lines for the different elemental orbitals shown in Fig. 3-2 are referred to as the binding energies; the kinetic energies of the Auger lines (or, more correctly, the collection of lines comprising the Auger peak) are also presented in terms of a binding energy. The binding energy is easily the most widely quoted experimental value in x-ray photoelectron spectroscopy. The core level binding energies of elements can many times be used to differentiate between several different oxidation (and thus chemical species) states of an element. These binding energy changes as a function of oxidation state can vary by a fraction of an electron volt or several electron volts over the entire range of oxidation states for an element. In the case of tin, for example, there is a binding energy difference of less than 2 eV between zerovalent, elemental tin and the tin(IV) state; for purposes of differentiating between tin(II) and tin(IV), the separation is sometimes almost negligible, depending on the species of the two oxidation states being studied. A broad range of binding energies, however, is exhibited by sulfur, ranging from ≈ 162 eV for the sulfide (S^{II-}) to ≈ 168 eV for the $S^{VI}O_4^{2-}$ anion.

Auger shifts, however, will often exceed photoelectron binding energy shifts for elements. One study of Auger and photoelectron lines of the oxides of a series of elements revealed that the shift of the Auger line could be from 2 to 10 times that of the shift of the photoelectron binding energy [43]. Two conditions must be met if this large ratio of Auger shift to photoelectron shift is observed. First, the element is a conductor, and second, the initial vacancy in the Auger process must be effected in the inner shell. Several elements such as sodium, cadmium, silver, zinc, magnesium, and copper display quite large Auger shifts compared to their relatively small photoelectron binding energy shifts. More detailed treatises on the relationship between the two parameters can be obtained from the literature [44, 45].

2.3.2. Spin-Orbit Splitting. During the ionization of a p, d, or f orbital in the photoelectron emission process, one of the direct results is the formation of two ionic states. These two states are represented by a so-called spin-orbit doublet in the x-ray photoelectron spectrum. For the p orbital, the two lines are designated the $p_{3/2,1/2}$ doublet, for the d lines, $d_{5/2,3/2}$, and for the f lines, $f_{7/2,5/2}$. The two major lines shown in the spectrum in Fig. 3-3, for example, represent the $2p_{3/2,1/2}$ spin-orbit doublet for chromium(III), with the values for the chromium $2p_{3/2}$ and chromium $2p_{1/2}$ being 577.0 and 586.4 eV, respectively. The spin-orbit splitting is merely the difference between these lines, or 9.4 eV. Many times, the spin-orbit splitting is coupled to the oxidation state of the metal ion and/or the electronic "spin state" of the ion. This is the case for cobalt, where the spin-orbit splitting difference can be used to differentiate between the paramagnetic "high-spin" cobalt(II) ion and the diamagnetic cobalt(III) "low-spin" species [46]. The reader is referred to other treatises [8, 9, 47] for a more comprehensive discussion on spin-orbit splitting of elements.

2.3.3. Multiplet Splitting. The process of emitting a core level photoelectron, such as from the chromium 2p level shown in Fig. 3-3, is not an isolated event. Rather, when the event is coupled by interaction with one or more valence shell electrons, a second phenomenon, that of multiplet splitting, is also observed. The concept of multiplet splitting [48] can be quite useful in studying paramagnetic metal ions, since the electronically unpaired "hole" created in the core level is interacting with unpaired electrons in the valence shell (in the case of chromium, the 3d electrons). The 3d metal ions that are paramagnetic, all of which are members of the first-row transition metal ions, exhibit broadening of their 2p photoelectron spectra. While not readily obvious due to overlap of the two doublets, the $2p_{3/2,1/2}$ lines attributable to the chromium(VI) species in Fig. 3-3 are narrower than the lines attributable to the chromium(III) species. The

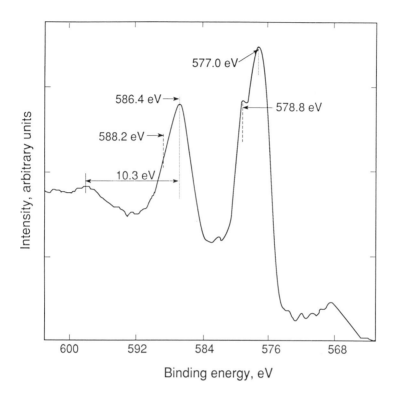

Fig. 3-3. The high-resolution chromium $2p_{3/2,1/2}$ x-ray photoelectron spectrum of galena, PbS, reacted with aqueous sodium dichromate, $Na_2Cr_2O_7$ (adapted from Ref. [103]).

reason for this is that the chromium(III) compound on the surface is an unpaired $3d^3$ electronic system and is thus exhibiting multiplet splitting; this paramagnetic configuration is in opposition to that of the chromium(VI) species, a $3d^\circ$ diamagnetic configuration. A second manifestation of multiplet splitting in paramagnetic 3d ions is an actual splitting of the 3s photoelectron level of the ions. Thus, the investigator can observe both different binding energies and different separations in the doublet peaks in this level [49] for various compounds. This can be used in conjunction with model compounds to identify a particular surface species or, conversely, to rule out a suspected species.

2.3.4. "Shakeup" and "Shakeoff" Satellites. Two more types of configurational interactions, similar in nature to that of multiplet splitting, are those of "shakeup" and "shakeoff." Under normal conditions, an x-ray photoelectron possesses a relatively large kinetic energy and is most often removed from its core orbital without causing additional excitation of other electrons. Some electronic relaxation always occurs that affects the kinetic energy of the photoelectron, but if additional (often less probable) interactions occur between the photoelectron and the valence electrons during the ionization event, the photoelectron will lose some of its energy, and additional peaks will be observed in the spectrum. These peaks will occur as satellite structure located to the high-binding-energy side of the main peak. Two possibilities are often discussed. First, another electron can be promoted to an excited state; this process is called "shakeup." If, however, that electron is promoted to a continuum state, the process is referred to as "shakeoff." The various energy-loss processes can be specifically identified for simple gases, but for solids this structure often exists as one or more broad bands.

The position and shape of these satellite structures can often be used as a fingerprint to identify the chemical species on the surface without having to assign the energy-loss mechanism. The most intense satellite structure is observed for the transition metals [50, 51] and rare earth compounds, which possess unpaired electrons and are therefore paramagnetic species. Thus, high-spin cobalt(II) (d^7), high-spin iron(III) (d^5), high-spin iron(II) (d^6), nickel(II) (d^8), chromium(III) (d^3), and copper(II) (d^9) complexes give very strong satellites with resolvable fine structures in most cases. In several of these metal ion systems, differentiation between different complexes of the same metal ion can be effected on the basis of the difference in their satellite structure. A good review of the rules describing the variation in the satellite structure as a function of the various electronic types of metal ions and their satellite patterns can be found in the literature [52].

2.3.5. Plasmon-Loss Lines. For solids another energy-loss mechanism, the plasmon loss, must also be considered. The observed satellite structure is then a combination of both intrinsic and extrinsic losses. The intrinsic losses would include the "shakeup" or any other relaxation mechanism similar to those described in Section 2.3.4. These are associated with the ionization event itself. However, an electron passing through a solid can interact with the valence or conduction electrons of the solid giving up some of its energy in discrete amounts, ΔE_L. This is then the extrinsic or plasmon-loss contribution. The energy-loss shift ΔE_L is measured simply by taking the difference in binding energy between the main photoelectron lines (Al 2s for the example shown in Fig. 3-4) and the first energy-loss peak. For the plasmon loss, in the simplest model, the photoelectron interacts with the free electron gas in a metal,

$$\Delta E_L = \frac{h}{2\pi} \, (4\pi e^2 n/m)^{1/2} = \frac{h}{2\pi} \omega_p \tag{3-2}$$

where e and m are the charge and mass of the electron, n is the density for the electrons, and *h* is Planck's constant. The loss is said to be that of a bulk plasmon if the excitation occurs within the solid. For some metals such as aluminum, a surface plasmon loss [53, 54] equal to $\Delta E_L / \sqrt{2}$ also is observed. The plasmon loss is proportional to the electron density, which is specific to each individual metal, so that E_L can be used to identify dif-

Fig. 3-4. High-resolution x-ray photoelectron spectra of the aluminum 2s region of a clean, evaporated aluminum film (lower) and the same film after exposure to oxygen at atmospheric pressure for 30 min (adapted from Ref. [57]).

ferent metals. It has also been shown that ΔE_L is independent of sample charging, and it has been argued that the plasmon-loss effect is independent of relaxation, or it represents changes in the initial electronic state [55, 56]. The intrinsic losses are due to relaxation during the photoionization process and thus represent changes in the final states. Plasmon interactions also occur for wide band-gap materials with the photoelectron interacting with the valance bands if they are highly delocalized and collectively shared. As the band gap increases, forming good insulators, it becomes more difficult to separate the intrinsic and extrinsic effects [55]. For many materials, the plasmon losses dominate the spectrum, and this is most likely true for the aluminum and silicon compounds. For compounds of the transition metals, as previously noted, the intrinsic loss can be very strong, obscuring the plasmon losses. The energy-loss shift has been shown to change significantly for many different compounds [55], and if the loss structure is comprised mainly of plasmon losses, changes in ΔE_L will indicate changes in the initial electron densities for the material.

Figure 3-4 illustrates a good example of using plasmon-loss lines to follow the oxidation of a clean aluminum surface to form an overlayer of alumina, Al_2O_3 [57]. In the bottom spectrum, one sees an evaporated film of aluminum metal with no oxidation; the bulk plasmon losses are separated by 15.2 eV, while the surface plasmon losses are separated by 10.7 eV. Upon exposure to oxygen at atmospheric pressure for 30 min, however, an overlayer of Al_2O_3 forms. As would be expected, surface plasmon losses attributed to the elemental, unoxidized aluminum are absent. Completely oxidized aluminum has a ΔE_L of about 24 eV.

3. Auger Parameters

The experimental binding energies are derived from Eq. (3-1) using the values of $h\upsilon$, ϕ_s, and the measured kinetic energy E_k. The measured binding energies thus obtained are equal to the difference in total energy of the final state ion and the initial state of the compound.

$$E_b = E_f - E_i \qquad (3\text{-}3)$$

It is emphasized that the difference in the measured binding energies has both changes in the initial state, due to chemical changes, and changes in the final ionized state, due to rearrangement of orbitals as a consequence of the ionization process. These two contributions cannot be separated by just measuring shifts in the core binding energies. Often, for similar compounds, the final state changes are similar, and one sees a predicted shift in binding energies with a change in chemistry; however, this is not observed for all compounds.

Another way of defining the binding energy is to set it equal to the orbital energy of an electron occupying the initial or un-ionized energy level minus a relaxation energy, R.

$$E_b(i) = \varepsilon(i) - R \tag{3-4}$$

The relaxation energy [58] can be loosely divided into two parts: intraatomic and extraatomic. Intraatomic relaxation includes the rearrangement of any orbital due to the ionization event, occurring for any atom, isolated or not, and is invariant to any changes in the environment surrounding the atoms. Extraatomic relaxation arises from the redistribution of electrons in neighboring atoms or in the conduction band of a metal. Extraatomic relaxation is also known as the polarization energy or screening energy, since the electrons rearrange themselves to screen the hole produced in the core orbital to obtain a minimum energy configuration.

Such a distinction may seem arbitrary but, as will be shown, there is a way to measure the extraatomic relaxation directly. The change in binding energy for various chemical compounds from Eq. (3-4) for the photoelectrons is

$$\Delta E_b(i) = \Delta\varepsilon(i) - \Delta R^{ea}(i) \tag{3-5}$$

The relaxation energy now is essentially extraatomic, since the intraatomic part will be about the same for different compounds containing the same atom. For the shift in the Auger line, three energy levels (ijk) must be considered.

$$E_k(ijk) = E_b(i) - E_b(j) - E_b(k) - e(jk) + R(jk) \tag{3-6}$$

The term $e(jk)$ is the interaction energy of the two holes in the final state. R, then, is the total relaxation energy with both intra- and extraatomic contributions. The change in the Auger kinetic energy from chemical changes would be

$$\Delta E_k(ijk) = \Delta E_b(i) - \Delta E_b(j) - \Delta E_b(k) + \Delta R^{ea}(jk) \tag{3-7}$$

For KLL Auger transitions, several approximations can be made.

$$\Delta R^{ea}(LL) \approx 2\Delta R^{ea}(K) \tag{3-8}$$

$$\Delta E_b(K) \approx \Delta E_b(L) \tag{3-9}$$

Equations (3-6), (3-7), and (3-8) can then be combined to give

$$\Delta E_k(KLL) \approx -\Delta\varepsilon(K) + 3\Delta R^{ea}(K) \tag{3-10}$$

By comparing Eqs. (3-5) and (3-10), one notes that in general the chemical shift for Auger lines will not be the same as for the photoelectron lines because of the significant difference in ΔR^{ea} associated with the two-hole Auger final state as compared to ΔR^{ea} for direct photoionization. In fact, the chemical shift for the Auger lines is often much larger.

If both the binding energy of the photoelectron and the kinetic energy of the Auger electron are measured, a new parameter, known as the Auger parameter (α), can be determined. Wagner originally defined the Auger parameter (α_w) [59] as the difference between the kinetic energy of the most intense Auger line, $[E_k(ijk)]$, and the most intense photoelectron line, $E_k(i)$.

$$\alpha_w = E_k(ijk) - E_k(i) \tag{3-11}$$

E_k, the kinetic energy, then equals

$$E_k(i) = h\upsilon - E_b(i) \tag{3-12}$$

In this case the reference is assumed to be made from the Fermi level rather than the vacuum level (Eq. 3-1), and $E_k(Fermi) = E_k(vacuum) + \phi_s$. The definition of the modified Auger parameter [60], as currently used in most papers, is obtained by combining Eqs. (3-11) and (3-12).

$$\alpha = \alpha_w + h\upsilon = E_k(ijk) + E_b(i) \tag{3-13}$$

$$\alpha + h\upsilon = E_k(Auger) + E_b(photoelectron) \tag{3-14}$$

The modified parameter as so defined is then independent of $h\upsilon$ and is always positive. As shown in Fig. 3-5, the actual measurement of the Auger parameter is simply the difference in the kinetic energy between two lines. This difference will remain the same independent of sample charging because any charging shifts will cancel. This is important for geologic materials, since many are insulators and thus exhibit charging. For the same reason, the Fermi level does not need to be precisely determined; data referenced to the vacuum level can also be directly compared. Often binding energy data and Auger kinetic energy data are combined, creating a two-dimensional plot as shown in Figs. 3-6 and 3-7. Of course the binding energy data must be referenced to the Fermi level and

Fig. 3-5. X-ray photoelectron spectrum of sodium hydrogen phosphate, Na_2HPO_4, showing the Auger parameter. (Adapted from Ref. [61].)

corrected for charging. Changes in the binding energies of photoelectron lines are taken from the x axis, and the changes in the kinetic energies of the Auger lines are determined from the y axis. As previously described, they are usually not the same. The plot then defines the Auger parameter (Eq. 3-14) as a series of diagonal lines. Any points lying on the diagonal would represent equal Auger parameters. By constructing such two-dimensional plots, much more chemical information can be extracted from the x-ray photoelectron spectra.

The Auger parameter of many different compounds has been measured and tabulated, showing that the parameter changes with the chemical environment [61, 62]. For KLL Auger transitions, Eqs. (3-5) and (3-10) can be combined, giving

$$\Delta\alpha = 2\Delta R^{ea}(i) \qquad (3\text{-}15)$$

This states that the change in the Auger parameter in the first approximation [63] is a direct measure of the change in the extraatomic relaxation.

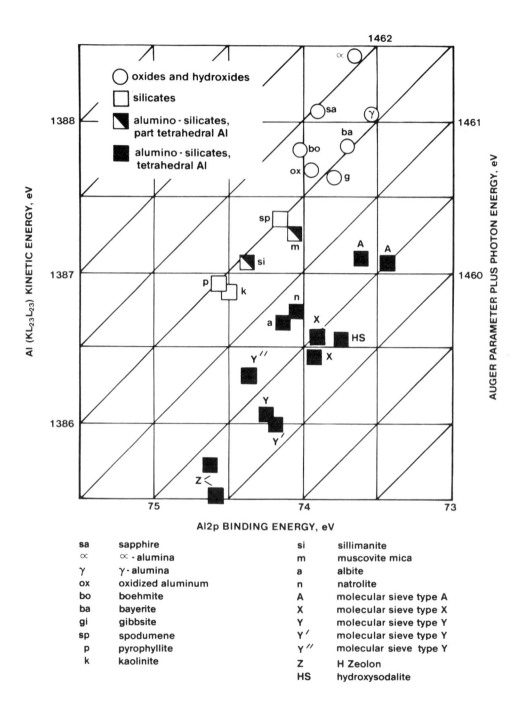

Fig. 3-6. Aluminum chemical state plot for aluminum/oxygen compounds. (Reprinted
from Ref. [67], with permission.)

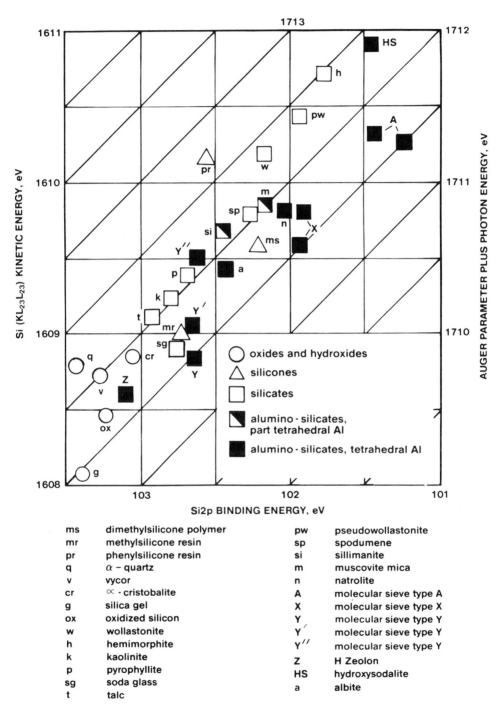

Fig. 3-7. Silicon chemical state plot for silicon/oxygen compounds. (Reprinted from Ref. [67], with permission.)

A fair question to ask is whether the extraatomic relaxation has any physical meaning. A number of studies have indicated that it is directly related to the polarizability of the material. The polarizability arises from various electronic interactions with the hole produced by the ionization, and these interactions slightly affect the kinetic energy of the escaping electron [64]. For good dielectrics a slight displacement of the negatively charged electron cloud of neighboring atoms occurs in response to the production of the positive hole. For ionic crystals, anion and cation interactions dominate [65]. If the material possesses a permanent electric dipole, this would be another source of interaction. For conductors, charge is no longer bound and thus is free to move within the conduction band to more completely screen the hole.

All of these interactions are of course additive, but often for a specific class of materials, one will dominate [65]. For example consider various sodium compounds [61]. A gaseous sodium atom possesses the smallest Auger parameter because the extraatomic relaxation energy would be zero. Metallic sodium, a good conductor, has the highest Auger parameter, because the polarization energy or extranuclear relaxation term is large. The difference between the Auger parameter of the gaseous atom and any other compound yields the extraatomic relaxation or polarization energy. The Auger parameter for the rest of the sodium ionic salts lies between these values, with NaI greater than most of the salts and NaF less than most of the other salts. Studies have also been conducted on frozen aqueous solutions of various sodium salts [66]. Results showed only small changes in the Auger parameters for all of the sodium salts in solution; a much greater change was observed for the solid ionic salts. For the frozen aqueous solutions, the Na^+ is surrounded by a similar environment of water molecules for each case. In general, the greater the polarizability of the compound, the greater the Auger parameter will be.

The application of the Auger parameter using x-ray-induced photoelectron and Auger lines in the same spectrum can be effected by considering what photoelectron and Auger lines are present for the elements of the material to be studied. Figure 3-5, for example, shows the lines used to obtain the Auger parameter for sodium hydrogen phosphate, Na_2HPO_4. Even in a pure compound, however, this can sometimes be difficult because of the close proximity of some of the lines to one another. In the case of uranium oxides, a detailed study of the uranium $4d_{5/2,3/2}$ photoelectron lines is difficult because of the complex oxygen KLL Auger peaks which occur in the same region. This also makes the study of related materials such as uranium aluminosilicates and other uranium/oxygen systems quite complex. The same situation exists for chromium and vanadium systems of oxygen, where the oxygen KLL Auger peaks overlap the chromium/vanadium LMM peaks. For cases such as these, the investigator must rely not only on the positions of photoelectron and Auger lines but also a quite detailed and careful lineshape analysis of all of the various components.

While these problems do hinder the use of the Auger parameter concept to geologic studies in several cases, there are indeed many applications for which it is a powerful tool. In the following discussion, its use is detailed in the study of such areas as bonding in geologic materials, sorption studies, and surface chemical modifications as they relate to such areas as mineral processing and flotation. Table 3-2 contains a representative sample of different applications of x-ray photoelectron and combined x-ray photoelectron/Auger spectroscopy to geologic materials, in both their pristine and their reacted states. While certainly not intended to be a comprehensive compilation, it presents a fair balance with respect to possible applications.

4. Applications of X-Ray Photoelectron and Combined X-Ray Photoelectron/Auger Spectroscopy

4.1. Structure

The first attempts to use x-ray photoelectrons to study the surfaces of the aluminosilicates and other similar materials proved to be disappointing. Because the binding energy shifts are small and further confused by sample charging, the results were often conflicting and little progress was made in relating different chemical structures to the x-ray photoelectron spectra. Other reasons for this are that the aluminum and silicon 2p lines are most often broad and weak. Differences in the oxygen 1s spectra, showing hydroxides and oxides (often present for metal oxides), are absent for the aluminosilicates. Differences between the binding energies of silicon 2p and oxygen 1s lines are very similar for many of these compounds. Thus, until recently, many investigators were content to examine the Si/Al ratio of an altered surface as compared with those for the bulk compounds. Over the last few years, significant progress has been made in relating the x-ray photoelectron spectra to the chemical structure of these compounds by carefully examining the Auger parameters, the Auger lineshapes, electron energy-loss spectra (plasmon loss), and shape of the

Table 3-2. Applications of X-Ray Photoelectron and Combined
X-Ray Photoelectron/Auger Techniques to Geologic Studies

Type of Study	System Studied	Technique(s)	Reference
Structure	Titanates	XPS	[75]
	Aluminosilicates	XPS	[67]
	Aluminosilicates	XPS	[68]
	Aluminosilicates	XPS	[72]
	Aluminosilicates	XPS	[73]
	Silicates	XPS	[74]
	Zeolites	XPS	[69]
	Micas	XPS	[71]
Analysis	Coal macerals	XPS	[76]
	Sediments	XPS	[79]
	Biological minerals	XPS	[12]
	Kerogens, coal	XPS	[77]
	Sediments, soils	XPS/AES	[78]
	Mt. St. Helens ash	XPS	[108]
Electronic state	Hollandites	XPS	[80]
	Forsterite	XPS	[81]
	Manganese oxides	XPS	[82]
	Lepidolite, manganese oxides	XPS	[83]
	Copper minerals	XPS	[84]
	Copper minerals	XPS/AES	[85]
	Copper minerals	XPS	[86]
	Vanadium aegirines	XPS	[87]
	Iron clay minerals	XPS	[88]
	Garnierite	XPS	[89]
	Lead minerals	XPS/AES	[90]
	Carbonate minerals	XPS	[91]
	Iron oxides	XPS	[92]
	Biological minerals	XPS	[121]
Reaction chemistry	Imines on mica	XPS	[109]
	Metal ions on sulfides	XPS	[93]
	Metal ions on sulfides	XPS/AES	[110]
	Oxidation of sulfides	XPS	[100]
	Organics on pyrite	XPS	[102]
	Reactions of asbestos	XPS	[107]
	Oxidation of manganese(II)	XPS	[111]
	Oxidation of bornite	XPS	[101]
	Cobalt on zeolites	XPS	[112]
	Seawater on basaltic glass	XPS	[105]
	Gold on sulfides	XPS	[94]
	Seawater on basaltic glass	XPS	[106]

Table 3-2. (Continued)

Type of Study	System Studied	Technique(s)	Reference
	Metal ions on clays	XPS/AES	[96]
	Flotation of sulfides	XPS	[113]
	Flotation of sulfides	XPS	[114]
	Hydrogen chloride/water on soils	XPS	[115]
	Chromium complexes on clays	XPS	[98]
	Uranium, thorium on basalt	XPS	[99]
	Flotation of sulfides	XPS	[116]
	Chromium on clays	XPS	[117]
	Mercury on sulfide minerals	XPS	[95]
	Cation exchange on silicates	XPS	[118]
	Copper on sulfide minerals	XPS/AES	[104]
	Cobalt on clays	XPS	[119]
	Chromium on galena	XPS	[103]
	Nickel, copper on clays	XPS	[97]
	Chromium, iron on clays	XPS	[120]
Review	Clay minerals	XPS	[122]
	Clay minerals	XPS	[123]
	Clay minerals and soils	XPS	[124]
	Geochemical processes	XPS	[125]
	Mineral surface chemistry	XPS	[126]

valence bands. This is all in addition to carefully measuring the binding energies.

The bulk structure of the aluminosilicates has been extensively studied by many different techniques. The silicate structure can be classified as chains, layers, and networks. Table 3-3 shows the stoichiometry and formal charge of the various silicate structures. Aluminum can be incorporated into the silicate structure as a cation or as a $[AlO_2]^-$ group tetrahedrally coordinated with the silicate unit. For example, mica, a complex layered structure, contains aluminum bonded both octahedrally as a cation and tetrahedrally with the silicate $Zn[Al_2][Si_3AlO_{10}](OH)_2$. The $[AlO_2]^-$ group can also be bonded to SiO_2 groups in the network systems. This adds a negative charge to the system, typical of the structure for zeolites. If the negative charge is viewed as distributed over the entire group, then the formal charge per silicon atom is proportional to the Al/Si ratio.

Table 3-3. Summary of Silicate Structures

Structure	Formal Charge	O/Si Ratio
Isolated silicates	$[SiO_4]^{4-}$	4.0
Dimeric ions	$[Si_2O_7]^{3-}$	3.5
Chains	$[SiO_3]^{2-}$	3.0
Layers	$[SiO_{2.5}]^{-}$	2.5
Network	$[SiO_2]^{0}$	2.0

Wagner et al. [67] and West and Castle [68] have made useful measurements of the Auger parameter for a number of aluminosilicates and zeolites, producing the Auger parameter plots being shown in Figs. 3-6 and 3-7. Good agreement was obtained for both sets of measurements. The silicon Auger parameters for most of the aluminosilicates are similar, but small measurable differences are found. Recall that the Auger parameter can be precisely measured. However, if the Auger parameter is approximately equal for a group of compounds, then the changes in the observed binding energies reflect changes in the electron densities of the initial states (see Eq. 3-10). By measuring the Auger parameter, such changes in the initial states may be inferred which cannot be made by just measuring the binding energies alone. By taking the silicates as a group, it is observed that the binding energy of the silicon 2p and oxygen 1s shift is about 0.5–0.6 eV for each unit of formal charge. Note that the difference between the silicon 2p and oxygen 1s is basically constant. The binding energy is highest for $[SiO_2]^{0}$ and lowest for $[Si_2O_7]^{-3}$. A similar shift for the zeolites is observed as the negative charge per silicon atom or Al/Si ratio increases. The Auger parameter for the aluminum can be roughly divided into two groups. The oxides and octahedrally bonded aluminum cations consist of one group with a slightly higher Auger parameter than for the second group, the zeolites, with aluminum tetrahedrally bonded with the silicate unit. The binding energy of the aluminum 2p line for the zeolites shifts consistently to higher binding energies as the Al/Si ratio decreases, similar to the shift observed for the silicon 2p line. In contrast the silicon oxides possess higher binding energies than the rest of the silicates, while the aluminum oxides possess lower binding energies than the rest of the other aluminosilicates.

Barr and Lishka [69] also observed similar shifts in binding energies for aluminum and silicon for the zeolites. They, like Wagner et al. [67], emphasize that the aluminosilicate species is better viewed as an entire unit with the formal charge distributed among the various atoms. For these compounds the binding energies exhibit shifts from a cluster of atoms rather than from the separate atoms themselves. As the number of neutral SiO_2 units increases, the negative charge per atom decreases, and the binding energy increases. By taking advantage of such shifts, these authors were able to conclude that especially for the zeolites with Si/Al ratios greater than 4, extensive amounts of alumina and sodium aluminate residues were found on the surface in addition to the zeolite structure. When the sodium was removed, mostly alumina was left on the surface. Obviously, it is quite important to characterize the structure of the surface before one can understand the significance of an adsorption experiment.

Barr et al. [70] have also studied the shape of the various valence bands with x-ray photoelectron spectroscopy for a number of the zeolites. They found that the shapes of valence bands for the zeolites differ significantly from those of alumina, silica, or sodium aluminate for Si/Al ratios of 1 to 3. As the amount of aluminum decreases, the valence bands resemble that of silica with small perturbations. Wagner et al. [67] also have compared the shapes of the KVV Auger transitions for oxygen for these compounds. Since the final ionic states have two vacancies in the valence levels, the shape should reflect changes in chemical structure. The shape can be divided into three groups: the aluminas, silicas, and zeolites (with Si/Al ratios less than 3). Zeolites with large Si/Al ratios are similar to silica. Barr and Lishka [69] also have mentioned that differences in the x-ray photoelectron energy-loss shift ΔE_L exist for the zeolites as compared to alumina and silica.

West and Castle [68] have greatly extended the analysis of the Auger parameter in their study of various aluminosilicates. It has already been stated that changes in the Auger parameter are related to changes in polarizability of the compounds. The index of refraction is also related to the bulk polarizability (P_b) by the Lorentz-Lorentz relationship,

$$\frac{n^2 - 1}{n^2 + 2} = \frac{4}{3} \pi \frac{N}{V} P_b \qquad (3\text{-}16)$$

where n is the index of refraction, N Avagadro's number, V is the molar volume. A plot of $(n^2 - 1)/(n^2 + 2)$ against the measured Auger parameters for silicon and aluminum demonstrated a good correlation. These authors then suggested that most of the polarizability was accounted for by the oxygen atoms. They also assumed that the Auger parameter was directly proportional to the oxygen polarizability P_o.

$$\Delta\alpha = \alpha - \alpha_o = KP_o \tag{3-17}$$

The zero point α_o can be estimated from the y intercept (Auger parameter) or by other means. The K was determined by measuring the Auger parameters for several standards and by using Eq. (3-16) to calculate P_b. They found that K was dependent upon the coordination of the oxygen: tetrahedrally bonded to silicon, tetrahedrally bonded to aluminum, or octahedrally bonded to aluminum. Excellent correlation was obtained between the oxygen polarizabilities calculated using the Lorentz-Lorentz equation and those derived from the Auger parameters. Many other interesting structural relationships can then be inferred once the polarizabilities have been determined.

For single-crystal substrates, photoelectron diffraction has also been used to study changes in the positions of atoms at the surfaces of quite complex minerals. The photoelectron diffraction pattern is obtained by measuring the intensity of a photoelectron line as the angle of the substrate is rotated. The change in intensity as a function of substrate angle can result from a number of different factors. Sample roughness and other instrumental changes can be factored out if the ratio of two lines of approximately the same binding energy is used rather than the absolute values. In this case the changes in the angular distributions will often be dominated by a diffraction effect. That is, a beam of electrons can act as a wave as it interacts with matter; the associated wavelength (λ, in nm) is proportional to its kinetic energy (E_k, in meV)

$$\lambda = \left[\frac{1.5}{E_k} \right]^{1/2} \tag{3-18}$$

If the electron waves are scattered by an ordered array of atoms, such as in a single crystal, they can be in phase in particular directions in the crystal, or they can be out of phase. If they are in phase, reinforcement will occur, and a strong, scattered beam will be present. If they are out of phase, they will interfere, and a weak beam will be observed. The angular resolution is usually poor in such experiments, so that quantitative information such as d spacings or a partial radial distribution function is not obtained. However, by comparing the angular distributions for a number of pairs of elements, one can say something about the sites they occupy without complicated computations.

Evans et al. [71] have used this technique to investigate the cleavage planes for micas, such as muscovite, lepidolite, phlogopite, and vermiculite. By closely examining the x-ray photoelectron spectral line intensities and the photoelectron angular intensities to obtain atomic ratios and the photoelectron angular distribution ratios of elemental pairs, they concluded that the structure of the cleavage planes for muscovite and lepidolite were similar to the accepted bulk structure as determined by x-ray diffraction and fluorescence. By comparing the angular distribution ratios for a pair of elements, the equivalency or near equivalency of atomic position can be inferred. Such analysis indicates that aluminum exists in both tetrahedrally and octahedrally coordinated positions, and potassium and other interlayer cations occupy quite different sites than the silicon. For muscovite, sodium, present in small amounts, was shown to occupy a slightly different position than the potassium and is not just randomly substituted. The potassium tends to reside in a central position with a coordination number of 12, while the sodium prefers an off-center position coordinating with six oxygen atoms. For phlogopite and vermiculite, the cleavage planes were found to be rich in aluminum and deficient in magnesium. Interestingly enough, they found that the excess aluminum is tetrahedrally coordinated, giving rise to a deficiency of octahedrally coordinated atoms. Such lattice strains may account for why these planes tend to be preferentially cleaved. For vermiculite the interlayer cations calcium or potassium were exchanged with lead. The potassium and calcium occupy different sites with potassium as anhydrous and calcium as hydrated. The lead after exchange is also hydrated, occupying a site similar to the calcium.

Quantitative applications of x-ray photoelectron spectroscopy also have been applied to aluminosilicates. Adams et al. [72] have used relative photoionization cross sections for the 1s, 2s, and 2p subshells obtained from model compound x-ray photoelectron line intensities. The data were then applied to freshly cleaved lepidolite, phlogopite, and muscovite samples to determine atomic ratios. This approach, when compared to more established bulk analytical methods, gave an accuracy of 5%.

Another group of researchers [73] studied the aluminum 2s and 2p photoelectron lines as a function of the aluminum/oxygen bond length. The samples were studied as powders using gold mesh to dissipate charging on the surface. The binding energies of the two main lines showed an increase of approximately 0.5 eV in going from albite and microcline (with Al/O bond distances of 175 pm) to grossularite (195 pm) and α-Al_2O_3 (192 pm). Adams and co-workers [74] also have made a study of the core binding energies of aluminum, along with those for iron, silicon, magnesium, and oxygen in a series of well-characterized silicate minerals.

X-ray photoelectron spectroscopy has been used to investigate a hollandite ($BaAl_2Ti_6O_{16}$), a zirconolite ($ZrCaTi_2O_7$), and two perovskites [75]. In addition to checking on the agreement of the elemental stoichiometries for the minerals, ion bombardment (sputtering) studies were conducted with respect to both reduction of the titanium and alterations of the titanium 2p photoelectron lineshapes. The same studies were conducted with the calcium 2p lines.

4.2. Analysis

X-ray photoelectron and Auger spectroscopy have played an important role in determining the analytical composition of many geologic surfaces. McIntyre et al. [76] have applied the techniques to both surface and bulk studies of trace elements in the coal macerals vitrinite, fusinite, and exinite. In the vitrinite sample, both aluminosilicate and organic-phase titanium components were found. The fusinite contained an organofluorine compound and calcium and magnesium dispersed with the fluorine over a large part of the maceral, while the exinite contained lower concentrations of elements than did the other macerals that were studied. Torbanite, gilsonite vitrinite, Kimmeridge kerogen, and brown coal also have been studied by other researchers [77].

One investigation that showed the strength of a combined x-ray photoelectron/Auger approach was one dealing with the surface composition of sediment and soil models [78]. The two model systems studied were ferric oxide, Fe_2O_3, and the clay mineral montmorillonite; the Fe_2O_3 was treated with sodium dihydrogen phosphate, $NaH_2PO_4 \cdot 2\,H_2O$, in order to approximate the phosphorus: iron atomic ratio of 1:100 found in soil. Figure 3-8, for example, shows the spectra for montmorillonite samples prepared by two different routes. The reader should note the difference in the intensity ratio of the Ca 2p/Na KLL lines for the two differently prepared samples. The increase in the calcium intensity in one spectrum was attributed to the formation of a $CaCO_3$-enriched surface during its preparation. The suspected reason for this reaction occurring was the dissolution of carbon dioxide in the water that evaporated from the clay, thus forming a surface-enriched carbonate. This hypothesis was confirmed by preparing the collodial clay sample in an inert (nitrogen)

Fig. 3-8. Calcium 2p and carbon 1s region of the x-ray photoelectron spectrum of powdered montmorillonite as received (top) and as a dried, aqueous colloidal suspension (bottom). (Adapted from Ref. [78].)

atmosphere. The calcium 2p line intensity was greatly diminished; this particular study is a good example of how changing the experimental aspects conform a reaction mechanism for the formation of a surface complex. This same group [79] also has published x-ray photoelectron studies related to the surfaces of sediment reference materials.

4.3. Electronic States

In light of the discussion above regarding the sensitivity of x-ray photoelectron and Auger spectroscopy to the electronic state of elements, it should not be surprising that the two techniques have proved to be powerful tools in expanding the detailed knowledge of geologic materials in this area, particularly in the role of electronic states in bonding. One group of researchers [80] used x-ray photoelectron spectra to determine the Ti^{3+}/Ti^{4+} ratio in synthetic hollandites. The results were compared to results reported previously for these materials using electron microprobe techniques. X-ray photoelectron spectroscopy has been used by Al-Kadier et al. [81] to study theoretical models for the SiO_4^{4-} species in forsterite, Mg_2SiO_4. This approach, coupled with the analysis of corresponding x-ray emission data for silicon, magnesium, and oxygen, showed the magnesium/oxygen bond. The spatial orientation of the magnesium cations and the SiO_4^{4-} anions appeared to effect sigma/pi mixing of orbitals, which could be viewed as either silicon/oxygen bonding or oxygen lone pairs.

One area of interest in electronic state studies that has been intensely investigated is that of manganese minerals. Sherman [82] used experimental x-ray photoelectron, x-ray emission, and optical spectra to study manganese oxide minerals. The data were used in conjunction with molecular orbital calculations to examine the clusters MnO_6^{10-}, MnO_6^{9-}, and MnO_6^{8-} corresponding to Mn^{2+}, Mn^{3+}, and Mn^{4+} in octahedral coordination with the O^{2-} oxide anion. The agreement between calculated and experimental spectroscopic transition energies was quite high, indicating that such isolated clusters are very good models for localized electronic structures in manganese oxides. This work also compared the degree of ionicity/covalency in the clusters and the possibility of various spin states for the Mn^{2+}, Mn^{3+}, and Mn^{4+} in manganese oxide minerals.

Evans and Raftery [83] also have used x-ray photoelectron spectroscopy to study the manganese oxides MnO, Mn_3O_4, Mn_2O_3, and MnO_2, along with Norwegian lepidolite. The oxidation state of manganese in the oxides was determined by use of the oxygen 1s and manganese 2p binding energies and the manganese 2p/oxygen 1s binding energy differences. The multiplet splitting of the manganese 3s photoelectron level for the oxides was compared; the multiplet splitting was observed to decrease as the oxidation state of the manganese increased. Two other trends also were identified, both related to the oxidation state of the manganese. First, as the oxidation state increased, the binding energy of the manganese 2p line increased. Second, the manganese 2p/oxygen 1s binding energy separation increased as the oxidation state of the manganese increased. By comparing all of these parameters derived from the model compounds, it was determined that the manganese in the lepidolite samples was present as manganese(II).

Another area in which x-ray-induced photoelectron and Auger spectroscopy has been of tremendous value in increasing the understanding of the electronic state of geologic materials is that of copper minerals. Nakai et al. [84] studied over two dozen copper sulfide minerals to determine the oxidation state of copper and sulfur. Perry and Taylor [85] were able to distinguish between covellite (CuS) and chalcocite (Cu_2S) by studying their core level binding energy differences, Auger parameters, and lineshape differences. Another study [86] of copper minerals used the extent of the chemical shift in the copper $2p_{3/2}$ binding energy to differentiate the copper(I) and copper(II) oxidation states, along with the absence or presence of satellite structure associated with the main copper $2p_{3/2}$ line.

Other transition-element-containing minerals have been studied, including some that have undergone weathering. One group [87] has studied vanadium-bearing aegirines in order to determine the oxidation state of the vanadium; the samples studied included both vanadium aegirites and a barium vanadyl silicate. Iron-containing clay minerals have been studied [88], both in their original state and after they had been reduced and oxidized. Studies [89] have been performed on garnierite which had been formed by the weathering of nickeliferous pyroxenite. Further weathering of the material was shown to induce distinct chemical changes, including the decrease of nickel and silica and an increase in iron(III) and aluminum.

Metal oxide and carbonate minerals have been studied extensively. Taylor and Perry [90] have studied the spectra (Fig. 3-9) of lead oxide, hydroxide, and carbonate minerals. By exposing a clean metallic lead surface to dry oxygen at temperatures below the melting point of lead, the orthorhombic form of PbO (massicot) could be formed as an overlayer mineral. The tetragonal form (litharge), however, was produced if the same experiment was conducted just at or above the melting point of lead. Air exposure of both the massicot and litharge resulted in carbonate/hydroxide overlayers. Sommer [91] has studied the carbon 1s and oxygen 1s photoelectron lines for a series of carbonate minerals. Allen and co-workers [92] made a comprehensive study of iron oxide and hydroxide minerals, including the effects of multiplet splitting and "shakeup" processes on the main iron $2p_{3/2,1/2}$ photoelectron lines.

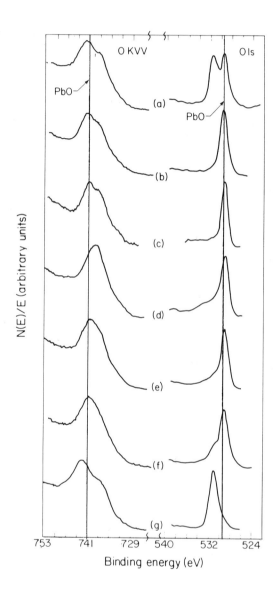

Fig. 3-9. Oxygen KVV and 1s lines for (a) PbO (massicot), (b) same PbO sample heated in situ in O_2, (c) clean, metallic lead exposed to O_2 at 150°C, (d) PbO_2 (plattnerite), (e) same PbO_2 heated to 320°C in vacuo, (f) Pb_3O_4 (minium), and (g) $2\,PbCO_3\cdot Pb(OH)_2$ (hydrocerussite). (Reprinted from Ref. [90], with permission.)

Chemisorption of water on the minerals was shown to alter the peak lineshapes, and the oxygen 1s peaks attributable to the oxide, hydroxide, and adsorbed water molecules were thoroughly characterized.

4.4. Reaction Chemistry

One of the most important roles of surface techniques in geologic chemistry is that of elucidating the reaction chemistry of species reacting with surfaces. While the number of chemical reaction possibilities is infinite, there are several general areas in which x-ray photoelectron and Auger spectroscopy have been used more than other areas. All of these reaction systems involve geologic substrate/fluid interactions, with the fluid usually being either a liquid such as water or an aqueous solution of an inorganic or organic species. Interactions involving a substrate/gas interface also are possible, but they are less studied; this is possibly due to a perception that such reactions are not as geochemically widespread as are the substrate/liquid reactions. (The reader should be aware of the extreme importance of zeolite/gas interactions in heterogeneous catalysis, for example; this extremely large subfield is beyond the scope of the present work and will not be discussed.) Three of the most common and representative areas of study are metal ion adsorption from aqueous solution, mineral flotation processing, and rock/water type reactions to simulate weathering processes. It is these areas that will be discussed in detail below.

Much work has been reported in the literature related to the adsorption of metal ions on sulfide minerals. Jean and his co-workers [93] have studied the aqueous adsorption of mercury, zinc, cadmium, and lead complexes on sulfide minerals as a function of the solution pH. The adsorption was shown to be heavily dependent on the pH, a factor also directly related to the hydrolysis of the metal ions in solution. Bancroft and Jean [94] have reported the low-temperature deposition of gold on pyrite, pyrrhotite, sphalerite, and galena; the deposition mechanism was shown to be the adsorption of the hydrated gold(III) species, followed by its reduction by the sulfide. Brown and his co-workers [95] have published a detailed study of the reaction of elemental mercury and mercury(II) salts with both pyrite and pyrrhotite.

Another area of study that has been extensively reported is that of the adsorption of metal ions on clays, non-sulfide minerals, and rocks. Dillard and co-workers [96] have investigated the surface chemistry of calcined cobalt/kaolinite materials in order to determine the chemical state of the cobalt. Figure 3-10, for example, shows spectra for a cobalt aluminate

Fig. 3-10. High resolution x-ray photoelectron cobalt $2p_{3/2,1/2}$ spectra for $CoAl_2O_4$ (top) and an air-calcined mixture of $Co(NH_3)_6Cl_3$ and kaolinite hydrite PX (bottom). (Adapted from Ref. [96].)

standard and a sample of a fired cobalt/clay mixture; the spectra are identical, showing $CoAl_2O_4$ to be the product of the solid-state reaction. This was confirmed by x-ray diffraction. Other studies of metal ion/clay interactions include those of nickel(II) and copper(II) on kaolinite, chlorite, and illite [97] and the adsorption of chromium(III)/amine complexes on kaolinite, chlorite, and illite [98]. Perry has reported the deposition of uranium and thorium hydrolysis products on the surface of basalt [99].

Surface techniques have also been able to yield extremely important data involving the chemistry of mineral flotation processing. Buckley et al. [100] have studied the surface oxidation of sulfide minerals in flotation. This same group has also studied the oxidation of natural bornite by the atmosphere [101]. Pillai et al. [102] have reported the adsorption of organic xanthates on pyrite, identifying the active collector formed on the surface as being dixanthogen.

Perry and co-workers [103] have reported the reaction of the dichromate ion, $Cr_2O_7^{2-}$, with galena surfaces and shown that both chromium(III) and chromium(VI) species are present on the surface (Fig. 3-3). This study used model compounds of both chromium(III) and chromium(VI) to determine the surface reaction products, including models such as Cr_2O_3, $Cr(OH)_3$, and $Cr_2O_3 \cdot n\ H_2O$. The carbon 1s region of the spectrum, in addition to the chromium 2p and oxygen 1s regions, was studied to document the formation of a mixed hydrated oxide/carbonate species as the surface product. Again, as with the case of the manganese oxide/lepidolite study discussed above, the splitting of the 3s photoelectron line was used to study both the model compounds and the reacted galena surface. The spin-orbit splitting differences for chromium(III,VI) that had been reported previously in the literature were used also to verify the two states. The reaction of aqueous copper(II) salts with galena and sphalerite also has been reported [104].

5. Summary

The application of x-ray-induced photoelectron and Auger spectroscopies to geologic surface chemistry already has proved to be a powerful tool in gaining a detailed understanding of interfacial reactions. A wide variety of chemical/geochemical systems has been studied, systems that are both naturally occurring (such as weathering) in the environment and systems created under laboratory conditions. Other studies, such as identifying electronic states of metal ions in minerals, have been invaluable in gaining an understanding of both their solid state chemistry and the reactions they undergo. The expansion of applications such as these, coupled with yet still to be realized applications, will continue to add to the body of knowledge concerning the field of geologic materials.

Acknowledgments

One of the authors (DLP) wishes to acknowledge support of his research by the U.S. Department of Energy (OBES and OCRWM) under Contract No. DE-AC03-76SF00098. Additional support has been provided by the Bureau of Mines, U.S. Department of Interior, and the U.S. Nuclear Regulatory Commission.

References

[1]   Einstein, A., Ann. Phys. (Leipzig), *1905*, 17, 132.

[2]   Auger, P., Comp. Rend., *1923*, 177, 169.

[3]   Lander, J. J., Phys. Rev., *1953*, 91, 1382.

[4]   Siegbahn, K., Ed., "Alpha-, Beta-, and Gamma-Ray Spectroscopy." North Holland Publishing, Amsterdam, 1965.

[5]   Harris, L. A., J. Appl. Phys., *1968*, 39, 1419.

[6]   Weber, R. E.; Peria, W. T., J. Appl. Phys., *1967*, 38, 4355.

[7]   Palmberg, P. W.; Bohn, G. K.; Tracey, J. C., Appl. Phys. Lett., *1969*, 15, 254.

[8]   Siegbahn, K.; Nordling, C. N.; Fahlman, A.; Nordberg, R.; Hamrin, K.; Hedman, J.; Johansson, G.; Bergmark, T.; Karlsson, S. E.; Lindren, I.; Lindberg, B., "ESCA, Atomic, Molecular, and Solid State Structure Studied by Means of Electron Spectroscopy." Almqvist and Wiksells, Upsala, 1967.

[9]   Siegbahn, K.; Nordling, C.; Johansson, G.; Hedman, J.; Heden, P. F.; Hamrin, K.; Gelius, U.; Bergmark, T.; Werme, L. O.; Manne, R.; Baer, Y., "ESCA Applied to Free Molecules." North-Holland, Amsterdam, 1969.

[10] Brundle, C. R.; Baker, A. D., Eds., "Electron Spectroscopy: Theory, Techniques, and Applications." Academic Press, New York, 1977.

[11] Windawi, H.; Ho, F., Eds., "Applied Electron Spectroscopy for Chemical Analysis." Wiley-Interscience, New York, 1982.

[12] Kane, P. F.; Larrabee, G. B., Eds., "Characterization of Solid Surfaces." Plenum Press, New York, 1974.

[13] Brigg, D.; Seah, M. P., Eds., "Practical Surface Analysis." John Wiley and Sons, New York, 1983.

[14] Carlson, T. A., "Photoelectron and Auger Spectroscopy." Plenum Press, New York, 1975.

[15] Czanderna, A. W., Ed., "Methods of Surface Analysis." Elsevier, New York, 1975.

[16] Thompson, M.; Baker, M. D.; Christie, A.; Tyson, J. F., Eds., "Auger Electron Spectroscopy." Wiley-Interscience, New York, 1985.

[17] Hercules, D. M., Anal. Chem., *1972*, 44, 106R.

[18] Hercules, D. M., Anal. Chem., *1974*, 46, 133R.

[19] Hercules, D. M., Anal. Chem., *1976*, 48, 294R.

[20] Kane, P. F.; Larrabee, G. B., Anal. Chem., *1977*, 49, 221R.

[21] Baker, A. D.; Brisk, M. A.; Liotta, D. C., Anal. Chem., *1978*, 50, 328R.

[22] Kane, P. F.; Larrabee, G. B., Anal. Chem., *1979*, 51, 308R.

[23] Baker, A. D.; Brisk, M. A.; Liotta, D.C., Anal. Chem., *1980*, 52, 161R.

[24] Larrabee, G. B.; Shaffner, T. J., Anal. Chem., *1981*, 53, 163R.
[25] Turner, N. H.; Colton, R. J., Anal. Chem., *1982*, 54, 293R.
[26] Bowling, R. A.; Larrabee, G. B., Anal. Chem., *1983*, 55, 133R.
[27] Turner, N. H.; Dunlap, B. I.; Colton, R. J., Anal. Chem., *1984*, 56, 373R.
[28] Bowling, R. A.; Shaffner, T. J.; Larrabee, G. B., Anal. Chem., *1985*, 57, 130R.
[29] Turner, N. H., Anal. Chem., *1986*, 58, 153R.
[30] McGuire, G. E., Anal. Chem., *1987*, 59, 294R.
[31] Turner, N. H., Anal. Chem., *1988*, 60, 377R.
[32] Riviere, J. C. In "Solid State Surface Science," Vol. 1, Green, M., Ed. Dekker, New York, 1969, p. 179.
[33] Wagner, C. D.; Riggs, W. M.; Davis, L. E.; Moulder, J. F.; Muilenberg, G. E., "Handbook of X-Ray Photoelectron Spectroscopy." Perkin-Elmer, Eden Prairie, MN, 1979.
[34] Reference [13], Chapter 5, pp. 181–214.
[35] Wagner, C. D.; Davis, L. E.; Zeller, M. V.; Taylor, J. A.; Raymond, R. H.; Gale, L. H., Surface Interface Analysis, *1981*, 3, 211.
[36] Brinen, J. S. In "Electron Spectroscopy," Candano, R.; Verbist, J., Eds. Elsevier, New York, 1974, p. 377.
[37] Klein, J. C.; Li, C. P.; Hercules, D. M.; Black, J. F., Appl. Spectrosc., *1984*, 38, 729.
[38] Frost, D. C.; Ishitani, A.; McDowell, C. A., Mol. Phys., *1972*, 24, 861.
[39] Sarma, D. D., Indian J. Chem., *1980*, 19, 1046.
[40] Wallbank, B.; Johnson, C. E.; Main, I. G., J. Electron Spectrosc. Relat. Phenom., *1974*, 4, 263.
[41] Yamada, H.; Shriaishi, K.; Watanibe, N., Nippon Kagaku Kaishi, *1981*, 8, 1245.
[42] Brundle, C. R.; Chuang, T. J.; Wandelt, K., Surf. Sci., *1977*, 68, 459.
[43] Wagner, C. D.; Biloen, P., Surf. Sci., *1973*, 35, 82.
[44] Wagner, C. D., J. Electron Spectrosc. Relat. Phenom., *1977*, 10, 305.
[45] Wagner, C. D. In "Handbook of X-Ray and Ultraviolet Photoelectron Spectroscopy," Briggs, D., Ed. Heyden, London, 1977, Chapter 7.

[46] Frost, D. C.; Ishitani, A.; McDowell, C. A., Mol. Phys., *1972*, 24, 861.

[47] Jolly, W. L., Coord. Chem. Rev., *1974*, 13, 47.

[48] Vernon, G. A.; Stucky, G.; Carlson, T. A. Inorg., Chem., *1976*, 15, 278.

[49] Carver, J. C.; Schweitzer, G. K.; Carlson, T. A., J. Chem. Phys., *1972*, 57, 973.

[50] Aarons, L. J.; Guest, M. F.; Hillier, I. H., J. Chem. Soc., Faraday Trans., *1972*, 68, 1866.

[51] Aberg, T., Phys. Rev., *1967*, 156, 35.

[52] Vernon, G. A.; Stucky, G.; Carlson, T. A., Inorg. Chem., *1976*, 15, 278.

[53] Riviere, J. C., In "Surface Analysis of High Temperature Materials: Chemistry and Topography," Kemeny, G., Ed. Elsevier Applied Science Publishers, London, 1984; Chapter 2.

[54] Ritchie, J. M., Phys. Rev., *1957*, 106, 874.

[55] Kubiak, C. J. G.; Aita, C. R.; Tran, N. C.; Barr, T. L., In "Defect Properties and Processing of High-Technology Nonmetallic Materials," Chen, Y.; Kingery, W. D.; Stokes, R. J., Eds. Materials Research Society, Pittsburgh, PA, 1984, p. 379–386.

[56] Barr, T. L.; Kramer, B.; Shah, I.; Ray, M.; Greene, J. E., In "Thin Films: The Relationship of Structure to Properties," Aita, C. R.; SreeHarsh, K. S., Eds. Materials Research Society, Pittsburgh, PA, 1985, p. 205–234.

[57] Barrie, A., Chem. Phys. Lett., *1973*, 19, 109.

[58] Kowalczyk, S.P.; Ley, L.; McFeely, F.R.; Pollak, R.A.; Shirley, D.A., Phys. Rev., B., *1974*, 9, 381.

[59] Wagner, C.D., Faraday Discuss. Chem. Soc., *1975*, 60, 291.

[60] Gaarenstroom, S.W.; Winograd, N., J. Chem. Phys., *1977*, 67, 3550.

[61] Wagner, C.D.; Joshi, A., J. Electron Spectrosc. Relat. Phenom., *1988*, 47, 283.

[62] Reference [13], Appendix 4.

[63] Thomas, T.D., J. Electron Spectrosc. Relat. Phenom., *1980*, 20, 117.

[64] Mott, N.F.; Gurney, R.W., "Electronic Processes in Ionic Crystals." Clarendon Press, Oxford, 1948.

[65] Citrin, P.H.; Thomas, T.D., J. Chem. Phys., *1972*, 57, 446.

[66] Kuroda, H.; Ohta, T.; Sato, Y., J. Electron Spectrosc. Relat. Phenom., *1979,* 15, 466.

[67] Wagner, C. D.; Passoja, D. E.; Hillery, H. F.; Kinsky, T. G.; Six, H. A.; Jansen, W. T.; Taylor, J. A., J. Vacuum Sci. Technol., *1982,* 21, 944.

[68] West, R. H.; Castle, J. E., Surf. Interface Anal., *1982,* 4, 68.

[69] Barr, T. L.; Lishka, M. A., J. Am. Chem. Soc., *1986,* 108, 3178.

[70] Barr, T. L.; Chen, L. M.; Mohsenian, M.; Lishka, M. A., J. Am. Chem. Soc., *1988,* 110, 7962.

[71] Evans, S.; Adams, J. M.; Thomas, J. M., Phil. Trans. Roy. Soc. London, *1979,* 292, 563.

[72] Adams, J. M.; Evans, S.; Reid, P.I.; Thomas, J. M.; Walters, M. J., Anal. Chem., *1977,* 49, 2001.

[73] Urch, D. S.; Murphy, S., J. Electron Spectrosc. Relat. Phenom., *1974,* 5, 167.

[74] Adams, I.; Thomas, J. M.; Bancroft, G. M., Earth Planet. Sci. Lett., *1972,* 16, 429.

[75] Myhra, S.; Bishop, H. E.; Riviere, J. C., Surf. Technol., *1983,* 19, 161.

[76] McIntyre, N. S.; Martin, R. R.; Chauvin, W. J.; Winder, C. G.; Brown, J. R.; MacPhee, J. A., Fuel, *1985,* 64, 1705.

[77] Clark, D. T.; Wilson, R.; Quirke, J. M. E., Chem. Geol., *1983,* 39, 215.

[78] Soma, M.; Seyama, H., Appl. Surf. Sci., *1981,* 8, 478.

[79] Soma, M.; Seyama, H.; Okamoto, K., Talanta, *1985,* 32, 177.

[80] Myhra, S.; White, T. J.; Kesson, S. E.; Riviere, J. C., Am. Mineral., *1988,* 73, 161.

[81] Al-Kadier, M. A.; Tolon, C.; Urch, D. S., J. Chem. Soc. Faraday Trans. II, *1984,* 80, 669.

[82] Sherman, D. M., Am. Mineral., *1984,* 69, 788.

[83] Evans, S.; Raftery, E., Clay Miner., *1982,* 17, 477.

[84] Nakai, I.; Sugitani, Y.; Nagashima, K.; Niwa, Y., J. Inorg. Nucl. Chem., *1978,* 40, 789.

[85] Perry, D. L.; Taylor, J. A., J. Mater. Sci. Lett., *1986,* 5, 384.

[86] Nakai, I.; Izawa, M.; Sugitani, J.; Niwa, Y.; Nagashima, K., Mineral. J., *1976,* 8, 135.

[87] Nakai, I.; Ogawa, H.; Sugitani, Y.; Niwa, Y.; Nagashima, K., Mineral. J., *1976,* 8, 129.

[88]   Stucki, J. W.; Roth, C. B.; Baitinger, W. E., Clays Clay Miner., *1976*, 24, 289.

[89]   Bosio, N. J.; Hurst, V. J.; Smith, R. L., Clays Clay Miner., *1975*, 23, 400.

[90]   Taylor, J. A.; Perry, D. L., J. Vacuum Sci. Technol. A, *1984*, 2, 771.

[91]   Sommer, S., Am. Mineral., *1975*, 60, 483.

[92]   Allen, G. C.; Curtis, M. T.; Hooper, A. J.; Tucker, P. M., J. Chem. Soc., Dalton Trans., *1974*, 14, 1525.

[93]   Jean, G. E.; Bancroft, G. M., Geochim. Cosmochim. Acta, *1986*, 50, 1455.

[94]   Bancroft, G. M.; Jean, G., Nature (London), *1982*, 298, 730.

[95]   Brown, J. R.; Bancroft, G. M.; Fyfe, W. S.; McLean, R. A. N., Environ. Sci. Technol., *1979*, 13, 1142.

[96]   Dillard, J. G.; Schenck, C. V.; Koppelman, M. H., Clays Clay Miner., *1983*, 31, 69.

[97]   Koppelman, M. H.; Dillard, J. G., Clays Clay Miner., *1977*, 25, 457.

[98]   Koppelman, M. H.; Dillard, J. G., Clay Clay Miner., *1980*, 28, 211.

[99]   Perry, D. L., In "NRC Nuclear Waste Geochemistry '83," Alexander, D. H.; Birchard, G. F., Eds. U.S. Nuclear Regulatory Commission Report No. NUREG/CP-0052, Washington, D.C., 1983, Chapter 2.2.

[100] Buckley, A. N.; Hamilton, I. C.; Woods, R., Dev. Miner. Process., *1985*, 6[Flotation Sulphide Miner.], 41.

[101] Buckley, A. N.; Woods, R., Aust. J. Chem., *1983*, 36, 1793.

[102] Pillai, K. C.; Young, V. Y.; Bockris, J. O'M., J. Colloid Interface Sci., *1985*, 103, 145.

[103] Perry, D. L.; Tsao, L.; Taylor, J. A., Inorg. Chim. Acta, *1984*, 85, L57.

[104] Perry, D. L.; Tsao, L.; Taylor, J. A., Proc. Electrochem. Soc., *1984*, 84-10[Proc. Int. Symp. Electrochem. Miner. Met. Process., 1984], 169.

[105] Crovisier, J. L.; Thomassin, J. H.; Juteau, T.; Eberhart, J. P.; Touray, J. C.; Baillif, P., Geochim. Cosmochim. Acta, *1983*, 47, 377.

[106] Crovisier, J. L.; Eberhart, J. P.; Thomassin, J. H.; Juteau, T.; Touray, J. C.; Ehret, G., C. R. Seances Acad. Sci., Ser. 2, *1982*, 294, 989.

[107] Thomassin, J. H.; Touray, J. C.; Baillif, P.; Jaurand, M. C.; Magne, L.; Goni, J., IARC Sci. Publ., *1980*, 30[Biol. Eff. Miner. Fibres, vl], 105.

[108] Brown, J. R.; Fyfe, W. S.; Bancroft, G. M., Appl. Surf. Sci., *1981*, 7, 419.

[109] Goelander, C. G.; Eriksson, J. C., J. Colloid Interface Sci., *1987*, 119, 38.

[110] Claesson, P. M.; Herder, P.; Stenius, P.; Eriksson, J. C.; Pashley, R. M., J. Colloid Interface Sci., *1986*, 109, 31.

[111] Murray, J. W.; Dillard, J. G.; Giovanoli, R.; Moers, H.; Stumm, W., Geochim. Cosmochim. Acta, *1985*, 49, 463.

[112] Lunsford, J. H.; Hutta, P. J.; Lin, M. J.; Windhorst, K. A., Inorg. Chem., *1978*, 17, 606.

[113] Predali, J. J.; Brion, D.; Hayer, J.; Pelletier, B., Dev. Miner. Process., *1981*, 2[Miner. Process., Part A, Volume Date 1979], 134.

[114] Nefedov, V. I.; Salyn, Y. V.; Solozhenkin, P. M.; Pulatov, G. Y., Surf. Interface Anal., *1980*, 2, 170.

[115] Kang, Y.; Pellett, G. L.; Skiles, J. A. Wightman, J. P., J. Colloid Interface Sci., *1980*, 75, 313.

[116] Predali, J. J.; Brion, D.; Hayer, J.; Pelletier, B., 13th Int. Mineral. Process. Congr., *1979*, 1, 55.

[117] Koppelman, M. H.; Emerson, A. B.; Dillard, J. G., Clays Clay Miner., *1980*, 28, 119.

[118] Adams, J. M.; Evans, S., Clays Clay Miner., *1979*, 27, 248.

[119] Koppelman, M. H.; Dillard, J. G., J. Colloid Interface Sci., 1978, 66, 345.

[120] Koppelman, M. H.; Dillard, J. G., ACS Symp. Ser., 1975, 18[Mar. Chem. Coastal Environ.], 186.

[121] Landis, W. J.; Martin, J. R., J. Vacuum Sci. Technol., A. *1984*, 2, 1108.

[122] Canesson, P., Dev. Sedimentol., 1982, 34[Adv. Tech. Clay Miner. Anal.], 211.

[123] Koppelman, M. H., NATO Adv. Study Inst. Ser., Ser. C, *1980*, 63[Adv. Chem. Methods Soil Clay Miner. Res.], 205.

[124] Defosse, C.; Rouxhet, P. G., NATO Adv. Study Inst. Ser., Ser. C, *1980*, 63[Adv. Chem. Methods Soil Clay Miner. Res.], 169.

[125] Bancroft, G. M.; Brown, J. R.; Fyfe, W. S., Chem. Geol., *1979*, 25, 227.

[126]   Koppelman, M. H.; Dillard, J. G., Dev. Sedimentol., *1979*, 27[Int. Clay Conf. Proc., 6th, 1978], 153.

4. AUGER ELECTRON SPECTROSCOPY AND MICROSCOPY

Raymond Browning
Center for Integrated Systems
Stanford University
Stanford, CA 94305 USA

Michael F. Hochella, Jr.
Department of Geology
Stanford University
Stanford, CA 94305 USA

1. Introduction

There is no question that the analysis of earth materials using surface-sensitive spectroscopies has revolutionized the field of mineral surface geochemistry. These spectroscopies have allowed earth scientists, for the first time, to elucidate directly the chemical and structural nature of mineral surfaces so that the important geochemical processes which occur there can be understood. In other fields, there has been little surprise to find that solid surfaces often do not reflect the chemistry, atomic structure, and properties of the bulk solid. Although still in its infancy, the study of mineral surfaces should yield similar trends in this respect. However, natural materials are generally much more heterogeneous than most other classes of materials, and the scale of grains and particles in geologic specimens can be extremely small. Therefore, surfaces and interfaces are exceptionally important in geologic systems, but unfortunately they can be very difficult to study.

Although there are probably over 50 surface-sensitive techniques currently in use, electron-stimulated Auger electron spectroscopy (AES) was one of the first developed and is today probably only surpassed in popularity by X-ray photoelectron spectroscopy (XPS). AES is highly surface specific, and for most materials it is only sensitive to the top few monolayers. In addition, very high spatial resolution can be easily obtained and, for most materials, interpreting AES spectra is relatively straightforward. This combination of surface sensitivity, spatial resolution, and its often unambiguous interpretation helps AES find enormous applicability in many fields, and particularly in metallurgy, semiconductor physics, and microelectronics. By comparison, however, AES has been used relatively sparingly in the analysis of earth materials. To date, it has mostly been used in the analysis of thin mineral accretions and mineral weathering processes (see below). Therefore, it is quite apparent that the geologic applications of AES have not been developed to the same degree as in other disciplines. To a large extent, this is because of the nature of earth materials (commonly oxides and insulators). The difficulty involved in using AES to analyze minerals was noticed early on, particularly in the study of mica catalytic supports [1-3] and lunar rock surfaces [3-7] in the 1970s. Two main problems were noted, namely the charging of the specimens by the exciting electron beam, and changes in the surface under the electron beam with time. We will discuss these problems in some detail as they determine which areas are appropriate for the use of AES analysis. We will show that although these problems are often present, they only occasionally are deterrents to analysis. These special problems posed by the nature of earth materials can, to a large degree, be removed by the design of appropriate instrumentation and the refinement of techniques. Ironically however, the success of Auger spectroscopy in fields such as semiconductor processing has tended to drive commercial instrument development in particular directions, and specifically toward the use of high primary beam energies and high beam currents with an emphasis on depth profiling using ion beam etching. As will be seen, these are not necessarily the best directions in instrument development for geologic specimens.

In this chapter we will discuss how Auger spectroscopy can be used to look at geologic systems and what answers can be found when the samples and/or experiments are carefully chosen. The first part of the chapter has been divided into sections concerning basic aspects of the Auger technique, including a short discussion on the interaction of an electron beam with a solid. This is followed by some specific geologic applications, with examples chosen to bring out a discussion of the benefits and the problems with AES. The second part of the chapter will discuss the current state of geochemical research using AES, its impact on geochemistry, and the future developments expected for the technique. We will also be arguing that the approach to microscopic analysis using Auger spectroscopy is at present very limited, and that there are more powerful ways to perform Auger data collection and analysis than are currently being used.

2. Electron-excited Auger Spectroscopy

Auger electron spectroscopy is based on the Auger effect, which was first described by Lisa Meitner and Pierre Auger, independently of one another, in 1923 [8]. They discovered that if an atom is ionized by the removal of an electron from an atomic core level, an electron with a characteristic energy can be ejected irrespective of the energy of the ionizing source. Thus, elements can easily be identified from these emitted electron energies. Any ionizing radiation with sufficient energy to ionize a core level will stimulate Auger emission, but by convention the term AES implies the use of an electron beam as the primary radiation, and it is in this sense that we are using it here.

The great advantage of AES over other surface-sensitive analysis techniques is its spatial resolution. The spatial resolution is largely determined by the spot size of the primary electron beam. The spatial resolution of commercial instruments at present is between 50 and 100 nm.

2.1. The Origin and Nature of Auger Electrons

The basic Auger process can be understood from a diagram of the electron energy levels in Fig. 4-1. Ionizing radiation ejects an electron from an atom in the solid leaving a hole in one of the atomic core levels. This core level is quickly filled by an electron from a higher level shell and energy is released. This energy can leave the atom either as a photon in the X-ray range (the characteristic X-ray used in electron microprobes and X-ray fluorescence instruments), or as another ejected electron from the same atom. This second ejected electron is the Auger electron, and its energy depends on the energy of the atomic levels involved in its production, not on the energy of the initial ionizing radiation. The energy of the Auger electrons can therefore be used directly in elemental analysis. Unfortunately, the Auger process, which involves three electrons, precludes the direct analysis of H and He, although indirect detection through plasmon energy loses is possible (see section 2.2). For heavier elements, there can be several Auger transitions seen as the higher level electrons cascade down toward the nucleus. As might be expected, this cascade can also cause the ejection of the atom itself from the surface as chemical bonds are disrupted.

The sequence of core hole creation, hole filling, and Auger electron ejection does not happen instantaneously and there is time for relaxation of the electrons surrounding the initial state hole. This relaxation process affects the energy of the Auger electron, and this

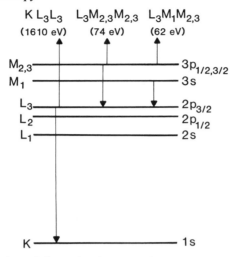

Fig. 4-1. Schematic diagram of the electron energy levels for Si showing the three Auger transitions that result in the most intense spectral lines. The transitions are labeled according to the electron level of the initial vacancy, the initial level of the hole-filling electron, and the level from which the Auger electron originates. The energies shown for these Auger electrons are taken from a quartz spectrum. (From Hochella et al., Am. Mineral., *1986*, 71, 1247, Fig. 1. Reprinted with permission from the Mineralogical Society of America.)

relaxation (or polarization) energy depends on the chemical environment of the atom. Therefore, the relaxation means that the energy level of the final Auger electron is not simply a result of the energy levels of the original atom, but it is material dependent. This so-called Auger chemical shift is therefore an indicator of chemical state and can provide very useful information. (A theoretical explanation of Auger chemical shifts can be found in Wagner [9].) The chemical shift of an Auger line from one material to another can be very large (up to 8-10 eV) as first reported by Wagner and Biloen [10]). However, because Auger transitions involve three atomic sublevels (as opposed to one for photoemission), and because electron energy analyzers typically used for Auger analysis are not optimized for energy resolution (see below), Auger lines are inherently quite broad and the advantage of large chemical shifts is somewhat negated. Also, any electron beam-induced surface damage (described in detail below) can easily change the shape and position of Auger lines. As a result, the use of Auger chemical shifts has never been extensively pursued for earth materials, although there is a limited amount of literature on the subject (e.g., Briggs and Riviere [11] and Remond et al. [12]).

2.2. *Electron/Solid Interactions*

The interaction between an electron beam and a solid is complex. Energetic primary electrons penetrate the solid surface and are scattered both elastically and inelastically by electrons in the solid. There are a variety of inelastic loss processes. With a primary electron beam energy more than a few keV, most of the energy of the incident electrons is dissipated by interactions with electrons in atomic core levels. These core electrons are scattered out of the atoms and add to the flux of scattering electrons. There are smaller loss processes, on the order of tens of eV, where a primary or scattered electron loses energy to coordinated oscillations of conduction and valence band electrons called plasmons. These losses do contain chemical-specific information, but they are difficult to interpret. However, they have been used successfully for detection and imaging of hydrogen in certain materials (Bevolo [13]), and there are clear differences in the plasmon loss features from certain polymorphs [14].

The energy distribution of electrons that emanate from a sample being struck by a primary electron beam is illustrated in Fig. 4-2. This particular example is for a 1-keV electron beam impinging on a Ag target. At the high-energy side there are the reflected (elastically backscattered) primary electrons. Just to the low-energy side of this peak are primary electrons that have suffered small energy losses to surface and bulk plasmons. There are also loss processes that entail transitions from bound atomic core states to unfilled band states. Below the loss electrons is a long tail of high-energy electrons that have lost energy in multiple collisions. Superimposed on this distribution of rediffused primary electrons are features due to the characteristic Auger electrons from Ag atoms. Below each Auger line is a tail of Auger electrons that have lost energy to plasmons and other processes. Finally, at the lowest energies are the so-called secondary electrons which are mostly ejected from the valence and conduction bands. It is

Fig. 4-2. Electronic spectrum generated by a 1-keV electron beam impinging on a Ag target. See text for details. (From Hochella, Rev. Mineral., *1988*, 18, 591, Fig. 11. Reprinted with permission from the Mineralogical Society of America.)

these electrons that are used to produce SEM images. However, electrons with energies in the range of 20 to 2000 eV (the energy range of Auger electrons used for practical surface spectroscopy) travel up to a few tens of monolayers before losing energy; therefore, characteristic Auger electrons originate in the very near surface of the sample.

Auger spectra can be displayed in one of two ways. The direct spectra can be plotted as shown in Fig. 4-2, but because of the high background and relatively small peaks, it is traditional to plot the first derivative of the direct spectra as shown in Fig. 4-3.

2.3. Auger Analysis of Insulating Surfaces

The vast majority of geologic materials are insulators. This leads to one of the difficulties of using AES on earth materials, that of sample charging. Slight negative charging of the surface during AES

Fig. 4-3. High resolution differentiated spectrum over the Ag MNN region of the spectrum in Fig. 4-2. The "V" in the label indicates the valence level. The position of each Auger line (by convention) is taken as the point of greatest negative excursion of the differentiated peak. (From Hochella, Rev. Mineral., *1988*, 18, 591, Fig. 11. Reprinted with permission from the Mineralogical Society of America.)

analysis will shift the Auger peaks to higher kinetic energy, but this does not preclude the collection of useful data. Excessive charging, however, results in unusable spectra, and this is generally the case unless charge neutralization measures are invoked. Coating the sample with a conductor, as is done for electron microprobe analysis, is obviously not an option because of the surface-sensitive nature of AES. Putting a conductor very close to the analytic surface, such as pressing grains of the sample into indium foil, also does not effectively reduce charging unless the grains are exceptionally small (less than a micron in size), allowing the majority of the primary electrons to penetrate the insulating grains and enter the conductor beneath. The only effective and practical method for charge neutralization is to set up the beam conditions and sample tilt such that the number of primary electrons entering the sample is roughly equal to the number of exiting electrons. (Bruining [15] was one of the first to show that this was possible and under what conditions this could be obtained.) The exiting electron current is made up of backscattered electrons (both elastic and inelastic), low-energy secondary electrons (defined above), and Auger electrons. The Auger current is exceptionally small compared to the backscattered and secondary currents, so it need not be considered in the rest of this discussion. It has been shown experimentally that the backscattered current has a relatively small dependence on the energy of the primary beam (e.g., Goldstein et al. [16] and Reimer [17]), but that the secondary current has a much greater dependence, especially for nonmetals (e.g., Dawson [18]). Because of the particular dependence of the secondary current on the primary beam energy, as the voltage of the primary beam is adjusted, there are two primary beam energies (called crossover voltages) where the number of entering and exiting electrons will balance and insulating surfaces will remain neutral. However, in order to increase the Auger signal, it is often desirable to increase the primary beam current at a given crossover voltage. By tilting the sample with respect to the primary beam, more electron/electron interactions occur nearer to the surface of the sample, and the backscattered and secondary emission efficiencies are increased. Therefore, neutral conditions can usually be found even at higher primary beam currents. This also explains why it is usually easier to find neutral conditions on flat surfaces rather than rough ones, where in the latter case the angle between the primary beam and surface changes over the area of analysis. In practice for the analysis of insulators, the primary beam voltage is set (usually around 3 keV), and then the primary beam current and sample tilt are adjusted until neutral conditions are found at an acceptable signal-to-noise ratio for Auger analysis.

2.4. Quantitative Analysis

Chemical quantification has been, and continues to be, one the most important areas of AES research. Qualitative analysis, using sensitivity factors derived from elemental and other chemically simple standards (e.g., Davis et al. [19]), has been available for some time and is very simple to perform. Semiquantitative to quantitative analysis is also straightforward if standards fairly similar to the unknown are available. In this case, quantification is routinely accurate to between 1 and 10% (relative) of the true composition. However, a more generally useful quantitative analysis scheme for Auger analysis is not yet available. One of the more promising approaches is to develop the Auger equivalent to the ZAF matrix correction method used for electron microprobe quantitative analysis [20]. For this to be fully successful, a better understanding of the material dependence of electron escape depths, ionization cross-sections, and Auger decay probabilities will be needed. There are other factors, no less important, which need to be considered. These include the dependence of Auger emission on crystallographic orientation (e.g., Armitage et al. [21] and LeGressus et al. [22]) and the angle between the sample surface and analyzer entrance direction (e.g., deBernardez et al. [23] and Wehbi and Roques-Carmes [24]). It is important also to remember that the usefulness of a quantitative analysis scheme will be diminished or negated if electron-induced surface damage, as described below, is taking place. Finally, an additional problem with Auger quantitative analysis involves Auger peak intensity measurement. Because Auger peaks can have complex shapes, and because they tend to be low in intensity on very intense and sloping backgrounds, accurate Auger peak intensity measurements are not trivial. Auger backgrounds can be removed by differentiating the spectra, either electronically (with a lock-in amplifier) or via computer. The peak intensity can then be measured by the peak to peak excursions of the differentiated Auger lines. Although this is very convenient and acceptable for semiquantitative analysis [25], there are hidden problems with this approach [26]. Especially for small signals, quantitative analysis requires a background to be fit and subtracted. Great strides have been made in background analysis since Sickafus first suggested that it could be described by a power law function [27-29]. Practical use of this model has been developed and reviewed by, among others, Prutton and El Gomati [30], Browning [31], and Mathew and others [32].

2.5. Instrumentation

The heart of any Auger spectrometer is the electron energy analyzer. The two most popular current analyzer designs that are used

ELECTRON ENERGY SPECTROMETERS

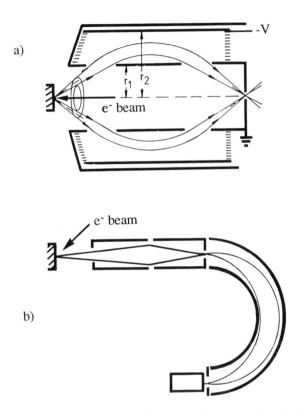

Fig. 4-4. Schematic diagrams of (a) a cylindrical mirror analyzer or CMA and (b) a concentric hemispherical analyzer or CHA. Electron trajectories are shown.

for AES are shown in Fig. 4-4. The cylindrical mirror analyzer (CMA) is most often used for routine AES. This analyzer consists of two concentric cylinders. The inner cylinder is typically at ground, and a negative voltage is applied to the outer cylinder which results in the electron trajectory through the analyzer as shown. Energy ranges are scanned by ramping the outer cylinder voltage. This analyzer has a well-defined focal point, so that the distance of the sample surface from the end of the analyzer is important for accurate electron energy measurement. Although the energy resolution of this particular analyzer design is only moderate, the entrance area into the analyzer is very large and the transmission of electrons through the analyzer is very high. Therefore, the data collection efficiency of this analyzer is excellent. Such an analyzer is important because of the inherently poor Auger signal-to-background ratio discussed above and also because of electron beam damage. With such an efficient analyzer, data collection times or beam currents, or both, can be reduced, thus minimizing the potential for damage. Another advantage of this analyzer is that the electron gun can be mounted inside the inner cylinder so that the gun and analyzer are co-axial with one another. This is extremely important for scanning Auger microscopy (see below) because it minimizes shadowing effects caused by blockage of electron-stimulated Auger electrons from the analyzer entrance because of rough surfaces.

On the other hand, superior energy resolution is sometimes required when one is studying, for example, Auger lineshapes. In this case, the concentric hemispherical analyzer (CHA) is usually the analyzer of choice. This analyzer consists of two concentric hemispheres with a retarding grid at the entrance slit. The purpose of the grid is to retard (or accelerate) the electrons one wishes to analyze to an energy (called the pass energy) most conducive to high energy resolution measurements while maintaining acceptable counting rates. Energy ranges are scanned by ramping the deflecting voltages on the hemispheres as well as ramping the voltage to the retarding grid. Because the acceptance solid angle into the analyzer is relatively small, this analyzer is also particularly well suited for angle-resolved AES studies.

3. Electron and Ion Beam Damage

Most materials can be easily damaged during Auger analysis unless special care is taken. This damage consists of structural and/or compositional modifications on the surface of the sample and can be caused by the electron beam used for Auger analysis and SEM imaging or the ion beam used in depth profiling experiments. We will discuss each in turn.

3.1. Electron Beam Damage

When energetic electrons impinge on a solid target, the most probable type of interaction is inelastic collisions with the valence electrons in the solid. Mostly through this mechanism, chemical bonds are disrupted. If these bonds are near the surface, atoms or small molecules can be lost to the vacuum, or residual gases in the vacuum can attach to the sample surface. These phenomena are called electron-stimulated desorption and absorption (ESD and ESA), respectively. These processes can significantly affect the composition and atomic structure of the area of analysis (see Carriere and Lang [33] and Pantano and Madey [34] for examples). In addition, electric-field-induced ion migration is likely if charging is allowed to occur during the analysis of insulators, and sample heating is a common problem for poor thermal conductors. These two effects can also modify, sometimes dramatically, the composition and structure of surfaces that are being analyzed. Examples of these degradational effects can be found in Dawson and others [35], Gossink and others [36], Yau and others [37], Clark and others [38], and Storp [39].

Electron-beam damage during Auger analysis, whether it takes the form of ESD, ESA, electric-field-induced ion migration, or thermal-induced disruption, can most easily be detected by the lack of analytic reproducibility. It is generally a good idea to test the integrity of the sample continually under the influence of the primary electron beam by collecting multiple sets of data over the same sample area under the same beam conditions as a function of time. If the results vary as a function of time, and especially if the variation shows a recognizable trend, electron-beam-induced damage is almost certainly taking place. This assumes, of course, that instrumental systematic errors have been eliminated. One caution is that beam damage, under certain conditions, can occur within a matter of seconds to minutes from the time of initial beam exposure. Therefore, Auger lines of the elements most likely to be affected should be measured quickly just after initial beam exposure.

There are a number of ways to minimize or altogether avoid electron-beam damage during Auger analysis. These have been discussed in an excellent review by Levenson [40], and they are summarized here. First, it is important to perform Auger analysis under the best vacuum conditions possible in order to reduce or eliminate ESA. Lower vacuum levels are acceptable if the residual gases in the vacuum are relatively inert. Second, the lowest accelerating voltage consistent with the needs of the analysis should be used in order to minimize potential sample heating and ion migration problems. Last, the beam current density (current per area) should also be reduced to achieve the minimum signal-to-noise level required for the analysis. The beam current density can be reduced by decreasing the beam current (typically down to the low nanoamp range for insulating samples), by defocusing the beam, and/or by rastering the beam over a larger area. As an example, let us assume that we wish to do an analysis with a 3-keV beam having a beam current of 20 nA, and that this produces a beam approximately 0.3 μm in diameter. If the beam is not rastered, the beam current density at the point of analysis would be 28 A/cm^2. Rastering this beam over an area of 4 μm^2 reduces the effective beam current density to a more acceptable 0.5 A/cm^2.

It should be remembered that all material surfaces are susceptible to electron-beam-induced damage and that care most be taken in all cases. Often, it may be found that the minimum beam conditions required to give an acceptable signal-to-noise ratio causes sample damage, but often the data can be collected before damage begins to occur.

3.2. Ion-beam Damage

Ion bombardment, besides removing material via the sputtering process, also can create unwanted damage in the near-surface region of samples that are undergoing depth profile analysis. (Storp [39] provides an excellent review of ion-beam damage for many classes of materials.) Inert gases are commonly used for sputtering to minimize the chemical interaction between the sputtering ions and the target, but the physical nature of sputtering can cause significant and unwanted alteration. An incoming ion can be scattered back into the vacuum, or it can remain implanted in the sample, thus changing the bulk composition of the sample and modifying its structure. During the interaction of a bombarding ion with the sample, the ion can eject atoms or molecules from the sample into the vacuum, but the sputtering process is inefficient, and much of the energy of the ion, on the average, is converted into heat. Besides implantation and heating, ion bombardment also affects the sample adversely by altering the atomic structure to a depth of the ion penetration, which can be tens of nanometers at energies typically used for ion bombardment (0.5-5.0 keV). Bombarding ions tend to mix atomic layers, and the near-surface regions of most crystalline materials will become amorphous during sputtering. Finally, ion bombardment can alter the near-surface composition irrespective of implantation by a process called preferential or differential sputtering. This occurs when the sputtering yields of two or more elements in the same material differ significantly. Pivin [41] gives a number of examples of this phenomenon for several types of materials, and Hochella and others [42] give several examples specifically involving minerals (Fig. 4-5). The element that sputters most rapidly becomes depleted in the near-surface region until its sputtering rate matches the rate of the slowest sputtering element. Therefore, as the material is sputtered away, a depletion front or fronts (depending on the number of constituent elements and their sputtering rates) advances through the material ahead of the retreating sample surface as the sputtering of all elements reaches a steady state. In the study of Hochella and others [42], using ion energies of 1 keV, steady state was reached within 3-4 minutes of sputtering for each mineral sample (Fig. 4-5).

There is one additional problem caused by the ion bombardment of insulating materials specifically when Auger spectra are being collected. The charge neutralization conditions set for Auger analysis of an insulating unsputtered material often changes after sputtering is started. This is most commonly due to the fact that sputtering often changes the composition and structure of the surface, which in turn changes its secondary electron emission characteristics. After sputtering commences, therefore, the sample tilt and instrument conditions may need to be adjusted to regain charge neutrality conditions.

4. Applications of AES to Mineral Surfaces

4.1. The Surface Composition of Lunar Samples

The surface of the moon is unusual in terms of mineralogy because, among other things, it is a system that has evolved in a vacuum. The processes of erosion have therefore been very different from terrestrial systems and certain resulting properties, such as the optical albedo, appear to be anomalous compared with terrestrial examples. The polarization law and optical reflectivity of the moon had been studied using ground-based observations. The conclusions from these studies suggested that the lunar surface was composed of a fine dark opaque powder. (There was some concern before the first landing that a lunar module might sink into this dust.) The production of a fine particulate surface from eons of micrometeorite bombardment is understandable, but there were very few terrestrial rocks dark enough in a powder form to resemble most of the moon. A typical highland albedo is 15%, and only 5% for the region sampled by Apollo 12. Before the Apollo missions, it was suspected that the lunar darkening was the result of excess iron at the soil surface from some mechanism such as preferential sputtering by the solar wind or micrometeorite bombardment. With the availability of lunar samples from the Apollo missions, it became possible to analyze the vacuum-eroded specimen surfaces using AES to try to cast some light on the darkening mechanism. It was also possible to try to reproduce in the laboratory the vacuum erosion processes on crushed lunar and terrestrial rocks and directly compare these with the surface lunar fines.

Gold and others [4] used a retarding field analyzer with a primary beam energy of 2 keV and current of up to 10^{-6} A over a 1 mm^2 area. The lunar specimens were taken from the recovered materials from four Apollo missions and were used as received from the

Fig. 4-5 (Opposite). Variation in atomic ratios, $\Delta\%$ (percent change relative to the composition of the unsputtered surface), in the near-surface regions of quartz, albite, calcite, and anhydrite as a function of sputtering time as determined by XPS. (From Hochella et al., Am. Mineral., *1988*, 73, 1449, Fig. 1. Reprinted with permission from the Mineralogical Society of America.)

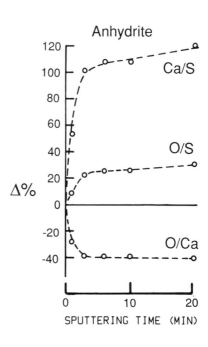

curator with no cleaning. The samples were pressed into a sample tray with no binder. The samples typically charged in the beam at all primary energies and currents. However, Gold and others found that if smooth flat surfaces were chosen for analysis, there was little if any charging. They initially investigated three samples of crushed lunar rocks and seven samples of fines. The ratio of the Auger Fe to O peak heights were normalized to the bulk iron concentration of one of the rock samples. Comparing the surface Fe/O ratios (from the fines) with the bulk Fe/O ratios (from the crushed rock) and the albedo, they showed that there is a strong correlation between surface concentration of Fe and light absorption (Figs. 4-6 and 4-7). As argued by Gold and others, these results clearly suggest a strong connection between the presence of surface Fe and the low albedo.

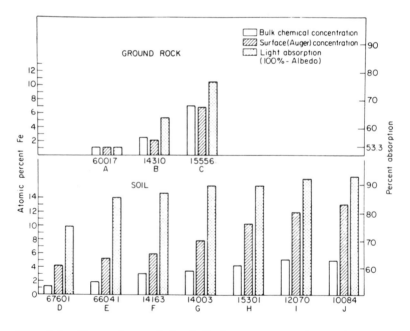

Fig. 4-6. Comparison of the bulk iron content, the surface iron content as measured by AES, and light absorption in lunar rock and soil samples. (From Gold et al., Proc. 5th Lunar Sci. Conf., *1974*, 3, 2413, Fig. 2. Reprinted with permission from Pergamon Press plc.)

Fig. 4-7. Albedo vs. atomic percent Fe for a number of lunar rock and soil samples. See text for details. (From Gold et al., Proc. 5th Lunar Sci. Conf., *1974*, 3, 2413, Fig. 3. Reprinted with permission from Pergamon Press plc.)

The use of one rock specimen as a standard and the normalization of the Auger peak heights to the measured bulk concentration seems well justified in this instance. Gold and others show that the Fe Auger peak intensity is proportional with bulk concentration from various rock samples and the extension to the darkened soil surfaces is reasonable. Simply using elemental standards for comparison in these circumstances would not be expected to be fruitful. There are several complications in the analysis that Gold and others avoided by using rock standards. First, the peak intensities and positions of the Fe, Ti, Ca, and Si Auger lines that were measured in these experiments are all sensitive to the oxidation conditions. For example, in the case of Si, there is a large peak shift in energy (up to 10 eV) which is dependent on its associated chemistry. Second, the Auger intensity is dependent on the matrix of the material and its average atomic number. This later effect is caused both by the changes in the backscattering of primary electrons which will add to the Auger yield, and also by the change in electron escape depth as a function of bulk composition and materials properties. Using closely related samples as self-consistent standards reduces the importance of these matrix effects. However, there are problems hidden in the analysis even with this method of using a closely related standard. There is both radiation damage from the electron beam and the depth distribution within the escape depth of the Auger electrons to consider. These problems become apparent when trying to compare results from different studies and with other surface-sensitive spectroscopies, such as XPS.

The darkening of the lunar soil by Fe enrichment might be caused by several mechanisms, the two most likely being solar wind sputtering and meteoritic evaporation and subsequent condensation. In fact, these might be the principle darkening processes on all airless bodies in the solar system. The differing albedo across the moon could be seen as aging differences under these processes. Terrestrial rocks containing Fe do darken under laboratory ion bombardment corresponding to solar wind sputtering of many thousands of years (Gold et al. [5] and Hapke [43]); therefore, there is very good circumstantial evidence for the solar wind explanation. However, the impact evaporation or some other mechanism could not be ruled out as a possible relevant mechanism.

With regard to this last point, Grant el al. [7] used Auger depth profiling to try to discriminate between the mechanisms of preferential sputtering by the solar wind, and melting or redeposition of material after micrometeorite bombardment. After lunar soil samples were cleaned in ethanol and fine grained material ultrasonically removed, the larger soil particles were pressed into a Pd tube. They found that carbon contamination and the nonuniform nature of the grains complicated the charging problem, but that charging was normally avoided at the low primary beam energy of only 1.4 keV and a 2×10^{-6} A beam current with a 0.3 mm primary beam diameter (2×10^{-3} A/cm^2). They picked out four anorthite grains, reasoning that any surface coating from the surrounding Fe-rich regolith would be distinctive. They showed by depth profiling that such a vapor deposited layer on these grains was unlikely, and that the very thin (0.3 nm) Fe-rich layer observed would not change the optical properties. They also showed by actually melting an edge of a particle in the electron beam that the composition and low optical transmission of the soil could not be reproduced by vacuum melting.

4.2 Depth Profiling Studies

Auger electron spectroscopy is widely used in conjunction with ion sputtering (etching) in order to obtain depth profiles of the elemental distribution into a material (in the geologic literature, see Morse et al. [44], Mucci et al. [45], Mucci and Morse [46], and Mogk and Locke [47]). An ion beam (typically Ar ions) with kinetic energy between 0.5 and 5.0 keV is used to sputter the specimen surface. Problems and artifacts that result from the ion bombardment of minerals have been discussed in detail above. What follows are descriptions of actual AES/depth profiling studies on minerals which demonstrate the attributes and uncertainties of such analyses.

The limitations of mineral depth profiling were first shown by Staib [48]. In this study, 0.5-keV argon ions were used to sputter muscovite for short periods alternating with AES analysis. The expected sequence of elemental concentrations through the muscovite structure (alternating K- and Al-rich layers approximately 0.5 nm apart) were not observed. Instead, the K level reached an equilibrium value at about half its initial value, suggesting that the layered structure was destroyed and mixed. In order to understand this result better, the structure of the mica surface was monitored using low-energy electron diffraction (LEED). It was observed that the LEED patterns indicative of a crystalline surface were rapidly destroyed by the ion beam before the Auger intensities showed any appreciable change. Clearly the structure under ion bombardment is very different from the original. Therefore, even though AES is only sensitive to the top few monolayers, the mixing of atomic layers by sputtering does not allow for profiling with anything near 0.5-nm depth resolution. On the other hand, profiling with depth resolution in the range of 3-5 nm has been routinely demonstrated.

An example of the AES/depth profiling technique used in an applied geologic study can be found in the work of Mucci and co-workers [44-46], who attempted to further the understanding of the growth of carbonate minerals in marine environments. In previous studies, the determination of a low apparent solubility of calcite in natural seawater has suggested that the pure calcite surface is stabilized by a layer of magnesian calcite. Such a mechanism would lead to considerable difficulties in modeling the accumulation of calcite in deep sea sediments. Initial experiments showed high Mg content in overgrowths on calcite using artificial seawater. But these experiments used supersaturations much higher than normally found, and this led to a controversy over the relative influences of the thermodynamic and kinetic factors determining the composition and stability of calcites and magnesian calcites. Using seawater very close to saturation with respect to the magnesium calcite phase in each solution, Mucci and co-workers were able to grow and characterize exceptionally thin (less than 0.5-μm) overgrowths. This allowed them to obtain previously unknown information on the parameters that control the composition of the solubility controlling overgrowth phase.

Their study used four carbonate minerals (calcite, aragonite, dolomite, and magnesite) as seed crystals in two synthetic seawaters with different saturation states. After extended exposure to the seawaters (times up to 5 months), the crystals were shaken to remove residual solution and dried in a vacuum dessicator before AES/depth profile analysis. They used a 60° sample tilt and a 3-keV primary beam energy to reduce charging effects. A 2-keV argon ion beam was used for sputtering.

The Auger results have to be interpreted with some care. The initial concentrations of the elements as measured by AES were very sensitive to electron-beam exposure, and a sharp decrease in the chlorine, sulfur, and sodium Auger signal was observed. The authors attribute this decrease to the electron-stimulated desorption of residual solution salts picked up from the seawater in the outermost layers. Similarly, although not mentioned explicitly, the beam damage was severe enough to desorb carbon (and probably a good deal of oxygen) from the surface of the analyzed surfaces, and the ion beam during sputtering would be expected to cause preferential loss of one or more of the components of interest.

Fig. 4-8. Sputter depth profile through an unrinsed calcite crystal after exposure to synthetic seawater. See text for details. (From Mucci et al., Am. J. Sci., *1985*, 285, 289, Fig. 3. Reprinted with permission from the American Journal of Science.)

Mucci and co-workers suggest that there are five layers distinguishable from the Auger depth profiles of pure calcite after exposure to seawater in their experiments (Fig. 4-8). The surface (I) and near-surface (II and III) layers are very high in magnesium and the Mg/Ca ratio at the surface is very dependent on the Mg^{2+}/Ca^{2+} concentration in solution. They point out that the assignment of layers II and III could either be due to a real concentration gradient between the surface and a solid overgrowth (layer IV, see below), or that this could be an artifact of ion mixing and topographic shadowing of the ion beam. We might also add the sample tilt with respect to the electron beam also causes shadowing affects on rough surfaces both before and after sputtering. No satisfactory conclusion could be reached from their data anyway, as layers II and III are not much thicker than the depth resolution of the AES/depth profiling technique. Below layers II and III, Mucci and others suggest that there are two additional layers. In the outer of these (IV), the Mg concentration is constant; in the inner region

(V), the concentration drops toward the seed bulk concentration. The presence of the constant concentration region (IV) means that despite the ambiguities caused by surface roughness, preferential sputtering, and mixing of the layers, it can be said that there is a Mg-rich overgrowth on calcite that is between 10 and 20 nm thick. It was also important to note that different Mg/Ca concentrations in the seawater produced different Mg/Ca ratios in this layer. Further, the surface layers (II and III) were apparently easily removed by rinsing the overgrowth in distilled water, leaving only layers IV and V. This result tends to confirm the idea that ion mixing is responsible for the near-surface Mg depth profile. There is also a fortunate agreement between the Mg concentration measured in the solid overgrowth (layer IV) after ion milling and after water rinsing. As it would not be expected that the results of preferential ion sputtering and preferential leaching would be the same, there is some confidence that the results are not overly distorted by these mechanisms.

Mucci and co-workers come to the conclusion that the overgrowth on calcite has a magnesium content that is dependent only on the seawater composition, and that the surface most Mg-rich layer (I) does not dictate the composition of the overgrowth. We conclude from their data that there are three layers: the top surface layer, the magnesian calcite overgrowth between 10 and 20 nm thick, and the underlying calcite substrate. It is clear that the understanding of this overgrowth mechanism and chemistry was greatly enhanced by the powerful combination of depth profiling and AES analysis.

5. Scanning Auger Microscopy

Scanning Auger microscopy (SAM) generally refers to Auger spectroscopy performed with high spatial resolution using a rastering electron beam. SAM was first performed on conventional scanning electron microscopes fitted with electron energy analyzers (Wells and Bremer [49] and MacDonald [50]). The best scanning Auger microprobes built today are not add-on systems, but are specifically designed for high spatial resolution surface analysis with secondary electron detectors added for SEM imaging. With these instruments, spatial resolutions in the 30-60 nm range have been achieved with conducting materials (e.g., Venables et al. [51] and Hochella et al. [52]), but sample damage at higher beam current densities (see above) has generally prevented useful Auger analysis of insulating materials at these very high spatial resolutions. Nevertheless, spatial resolutions of slightly less than 1 µm have been demonstrated on insulating mineral surfaces without significant beam damage (Hochella et al. [53]).

Because most geologic materials are nonhomogeneous, the spatial resolution advantage of SAM qualifies it as a very important technique for most problems in surface geochemistry. For example, a number of different surface analysis techniques, such as X-ray photoelectron spectroscopy (XPS) (e.g., Berner et al. [54] and Hellmann et al. [55]), secondary ion mass spectroscopy (SIMS) (Nesbitt and Muir [56] and Muir et al. [57]), resonant nuclear reaction (RNR) analysis (Petit et al. [58]), and Rutherford backscattering (RBS) (Casey et al. [59]) have been critical in developing an understanding of mineral dissolution. However, SAM, specifically because of its spatial resolution capabilities, has added and will continue to add a great deal to the understanding of how minerals dissolve. The fact that mineral dissolution is chemically inhomogeneous across the surface of a dissolving grain was first shown by Hochella and others [60] using SAM. It had been previously suggested that leached surface zones would develop near high energy surface sites such as crystal defect and microcrack outcrops (Berner et al. [54], Chou and Wollast [61], Holdren and Speyer [62]). Although in the study by Hochella and others [60] it was impossible to observe the surface chemistry directly on the scale of crystal defects, the spatial resolution (2-10 µm) was sufficient to analyze the surface chemistry in and around etch pits, as well as on surfaces with various dissolution-induced morphologies. The material analyzed in this study was labradorite powder that had been dissolved under hydrothermal conditions (300°C and 300 bars) in neutral to slightly basic solutions for 2 months. The reacted surfaces were generally enriched in Al and depleted in Si, Ca, and Na relative to the starting composition. Depth profiling was not performed, so the thickness of these leached layers was not determined. However, from XPS depth profiling studies of albite dissolution under hydrothermal conditions (Hellmann et al. [55]), the labradorite leached layers are expected to be several hundreds of angstroms thick. Most importantly, however, the SAM analysis clearly shows the complexity of mineral dissolution. In the labradorite hydrothermal experiment, the surface composition changed systematically down the length of the core of powder in the reactor, but also locally on individual grains. In fact, it was shown that the surface chemistries both within etch pits and on smooth areas outside the etch pits were essentially the same. However, certain surfaces that showed a distinctive roughened and fragmented structure were severely depleted in Ca. In some of these areas, Ca was entirely removed from the near surface region. Na was not analyzed with SAM because of its well-known mobility under the influence of electron beams, especially near surfaces. However, XPS analysis of the same powder, although it lacked spatial information, showed that Na was also depleted in the near-surface region, and it is probably reasonable to assume that Ca and Na depletions are roughly parallel in a spatial sense. The upshot of this study is that mineral dissolution, as can be studied in the laboratory in a flow-through system, is a complex and inhomogeneous process. The variability of surface chemistry probably depends on the local density and spatial distribution of microcracks and other surface defects or high-energy sites, as well as the local flow patterns of the reacting solutions through the rock core.

6. Future Trends

There have only been a small number of effective studies in the earth sciences using AES. In these studies, AES has been used to add significantly to the understanding of certain aspects of mineral weathering and accretion mechanisms. However, the unique attributes of AES have not been exploited on a regular basis to solve problems in the earth sciences. AES has been used more as a novel probe rather than an integral part of the methodology of the geochemist.

There are both practical and scientific reasons for the relatively small number of AES studies in geochemistry. The most obvious reason is the effect of the electron beam on the samples, both in radiation damage and in charging. However as has been shown here the problems of charging can be overstated, and very often high spatial resolution Auger microprobe information is readily obtainable. Further, the quality of analysis of most elements can be relatively insensitive to beam-induced surface damage. Although it may seem surprising given the intensive commercial development of scanning Auger microscopes, a large part of the problem of applying AES to geochemical problems lies with the instrument design and data reduction techniques, which are really targeted for the semiconductor industry. This has lead to the development of instruments that are not optimum for geologic samples. First, the very heterogeneous nature of earth materials means that spatial sampling rather than imaging is a better technique for determining mineral and surface phase distributions. Second, the sensitivity of the electron spectrometers is two orders of magnitude below feasible practical limits. Third, the spectral data reduction techniques, and thus the limits of detection, are still developing, and again are not tuned to the needs of the geology community. Unfortunately, with the cost of these instruments being an order of magnitude more than a conventional SEM, it is unlikely that the geological applications of AES will ever do more than follow the trends of instrument development for other fields.

Signal processing of images and spectra is making considerable gains, and in directions that will be of use on geological specimens. Treating Auger imaging as a spatial sampling technique has significant benefits leading to the identification and quantification of chemical phase distributions at very high resolutions. For example, this approach has been used in the analysis of sintered SiC ceramics in a study by Browning and others [63]. The type of analysis has direct bearing on very high spatial resolution surface analysis of rocks and minerals. The SiC specimens were formed by the high-temperature sintering of SiC powder using B_4C as the sintering agent. Two specimens with different sintering histories were spatially sampled by stepping an electron beam in a raster across the sample in the region of triple points between three SiC grains. The peak heights of the Auger lines for C, B and N were collected at each point in the scans. The scans were 32 by 32 points over 7 by 7 μm areas using a 50-nm beam diameter. The dwell time per point was 600 ms with a beam current of 3×10^{-9} A (10^2 A/cm^2). While this dose rate is 10^4 times the rate used, for example, by Grant and others [7] on the lunar sample described above, the total dose per point is similar as for a spectrum over 0.1 mm in diameter. The

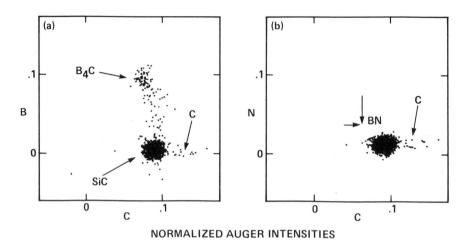

NORMALIZED AUGER INTENSITIES

Fig. 4-9. Scatter diagrams in two-dimensional composition space for (a) B vs. C and (b) N vs. C. See text for details. (From Browning et al., Advan. Ceram. Mater., *1987*, 2, 773, Fig. 7. Reprinted with permission from the American Ceramic Society.)

spatially sampled Auger data can be displayed as scatter diagrams, where chemically distinct areas on the sample bunch in different areas of these plots. For example, in Fig. 4-9 the Auger signals of B and C form a cluster of points with a high boron content corresponding to B_4C and a cluster with low B corresponding to SiC. This sample had been sintered in an argon atmosphere. In Fig. 4-10, the scatter diagrams of B, N, and C are shown for the SiC sampled sintered in a N_2 atmosphere. It can be seen from the N/B scatter diagram that the B and N signals are highly correlated, and from the B/C scatter diagram that there is no B_4C cluster. From these results, the conclusion that the B_4C had reacted with the N_2 sintering atmosphere to form BN and C is obvious. However, this was the first observation of BN in these systems; it had been missed using other techniques. Browning and others [63] compare the images collected using this method of AES microscopy with conventional backscattered images and carbon X-ray maps by wavelength dispersive analysis. The spatial resolution is greater than 30 times higher than the wavelength dispersive C map.

Spatial sampling can also be used to discriminate against charging over rough surfaces. Fig. 4-11a shows this type of result from

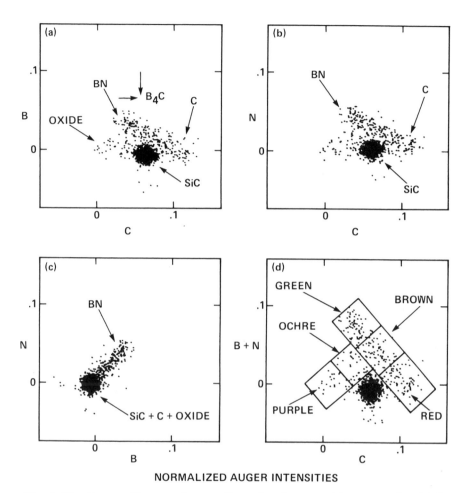

Fig. 4-10. Scatter diagrams in two-dimensional composition space for (a) B vs. C, (b) N vs. C, (c) N vs. B, and (d) B+N vs. C, plus color assignment windows. (From Browning et al., Advan. Ceram. Mater., *1987*, 2, 773, Fig. 11. Reprinted with permission from the American Ceramic Society.)

an uncoated plagioclase grain used in a hydrothermal dissolution experiment (Hochella et al. [60] and Browning [64]). The partially dissolved grain is fairly rough due to heterogeneous dissolution and secondary phase formation. As the electron beam scans across the sample surface there are some areas, generally indentations, that charge rapidly. If these areas are removed from the data set, the rest of the information can be put to good use. This can be achieved rather simply

a

b

Σ Backgrounds

Fig. 4-11. (a) SEM and (b) scatter diagram of hydrothermally altered plagioclase grain. Scatter diagram is from a 64 x 64 point Al-Si-Ca Auger image over the region marked on the SEM. See text for details. (From Browning, "Analytical Electron Microscopy", Joy, D.C., Ed. San Francisco Press, San Francisco, 1987, p. 311, Fig. 5. Reprinted with permission from the San Francisco Press.)

by using the fact that the background level changes dramatically during charging and far more than would be expected due to topographic effects. The Al Auger peak intensity is plotted against the sum of the backgrounds of all measured peaks, in this case Auger peaks for Al, Si, and and Ca, in Fig. 4-11b. The points that are outside the central distribution can be removed by windowing the data. Other signals such as the SEM signal could be used in a similar manner, dependent on the specimen and the measurement intended. Unfortunately, there are no such data manipulation routines in commercial instruments at present.

In addition to new techniques in data reduction of spectral information as described above, new AES instruments are currently being developed that will make this spectroscopy even more useful for earth scientists. One development is in the area of electron energy analyzers which will have a higher collection and transmission efficiency. These new analyzers will allow the operator to collect data with suitable signal-to-noise ratios while using lower beam currents over shorter times, and thus more effectively minimizing electron beam-induced surface damage and charging than is possible today. Another development is in the use of field-emission electron sources in beam columns. The advantage of these sources, besides the much higher brightnesses attainable compared to conventional thermoemission sources, is that high spatial resolutions are possible with low primary beam energies, again reducing the potential of surface damage and charging problems. The SEM photomicrograph shown in Fig. 4-11a is of an *uncoated* plagioclase grain imaged at a primary beam energy of 5 keV using a field-emission source. Another development that will make a major impact in the analysis of the surfaces of geologic samples is projection microscopy. A variety of instruments are being developed, and although at present there is no clear indication which design will be at the leading edge, sub-100-nm analysis of insulating surfaces appears feasible.

Therefore, the perceivable trends in the development of AES instrumentation are toward lower primary energies, higher collection efficiencies, and sophisticated data reduction techniques. All of these will make AES a more valuable technique for the earth sciences and will open up still newer frontiers in the study of surface-related processes.

Acknowledgments

Our Auger spectroscopic work has been generously supported by the Center for Integrated Systems and the Center for Materials Research at Stanford University.

References

[1]   Poppa, H.; Elliot, A.G., Surf. Sci., *1971*, 24, 149.

[2]   Carriere, B.; Deville, J.-P.; Goldsztaub S., C. R. Acad. Sci. Paris, *1970*, 271, 796.

[3]   Carriere, B.; Deville, J.-P.; Goldsztaub S., C. R. Acad. Sci. Paris, *1971*, 272, 951.

[4]   Gold, T.; Bilson, E.; Baron R.L., Proc. 5th Lunar Sci. Conf., *1974*, 3, 2413.

[5]   Gold, T.; Bilson, E.; Baron R.L., Proc. 6th Lunar Sci. Conf., *1975*, 3285.

[6]   Baron, R.L.; Bilson, E.; Gold, T.; Colton, R.J.; Hapke, B.; Steggert, M.A., Earth Planet. Sci. Lett., *1977*, 37, 263.

[7]   Grant, R.W.; Housley, R.M.; Szalkowski, F.J.; Marcus, H.L., Proc. 5th Lunar Sci. Conf., *1974*, 3, 2423.

[8]   Meitner, L., Zeits. Physik., *1923*, 17, 54; Auger, P., Comp. Rend., *1923*, 177, 169.

[9]   Wagner, C.D., Disc. Faraday Soc., *1975*, 60, 291.

[10]   Wagner, C.D.; Biloen P., Surf. Sci., *1973*, 35, 82.

[11]   Briggs, D.; Riviere, J.C., In "Practical Surface Analysis by Auger and X-ray Photoelectron Spectroscopy," Briggs, D., Seah, M.P., Eds. Wiley, New York, 1983, Chapter 3.

[12]   Remond, G.; Holloway, P.H.; LeGressus, C., Scanning Electron Microsc., *1981*, 1981/I, 483.

[13]   Bevolo, A.J. "Analytical Electron Microscopy," Joy, D.C., Ed. San Francisco Press, San Francisco, 1987, p. 340.

[14]   Lurie, P.G.; Wilson, J.M., Surf. Sci., *1977*, 65, 476.

[15]   Bruining, H. "Physics and Applications of Secondary Electron Emission," Pergamon Press, London, 1954.

[16]   Goldstein J.I.; Newbury, D.E.; Echlin, P.; Joy, D.C.; Fiori C.; Lifshin E., "Scanning Electron Microscopy and X-ray Microanalysis," Plenum Press, New York, 1981.

[17]   Reimer, L. "Scanning Electron Microscopy," Springer-Verlag, New York, 1985.

[18]   Dawson, P.H., J. Appl. Phys., *1966*, 37, 3644.

[19]   Davis, L.E.; MacDonald, N.C.; Palmberg, P.W.; Riach, G.E.; Weber, R.E., "Handbook of Auger Electron Spectroscopy," 2nd edition, Physical Electronics, Eden Prairie, MN, 1976.

[20] Sekine, T.; Hirata, K.; Mogami, A., Surf. Sci., *1983*, 125, 565.

[21] Armitage, A.F.; Woodruff, D.P.; Johnson P.D., Surf. Sci., *1980*, 100, L483.

[22] LeGressus, C.; Duraud, J.P.; Massignon, D.; Lee-Deacon, O., Scanning Electron Microsc., *1983*, 1983, 537.

[23] deBernardez, L.S.; Ferron, J.; Goldberg, E.C.; Buitrago, R.H., Surf. Sci., *1984*, 139, 541.

[24] Wehbi, D.; Roques-Carmes C., Scanning Electron Microsc., *1985*, 1985/I, 171.

[26] Anthony, M.T.; Seah, M.P. J. Electron Spectros. Relat. Phenom., *1983*, 32, 73; 32, 87.

[27] Sickafus, E.N., Phys. Rev., B, *1977*, 16, 1436.

[28] Sickafus, E.N., Phys. Rev., B, *1977*, 16, 1448.

[29] Sickafus, E.N.; Kukla, B., Phys. Rev., B, *1979*, 19, 4056.

[30] Prutton, M.; El Gomati, M.M., Surf. Interf. Anal., *1986*, 9, 99.

[31] Browning, R., Mater. Res. Soc. Bull., *1987*, Aug./Sept., 75.

[32] Matthew, J.A.D.; Prutton, M.; El Gomati, M.M.; Peacock, D.C., Surf. Interf. Anal., *1988*, 11, 173.

[33] Carriere, B.; Lang, B., Surf. Sci., *1977*, 64, 209.

[34] Pantano, C.G.; Madey, T.E., Appl. Surf. Sci., *1981*, 7, 115.

[35] Dawson, P.T.; Heavens, O.S.; Pollard, A.M., J. Phys. C, Solid State Phys., *1978*, 11, 2183.

[36] Gossink, R.G.; van Doveren, H.; Verhoeven, J.A.T., J. Non-Cryst. Solids, *1980*, 37, 111.

[37] Yau, Y.W.; Pease R.F.W.; Iranmanesh, A.A.; Polasko, K.J., J. Vacuum Sci. Technol., *1981*, 19, 1048.

[38] Clark, D.E.; Pantano C.G.; Hench, L.L. "Corrosion of Glass," Books for Industry and the Glass Industry, New York, 1979, pp. 15-19.

[39] Storp, S., Spectrochim. Acta, *1985*, 40B, 745.

[40] Levenson, L.L., Scanning Electron Microsc., *1982*, 1982/III, 925.

[41] Pivin, J.C., J. Mater. Sci., *1983*, 18, 1267.

[42] Hochella, M.F.; Lindsay, J.R.; Mossotti, V.G.; Eggleston, C.M., Am. Mineral., *1988*, 73, 1449.

[43] Hapke, B.W., The Moon, *1973*, 7, 342.

[44] Morse, J.W.; Mucci, A.; Walter, L.M.; Kaminsky, M.S., Science, *1979*, 205, 904.

[45] Mucci, A.; Morse, J.W.; Kaminsky, M.S., Am. J. Sci., *1985*, 285, 289.

[46] Mucci, A.; Morse, J.W., Am. J. Sci., *1985*, 285, 306.

[47] Mogk, D.W.; Locke, W.W., Geochim. Cosmochim. Acta, *1988*, 52, 2537.

[48] Staib, P., Rad. Effects, *1973*, 18, 217.

[49] Wells, O.C.; Bremer, C.G., J. Sci. Instrum. (J. Phys. E), *1969*, 2, 1120.

[50] MacDonald, N.C., Appl. Phys. Lett., *1970*, 16, 76.

[51] Venables, J.A.; Janssen, A.P.; Harland, C.J.; Joyce, B.A., Phil. Mag., *1976*, 34, 495.

[52] Hochella, M.F.; Turner, A.M.; Harris, D.W., Scanning Electron Microsc., *1986*, 1986/II, 337.

[53] Hochella, M.F.; Harris, D.W.; Turner, A.M., Am. Mineral., *1986*, 71, 1247.

[54] Berner, R.A.; Holdren G.R.; Schott, J., Geochim. Cosmochim. Acta, *1985*, 49, 1657.

[55] Hellmann, R.; Eggleston, C.M.; Hochella, M.F.; Crerar, D.A. Geochim. Cosmochim. Acta, *1990*, in press.

[56] Nesbitt, H.W.; Muir, I.J., Nature, *1988*, 334, 336.

[57] Muir, I.J.; Bancroft, G.M.; Nesbitt, H.W., Geochim. Cosmochim. Acta, *1989*, 53, 1235.

[58] Petit, J.-C.; Mea, G.D.; Dran, J.-C.; Schott, J.; Berner, R.A., Nature, *1987*, 325, 705.

[59] Casey, W.H.; Westrich, H.R.; Arnold, G.W.; Banfield, J.F., Geochim. Cosmochim. Acta, *1989*, 53, 821.

[60] Hochella, M.F.; Ponader, H.B.; Turner A.M.; Harris, D.W., Geochim. Cosmochim. Acta, *1988*, 52, 385.

[61] Chou, L.; Wollast, R., Am. J. Sci., *1985*, 285, 963.

[62] Holdren, G.R.; Speyer, P.M., Geochim. Cosmochim. Acta, *1985*, 49, 675.

[63] Browning, R.; Smialek, J.L.; Jacobson, N.S., Advan. Ceram. Mater., *1987*, 2, 773.

[64] Browning, R., "Analytical Electron Microscopy," Joy, D.C., Ed. San Francisco Press, San Francisco, 1987, p. 311.

5. FOURIER TRANSFORM INFRARED AND RAMAN SPECTROSCOPY

Clifford T. Johnston
Department of Soil Science
University of Florida
Gainesville, FL 32611-0150

1. Introduction

Vibrational spectroscopy has contributed more direct information about the structure and bonding of mineral-adsorbate complexes than any other experimental technique. Raman and infrared (IR) modes are intrinsically sensitive to small changes in the molecule's local environment. In surface studies, these perturbations can serve as diagnostic properties that may be used to explore surfaces and to gain insight into the interaction between adsorbates and mineral surfaces. As changes occur at the interface, the molecular environment around those constituent atoms of the molecule which execute periodic motion at a particular frequency (i.e., the vibrational chromophore) is perturbed. These perturbations are often represented as the appearance or loss of a particular set of bands, a change in relative intensity, a shift in frequency, or a change in the lineshape of a particular vibrational band. Correlation of these changes with data obtained using different approaches, and especially with macroscopic properties of the system such as thermodynamic stability constants, reaction kinetics, or dissolution phenomena, provides a powerful approach to characterize the behavior of mineral surfaces [1, 2]. A large body of experimental and theoretical evidence has established a strong correlation between vibrational spectra and surface structure and/or reactivity.

Although vibrational spectroscopy has long been recognized as a sensitive probe for surface applications, severe interferences from water, impurities, and the bulk adsorbent have restricted their application. Water, for example, presents a strong interference in IR spectroscopy because of its large IR absorption coefficient; the presence of even trace amounts of water can mask a significant portion of the IR spectrum. Consequently, many of the earlier vibrational studies were restricted to "dry" systems where removal of water was dictated by the sample presentation method. Nonetheless, these surface spectroscopic studies have provided a tremendous amount of mechanistic information. Since the first reported IR absorption study of hydrogen bonding between water and clay surfaces by Buswell et al. [3] in 1937, IR and Raman methods have contributed steadily to our understanding of the behavior of adsorbate interactions near mineral surfaces. Several excellent reviews of surface studies using conventional IR [4-8] and Raman [9, 10] techniques, provide a thorough introduction to the earlier work in this area.

The focus of this chapter will be to introduce Raman and Fourier transform infrared spectroscopy (FTIR) briefly, and to discuss recent applications to the study of mineral surfaces and adsorbed species. Although most of applications are confined to systems related to aqueous geochemistry, the techniques presented have been successfully used in a wide diversity of applications [11, 12]. The emphasis given here has been selected to illustrate the complementary relationship between Raman and FTIR spectroscopy in their application to surface chemistry, as well as their relationship to data obtained using different experimental methods.

Among the current trends in vibrational spectroscopy of mineral surfaces, three in particular stand out. The first is toward the application of in situ techniques which can provide nondestructive information about the sample [13,14]. There is a chronic need to obtain molecular information about the surface without disturbing or altering the interface. Many of the conventional surface methods involve harsh measurement techniques. Electron microscopy and microprobe analysis, inelastic electron tunneling (IETS), electron energy-loss (EELS) and x-ray photoelectron spectroscopy, for example, all require that the sample be subjected to a high vacuum. Although these methods have contributed significantly to our understanding of mineral surfaces, the invasive nature of the sample presentation technique often significantly perturbs the interfacial species of interest. Mineral surface chemistry reactions generally occur at hydrated interfaces; thus, there is a need for techniques that can observe surface complexes in the presence of water.

Another area of current interest results from the significant improvements in sensitivity and spectral discrimination of Raman and FTIR methods over the past few years. The recent introduction of continuous wave (cw) near-IR lasers and the application of quantum mid-IR detectors to surface studies, for example, are providing researchers with powerful tools with which to detect adsorbates at submonolayer coverages on difficult to analyze substrates previously. Two approaches are typically invoked to achieve this objective. The first attempts to enhance the spectroscopic signature of the adsorbate or surface functional group of interest, whereas the second involves minimizing the predominant interferences from the solvent, usually water, and from the bulk adsorbent.

Finally, a third area that has received considerable attention in recent years is the relationship of vibrational data to those obtained using experimental methods which are characterized by different time and energy scales [1, 2-15]. Structural information suggested by a particular method is intimately related to the time and energy scales associated with that approach. In addition, each technique is characterized by a particular set of strengths and limitations for a given application. Thus, a multiple-technique approach can generally provide a more coherent picture of the system of interest [16, 17]. This is especially true for the combined application of Raman and infrared methods, because the two methods are fundamentally different. Vibrational state transitions probed by IR spectroscopy must involve changes in dipole moment, whereas those probed by Raman spectroscopy must involve changes in polarizability [24]. Thus, each technique can, in principle, give a different picture of the vibrational spectrum. Combined application of the two techniques is often required in order to observe all of the available vibrational bands. Correlation of the spectroscopic data to macroscopic observables, however, remains the preeminent goal for molecular studies of mineral surfaces [18, 19].

2. Techniques

2.1 Raman Spectroscopy

When photons from a monochromatic light source interact with a sample, most of the scattered photons are not perturbed; i.e., they have the same energy as the incident photons. However, a small minority of the

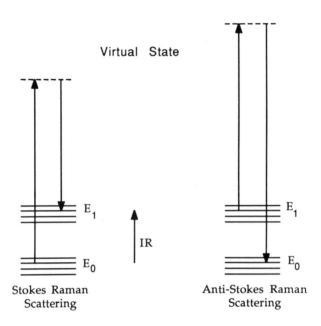

Fig. 5-1. Schematic of Stokes and anti-Stokes Raman scattering.

incident photons, roughly on the order of one out of every 10^8, interact with (are inelastically scattered) by the sample. Depending on the vibrational states of the sample, the incident photon can either gain energy from $(S_1 \rightarrow S_0)$, or lose energy to $(S_0 \rightarrow S_1)$ the sample (Fig. 5-1).

The latter process, known as anti-Stokes scattering, is negligible for vibrational energy transitions greater than a few hundred wavenumbers because essentially all of the molecules will be in their ground state. Conventional Raman spectroscopy is based on the former process which is termed Stokes scattering. In this case, a small amount of energy is transferred from the photon to the sample resulting from a vibrational transition from the ground state to an excited level. The energy of this inelastically scattered photon will be less than that of the incident photons, with the difference in energies corresponding to the energy of the vibrational transition. In other words, a small portion of the light scattered by the sample will be shifted to lower frequencies by discrete amounts corresponding to the vibrational energy levels encountered. To illustrate these processes, the Stokes and anti-Stokes Raman spectra of acetanilide are shown in Fig. 5-2. The strong, narrow spike at $\Delta v = 0$ cm^{-1} corresponds to the laser excitation frequency.

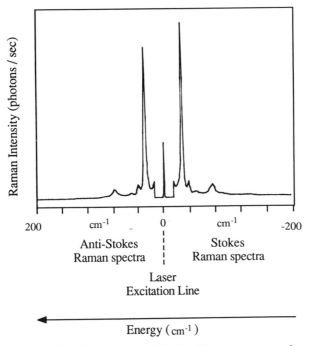

Fig. 5-2. Stokes and anti-Stokes Raman spectra of crystalline acetanilide in the -200–200 cm^{-1} region.

Since its discovery in 1928 by Prof. C. V. Raman [20], the Raman effect has proved to be a powerful method for studying vibrational and rotational transitions of molecules. Although Raman methods were restricted prior to the mid 1960s by the low intensity of gas discharge lamps (e.g., Hg-arc lamps), the use of lasers as illumination sources has removed this limitation. An illustration of this point is a comparison of the photon flux of a typical ion laser to that of a gas discharge lamp. The radiance from one of the most intense super-high pressure mercury lamps is about 250 W/(cm^2 sr); in contrast, the radiance of an Ar$^+$ ion laser is orders of magnitude greater. For example, consider an Ar$^+$ ion laser emitting light at 514.5 nm at 200 mW with a collimated beam diameter of 1.5 mm and a divergence of 0.5 mrad. The beam can be focused into a spot of r = 50 μm with the beam making a cone of half-angle θ = 2 mrad. The solid angle that the cone makes is $d\Omega \cong (\theta R)^2 \pi / R^2 \cong \theta^2 \pi \approx 1.3 \times 10^{-5}$ sr. The cross-sectional area of the laser beam $dA = 7.8 \times 10^{-5}$ cm^2. The radiance is defined by the incident power divided by the product (dA dΩ) or about $\approx 2.0 \times 10^8$ W/(cm^2 sr). This tremendous increase in intensity of the laser in comparison to the mercury arc lamp has resulted in the application of the Raman effect to a wide variety of applications in physics and chemistry, which have been reviewed extensively by Long [21], Turrell [22], Herzberg [23], and Wilson et al. [24].

As a surface technique, Raman spectroscopy provides a number of advantages over infrared absorption spectroscopy [25]. In addition to the high photon flux of the laser, the small laser spot size can be focused down to $\approx 10 \, \mu m$ in diameter. This can provide high spatial discrimination both of heterogenous surfaces and of mixtures of different solid phases [26]. Laser excitation sources can also provide detailed information about the molecular symmetry of Raman-active modes of crystalline materials when used with an optical polarizer and analyzer, because the electric vector of the light emitted from the laser is highly polarized. Sorption processes often result in a reduction in symmetry of the adsorbate at the interface, which may result in a breakdown in the selection rules for a particular species relative to its symmetry in solution.

Resonance Raman methods have recently been shown to offer considerable promise for interfacial studies. When the excitation wavelength of the laser coincides with an electronic transition of the sample, the vibrational modes associated with the electronic chromophore (that part of the molecule responsible for the electronic transition) can be resonantly enhanced by a factors of up to 10^3–10^6. Although resonance Raman spectra of biological materials have been obtained routinely over the past 10 years [27, 28], relatively few applications of this technique to surface studies have been reported [29–32]. There must be substantial overlap between the electronic transition of a particular chromophore and the laser excitation frequency in order for resonance Raman scattering to occur. Recent technological advances have resulted in the extension of laser excitation frequencies into both the ultraviolet (UV) and the near-IR regions. Thus, a greater number of surface complexes showing resonantly enhanced Raman spectra are expected, due to the extended range of laser frequencies.

2.2 Fourier Transform Spectroscopy

Perturbations of IR-active modes that accompany sorption processes have provided detailed information about the structure and reactivity of mineral surfaces. The application of IR methods to the study of adsorbed species has a long history and has been the subject of numerous reviews [4, 5, 7, 33, 34]. The majority of surface IR studies have employed transmission sample presentation using a dispersive-IR spectrometer. Fourier transform infrared spectrometers, however, have emerged over the past 20 years as being superior to dispersive-IR spectrometers for most applications. Dispersive spectrometers utilize gratings and slits to pass a nearly monochromatic component to the detector. Consequently, most of

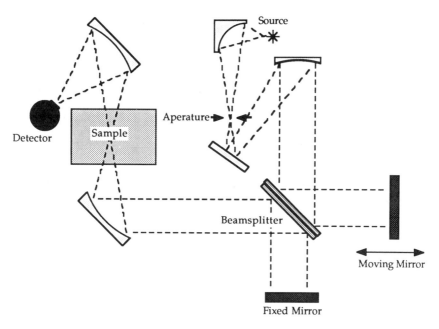

Fig. 5-3. Schematic of an interferometer showing the source, moving mirror, fixed mirror, beam splitter, and detector.

the incident energy from the broadband IR source is discarded by the monochromator; only a narrow spectral element is transmitted through the instrument. FTIR spectrometery, in contrast, is based upon optical interference between the two beams. The incident light coming into an interferometer is divided into two components by a beam splitter, and a phase difference between the two beams is introduced by changing the optical pathlength traveled by the two beams. Optical interference is achieved by subsequent superposition of the two beams. A thorough comparison of the performance and optical efficiency of FTIR to that of dispersive IR spectrometers is given by [7, 11, 35]. For comparable resolution and sensitivity in the midinfrared region, the overall efficiency advantage of FT over dispersive spectrometers has been estimated to be between a factor of 10 and 200 depending on the spectral region [11, 35].

The three principal components of a Fourier transform spectrometer are the source, interferometer, and detector. A schematic of a FT spectrometer incorporating a Michelson interferometers is shown in Fig. 5-3. Although many spectrometers use the Michelson interferometer, several commercial spectrometers use other types of two-beam interferometers. Most of the published work for chemical applications using Fourier transform infrared methods have been restricted to the mid- and far-IR spectral regions (i.e., 20 to 4000 cm^{-1}). This restriction no longer applies, as single FT instruments are currently available that can cover a spectral range extending from the far-IR through the entire visible portion of the spectrum [36]. The heart of the FT spectrometer is the interferometer, which consists of a beam splitter, a moving mirror, and a fixed mirror. A collimated beam from the source is divided into two components by the beam splitter. The portion of the incident beam that is transmitted through the beam splitter is passed directly onto the fixed mirror (though some interferometer designs switch the moving and fixed mirrors), whereas the reflected portion is incident upon the moving mirror. A modulated beam results from the optical interference from recombination of the transmitted and reflected portions (i.e., superposition of the two components) at the beam splitter surface. The modulated beam from the interferometer then passes through the sample compartment and on to the detector (Fig. 5-3). The emissivity of the source and the frequency responses of the beam splitter and the detector must be given careful consideration in order to obtain optimum sensitivity for a given spectral region of interest.

A major development which has improved the sensitivity of FTIR spectroscopy significantly has been the development of quantum mid-infrared detectors. A common example is the liquid nitrogen cooled, mercury/cadmium/telluride (MCT) photoconductive detector. These detectors provide improved sensitivity over the conventional pyroelectric deuterated triglycine sulfate (DTGS) detector by a factor of 4 to 50, depending on the composition of the MCT detector. In addition to increased sensitivity, these detectors are characterized by a much faster response time than the DTGS detectors. Hence, whereas relatively slow mirror velocities are required for DTGS detectors, considerably higher mirror velocities can be used with MCT detectors. This produces rapid-scan capabilities; scan collection rates of up to 60 scans per second have been reported using an MCT detector. The spectral resolution of FTIR spectra collected using rapid-scan methods, however, is very low (> 16 cm^{-1}).

In comparing the performance of FT to dispersive spectrometers, two main differences stand out [7,11,35]. First, the optical throughput is greater (referrred to as the Jacquionot advantage) for an interferometer than for a monochromator of comparable size operating in the mid-IR region. The optical throughput advantage of a FT spectrometer is particularly high compared to that of a dispersive instrument in the higher wavenumber region; however, this advantage diminishes in the lower frequency ranges. A second main distinction between the two types of spectrometers is related to the fact that radiation at all frequencies are measured simultaneously with a FTIR spectrometer versus sequentially with a grating monochromator. All of the energy remains incident upon the detector at all times for a FT spectrometer, whereas only a small fraction of the energy is incident upon the detector at any particular time for a dispersive instrument. This advantage, called the multiplex or Fellgett's advantage, is directly proportional to the square root of the number of spectral elements obtained. For a typical measurement from 400 to 4000 cm^{-1} at 1 cm^{-1} resolution, the theoretical multiplex advantage of the FT method is 84.

The most common method of sample presentation for both IR and FTIR studies of mineral surfaces is transmission spectroscopy. Included in this category are methods using KBr pellets, self-supporting clay films, deposits on IR windows, and compressed wafers (e.g., Si wafers [77]). Surface studies and descriptions of transmission cells using this technique have been reviewed extensively [4,35] and will not be discussed here. Diffuse-reflectance (DR) FTIR methods have been used recently, however, to provide a sensitive tool for studying powders, such as clay minerals and hydrous metal oxides [37–39]. The application of DR methods to the study of adsorbed species has been extended recently by the development of several controlled-environment diffuse reflectance cells [40]. These cells allow the composition of the vapor phase in contact with the adsorbent to be controlled. A limitation of diffuse-reflectance methods has been the lack of spectral reproducibility. This is because the observed signal depends largely on how the sample is placed in the sample cup; subtle changes introduced by repacking or otherwise changing the sample typically result in large spectral artifacts. Controlled-environment (CE) DR cells minimize this problem by allowing spectra of the solid/adsorbate complex to be collected in-situ; the surface loading of the sample can be changed without disturbing the sample in the spectrometer. In particular, CEDR methods should provide a useful approach for examining vapor-phase sorption and desorption onto and from mineral surfaces. In addition to controlling the composition of the vapor phase, such cells also permit heating of the sample to approximately 600°C, and evacuation of the cell to 10^{-6} Torr [40].

3. Applications

3.1. Observation of Low-Frequency Adsorbate Modes

A major disadvantage of infrared absorption techniques for surface studies is the opacity of the adsorbent in the low-frequency region below ≈ 1200 cm^{-1} [4,41-43]. The strong IR-active absorption bands of the adsorbent block out a significant portion of the incident radiation in this region. This problem is enhanced further by the fact that a significant amount of adsorbent is often required in order to detect the adsorbate. FTIR spectra of surface complexes are frequently obtained by obtaining absorbance or transmittance spectra of the adsorbent prior to treatment and of the reacted adsorbate/adsorbent complex. The spectrum of the adsorbent is then typically subtracted from that of the adsorbate/adsorbent complex in order to observe the perturbed adsorbate bands. A subjective judgment is often required to decide how much of the adsorbent spectrum of the substrate to subtract from the spectrum of the surface complex. This makes the unambiguous observation of adsorbate vibrational modes based upon spectral subtractions difficult in the lower wavenumber region [44]. In contrast, the corresponding Raman-active structural modes of the adsorbent are poor Raman scatterers and present minimal interference [9, 10]. Thus, Raman methods are particularly well suited for studying vibrational modes of surface/adsorbate complexes below 1200 cm^{-1}, provided that background fluorescence of the sample can be minimized [29 ,30 ,43 ,45–47].

To illustrate this point, compare the Raman spectra of the kaolinite-dimethyl sulfoxide complex [45] and of kaolinite (Fig. 5-4) to their corresponding IR spectra (Fig. 5-5). Although the amount of dimethyl sulfoxide (DMSO) intercalated by the kaolinite sample is high ($\approx$2 DMSO molecules per unit cell of kaolinite), the IR spectrum of the clay/organic complex is similar to that of the clay itself. The most significant differences apparent between the two spectra are some weak bands at 1320, 1110, and 960 cm^{-1}, and a slight reduction in the intensity of the 940 cm^{-1} band in the IR spectrum of the kaolinite/DMSO intercalate (Fig. 5-5). A significantly different result for these samples, however, is obtained using Raman spectroscopy (Fig. 5-4). The spectrum of the adsorbent kaolinite is characterized by one strong feature at 141 cm^{-1} and by several weak features in the 100 to 1000 cm^{-1} region [14]. In contrast, the Raman spectrum of the kaolinite/DMSO intercalate (Fig. 5-4) is dominated by Raman-active modes of the adsorbate DMSO. Two observations appear evident from these data: (1) the adsorbate vibrational bands can readily be observed, and (2) the structural framework bands of the substrate kaolinite present virtually no interference in this region [45].

The diagnostic vibrational modes of the adsorbate, those bands which provide the most valuable information about their local environment, are often found below 1200 cm^{-1}. Thus, mechanistic information about the interaction between the guest adsorbate and the surface often relies upon observations in this region. For the kaolinite/DMSO intercalate, an important diagnostic vibrational band is the sulfonyl stretching mode. This band occurs at 1042 cm^{-1} in pure DMSO. The DMSO molecule has been observed to coordinate to surfaces via both the oxygen and the sulfur atoms. On the basis of previous spectroscopic investigations [48], a shift to lower

Fig. 5-4. Raman spectra of kaolinite and of the kaolinite/Me$_2$SO (DMSO) complex, in the 100 to 1000 cm^{-1} region. Reproduced from Ref. [45] with permission .

wavenumber of the ν(S=O) band is indicative of coordination via the oxygen atom, whereas a shift to higher wavenumber is attributed to coordination through the sulfur atom. Upon examination of the ν(S=O) band of the kaolinite-DMSO complex in the 900–1150 cm^{-1} spectral region, a shift to lower wavenumber was observed. This indicated that DMSO coordinated to the kaolinite surface through the oxygen atom in the interlamellar region (Fig. 5-6). This result was later confirmed by combined neutron and x-ray powder diffraction [49] and solid state ^{13}C and ^{29}Si magic angle spinning nuclear magnetic resonance (MAS NMR) [50] studies of the kaolinite-DMSO intercalate where conclusive structural data were obtained which indicated coordination through the oxygen atom.

Fig. 5-5. Diffuse reflectance FTIR spectra of kaolinite and of the kaolinite/Me$_2$SO (DMSO) complex.

Observation of the $\nu(S=O)$ band was not possible in the IR spectrum because of the intense interference from the lattice vibrational modes in the 1125–975 cm^{-1} region (Fig. 5-5).

Discrimination of surface species using Raman spectroscopy in regions virtually inaccessible to IR absorption also has been reported for a number of sorption studies using hydrous metal oxides as adsorbents. Knözinger and Jeziorowski [46, 51], for example, discriminated against the paramolybdate ion Mo$_7$O$_{24}^{2-}$ in favor of the MoO$_4^{2-}$ ion as the predominant surface species on η- and γ-Al$_2$O$_3$ on the basis of in situ Raman data in the low-frequency region. A related study which followed the calcination of aluminum and titanium oxide supported WO$_3$, MoO$_3$, and V$_2$O$_5$ using in situ Raman techniques [43]. Low-frequency Raman methods have also been used to study the interaction between chemisorbed hydrogen-sequestering agents and the surface hydroxyl groups of silica

Fig. 5-6. (Opposite page) Projection of four half-unit cells of (a) kaolinite and (b) a kaolinite/Me$_2$SO (DMSO) complex onto the (100) plane. Reprinted from Ref. [45] with permission.

[42]. In each case, the observed perturbations of low-frequency Raman bands contributed significantly to our overall understanding of the interfacial chemistry. Similar observations using IR methods would be difficult if not impossible in this spectral region, due to strong interference from the adsorbent.

Although Raman techniques have an advantage in the spectral region below 1200 cm^{-1} for systems with tolerable fluorescence, IR absorption methods have often been used to examine adsorbate modes in this region [52]. Despite the fact that the strong IR-active modes of the substrate block out certain spectral regions, spectral "windows" of suitable optical transmission to observe adsorbate modes can still be found. For example, IR absorption bands of the C–D deformation and the C–O stretching bands deuterated dimethyl ether adsorbed on alumina in the 1150–1000 cm^{-1} regions were studied as a function of temperature [90]. Upon heating the complex from 150 K to 400 K the spectral perturbations of these two bands indicated clearly that a surface methoxy species was formed above 250 K at the expense of the dimethyl ether. Similar IR methods were used to study the Claus reaction

$$2\,H_2S \quad + \quad SO_2 \quad \xrightarrow{\quad Al_2O_3 \quad} \quad 2\,H_2O \quad + \quad (3/n)\,S_n \quad (5\text{-}1)$$

by examining the low-frequency IR-active modes of SO_2 and H_2S adsorbed on γ-Al_2O_3 [53–55]. A recent review of spectroscopic techniques used to characterize systems with respect to catalysis, including FTIR and Raman methods, has appeared [16].

3.2 Surface Resonance Raman Studies

One of the greatest limitations of Raman spectroscopy as a surface technique is the intrinsic inefficiency of the scattering process itself. For a typical surface complex, on the order of only one out of every 10^{15} photons incident upon the sample will be Raman scattered. This extremely low flux of Raman-scattered photons has severely restricted application of the technique to surface studies. Photon contributions from background fluorescence, a major interference in Raman spectroscopy, enhance the problem further by swamping the Raman signal with a strong emission background [9, 10]. Consequently, the more successful Raman surface studies have utilized background substrates which are characterized by (1) low fluoresence levels, and (2) high surface areas in order to increase the number of adsorbate Raman scatterers (e.g., silica). Ideal adsorbates, in this respect, are those molecules which have large Raman-scattering cross-section values (i.e., strong Raman scatterers). Pyridine, for example, has frequently been used as a probe molecule to explore active sites on surfaces.

The symmetric ring-breathing mode at 991 cm^{-1} appears as a strong band in the Raman spectrum of pyridine. The position of this band is strongly correlated to the mechanism of adsorption; bonding mechanisms involving physical adsorption, hydrogen bonding, proton transfer, or coordination to electron-deficient Lewis acid sites have been distinguished depending on the shift in frequency of this band.

The Raman-scattering cross-section of a particular adsorbate vibrational mode can be enhanced tremendously if the mode is coupled to an electronic transition that can be stimulated by the excitation wavelength of the laser. This effect is called resonance Raman scattering and can result in enhanced signals that are on the order of 10^3–10^6 greater than those observed for the non-resonance case [28, 56]. Resonantly enhanced Raman-scattering for adsorbed species can be obtained if the adsorbate has an electronic transition(s) which has significant overlap with the excitation frequency of the laser. A continuum of highly monochromatic laser-excitation frequencies is currently accessible through the combined application of ion, solid-state Nd:YAG, diode, diode-pumped, and dye lasers in extending the range of this technique from the ultraviolet region through the visible into the near-IR region [57]. Thus, a greater number of applications showing resonantly enhanced Raman spectra of surface species are being reported as technological advances continue to expand the range of laser-excitation frequencies. As a protoypical example of this method, the application of resonance Raman spectroscopy (RR) to the study of aromatic hydrocarbon interactions with clay mineral surfaces will be considered.

Aromatic compounds have been observed to undergo chemical transformations on a variety of 2:1 layer-silicate surfaces. Under dry conditions, an electron-transfer mechanism has been proposed to account for the chemisorption of aromatic compounds (e.g., benzene, toluene, phenol, and anisole) with transition-metal-exchanged montmorillonite clays. Transition metal cations that have been found to promote such reactions include Cu(II), Fe(III), Ru(III) and VO^{2+}. To illustrate the steps involved in the proposed electron-transfer mechanism, the Cu^{2+}-montmorillonite/benzene complex (represented by X=H) will be considered below.

The aromatic hydrocarbon (benzene in this example) is adsorbed on the interlamellar surface of the hydrated clay (Step 1). The d$_{001}$ spacing of the Cu-montmorillonite at 50% RH increases from 1.24 nm to 1.50 nm for the Cu-montmorillonite/benzene complex [58]. The clay must be hydrated for sorption to occur; otherwise, the interlamellar surfaces of the clay are not accessible. The color of the clay film changes from pale blue to light yellow upon adsorption of benzene, with the electron spin resonance (ESR) spectrum of this complex being dominated by the Cu^{2+} signal (d^9 electron configuration). Removal of water from the hydration sphere of the interlamellar transition metal cation (e.g., Cu^{2+}) makes possible the donation of π electrons from the aromatic adsorbate to the unfilled d orbitals of the metal cation (Step 2). This can be accomplished by desiccation using P_2O_5, by use of a vacuum, or by refluxing in an anhydrous solvent such as hexane.

Step 1: Intercalation into hydrated montmorillonite

Step 2: Dehydration-induced electron transfer reaction

The single electron-transfer reaction results in reduction of the metal cation (e.g., $Cu^{2+} \rightarrow Cu^{1+}$) along with concurrent formation of a radical organic cation. Step 2 is characterized by several prominent physical changes. First, the ESR signal from the Cu^{2+} (d^9 electron configuration) ion is strongly reduced with a concomitant increase in intensity of a new band having a g value of 2.0024 [59, 60]. Such observations indicate that the Cu^{2+} metal cation is reduced to Cu^+, with an organic free radical being formed concurrently. Second, the color of the Cu-montmorillonite/benzene complex changes from light yellow to dark red, suggesting that a polymerization or electron-transfer reaction has occurred.

Step 3a: Polymerization of benzene to poly(p-phenylene)

Further polymerization reactions can occur if the para positions on the ring are open, as in the case of benzene (Step 3a). Evidence for the polymerization of benzene is provided by the fact that a strong band grows in intensity at $\lambda_{max} \approx 500$ nm and shifts into near-IR region as the reaction proceeds. Benzene itself does not have an absorption feature in this region. Severe attenuation of the IR signal in the 4000–2500 cm^{-1} region suggests that this low-energy electron-transfer adsorption band extends into the IR region as well [58, 60, 61]. Additional evidence was provided by a mass spectroscopic study of the Cu-montmorillonite/benzene complex [62]. Although no high molecular weight materials from the complex were detected directly, such materials were identified in methanol extracts. This suggests that polymerization of benzene had occurred.

If, on the other hand, the para positions are occupied by a substituent, as in the case of p-xylene and p-dimethoxybenzene (Step 3b), then polymerization of the adsorbate either does not occur or is strongly reduced. In this case, the radical organic cation is itself "trapped" on the surface. It has been suggested by several researchers that 2:1 clays may provide a novel template for the "trapping" radical organic cations [59,61]. Upon rehydration, the ESR Cu^{2+} signal reappears and the color changes back to its original color prior to dehydration (Step 4). This indicates that the electron-transfer reaction is reversible. This drying/wetting cycle can be repeated several times.

Step 3b: Polymerization hindered if para positions are occupied

Step 4:   Rehydration of clay/arene complex

Several significant changes in the IR spectra of these surface complexes have been observed to accompany the above-mentioned color changes [58, 60, 63]. The IR-active C—C stretch and the C—H out of plane deformation vibrational modes, in particular, are sensitive to the apparent changes reflected by the changes in color. The C—C stretching band has been observed to decrease in frequency by 8–15 cm^{-1}, and the C—H out of plane deformation band to increase in frequency by 23–53 cm^{-1}, upon desiccation (Step 2) [58]. These changes indicated a loss in aromatic character of the adsorbate upon dehydration, similar to the changes induced in p-disubstituted benzene compounds by strongly electron-withdrawing substituents. Although the limited IR data indicate that a chemical transformation of the adsorbate occurs, the identity of the new surface species remains unknown. Detailed observation of additional IR modes, such as the ν(C—H) bands, was not possible. This was because of the strong scattering of radiation by the substrate, along with the limited sensitivity of the dispersive IR spectrometer used for these studies.

Resonance Raman (RR) spectroscopy provided a solution to this interference problem by increasing the effective Raman cross-section value of the adsorbate through resonance Raman enhancement. Soma et al. [31, 32] utilized the excitation frequencies of an ion laser, and of a dye laser pumped by an ion laser, to collect resonantly enhanced Raman spectra of aromatic hydrocarbons sorbed on Cu(II)- and Fe(III)-exchanged montmorillonites. Upon dehydration, the large color changes reflected in the visible absorption spectra provided a novel means with which to study the chemisorbed species responsible for the color change. The laser excitation frequency was selected to overlap significantly with the "new" UV/visible adsorption band of the clay/arene complex. The resulting RR spectrum of the chemisorbed species was quite different from the spectrum of free benzene and similar to that of p-terphenyl. On the basis of the RR spectra, these workers proposed the formation of poly(p-phenylene) chains. More recent ESR evidence, however, has indicated that a variety of reaction products are formed in the interlamellar region, depending on the type of cation, water content, temperature, and reaction time [64].

In the case of transition metal/clay/arene complexes, the principle electronic absorption bands in the UV-visible spectrum are those of the adsorbate. More importantly, the chemisorption of the aromatic adsorbate is solely responsible for the significant change in color. Thus, the RR spectrum of the chemisorbed species can be obtained by appropriate selection of a laser excitation frequency. Although surface and background fluorescence still contribute unwanted photons to the signal, the greatly enhanced signal of the adsorbate permits discrimination of the surface species. Resonantly enhanced Raman spectra also have been observed for a select group of aromatics which includes biphenyl and terphenyl [32], monosubstituted benzenes [30], anisole, p-dimethoxybenzene [29, 65], thiophene, and methylthiophene [61].

There are recent FTIR, ESR, sorption, and kinetic data which suggest that other adsorbates as well may be susceptible to similar electron-transfer reactions; these include 4-chloroanisole [66], dibenzo-p-dioxin, and 1- and 2-chlorodibenzo-p-dioxin [67], dimethylaniline [68], benzidine [69,70], aniline and p-chloroaniline [71]. A comprehensive review of abiotic organic reactions at mineral surfaces, which include single electron-transfer reactions, has been made recently by Voudrias and Reinhard [72]. The growing list of aromatic and other conjugated compounds that have been shown to participate in these surface-induced electron-transfer reactions attests to their importance in natural systems.

Unfortunately, many chemisorption reactions that occur in nature do not provide the dramatic color changes necessary to exploit the resonance Raman effect. Technological advances, however, are resulting in the continual expansion of laser-excitation frequencies. This, in turn, should allow resonantly enhanced Raman spectra to be obtained in spectral regions previously unaccessible to conventional laser excitation frequencies. A recent example is the commercial development of cw Nd:YAG lasers which emit at 1.064 μm (9398 cm^{-1}) [73–75]; this "new" excitation frequency will undoubtedly result in the observation of near-IR resonantly enhanced Raman scattering. Application of solid-state diode lasers has even greater potential, because such units are tunable over a significant portion of the near-IR regions. Few surface Raman studies have been reported to date that have used these excitation lines.

3.3. Interference from Fluorescence Background Scattering

The strong background scattering known as fluorescence is one of the greatest obstacles to the application of Raman spectroscopy for surface studies. Raman studies of high surface area materials, in particular, are plagued by high fluorescence levels [9, 10]. Because fluorescence is an inherently efficient resonance phenomenon, it is often much stronger than the intrinsically weak Raman signal. Detection of Raman-scattered photons for a highly fluorescent sample, which includes most surface/adsorbate complexes, is extremely difficult because the photomultiplier tube is a shot-noise limited device. In this case, the noise of the photomultiplier tube is proportional to the square root of the intensity of the incident light. Thus, as the light intensity increases due to background fluorescene, there is a corresponding increase in the noise level of the detector. For highly fluorescent samples, the shot noise generated from fluorescence can completely mask the Raman signal.

Organic impurities, surface hydroxyl groups, and transition metal impurities such as iron have all been shown to contribute to sample fluorescence. Several experimental approaches have been proposed to reduce the amount of fluorescence [9, 10, 34]. Among the least disruptive (invasive) of these is the use of a pulsed excitation source, such as a mode-locked laser. This method takes advantage of the fact that vibrational transitions occur over a considerably shorter time scale (between 10^{-12} and 10^{-15} s) than that of the fluorescence process ($\approx$ 10^{-9} s) [34]. For each pulse from the mode-locked laser, the detector is turned on to count the Raman-scattered photons and then turned off before the "fluorescent" photons contribute significantly to the signal. An associated disadvantage of this method, however, is that sufficient time must be allowed between pulses to allow the fluorescence signal to decay once more to a low enough level; thus, the effective duty cycle can be very low. A simpler, but less effective, procedure is the "drench/quench" method. The fluorescence level has been observed to decrease slowly to a tolerable level after the sample has been exposed to the laser beam over an extended period of time (several hours). Little is known about the mechanism of this photobleaching process [10].

Numerous invasive approaches also have been used to reduce fluorescence levels. For example, several researchers have used "cleaning" procedures to modify the surfaces of interest. For studies involving silica gel, porous Vycor glass, and Cab-O-sil, fluorescence levels were reduced by heating the samples in an oxygen atmosphere at 500°C and, subsequently, keeping the samples under vacuum. Additional fluorescence-suppression methods include washing the adsorbent in 30% hydrogen peroxide at 100°C, in concentrated nitric acid laced with potassium perchlorate, or in 0.2 mol/L NaOH followed by calcining at 400°C [76]. For surface studies of adsorbed species, however, these invasive methods are clearly undesirable.

A practical solution for the suppression of fluorescence is to choose an excitation frequency which reduces the resonance effect. This has typically required the selection of longer excitation wavelengths, such as the 647.1 nm line of the Kr^+ ion laser. As Chase [74] has recently pointed out, however, this excitation frequency reduces but does not eliminate fluorescence for most samples. As mentioned previously, cw Nd:YAG lasers that emit at 1064 nm have recently become commercially available. The real advantage that these lasers provide in surface Raman applications is a solution to the fluorescence problem. Fluorescence is strongly reduced at 1 μm; thus, the cw Nd:YAG excitation source is ideal for samples with high fluorescence backgrounds. This benefit is not without penalty, however, since Raman-scattering cross-section is sharply reduced at longer wavelengths. Raman-scattering intensity is inversely proportional to the fourth power of the laser wavelength [21]. Thus, by using the 1064-nm line of the Nd:YAG laser, a significant loss in Raman intensity resulted from the v^4 dependence. For comparison, the Raman intensity obtained using the 1064-nm line of the Nd:YAG laser would be approximately 30 times less than that obtained using the 457-nm line of the Ar^+ ion laser at the same incident laser power. Higher laser intensities thus must be used for near-IR Raman-scattering studies to compensate for the reduced cross-section due to the v^4 dependence. Care must then be taken to avoid unwanted sample heating and photodecomposition of the sample which can result from high incident power levels. Low-energy surface/adsorbate interactions are particularly susceptible to these undesirable effects.

Fourier transform Raman (FT-Raman) spectroscopy has received considerable attention recently [73, 74]. The multiplexing advantage of the Fourier transform spectrometer makes up for some of the loss in intensity due to the longer wavelength excitation frequency of the Nd:YAG laser. It has been demonstrated that a conventional FTIR spectrometer can even be used for FT-Raman measurements if the proper beam splitter and detector are chosen. The FT spectrometer is used in the emission mode, with the Rayleigh- and Raman-scattered light from the sample serving as the light source. The typical frequency range for FT-Raman studies using the 1064 nm excitation frequency is from approximately 5500 to 9500 cm^{-1}, a region which is accessible to most commercial FTIR spectrometers. In a comparison of FT- and conventional dispersive Raman methods, the quantum efficiency of the detector and the optical throughput of the spectrometer were greater for the FT instrument than for the scanning instrument [74]. A schematic of a FT-Raman experiment is shown in Fig. 5-7. Typically, a quartz or CaF$_2$ beam splitter is used in conjunction with an indium gallium arsenide or germanium photodiode detector.

The laser excitation line is many orders of magnitude more intense than the weak Raman lines. Thus, the laser line and the accompanying Rayleigh wings must be attenuated. Dielectric interference filters have been used that have high extinction at 1064 nm; these filters, however, prevent observation of low-frequency Raman-active modes below 200–300 cm^{-1}. A number of research groups are currently working on improving the collection optics and sample presentation geometry, removal of the quasi-elastic scattered light, and improving the detector for FT-Raman spectroscopy.

Although no surface studies of adsorbed species have been reported to date that utilize FT-Raman spectroscopy, this method has been applied successfully to biological molecules [75]. The principle limitations of Raman techniques for surface studies has been the high levels of background fluorescence and the low efficiency of Raman scattering. The application of FT-Raman methods using the 1064 nm line of the cw Nd:YAG laser addresses both of these restrictions. First, fluorescence is eliminated by using the 1064 nm excitation frequency. Samples that were previously considered impossible to analyze using conventional laser excitation sources can thus be examined. Second, the optical throughput and efficiency of the Fourier transform spectrometer are greater than for a conventional scanning monochromator. However, this second advantage only partially compensates for the corresponding large loss in intensity due to the v^4 dependence.

3.4 Aqueous Studies of Colloidal Suspensions

The infrared intensity of a particular mode depends on the square of the variation of the dipole moment, μ_i, with respect to the normal coordinate Q_i, $(\partial\mu_i/\partial Q_i)^2$. Raman band intensities, however, depend on the square of the change in polarizability, α_i, with normal coordinate, $(\partial\alpha_i/\partial Q_i)^2$. This difference in selection rules results in the fact that liquid water is less Raman active than IR-active. The O—H stretching and bending modes of water induce a large change in the dipole moment, and thus absorb radiation

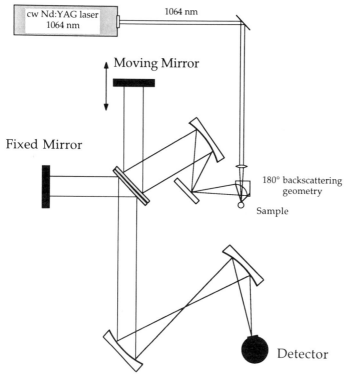

Fig. 5-7. Schematic of an FT Raman spectrometer.

strongly in the 3100–3500 cm^{-1} (v_1 and v_3; v[O—H]) and 1600–1700 cm^{-1} (v_2; δ[O—H]) regions. These same vibrational modes, however, produce a smaller change in polarizability, $(\partial\alpha_i/\partial Q_i)^2$. As a result, these modes have lower Raman intensities. A direct comparison between IR and Raman band intensities of the water molecule is not possible because the vibrational excitation mechanisms are fundamentally different. Infrared transitions involve the direct absorption of light, whereas Raman transitions result from the inelastic scattering of light by the sample. In general, however, water produces less interference in Raman than in IR spectroscopy. Consequently, Raman methods are well suited for studying surface reactions in aqueous suspension with minimal interference from the solvent, water.

Raman spectra of kaolinite in aqueous suspension compared to dry kaolinite are shown in Figs. 5-8, and 5-9 for several different spectral regions [14]. The presence of water did not interfere significantly with observation of the Raman-active bands of kaolinite. This is true even in the

Fig. 5-8 Raman spectra of (a) KGa-1 kaolinite in aqueous suspension, (b) dry KGa-1 kaolinite, and (c) dry Mesa Alta kaolinite in the 100–1000 cm^{-1} region (left side) and in the 200–1000 cm^{-1} region (right side) Reprinted from Ref. [14] with permission.

hydroxyl stretching region (3600–3700 cm^{-1}), which is close to the v_1 and v_3 water bands (Figs. 5-9, 5-10). Raman spectra of the v(O-H) bands of kaolinite exhibit a strong pH dependence, showing significant enhancement at higher pH values (Fig. 5-10); e.g., a fourfold increase in intensity was observed at pH 9.5 in comparison to pH 5. It is well established that the interparticle association of kaolinite is highly pH dependent. Thus, the observed pH dependence of the Raman data may provide a useful approach for studying interparticle colloidal interactions in aqueous suspension [14].

Adsorption isotherms for benzene adsorption onto silica in aqueous suspension, and from the vapor phase, have been determined using Raman methods [77]. The integrated intensity of a specific vibrational mode, the ring-breathing mode, of adsorbed benzene was determined as a function of benzene solution/vapor concentration. Although the vapor-phase adsorption isotherms were clearly resolved, the solution spectra were complicated by the presence of a strong benzene component in solution. A major difficulty that is frequently encountered with spectroscopic surface studies is associated with the unambiguous assignment of spectral features to a surface species. Discrimination of spectral features for an adsorbate on the surface from those of the adsorbate in bulk solution is an especially difficult

Fig 5-9 Raman spectra of (a) KGa-1 kaolinite in aqueous suspension, (b) dry KGa-1 kaolinite, and (c) dry Mesa Alta kaolinite in the 3600–3725 cm^{-1} region Reprinted from Ref. [14] with permission.

task for physical adsorption processes. An example of such processes is the adsorption of benzene onto silica, where the surface interaction is relatively weak [77] and hence the observed vibrational pertubations are small. In the specific case of the ring-breathing mode of benzene adsorbed onto silica, adsorbed benzene was identified at 989 cm^{-1} in comparison to the solution benzene value at 992 cm^{-1}. Given that the full-width at half-maximum value for this band is 10 cm^{-1}, assigning physical meaning (surface vs. bulk solution) to two bands separated by 3 cm^{-1} may not be justified [78]. As the strength of the surface interaction increases (e.g., physisorption $\rightarrow$ chemisorption), larger vibrational perturbations occur. These permit spectral features of surface components to be assigned with greater confidence.

Raman methods have also been used to examine solute/solvent interactions in aqueous suspension. Hydrogen-bonding interactions of

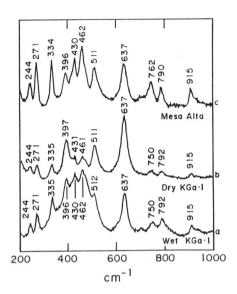

Fig. 5-10. pH dependence of Raman spectra of
kaolinite in aqueous suspension. Reprinted from
Ref. [14] with permission.

dilute polar solutes, such as DMSO, have been studied in aqueous solution
[48, 79]. One area that has not been explored fully is the potential use of
Raman methods to probe changes in aqueous solution for a particular
component which is associated with an interfacial surface. Raman and IR
methods have been used to study of the aqueous surface chemistry of
soluble silicates over a wide range of pH [80]. The Raman spectra provide
direct information about the chemical speciation of the silicate ions as a
function of pH. Although the study in question was restricted to soluble
silicates, this type of approach should have direct application to the study of
mineral formation, stability, and dissolution reactions.

As noted above, water strongly interferes with IR spectroscopy
because of its large absorption (i.e., extinction) coefficient. For effective
cell pathlengths greater than 6 μm, virtually all of the incident IR radiation
in these regions is attenuated by the sample [11, 81, 82]. Thus, cells of
extremely short pathlength (< 6 μm) must be used for IR measurements in
these regions. As the pathlength of the cell increases, the optical "window"
where sufficient energy is transmitted through the sample is severely
constricted due to beam attenuation associated with water absorption.

Short-pathlength IR transmission and reflection cells have now been
designed for studies in aqueous solution [81, 83]. The band intensity for a
particular component is directly proportional to the effective pathlength of
the cell according to the Beer-Lambert law.

$$\tau(v) = \exp(-\varepsilon c d) = \left(\frac{I(v)}{I_o(v)}\right) \qquad (5\text{-}2)$$

$$A(v) = (\varepsilon c d) = \log_{10}\left(\frac{I_o(v)}{I(v)}\right) \qquad (5\text{-}3)$$

$$\overline{A} = \int_{v_1}^{v_2} A(v)\, dv = \int_{v_1}^{v_2} \log_{10}\left(\frac{I_o(v)}{I(v)}\right) dv \qquad (5\text{-}4)$$

where the following definitions apply:

$\tau(v)$ sample transmittance at v

$A(v)$ sample absorbance at v

$\overline{A}$ integrated absorbance

ε molar absorptivity ($dm^2\,mol^{-1}$)

d effective pathlength (dm or cm)

c sample concentration ($mol\,dm^{-3}$)

$I_o(v)$ intensity of incident beam

$I(v)$ intensity of transmitted beam

As shown above, the integrated absorbance for a particular component is directly proportional to the pathlength, d (Eqt. 5-4). The short pathlength of the above-mentioned cells generally restricts their application to concentrations greater than 10^{-3} mol/L. More dilute solutions can be examined as well if the pathlength of the cell is increased; however, this also increases the unwanted interference from water and reduces the energy transmitted by the sample. The optical materials used to construct these cells have characteristic cutoff frequencies that further restrict the optical "window" available for spectral observations.

    Transmission IR cells with pathlengths on the order of 5 μm are not suitable for studies of most aqueous colloidal suspensions because of clogging and particle exclusion problems. Internal-reflection spectroscopy (IRS) has been used successfully as an alternative approach for aqueous studies. Rather than passing the incident beam directly through the sample, in this case, the beam is passed through an internal-reflection element in contact with the sample. The effective pathlength of the cell can be varied by changing the number of reflections. Internal-reflection elements, prism designs, sample presentation geometries, and numerous applications have been reviewed [81, 82]. Although IRS-based methods were used even prior to the introduction of FTIR spectrometers, the higher optical throughput and efficiency of interferometer-based techniques has significantly expanded their application to surface studies.

The majority of IRS aqueous studies have been oriented toward interfacial reactions involving biological or polymeric materials [84, 85]. A recent application of IRS to the study of goethite in aqueous solution [13] has demonstrated that the technique can also provide useful information about the aqueous surface chemistry of minerals. In situ FTIR spectra of goethite were studied in aqueous (H_2O and D_2O) suspensions using a cylindrical internal-reflection cell. The infrared-active modes of goethite were shown to be perturbed by changes in pH, ionic strength, and phoshphate concentration.

Reflection-based techniques are characterized as well by several disadvantages. First, internal-reflection spectra of dilute solutes rely heavily upon spectral subtraction techniques, where a subjective judgment is required to determine how much of the reference spectrum must be subtracted from the sample spectrum. Spectral subtraction techniques can often introduce spectral artifacts near the regions where water or other components absorb radiation strongly. Second, the short pathlength of the IRS cell limits the sensitivity for detection of dilute species. An additional complication for adsorbate/surface studies is that the colloid in aqueous suspension can interact directly with the internal reflection element. For a charged colloid in aqueous suspension such as a hydrous metal oxide or clay mineral the interaction between the colloid and the IRE will also be pH dependent. Changes due to surface coating of the IRE by the colloid can be difficult to separate from changes taking place at the colloid/solution interface. Conversely, there is evidence that the behavior of the adsorbates themselves can be influenced by the IRE.

3.5. Intercalation Studies of Mineral/Organic Complexes

Vibrational spectroscopy has been used extensively in conjunction with powder x-ray diffraction techniques to study the mechanism of insertion and intercalation of organic molecules into the interlamellar region of kaolin subgroup minerals. Kaolinite, the most important member of this group of 1:1 layer aluminosilicates, possesses both a gibbsite-like aluminum hydroxyl layer containing inner-surface hydroxyl groups, and an opposing siloxane surface. The siloxane surface is characterized by a negative surface charge. Therefore, moderate strength electrostatic bonds are formed between the inner-surface hydroxyl groups of the aluminum octahedral layer and the siloxane oxygen atoms in the interlamellar region. Intercalation of kaolinite is restricted to a select group of small, highly polar, organic molecules which include dimethyl sulfoxide (DMSO), hydrazine, formamide, and N-methyl formamide. Each is capable of disrupting the cohesive interlayer electrostatic bonds, permitting insertion of the guest species into the interlamellar region.

Upon intercalation, the effective surface area of kaolinite increases from ≈ 20 m^2/g to a value in excess of 800 m^2/g, a 40-fold increase. The structure of the kaolinite/DMSO intercalation complex is shown in Fig. 5-6. Once the kaolinite structure has been expanded, additional guest species, which are not capable of direct intercalation reactions by themselves, can then be inserted into the expanded structure. Because the number of adsorbers (IR) or scatterers (Raman) increases substantially upon intercalation as a result of the increased surface area, a concomitant increase in the signal from the adsorbed species is achieved. Detection levels of adsorbed intercalates at a fraction of monolayer coverage (< 0.05) can therefore be obtained.

A major problem in surface spectroscopy is the discrimination of spectral contributions of the surface species from those present in the surrounding liquid or vapor phase. The vibrational modes of intercalated species can often be discriminated from those of nonintercalated species because the intercalated species form moderate to strong hydrogen bonds with inner-surface hydroxyl groups which induce shifts in frequencies of the adsorbate [86, 87]. As a result, the intercalated species is characterized by a set of vibrational bands that are distinct from those of the non-interfacial species. The intercalation induced frequency shifts of the adsorbate serve as a sensitive diagnostic property of the probe intercalate molecules.

3.6. Vibrational Studies of Surface Hydroxyl Groups

Another approach to studying surface adsorbate complexes involves the use of surface hydroxyl groups as stationary reporter groups. These functional groups have been utilized extensively as reporter groups, with the shifts in position of the hydroxyl stretching and deformation bands being used to determine the bonding mechanisms of the adsorbate to the interlamellaer surface. Structural hydroxyl groups, in general, are characterized by large absorptivities and therefore provide a useful tool with which to examine the behavior of adsorbed species in the interfacial region.

The interaction of phosphate with the surface hydroxyl groups of goethite has received considerable attention using IR methods as a prototypical oxyanion/hydrous metal oxide system. Goethite is a model adsorbent for oxyaninon sorption studies in at lease two respects; first, the vibrational modes associated with the surface hydroxyl groups have been well characterized; and, second, this oxide has a high adsorption capacity for oxyanions such as phosphate. At least three types of surface hydroxyl groups (A, B, and C) occur on the exposed surface of goethite [88, 89].

Parfitt observed that after goethite was reacted with 205 μmol/g of HPO_4^{2-} the intensities of the type B and C bands were only slightly perturbed, but the relative intensity of the type A hydroxyl was strongly reduced. This result suggested strongly that the type A hydroxyl groups were considerably more reactive to phosphate than were the B and C groups.

4. Summary

Raman and FTIR spectroscopy are versatile techniques for studying mineral surfaces and their interaction with adsorbed species at the solid interface. Very few restrictions apply with respect to the type of sample that can be analyzed. High surface area materials, single crystals, composite mixtures, and colloids in aqueous suspsension have all been subjected to vibrational analysis. A general trend in vibrational spectroscopy has been toward the development of nondestructive sampling techniques. Considerable progress has been made in this area, so in situ surface studies are being reported with increasing frequency in the literature. Of particular interest for aqueous geochemistry applications has been the development and application of methods that can observe colloids in aqueous suspension.

Recent technological advances have extended the range of laser-excitation frequencies available for Raman studies. The two most noteworthy advances in this regard have been (1) the recent application of cw Nd:YAG lasers to Raman studies of surfaces, which eliminates background fluorescence; and (2) the use of multiple laser-excitation frequencies to obtain resonantly enhanced Raman spectra of adsorbed species. In addition, the development of quantum midinfrared detectors has greatly increased the sensitivity of FTIR methods. All of these improvements have resulted in increased ability to analyze difficult samples. Many surface/adsorbate complexes that were once considered to be virtually impossible to analyze using Raman or FTIR methods can now be examined quite readily.

Each of the physical methods and sampling techniques discussed herein is characterized by a distinct set of advantages and limitations. An attempt has been made to identify each of these areas, so that more judicious application of these techniques to a particular problem can be made. Raman and FTIR methods are complementary methods; combined application of the two techniques always results in a more complete picture of the surface or of the surface complex of interest. Since each technique is also restricted to a fundamental time scale, experimental techniques that are characterized by a range of time scales must often be applied in order to develop a more coherent picture of the system. The time scale associated with vibrational transitions is very short. Therefore, combined application using medium time-scale techniques (NMR, ESR, dieletric relaxation) and long time-scale measurements (x-ray powder diffraction, thermodyanmic properties, reaction kinetics) will provide more complete information about the static and dynamic properties of the surface complex.

Acknowledgments

This work was supported, in part, by the Environmetnal Sciences Branch of the United States Air Force Contract No F0835-89-C-0152.

References

[1]   Rudin, M.; Motschi, H., J. Colloid. Interf. Sci., *1984*, 98, 385.

[2]   Motschi, H., Colloids and Surfaces, *1984*, 9, 333.

[3]   Buswell, A.M.; Krebs, K.; Rodebush, W.H., J. Am. Chem. Soc., *1937*, 59, 2603.

[4]   Little, L.H., Kiselev, A.V. and Lygin, V.I., "Infrared Spectra of Adsorbed Species." Academic Press, London, 1966.

[5]   Hair, M.L., Transmission infrared spectroscopy for high surface area oxides. In "Vibrational Spectroscopies for Adsorbed Species", Bell, A.T.; and Hair, M.L., Eds. Washington D.C., ACS Symposium Series No. 137, 1980.

[6]   Hair, M.L., "Infrared Spectroscopy in Surface Chemistry," Dekker, New York, 1967.

[7]   Bell, A.T., Applications of Fourier transform infrared spectroscopy to studies of adsorbed species. In "Vibrational Spectroscopies for Adsorbed Species", Bell, A.T.; and Hair, M.L., Eds. Washington D.C., ACS Symposium Series No. 137, 1980.

[8]   Mortland, M.M., Advan. Agron., *1970*, 22, 75.

[9]   Egerton, T.A., Catal. Rev. Sci. Eng., *1975*, 11, 71.

[10]   Cooney, R.P.; Curthoys, G.; Tam, N.T., Advan. Catal. *1975*, 24, 293.

[11]   Griffiths, P.R. and deHaseth, J.A., "Fourier Transform Infrared Spectrometry," John Wiley & Sons, New York, 1986, pp. 1-656.

[12]   Willis, R.F., Ed., "Vibrational Spectroscopy of Adsorbates," Springer-Verlag, Berlin 1980, p. 1.

[13]   Tejedor-Tejedor, M.A.; Anderson, M.A., Langmuir, *1986*, 2, 203.

[14]   Johnston, C.T.; Sposito, G.; Birge, R.R., Clays and Clay Miner., *1985*, 33, 483.

[15]   Sposito, G.; Prost, R., Chem. Rev., *1982*, 82, 553.

[16]   Austermann, R.L.; Denley, D.R.; Hart, D.W.; Himelfarb, P.B.;
       Irwin, R.M.; Narayana, M.; Szentirmay, R.; Tang, S.C.; Yeates,
       R.C., Anal. Chem., *1987*, 59, 68R.

[17]   Hipps, K.W.; Dunkle, E.A.; Mazur, U., Langmuir, *1986*, 2, 528.

[18]   Sposito, G., Distinguishing adsorption from surface precipitation.
       In "Chemical Processes at Mineral Surfaces," Davis, J.A. and
       Hayes, K.F. Eds., American Chemical Society, Washington D.C.
       1986, p. 217.

[19]   Johnston, C.T.; Sposito, G., Disorder and early sorrow: progress
       in the chemical speciation of soil surfaces. In "Future Developments
       in Soil Science Eesearch," Boersma, L.L., Ed., Soil Science
       Society of America, Madison, Wisconsin, 1987, p. 89.

[20]   Krishnan, R.S.; Shankar, R. K., J. Raman Spectrosc., *1981*,
       10, 1.

[21]   Long, D.A., "Raman Spectroscopy." McGraw-Hill, New York,
       1977, pp. 1-276.

[22]   Turrell, G., "Infrared and Raman Spectra of Crystals." Academic
       Press, London, 1972. pp. 1-384.

[23]   Herzberg, G., "Molecular Spectra and Molecular Structure: II.
       Infrared and Raman Spectra of Polyatomic Molecules." Van
       Nostrand Reinhold Company, New York, 1945, pp. 1-632.

[24]   Wilson, E.B., Decius, J.C. and Cross, P.C., "Molecular
       Vibrations: The Theory of Infrared and Raman Vibrational Spectra."
       Dover Publications, Inc., New York: 1955. pp. 1-388.

[25]   Morrow, B.A., Raman spectroscopic studies of surface species. In
       "Vibration Spectroscopies for Adsorbed Species," Bell, A.T. and
       Hair, M.L., Eds., ACS Symposium Series No. 137, Washington,
       D.C., 1980, p. 119.

[26]   Haley, L.V.; Wylie, I.W.; Koningstein, J.A., J. Raman Spectrosc.,
       1982, 13, 203.

[27]   Spiro, T.G.; Stein, P., Ann Rev. Phys. Chem., *1977*, 28, 501.

[28]   Clark, R.J.H.; Stewart, B., Struct. Bond., *1979*, 36, 1.

[29]   Soma, Y.; Soma, M., Chem. Phys. Lett., *1983*, 94, 475.

[30]   Soma, Y.; Soma, M.; Harada, I., J. Phys. Chem., *1985*, 89, 738.

[31]   Soma, Y.; Soma, M.; Harada, I., Chem. Phys. Lett., *1983*, 99,
       153.

[32]   Soma, Y.; Soma, M.; Harada, I., J. Phys. Chem., *1984*, 88, 3034.

[33]   Kiselev, A.V. and Lygin, V.I., "Infrared Spectra of Surface
       Compounds.", Wiley, New York, 1975.

[34]   Delgass, W.N., Haller, G.L., Kellerman, R.; Lunsford, J.H., "Spectroscopy in Heterogeneous Catalysis." Academic Press, New York, 1988, pp. 1-341.

[35]   Bell, A.T., Infrared spectroscopy of high-area catalytic surfaces. In "Vibrational Spectroscopy of Molecules on Surfaces," Yates, J.T. and Madey, T.E., Eds., Plenum, New York, 1987, p. 105.

[36]   Baiardo, J.; Mukherjee,R.; Vala, M., Appl. Spectrosc., *1981*, 35, 510.

[37]   Fuller, M.P.; Griffiths, P.R., Anal. Chem., *1978*, 50, 1906.

[38]   Fuller, M.P.; Griffiths, P.R., Appl. Spectrosc., *1980*, 34, 533.

[39]   Hamadeh, I.M.; Yeboah, S.A.; Trumbull, K.A.; Griffiths, P.R., Appl. Spectrosc., *1984*, 38, 486.

[40]   Hamadeh, I.M.; King, D.; Griffiths, P.R., J. Catal., *1984*, 88, 264.

[41]   Morrow, B.A., J. Phys. Chem., *1977*, 81, 2683.

[42]   Morrow, B.A.; Hardin, A.H., J. Phys. Chem., *1979*, 83, 3135.

[43]   Chan, S.S.; Wachs, I.E.; Murrell, L.L.; Wang, L.; Hall, W. K. J., Phys. Chem., *1984*, 88, 5831.

[44]   Morrow, B.A.; Devi, A., J. Chem. Soc. Faraday Trans. 1, *1972*, 68, 403.

[45]   Johnston, C.T.; Sposito, G.; Bocian, D.F., J. Phys. Chem., *1984*, 88, 5959.

[46]   Jeziorowski, H.; Knozinger, H., J. Phys. Chem., *1979*, 83, 1166.

[47]   Reichmann, M.G.; Bell, A.T., Langmuir, *1987*, 3, 563.

[48]   Bertoluzza, A.; Bonora, S.; Battaglia, M.A.; Monti, P., J. Raman Spectrosc., *1979*, 8, 231.

[49]   Thompson, J.G.; Cuff, C., Clays Clay Miner., *1985*, 33, 490.

[50]   Thompson, J.G., Clays Clay Miner., *1985*, 33, 173.

[51]   Knozinger, H.; Jeziorowski, H., J. Phys. Chem., *1978*, 82, 2002.

[52]   Morrow, B.A.; Cody, I.A., J. Phys. Chem., *1976*, 80, 1998.

[53]   Datta, A.; Cavell, R.G.; Tower, R.W.; George, Z.M., J. Phys. Chem., *1985*, 89, 443.

[54]   Datta, A.; Cavell, R.G., J. Phys. Chem., *1985*, 89, 450.

[55]   Datta, A.; Cavell, R.G., J. Phys. Chem., *1985*, 89, 454.

[56]   Drissler, F., Collective Phenomena, *1981*, 3, 147.

[57]   Lafferty, H.W., Spectroscopy, *1988*, 2, 23.

[58]   Pinnavia, T.J.; Mortland, M.M., J. Phys. Chem., *1971*, 75, 3957.

[59] Rupert, J.P., J. Phys. Chem., *1973*, 77, 784.

[60] Pinnavia, T.J.; Hall, P.L.; Cady, S.S.; Mortland, M.M., J. Phys. Chem., *1974*, 78, 994.

[61] Soma, Y.; Soma, M.; Furukawa, Y.; Harada, I., Clays Clay Miner., *1987*, 35, 53.

[62] Mortland, M.M.; Halloran, L.J., Soil Sci. Soc. Am. J., *1976*, 40, 367.

[63] Stoessel, F.; Guth, J.L.; Wey, R., Clay Min., *1977*, 12, 255.

[64] Eastman, M.P.; Patterson, D.E.; Pannell, K.H., Clays and Clay Min., *1984*, 32, 327.

[65] Fenn, D.B.; Mortland, M.M.; Pinnavaia, T.J., Clays Clay Min., *1973*, 21, 315.

[66] Govindaraj, N.; Mortland, M.M.; Boyd, S.A., Environ. Sci. Technol., *1987*, 21, 1119.

[67] Boyd, S.A.; Mortland, M.M., Nature, *1985*, 316, 532.

[68] Vansant, E.F.; Yariv, S., J. Chem. Soc. Faraday Trans., *1977*, 73, 1815.

[69] Tennakoon, D.T.B.; Thomas, J.M.; Tricker, M.J., J. Chem.Soc. Dalton Trans., *1974*, 20, 2207.

[70] McBride, M.B., Clays Clay Miner., *1979*, 27, 224.

[71] Cloos, P.; Moreale, A.; Broers, C.; Badot, C., Clay Miner., *1979*, 14, 307.

[72] Voudrias, E.A.; Reinhard, M., Abiotic organic reactions at mineral surfaces. In "Geochemical Processes at Mineral Surfaces," ACS Symposium Series Vol 323, Davis, J. A. and Hayes, K.F., Eds., American Chemical Society, Washinton,D.C., 1986, p. 462.

[73] Hallmark, V.M.; Zimba, C.G.; Swalen, J.D.; Rabolt, J.F., Spectroscopy, *1987*, 2, 40.

[74] Chase, B., Anal. Chem., *1987*, 59, 881.

[75] Lewis, E.N.; Kalasinsky, V.F.; Levin, I.W., Anal. Chem., *1988*, 60, 2306.

[76] Saperstein, D.D.; Rein, A.J., J. Phys. Chem., *1977*, 81, 2134.

[77] Sayed, M.B.; Cooney, R.P., J. Colloid. Interf. Sci., *1983*, 91, 552.

[78] Cameron, D.G.; Kauppinen, J.K.; Moffatt, D.J.; Mantsch, H.H., Appl. Spectrosc., *1982*, 36, 245.

[79] Kelm, H.; Klosowski, J.; Steger, E., J. Mol. Struct., *1975*, 28, 1.

[80] Marinangeli, A.; Morelli, M.A.; Simoni, R.; Bertoluzza, A., Can. J. Spectrosc., *1978*, 23, 173.

[81] Harrick, N.J., "Internal Reflection Spectroscopy, 2nd ed", Harrick Scientific Corp., Ossining, New York, 1979, pp. 1-327.

[82] Mirabella, F.M. and Harrick, N.J., "Internal Reflection Spectroscopy: Review and Supplement." 1985, pp. 1-195.

[83] Bertie, J.E.; Eysel, H.H., Appl. Spectrosc., *1985*, 39, 392.

[84] Knutson, K.; Lyman, D.J., Surface infrared spectrsocopy. In "Surface and Interfacial Aspects of Biomedical Polymers. Vol. 1. Surface Chemistry and Physics," Andrade, E., Ed. Plenum Press, New York, 1985 p. 197.

[85] Koenig, J.L., Anal. Chem., *1987*, 59, 1141A.

[86] Olejnik, S.; Aylmore, L.A.G.; Posner, A.M.; Quirk, J.P., J. Phys. Chem., *1968*, 72, 241.

[87] Olejnik, S.; Posner, A.M.; Quirk, J.P., Clays Clay Miner., *1971*, 19, 83.

[88] Parfitt, R.L., Adv. Agron. *1978*, 30, 1.

[89] Parfitt, R.L.; Russell, J.D., J. Soil Sci. *1988*, 28, 297.

[90] Chen, J.G.; Basu, P.; Ballinger, T. H.; Yates, J. T., Langmuir, *1989*, 5, 352.

6. MÖSSBAUER SPECTROSCOPY

Frank J. Berry
School of Chemistry
University of Birmingham
Birmingham B15 2TT, United Kingdom

1. Introduction

In recent years mineralogy has developed even stronger links with solid-state chemistry and physics. Central to this field of study has been an understanding of chemical bonding or electronic structure in minerals which has been developed both theoretically and by the use of spectroscopic techniques. Mössbauer spectroscopy, mainly in conventional transmission geometry, has been used for many years to examine the bulk bonding, structural, and magnetic properties of naturally occurring solids, and the subject has been thoroughly reviewed in the recent past [1].

As mentioned in the preface to this book, current interest in the nature of solid surfaces is, to some extent at least, a reflection of recent developments in techniques with a capacity to examine those superficial regions of solids that have hitherto resisted investigation. Such developments have enabled more accurate assessments to be made of the bonding characteristics, structural properties, and elemental compositions of solid surfaces. Mössbauer spectroscopy has also been developed in recent years to enhance its capacity to examine the fundamental properties of solid surfaces and molecular processes at the solid/gas and solid/liquid interfaces. This chapter therefore attempts to outline the principles of Mössbauer spectroscopy and to give some indication of the apparatus involved and how the technique may be used for the examination of solid surfaces. The remainder of the chapter is devoted to examples of how Mössbauer spectroscopy can be used to study the surface properties of mineral and geologic materials and seeks to illustrate how the Mössbauer spectra can be interpreted in terms of the composition, bonding, and structural properties of the solid surfaces.

2. Mössbauer Spectroscopy

The theory and practice of Mössbauer spectroscopy are described in detail elsewhere [2]. The reader requiring a less detailed summary of the fundamentals of the subject is referred to an article written to celebrate the twenty-fifth anniversary of the Mössbauer effect [3], while the use of Mössbauer spectroscopy to elucidate information of a chemical, structural, magnetic, and dynamical nature has been comprehensively surveyed in a recently published book [4]. The following is a brief account designed to give the reader an outline appreciation of the general principles of the technique, the instrumentation involved, and the interpretation of the results.

2.1. The Mössbauer Effect

The Mössbauer effect is the recoilless emission of gamma radiation from a nucleus in an excited state and its subsequent recoil-free resonant absorption and reemission, i.e., resonant fluorescence, by a nucleus of the same isotope in the ground state.

The decay of many radioactive nuclei in an excited state occurs by the emission of gamma rays, and the possibility that these might excite other stable nuclei of the same isotope and give rise to nuclear resonant absorption and fluorescence was recognized earlier this century. The initial attempts to detect these resonant processes were unsuccessful mainly because of the nuclear recoil which accompanies both the emission and absorption of the γ-ray by a free nucleus. As a result of this nuclear recoil, the energy of the emitted γ-ray is less than the energy difference between the two nuclear levels and consequently is insufficient to be resonantly absorbed by another nucleus in the ground state. Hence, for free nuclei, the recoil energy prevents resonant absorption of γ-rays under normal circumstances.

In 1957 Mössbauer showed that a nucleus in a solid can sometimes emit and absorb γ-rays without recoil because, when embedded in a solid matrix, the nucleus is no longer isolated but is fixed within the lattice. In this situation the recoil energy may be less than the lowest quantized lattice vibrational energy, such that the γ-ray may be emitted without any loss of energy due to the recoil of the nucleus. Since the probability of these recoil-free events depends on the energy of the γ-ray, the Mössbauer effect is restricted to certain isotopes with low-lying excited energy states. The probability is also optimized at low temperatures and depends on the vibrational properties of the solid in which the Mössbauer nucleus is situated. It is also important to appreciate that the Mössbauer γ-ray has a finite spread in energy, which is quantified by a linewidth and is related to the lifetime of the nuclear excited state. Hence, whether or not an isotope of an element exhibits the Mössbauer effect, and whether or not an isotope will give chemically significant data, depends on inherent properties of the nucleus that cannot be changed. In general, the Mössbauer effect is optimized for low-energy γ-rays associated with nuclei strongly bound in a crystal lattice at low temperatures.

Although the Mössbauer effect has been detected in over 100 isotopes, most of the practical applications of the effect in the technique known as Mössbauer spectroscopy have involved a smaller number of isotopes in which the conditions for observing the Mössbauer effect and obtaining useful information are particularly favorable. Indeed the majority of Mössbauer spectroscopic studies have involved the use of iron-57 as this has the most advantageous combination of properties for Mössbauer spectroscopy. However, similar and equally valuable information can be obtained from isotopes of many other elements. For example, the author of this article has examined solids containing Mössbauer active isotopes of iron, tin, antimony, tellurium, iodine, iridium, and europium, and the list may readily be extended to include others such as tungsten, ruthenium, and gold.

2.2. Mössbauer Spectroscopy

2.2.1. Principles. The energy levels of a nucleus of an atom in a solid are modified by the electronic and lattice environment of the nucleus. Mössbauer spectroscopy enables these nuclear energy levels to be investigated and is therefore a means by which the nucleus can be used as a probe of its environment.

The most usual experimental arrangement for Mössbauer spectroscopy involves a radioactive source containing the Mössbauer isotope in an excited state together with the material to be investigated, which contains the same isotope in the ground state and which is called the absorber. For example, the source for ^{57}Fe Mössbauer spectroscopy is normally radioactive ^{57}Co, which undergoes a spontaneous electron capture transition to give a metastable state of ^{57}Fe; this then decays to the ground state via a γ-ray cascade which includes the 14.4-keV Mössbauer γ-ray.

The radioactive source nuclei are usually embedded in a solid matrix to provide the required rigid environment. In conventional transmission mode Mössbauer spectroscopy, the γ-rays emitted by the source pass through the absorber, where they may be partially absorbed, and then pass to a suitable detector. The examination of the energy levels of the Mössbauer nuclei in the absorber requires the modification of the energy of the γ-rays emitted by the source so that they can have the correct energy for resonant absorption. This is usually achieved by moving the source relative to the stationary absorber and thereby giving the γ-rays an energy shift as a result of the first-order relativistic Doppler effect. Resonant absorption occurs when the energy of the γ-ray just matches the nuclear transition energy for the Mössbauer nucleus in the absorber. The situation is described schematically in Fig. 6-1. The Mössbauer spectrum therefore consists of a plot of γ-ray counts against the velocity of the source with respect to the absorber and is measured in millimeters per second. A Mössbauer spectrum is illustrated in Fig. 6-2 and shows that, in the simplest case where the source and absorber contain the Mössbauer isotope in the same cubic environment, a single line spectrum at zero velocity is obtained.

2.2.2. Mössbauer Spectrometer. A schematic representation of a Mössbauer spectrometer is illustrated in Fig. 6-3. The Doppler velocities of ca. ±10 mm/s are normally obtained by an electromechanical trans-

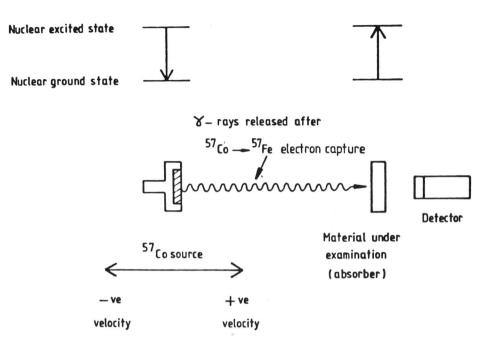

Nuclear excited state

Nuclear ground state

γ- rays released after

$^{57}Co \longrightarrow {}^{57}Fe$ electron capture

Detector

Material under
examination
(absorber)

^{57}Co source

— ve
velocity

+ ve
velocity

Fig. 6-1. Schematic representation of Mössbauer spectroscopy. The energy of the γ-ray emitted by the source is modified by oscillating the source backwards and forwards with respect to the stationary absorber. (Reproduced from [3] with permission.)

ducer, similar to a loudspeaker, and driven by a suitable electronics system. The detection of the γ-radiation is normally achieved by a gas-filled proportional counter, a scintillation detector, or a high-resolution germanium or lithium-drifted germanium detector. The single-channel analyzer selects the counts corresponding to the Mössbauer γ-ray. The synchronization of the γ-ray counting and the source motion is now generally achieved by a microprocessor system in which the counts are accumulated in channels corresponding to the velocity of the source relative to the absorber. The spectrum is accumulated over periods of hours or days, during which time the spectrum may be monitored on a display screen. When a spectrum with a satisfactory signal-to-noise ratio has been obtained, it is normally stored on magnetic tape or disc for subsequent computer analysis.

The quality of the Mössbauer spectrum and the information that can be obtained from it can frequently be enhanced by making measurements with the absorber, and sometimes the source, at low temperatures. Sometimes additional information can be obtained by recording the spectra in

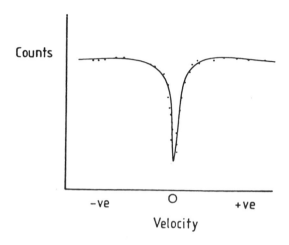

Fig. 6-2. Simple representation of a Mössbauer spectrum which is composed of a plot of γ-ray counts against the velocity of the source with respect to the stationary absorber measured in millimeters per second.

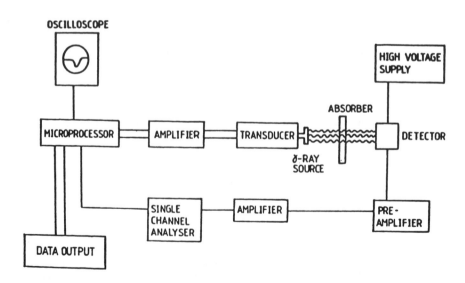

Fig. 6-3. Schematic representation of a microprocessor-controlled Mössbauer spectrometer.

applied magnetic fields, and many of the experimental aspects of Mossbauer spectroscopy concern the provision of the required sample environment. The sample under investigation is important and must contain a sufficient amount of the Mössbauer isotope to provide a measurable resonant absorption.

2.2.3. Surface Studies by Mössbauer Spectroscopy

Transmission Mössbauer spectroscopy. The experimental aspects of Mössbauer spectroscopy that have been outlined above correspond to the most usual case of an absorption experiment involving transmission geometry and represent the way in which most Mössbauer spectroscopic studies of minerals have been performed in the past. Such methods can be an important means of examining surface effects if the particles are small so that the high surface area of the mineral contains a large number of Mössbauer-active species in surface sites.

Source experiments. The Mössbauer technique has also been used in other ways for the study of mineralogical materials. For example, it is possible to perform Mössbauer spectroscopic measurements in which the sample under investigation forms the radioactive source; this allows measurements involving elements that form the parent nuclei in a Mössbauer decay scheme. In this method the Mössbauer nucleus can be incorporated within the sample surface, and the spectrum is recorded using the radioactive sample as a source against a standard absorber to give a Mössbauer emission spectrum. In the case of ^{57}Co Mössbauer emission spectroscopy, for example, 1 mCi of ^{57}Co is sufficient to permit the acquisition of good quality data, and the technique is therefore capable of examining systems containing a very dilute concentration of the Mössbauer nucleus in the surface. Although caution must be exercized when interpreting Mössbauer emission spectra in terms of the oxidation states of cations and other phenomena that may be influenced by nuclei transformation processes, the data often provide a good means of probing the local structural environment of the Mössbauer nucleus.

Backscattering techniques. The acquisition of Mössbauer spectra from the surfaces of large samples is best achieved by using the technique in modes of operation that involve the direct detection of radiation emanating from the surface of the solid material. In such experiments the γ- and x-rays that arise as a result of internal conversion processes and that are resonantly emitted from the surface may be counted with apparatus similar to that normally used in transmission mode Mössbauer spectroscopy.

A significant advance in the use of Mössbauer spectroscopy for the study of solid surfaces has involved the development of conversion electron Mössbauer spectroscopy (CEMS), the detailed principles and applications of which have been described in other articles [5–7]. Iron-57 CEMS has been the most commonly applied form of this technique and, in brief, involves the absorption of γ-rays emitted from the Mössbauer source by ^{57}Fe nuclei in the sample so as to give excited nuclei, the majority (ca. 91%) of which decay by emitting internally converted electrons that can be detected by the apparatus described later. The ability of CEMS to investigate the superficial regions of solids is vested in the shallow escape depth of the internally converted electrons; indeed 95% of the 7.3-keV electrons ejected from the K shell originate from within 300 nm of the surface and 60% emanate from within 54 nm. Following the emission of these conversion electrons, L Auger electrons are also emitted in about 70% of cases, and these also contribute to the Mössbauer spectrum. In the remaining 30% of cases, a 6.3-keV x-ray is emitted which has an escape depth of about 10 μm. Although this is within the escape depth of a scattered 14.4-keV γ-ray, the x-ray has a higher emission probability. Hence, by monitoring electrons, x-rays, and γ-rays, it is possible to examine solid surfaces sometimes to a depth of about 20 μm. This means that grossly modified surfaces, as well as those in which only the first few layers are of interest, are amenable to examination by these methods. It is also important to note the high sensitivity of CEMS, which is another favorable feature of the technique. For example, when ^{57}Fe, ^{119}Sn, ^{151}Eu, or ^{197}Au is being used, the concentration of the isotope in the sample can be up to 100 times smaller than that which is required for conventional transmission mode Mössbauer measurements.

The greatest restriction on the general application of CEMS results from the relatively few elements that are amenable to practical examination by the method. Indeed, although CEMS studies have been recorded [8] from a number of isotopes, only a few are, in reality, suitable for extensive use in the examination of solid surfaces. This is because the practical application of CEMS in surface analysis requires the Mössbauer isotope to have a favorable conversion factor and the energy of the conversion electron to be such that it corresponds to a meaningful escape depth. The characteristics of the various isotopes have been tabulated in a recent review [8] of the use of CEMS for the structural microanalysis of solid surfaces and thin films. Most of the chemically significant CEMS experiments have involved studies of iron-bearing solids and, although investigations of tin compounds are now being successfully conducted and the examination of materials containing tellurium, europium, ytterbium, tungsten, iridium, and gold are possible, the total number of isotopes that may be developed for use in CEMS in the future is unlikely to be large.

The experimental features of CEMS are similar to those used in conventional transmission mode Mössbauer experiments. Although the detection of internally converted electrons instead of γ-rays is, in principle, a reason why CEMS should be a more efficient method of Mössbauer spectroscopy, the low energy of the electrons inhibits their ability to penetrate a detector window. Hence, the sample needs to be inside the detector, which is called a resonance counter. Instruments similar to that depicted in Fig. 6-4 have been successfully developed by several workers including the author of this chapter. The apparatus is essentially an 8-mm-thick flat cylindrical flow counter with mylar windows. The sample is fixed onto one of these windows, and the pair is set into aluminum rings which are screwed onto a perspex block containing three parallel 0.1-mm-thick stainless steel anode wires. The inlet and outlet tubes permit the flow of a counting gas composed of 90% helium and 10% methane. The backscattered internally converted electrons are detected as a function of the Doppler modulated incident γ-ray energy and, in contrast to transmission mode Mössbauer spectra, maxima in electron counts occur at velocities when resonant absorption occurs.

In more recent times, the CEMS technique has been developed to facilitate the examination of the solid as a function of depth below the surface. This modification of CEMS, which is known as depth-selective conversion electron Mössbauer spectroscopy (DCEMS), permits the investigation of different surface layers without disturbing the intrinsic nature of the regions as sometimes occurs when etching or sputtering techniques are used. The basic principle is that the conversion electrons (e.g., 7.3-keV K-conversion electrons in the case of ^{57}Fe) suffer an energy loss as they escape from the deeper regions as a result of inelastic collisions. Hence the energy of the electrons when they emerge from the solid can be related to their depth of origin below the surface. Information on the depth dependence of structure and composition may therefore be obtained, and the technique can, in principle, be applied to any sample that contains a sufficient amount of the isotope of interest.

The initial DCEMS studies were performed with the basic resonance counters described earlier [5, 7]. One method involved the variation of the incident γ-ray angle through 90° from normal to give both increased count rates and the detection of effects from different surface regions in materials enriched in iron-57. Attempts to limit the probing depth by

Fig. 6-4. A conversion electron Mössbauer resonance counter. (Reproduced by permission from J. Fenger, Danish Atomic Energy Commission, Riso, Denmark.)

either altering the distance between the counter wires and the sample or by exploiting the poor resolution of the helium/methane counter illustrate other methods by which the sensitivity of the basic instrument to surface effects in bulk solids may be refined. Subsequent successful depth-resolved CEMS studies entailed the use of β-ray spectrometers [5]. The basic principle involves the detection of electrons emerging from the surface with an electron spectrometer of several percent energy resolution so that DCEM spectra can be obtained at various selected electron energies. More recent developments in DCEMS [8, 9] have involved measurements at different electron emission angles, i.e., angle resolved DCEMS, to give additional information on depth-dependent properties. A description and comparison of various electron spectrometers now being used for DCEMS has recently been published [8].

Since measurements at low temperatures are very important in Mössbauer spectroscopy, there have been many attempts to develop experimental facilities to record CEM spectra at temperatures down to 4.2 K. The first CEMS experiments at liquid helium temperatures involved the use of complex β-spectrometers. The low counting efficiency as well as the complexity of these instruments precluded their facile application. The high-resolution electron spectrometers developed for DCEMS studies were found to be difficult to adapt for low-temperature studies, and proportional helium counters were found to be unsuccessful at temperatures below ca. 30 K. More recent methods for the acquisition of CEM spectra at low temperatures have entailed the use of channel electron multipliers [8]. By these methods the CEM spectra from ^{57}Fe, ^{119}Sn, ^{125}Te, ^{151}Eu, ^{170}Yb, ^{182}W, ^{193}Ir, and ^{197}Au have been recorded.

2.3. Interpretation of the Mössbauer Spectra

A Mössbauer spectrum is characterized by the number, shape, position, and relative intensity of the various absorption lines. These features result from the nature of the various hyperfine interactions, their time dependence, and their dependence on any motion of the Mössbauer nuclei. The total absorption intensity of the spectrum is a function of the concentration of Mössbauer nuclei in the absorber and the cross-sections of the nuclear processes involved. This absorption intensity, together with the signal-to-noise ratio of the detection system and the total number of counts, determines the quality of the Mössbauer spectrum and the accuracy with which information can be derived from it.

The three main hyperfine interactions that determine the nature of the Mössbauer spectrum are primarily reflected in the chemical isomer shift, δ, the quadrupole splitting, Δ, and the magnetic splitting. Further deductions can be made from linewidth data, the sign of the quadrupole interactions, and the influence of relaxation times on the Mössbauer parameters.

2.3.1. Chemical Isomer Shifts. The chemical isomer shift of the absorption lines in the Mössbauer spectrum is a result of the electric monopole interaction between the nuclear charge distribution over the finite nuclear volume and the electronic charge density over this volume. This shift arises because of the difference in the nuclear volume of the ground and excited states and the difference between the electron densities at the Mössbauer nuclei in different materials. In systems where this electric monopole interaction is the only hyperfine interaction affecting the nuclear energy levels, the nuclear ground and excited states are unsplit, but their separation is different in the source and absorber by an amount given by the chemical isomer shift, δ. This situation is shown schematically in Fig. 6-5. In this diagram the transition energy between the nuclear ground and excited states in an isolated nucleus, which is illustrated on the left of the diagram, is modified by the different electronic environments of the nucleus in the source and in the absorber. The application of a Doppler velocity to the source is therefore necessary to attain resonance and detect the shift in the position of the resonance line from

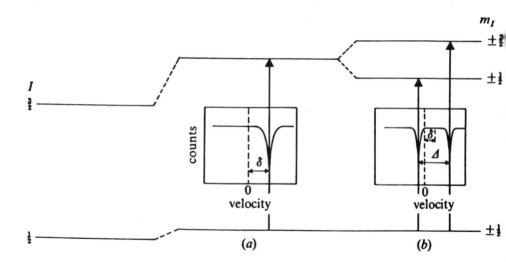

Fig. 6-5. The effects on the nuclear energy levels of ^{57}Fe of (a) the chemical isomer shift and (b) the quadrupole splitting. The Mössbauer absorption transitions and the resulting spectra which give the chemical isomer shift, δ, and the quadrupole splitting, Δ, are also shown. (Reproduced from [4] with permission.)

zero velocity (see Fig. 6-2), and the resulting Mössbauer spectrum consists of a single absorption line at a position determined by the chemical isomer shift (Fig. 6-5a). When additional hyperfine interactions are present, the chemical isomer shift sets the position of the centroid of the whole Mössbauer spectrum. The chemical isomer shift is not an absolute quantity since it represents the difference between the electric monopole interactions in the source and the absorber. Hence, in order to make comparisons of the chemical isomer shifts obtained from different absorbers, the chemical isomer shift data are generally expressed relative to a standard absorber which is also used to determine the zero of the velocity axis of the spectrum. These standard absorbers are specific to the particular isotope used and for the common Mössbauer isotopes there are universally recognized reference standards for chemical isomer shifts. Since temperature-dependent thermal vibrations of the nuclei will shift the γ-ray energies, it is important that the temperature of both the source and absorber be considered when quoting and comparing chemical isomer shift data.

The chemical isomer shift is, as shown above, a measure of the difference in the electronic environments of the source and absorber nuclei, particularly the difference in s electron densities. Hence Mössbauer spectroscopy provides a means of monitoring s electron density at the nucleus which is dependent on p, d, and f electron disposition. Therefore, the chemical isomer shift δ is an important means by which atomic oxidation states, which have sometimes been difficult to determine in the past, may be investigated directly. Similarly, covalency effects and the shielding of one set of electrons by another, which also influence the electronic environment of the nucleus, may be reflected in changes in δ. The chemical isomer shift data can also be used to assess quantitatively the electron-withdrawing power of substituent electronegative groups as well as the degree of π-bonding and back donation from metal atoms to substituent groups.

2.3.2. Quadrupole Splittings. Nuclei in states with a nuclear spin $I > 1/2$ have nonspherical charge distributions which are characterized by a nuclear quadrupole moment. When the nuclear quadrupole moment experiences an asymmetric electric field produced by an asymmetric electronic charge distribution or ligand arrangement and characterized by a tensor quantity called the electric-field gradient, an electric quadrupole interaction occurs that gives rise to a splitting of the nuclear energy levels corresponding to different alignments of the quadrupole moment with respect to the principal axis of the electric-field gradient. In the case of ^{57}Fe, the excited state has I = 3/2 and, in the presence of a nonzero electric-field gradient, this splits into two substates. This situation leads to a two-line spectrum with the two lines separated by the quadrupole splitting Δ as shown in Fig. 6-5b.

Since the value of the nuclear quadrupole moment is fixed for a given nuclide, the quadrupole split Mössbauer spectrum gives details of the electric-field gradient, which contains a contribution from any asymmetry in the electronic structure of the Mössbauer atom itself, called the valence contribution, and a lattice contribution arising from any asymmetry in the atomic or ligand array around the Mössbauer atom.

The quadrupole splitting therefore reflects the symmetry of the bonding environment and the local structure in the vicinity of the Mössbauer atom and can give information on the electronic populations of orbitals, isomerization phenomena, and the defect structure of solids.

2.3.3. Magnetic Splittings. When a nucleus is placed in a magnetic field, there is a magnetic dipole interaction between any nuclear magnetic moment and the magnetic field. This interaction completely removes the degeneracy of a nuclear state with an angular momentum quantum number I > 0 and splits it into 2I + 1 substates. For example, in the case of ^{57}Fe the ground state with I = 1/2 splits into two substates, and the excited state with I = 3/2 splits into four substates such that the six transitions appropriate to the Δm_I = 0, ±1 selection rule give a six-line Mössbauer spectrum as depicted in Fig. 6-6. The splitting of the spectral lines is directly proportional to the magnetic field experienced by the nucleus; hence, Mössbauer spectroscopy provides an effective means by which the magnetic field can be measured and the intensities of the lines can give information on the angle between the magnetization direction and the incident γ-ray.

The effects of magnetic and quadrupole splitting on the Mössbauer spectrum may be considerably more complex when they are present together and the observed spectrum is then strongly dependent on their relative magnitudes and orientations. The application of a magnetic field to a system with no unpaired spins, and therefore no magnetic hyperfine interaction, leads to a magnetic splitting in addition to the quadrupole splitting, and this can provide information on the sign of the electric field gradient at the nucleus. The data obtained from magnetically split Mössbauer spectra can be used to investigate the magnetic ordering and structure of magnetically ordered systems, the nature of the magnetic

Fig. 6-6. The effect of magnetic splitting on the nuclear energy levels of ^{57}Fe showing the allowed transitions and the resulting spectrum in the absence of quadrupole splitting. (Reproduced from [4] with permission.)

interactions, the size of the magnetic moment on particular atoms, and details of the electronic structure of the atom which relates to the magnetic hyperfine field at the Mössbauer nucleus.

2.3.4. Time-Dependent Effects, Relaxation, and Dynamics. The Mössbauer process and the hyperfine interactions have characteristic times, and the spectrum observed in any situation depends on whether the properties of the nuclear environment or the position of the nucleus are changing relative to these times. These time-dependent effects can influence both the spectral lineshapes and the values of the Mössbauer hyperfine parameters. Time-dependent changes in the nuclear environment, often referred to as relaxation processes, can relate to structural changes in a system as a function of time as well as to changes involving the electronic configuration. Any nuclear motion in a system can also be examined by Mössbauer spectroscopy, and the results provide a means by which nuclear dynamics can be investigated. These matters are fully discussed elsewhere [4].

3. Application of Mössbauer Spectroscopy to the Study of the Surface Chemistry of Geologic Materials

Conversion electron Mössbauer spectroscopy has shown itself to be the most powerful mode of Mössbauer spectroscopy when used for the examination of solid mineral surfaces, and it is therefore appropriate that this technique be given first consideration in this chapter.

3.1. Conversion Electron Mössbauer Spectroscopy

One of the first applications of conversion electron Mössbauer spectroscopy involved the examination of lunar materials [10]. The CEM spectrum of a lunar fine with particles of ca. 147 μm diameter, i.e., greater than the escape depth of the conversion electrons, was compared with the CEM spectrum of an unoxidized simulated lunar material and to the transmission mode Mössbauer spectra of the lunar and simulated samples. The CEM spectrum from the lunar fine (Fig. 6-7) showed a signal close to zero velocity which was associated with the unexpected presence of superparamagnetic iron in the outermost regions of the material.

It is also relevant to cite a more recent CEMS study of an extraterrestrial solid. The work involved the examination of iron-nickel alloys in lamella found in meteorites [11]. It is well known that many meteorites

Fig. 6-7. Conversion electron Mössbauer spectrum recorded from a lunar fine. (Reproduced from [10] with permission.)

contain two phases, one called kamacite, which is an iron-nickel alloy, and the other taenite, which often occurs as lamellae of about 100 μm thickness lying between the thicker (~1 mm) plates of kamocite. The transmission mode Mössbauer spectrum recorded from an untreated taenite lamella of 63 μm thickness from the iron meteorite Cape York is shown in Fig. 6-8a and may be interpreted in terms of the superposition of three spectra:

1. A central line from a paramagnetic atomically disordered iron-nickel alloy containing about 25% nickel (Pa)

2. An asymmetric six-line spectrum showing a magnetic field of 287 kG and characteristic of a ferromagnetic atomically ordered iron-nickel alloy containing ~50% nickel (O)

3. A symmetric six-line spectrum showing a magnetic hyperfine field of 342 kG corresponding to martensite (M).

The conversion electron Mössbauer spectrum recorded from the lamella is shown in Fig. 6-9a and, by direct comparison with the transmission mode Mössbauer spectrum (Fig. 6-8a), demonstrates that a clear difference exists between the surface layer of the lamella and its bulk composition. The transmission mode Mössbauer spectrum is dominated by the martensite contribution and the central peak from the paramagnetic iron-nickel alloy phase, while the CEM spectrum is dominated by the component from the ordered iron-nickel alloy.

The transmission and conversion electron Mössbauer spectra recorded after etching the lamella in a solution of 5% hydrochloric acid and 5% nitric acid in water are shown in Figs. 6-8(b) and (c) and Figs. 6-9(b)–(e), respectively. The transmission mode Mössbauer spectra show that the martensite is mainly present in the inner parts of the lamella, indeed even after 580 min of etching, which corresponds to the removal of 25 mg/cm of lamella in total, the martensite dominates the spectrum (Fig. 6-8c). However, the conversion electron Mössbauer spectra show that the area of the central peak representing the paramagnetic atomically disordered iron-nickel alloy increases relative to the area of the spectrum corresponding to the ferromagnetic atomically ordered iron-nickel alloy as more and more of the lamella is etched away. The results are consistent with the suggested structure of taenite lamellae as illustrated schematically in Fig. 6-10.

The results show that the combined use of etching, transmission mode Mössbauer spectroscopy, and conversion electron Mössbauer spectroscopy can give essential information on the distribution of the different phases in taenite lamellae of iron meteorites. The information is also

Fig. 6-8. Transmission mode Mössbauer spectrum recorded at 298 K from a thick lamella from the iron meteorite Cape York after different etching times: M, martensite; O, ordered phase; and Pa, paramagnetic phase; (a) lamella as found, (b) after 290 min of etching, (c) after 580 min of etching. (Reproduced from [11] with permission.)

important because a more detailed knowledge of the structure of taenite lamellae may lead to a better understanding of how these lamellae have

Fig. 6-9. Conversion electron Mössbauer spectra recorded at 298 K from a thick taenite lamella from the iron meteorite Cape York after different etching times: O, ordered phase; Pa, paramagnetic phase; M, martensite; (a) lamella as found, (b) after 25 min of etching, (c) after 80 min of etching, (d) after 155 min of etching, (e) after 580 min of etching. (Reproduced from [11] with permission.)

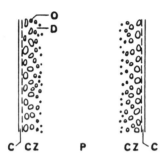

Fig. 6-10. Representation of the composition of a
thick taenite lamella (thickness 70 μm): C, ordered
FeNi alloy; CZ, mixture of ordered FeNi alloy and
disordered paramagnetic FeNi alloy; O, domains of
ordered FeNi alloy; D, disordered FeNi alloy filling
space between the ordered domains; P, martensite.
(Reproduced from [11] with permission.)

been formed and thereby a better knowledge of the iron-nickel phase
diagram and of the thermal history of iron-containing meteorites.

A study that illustrates the capacity of CEMS to identify changes in
the cationic oxidation states in the surface layers of minerals as a result of
weathering has been performed on siderite [5]. The CEM spectra
recorded from two different regions of a large sample are shown in Fig.
6-11. The spectrum in the top part of the figure shows two peaks emanating from the Fe^{2+} ions in $FeCO_3$. The other spectrum shows a doublet
characteristic of Fe^{3+}, which is partially superimposed on the Fe^{2+} signal
that is characteristic of siderite. The Fe^{3+} component was attributed to the
presence of small superparamagnetic particles of geothite (α-FeOOH),
which is well known to be a product of the weathering of siderite.

The amenability of layered sheet silicate materials, which readily
give thin plate-type crystals that can be easily mounted in the resonance
counter, to examination by CEMS is well illustrated by several studies of
biotite. The CEM spectrum shown in Fig. 6-12 recorded from the surface
of the freshly cleaved basal face of a single crystal of biotite [12] shows a

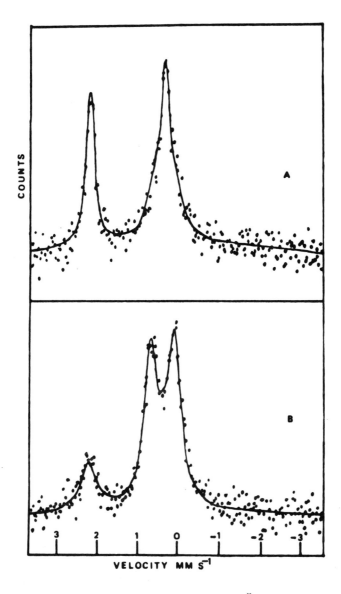

Fig. 6-11. Iron-57 conversion electron Mössbauer spectra recorded from two different regions of a siderite surface. (Reproduced from [5] with permission.)

large quadrupole split absorption with a positive chemical isomer shift characteristic of the presence of high-spin Fe^{2+} ions in distorted octahedral sites. The asymmetry of the peaks reflects the orientation of the principal

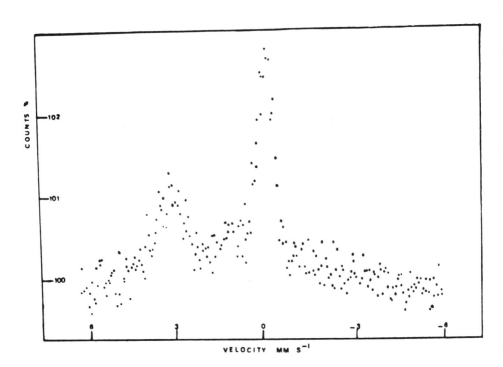

Fig. 6-12. CEM spectrum of freshly cleaved (OOOl) face of a single crystal of biotite showing the presence of high-spin Fe^{2+}. (Reproduced from [12] with permission.)

component of the electric-field gradient tensor with respect to the incident γ-radiation. This observation was given further attention in a subsequent study [13] of the dependence of the spectral peak areas on the angle of incidence of the γ-ray beam and is important since these types of investigations are not easy to perform by conventional transmission mode Mössbauer spectroscopy.

A later and more comprehensive investigation of biotite [14] is also worthy of note since it demonstrates the capacity of ^{57}Fe conversion electron Mössbauer spectroscopy to monitor the early stages of solid-state reactions which begin in the regions close to mineral surfaces. The transmission mode Mössbauer spectrum (Fig. 6-13a) and conversion electron Mössbauer spectrum (Fig. 6-14a) recorded from the freshly cleaved, unheated biotite crystals were found to be essentially similar and associated with the presence of Fe^{2+} ions in two types of similar oxygen/hydroxy coordination within the biotite crystal. The presence of a

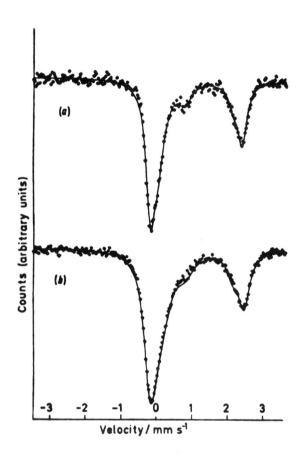

Fig. 6-13. Transmission ^{57}Fe Mössbauer spectra of biotite (a) before and (b) after heating at 550 K for 700 h. (Reproduced from [14] with permission.)

small peak at ca. 1 mm/s corresponding to one component of a doublet emanating from a small quantity of Fe^{3+} in the unheated samples was also clearly visible. The increase in the Fe^{3+} component detected by conversion electron Mössbauer spectroscopy when the biotite was heated at 550 K for 100 h (Fig. 6-14b), and which was not observed in the transmission Mössbauer spectrum even after calcination at 550 K for 700 h (Fig. 6-13b), showed that oxidation is confined to the outermost regions of the biotite crystal. Crystals that were subjected to longer heat treatments at 500 K gave CEM spectra which showed an increase in the Fe^{3+} content at

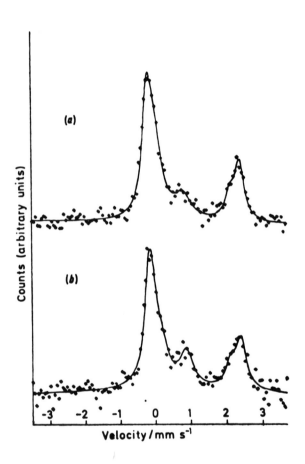

Fig. 6-14. Conversion electron Mössbauer spectra recorded from biotite (a) before and (b) after heating at 550 K for 100 h. (Reproduced from [14] with permission.)

the biotite surface, while the transmission mode Mössbauer spectra showed the $Fe^{2+}:Fe^{3+}$ ratio in the bulk to remain unaltered. Careful analysis of the ^{57}Fe CEMS chemical isomer shift and quadrupole splitting data enabled the surface Fe^{3+} species to be associated with $Fe^{3+}(O^{2-})_6$ and/or similar entities described as $Fe^{3+}(O^{2-})_5(OH^-)$ and $Fe^{3+}(O^{2-})_4(OH^-)_2$. The process of oxidation as detected by CEMS was associated with the occurrence of three reactions:

Dehydroxylation:

$$2\,OH^- \rightarrow O^{2-} + H_2O \qquad (6\text{-}1)$$

Oxygenation:

$$4\,Fe^{2+} + O_2 \rightarrow 4\,Fe^{3+} + 2\,O^{2-} \qquad (6\text{-}2)$$

Dehydrogenation:

$$4\,Fe^{2+} + 4\,OH^- + O_2 \rightarrow 4\,Fe^{3+} + 4\,O^{2-} + 2\,H_2O \qquad (6\text{-}3)$$

The formation of Fe^{3+} at the surface was associated with reactions (6-2) and (6-3) with reaction (6-3) being envisaged as leading to the production of $Fe^{3+}(O_2^-)_5(OH^-)$ and $Fe^{3+}(O_2^-)_4(OH)_2$ as indicated by the CEMS results. The favorable oxygenation reaction at the surface, in contrast to the bulk, was attributed to the direct reaction of Fe^{2+} with oxygen at basal surfaces and at exposed prismatic edges on the steps of the basal surfaces. Such processes were considered to be particularly favored by the ability of oxygen to gain access to the internal superficial structure by diffusing into cracks associated with grain and subgrain boundaries close to the surface. The dehydroxylation reaction (6-1) was also envisaged as occurring in the surface regions and to involve the initial removal of one OH^- ion followed by rapid oxygenation to complete the vacancy in the coordination shell. The broadening of the Fe^{3+} resonance in the transmission mode Mössbauer spectrum (Fig. 6-13b) was associated with a distortion of the lattice arising from dehydroxylation in the bulk (reaction 6-1) and the consequent formation of O^{2-} and trapped water molecules. Hence the study is a very good illustration of the power of CEMS for the elucidation of the early stages of various solid-state reactions in mineral surface regions. In particular the report demonstrates how information relating to the nature of the surface product and its rate of growth may be obtained.

The power of CEMS for elucidating unexpected phenomena at mineral surfaces is well illustrated by a study [15] of the naturally occurring hydrated iron phosphate called vivianite of formula $Fe_3(PO_4)_2 \cdot 8H_2O$. The fresh mineral is colorless but turns blue on exposure to air as a result of the partial oxidation of Fe^{2+} ions to Fe^{3+} and the consequent charge-transfer interactions via bridging O^{2-} ions. The transmission mode Mössbauer spectrum (Fig. 6-15a) showed a broad relatively intense absorption with a pronounced shoulder at about 0.15 mm/s and a weaker structured absorption at about 2.6 mm/s. The spectrum may be interpreted in terms of the superposition of two large quadrupole split absorptions

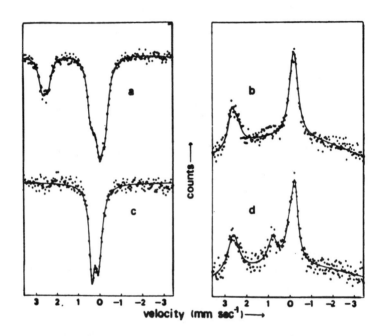

Fig. 6-15. (a) and (c) Transmission mode ^{57}Fe Mössbauer spectra recorded from vivianite before and after heating at 120°C for 1 h, (b) and (d) the CEM spectra recorded simultaneously from the sample. (Reproduced from [15] with permission.)

attributable to Fe^{2+} ions in two inequivalent sites and a narrow doublet centered at about 0.43 mm/s, which is characteristic of the Fe^{3+} ions which originate from the partial oxidation of the sample. In marked contrast to the transmission mode Mössbauer spectrum, the conversion electron Mössbauer spectrum recorded simultaneously from the same sample (Fig. 6-15b) showed the Fe^{3+} component at about 0.43 mm/s to be essentially absent in the surface regions of the crystal. The transmission mode Mössbauer spectrum (Fig. 6-15c) recorded from the sample following treatment in air at 120°C for 1 h showed the Fe^{2+} species to be gradually converted to Fe^{3+}. In marked contrast, no facile growth of Fe^{3+} was observed in the CEM spectrum (Fig. 6-15d) showing that the surface regions are only slowly and partially oxidized to Fe^{3+}. The result is a clear demonstration of the unexpected stability of the surface regions of vivianite to oxidation, and the phenomenon has also been shown to occur in iron(II) phosphate octahydrate formed by chemical treatment on the surface of metallic iron [16]. The formation of Fe^{3+} in the bulk of vivianite by oxidation of Fe^{2+} may occur by the process

$$4\,Fe^{2+}(H_2O)_n + O_2 \rightarrow 4\,Fe^{3+}(OH^-)(H_2O)_{n-1} + 2\,H_2O \qquad (6\text{-}4)$$

which eventually leads to the formation of a basic iron(III) phosphate as a final product. The relatively inert nature of the surface region of vivianite may therefore result from a preferential dehydration of these regions to an oxidatively more stable lower hydrate of formula $Fe_3(PO_4)_2 \cdot nH_2O$, where $n < 8$.

Conversion electron Mössbauer spectroscopy has also been used to complement nuclear microprobe analysis in an investigation [17] of the desert varnish surface coatings, which is the name given to the reddish brown weathered iron oxide surface coatings that are formed on rocks in desert regions, which have been found on samples of dolerite and sandstone in Antarctica. The nuclear microprobe analysis was used to provide a direct measurement of the oxygen, iron, calcium, and sulfur content of the varnish coating and underlying bulk rock. The concentration profiles of these elements showed that in the iron-containing dolerites, the desert varnish is formed by subsurface leaching of the iron from the bulk rock which segregates to the surface to form a hydrous coating of iron(III) oxide. The iron formed in the surface varnish coating of the essentially iron-deficient sandstone was envisaged as being derived externally from the overlying dolerite sills.

The conversion electron- and $Fe\text{-}K_\alpha$ x-ray Mössbauer spectra recorded from a coarse-grain dolerite DN 48, a fine-grain dolerite MM69, and sandstone WB62 are shown in Fig. 6-16. The doublet resonance with peaks around 0.1 mm/s and 2.3 mm/s is characteristic of Fe^{2+}, while the doublet with peaks at 0.1 mm/s and 0.8 mm/s is indicative of the presence of Fe^{3+}. The spectrum of the fine-grained dolerite MM69 that was recorded by monitoring the backscattered x-rays, which have a greater escape depth than the conversion electrons, shows that the iron at a depth of about 10 μm from the surface is predominantly in the Fe^{2+} state with only a minor Fe^{3+} component being observed. The conversion electron Mössbauer spectrum shows that at shallower depths of about 0.1 μm the greater proportion of the iron is in the Fe^{3+} state. In contrast to this, the x-ray Mössbauer spectrum recorded from the coarse-grained dolerite DN48 shows that at the greater depth, Fe^{3+} is predominant. The results were correlated with the texture of the varnish coatings which on the fine-grained dolerites were usually smooth and often polished but on the coarse-grained dolerites were flaky and of less uniform appearance. The

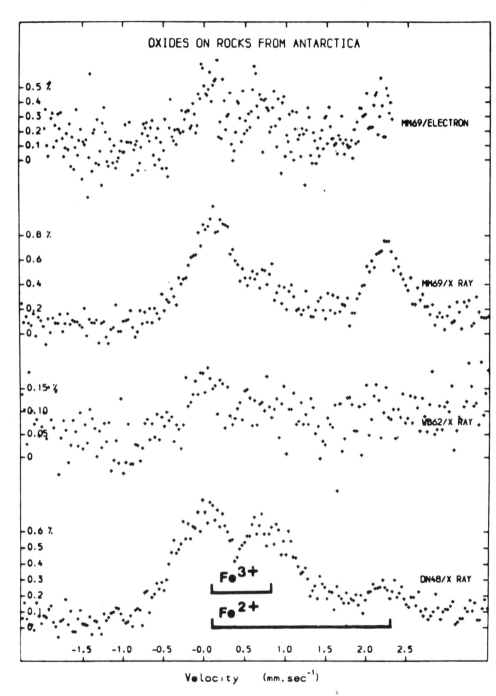

Fig. 6-16. Backscattered conversion electron and Fe-K_a X-ray Mössbauer spectra recorded from DN48, WB62, MM69. (Reproduced from [17] with permission.)

boundaries between the coarser crystallites were envisaged as readily facilitating the passage of water containing dissolved oxygen into the rock, thereby forming the desert varnish and consequently higher Fe^{3+} contents at greater depths as compared to the fine-grained dolerites.

The poor quality of the backscattered x-ray Mössbauer spectrum recorded from the sandstone WB62 was associated with the relatively low iron content. However, the discernable Fe^{3+} resonance is indicative of the presence of oxidized iron at a greater depth in this varnish coating than in that formed in the dolerites and is consistent with the postulation that the iron originates from the overlying dolerite layer. The study illustrates the depth selectivity that backscattered Mössbauer spectra involving the detection of x-rays on the one hand and conversion electrons on the other can provide. It also demonstrates how backscattered Mössbauer spectroscopy can be used as a complementary technique to analytical methods for the characterization of the surface regions of natural materials.

3.2. Mössbauer Emission Spectroscopy

The use of Mössbauer emission spectroscopy in the study of the surface properties of natural materials is well illustrated by an in situ investigation of ^{57}Co and ^{119}Sb ions, which decay to the Mössbauer nuclides ^{57}Fe and ^{119}Sn when adsorbed at the iron(III) oxide (hematite)/aqueous solution interface [18]. The Mössbauer emission spectra of ^{57}Fe arising from ^{57}Co at the interface between hematite particles and 0.1 mol/dm^3 NaCl solution at various pH values of the aqueous phase are collected in Fig. 6-17 and show a marked dependence on the pH of the aqueous phase. The chemical isomer shifts are characteristic of the presence of Fe^{3+}. At pH 5.7 the spectrum (Fig. 6-17a) consisted of a partly resolved sextet indicative of a hyperfine magnetic field distribution presumably reflecting a magnetic interaction between the adsorbed Fe^{3+} and the ferric ions in the substrate. These spectra also show that an increase in pH of the solution gives rise to an increase in the splitting of the sextet and a decrease in linewidth and is thereby indicative of a pH dependence of the interaction (Fig. 6-17b–e). The readjustment of the pH from a high value of 12.7 to a low value of 3.0 gave slow desorption of the ^{57}Co ions without significant change in the Mössbauer spectrum (Fig. 6-17f) and thereby demonstrated that the chemical form of $^{57}Co^{2+}$ adsorbed from alkaline solution is retained when the pH of the solution is lowered to an acidic value.

The Mössbauer emission spectra of ^{119}Sn arising from the adsorption of $^{119}Sb^{5+}$ ions on hematite particles in 0.25 mol/dm^3 LiCl solutions of different pH values at room temperature are shown in Fig. 6-18. The

Fig. 6-17. Mossbauer emission spectra recorded from ^{57}Fe arising from divalent ^{57}Co at the hematite/0.1 mol/dm^3 NaCl solution interface at room temperature. (a) pH 5.7, (b) pH 7.4, (c) pH 9.6, (d) pH 11.0, (e) pH 12.7, (f) pH 3.0. (Reproduced from [18] with permission.)

spectra were composed of essentially symmetric lines centered around zero velocity relative to the barium stannate absorber and demonstrated the presence of tetravalent ^{119}Sn. The magnetically split spectrum, which

Fig. 6-18. Mossbauer emission spectra of ^{119}Sn arising from pentavalent ^{119}Sb at the hematite/0.25 mol/dm^3 LiCl solution interface at room temperature. (a) pH 8.5, (b) pH 6.6, (c) pH 4.6, (d) pH 3.4, (e) pH 2.5, (f) pH 8.6. (Reproduced from [18] with permission.)

appears as a single broad band, may be associated with magnetic interactions with ordered Fe^{3+} ions in the substrate. The linewidths showed a strong dependence on pH, suggesting that the chemical structure of adsorbed pentavalent ^{119}Sb is considerably dependent on the pH of the solution. The decrease in pH of the aqueous phase (Fig. 6-18b–e) gave rise to an increase in linewidth, suggesting stronger interactions of ^{119}Sn with the Fe^{3+} ions of the substrate in weakly acidic regions. The chemical structure of the pentavalent ^{119}Sb adsorbed from acidic solution was retained when the pH of the solution was raised above 7.

The hyperfine magnetic fields on trivalent ^{57}Fe and tetravalent ^{119}Sn were interpreted in terms of the nature of the metal ions adsorbed at the interface between hematite and the aqueous solution. For example, in the slightly acidic region, the adsorbed ^{57}Co ions were considered to be distributed between at least two chemical forms, one attributable to the ^{57}Co ions coordinatively bound to surface sites and the other to ^{57}Co weakly bound to the hematite surface. In the alkaline region, most of the adsorbed ^{57}Co ions were shown to be within the first metal ion layers of the substrate forming Co—O—Fe bonds. The desorption of ^{57}Co^{2+} from strongly coordinated surface complexes was shown to occur when the pH was lowered from alkaline to acidic values. In contrast, the pentavalent ^{119}Sb ions at the interfaces were found to be weakly bonded to the oxide ion layer of the hematite surfaces in neutral and slightly acidic regions, while in the acidic region, most of the adsorbed ^{119}Sb ions were envisaged as being held within the first metal ion layers of the substrate forming Sb—O—Fe bonds.

It is clear that in situ Mössbauer emission measurements such as these are applicable to a variety of mineral/aqueous solution interfaces, and future developments in these methods offer considerable scope for the study of surface phenomena at such boundaries.

4. Conclusions

The study of mineral surfaces by Mössbauer spectroscopy has much potential, especially when the technique is used in backscattered or emission geometry. The technique has a capacity to examine oxidation states, local structure, and magnetic effects within the surface and subsurface regions and is well disposed to studies of the effects of weathering, adsorption, and interfacial phenomena. Although the full capacity of the technique for the study of naturally occurring systems has yet to be completely explored, the work reported to date shows that the field is fertile and is clearly open for much imaginative and fruitful activity.

References

[1] Maddock, A.G., In "Chemical Bonding and Spectroscopy in Mineral Chemistry," Berry, F.J.; Vaughan, D.J., Eds. Chapman and Hall, London, 1985, p. 141.

[2] Greenwood, N.N.; Gibb, T.C., "Mössbauer Spectroscopy." Chapman and Hall, London, 1971.

[3] Berry, F.J., Phys. Bull., 1983, 34, 517.

[4] Berry, F.J.; Dickson, D.P.E., "Mössbauer Spectroscopy." Cambridge University Press, Cambridge, 1986.

[5] Tricker, M.J., In "Surface and Defect Properties of Solids," Vol. 6, Roberts, M.W.; Thomas, J.M., Eds. Specialist Periodical Reports, The Chemical Society, London, 1977, p. 106.

[6] Berry, F.J., Transition Metal Chem., *1979*, 4, 209.

[7] Tricker, M.J., In "Mössbauer Spectroscopy and its Chemical Applications," Stevens, J.G.; Shenoy, G.K., Eds. Advances in Chemistry Series, American Chemical Society, Washington, D.C., 1981, p. 63.

[8] Sawichi, J.A., In "Industrial Applications of the Mössbauer Effect," Long, G.J.; Stevens, J.G., Eds. Plenum Press, New York, 1986, p. 83.

[9] Saneyoshi, K.; Debusmann, K.; Keune, W.; Brand, R.A.; Liljequist, D., In "Industrial Applications of the Mössbauer Effect," Long, G.J.; Stevens, J.A., Eds. Plenum Press, New York, 1986, p. 121.

[10] Forester, D.W., Proc. Fourth Lunar Sci. Conf. Suppl. 4, Geochim. Cosmochim. Acta. *1973*, 3, 2697.

[11] Christiansen, A.; Larsen, L.; Roy-Poulson, H.; Roy-Poulson N.O.; Vistisen, L.; Knudsen, J.M., Phys. Scripta, *1984*, 29, 94.

[12] Tricker, M.J.; Thomas, J.M.; Winterbottom, A.P., Surf. Sci., *1974*, 45, 601.

[13] Tricker, M.J.; Freeman, A.G., Surf. Sci., *1975*, 52, 549.

[14] Tricker, M.J.; Winterbottom, A.P.; Freeman, A.G., J. Chem. Soc., Dalton Trans., *1976*, 1289.

[15] Tricker, M.J.; Ash, L.A.; Jones, W., J. Inorg. Nucl. Chem., *1979*, 41, 891.

[16] Berry, F.J., J. Chem. Soc., Dalton Trans., *1978*, 308.

[17] Johnston, J.H.; Cardile, C.M.; Coote, G.E.; Sparks, R.J.; Wallace, G.; McMillan, J.W.; Pummery, F.C.; Longworth, G., Chem. Geol., *1984*, 42, 189.

[18] Ambe, F.; Ambe, S.; Okada, T.; Sekizawa, H., In "Geochemical Processes at Mineral Surfaces," Davis, J.A.; Hayes, K.F., Eds. ACS Symposium Series 323, American Chemical Society, Washington, D.C., 1986, p. 403.

7. NUCLEAR MAGNETIC RESONANCE SPECTROMETRY

Cecil Dybowski
Department of Chemistry and Biochemistry
and
Center for Catalytic Science and Technology,
University of Delaware
Newark, Delaware 19716

1. Introduction

Nuclear magnetic resonance (NMR) spectrometry is not usually thought to be a technique for the examination of species at surfaces. Indeed, the technique has been generally restricted to use in the liquid state, with recent developments making it more generally applicable to solid and solid-like materials [1]. It has a rather high limit of detection compared to analytical techniques such as mass spectrometry and visible spectrometry, but it has become an important tool for analysis of organic materials in solution [2]. The major reasons for this importance are easily seen: (1) The spectrometry is conveniently related to *chemical structure* through the dependence of such parameters as the chemical shift on local electronic environment; (2) the *connectivity* of functional groups is readily apparent in the NMR spectrum through parameters such as the scalar and dipolar couplings; (3) the relative *proportions* of nuclei in various environments are straightforwardly determined; and (4) the ubiquity of elements with nuclear spins makes the technology applicable to a wide variety of problems in chemistry and materials science. It is in this context that NMR spectrometry has become a standard analytical tool in chemistry and chemistry-related areas.

Many events that determine chemical processes occur at interfaces [3]. Melting and freezing are two common examples, but the production of materials with heterogeneous catalysts and the rectification of electric current in a semiconducting diode represent surface-mediated processes that are technologically significant in our daily lives. To control and use surface-related technologies, one must understand surface processes in detail, and that requires a determination of the type and number of chemical species present at a surface under various conditions.

While the major examples of its use are in solution-state chemistry, and to a lesser degree in the solid state, NMR spectrometry should be applicable to problems in surface chemistry. The fact that only limited utility of this tool has been made requires an answer to least the following two questions:

1. What are the characteristics of NMR spectrometry that limit its use in analyzing surface species?

2. What processes in surface chemistry might be susceptible to analysis with the technique?

In this chapter, I propose to describe some answers to the first question and, by selected examples, point the way to a partial answer to the second.

2. The Theory of NMR Spectrometry

2.1. The Zeeman Effect

Every nucleus having an odd atomic mass and every nucleus with an even atomic mass and an odd nuclear charge has, in its ground state, the property known as intrinsic angular momentum, or *spin* [4]. Like other angular momenta, the nuclear spin is quantized. For a nucleus with spin I, there is only a denumerable set of substates, 2I + 1, that have different azimuthal quantum numbers, m, where

$$-I \leq m \leq +I \tag{7-1}$$

In free space, all these substates have the same energy. However, a nucleus in a magnetic field is subject to a coupling to the field that removes the degeneracy, the Zeeman interaction, as shown in Fig. 7-1. In thermal equilibrium in a magnetic field, the populations of the magnetic substates are then slightly different. Thus, the application of a radiation field that induces transitions among the substates results in a net absorption of energy:

$$\Delta E = \gamma h B_0 / 2\pi \tag{7-2}$$

where B_0 is the magnetic field strength, h is Planck's constant, and the parameter γ is known as the *gyromagnetic ratio* of that particular nuclear

No   Zeeman   Chemical
Field   Only   Shift

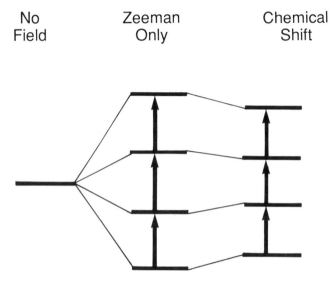

Fig. 7-1. Grotrian diagrams, showing the effects of the Zeeman interaction and the chemical shift on the nuclear spin levels for a spin of 3/2. The effects on the energy differences due to the chemical shift are exaggerated for purposes of illustration.

species. Table 7-1 gives the gyromagnetic ratios for several common nuclear spins. For magnetic field strengths routinely available in the laboratory, the transition energies for nuclei fall in the range of radiofrequency excitation. Thus, the NMR spectrometer is very much like a small television station or a "ham" radio station.

2.2. The Chemical Shift

Were the Zeeman coupling to the magnetic field the only effect in a nuclear-magnetic system nuclear magnetic resonance would be an interesting phenomenon but its utility in the analysis of materials would be quite limited. The presence of electrons in the environment of the nucleus, however, has a detectable effect on the relative energies of the magnetic substates. In particular, the exact transition energy for a nucleus in a particular environment depends on the local electronic environment which gives rise to the phenomenon known as *chemical shift* [1, 4].

$$\Delta E_i = \gamma h B_0 (1 - \sigma_i)/2\pi \qquad (7\text{-}3)$$

Table 7-1. The Magnetic Properties of Selected Isotopes

Isotope	Spin	Gyromagnetic Ratio (Hz/Gauss)
^{1}H	1/2	4257.7
^{2}H	1	653.6
^{13}C	1/2	1070.7
^{14}N	1	307.6
^{23}Na	3/2	1126.2
^{27}Al	5/2	1109.4
^{29}Si	1/2	845.8
^{31}P	1/2	1732.5
^{57}Fe	1/2	137.6
^{63}Cu	3/2	1128.5
^{95}Mo	5/2	277.4
^{129}Xe	1/2	1177.7
^{133}Cs	7/2	558.5
^{195}Pt	1/2	915.3
^{207}Pb	1/2	890.8

where σ_i is the chemical shift of a nucleus at site, i. For a sample containing nuclei in several different electronic environments, the chemical shift effects make the transition energies slightly different. Thus, the NMR spectrum of a particular nuclear species has a series of

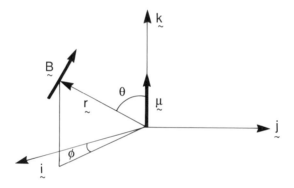

$$B(r) = \frac{3(\underset{\sim}{\mu}\cdot\underset{\sim}{r})}{|r|^5}\,\underset{\sim}{r} - \frac{\underset{\sim}{u}}{|r|^3}$$

$\underset{\sim}{\mu} \equiv$ magnetic moment

$\underset{\sim}{r} \equiv$ position relative to moment

Fig. 7-2. The dipolar local field at a site **r** relative to a magnetic moment, *μ*, at the origin. The dipolar local field depends on all three polar co-ordinates.

transitions corresponding to the number of different environments present in the sample. In addition, the transition moments for the nuclei in different environments can be made identical in carefully crafted experiments, allowing one to determine the relative number of nuclei in each environment in a complex sample. Thus, chemical type and relative quantity are readily available in the NMR experiment.

2.3. The Dipolar Interaction

The presence of other nuclei in the neighborhood of a nucleus also has an effect on the nuclear magnetic absorption spectrum. Because each nuclear magnetic moment is like a small bar magnet, it too produces a component of the magnetic field that can be felt at nearby points in space. This part of the magnetic field is often called the *dipolar local field* [4] (see Fig. 7-2). Any nearby nucleus is affected by this field, resulting in a coupling of the two nuclei. This is the *direct dipole-dipole interaction.* For spin-1/2 nuclei held fixed as in a solid, this interaction is frequently the dominant effect in the NMR spectrometry, being many orders of magnitude larger than the chemical shift interaction with the electrons. In liquids and solutions, the rapid, random isotropic tumbling of molecules (and their associated nuclei) causes the average effect of this interaction to be reduced such that, in first approximation, it is negligibly small.

2.4. The Indirect Coupling

Nuclei also interact with each other through an electron-mediated coupling, called the *scalar* or *indirect* coupling. Unlike the direct dipolar coupling, the rapid motion characteristic of the liquid state does not eliminate the effect of this coupling. It generally is short range, being important only for nuclei separated by a few bonds. The effects of the indirect couplings can give the spectrometrist a picture of the connectivity of chemical groups in a molecule and the numbers of spins in the groups. Such connectivity information is a major reason NMR spectrometry is used whenever the identification of the chemical structure of a new material is required, such as in organic and inorganic syntheses. The NMR spectrum is reasonably easily interpretable to give such chemical information.

2.5. Quadrupolar Interactions

For nuclei of spin greater than 1/2, the nuclear charge distribution in the ground state is not spherically symmetric. Thus, the energies of the various substates will depend on the existence of electric-field gradients to which the nucleus is subject through the coupling of the quadrupole moment of the nuclear charge distribution to the electric-field gradient [1]. Of course, it will also depend on the Zeeman coupling to a magnetic field. Depending on the relative magnitude of these two effects, the spectrometry falls into one of three cases:

1. For much stronger coupling to the electric-field gradient than to the magnetic field, the spectrometry is Zeeman-

perturbed nuclear quadrupole resonance spectrometry. For some nuclei with very large quadrupole moments, even the strongest magnetic fields do not produce an effect sufficiently strong to dominate the electric-field coupling. Even though one may not detect the entire spectral structure, magnetic resonance can be used in some cases.

2.   When the two couplings are roughly equal, the most complex case arises; one has neither pure NQR nor pure NMR spectrometry.

3.   When the coupling to the electric field is weak compared to the Zeeman interaction with the magnetic field, the spectrum is a perturbed magnetic resonance spectrum, with the quadrupolar interaction causing splitting of the NMR transitions.

2.6. *Orientational Effects*

The interactions among spins and of spins with the magnetic field not only depend on the magnitudes of couplings; they depend on the orientation of the local molecular structure relative to the magnetic field, as can be seen for the dipolar coupling in Fig. 7-2. This orientational dependence exists for most interactions experienced by spins, including the chemical shift, quadrupolar coupling to an electric field gradient, and the indirect coupling. For a single-crystal sample, this orientation dependence makes the spectrum a strong function of how the sample finds itself situated in the magnetic field [4]. For a sample that is a powder of many microcrystallites, the orientational effects produce a disperse resonance band - a composite of the resonances of the nuclei in all the various orientational environments in the various microcrystallites in the sample.

Orientational effects can be incorporated into the theoretical description of magnetic resonance by couching the interaction Hamiltonians as products of the basis operators for the spins, I_i, with tensors, A, that specify the magnitudes and the orientational effects in a single entity. For example, any bilinear interaction in spin operators can be written as: [4]

$$H_{Aij} = I_i \cdot A \cdot I_j \qquad (7\text{-}4)$$

The elements of the tensor **A** depend on the definition of the coordinate system. It is conventional in magnetic resonance to choose an axis of the system (usually the z axis) parallel to the magnetic-field direction.

There is generally a specific orientation of the molecular framework relative to the magnetic field for which the tensor is diagonal, A_p. Referred to this special coordinate system, the interaction contains no cross terms in the basis operators. This particular system is called the *principal-axis system* (PAS) of the tensor. The form of the tensor for any other orientation of the molecular system relative to the magnetic field can be expressed in the terms of the tensor in the PAS by a transformation that depends on the orientation of the PAS and the co-ordinate system in which one must specify the tensor:

$$A = T^{-1} \cdot A_p \cdot T \qquad (7\text{-}5)$$

The transformation matrix, **T**, depends only on the orientational relationship of the two coordinate systems. The components of A_p, called the principal elements of the tensor, give the details of the interaction and are often reported from the analyses of spectra. [4] They define the "magnitude" of the interaction.

2.7. Magic-Angle Spinning

The orientational dependence of the chemical shift limits the resolution of spectra of complex materials because of substantial overlap of resonances from nuclei in different environments, even when the effects of dipole-dipole coupling are suppressed by high-power decoupling, as is done for rare spins, such as carbon, by irradiation of the interfering protons. Since many samples of chemical interest cannot be obtained as single crystals, the loss of resolution would limit the utility of these techniques in analysis of such disordered systems as powdered catalysts.

It is well known that rapid isotropic random motion of the molecular framework relative to the magnetic field results in a situation in which the resonance occurs only at the average shift for each nuclear site, rather than having the intensity spread throughout a band, as

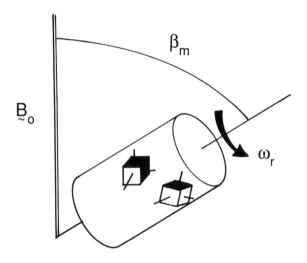

Fig. 7-3. The geometry of magic-angle spinning. When each crystallite, indicated by a small cube, is rotated rapidly about an axis inclined at β_m to the magnetic field, the average chemical shift interaction for every spin reduces to the isotropic value.

happens when one takes the spectrum of a powder. Early on, Lowe [5] and Andrew [6] independently discovered that coherent motion can produce a situation in which a similar effect is observed in the NMR spectrum of a powder of small crystallites. Under rapid rotation of the sample (all crystallites) about an axis inclined at a special angle, β_m (the magic angle), to the field direction, the resonance consists of a series of absorption lines at $\omega_o \pm m\omega_r$, where ω_r is the frequency of rotation [7]. (See Fig. 7-3.) The intensities of the lines at positions with m not equal to zero relative to that with m = 0 (the centerband) are strong functions of the spinning rate. In the limit of very rapid spinning, the sidebands are completely suppressed. Thus, a spectrum taken under conditions of rapid spinning about an axis inclined at the magic angle consists of only resonances at the isotropic positions for the nuclei in the various environments in the sample, in essence giving similar information to the solution-state spectra so familiar to the chemist.

The specific value of the magic angle is determined by the

manner in which the tensor for the chemical shift transforms under rotations. [4] In particular, the static term in the anisotropic part of the chemical shift interaction varies as $3 \cos^2 ß - 1$, where ß is the angle between the spinning axis and the magnetic field. When that term is zero, as it is for the magic angle ($ß_m = \arccos 3^{-1/2}$), the anisotropic terms do not contribute to the rapid-spinning spectrum and the result is a spectrum of the isotropic chemical shifts. For other interactions with more complex orientational dependences, to achieve averaging of the interaction by coherent motion of the sample may require rapid rotation about another axis or modulation by simultaneous reorientations about several axes.

2.8. Slow Spinning to Regain Chemical-Shift Parameters

While magic-angle spinning (MAS) suppresses some information to permit access to other average information, the knowledge of all tensor elements gives a more nearly complete picture of the electronic structure around each nuclear site. Thus, several techniques have been developed to retrieve this information from the spectrum [8-10]. The most commonly used technique is the analysis developed by Herzfeld and Berger [8] of the sideband pattern in an experiment in which the rotation frequency is not fast compared to the characteristic parameters of the resonance. In that case, sidebands appear. The relative intensities of the sidebands are complex functions of the chemical-shift tensor elements, the Larmor frequency, and the rotation frequency. One can use a series of graphs of these functions given by Herzfeld and Berger [8] to determine the values of two parameters, ρ and μ, from the sideband intensity pattern at a given spinning rate or at several rates. Once known, these two parameters and the isotropic chemical shift can be used to calculate the principal elements of the chemical-shift tensor. The principal elements are, in fact, characteristic of the particular environment of the nucleus and may be used as an even more definitive identifier of chemical type than the isotropic chemical shift alone. A compendium of carbon-13 chemical-shift tensor principal elements has recently been published [11] from which one may discern relationships of these parameters to chemical structure.

3. Problems of Applying NMR Spectrometry to Surfaces

Magnetic resonance spectrometry is an extremely accurate tool for identifying chemical species. In solution the resonances are rather sharp and the subtle differences in environment among various nuclei of the same type can be easily discriminated through the NMR spectrum. For solid-like materials, particularly rather regular crystalline materials, one can apply techniques to suppress interactions that cause loss of resolution to obtain a similar, though generally less well resolved, spectrum of the material [4]. To obtain spectra of species at surfaces, one must overcome several severe problems inherent to the NMR technique to obtain useful information on function and structure from the spectrum.

3.1. Sensitivity

NMR spectrometry is useful because the information is easily related to chemical structure and function, and because it can be made relatively quantitative. A major problem for NMR spectrometry when applied to surfaces is the lack of sensitivity of the technique. For example, pure water is 55 mol/L, and 1 cm^3 contains $\approx 6.62 \times 10^{22}$ protons. From this amount of material, one can easily obtain a strong NMR signal. A typical catalyst containing metal will have a metal surface area of ~ 2 m^2/g. If one assumes that one can easily get 1 g of catalyst into the resonance volume of the NMR spectrometer and that all metal sites can be saturated with benzene, then the signal will arise from about 7.5×10^{19} protons. While the signal from such a sample is observable, it is three orders of magnitude smaller than the typical signals with which a spectrometrist usually deals. If one wishes to examine with NMR spectrometry materials loaded at submonolayer coverages, the problem becomes just that much more difficult to master as the coverage is lowered. Thus, most studies have tended to concentrate on the NMR properties of saturated surfaces or of materials in multilayers above the surface, and those that have addressed monolayer and submonolayer coverages have been carried out by averaging over many hours or days.

3.2. Interferences from the Bulk

NMR spectrometry is not a surface-sensitive technique.

Whatever resonant nuclei are in the resonance region of the spectrometer will contribute to the spectrum. Thus, if the bulk of the material in the resonance region contains a nucleus like the one being detected at the surface, the spectrum will be a composite spectrum, with both the surface and the bulk phases contributing. By virtue of the usual property of such systems, the signal from the bulk material, being here in much larger abundance, will swamp the signal from the surface species. Thus, it is not generally likely that NMR spectrometry will contribute much to the understanding of the difference between molecules at the surface and in the bulk of a material such as a polymer. It may however be of immense value in studying overlayers of polymers on other materials, and may provide information in such areas as coatings technology.

In a related problem, even though the bulk of the sample itself may not contain nuclei that interfere with the detection of the NMR spectrum of the surface material, the NMR spectrometer may contain nuclei in sufficient numbers to be a contributor to the spectrum. Many parts in the resonance region contain materials that have NMR-active nuclei. These may be at low levels as compared to the usual liquid or solid sample and do not contribute to the overall spectrum in those cases. However, with surface samples, the number of nuclei in the sample may be so small that the intrinsic resonance of the NMR probe becomes the dominant characteristic of the spectrum. For example, MAS rotors are often made of alumina or of some polymer material that gives a resonance. Cables contain dielectric materials that are polymers of carbon and hydrogen. Care must always be taken in these circumstances to ensure that the spectra are spectra not of the probe, but rather of the surface sample.

3.3. Sample Integrity

Surfaces are very reactive chemical systems. It is essential that, once the sample is prepared, it be kept away from contamination through exposure to the atmosphere. NMR sample-handling procedures must be carefully scrutinized to ensure that inadvertent contamination of the surface does not occur during the spectrometric experiment. Generally this means that the sample must be contained in a sealed sample tube at all times. Plastic caps are not sufficient to maintain sample integrity for long periods, and flame seals are to be preferred. For magic-angle spinning, two approaches seem to be in practice. First, one prepares the sample in a glass insert that is tailored to the MAS rotors of the spectrometer. It is carefully sealed under flame, often with rotation in a lathe to maintain the delicate balance needed for magic-angle spinning. The second approach, used by both Gay [12] and Apple [13], is to build a probe system with a special collar that fits around a standard NMR sample tube, to produce a spinning system much like the original spinners for liquid samples. It is more tolerant of weight disbalance than the insert and, hence, the process of sealing under flame need not be so carefully controlled in this latter process. The disadvantage (at the present) is that such technology is not commercially available, whereas inserts for some spinners are [14].

3.4. *Spectrometry at Catalytic Conditions*

Most NMR spectrometers are equipped with a probe whose temperature can nominally be varied from about -100°C to about 150°C. There is reason to investigate the NMR spectrometry of catalytic systems at high temperatures where the catalytic processes are carried out. Haddix et al. [15] have described a probe that they have designed to allow the measurement of NMR spectra in a flowing system at temperatures comparable to those used in catalytic reactions. The advantage of such a system is that NMR spectrometry can be used to probe the steady-state populations of various surface species under catalytic conditions. A kind of temperature-programmed desorption spectrometry with NMR detection can be carried out with such a probe system. The difference spectrometry that is the basis of this technique will perhaps provide an improvement in resolution of the spectrometry by the thermal changes in concentrations.

4. *Selected Applications of NMR Spectrometry to Problems in Surface Chemistry*

The use of NMR spectrometry to investigate surface chemical properties of materials is restricted to cases in which the surface layer is richer in a nuclear species by orders of magnitude than the bulk - quite a severe restriction [16 - 18]. One may, in some cases, exploit differences in relaxation rates between surface and bulk to produce a selective detection of the surface phase, but most published experiments have concentrated on systems in which the surface layer is enriched, either by choice of system or by artificial enrichment, in the nuclear spin. Thus, most published reports tend to emphasize materials for which the surface and the bulk are chemically dissimilar.

One area of study where there is usually a dissimilarity of the surface phase and the bulk is heterogeneous catalysis. Many of the catalysts used to transform organic materials are inorganic, geologic materials. It should come as no surprise that NMR spectrometry, which has been used to identify organic materials in solution for almost 40 years, has been applied to the study of organic materials in surface phases related to heterogeneous catalysis. The major problem for this kind of investigation (and, in a way, a definite advantage) is the low natural abundance of carbon-13. This low natural abundance, along with the low concentration of surface species, would make it exceedingly difficult to investigate these kinds of system, were it not for the availability of ^{13}C-enriched materials. Thus, most experiments have been carried out with enrichment of the surface species. Of course, many catalysts have NMR-active nuclei in them. Thus, aluminum-27 and silicon-29 nuclei have been the subject of a number of NMR experiments which address the structure of materials such as the aluminosilicates, both under catalytic conditions and under a variety of preparations. Similarly, counterions can be investigated by NMR spectrometry, e.g., in sodium-23 or ^{133}Cs NMR spectrometry of the counterions in zeolites. When metals are present at the surface, the NMR spectrometry of the nuclei in the metal (e.g., platinum-195) may be investigated to determine the chemical state of the metal under catalytic conditions [18]. In some cases it is possible to introduce a material that is NMR-active which acts as a "reporter" on the actual geologic material. Such a case is seen in the adsorption of xenon gas by porous materials. The NMR spectrometry of xenon-129 (and, perhaps, xenon-131) is relatively easy to carry out and the NMR parameters give detailed, indirect information on the structures in which the xenon is adsorbed [19].

The representative problems presented below tend to come from experiments in heterogeneous catalysis, a reflection of my own bias and interests. One should, however, be aware that there are other interesting applications of NMR spectrometry to surface chemistry being developed constantly. The discussion is meant only to be exemplary, and not exhaustive.

4.1. Surface Hydroxyl Groups

Because the proton has the largest gyromagnetic ratio of any nucleus other than the triton, the problem of low signal-to-noise ratio

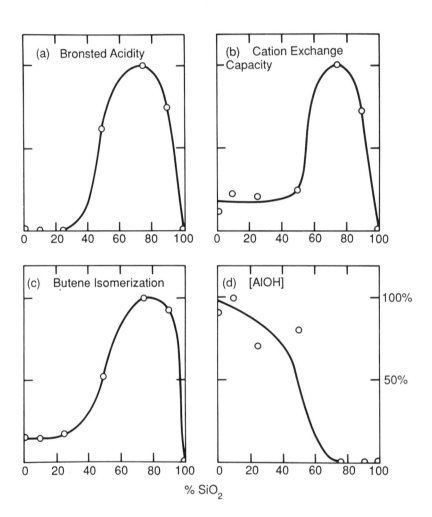

Fig. 7-4. Comparison of Gaussian fraction in the NMR spectrum of mixed alumina-silica catalysts with other measures of the catalytic activity. (From reference [20])

will be most easily overcome by study of proton-containing surface pecies. One major problem with study of the proton resonance is that many surfaces are coated with large numbers of hydroxyl groups that can obscure the resonances of other important surface species in a manner analogous to shrouding by resonances of the bulk. With proper choice of the problem, however, one may find many cases where proton NMR spectrometry will give information about the nature of surfaces.

In pioneering work applying NMR spectrometry to a catalytic problem, Schreiber and Vaughan [20] investigated the nuclear magnetic resonance of surface hydroxyls on mixed silica/alumina materials, which are used in a variety of catalysts either as the catalyst or as a support for some surface phase that is the catalytic material. The proton NMR spectra of such materials exposed to ambient air will show the large amount of superficial water that is present as a sharp liquid-like resonance. After careful drying at elevated temperatures and sealing to prevent exposure, the spectra of the mixed silica/aluminas show the surface hydroxyls to be a two-phase system. In one phase the surface hydroxyls exhibit a rather sharp Lorentzian resonance, while in the second phase the resonance is broader and Gaussian in shape. The two phases can be associated with the surface silanol groups and the surface aluminum-associated hydroxyl groups, respectively. Thus, through NMR spectrometry, one is able to quantify the fraction of exposed aluminum atoms at the surface of the mixed catalyst. The relative amount and distribution of the sites for interaction of reactants at the surface have been thought, for a long time, to determine the efficacy of a catalyst. What Schreiber and Vaughan demonstrated is that the fraction of surface hydroxyls in one phase, as measured by NMR spectrometery, is correlated with other measures of the catalytic activity. In Fig. 7-4 are shown the results of that experiment compared to other measures of catalytic activity of the same materials. This elegant experiment shows a major strength of NMR spectrometry for identifying surface phases - quantification of sites.

4.2. Benzene Sorbed on Alumina

Sorbed molecules may also be analyzed with proton NMR spectrometry. When examined, the surface molecules may have different properties from the same material in the bulk phase. A very good example of this difference can be seen in the NMR spectra of benzene sorbed on materials such as charcoal [21] or alumina [22]. Pure benzene exhibits a transition from the liquid state to a solid phase

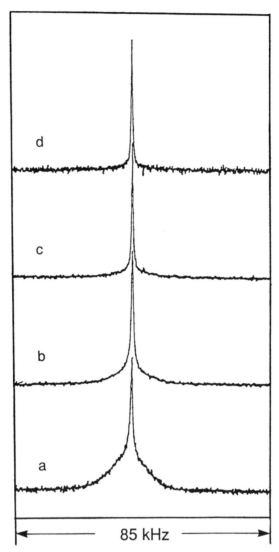

Fig. 7-5. Proton NMR spectra of benzene sorbed in alumina as a function of coverage. a) 3.3 statistical monolayers; b) 2.4 statistical monolayers; c) 1.3 statistical monolayers; d) 0.9 statistical monolayers. T = 173K. State always obtained by desorption from the saturated surface. (From reference [22])

in which the molecules rotate rapidly around the sixfold axes while remaining localized at lattice sites [23]. At temperatures below about

90 K, the rotation is quenched to give a rigid solid between these two temperatures, the spectrum is a composite of both a Lorentzian phase and a Gaussian phase. Both of these dynamic transitions are observed by their effects on the NMR lineshape as a function of temperature [23]. When benzene is sorbed by a material such as alumina, the temperature dependence of the NMR spectrum is rather different from that of the bulk. First, the spectrum at temperatures below the freezing point of bulk benzene is rather narrow. At a temperature of about 185 K, one begins to detect a broad resonance that grows in as the temperature is decreased. Only at temperatures below about 125 K does the spectrum of sorbed benzene exhibit a single Gaussian form. At temperatures intermediate to these limits it is a composite of a Lorentzian and a Gaussian line. Because of the manner in which the experiment is carried out, only the resonance of the benzene is detected [20, 22]. Such a temperature dependence of the NMR spectral parameters demonstrates the existence of a distribution of binding energies of benzene molecules on alumina [21]. What is at first puzzling is the fact that those molecules which are most easily removed by outgassing at 295 K are the ones that are least mobile in the intermediate temperature range, i.e., the ones that contribute to the Gaussian part of the bifurcated resonance at the highest temperature in this two-phase range (see Fig. 7-5). We explain this fact by assuming that the materials most difficult to desorb at 295 K are, in fact, sorbed in regions of the sample in which they have greater free volume than do the ones that are most easily desorbed. In the intermediate-temperature region these molecules still have sufficient freedom to move, whereas those easily removed by desorption are in regions in which they have essentially bulk properties and thus are "frozen" at these temperatures much like bulk materials. Some support for this explanation can be seen in the NMR spectra taken at various coverages at a constant temperature. The molecules that contribute to the narrow resonance are those left when only a statistical monolayer of material is on the alumina surface after desorption. If these molecules are located in places where molecular packing is hindered by interaction with the walls, e.g. in small pores, one can qualitatively understand the results because these molecules will have more free volume than those in large pores which are very much like bulk benzene. Thus, careful studies of the NMR spectrometry as a function of the coverage and temperature may unravel the complex dynamic structure of these heterogeneous materials.

4.3. *Silanized Surfaces*

The adsorption of material such as benzene presents one kind of surface-bound material that can be investigated with NMR spectrometry. Other systems amenable to investigation with NMR spectrometry are surface phases that are chemically bound to the substrate [24]. For example, to change the properties of silicas for chromatographic purposes, one often adds a surface-attached layer by reaction of pendant silane groups. One such treatment involves the addition of silanizing agents such as trimethylchlorosilane to produce a more uniform surface layer for the chromatographic process. Such surface materials can be investigated with NMR spectrometry of the silicon, as is shown in Fig. 7-6, or the carbon in the chains. In this case, if the bulk material contains silicon, the bulk material may also show a signal, but it can be ascertained which unique signals arise from surface species by examining the material before and after silanation [24]. The resulting spectra of the surface can be used to determine the various environments produced by this process - a useful bit of information in determining the chromatographic effects of the process. Spectroscopic changes under chromatographic conditions may also be important in specifying the nature of surface interactions that result in the separation process.

One difficulty with the cross-polarization experiment is that the intensities in a spectrum like that of Fig. 7-6 do not directly reflect the relative concentrations of nuclei in the various environments. This effect results from the manner in which rare-spin magnetization is generated for a nuclear species such as silicon. The process, known as *cross polarization*, results in a transfer of polarization from another nuclear species, often protons, through the intermediacy of the dipole-dipole coupling of the rare spin with the protons. Rare spins in various environments have different rates for this transfer. Thus, transferring for a time, called the *contact time*, may result in different relative amounts of magnetization for the different resonances. An additional complication is the depletion of magnetization due to relaxation that may also destroy various components of the magnetization with different rates. These two competing processes produce a complicated time dependence for the relative magnetizations that can only be sorted out by a detailed study of the dependences of the relative intensities on contact time. An example is shown in Fig. 7-7 [24]. The three lines indicate how the intensities of the three silicon resonances of silica gel

Fig. 7-6. The MAS silicon-29 spectrum of silica gel (A) and of the reaction product formed by reaction of the silica gel with trimethylchlorosilane (B). The chemical shifts are reported relative to the resonance of silicon-29 in tetrachlorosilane. Note the additional resonance due to the presence of the bonded groups. (From reference [24])

Fig. 7-6) vary as the contact time for transfer of magnetization is systematically changed. As one can see, the optimal contact time for each of the three resonances is different and the relative intensities of two of the lines are nearly equal for contact times of less than 5 milliseconds, but they become unequal at longer contact times due to the kinetics of the transfer and relaxation processes. The detailed analysis of the time dependence is shown by the solid lines, from which one can extract the relative numbers of the three species.

4.4. *Surface Complexes with Hydroxyls*

It is well known that the binding of bases with different acid sites at the surface of alumina produces several different adducts [25]. For example, the attachment of pyridine to Brønsted acid sites produces a pyridinium ion, whereas the attachment to a Lewis acid site produces the Lewis acid/pyridine adduct. The chemical shifts of carbon in the material made by adsorbing butyl amine on alumina show that there are two species formed, which can easily be distinguished in the ^{13}C NMR spectrum [26]. Similar results have been seen in the nitrogen-15 [27] spectrum of adsorbed amines and in the P-31 NMR spectrum of

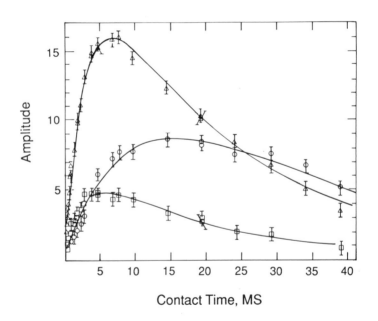

Fig 7-7. The dependence of the intensities of the three silicon-29 resonance lines on contact time for the resonances of silica gel. Note that the relative intensities are rather strong functions of the contact time. (From reference [24])

adsorbed phosphines [28]. With care, one can quantify the amount of each site by a series of NMR experiments that accounts for the relative cross-polarization efficiency of the various sites, as was done for the silicon-29 spectrum of silica gel.

4.5. Metal-containing Surfaces

A typical cracking catalyst nowadays contains a metal component as well as an oxide such as alumina or silica [29]. The presence of this additional phase at the surface of a heterogeneous catalyst provides additional sites for the formation of complexes that can be investigated with NMR spectrometry. Indeed, the metal itself may be the focus of NMR investigation. For example, in a classic series of experiments, Slichter and his co-workers [18, 30] investigated the resonance of platinum-195 in small crystallites supported on the surface of alumina to understand how the local susceptibility of the conduction electrons varies in the region of the surface. The range of shifts of the platinum resonance in these small particles is particularly large, and the spectra were gathered in a point by point fashion over a much larger change of magnetic field than is done in typical NMR experiments [30]. Examples of Pt-195 spectra obtained from platinum particles on an alumina support are shown in Figure 7-8 [30]. The spectrum is plotted as a function of the field to frequency ratio, H_0/f, rather than by the conventional definition. The range of resonance intensity is extremely large compared to most spectra. Note how the relative intensity in the range around $H_0/f = 1.09$ changes as the dispersion changes. As dispersion becomes higher, the intensity in this region becomes greater, indicating that surface atoms resonate in this region.

Because the metal particles have nuclei with spin, it is also possible to perform a double-resonance experiment on an adsorbed molecule in contact with the surface to locate the resonance of the platinum nuclei that are at the surface. Slichter and his co-workers [31] used adsorbed ^{13}C-enriched CO to make such an identification for platinum particles with different dispersions.

King et al. [32] did a similar experiment to identify the surface atoms of a supported copper catalyst. Copper-63 is a quadrupolar nucleus which, when subject to a strong electric field gradient, is broadened beyond detection with typical NMR spectrometric instrumentation. Such is the case for those copper nuclei at the surface

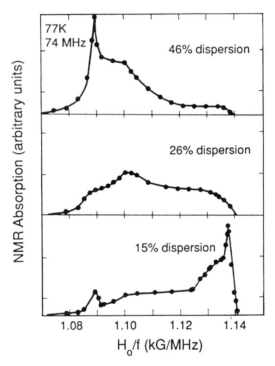

Fig. 7-8. Platinum-195 spectra of supported platinum particles with varying dispersions. Note the manner in which shift is plotted and how the intensity changes with dispersion. (From reference [30])

of the supported particle. However, treatment of the catalyst with HCl chlorinates the surface to produce a much weaker electric field gradient. Thus, the NMR spectrum after such a treatment contains a component for these nuclei as well as those in the bulk of the copper particle. The results showed that by comparison of the spectra before and after such a treatment, one could determine the number of copper atoms at the surface of the particle - that is, NMR spectrometry could be used to determine the dispersion of the metal in this catalyst.

7.6. Investigation of Nuclei in a Catalyst

As one can discern from the previous examples, the structure of

Fig. 7-9. Silicon-29 MAS NMR spectra of faujasites of varying silicon-to-aluminum (Si/Al) ratio and the theoretical spectra generated from the known silicon-to-aluminum ratio, assuming Lowenstein's rule to be valid. (From reference [34])

the catalyst, be it the hydroxyl groups or the metal placed at the surface of an oxide, can be addressed with NMR spectrometry. In early work to apply NMR spectrometry in catalysis, the structures of microporous zeolites were investigated with silicon-29 and aluminum-27 NMR spectrometry [33, 34]. Zeolites are microporous aluminosilicate structures, whose properties depend, in part, on the relative numbers of silicon and aluminum atoms in the framework and the distribution of these atoms [29]. It was discovered that the silicon-29 MAS NMR spectrum of such materials gives detailed information on the structure and distribution of silicon and aluminum in these materials. For example, Fig. 7-9 shows the silicon-29 MAS NMR spectra of several faujasites (Y zeolites) of varying silicon/aluminum ratio. One identifies five resonances in the spectrum, showing that the silicon spectrometry is sensitive to whether the atoms in positions two bonds away are silicons or aluminums. From most shielded to most deshielded these correspond, respectively, to silicon with no, one, two, three, or four aluminum atoms in the positions. The knowledge that these systems obey Lowenstein's rule allows one to estimate the silicon/aluminum ratio from the intensity pattern of these five lines, as shown in the figure. Thus, an important structural parameter is directly available from such spectra.

Aluminum present in zeolites can also be detected with MAS NMR spectrometry of the aluminum-27 nucleus. Even though it has a quadrupole moment, when it is present in nearly cubic symmetry the effect of the quadrupole coupling is small because the electric field gradients are small, such as the tetrahedral sites in the faujasites or when it is in its usual octahedral environment when it is hydrated. The octahedral and tetrahedral coordinations have different chemical shifts and can be easily distinguished, as can be seen in Fig. 7-10. This fact was the basis for investigation of the effects of dealumination [34] with NMR spectrometry. The existence of a resonance for octahedral aluminum after dealumination and before washing showed that nonframework aluminum had been produced by dealumination. Subsequent washing removed most of the nonframework aluminum, as can be seen with the NMR spectrometry.

4.7. Investigation of Small Molecules on Catalysts

Inorganic materials such as alumina, silica and other oxides are used as catalysts, either separately or in combination, and sometimes with small metallic crystallites on the surface, particularly for transformations of organic materials. Thus, spectrometrists first investigated the properties of the surface-bound organic species with the goal of determining the structures that exist in a catalytic complex on such a surface. Perhaps the most studied small molecule at a surface

Fig. 7-10. The aluminum-27 MAS NMR spectrometry of a faujasite at several stages of dealumination with $SiCl_4$. a) the starting material; b) after dealumination but before washing; c) after washing to remove occluded aluminum species; d) after several washings. (From reference [34])

is carbon monoxide [35]. On supported metals there are at least three forms for the bonding of CO: linear, bridged, and <u>gem</u>-dicarbonyl forms [36] (Fig. 7-11). Because of the differences in bonding arrangements, each form has a unique signature in the carbon NMR spectrum, and is readily distinguishable by carbon spectrometry. As one can see in Fig. 7-11, the relative abundances of the various species on catalysts from different preparations gives an indication of how the spectrometry and the changes in these materials can be studied in a stopped-flow analysis [37]. Again, these experiments are carried out using enriched material. The results have importance for the studies of these model reactions, but it will probably not be possible to investigate on-line operation without enrichment using carbon-13 NMR spectrometry.

The interaction of complicated materials such as ethylene [38] or benzene [39] with surfaces has also be addressed with NMR spectrometry. For example, ethylene bound to platinum on a supported-platinum catalyst was investigated by measurement of the dipolar coupling of carbon to the attached hydrogens. Wang et al. [38] conclude that, on the surface, attachment of ethylene results in the formation of the CCH_3 fragment. Similarly, the presence of binding to platinum is observed in proton spectrometry of benzene by the presence of proton/proton dipolar couplings which can be suppressed by multiple-pulse decoupling sequences [39]. Acetylene is found to exist in two forms on several different metals: CCH_2 and HCCH. [40] More recently, the reactions of various osmium carbonyls with dehydroxylated magnesia have been investigated by carbon NMR spectrometry. The results indicate how the various complexes tether to the surface to form catalytic centers of unique stoichiometry [41]. The comparison of the ^{13}C MAS NMR spectra of the species formed at the surface with solution spectra of model compounds made it possible to determine the kinds of osmium surface complexes formed in various reactions. Again, enrichment of the precursors was necessary to study these model catalysts.

Pfeifer and the Leipzig group [42] have examined the ^{13}C NMR spectra of a wide variety of small organic molecules sorbed in zeolites and other oxides. They observe slight effects of the local environment through a dependence of the NMR chemical shift on coverage. These shifts are due to intermolecular and molecule/wall interactions for such molecules as ethylene, the butenes and acetone. Similar interactions also appear in the ^{15}N spectrum of ammonia sorbed in a Na Y zeolite.

The major difficulty in all of these experiments is the insensitivity of the NMR technique. While it may be extremely useful for these model systems, there seems to be very little doubt that it cannot serve as a routine monitor of the state of a system containing materials of natural abundance in carbon-13.

One may use proton NMR spectrometry to investigate these systems. Supported-metal catalysts frequently take up hydrogen to form

CO Adsorbed on
Ru/Silica

Sample (A)
CO/Ru = 1.54

Sample (B)
CO/Ru = 1.66

Frequency, in ppm, Relative to TMS

Fig. 7-11. The carbon-13 NMR spectrum of the material formed by the adsorption of carbon monoxide on a silica-supported ruthenium catalyst. There are four species whose spectra appear, the three described in the text and a small amount of gaseous carbon dioxide, which gives the sharp resonance at 124 ppm. (From reference [36])

preparation has changed the number of each binding site and how each is filled as the total amount of carbon monoxide is changed. A n important point is that these experiments are carried out with enriched carbon monoxide and thus represent experiments on model systems. At natural abundance, such spectra would be difficult to detect. The quantification of these three species is important in being able to specify the kinds of binding sites available in a catalyst. It is also important to understand the role of carbon monoxide and the surface species derived from it in "syngas" reactions. For example, the various carbonaceous deposits formed by the reaction of the carbon monoxide with hydrogen over a supported-metal catalyst can be identified with carbon NMR supported-metal hydrides [43]. The signatures of these hydrides are easily discernible in the proton NMR spectrum of the material without decoupling experiments. It has been shown for palladium "hydride" that the NMR chemical shift can be used as a measure of the Pd/H ratio in

Fig. 7-12. The proton NMR spectra of hydrogen absorbed by bulk palladium and alumina-supported palladium at two overpressures of hydrogen. The change in shift of the resonance is related to the H/Pd ratio in the system, as determined from the uptake isotherms. (From reference [44])

the bulk and in the oxide supported palladium. Figure 7-12 [44] shows a comparison of the resonance of protons absorbed by a palladium powder and an alumina-supported palladium catalyst. As the pressure of hydrogen is changed, the Pd/H ratio is changed, with a corresponding change in the NMR shift.

These hydrides participate in chemical reactions such as the hydrogenation of olefins. Figure 7-13 [44] shows a series of spectra of the protons in a sample containing the catalysts and a mixture of hydrogen and ethylene. Figure 7-13 shows the hydride resonance before addition of ethyelene. After addition of ethylene, the reaction of hydrogen with ethylene produces physisorbed ethane, which is detected in the proton spectrum of Fig. 7-13. Importantly, the amount of hydride present is decreased by the reaction, resulting in a shift of the hydride resonance. Addition of more ethylene continues this process, as can be seen in Fig. 7-13. Such studies may add information on the participation of species in reaction networks at the surfaces of catalysts.

4.8. Relaxation Studies of Molecules at Surfaces

One of the first NMR techniques applied to molecules sorbed at a catalytic surface is the characterization of relaxation processes that return the nuclear magnetic system to equilibrium [45]. The proximity of the surface atoms with a variety of binding sites results in a more complex dependence of relaxation rates on system variables such as temperature and concentration than is seen for bulk liquids or solids, principally because the mobility of the material at the surface may be changed dramatically by the surface. In one study, Bronniman and Maciel [46] investigated the properties of methanol sorbed into a H Y-zeolite dehydrated at 400°C. They measured the spin-lattice relaxation time of ^{13}C in the rotating frame as a function of temperature at a fixed loading of 10 wt. %. From these measurements, they were able to estimate the correlation time at each temperature for the motion inducing relaxation. They suggest that this motion is a proton-exchange process that involves two methanol molecules. The spin-lattice relaxation of ^{13}C in the laboratory frame is at least biexponential, from which it is concluded that there are two types of methanol, a mobile material much like a liquid, and a more rigidly held material. A third resonance slightly shifted from the others was attributed to methanol molecules that find themselves in the smaller ß cages.

In a similar study by Tsiao [47], carbon-13 MAS NMR

in contact with H-ZSM-5, a catalyst known to produce a distribution of products akin to gasoline [48]. The fully relaxed ^{13}C MAS NMR spectra for methanol sorbed in a series of zeolites that differ only in the silicon/aluminum are given in Fig. 7-14. Each material was treated by heating for a short while (15 min) at 100°C and then at 200°C to produce reaction. The appearance of a resonance attributable to the presence of dimethyl ether shows that the first step in the reaction is the dehydration of methanol. No resonances attributable to subsequent steps in the reaction are seen for the conditions of treatment given here, as has been seen by Derouane et al. [49]. The difference in reactivity as a function of the silicon/aluminum ratio can be clearly seen in the production of dimethyl ether as a function of the temperature of pretreatment. The catalyst with silicon/aluminum ratio of 67 produces dimethyl ether by contact at room temperature and is the most active of the ones Tsiao investigated [47]. The results also indicate that the conversion of methanol to dimethyl ether is rather facile and that the subsequent steps require higher thermal energy to occur.

Tsiao also investigated the carbon-13 relaxation behavior of the methanol-derived material [47]. He found that there are two types of species present not counting the dimethyl ether that could be distinguished on the basis of spin-lattice relaxation times. The temperature dependence of the relaxation times allowed him to estimate the activation energy of the process that affect the rotational mobility of these species. One appears to be surface-bound methoxyl group and the other is a more liquid-like material.

4.9. The NMR Properties of Xenon Gas Adsorbed in Microporous Materials.

The structure and chemical properties of a system may sometimes be probed indirectly with NMR spectrometry. One interesting use related to catalysis and surfaces is the application of xenon-129 NMR spectrometry to the investigation of microporous materials [50]. Xenon-129 is reasonably abundant and the gyromagnetic ratio is close to that of carbon-13. Its receptivity relative to carbon-13 is about 32, making it easier to detect than carbon-13 at equivalent conditions. Like other properties of gases, the NMR parameters are strong functions of density. This results from the strong perturbations that occur when a gas molecule collides with another molecule or any other object. One typically uses a virial expansion to

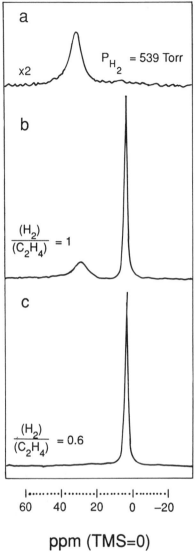

Fig. 7-13. Proton NMR spectra of materials produced by mixing hydrogen and ethyelene over an alumina-supported palladium catalyst. A) The "hydride" before addition of ethylene; B) After subsequent addition of ethylene; C) After addition of more ethylene. The sharp resonance is that of ethane on the surface. Note how the "hydride" resonance shifts as it is presumably depleted. (From reference [44])

spectrometry was the primary tool to investigate the dynamics of methanol and the surface species derived from methanol when it is put

Fig. 7-14. The carbon-13 MAS NMR spectra of materials made by adsorption of C-13-enriched methanol on H-ZSM-5 of varying silicon-to-aluminum ratio. The presence of the resonance of dimethyl ether indicates that dehydration of methanol has occurred on some catalysts. (From reference [47])

express this dependence on density, e.g. for the chemical shift:

$$\sigma(\rho, T) = \sigma_0 + \sigma_1\rho + \sigma_2\rho^2 + \sigma_3\rho^3 + \dots \qquad (7\text{-}6)$$

where σ_i is the ith virial coefficient of the chemical shift. In a bulk gas, the dominant effects arise from gas/gas collisions, and each virial term represents the effects of wall collisions, binary, ternary, quarternary and higher order collisions, respectively. Much of the attention in gas-phase NMR spectrometry has focused on measuring and calculating the coefficients σ_1, σ_2, etc. [51].

For a gas sorbed in a microporous material such as a zeolite, collisions with the wall, both elasitc and inelastic, can have a significant effect on the observed NMR properties. This dependence allows one to determine properties of the "walls" indirectly from the NMR spectrometry of sorbed xenon [50]. In one restricted sense, the collisions with walls are similar to collisions with other gas molecules in a mixture [52]. For example, the intercept at $\rho = 0$, σ_0, for xenon adsorbed in sodium or hydrogen Y zeolites is approximately 60 ppm deshielded relative to that of bulk xenon gas. This 60-ppm shift is a measure of the effect of the zeolite structure on the xenon.

The presence of ions exchanged into the structure, metal particles and other adsorbed materials can be detected by the effects interaction with these materials have on the xenon chemical-shift parameters [50, 53]. Metallic particles in the zeolite for catalytic enhancement, for example, have a pronounced effect on the resonance of sorbed xenon. The presence of material such as hydrogen or carbon monoxide adsorbed on the metal substantially changes this effect such that one can detect the presence of these coadsorbates easily with xenon NMR spectrometry [50]. Such indications in the xenon NMR spectrum have been used to distinguish inhomogeneous adsorption of hydrogen and carbon monoxide in zeolite-supported platinum catalysts.

One effect observed in NMR spectrometry of xenon in these microporous environments not seen for the bulk gas is the low-density behavior of the chemical shift for xenon in contact with zeolites having certain counterions [50, 53]. For multiply-charged counterions such as calcium and nickel, the chemical shift becomes strongly deshielded at low densities and, in the limit of low density, is inversely proportional to density [50,53]. Fraissard and Ito [50] attribute this to an "electric-field" effect. Cheung [54] argues that such a dependence on density can be understood if the xenon is adsorbed in two ways in these catalysts, one to form a phase similar to that of xenon in H Y zeolite and a second in which the xenon is bonded (presumably in the proximity of the ion). He assumes that rapid exchange is occurring between these two sites, the result of which is the observed dependence of chemical shift on density. Measurements of the chemical shift of xenon on Y zeolites with different levels of exchange of nickel ion [53] shows that, at low nickel loading, the inverse dependence on density is not required to explain the data and is only required at nickel loadings above 15% exchange of the sodium ions to explain the dependence of xenon chemical shift on density. (Fig. 7-15) Thus the first exchanged nickel ions do not affect the xenon by this mechamism. One could understand

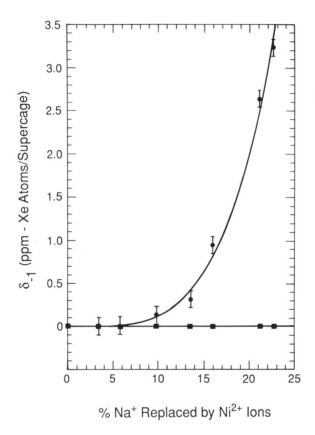

% Na⁺ Replaced by Ni²⁺ Ions

Fig. 7-15. The coefficient for the term inverse in the density in the chemical-shift expansion for xenon sorbed in nickel-sodium Y zeolites as a function of the amount of sodium removed. Note that a low concentration there seems to be no dependence on this term. (From reference 47)

how this might happen if (1) direct contact between the ion and the xenon were required to show the effect and (2) if the first nickel ions were exchanged into sites in which the xenon could not have direct access to them. Such a situation would occur if the nickel were first to be exchanged into sites in the hexagonal prisms or the sodalite cages, with occcupation of cation sites in the supercages by nickel occurring only after these are filled. This seems to be the case for nickel ions [55]. If that is correct, it is likely that direct contact with the ion by xenon is required to show this effect, and this could be interpreted as evidence of the formation of a transient adsorption bond between xenon and the site created by the exchange of cations.

The wall collision term is affected by the rate of collision with the "wall" of the zeolite. The more frequent the collisions, the stronger is the time-averaged effect on the xenon chemical shift. This property has been exploited [56] to estimate the size of pores and channels in zeolites. A correlation has been found between the chemical shift and the "mean free path in the structure," which can be related to dimensions of the zeolite environment. Low-temperature xenon NMR spectrometry may also be useful in estimating available volume in that it can be used as a detected of the filling of the void and the condensation of solid xenon in a kind of titrimetric experiment [54].

5. Epitome

NMR spectrometry has been, and continues to be, an excellent tool for identifying chemical structures in liquids and solutions. The ability to quantify the resonance allows one to ascertain the relative numbers of the various functional groups in a sample easily and, hence, straightforwardly specify the chemical constituents. More recently, sophisticated techniques for determining the spectra of solids and solid-like materials have expanded the utility of the technique into spheres in which chemical information was previously limited. One major drawback that remains is the relative insensitivity of NMR spectrometry when compared to visible and other forms of detection -- a particularly acute problem when studying surface phases. Another major difficulty of application to surfaces is the potential of interferences from the bulk that may obscure surface resonances or, worse, be misinterpreted as arising from surface phases. The technique of NMR spectrometry will most likely find use when applied to materials in which the surface has been treated to have an enhanced NMR signal, for example by the deliberate sorption of an NMR-unique material on one that is transparent to the NMR technique. A paragon of that type of problem is the investigation of surface states of organic materials on catalysts made of refractory oxides.

I have presented a few selected examples of the application of NMR spectrometry to surface studies to indicate some of the problems and some of the very interesting results one obtains from surface systems. These are, by no means, exhaustive; there are many more in the literature, and newer applications are seen every day. NMR spectrometry has now reached the stage of wider application in surface chemistry, and it should be recognized by surface chemists as a major tool for these types of investigations, in a manner analogous to its use in solution, provided one takes the proper steps to ensure that spectrometric interferences, low concentrations, and loss of signal due to strong magnetic interactions at the surface do not limit the spectrometry.

Acknowledgments

I have been fortunate to have been supported financially over the years by various institutions: the National Science Foundation, the Department of Energy, the Petroleum Research Fund, the Research Corporation, Sun Refining and Marketing Company, Hercules, Inc., and the University of Delaware Research Foundation. To these institutions who have placed trust in me and to my many students, I am grateful for the ability to explore new venues for the use of NMR spectrometry.

References

[1]   Abragam, A., "The Principles of Nuclear Magnetism," Oxford University Press, London, 1961.
[2]   Becker, E. D., "High Resolution NMR," Academic Press, Orlando, FL, 1980.
[3]   Adamson, A. W. "Physical Chemistry of Surfaces," Wiley, New York, 1982.
[4]   Gerstein, B. C.; Dybowski, C. R., "Transient Techniques in NMR of Solids," Academic Press, Orlando, Fl, 1985.
[5]   Lowe, I. J., Phys. Rev. Lett., *1959*, 2, 285.
[6]   Andrew, E. R.; Bradbury, A.; Eades, R. G., Nature, *1958*, 182, 1659.
[7]   Schaefer, J.; Stejskal, E. O.; Buchdahl, R., Macromolecules, *1975*, 8, 291.
[8]   Herzfeld, J.; Berger, A., J. Chem. Phys., *1980*, 73, 6021.

[9]   Gerstein, B. C.; Pembleton, R. G.; Wilson, R. C.; Ryan, L., J. Chem. Phys., *1977*, 66, 362.
[10]   Bax, A.; Szeverenyi, N.; Maciel, G., J. Magn. Reson., *1983*, 55, 494.
[11]   Duncan, T. M., J. Phys. Chem. Ref. Data, *1987*, 16, 125.
[12]   Gay, I. D., J. Magn. Reson., *1984*, 58, 413.
[13]   Shoemaker, R. D.; Apple, T. M., J. Magn. Reson., *1986*, 67, 367.
[14]   Inserts for some commercial spinners are available from Wilmad Glass Company, Buena, New Jersey 08301.
[15]   Haddix, G. W.; Reimer, J. A.; Bell, A. T., J. Catal., *1987*, 106, 111.
[16]   Tabony, J. B., Progr. NMR Spectrosc., *1982*, 14, 1.
[17]   Duncan, T. M.; Dybowski, C. R., Surf. Sci. Rep., *1981*, 1, 157.
[18]   Wang, P., Ansermet, J. P.; Wang, Z.; Shore, S.; Slichter, C. P.; Sinfelt, J. H., Science, *1986*, 234, 35.
[19]   Fraissard, J. P.; Ito, T., ZEOLITES, *1988*, 8, 350.
[20]   Schreiber, L. B.; Vaughan, R. W., J. Catal., *1975*, 40, 226.
[21]   Resing, H. A., J. Phys. Chem., *1965*, 43, 669.
[22]   Tirendi, C. F.; Mills. G. A.; Dybowski, C., J. Phys. Chem., *1989*, 93, 3282.
[23]   Andrew, E. R.; Eades, R. G., Proc. Roy. Soc. (London), *1953*, A218, 537.
[24]   Maciel, G. E.; Sindorf, D. W., J. Am. Chem. Soc., *1981*, 103, 4263; Maciel, G. E.; Sindorf, D. W., J. Am. Chem. Soc., *1980*, 102, 7606.
[25]   Peri, J. B., J. Phys. Chem., *1974*, 78, 588.
[26]   Dawson, W.; Ellis, P. D.; Kaiser, S.; Inners, R. R., J. Phys. Chem., *1982*, 86, 867.
[27]   Haw, J. F.; Chuang, I. S.; Hawkins, B. L.; Maciel, G. E., J. Am. Chem. Soc., *1983*, 105, 7206.
[28]   Rothwell, W. P.; Shen, W. X.; Lunsford, J., J. Am. Chem. Soc., *1984*, 106, 2452.
[29]   Gates, B. C.; Katzer, J. R.; Schuit, G. C. A., "Chemistry of Catalytic Processes," McGraw-Hill, New York, 1979.
[30]   Rhodes, H. E. Wang, P.-K.; Stokes, H. T.; Slichter, C. P.; Sinfelt, J. H., Phys. Rev., B, *1982*, 26, 3559.
[31]   Makowka, C. D.; Slichter, C. P.; Sinfelt, J. H., Phys. Rev. Lett., *1982*, 49, 379.
[32]   King, T. S.; Goretzke, W. J.; Gerstein, B. C., J. Catal., *1987*, 107, 583.

[33] Engelhardt, G.; Lohse, U.; Lippmaa, E.; Tarmak, M.; Magi, M., Z. Anorg. Allg. Chem., *1981*, 482, 49.

[34] Klinowski, J.; Ramdas, S.; Thomas, J. M.; Fyfe, C. A.; Hartman, J. S., Faraday Soc. Trans. 2, *1982*, 78, 1025; Fyfe, C. A.; Klinowski, J.; Gobbi, G. C., Angew. Chem., Int. Ed. Engl., *1983*, 22, 259.

[35] Duncan, T. M.; Yates, J. T.; Vaughan, R. W., J. Chem. Phys., *1980*, 70, 1219.

[36] Duncan, T. M.; Root, T. W., J Phys. Chem., *1988*, 92, 4426.

[37] Duncan, T. M.; Winslow, P.; Bell, A. T., Chem Phys. Lett., *1983*, 102, 163.

[38] Wang. P. K.; Slichter, C. P.; Sinfelt, J. H., J. Phys. Chem., *1985*, 89, 3606.

[39] Tirendi, C. F.; Mills, G. A.; Dybowski, C., J. Phys. Chem., *1984*, 88,5765.

[40] Wang, P. K., Ph.D. dissertation, University of Illinois at Urbana, Urbana, Illinois, 1984.

[41] Hasselbring, L. C., Ph.D. dissertation, University of Delaware, Newark, Delaware, 1987.

[42] Pfeifer, H.; Meiler, W.; Deininger, D., Ann. Rept. NMR Spectrosc., *1983*, 15, 291.

[43] Sheng, T.-C.; Gay, I. D., J. Catal., *1982*, 77, 53.

[44] Barabino, D. J., Ph.D. dissertation, University of Delaware, Newark, Delaware, 1988.

[45] Zimmerman, J. R.; Brittin, W. E., J. Phys. Chem., *1957*, 61, 1328.

[46] Bronniman, C. E.; Maciel, G. E., J. Am. Chem. Soc., *1986*, 108, 7154.

[47] Tsiao, C. J., Ph.D. dissertation, University of Delaware, Newark, Delaware, 1989.

[48] Kaeding, W. W.; Butter, S. A., J. Catal., *1980*, 61, 155.

[49] Derouane, E. G.; Nagy, J. G.; Dejaifve, P.; van Hooff, J. H. C.; Spekman, B. P.; Vedrine, J. C.; Naccache, C., J. Catal., *1978*, 53, 40.

[50] Fraissard, J.; Ito, T., ZEOLITES, *1988*, 8, 350.

[51] Jameson, A. K.; Jameson, C. J.; Gutowsky, H. S., J. Chem. Phys., *1970*, 53, 2310.

[52] Jameson, C. J.; Jameson, A. K.; Cohen, S. M., J. Chem. Phys., *1973*, 59, 4540.

[53] Bansal, N.; Dybowski, C., J. Phys. Chem., *1988*, 92, 2333.

[54] Cheung, T. T. P.; Fu C. M.; Wharry, S., J. Phys. Chem., *1988*, 92, 5170.

[55] Jablonski, J. M.; Mulak, J. Romanowski, W., J. Catal., *1977*, 47, 147.

[56] Demarquay, J.; Fraissard, J. P., Chem. Phys. Lett., *1987*, 136, 314.

8. ELECTRON SPIN RESONANCE SPECTROSCOPY

Murray B. McBride
Dept. of Soil, Crop and Atmospheric Sciences
Cornell University, Ithaca, NY 14853

1. Introduction

Electron spin resonance (ESR), also termed electron paramagnetic resonance (EPR), can be an informative method in the study of bulk and surface properties of geologic materials. ESR, detecting unpaired electrons within materials, is not strictly a surface method. When applicable, the method has inherent advantages over most other analytical techniques because:

1. It allows investigation of samples under near-ambient conditions, with little disruption of surface structures or chemical equilibria. Studies of mineral surfaces under environmentally realistic conditions are possible.
2. It is not confounded by the sample matrix regardless of its chemical complexity (assuming the matrix is nonmagnetic). Therefore all diamagnetic materials with paramagnetic centers are accessible to study.
3. It is inherently sensitive, able to detect concentrations of paramagnetic centers as low as 10^{-9} mol/L [1].

The last advantage permits the study of trace metals and rather unstable electronic defects to be investigated in minerals. These paramagnetic centers may be sources of chemical reactivity, as will be discussed later in this chapter. Thus, ESR often selectively "sees" reactive species, since these species tend to possess unpaired electrons, and is "blind" to the diamagnetic matrix.

Despite the potential utility of ESR, the method has not found widespread use in mineralogical and surface chemical studies. In part, this may be due to investigators' unfamiliarity with the theory and practice of ESR. This chapter presents sufficient theory to enable the main features of ESR spectra to be interpreted and demonstrates, using examples, how the method can be applied to the study of geologic materials. Particular emphasis is placed on the use of ESR as a surface spectroscopic method.

2. Theory

2.1. The ESR Experiment

When an external magnetic field H is applied to a sample having a magnetic moment μ_e, the electron spin S is quantized in that direction. Defining the magnetic field to be in the Z-axis direction, there are two allowed orientations of the spin, parallel and antiparallel to H, with spin quantum numbers $m_s = 1/2$ and and $-1/2$. These quantized spin states are depicted in Fig. 8-1a. The energy of interaction E between the applied field

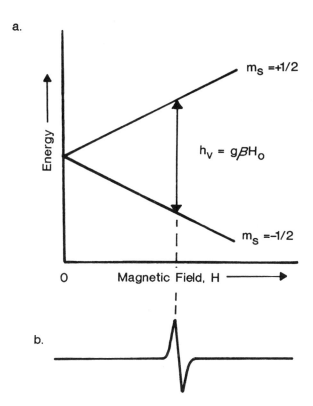

Fig. 8-1. (a) Zeeman energy levels of an unpaired electron in an applied magnetic field, H. (b) Signal arising from electron spin resonance at $H = H_0$.

and the electron is given by:

$$E = -\mu_e \cdot H = g\beta H m_s \qquad (8\text{-}1)$$

where g is a dimensionless constant called the g factor and β is the Bohr magneton constant. The +1/2 and -1/2 spin states, denoted as eigenfunctions of the Schroedinger equation by $|+1/2\rangle$ and $|-1/2\rangle$, are then operated on by the spin hamiltonian operator, H, to obtain the allowed energy states E_i:

$$H|+1/2\rangle = E_{+1/2}|+1/2\rangle \qquad (8\text{-}2)$$

$$H|-1/2\rangle = E_{-1/2} \cdot |-1/2\rangle \qquad (8\text{-}3)$$

The operator, H, for this simple $S = 1/2$ spin system, based upon Eq. (8-1), is:

$$H = g\beta H S_z \qquad (8\text{-}4)$$

where S_z is the spin operator. Then:

$$H|+1/2\rangle = g\beta H S_z |+1/2\rangle = 1/2\ g\beta H\ |+1/2\rangle \qquad (8\text{-}5)$$

$$H|-1/2\rangle = -1/2\ g\beta H\ |-1/2\rangle \qquad (8\text{-}6)$$

These equations, compared with Eq. (8-2) and (8-3), reveal that the allowed energy levels, termed the Zeeman energies, are $E_{+1/2} = 1/2 g\beta H$ and $E_{-1/2} = -1/2 g\beta H$, so that the energy difference between the spin states in Fig. 8-1 is $\Delta E = g\beta H$.

If electromagnetic radiation of the proper frequency, ϑ, is applied perpendicular to H, transitions between these two Zeeman energy levels can be induced. The resonance condition:

$$h\vartheta = \Delta E = g\beta H_0 \qquad (8\text{-}7)$$

where H_0 is the applied field at which resonance occurs, must be met for transitions to be possible. In X-band ESR spectrometry, the frequency of radiation is fixed near $\vartheta = 9.5$ GHz, and all of the spectra displayed in this chapter were generated at x band. Spectra are obtained by scanning through a range of magnetic field strength, H, until absorption of radiation is detected. The field strength at which absorption occurs must then correspond to H_0. For a "free" electron (i.e., electron without orbital contributions to its magnetic moment), the g value is 2.0023, and H_0 is about 0.33 Tesla (3300 Gauss) for $\vartheta \approx 10^{10}$ Hz. Net absorption of electromagnetic radiation (and therefore, signal detection) occurs because there is a slight excess of spins in the lower ($m_s = -1/2$) energy level. This signal is recorded as the derivative of the absorption peak, as shown in Fig. 8-1b.

2.2. *The Electron Spin and g factor*

In most solids, electrons are paired in orbitals and generate no magnetic moments from their spins. Such solids are termed diamagnetic and fail to generate ESR signals. However, minerals commonly incorporate unpaired electrons into their structure in the form of 1.transition, rare earth, or other elements having partially filled electron shells; or 2. electronic defects in which an excess or deficit of electrons resides at a normally diamagnetic lattice site. In general, an unpaired electron has spin angular momentum, h $S/2\pi$, where h is Planck's constant and $S = 1/2$. The magnetic moment of the electron μ_e, is inferred from Eq. (8-1) to be given by:

$$\mu_e = -g\,\beta S \tag{8-8}$$

The g value varies for unpaired electrons in different atoms, determined by the extent to which orbital magnetic momentum contributes to the total magnetic moment, μ_e. In the general case, where spin/orbit coupling results in an orbital as well as spin contribution to the magnetic moment, the S quantum number is inadequate to describe the system, and the J quantum number must be introduced. The value of J ranges from $L + S$ to $|L - S|$ in integral steps, where L and S are the total orbital and spin quantum numbers for the atom. The g value is then given by the Landé formula:

$$g_J = 1 + \frac{[J(J + 1) + S(S + 1) - L(L + 1)]}{2J(J + 1)} \tag{8-9}$$

For transition metals in which the d shell is less than half full, the quantum state with minimum J value is lowest in energy, while the reverse is true for metals with d shells more than half full. for example, the VO^{2+} ion has one d electron, so that $S = 1/2$ and $L = 2$. J can then have the values 5/2 or 3/2, but the $J = 3/2$ state is the ground state. Thus, VO^{2+} in the ground state would be expected to have $g_J = 0.8$. By comparison, the Cu^{2+} ion, with only one unpaired d electron, also has the quantum numbers $S = 1/2$ and $L = 2$. However, in this case, the d shell is nearly filled, and the ground state has $J = 5/2$. The Landé formula then predicts that $g_J = 1.2$. These predicted g values are rarely correct for the first transition series metals in solids, because the crystal fields of these solids are usually of medium strength. Because the d electrons of the atom interact strongly with the crystalline electric field, the crystal-field energy exceeds the spin/orbit coupling energy. Some d electrons interact more strongly with the field than others and the energies of the d orbitals are split apart, producing *crystal-field splitting*. The energy separation of d orbitals is usually sufficiently large to allow only the lowest lying states to be thermally populated. If the lowest state has no orbital degeneracy, the magnetic moment of the atom will then be close to the contribution from the spin angular momentum only; the orbital angular momentum is effectively "quenched" by the crystal field, having an average value of zero. With crystal-field energy exceeding spin/orbit coupling energy, the J quantum number is no longer valid. Thus, for an orbitally nondegenerate ground-state, L = 0 and J is replaced by S in the Landé formula:

$$g_s = 1 + \frac{[S(S + 1) + S(S + 1)]}{[2S(S + 1)]} = 2 \qquad (8-3)$$

The g value of 2.00 is indicative of "spin-only" magnetic moment and is expected for unpaired electrons in s orbitals (L = 0) of solids. However, unpaired electrons in p orbitals (L = 1) are expected to have g values that deviate from 2 because of spin/orbit coupling. Similarly, spin/orbit coupling usually causes first-row transition metals to have g values that differ somewhat from 2 despite the quenching effect of the crystal field.

Since spin/orbit coupling is direction dependent, the experimentally determined g value of unpaired electrons in p or d orbitals of atoms in crystals varies with crystal orientation in the ESR spectrometer. The theoretical importance of this g-value anisotropy and its utility in extracting chemical information from ESR spectra will be discussed later in this chapter. However, a few general points about the relationship of g value anisotropy to the symmetry of paramagnetic species should be made. The anisotropy is mathematically expressed as a second-rank tensor; for all such g tensors there exists a principal axis system (x,y,z) which permits complete description of the tensor by only three principal values (g_{xx}, g_{yy}, g_{zz}). The principal axes are generally coincident with molecular symmetry axes. Thus, cubic symmetry (perfect octahedral and tetrahedral) produces an isotropic g factor ($g_{xx} = g_{yy} = g_{zz}$). Paramagnetic centers with one threefold or higher axis of symmetry (labeled the z axis by convention) have $g_{xx} = g_{yy} \neq g_{zz}$, since the x and y axes are equivalent. The g factors in such axially symmetric centers are usually labeled $g_{\parallel}$ (the g factor parallel to the symmetry axis; i.e., $g_{\parallel} = g_{zz}$) and $g_{\perp}$ (the g factor perpendicular to the symmetry axis, i.e., $g_{\perp} = g_{xx} = g_{yy}$). Centers with symmetry lower than axial are termed orthorhombic, having three different principal g factors ($g_{xx} \neq g_{yy} \neq g_{zz}$).

Anisotropy of the g tensor is averaged for paramagnetic species undergoing rapid thermal motion, as is the case for transition metal ions in aqueous solution at room temperature, and a single isotropic g factor is exhibited, given by $g_0 = 1/3 (g_{xx} + g_{yy} + g_{zz})$.

2.3. Hyperfine Splitting

The different g factors arise from local fields induced by the applied external field. However, other local fields exist which do not depend on the presence of the external field. These permanent local fields usually arise from magnetic nuclei in the vicinity of the unpaired electron and vary depending upon the spin quantum number, I, and nuclear g factor (g_N) of the nucleus. The interaction of the magnetic moment of an unpaired electron in an atom, μ_e, with the magnetic moment of the nucleus in the same atom, μ_N, is called the *nuclear hyperfine interaction*, generating *hyperfine splitting* of ESR spectra. As with the electron spin, the z-axis component of the nuclear spin is quantized in the applied magnetic field and allowed to have only certain values defined by the quantum number m_I. For a nucleus with $I = 1/2$, m_I can have the values $+1/2$ or $-1/2$ only. The local magnetic field generated by these two possible orientations of the nuclear spin will add to or subtract from H, splitting each electron spin state into two possible energy levels depicted in Fig. 8-2a. For an $S = 1/2$, $I = 1/2$ system (e.g., the H atom), the result is that two resonance lines appear as shown in Fig. 8-2b, split by A Tesla (or Gauss), where A is the hyperfine splitting constant. In general, for a nuclear spin I, there are $(2I + 1)$ nuclear spin states and therefore $(2I + 1)$ resonance lines due to hyperfine splitting in the $S = 1/2$ system. Different isotopes of the same element may have different nuclear spins, so that hyperfine splitting can be affected by isotope ratios.

The hyperfine interaction consists of an isotropic part, $hA_0 S \cdot I$, arising from the Fermi contact interaction between the electron and nucleus, and an anisotropic part, $h S \cdot T' \cdot I$, arising from the electron/nuclear dipolar interaction. A_0 is called the isotropic hyperfine coupling constant, while the hyperfine tensor, T', describes the dependence of hyperfine splitting on the orientation of a molecule or crystal relative to the external field, H. The spin Hamiltonian operator, assuming the hyperfine interaction is small relative to the electron Zeeman interaction, becomes:

$$H = \beta S_z \cdot g \cdot H + h\, A_0 S_z I_z + h S_z \cdot T' \cdot I_z \qquad (8\text{-}10)$$

for the general case of anisotropic g and hyperfine splitting.

In general, electrons in s orbitals have isotropic hyperfine interactions, while electrons in p, d and f orbitals have anisotropic interactions. Rapid thermal rotation of atoms or molecules can average the dipole/dipole (anisotropic) interaction to zero, but the isotropic hyperfine splitting always remains. For paramagnetic species rigidly adsorbed on or embedded in minerals, anisotropic hyperfine splitting is likely to be observed.

Hyperfine splittings can range from very large values for electrons in s orbitals, as is the case for the H-atom (510 Gauss), to smaller values for electrons in orbitals that interact less directly with the nucleus (e.g., 95 Gauss for Mn^{2+}).

Superhyperfine splitting, like hyperfine splitting, originates from electron/nuclear interaction but is differentiated by the fact that the interacting nuclei do not belong to the paramagnetic atom. For example, it is common for unpaired electrons in minerals to interact with magnetic nuclei of neighboring atoms, producing additional splitting in the ESR spectrum. Typically, this form of splitting is smaller in magnitude than

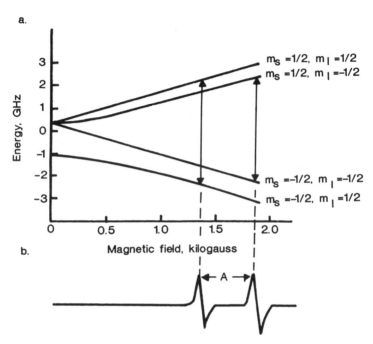

Fig. 8-2. (a) Energy levels of the H atom (S=1/2, I = 1/2) in an applied magnetic field. (b) Spectrum of the H atom, showing the hyperfine splitting, A. (Adapted from [1]).

hyperfine splitting because the electron/nuclear interaction is even less direct.

2.4. Fine Structure

The discussion above assumed that only one unpaired electron resided in the orbitals of an atom. It is clear from Table 8-1, however, that most first-row transition metals have S greater than 1/2. Taking, for example, the common case of S = 5/2, exemplified by Mn^{2+} and Fe^{3+} in high-spin states, all five d electrons are unpaired. These ions are said to be in a sextet state, because six spin states (m_s = +5/2, +3/2, +1/2, −1/2, −3/2, −5/2) are allowed. However, electron spin/orbit and spin/spin coupling mechanisms split the sextet into a quartet and a doublet. As demonstrated in Fig. 8-3, this splitting occurs even in the absence of an applied magnetic field and is termed *zero field splitting* (ZFS).

Table 8-1. Ground-State Properties of Common First-Row Transition Metal Ions[a]

No. of d Electrons	Quantum Numbers (Ground State)			Examples
	S	L	J	
1	1/2	2	3/2	Ti^{3+}, VO^{2+}
2	1	3	2	V^{3+}
3	3/2	3	3/2	Cr^{3+}, Mn^{4+}
4	2	2	0	Mn^{3+}
5	5/2	0	5/2	Mn^{2+}, Fe^{3+}
6	2	2	4	Fe^{2+}
7	3/2	3	9/2	Co^{2+}, Ni^{3+}
8	1	3	4	Ni^{2+}
9	1/2	2	5/2	Cu^{2+}

[a]adapted from [1]

The parameter D arises for the sextet state in axially symmetric ligand fields, while an additional parameter, E, is necessary to describe the less symmetrical rhombic ligand field [2]. The ZFS interactions are introduced into the Hamiltonian operator as the $S{\cdot}D{\cdot}S$ term, where **D** is a tensor. This term can be rewritten [1]:

$$S{\cdot}D{\cdot}S = D[S_z^2 - 1/3S(S + 1)] + E (S_x^2 - S_y^2) \qquad (8\text{-}11)$$

However, even in cubic symmetry, a small zero field splitting is predicted.

The consequence of zero field splitting, evident from Fig. 8-3, is that a multiplicity of lines, called the fine structure of the spectrum, may appear. In the case of S = 5/2, five resonance lines are possible. For octahedral ions such as $Mn(H_2O)_6^{2+}$ and $Fe(H_2O)_6^{3+}$, the ZFS is small and these five $\Delta m_S = \pm 1$ transitions contribute to each hyperfine line but are unresolved. However, metals in more distorted crystal fields have large ZFS terms which often result in only the middle electronic transition ($+1/2 \leftrightarrow -1/2$) being observable within the range of the applied magnetic field. In these cases, the fine structure is more important than hyperfine splitting in determining the spectral lineshape, often producing shifts in line positions of >1000 Gauss and very large changes in the apparent g values.

Zero field splitting generates paired spin states (e.g., $m_S = \pm 5/2, \pm 3/2, \pm 1/2$), called Kramers' doublets, for transition metals with odd numbers of unpaired electrons (see Fig. 8-3). Consequently, the lowest energy spin state is at least doubly degenerate, and in an applied magnetic field, resonance between these two states is always possible. The practical significance of this principle (Kramers' rule) is that metal ions with S = 1/2,

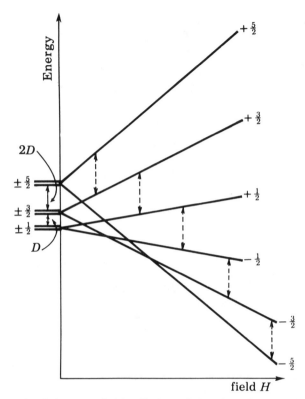

Fig. 8-3. Zero field splitting of the three Kramers'
doublets of S = 5/2 metal ions in a weak tetragonal
crystal field. (Reproduced from [79] with permission.)

3/2, and 5/2 are typically the most accessible to detection by ESR (see Table
8-1).

2.5 Powder Spectra

Most mineral samples of interest to geochemists and soil scientists (e.g.,
layer silicates, oxides) do not occur as large single crystals but rather as clay
sized particles. At best, these particles may adopt a preferred orientation in
one plane when allowed to deposit on a flat surface from suspension.
Thus, the c axes of layer silicate clay particles align perpendicular to the
surface, as the platelets lie flat on the surface upon drying. Nevertheless,
the a and b axes remain randomly oriented within the plane. In the general
case of powder samples, all orientations of the crystallite axes will be
equally probable. Considering a simple axially symmetric S = 1/2 spin

system in a powder, resonance can occur at all fields between $H_\perp$ (the field corresponding to $g_\perp$) and $H_\parallel$ (the field corresponding to $g_\parallel$). From the resonance condition (Eq. 8-7), it can be shown that in a powder, H_0 depends on θ, the angle between a given symmetry axis and the magnetic field direction [1]:

$$H_0 = \frac{h\vartheta}{g\beta} = \frac{h\vartheta}{\beta} [g_\parallel^2 \cos^2\theta + g_\perp^2 \sin^2\theta]^{-1/2} \qquad (8\text{-}12)$$

The probability $P(H)dH$ that the resonant position is between H_0 and $H_0 + dH_0$ can be shown by geometry to be proportional to $\sin\theta \, d\theta$ [1]. Since $P(H)$ is proportional to resonance absorption intensity, the $\sin\theta$ dependence indicates that a large number of particles with symmetry axes nearly perpendicular to the field direction contribute to the signal. In contrast, few particles have their symmetry axes aligned nearly parallel to the field direction. Consequently, more intense absorption is found at $H_\perp$ than at $H_\parallel$. In addition, since $P(H) \propto \sin\theta/(dH/d\theta)$, absorption will be most intense for small $dH/d\theta$, an indication that resonance lines are only likely to be detected near turning points (minima and maxima in H_0). Because $H_\perp$ and $H_\parallel$ represent extremes in the value of H_0, they are turning points, and the powder absorption lineshape will take the form depicted in Fig. 8-4a, with the most obvious features being inflections at $H_\parallel$ and $H_\perp$. Since the expression for $dH/d\theta$ can be obtained by taking the derivative of Eq. (8-12), an equation for $P(H)$ can be obtained which describes the absorption lineshape [1]:

$$P(H) \propto \left(\frac{h\vartheta}{\beta}\right)^2 \frac{1}{H_0^3 \, (g_\parallel^2 - g_\perp^2) \cos\theta} \qquad (8\text{-}13)$$

Because ESR spectra are recorded as derivative spectra, the actual powder spectrum of an axially symmetric $S = 1/2$ spin system will appear as shown in Fig. 8-4b.

For the less symmetrical orthorhombic spin system, three turning points, H_x, H_y and H_z exist, corresponding to the three principal g values. The outer two peaks of the derivative spectrum (g_{xx}, g_{zz}) will display an absorption-type lineshape, while the central peak (g_{yy}) has a derivative-type lineshape. The different lineshapes can be used to identify the extreme (g_{xx}, g_{zz}) and intermediate (g_{yy}) principal values of g for a given paramagnetic center.

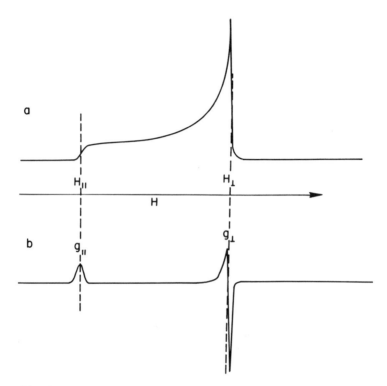

Fig. 8-4. Absorption (a) and derivative (b) lineshapes of the powder ESR spectrum, assuming an axially symmetric S = 1/2 paramagnetic species with $g_\parallel > g_\perp$.

2.6. Ferromagnetic Resonance

Generally, ESR is applied to the analysis of paramagnetic systems, that is, materials in which unpaired electrons are isolated in a diamagnetic matrix. However, ferromagnetic and ferrimagnetic minerals possess spin magnetic moment due to cooperative interaction among unpaired electrons in atoms of the solid. The phenomenon of microwave radiation absorption by such strongly magnetic materials is referred to as *ferromagnetic resonance*. Numerous common magnetic minerals generate strong ESR absorptions attributable to ferromagnetic resonance.

Below their characteristic Curie temperatures, the spin magnetic moments of lattice ions in, ferromagnetic solids are aligned parallel to one another. The electrostatic exchange interaction creating this magnetic ordering has no direct effect on the resonance but generates a very large net magnetization, M, as well as a large internal magnetic field. The external field, H, is no longer the field experienced by electrons, and an effective field, H_{eff}, is defined such that:

$$h\vartheta = g\beta H_{eff} = g\beta(H + 2K_1/M_s) \tag{8-14}$$

where K_1 is an anisotropy coefficient and M_s is the spontaneous magnetization of the crystal. The K_1 term in this equation can produce apparent experimental g values that are orientation dependent and very different from the spin-only value of 2. The "true" g value defined by the equation is generally very nearly 2 for ferromagnetics, indicating that the crystal field of the lattice "quenches" the orbital magnetic moment almost completely.

Ferromagnetics (and ferrimagnetics) have high spontaneous magnetizations (≈ 0.1 Tesla), and since magnetic resonance absorption is proportional to magnetic susceptibility, ESR absorption peaks due to ferromagnetics can be more than three orders of magnitude more intense than peaks due to paramagnetics. For this reason, ferrimagnetic impurities of mineral samples in the form of separate particles, coating, or inclusions can produce intense, usually broad, signals that obscure the paramagnetic signals of interest. A listing of magnetic properties of some common minerals in Table 8-2 indicates that iron oxides in particular are likely to generate ferromagnetic resonance signals. Those natural minerals that are ideally diamagnetic commonly possess paramagnetic properties because of structural defects and isomorphous substitutions by magnetic ions.

Above a critical temperature, termed the Curie point (T_C), ferromagnetic and ferrimagnetic minerals lose long-range magnetic ordering and behave as paramagnetics. The resultant effect on their ESR spectra is a drastic reduction in resonance absorption intensity.

It is evident from Table 8-2 that many of the secondary Fe-rich minerals are antiferromagnetic. These minerals are magnetically ordered, but the spin magnetic moments are aligned in such a way that the resultant magnetization is zero. In theory, antiferromagnetic crystals should not produce an ESR signal. However, above a critical temperature, termed the Néel temperature (T_N) for antiferromagnetics, the magnetic ordering is thermally destroyed and paramagnetic resonance is induced. Thus, ESR absorption by antiferromagnetic minerals such as hematite and goethite can be weak or moderately strong, depending upon the temperature at which absorption is measured. Absorption intensity is proportional to the magnetic susceptibility of the mineral. Table 8-3 provides a summary of magnetic properties and ferromagnetic resonance behavior of selected iron oxides.

Table 8-2. *Magnetic Properties of Some Common Minerals*[a]

Mineral	Magnetic Susceptibility $\times 10^6$ (cgs units)	Magnetism Class	Critical Temperature (°K)
Magnetite	$\leq +800,000$	Ferrimagnetic	$T_C = 858$
Hematite	$+20$	Antiferromagnetic	$T_N = 953$
Maghemite	$\leq +30,000$	Ferrimagnetic	
Goethite	$+42$	Antiferromagnetic	$T_N = 390$
Lepidocrocite	$+42$	Antiferromagnetic	$T_N = 70$
Vermiculite	$+12$	Paramagnetic	
Montmorillonite	$+2.2$	Paramagnetic	
Biotite	$+12 - 52$	Spin-glass[b]	$T_M = 7$
Dolomite	$+0.9$	Paramagnetic	
Quartz	-0.46	Diamagnetic	
Orthoclase, calcite	-0.38	Diamagnetic	
Muscovite	$+1.0 - 12$	Paramagnetic	
Kaolinite	-1.5	Diamagnetic	

[a]-Adapted from [3]
[b]-see Ref. [4].

Unfortunately, resonance in these minerals is very sensitive to their history of synthesis, presumably a manifestation of variable particle size and crystallinity, which strongly affects magnetic ordering. Consequently, while ferromagnetic signals of Fe oxides and other transition metal oxides are virtually ubiquitous in natural samples from soils, sediments, and other geologic materials, they have not proved to be reliable indicators of the presence of particular mineral phases. Nevertheless, for controlled reactions such as the thermally- induced transition of Fe_3O_4 to γ -Fe_2O_3 to α-Fe_2O_3, ESR can be a useful indicator of phase changes [5].

Table 8-3. Magnetic and Ferromagnetic Resonance Properties of Iron Oxide Minerals.

Mineral	Magnetic Properties	ESR Behavior at X Band
Ferrihydrite ($Fe(OH)_3$)	Paramagnetic	Moderately strong, broad ($\Delta H \approx 1000$ Gauss) signal near $g = 2$
Hematite (α-Fe_2O_3)	Weakly ferromagnetic above $-10°C$	Weak, broad low-field signal ($g > 2$) above $-60°C$
Goethite (α-FeOOH)	Antiferromagnetic with Néel temperature of 120°C	Weak, broad $g = 2$ signal above $+80°C$
Akaganeite (β-FeOOH)	Antiferromagnetic with Néel temperature of 20°C	
Maghemite (γ-Fe_2O_3)	Ferrimagnetic	Strong signal near $g = 2$
Lepidocrocite (γ-FeOOH)	Antiferromagnetic with Néel temperature of $-200°C$	Medium signal near $g = 2$
Magnetite (Fe_3O_4)	Ferrimagnetic	Strong, broad ($\Delta H \sim 3000$ Gauss) asymmetric low-field ($g > 2$) signal, with linewidth decreasing at very low and very high temperature
δ-FeOOH	Ferrimagnetic	Strong, broad signal at $g \geq 2$

3. Applications

3.1. Structural Paramagnetic Centers

Two common types of ESR-detectable centers occur in natural minerals: (1) those which are not inherently paramagnetic but generate unpaired electrons upon the absorption of energy, stabilized by isomorphous substitution in the structure, and (2) those which are paramagnetic by virtue of isomorphous substitution of metal ions having unpaired electrons in d or f orbitals into the diamagnetic structure.

The first type is typified by the $[AlO_4]^0$ center of quartz [6, 7], arising from a normally diamagnetic state in which Al^{3+} substitutes for Si^{4+} in a tetrahedral site, and the charge deficit is compensated locally by a nearby metal ion (usually H^+, Li^+, or Na^+). Upon irradiation with x-rays or γ-rays, for example, the paramagnetic site is created as an electron is removed from an oxygen ion adjacent to Al. The charge-compensating metal ion can then diffuse away, leaving the paramagnetic $[AlO_4]^0$ site. This electron "hole" in the 2p orbital of a structural oxygen magnetically interacts with the neighboring Al nucleus, since the only isotope of Al (^{27}Al) has a nuclear spin of I = 5/2. One would then expect the ESR spectrum to consist of 2I + 1 = 6 equally spaced resonance lines. In actual fact, single quartz crystals generate more complex spectra than this due to the nuclear Zeeman and quadrupole effects of ^{27}Al [6]. The former arises from the alignment of the nuclear magnetic moment, I, of Al in the external magnetic field and becomes important if the hyperfine splitting is relatively small. The latter only occurs if the nuclear spin I > 1/2, and is a result of the alignment of the nuclear spin by the local electric-field gradient.

The electron hole has some mobility at room temperature, jumping among the four oxygen atoms coordinated to Al. However, the presence of the hole may sufficiently distort the immediate structure to stabilize itself on one oxygen, causing jumps to other oxygens to require the absorption of energy. Certain visible absorption bands of smoky quartz have been attributed to transitions of this type. In fact, it is common for paramagnetic centers in crystals to be detected optically as well as by ESR.

The positive hole should not be construed as merely an artificial product of radiation damage, since there is evidence that such defects occur naturally in various silicates and oxides. They are undoubtedly stabilized by a structural vacancy or isomorphous substitution, allowing the positive charge to be neutralized locally.

One example of an electronic defect occurring naturally in layer silicate clays is that detected commonly in kaolinites by ESR. The spectrum, shown in Fig. 8-5, is characteristic of an axially symmetric centre in a powder, having $g_{\parallel} = 2.049$ and $g_{\perp} = 2.003$ [8]. The relative intensities of the $g_{\parallel}$ and $g_{\perp}$ components of the spectrum vary with the alignment of the ab plane of the layer silicate relative to the applied magnetic field. This can be verified with kaolinite films in which the platelets are aligned in the plane of the film either by deposition from suspension on a flat surface, or by unidirectional compression. Such films reveal the $g_{\parallel}$ signal to be more intense when the plane of the film is oriented normal to the magnetic field, and the $g_{\perp}$ signal to be more intense when the plane is parallel to the magnetic field (Fig. 8-5). Consequently, the symmetry axis of the defect must be approximately perpendicular to the ab plane of the kaolinite platelets. This defect may arise from an electron hole on an O atom adjacent

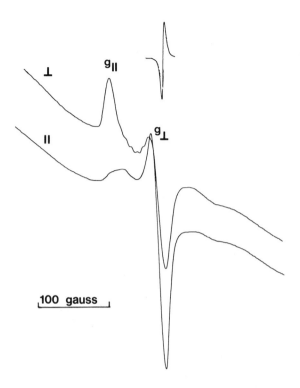

Fig. 8-5. ESR spectrum of the lattice defects in a kaolinite pellet oriented perpendicular ($\perp$) and parallel (||) to H. The strong pitch signal in this and subsequent spectra marks the $g = 2.0027$ position.

to a Mg^{2+} ion isomorphously substituting for Al^{3+} [8]. The g values of this site are similar to the principal g values for the axially symmetric O^- radical formed on a MgO surface ($g_{||} = 2.0013$, $g_\perp = 2.042$) [9]. In the latter case, the fact that $g_\perp$ (identified as the most intense of the two main resonances in powder spectra of axial species) is higher than g_e (g factor for the free electron) implies that the unpaired electron resides in the $2p_z$ orbital of O^-, as crystal field splitting has raised the energy of the $2p_z$ orbital relative to the $2p_x$ and $2p_y$ orbitals. In kaolinite, the crystal field of O^- presumably has a different symmetry, resulting in $g_{||} > g_\perp$, where the electron hole is assumed to reside in a nonbonding sp_z or p_x orbital of the oxygen atom [10].

The lack of superhyperfine structure in the kaolinite defect spectrum is consistent with the assignment of the defect to an O^- coordinated to Mg, since the commonest isotopes of Mg (^{24}Mg, ^{26}Mg) do not have nuclear spin. The ^{25}Mg isotope ($I = 5/2$) is 10.0% abundant and may account for weak hyperfine lines superimposed on the defect spectrum [10] (see Fig. 8-5). However, these features, which sometimes occur as multiplets of six, are believed to arise from a separate defect associated with ^{27}Al ($I = 5/2$) [10], which is 100% abundant and should generate $2I + 1 = 6$ lines. This second defect may be analogous to the $[AlO_4]^0$ center of quartz, where the electron hole is stabilized by isomorphous substitution of Al^{3+} for Si^{4+}. The quartz defect, like that shown in Fig. 8-5, is nearly axial, with principal g values of 2.0602, 2.0085, and 2.0019 [6].

Besides kaolinite, dickite, pyrophyllite, and talc also possess the spectrum of an axially symmetric defect devoid of superhyperfine structure [11, 12]. The powder spectrum of a pyrophyllite, displayed in Fig. 8-6a, reveals an axial species with $g_{\parallel} = 2.048$ and $g_{\perp} = 2.008$, presumably associated with Mg^{2+} substituted in the octahedral layer of this 2:1 layer silicate. The spectrum of an oriented crystal of this same pyrophyllite (Fig. 8-6b) indicates enhancement of the $g_{\parallel}$ component when the basal plane of the crystal is normal to the applied field, and enhancement of the $g_{\perp}$ component when the basal plane is parallel to the field. Therefore, the symmetry axis is oriented perpendicular to the basal plane in both pyrophyllite and kaolinite.

Weaker features are also noted in the pyrophyllite spectra (Fig. 8-5), presumably originating from a second electron hole defect stabilized by Al^{3+} for Si^{4+} substitution. A multiplet of 11 lines (partially obscured by the main defect spectrum) is most evident in the single-crystal spectrum (Fig. 8-6b), interpreted as arising from the interaction of the electron hole with two magnetically equivalent ^{27}Al nuclei. Such an electron hole could exist on an (inner) oxygen atom linking tetrahedral and octahedral Al atoms [11] or an (outer) oxygen atom linking two adjacent tetrahedral atoms. Although the latter possibility would require isomorphous substitution of Al^{3+} for Si^{4+} in adjacent tetrahedral sites, a situation that is generally believed to be unstable because of the positive charge developed in the structure, the 11-line spectrum of Al^{3+} $-O^-$ $-Al^{3+}$ sites in feldspars has been reported [13]. Furthermore, interaction of the hole with one tetrahedral and one octahedral Al is not likely to produce a simple 11-line spectrum, because these Al atoms are unlikely to be magnetically equivalent.

The ^{27}Al superhyperfine splitting constant in pyrophyllite (~21 Gauss) is much larger than that in feldspar [13] or kaolinite [11]. Since the isotropic component of the splitting arises from unpaired electron density in an s orbital of Al, the ESR spectrum then indicates a considerable degree of electron delocalization from the O^- atom to the two coordinated Al atoms.

Although much study of electronic defects in minerals has been done, the question of whether these defects have any manifestation in surface reactivity must be raised. It is quite evident that the O^- state (positive hole) is a powerful oxidizing agent. For example, in amethyst (quartz), Fe^{3+}

Fig. 8-6. Powder (a) and oriented crystal (b) spectra of the lattice defects in a pyrophyllite (Wards #49, Robbins, North Carolina).

substitutes for Si^{4+} to stabilize the electron hole, but the hole readily oxidizes Fe^{3+} as follows [14]:

$$Fe^{3+} \longrightarrow O^- \longrightarrow Si \quad \rightleftarrows \quad Fe^{4+} \longrightarrow O^{2-} \longrightarrow Si$$

Clearly, then, such electronic defects could not persist within the surface layer of minerals under conditions of interest to soil chemists or geochemists, even if high-energy radiation were available to form such defects. This may explain why the expanding clays (i.e., smectites) reveal little or no evidence of the electron hole defect [15], with each silicate sheet exposed to the ambient environment. Defects that are present in smectite samples may actually reside in impurity phases (e.g., quartz, anatase) Of course, under extremely dry, near-vacuum conditions, such as that found in Martian soil, high-energy UV radiation could generate the strongly oxidizing O^- centers on mineral surfaces.

The electron holes such as those buried in kaolinite are trapped coulombically at the site of isomorphous substitution. As the temperature is raised, broadening in the ESR spectra of these defects suggests that the holes jump among equivalent or nearly equivalent O atoms [13], but migration to distant surface sites is not possible. Statistically, the defect spectrum is most intense in well-crystalized kaolinites [16], a further indication that the electron hole can persist only in an electronically insulating matrix.

These internal defect centers arise from electromagnetic energy that has been stored electronically and may be responsible for certain energy-absorbing and emitting phenomena of layer silicates, such as luminescence [17–19] and optical absorption [20].

The fact that electronic defects tend to accumulate in proportion to the intensity of incident x-rays or other high-energy radiation has been utilized in attempts to date minerals by ESR. For example, the ESR signals arising from electronic defects in the structure of natural quartz [21, 22] and calcite [23] may result from natural radioactivity in the surrounding rock or sediment.

The second type of paramagnetic center in minerals, arising from isomorphously substituted metals with unpaired electrons in their valence shell, is extremely common and has been reported for many diamagnetic minerals. Particulary numerous are reports of Fe^{3+}, Mn^{2+}, Cr^{3+}, and V^{4+} ESR spectra in silicates, oxides, phosphates and carbonates [10, 11, 24–35]. Other less stable oxidation states reported include Ti^{3+} and Mn^{4+} in TiO_2 [34, 36]. Notably, all of these transition metal ions possess one, three, or five d electrons (see Table 8-1) and are therefore Kramers' ions (see Section 2, Theory). Non-Kramers' ions such as Fe^{2+} and Ni^{2+} are often much less easily detected by ESR.

An inherent difficulty in applying ESR to mineral studies is the fact that transition metal ions may not generate resonances because of strong interaction of the spin system with the lattice via spin/orbit coupling. Those metals having low-lying excited electronic states are prone to strong spin/orbit coupling, with the result that spin/lattice relaxation is rapid and the ESR signal is sufficiently broadened to prevent detection. In such unfavorable cases detection may require cooling the sample to temperatures near absolute zero in order to suppress relaxation. Mn^{2+} and Fe^{3+}, both d^5 ions, are examples of metals with very small spin/orbit coupling with excited electronic states, a consequence of the large energy separation between the ground state and lowest lying excited state. These ions are readily detected in minerals over a wide range of temperatures and lattice environments. Conversely, structural Fe^{2+} cannot be detected except possibly at temperatures below 20°K [1]. Time-dependent phenomena can profoundly affect the linewidths and detectability of ESR signals. As with NMR, these phenomena are quantified by the spin/lattice relaxation time (T_1) and the spin/spin relaxation time (T_2). Analysis of line-broadening effects can yield information about kinetics and structure [1], but this subject will not be developed any further here.

Useful chemical information can be gained from the ESR spectra of structural transition metals. Besides the obvious direct confirmation of the oxidation state of the metal, which can be useful in studies of surface electron transfer reactions (see later section), the ESR spectra provide:

1. g values. For a given transition metal ion, the g values can be correlated to the orbital energy levels, E, in the crystal field. V^{4+} is a particularly good "probe" ion for this purpose [35–37], with g values equated to:

$$g_{\parallel} = 2 \left[1 - \left(\frac{4\,\zeta}{\Delta_1} \right) \right]$$

$$g_{\perp} = 2 \left[1 - \left(\frac{\zeta}{\delta} \right) \right]$$

$$(8\text{--}15)$$

Where $\Delta_1 = E_{d_{x^2-y^2}} - E_{d_{xy}}$, $\delta = E_{d_{xz}} - E_{d_{xy}}$ (tetragonal distortion), and ζ (spin/orbit coupling) is about 150 cm^{-1}. The g values are sensitive to the chemical bonding properties as well as the symmetry of the anions coordinated to the metal. For example, g values of the tetragonal Cu^{2+} ion decrease with increasing covalency of the Cu^{2+}/ligand interaction [38], while Cu^{2+} substituted into a tetrahedral site of a phosphate mineral possesses higher g values than Cu^{2+} in tetragonal symmetry [39]. Interpretation of g values in chemical terms is problematical for ill-defined crystal fields because effects of covalency and crystal field symmetry cannot be separated.

2. Hyperfine splitting. This is observed for metals possessing nuclear spin (e.g., ^{57}Fe, 2.2% abundance, I = 1/2; ^{55}Mn, 100% abundance, I = 5/2; ^{51}V, 99.8% abundance, I = 7/2). Because the isotropic hyperfine interaction arises from unpaired electron spin density at the nucleus, the splitting constants are sensitive to electrostatic and covalent metal/ligand interactions. For example, the hyperfine splitting of V^{4+} substituted in various MO_2 crystals with the rutile structure (M = metal) varies considerably [40]. Small differences in the hyperfine splitting for Mn^{2+} substituted in dolomite, calcite, and other carbonates with the calcite structure reveal that, for this d^5 metal ion, hyperfine splitting is a rather insensitive indicator of small differences in the immediate ligand environment [25, 41]. Nevertheless, manganese located adjacent to gross lattice defects such as anion vacancies has clearly altered splittings [28]. For Mn^{2+}, the splitting parameter, A, tends to decrease in proportion to the covalent character of its bond to coordinating anions [34].

3. ZFS values . For metal ions with S > 1/2, zero field splitting (ZFS) parameters can determine the field positions (i.e., apparent g values) of the resonances and are often useful in determining the symmetry of the crystal field.

The next several pages describe some applications of the ESR spectra of Fe^{3+}, Mn^{2+}, and V^{4+} substituted into mineral structures.

3.1.1. Fe^{3+} . Crystal field parameters can be calculated from the ESR spectra of high-spin Fe^{3+} (S = 5/2) in minerals [42]. In cubic or nearly cubic crystal fields, the resonances should appear near g = 2.0, displaying a maximum of five fine-structure components from the five allowed ($\Delta m_s = \pm 1$) spin transitions. This is equivalent to the situation depicted in Fig. 8-3, but the ZFS term is very small for cubic symmetry. For axial symmetry with small ZFS terms, the resonances again appear near g = 2.0, but $g_{\parallel}$ and $g_{\perp}$ components are observed. Frequently, for ESR spectra of Fe^{3+} in rhombic and axial crystal fields, the ZFS terms are comparatively large, and it is necessary to describe the spectra by means of a spin Hamiltonian with terms in addition to that for the Zeeman interaction [11]:

$$H = g\beta H \cdot S + D \left[S_z^2 - 1/3\, S\,(S + 1) \right] + E \left[S_x^2 - S_y^2 \right] \qquad (8\text{-}16)$$

The second and third terms are the axial and rhombic operators of the ZFS interaction. These partly remove the spin degeneracy, leaving three Kramers' doublets (Fig. 8-3). Depending on the magnitude of D relative to $h\vartheta$ (the resonant energy) and E relative to D (expressed as $\lambda = E/D$), the apparent g value, g_{app}, as determined from the field position of the resonance, H_o, and the resonance condition ($h\vartheta = g_{app}\beta H_o$), commonly deviates greatly from 2.00. Nevertheless, the true g value as defined in Eq. (8-16) remains very near g = 2.00 for Fe^{3+} and is isotropic, a result of the weak spin/orbit coupling for the d^5 electronic configuration. Also, large ZFS terms (D >> $h\vartheta$) often result in only the +1/2 $\leftrightarrow$ −1/2 electronic transition producing a resonance, since the field positions of the other transitions are not experimentally accessible.

The parameter λ describes the symmetry of the crystal field of Fe^{3+}, with three cases possible [11]:

1. $\lambda = 0$. This represents axial symmetry. If $D > h\vartheta$, resonances at $g_{\parallel} = 2.0$ and $g_{\perp} = 6.0$ are expectedfor the $+1/2 \leftrightarrow -1/2$ transition.

2. $\lambda = 1/3$. This is the purely orthorhombic case in which the three Kramers' doublets are split equally by the crystal field. An isotropic resonance at $g = 4.3$ is predicted for the central doublet, and highly anisotropic resonances for the upper and lower doublets are expected. Consequently, only the $g = 4.3$ resonance is evident in powder spectra.

3. $1/3 > \lambda > 0$. This encompasses all intermediate symmetries. In powders, a triplet of resonances in the $g = 4$ region is expected [10]. Exact positions of Fe^{3+} resonances in powders as a function of λ and D have been calculated [43, 44]. It is beyond the scope of this chapter to detail the methods and results of these calculations.

Many clay minerals produce rather broad resonances near $g = 4.3$, attributable to Fe^{3+} in orthorhombic symmetry [15]. These resonances are only possible if paramagnetic Fe^{3+} ions are isolated in a diamagnetic structure. Commonly, clays also possess broad, sometimes intense ESR signals centered near $g = 2$, which are attributed to Fe^{3+} "clustered" in separate Fe oxide phases or structural inclusions [45, 46] (see Section 2.6).

To understand the nature of the Fe^{3+} paramagnetic resonance in minerals and interpret any useful information it may contain about the structure and surface reactivity of clays, it is useful first to consider the case of Fe^{3+} isomorphously substituted into a simple mineral, gibbsite. Figure 8-7 displays the low-field spectrum of a gibbsite film consisting of microcrystalline pseudohexagonal platelets with their (001) crystal planes aligned (imperfectly) in the plane of the film. Resonances at $g = 3.17$, 4.09, 4.14, 4.27, 5.11, and 8.44 are noted, clearly revealing Fe^{3+} to be in several different crystal fields. The multiplet of resonances in the vicinity of $g = 4$ can be assigned to transitions between the middle pair of spin states, labeled the $3 \leftrightarrow 4$ transition (Fig. 8-3). By comparing theoretical positions of powder lines [44] with these observed resonances, and taking into account the obvious direction dependence of the resonance intensities, the following two Fe^{3+} sites can be described:

Site I:

Parameters: $E/D \approx 0.125$, $\dfrac{h\vartheta}{D} \leq 2$ $(D \geq 0.15 \text{ cm}^{-1})$

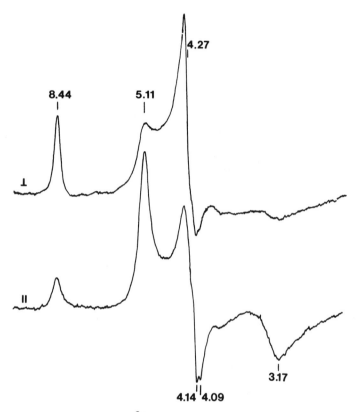

Fig. 8-7. Low-field Fe^{3+} spectrum of a synthetic gibbsite (<2 μm) oriented as a self-supporting film in the magnetic field. The main Fe^{3+} resonances are denoted by their g values.

Detected transitions: $x(1 \leftrightarrow 2) = 8.44$, $z(3 \leftrightarrow 4) = 5.11$, $x(3 \leftrightarrow 4) = 3.17$

Site II:

Parameters: $E/D \approx 0.30, \dfrac{h\vartheta}{D} \leq 2$

Detected transitions: $z\,(3 \leftrightarrow 4) = 4.27$, $x(3 \leftrightarrow 4) = 4.14$, $y(3 \leftrightarrow 4) = 4.09$

　　　Site II is almost completely orthorhombic, while Site I can be described as distorted axial. The principal axes, x, y, and z, of the fine-structure tensor can be located using the orientation dependence of the resonance

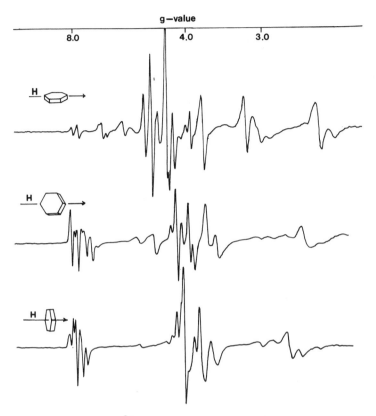

Fig. 8-8. Low-field Fe^{3+} spectra of a single hexagonal gibbsite crystal
(>1 mm) oriented with the crystal axes relative to the magnetic field
as shown diagrammatically.

intensities. Assuming that the above assignments of transitions are correct,
Sites I and II have different axis orientations relative to the basal plane of
gibbsite. A complete description of the fine-structure tensor and its
alignment relative to crystal axes can be obtained from spectra of single
crystals. In the case of Fe^{3+}-doped gibbsite, as shown in Fig. 8-8, the
spectrum is highly sensitive to the orientation of the crystal axes relative to
the applied magnetic field, H; the fine structure tensor is evidently very
anisotropic. Also, many more transitions are detected in the single crystal
than in a powder because the random orientation of crystallites in a powder
results in resonance lines being observed only for those orientations in
which H is parallel to any of the principal (x,y,z) axes or lies in any of the
principal (xy, xz, yz) planes of the fine structure tensor [44] (see Section

2.5). Thus, in Fig. 8-8, the large number of transitions is partly a result of the detection of nonprincipal values of the tensor. In addition, Fe^{3+} in the single crystal is likely to occupy physically identical but differently oriented sites. Each of these symmetry-related sites produces a distinct spectrum, multiplying the apparent complexity of the single-crystal spectrum. In the powder spectrum, physically identical sites are indistinguishable, since the principal values of their magnetic parameters are identical. It is clear from Fig. 8-8, however, that many of the resonances occur near g = 4 and g = 8, especially those whose field positions are relatively insensitive to crystal alignment in the field. It is these resonances when then dominate the powder spectrum.

Kaolinites possess Fe^{3+} spectra similar to those of gibbsite, interpreted in terms of an orthorhombic site (g = 4.2) and a partially orthorhombic site (g_z = 4.9, g_x = 3.7, g_y = 3.5) located in the octahedral layer of the mineral [11, 47]. The orthorhombic site is more prevalent in poorly crystalline kaolinites and halloysites [48, 49] and is therefore attributed to Fe^{3+} located adjacent to layer stacking disorders. In fact, intercalation of kaolinite with dimethylsulfoxide and other reagents causes the partially orthorhombic site to change symmetry such that only the g = 4.2 signal is apparent [11]. Evidently, expansion of the interlayer distorts the layered structure, reducing the symmetry of the crystal field around Fe^{3+}.

Smectites, like kaolinite, reveal a nearly isotropic resonance near g = 4.27, and anisotropic resonances at lower and higher field (Fig. 8-9). The Fe^{3+} resonances are broader than those for kaolinite and gibbsite, presumably because of the lack of stacking order (turbostratic behavior) and structural heterogeneity typical of these swelling clays. The anisotropic portion of the spectrum is sensitive to the exchange cation and hydration or solvation of the smectite [50–52]. Dehydration of the interlayers collapses the anisotropic resonance into the central g = 4.27 peak, suggesting that Fe^{3+} symmetry is lowered as the platelets are forced into more rigid alignment. Assignment of the isotropic and anisotropic Fe^{3+} signals to specific structural sites has not been made, and some ESR studies have suggested the existence of tetrahedrally as well as octahedrally coordinated Fe^{3+} [53]. If Fe^{3+} resides completely in octahedral sites, different symmetries can arise from *cis* and *trans* configurations of hydroxyls in the $FeO_4(OH)_2$ octahedra, or from the charge and size of the neighboring octahedral cations. The latter suggestion is supported by the observation that the principal axes of the ZFS tensor are oriented differently in montmorillonite (dioctahedral) relative to hectorite (trioctahedral) [8].

Fe^{3+} is commonly substituted in natural crystalline minerals such as micas, quartz, and feldspars, providing opportunity for detailed determination of the location of crystal field symmetry axes of Fe^{3+} relative to the mineralogical axes of single crystals. The details of these spectra will not be discussed here, but a common characteristic for micas, feldspars, and

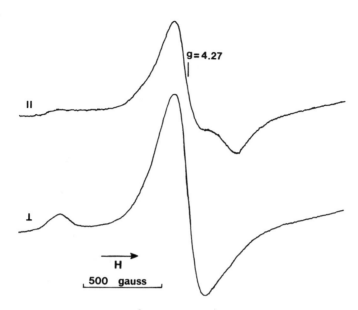

Fig. 8-9. Low-field Fe^{3+} spectrum of an oriented film of Umiat bentonite.

quartz is a low-symmetry crystal field (nearly orthohombic) and relatively large ZFS terms [26, 29, 30, 54–58]. Consequently, the most conspicuous resonances in powder samples have effective g values near 4.3. Synthetic quartz, for example, contains several slightly different Fe^{3+} sites, having g = 2.0032, $h\vartheta/D \approx 1$ and E/D = 0.17 – 0.21 [59]. Micas such as phlogopite and muscovite possess several Fe^{3+} centers with varying degrees of orthorhombic character [54, 55], some of which are attributed to tetrahedrally substituted Fe^{3+} [56]. However, these magnetically nonequivalent Fe^{3+} centers generally cannot be assigned with certainty to particular substitution sites in the minerals, and in this respect, ESR of structural Fe^{3+} has been of limited value.

The Fe^{3+} signal is a sensitive internal indicator of phase transitions in minerals [31] and is also valuable in monitoring oxidation/reduction reactions at mineral surfaces, as will be discussed later. The g = 4.3 resonance of illite, for example, increases in intensity upon heat treatment [60], an apparent consequence of the oxidation of structural Fe^{2+} by O_2. The weathering of phlogopite mica by the removal of K^+ weakens the g = 4.3 signal of structural Fe^{3+} [8], an unexpected consequence, since structural Fe^{2+} is oxidized during weathering. However, the development of an intense broad resonance centered near g = 2 accompanies the weathering process, suggesting the conversion of paramagnetic Fe^{3+} into ferromagnetic Fe^{3+}. Given the elemental composition of the octahedral sheet in the unweathered mica, $Mg_{2.32}$ $Al_{0.38}$ $Fe^{2+}_{0.13}$ $Fe^{3+}_{0.01}$, oxidation of Fe^{2+} must produce more Fe^{3+}/Fe^{3+} or Fe^{3+}/Fe^{2+} magnetic interaction, which may intensify the ferromagnetic resonance. It is not certain whether the ferromagnetism arises from Fe within the octahedral layer or Fe hydroxides formed as Fe is expelled from the structure to compensate charge imbalance created by oxidation. This type of result demonstrates the potential value of ESR in mineral weathering studies.

Fe^{3+} is readily doped into laboratory-synthesized minerals such as boehmite and amorphous aluminosilicates [61, 62], its ESR spectrum reporting the local structure in terms of the value of E/D. The Cr^{3+} ion similarly has potential as an ESR probe of aluminous minerals [63]. ESR clearly demonstrates the strong tendency of these metals to "cluster," forming ferromagnetic phases within the diamagnetic matrix. As a result, doping levels must be quite low to generate spectra of paramagnetic species.

3.1.2 Mn^{2+}. High-spin Mn^{2+} (S = 5/2) has the same electron spin quantum states as Fe^{3+}, which are separated in an applied magnetic field as depicted in Fig. 8-3. In a cubic (i.e., perfect octahedral or tetrahedral) lattice site, the five allowed transitions occur at nearly the same field position corresponding to g = 2.00. However, only the central transition ($m_S = -1/2 \leftrightarrow +1/2$)) is detectable in powder spectra . Interaction with the ^{55}Mn nucleus (I = 5/2) splits the transition into six equally intense hyperfine lines. For example, Mn^{2+} substituted into the cubic CaO lattice produces a simple sextet in which the hyperfine lines are separated approximately by A, the hyperfine splitting constant [64]. Exact field positions of the hyperfine lines in the powder spectrum of cubic Mn^{2+} are given by [65]:

$$H = H_0 - Am_I - \frac{A^2}{2H_0} \left(\frac{35}{4} - m_I^2 \right) \tag{8-17}$$

where $H_0 = \eta\vartheta/g\beta$ and m_I is the nuclear quantum number (m_I = 5/2, 3/2, 1/2, -1/2, -3/2, -5/2). Because of the last term in Eq. (8-17), the hyperfine lines are not exactly equally spaced.

A more complex powder spectrum is typical for Mn^{2+} in natural minerals, since the ZFS term, D, which describes the axial distortion of the lattice site, is not zero as it is for cubic sites. Consequently, terms in D^2/H_0 and D^2A/H_0^2 are added to Eq. (8-17), splitting each hyperfine line into a doublet. In addition, forbidden resonances appear as doublets between the main resonances, arising from $\Delta m_S = 2$ transitions [64, 65]. These features are useful in that D can be determined from the magnitude of splitting and intensity of forbidden lines. Mn^{2+} substituted into calcite generates this

Fig. 8-10. Spectrum of Mn^{2+} coprecipitated into $CaCO_3$.

distorted-site spectrum, and a typical example is shown in Fig. 8-10, in which Mn^{2+} was coprecipitated with $CaCO_3$.

Because the magnitude of D in most minerals is much smaller for Mn^{2+} than for Fe^{3+} (i.e., D << $h\vartheta$), Mn^{2+} spectra are centered at the field position corresponding to g = 2.00. The g factor is very close to the free electron value (e.g., g = 2.0010 – 2.0030 for Mn^{2+} in carbonates) because of the remoteness of excited electronic states for high-spin Mn^{2+}.

In oxyanion-based natural minerals, whose lattice structure is based on the oxyanion, the g values and hyperfine splittings of Mn^{2+} are not very sensitive to crystal-field differences. Thus, Mn^{2+} in carbonates, phosphates, and layer silicates has spectra with comparable g and A values. In contrast, the axial distortion parameter, D, is much more variable. For example, with carbonate minerals, D has the value 27.7 Gauss for calcite, 85.5 for magnesite, 3.2 for the Ca site in dolomite, and 152 for the Mg site in dolomite [66]. Mn^{2+} substituted into other metal carbonates can produce quite different values of D. As a magnetic probe, then, Mn^{2+} can identify mineral phases and signal phase transitions [41, 67].

In a few minerals, the Mn^{2+} lattice site is sufficiently distorted that D > A, and the single-crystal ESR spectrum consists of five sets of six hyperfine lines arising from the five allowed electronic transitions. In this case, the apparent g values of the resonances can deviate widely from 2.00. A case in point is the substitution of Mn^{2+} in one of the Ca sites of apatite, generating a spectrum with magnitudes of D (504 Gauss) and E (114 Gauss) that indicate a highly distorted (orthorhombic) site [28].

3.1.3. VO^{2+}. Like Fe^{3+} and Mn^{2+}, VO^{2+} (or V^{4+}) is commonly detected in clays and other minerals by ESR, complexed to organic matter in the samples or isomorphously substituted into the minerals [10, 11, 68]. With a d^1 (S = 1/2) electronic configuration, the spin Hamiltonian of the axially symmetric VO^{2+} oxycation is described by the electron Zeeman and hyperfine interaction [69]:

$$H = \beta [g_{\parallel} H_z S_z + g_{\perp} (H_x S_x + H_y S_y)] + A_{\parallel} I_z S_z + A_{\perp} (I_x S_x + I_y S_y) \quad (8\text{-}18)$$

where $g_{\parallel}$, $g_{\perp}$ and $A_{\parallel}$, $A_{\perp}$ are the principal values of the anisotropic g and A tensors. Hyperfine splitting arises from electron spin interaction with the ^{51}V (I = 7/2) nucleus, producing eight hyperfine lines. Consequently, the powder spectrum of axial VO^{2+} substituted in mineral structures possesses 16 lines, an octet of hyperfine lines from each of the $g_{\parallel}$ and $g_{\perp}$ resonances (distorted axial would produce $3 \times 8 = 24$ lines). The powder VO^{2+} spectrum of Llano vermiculite is displayed in Fig. 8-11a, revealing the extreme anisotropy of the axially symmetric A tensor ($A_{\parallel}$ = 181 Gauss, $A_{\perp}$ = 71 Gauss). The g tensor is more nearly isotropic, with $g_{\parallel}$ and $g_{\perp}$ values near 1.95 and 1.98, respectively. The orientation dependence of the VO^{2+} spectrum for a single vermiculite crystal, depicted in Figure 11b, reveals that the symmetry axis is aligned fairly close to the basal plane because A is largest for the basal plane oriented parallel to the magnetic field. Several symmetry-related V^{4+} sites appear to complicate the single-crystal spectrum; otherwise only eight resonance lines would be detected for each orientation of the crystal. Splitting of the high-field resonances into triplets, observed for the parallel orientation of the crystal (Fig. 8-11b), is probably a consequence of three equivalent V^{4+} sites rotated by 120°, as would be expected for tetrahedral (or octahedral) coordination sites in mica.

Closer inspection of this spectrum indicates that the symmetry is lower than axial, because more than eight hyperfine lines are detected in the $g_{\perp}$ region. Presumeably then, $A_{zz} >> A_{xx} \approx A_{yy}$, as might be expected for V^{4+} in a distorted axial site of the mica. Chemical evidence suggests that V^{4+} in this vermiculite, and in other layer silicates such as kaolinites and halloysites [10, 46] is structural rather than surface adsorbed. In kaolinite, $A_{\parallel}$ and $A_{\perp}$ have been estimated at 170 and 60 Gauss, respectively, with the symmetry axis aligned close to the basal plane of the layers (McBride, unpublished data).

The VO^{2+} ion is an excellent probe of its chemical environment, and the hyperfine splitting is particularly sensitive to the bonding properties of

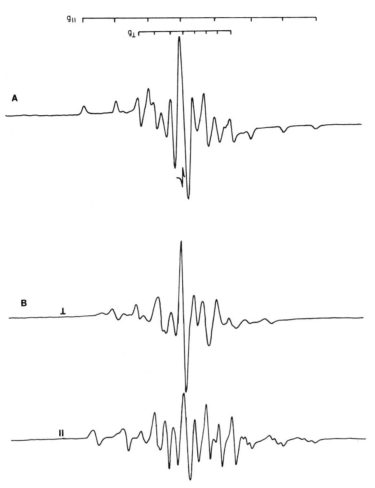

Fig. 8-11. Powder (a) and single crystal (b) spectra of VO^{2+} occurring naturally in Llano vermiculite. The single crystal was oriented with its basal plane parallel (∥) and perpendicular (⊥) to the field.

coordinating atoms as demonstrated by the principal values of A for VO^{2+} in various surface and structural sites (Table 8-4).

In all of these solids, the V^{4+} cation is coordinated to oxygen ions (O^{2-}, OH, or H_2O), yet considerable variation in A is noted. A larger range of hyperfine splittings is expected when comparing chemically dissimilar ligands coordinating to V^{4+} [73]. Coordinating atoms that produce strong

Table 8-4. *Principal Hyperfine Splitting Values (Gauss) for* VO^{2+}
in Selected Environments[a].

Environment	$A_{\parallel}$	$A_{\perp}$
Aqueous	203	78
Complexed in humic acids	185–193	67–75
Adsorbed on smectite	204	81
Adsorbed on $Al(OH)_3$	186	62
Substituted in $Al(OH)_3$	187	67
Substituted in kaolinite	~170	~60
Substituted in vermiculite	181	71
Substituted in TiO_2	156	42

[a] Data from [35, 70-72]

crystal fields (more covalent bonding) generate smaller values of $A_{\parallel}$ and $A_{\perp}$. Thus, the ligand environment of VO^{2+} substituted in Al oxides and layer silicates may be somewhat more covalent than the aqueous environment of hydrated VO^{2+} adsorbed on clay surfaces. The value of VO^{2+} as a paramagnetic probe in studies of metal ion adsorption on oxide and layer silicate surfaces will be discussed further in the next section.

3.2. Adsorption of Ions on Surfaces

3.2.1. Layer Silicate Clays and Zeolites. These minerals, possessing structural negative charge, retain cations rather loosely in exchangeable form. Numerous models have been developed to predict the behavior of such cations at the mineral/water interface, but ESR has provided a direct view of the orientation and mobility of these ions retained electrostatically at the hydrated surfaces. The subject has been reviewed elsewhere [8, 74, 75], but a brief summary of the behavior of the most informative paramagnetic probe cations will be presented here.

Cu^{2+} and VO^{2+}. Both ions are axially symmetric $S = 1/2$ spin systems. On fully wet clays and zeolites, the hydrated $Cu(H_2O)_6^{2+}$ ion produces a single broad resonance, as the expected four lines from hyperfine interaction with the Cu nucleus ($I = 3/2$) are obscured by line-broadening effects at room temperature [76, 77]. The $VO(H_2O)_5^{2+}$ ion similarly produces an isotropic spectrum, composed of eight hyperfine lines, as shown in Fig. 8-12. For both ions, the anisotropy of g and A tensors is averaged out by rapid molecular rotation, and isotropic g and hyperfine values, g_0 and A_0, appear in the spin Hamiltonian:

$$H = g_0 \beta_0 \, H{\cdot}S + A_0 \, S{\cdot}I \qquad (8\text{-}19)$$

where $g_0 = 1/3 \, g_{\|} + 2/3 \, g_{\perp}$ and $A_0 = 1/3 \, A_{\|} + 2/3 \, A_{\perp}$. Consequently, the transition energies for the isotropic spectrum are given by [78]:

$$\omega_0 = 2\pi \, g_0 \, \beta_0 \, H/h \; + A_0 m_I + \frac{A_0{}^2}{4\pi} \left(\frac{I(I+1) - m_I{}^2}{g_0 \beta_0 H/h} \right) \qquad (8\text{-}20)$$

where ω_0 is the microwave frequency (rads/s) of the resonances. The last term in this equation accounts for the unequal spacing of the hyperfine lines, but it is easily demonstrated that the outer hyperfine lines ($m_I = +7/2, -7/2$) are separated by $7A_0$, facilitating the measurement of A_0 (in Gauss or Tesla) directly from the spectrum. Notable also in the spectrum (Fig. 8-12) is the unequal linewidth of the resonances, a consequence of the g and A tensor anisotropy. Linewidths, ΔH, vary systematically as a function of the m_I quantum number according to [79]:

$$\Delta H(m_I) = a + bm_I + cm_I{}^2 \qquad (8\text{-}21)$$

where a, b, and c are constants, The term in b causes asymmetric broadening, while the term in c causes the outer resonances to be broader than the inner ones. For rapidly tumbling ions, the b and c terms are proportional to τ_c, the time required for molecular rotation. Thus, line broadening effects have been used to estimate mobility of adsorbed VO^{2+} on smectites, indicating at most a 50% reduction in rotational mobility relative to VO^{2+} in the aqueous phase [70]. Although the interlayer regions accommodating these cations are only about four water molecules thick (~10A), this narrow water layer is nearly solution-like.

As clays and zeolites are partially dehydrated by mild heating, freezing, or other methods, the isotropic spectra of Cu^{2+} and VO^{2+} are replaced by the anisotropic, solid-state spectra typical of axially symmetric hydrated ions [70, 76, 77, 80]. For example, the isotropic signal is totally replaced by the anistropic signal once the clay temperature is lowered to $-50°C$ [80]. The symmetry (z) axes of the ions are generally aligned perpendicular to the plane of the silicate sheets on partially dehydrated smectites, as indicated by the orientation dependence of the ESR spectrum in clay films aligned parallel and perpendicular to the applied field. This anisotropy is displayed for a Ca^{2+}-saponite film in Fig. 8-13, where a fraction of the Ca^{2+} exchange ions are replaced by Cu^{2+}. The four hyperfine lines of the $g_{\|}$ resonance are detected with the film perpendicular to the field, while the $g_{\perp}$ resonance (with unresolved hyperfine structure) is detected for the parallel alignment.

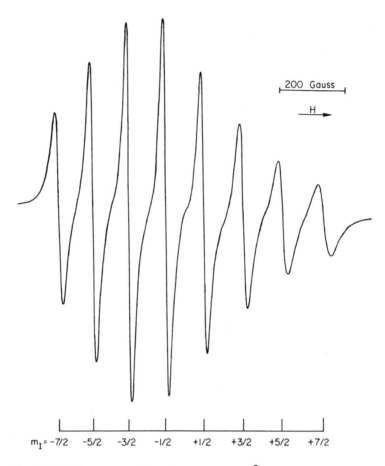

Fig. 8-12. Spectrum of exchangeable VO^{2+} on fully hydrated hectorite.

The preferred orientation of molecular axes relative to the clay sheets is based on the tendency of planar $Cu(H_2O)_4^{2+}$, or even $Cu(H_2O)_5^{2+}$, to lie "flat" as layer collapse forces them into rigid alignment. Thus, dehydrated clay interlayers are seen by ESR to be quasicrystalline.

The time scale of molecular motion separating "mobile" from "rigid" depends on the degree of anisotropy of the spectrum. Taking hydrated Cu^{2+}, for example, the anisotropy of g can be expressed as $|g_\| - g_\perp| = 2.38 - 2.07$. In frequency units, this is $\approx 1.5 \times 10^9$ s^{-1}; therefore, rigid-limit

Fig. 8-13. Spectra of Cu^{2+} adsorbed on an air-dry Ca^{2+} saponite film. The orientation of the platelets used to obtain the two spectra are depicted.

spectra (as in Fig. 8-13) replace solution-like spectra once the adsorbed ions have rotational frequencies lowered to 10^9 s^{-1} or less, the range of incipient slow motion. Since hydrated VO^{2+} and Cu^{2+} ions have rotational correlation times, τ_c, of about 5×10^{-11} s in water at room temperature [78], their rotational frequencies are about $1/\tau_c \approx 2 \times 10^{10}$ s^{-1}, approximately an order of magnitude more rapid than incipient slow motion. Conversely, the metal ions in collapsed interlayers must have rotational frequencies slower than 1.5×10^9 s^{-1}, but it is possible that ions such as $Cu(H_2O)_4^{2+}$ spin rapidly about their symmetry axis while remaining flat on the surfaces, motion which does not average the spectral anisotropy.

Alignment of hydrated Cu^{2+} ions in air-dry vermiculite with two monolayers of water molecules in the interlayer is different from the alignment in smectites, with the symmetry axis inclined nearly 45° away from the plane of the silicate sheets [76, 81]. This fact, combined with the higher $g_{\parallel}$ and lower $A_{\parallel}$ values for Cu^{2+} adsorbed on vermiculite, suggests a significantly different ligand environment for Cu^{2+} in the hydrated interlayers of vermiculite. The inclination of the $Cu(H_2O)_6^{2+}$ symmetry axis is consistent with the known hydration structure of vermiculite and may be a consequence of maximized H bonding between silicate O atoms and coordinated H_2O. The magnetic parameters of Cu^{2+} are consistent with a weaker crystal field or less covalent Cu ligand bond in comparison with montmorillonite [38]. In fact, the ESR parameters of Cu^{2+} on vermiculites and other tetrahedrally charged clays are similar to those of $Cu(H_2O)_6^{2+}$ frozen in water, while the parameters of Cu^{2+} on the octahedrally charged smectites suggest a more covalent (i.e., stronger) ligand field [82]. The principal components of the g tensor for axially symmetric Cu^{2+} are related to the crystal-field splitting energies, Δ_1 and Δ_2 (where Δ_1 is the separation between $d_{x^2-y^2}$ and d_{xy} orbitals, and Δ_2 is the separation between $d_{x^2-y^2}$ and d_{xz}, d_{yz}) by:

$$g_{\parallel} = 2\left(1 + \frac{4\zeta}{\Delta_1}\right) \qquad g_{\perp} = 2\left(1 + \frac{\zeta}{\Delta_2}\right) \tag{8-22}$$

The spin/orbit coupling constant, ζ, is 829 cm^{-1} for the free Cu^{2+} ion [1]. These equations predict that stronger ligand fields produce lower g values, as is observed for the octahedrally charged smectites. It is possible that monomeric Cu/hydroxy complexes form on these clays upon drying, since these complexes have lower g and greater A values than $Cu(H_2O)_6^{2+}$ [83]. The well-known Brønsted acidity of the octahedrally charged clays arises from hydrolysis reactions of the exchange cations [84]. If Cu^{2+} hydrolysis explains the difference in ESR spectra, the hydrolysis reaction must be less favorable on tetrahedrally charged clays.

More direct evidence for metal hydrolysis has been detected from VO^{2+} spectra, where small quantities of VO^{2+} adsorbed on moist smectites generated a rigid-limit spectrum suggesting that VO^{2+} was coordinated to OH^-, possibly as physically adsorbed $VO(OH)_2$ monomers [70]. This signal was obscured by the isotropic spectrum of hydrated VO^{2+} at higher adsorption levels, possibly because the surface acidity generated by VO^{2+} hydrolysis limited further hydrolysis.

ESR proves very useful in distinguishing different chemical forms of adsorbed metals. With Cu^{2+}, the ESR signal intensity is proportional to the hydrated Cu^{2+} content of the clay. As the clay pH is raised, this signal diminishes because hydrolysis and polymerization of Cu-hydroxy species occurs [85]. Polynuclear structures of Cu^{2+} and VO^{2+} are not usually detectable by ESR because of metal/metal interactions. The ESR signal of hydrated Cu^{2+} also diminishes at high temperatures ($\geq 170°C$) because the cation dehydrates. This signal loss is enhanced if the clay pH is initially high, suggesting that dehydration promotes the hydrolysis of Cu^{2+} [85].

Complexation chemistry of clay-adsorbed VO^{2+} and Cu^{2+} is readily followed by ESR because the g and A values of the complexes are often quite different from those of the hydrated metals. The ESR spectra of adsorbed Cu-amine, Cu-alcohol and various other cationic complexes have been reported [52, 86-88]. Not surprisingly, orientation studies of dried clay films typically reveal bulky complexes to lie "flat" in the constraining interlayer [89], or even to dissociate partially to minimize their molecular dimensions [86]. In ESR studies of Cu^{2+}/ligand complexes in zeolites, the size of the complex relative to that of the zeolite cage determines whether the complex can form [75].

Mn^{2+} and Fe^{3+}. The essentially cubic symmetry of $Mn(H_2O)_6^{2+}$ and $Fe(H_2O)_6^{3+}$ is expected to result in very small ZFS terms for the aqueous ions, producing isotropic ESR signals at g = 2.00. In fact, aqueous solutions of monomeric Fe^{3+} have extremely broad signals centered at g = 2.00 [69], as do Fe^{3+} cations adsorbed as exchange ions on hydrated zeolites and layer silicates [90, 91]. The broadening may be related to lowered symmetry arising from OH^- coordinated to Fe^{3+}. Aqueous Mn^{2+} generates a sextet of rather narrow hyperfine lines, each of which envelopes all five allowed electronic transitions for the S = 5/2 spin system. Line broadening in Mn^{2+} spectra is related to spin relaxation processes and can be used to estimate a rotational correlation time, τ_c, for the $Mn(H_2O)_6^{2+}$ ion [92]. Fig. 8-14a displays the spectrum of exchangeable Mn^{2+} on a fully wet smectite. This spectrum is only slightly broadened relative to that of aqueous Mn^{2+}, indicating an approximately 30% increase in τ_c for the adsorbed ion. More severe broadening results from air drying the clay (Fig. 8-14b), a consequence of motional restriction. Dehydration at 200°C actually narrows the spectrum again (Fig. 8-14c), but this is the consequence of converting from a solution to a solid-state powder spectrum in which only the +1/2 $\leftrightarrow$ –1/2 electronic transition is detected. (Similar line narrowing occurs when aqueous Mn^{2+} is frozen.) The solid-state spectrum reveals forbidden transitions and anisotropic splitting of the six principal resonances [93], features which were described earlier to be the result of a distorted axial crystal field. Evidently, dehydrated Mn^{2+} enters a low-symmetry hexagonal cavity site in the silicate. In zeolites, ESR detects a transition from the solution-like to the solid-state spectrum at –40 to –60°C, denoting the temperature at which zeolite cavity water is frozen out [94].

The Mn^{2+} ion is clearly a useful probe for surface studies of colloidal materials, reporting local viscosity and detecting temperatures of near-surface phase transitions [95].

Cr^{3+}. The $Cr(H_2O)_6^{3+}$ ion (S = 3/2) produces a single symmetrical resonance near g = 1.98 in aqueous solution [69]. The spectrum broadens if the symmetry of the ion is lowered (e.g., by entry of anions into the hydration sphere) or the rotational mobility is reduced, a consequence of ZFS effects. Thus, Cr^{3+} adsorbed on fully hydrated smectites has a greater ESR linewidth than aqueous Cr^{3+} [96]. In zeolites, $Cr(H_2O)_6^{3+}$ can persist to high temperatures in the structural cages according to ESR [97], but the

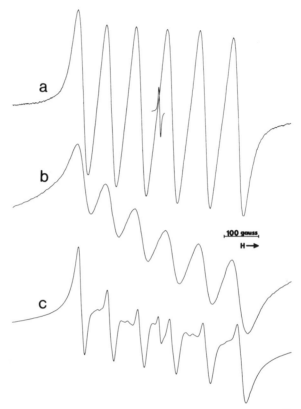

Fig. 8-14. Room temperature spectra of Mn^{2+} adsorbed on Mg^{2+}-hectorite powder samples, (a) Fully hydrated, (b) air-dried, and (c) thermally dehydrated at 200°C.

reduction in symmetry and mobility upon partial dehydration of Cr^{3+} produces the more complex powder spectrum [98].

3.2.2. Oxides and Noncrystalline Aluminosilicates . The ESR spectra of metals adsorbed on these minerals have two consistent characteristics which distinguish them from the spectra of exchangeable metal ions adsorbed on silicates possessing structural charge (see previous section). These are: (1) a rigid-limit line shape, even when the solid is fully wet, indicating low mobility; and (2) altered magnetic parameters, indicating ligand exchange within the coordination sphere of the hydrated metal. Both characteristics point to chemisorption, in which the metal ion coordinates directly to surface oxygen atoms of the oxide or aluminosilicate. ESR studies of chemisorption have been done with Cu^{2+} and VO^{2+} on diamagnetic minerals including Ti oxides, Al oxides and hydroxides, and allophanes (noncrystalline aluminosilicates). Although Cr^{3+} and Fe^{3+} adsorption has also been studied by ESR [99], the strong tendency of these metals to form hydroxide "clusters" at surfaces often produces broad ferromagnetic (or undetected) resonances rather than the spectra of isolated paramagnetic ions. In addition, ZFS effects dominate the paramagnetic spectra of Cr^{3+} and Fe^{3+} in rigid environments, effects that are not readily correlated to structural details [61]. Chemisorbed Mn^{2+} has broadened or undetectable resonances [100], a likely consequence of low symmetry and reduced mobility of the bonded metal. In contrast, VO^{2+} and Cu^{2+} report detailed chemical information about the metal/surface complex via the hyperfine splittings and g values [101–104]. For axially symmetric VO^{2+}, strong crystal fields generate smaller hyperfine splittings and higher g values [73], while for axially symmetric Cu^{2+}, strong crystal fields produce larger hyperfine splittings and lower g values [38]. These relationships are quantitatively described in Eq. (8-22) and (8-7).

Adsorption of $VO(H_2O)_5{}^{2+}$ and $Cu(H_2O)_6{}^{2+}$ on $Al(OH)_3$, allophane, and related minerals in the wet state produces a rigid-limit spectrum with parameters indicating enhanced covalent bonding (see Table 8-3). An example of this type of powder spectrum is shown in Fig. 8-15, with evidence for two forms of chemisorbed Cu^{2+}. The chemical environment at Al hydroxide surfaces as defined by these spectra is similar to that of VO^{2+} and Cu^{2+} coprecipitated in Al oxides; in fact, the coprecipitated metals reside largely at the hydroxide surfaces [105]. It is possible to locate metals in solids using ESR. If metals are at surfaces, their spectra are magnetically broadened when metals such as Ni^{2+}, which do not themselves generate resonances, are introduced. Also, surface-bound metals such as Cu^{2+} and VO^{2+} readily form ternary (surface/metal/ligand) complexes when ligands are introduced, identified by shifts in the magnetic parameters. Ternary surface complexes that have been identified by ESR include boehmite/VO^{2+}/phosphate [104], gibbsite/Cu^{2+}/glycine [106], allophane/Cu^{2+}/NH_3 [102], and imogolite/Cu^{2+}/glycine [107]. Metals inaccessible to complexing ligands cannot form ternary complexes, as is the case for some fraction of Cu^{2+} ions chemisorbed on imogolite when the bulky o-phenanthroline ligand is introduced [96]. Thus, ESR can indicate the entry of metals into very small pores or solid materials.

A description of the d orbital energies of Cu^{2+} in the free (aqueous) and adsorbed states, employing Eq. (8-22) to convert ESR parameters to crystal field splitting energies, is diagrammed in Fig. 8-16. Unfortunately, ESR provides an equivocal description of bonding at the surface. The number of ligating atoms is uncertain, since mononuclear or binuclear complexes with surface oxyanions could both generate g and A values in the range observed. In fact, two distinguishable chemisorption sites have commonly appeared in ESR spectra as exemplified by Fig. 8-15 [102, 105]. More

Fig. 8-15. Observed (solid line) and simulated (broken line) powder spectrum of Cu^{2+} adsorbed on alumina (10 mmol/kg). The simulation assumed two species with the g values shown.

detailed investigations of Cu^{2+} and VO^{2+} chemisorbed on Al oxide have been conducted by the related spectroscopic technique of ENDOR (electron/nuclear double resonance), confirming that the metals are coordinated directly to surface O atoms but failing to prove that the complexes are binuclear [108, 109].

ESR investigations have substantiated the concept of a chemisorption/precipitation continuum in sorption processes on oxides, distinguishing metals adsorbed at isolated sites from those clustered into small nuclei [100, 101]. Isolated metal ions generate paramagnetic spectra, while clustered metals are likely to display very broad or undetectable spectra as a result of metal/metal electronic and magnetic interactions. Studies with microcrystalline gibbsite have determined that the dominant (001) planar surface is unable to chemisorb Cu^{2+}, but crystallite edge or step sites possess reactive OH groups which bond Cu^{2+} [101]. Sorption beyond the limited capacity for chemisorption afforded by these bonding

Fig. 8-16. Approximate effect of chemisorption on the d orbital energies of Cu^{2+}, as deduced from ESR parameters.

sites involves the formation of $Cu(OH)_2$ nuclei. This example underscores one of the important advantages of ESR in surface studies: its ability to distinguish magnetically as well as chemically different environments of paramagnetic ions.

3.3. Chemical Reactions Involving Electron Transfer

Reactions at mineral surfaces that involve changes in oxidation state of the surface and/or the reactant may be accessible to study by ESR, either through changes in paramagnetic or ferromagnetic signals of the solid, or through the appearance of organic radicals or other paramagnetic species as reaction intermediates or products. Because of interest in their catalytic properties, several major classes of mineral materials have been investigated by ESR:

3.3.1. Layer Silicate Clays.

Potential electron donor or acceptor sites in these aluminosilicates are associated with structural transition metals (e.g., Fe^{3+}, Mn^{2+}, V^{4+}), adsorbed transition metal cations, and lattice defects (e.g., electron holes). However, there is no evidence to suggest that lattice defects react with surface-adsorbed molecules, probably because they are trapped deep below the surface in an insulating matrix. On the other hand, numerous cases of electron transfer between adsorbed or structural transition metals and adsorbed organics have been documented using ESR. Example redox reactions of this type are:

1. Oxidation of benzidine, aniline, and related aromatic amines on smectites, where adsorbed and structural Fe^{3+} accept an electron to form the radical cation (detected by ESR) and the intensity-colored benzidine-blue complex [110–112]:

$$Fe^{3+} \text{ (structural)} + H_3N(C_6H_4)_2NH_3{}^{2+} \rightleftarrows Fe^{2+}$$

$$\text{(structural)} + H_3N(C_6H_4)_2NH_3{}^{3+} \tag{8-23}$$

2. Oxidation of benzene and substituted benzenes, where VO^{2+}, Cu^{2+}, and Fe^{3+} adsorbed on dessicated smectites oxidize aromatics to radical cations [113–115]. ESR confirms both the disappearance of the oxidized forms of the metals and the formation of organic radicals.

3. Oxidation of dioxin and chlorinated unsaturated hydrocarbons by Cu^{2+} adsorbed on dessicated smectites [116].

Usually, the adsorbed organic radical intermediates produce structureless, narrow ESR signals near the free electron position (g = 2.0023), a possible consequence of rapid electron exchange between radicals or between a radical and a parent molecule [117]. Exchange processes can collapse the hyperfine structure expected from interaction of the unpaired electron with structural protons of the radical molecule. Furthermore, anisotropy of the g and A tensors in these adsorbed radicals may not be averaged out because of restricted motion [113], and magnetic interactions with the electron dipoles of adsorbed or structural transition metals may further deteriorate spectral resolution. In any event, the ESR-detectable radical concentration is only a small fraction of that expected from the chemical reaction [113]. This may reflect the tendency of radical cations to pair on clay surfaces to form essentially diamagnetic charge-transfer complexes with weak or nonexistent ESR signals. Pairing of benzidine, methylene blue, and methyl viologen radicals on layer silicates has been confirmed [110, 118].

3.3.2. Oxides of Fe and Mn. Some of the electron-acceptor properties of layer silicate clays outlined above may actually originate from Fe oxide and Mn oxide impurities. The clays commonly possess broad ESR signals, which can be at least partially removed by chemical treatments that dissolve oxides [111]. ESR has demonstrated the production of semiquinone radicals when Fe and Mn oxides are introduced into aqueous solutions of diphenols and polyphenols [119, 120]. ESR detects these anion radicals once they are released into aqueous solution but cannot detect them at the surfaces because of strong magnetic interaction with Fe and Mn. Magnetic interactions may also explain the failure to detect radical intermediates when phenol vapors are oxidized to quinone products on dry smectite surfaces [121].

ESR confirms the dissolution of Mn^{2+} by the appearance of the typical six-line resonances for $Mn(H_2O)_6^{2+}$ when Mn oxides are reacted with solutions of diphenols and polyphenols. Similarly, reactions of benzidine with Mn oxides release Mn^{2+} to solution, confirming the electron transfer (unpublished data). In this case, no radical product is detected by ESR despite the appearance of an intensely blue reaction product in solution, an indication that a charge-transfer complex is formed rather than a monomeric radical.

The potential for redox reactions by oxides with mixed oxidation states (e.g. Fe^{2+}/Fe^{3+}, Mn^{3+}/Mn^{4+}) may be high. Studies of model Fe-containing minerals suggest that electrons are quite mobile between crystallographically equivalent neighboring sites containing Fe^{2+} and Fe^{3+} [122]. Ferromagnetic resonance may play a key role in understanding electron transfer between these oxides and adsorbed organics or metals.

3.3.3. Diamagnetic Oxides. Oxides of Si, Al, and Ti have catalytic properties that are enhanced by thermal treatment and/or irradiation. Titanium oxides can be activated by heating in reducing atmospheres and generating a reduced form of the oxide, which possesses an ESR signal attributed by some workers to Ti^{3+} centers [123, 124]. The Ti^{3+} centers may then revert to Ti^{4+}, providing an electron for subsequent reactions involving adsorbed molecules [125, 126], eg.:

$$e^- + N_2O \rightarrow N_2 + O^- \qquad (8\text{-}24)$$

$$e^- + O_2 \rightarrow O_2^- \qquad (8\text{-}25)$$

The semiconductor properties of Ti oxides imply that these electron transfer reactions could be from the conduction band rather than from a localized surface site [125]. Yet, ESR detects localized centers of electron excess or deficiency (on the ESR time scale), supporting instead a model of catalysis which emphasizes the role of these centers. In fact, band theory with its depiction of delocalized electrons in the solid is viewed by some to be completely inadequate in explaining catalysis [127].

Aluminas, silicas, and amorphous aluminosilicates have long been known to develop catalytic activity following extreme dehydration. ESR can gauge the reducing or oxidizing capacity of these activated surfaces, since molecules can be introduced that are either reduced (e.g., tetracyanoethylene) or oxidized (e.g., benzene) to detectable radicals [128,

129]. The proposed reducing sites consist of the unsolvated hydroxyl ion site (<250°C) and the oxide anion site (>350°C), defects which are created by dehydration and dehydroxylation reactions at the surface [128]. On the other hand, exposed Al^{3+} ions are created by dehydroxylation (>250°C), and these catalyze oxidation when molecular oxygen is present, forming the cation radicals detected at the surface by ESR. In summary, reducing activity is attributed to cation vacancies, while oxidizing activity is associated with anion vacancies (Lewis acid sites). Aluminosilicates with low alumina content are poor reduction catalysts, tending to promote only oxidation. Pure aluminas possess redox activity, potentially catalyzing oxidation and reduction simultaneously.

One point becomes clear from these principles: The surface defects responsible for redox catalysis in aluminosilicates cannot exist at normal earth-surface conditions. Catalysis of redox processes by minerals in soils and sediments cannot be attributed to such high-energy defects.

4. Conclusion

It is evident from the examples given in this chapter that ESR has a unique role to play in the detailed study of adsorption and electron transfer reactions at surfaces, especially in "wet" systems, where other surface analytical techniques cannot be applied. Many of the future applications of ESR are likely to be found in environmental chemistry, for example in detailing decomposition reactions of organic compounds at mineral surfaces and in describing metal sorption processes. Combined with other methods such as diffuse reflectance FTIR and UV/visible spectroscopy, ESR can provide a detailed description of the surface-adsorbate complex on the molecular scale. A deeper understanding of the adsorptive and catalytic functions of mineral surfaces will undoubtedly emerge as a result.

References

[1] Wertz, J.E.; Bolton, J.R. "Electron Spin Resonance." McGraw-Hill, New York, 1972.
[2] Scullane, M.I.; White, L.K.; Chasteen, N.D., J. Magn. Reson., *1982*, 47, 383.
[3] McBride, M.B.In "Methods of Soil Analysis, Part 1. Physical and Mineralogical Methods"; Klute, A., Ed.; American Society of Agronomy: Madison, WI, *1986*; Chapter 9.
[4] Coey, J.M.D., In "Proceedings of the International Clay Conf.," Schultz, L.G.; van Olphen, H.; Mumpton, F.A., Eds. Clay Minerals Society; Bloomington, Indiana, 1987, p. 261.
[5] Klimaszewski, B., Poznan. Tow. Przyj. Nauk, Pr. Kom. Mat.-Przyr., Fiz. Dielek. Radiospektrosk., *1973*, 6, 217.

[6] Weil, J.A., Phys. Chem. Miner., *1984*, 10, 149.
[7] Nuttall, R.H.D.; Weil, J.A., Can. J. Phys., *1981*, 59, 1696.
[8] Pinnavaia, T.J. In "Advanced Chemical Methods for Soil and Clay Minerals Research," Stucki, J.W.; Banwardt, W.L., Eds. Reidel Publishing Co: Boston, 1980; Chapter 8.
[9] Wong, N.-B.; Lunsford, J.H., J. Chem. Phys., *1971*, 55, 3007.
[10] Meads, R.E.; Malden, P.J., Clay Miner., *1975*, 10, 313.
[11] Hall, P.L., Clay Miner., *1980*, 15, 321.
[12] McBride, M.B., Clays Clay Miner., *1976*, 24, 88.
[13] Speit, B.; Lehmann, G., Phys. Chem. Miner., *1982*, 8, 77.
[14] Freund, F.; Batllo; Freund, M.M. In "Spectroscopic Characterization of Minerals and their Surfaces," Coyne, L.M; McKeever, S.W.S; Blake, D.F., Eds.; American Chemical Society: Washington, D.C., 1990, p. 310.
[15] Craciun, C.; Meghea, A., Clay Miner., *1985*, 20, 281.
[16] Mestdagh, M.M.; Vielvoye, L.; Herbillon, A.J., Clay Miner., *1980*, 15, 1.
[17] Coyne, L.M.; Banin, A., Clays Clay Miner., *1986*, 34, 645.
[18] Coyne, L.M.; Sweeney, M.; Hovatter, W., J. Luminesc., *1983*, 28, 395.
[19] Calderon, T.; Aguilar, M.; Jaque, F.; Coy-yll, R., J. Phys. C., *1984*, 17, 2027.
[20] Avezov, A.D.; Vakhidov, S.A.; Gasanov, E.M.; Novozhilov, A.I.; Samoilovich, M.I., Phys. Status Solidi. A., *1977*, 44, K87.
[21] McMorris, D.W., Nature, *1970*, 226, 146.
[22] Shimokawa, K.; Imai, N., Geochim. Cosmochim. Acta, *1987*, 51, 115.
[23] Skinner, A.F., Nature, *1983*, 304, 152.
[24] Kemp, R.C., J. Phys.C, *1971*, 4, L11.
[25] Prissok, F.; Lehmann, G., Phys. Chem. Miner., *1986*, 13, 331.
[26] Scala, C.M.; Hutton, D.R.; McLaren, A.C., Phys. Chem. Miner., *1978*, 3, 33.
[27] Marshall, S.A.; Marshall, T.; Zager, S.A.; Zimmerman, D.N., Physica Status Solidi (B), *1979*, 91, 275.
[28] Warren, R.W.; Mazelsky, R., Phys. Rev., B, *1974*, 10, 19.
[29] Hofmeister, A.M.; Rossman, G.R., Phys. Chem. Miner., *1984*, 11, 213.
[30] Mombourquette, M.J.; Tennant, W.C.; Weil, J.A., J. Chem. Phys., *1986*, 85, 68.
[31] Lang, R.; Calvo, C.; Datars, W.R., Can. J. Phys., *1977*, 55, 1613.
[32] Lichtenberger, G.J.; Addison, J.R., Phys. Rev., *1969*, 184, 381.
[33] Gerritsen, H.J.; Harrison, S.E.; Lewis, H.R., J. Appl. Phys., *1960*, 31, 1566.
[34] Andresen, H.G., Phys. Rev., *1960*, 120, 1606.

[35] Gallay, R.; van der Klink, J.J.; Moser, J., Phys. Rev., B, *1986*, 34, 3060.
[36] Boudeville, Y.; de Montgolfier, P., Chem. Phys. Lett., *1975*, 30, 469.
[37] Jaroš , M.; Zítková, J., Phys. Status Solidi, *1967*, 20, K99.
[38] Kivelson, D.; Neiman, R., J. Chem. Phys., *1961*, 35, 149.
[39] Clark, C.J.; McBride, M.B., J. Solid State Chem., *1984*, 55, 106.
[40] Shimizu, T., Phys. Lett., *1966*, 23, 20.
[41] Barnett, J.D.; Nelson, H.M.; Tyagi, S.D., Phys. Rev., B, *1985*, 31, 1248.
[42] Ghose, S.; Tsang, T., Am. Miner., *1973*, 58, 748.
[43] Scullane, M.I.; White, L.K.; Chasteen, N.D., J. Magn. Reson., *1982*, 47, 383.
[44] Aasa, R., J. Chem. Phys., *1970*, 52, 3919.
[45] Goodman, B.A., Clay Miner., *1978*, 13, 351.
[46] Angel, B.R.; Vincent, W.E.J., Clays Clay Miner., *1978*, 26, 263.
[47] Hall, P.L.; Angel, B.R.; Jones, J.P.E., J. Magn. Reson., *1974*, 15, 64.
[48] Mestdagh, M.M.; Vielvoye, L.; Herbillon, A.J., Clay Miner., *1980*, 15, 1.
[49] Nagasawa, K.; Noro, H., Chem. Geol., *1987*, 60, 145.
[50] McBride, M.B.; Pinnavaia, T.J.; Mortland, M.M., Clays Clay Miner., *1975*, 23, 103.
[51] McBride, M.B.; Pinnavaia, T.J.; Mortland, M.M., Clays Clay Miner., *1975*, 23, 162.
[52] Berkheiser, V.; Mortland, M.M., Clays Clay Miner., *1975*, 23, 404.
[53] Olivier, D.; Vedrine, J.C.; Pezerat, H., Bull. Grpe Fr. Argiles, *1975*, 27, 153.
[54] Kemp, R.C., J. Phys. ,C., *1972*, 5, 3566.
[55] Kemp, R.C., Phys. Status Solidi (B), *1973*, 57, K79.
[56] Olivier, D.; Vedrine, J.C.; Pezerat, H., J. Solid State Chem., *1977*, 20, 267.
[57] Matarrese, L.M.; Wells, J.S.; Peterson, R.L., J. Chem. Phys., *1969*, 50, 2350.
[58] Rager, H.; Schneider, H., Am. Mineral., *1986*, 71, 105.
[59] Choi, D.; Choh, S.H., Bull. Magn. Reson., *1986*, 8, 192.
[60] Matyash, I.V.; Polshin, E.V.; Kalinchenko, A.M., Geokhimia, *1969*, 7, 840.
[61] Hare, A.S.; Vickerman, J.C., J. Chem. Soc., Faraday Trans. I, *1981*, 77, 1103.
[62] McBride, M.B.; Farmer, V.C.; Russell, J.D.; Tait, J.M.; Goodman, B.A., Clay Miner., *1984*, 19, 1.
[63] van Dijk, J., J. Inorg. Nucl. Chem., *1978*, 40, 1609.

[64] Angus, J.G.; Beveridge, D.; Raynor, J.B., Chem. Geol., *1984*, 43, 331.

[65] Wildeman, J.R., Chem. Geol., *1970*, 5, 167.

[66] Schindler, P.; Ghose, S., Am. Mineral., *1970*, 55, 1889.

[67] Barnett, J.D.; Nelson, H.M., Phys. Rev., B, *1986*, 33, 4464.

[68] Hall, P.L.; Angel, B.R.; Braven, J., Chem. Geol., *1974*, 13, 97.

[69] McGarvey, B.R., J. Phys. Chem, *1957*, 61, 1232.

[70] McBride, M.B., Clays Clay Miner., *1979*, 27, 91.

[71] McBride, M.B., Can. J. Soil Sci., *1980*, 60, 145.

[72] McBride, M.B., J. Colloid Interf. Sci., *1987*, 120, 419.

[73] Boucher, L.J.; Tynan, E.C.; Yen, T.F., In "Electron Spin Resonance of Metal Complexes," Yen, T.F., Ed. Plenum, New York, 1969; p. 111.

[74] Hall, P.L., Clay Miner., *1980*, 15, 337.

[75] Lunsford, J.H. In "Magnetic Resonance in Colloid and Interface Science," Fraissard, J.P.: Resing H.A., Eds. Reidel Publishing Co., Boston, 1980, p. 67.

[76] Clementz, D.M.; Pinnavaia, T.J.; Mortland, M.M., J. Phys. Chem., *1973*, 77, 196.

[77] Nicula, A.; Stamires, D.; Turkevich, J., J. Chem. Phys., *1965*, 42, 3684.

[78] Chasteen, N.D.; Hanna, M.W., J. Phys. Chem., *1972*, 76, 3951.

[79] Carrington, A.; McLachlan, A.D. "Introduction to Magnetic Resonance," Harper and Row, New York, 1967.

[80] McBride, M.B., J. Phys. Chem., *1976*, 80, 196.

[81] McBride, M.B., Clays Clay Miner., *1976*, 24, 211.

[82] McBride, M.B. In "Geochemical Processes at Mineral Surfaces," Davis, J.A.; Hayes, K.F., Eds. American Chemical Society, Washington, D.C., 1986, p. 362.

[83] Ottaviani, M.F.; Martini, G., J. Phys. Chem., *1980*, 84, 2310.

[84] Mortland, M.M.; Raman, K.V., Clays Clay Miner., *1968*, 16, 393.

[85] McBride, M.B., Clays Clay Miner., *1982*, 30, 200.

[86] Berkheiser, V.; Mortland, M.M., Clays Clay Miner., *1977*, 25, 105.

[87] Brown, D.R.; Kevan, L., J. Phys. Chem., *1988*, 92, 1971.

[88] Nagai, S.; Ohnishi, S.; Nitta, I.; Tsunashima, A.; Kanamaru, F.; Koizumi, M., Chem. Phys. Lett., *1974*, 26, 517.

[89] Velghe, F; Schoonheydt, R.A.; Uytterhoeven, J.B.; Peigneur, P.; Lunsford, J.H., J. Phys. Chem., *1977*, 81, 1187.

[90] Evmiridis, N.P., Inorg. Chem., *1986*, 25, 169.

[91] Coyne, L.M.; Banin, A., Clays Clay Miner., *1986*, 34, 645.

[92] McBride, M.B.; Pinnavaia, T.J.; Mortland, M.M., Am. Mineral., *1975*, 60, 66.

[93] McBride, M.B.; Pinnavaia, T.J.; Mortland, M.M., J. Phys.

Chem., *1975*, 79, 2430.
[94] Tikhomirova, N.N.; Nikolaeva, I.V.; Demkin, V.V.;
      Rosolovskaya, E.N.; Topchieva, K.V., J. Catal., *1973*, 29, 105.
[95] Martini, G., J. Colloid Interf. Sci., *1981*, 80, 39.
[96] McBride, M.B., unpublished data.
[97] Rubinshtein, A.M.; Slinkin, A.A.; Loktev, M.I.; Fedorovskaya,
      E.A.; Bremer, H.; Vogt, F., Z. Anorg. Allg. Chem., *1976*, 423,
      164.
[98] Hemidy, J.F.; Goupil, J.M.; Cornet, D., J. Chim. Phys. Phys. -
      Chim. Biol., *1977*, 74, 74.
[99] Hubald, J., Z. Chem., *1977*, 17, 273.
[100] Bleam, W.F.; McBride, M.B., J. Colloid Interf. Sci., *1985*, 103,
      124.
[101] McBride, M.B.; Fraser, A.R.; McHardy, W.J., Clays Clay Miner.,
      1984, 32,12.
[102] Clark, C.J.; McBride, M.B., Clays Clay Miner., *1984*, 32, 300.
[103] Bleam, W.F.; McBride, M.B., J. Colloid Interf. Sci., *1986*, 110,
      335.
[104] McBride, M.B., J. Colloid Interf. Sci., *1987*, 120, 419.
[105] McBride, M.B., Clays Clay Miner., *1982*, 30, 21.
[106] McBride, M.B., Clays Clay Miner., *1985*, 33, 397.
[107] Goodman, B.A.; Green, H.L.; McPhail, D.B., Geochim.
      Cosmochim. Acta, *1984*, 48, 2143.
[108] Motschi, H.; Rudin, M., Colloid Polym. Sci., *1984*, 262, 579.
[109] Rudin, M.; Motschi, H., J. Colloid Interf. Sci., *1984*, 98, 385.
[110] McBride, M.B., Clays Clay Miner., *1985*, 33, 510.
[111] McBride, M.B., Clays Clay Miner., *1979*, 27, 224.
[112] Cloos, P.; Moreale, A.; Broers, C.; Badot, C., Clay Miner., *1979*,
      14, 307.
[113] Pinnavaia, T.J.; Hall, P.L.; Cady, S.S.; Mortland, M.M., J. Phys.
      Chem., *1974*, 78, 994.
[114] Mortland, M.M.; Pinnavaia, T.J., Nature, *1971*, 229, 75.
[115] Boyd, S.A.; Mortland, M.M., Environ. Sci. Technol., *1986*, 20,
      1056.
[116] Boyd, S.A.; Mortland, M.M., Nature, *1985*, 316, 532.
[117] Rupert, J.P., J. Phys. Chem., *1973*, 77, 784.
[118] White, J.R.; Bard, A.J., J. Electroanal. Chem., *1986*, 197, 233.
[119] Kung, K.-H.; McBride, M.B., Clays Clay Miner., *1988*, 36, 297.
[120] Kung, K.-H.; McBride, M.B., Clays Clay Miner., *1988*, 36, 303.
[121] Sawhney, B.L.; Kozloski, R.K.; Isaacson, P.J.; Gent, M.P.N.,
      Clays Clay Miner., *1984*, 32, 108.
[122] Amthauer, G.; Rossman, G.R., Phys. Chem. Miner., *1984*, 11,
      37.
[123] Hasiguti, R.R.; Iguchi, E.; Takahashi, S., "Proc. 9th Int. Conf. on

the Physics of Semiconductors," 1968, 2, 1142.
[124] Iyengar, R.D.; Codell, M.; Karra, J.S.; Turkevich, J., J. Am. Chem. Soc., *1966*, 88, 5055.
[125] Cunningham, J.; Penny, A.L., J. Phys. Chem., *1974*, 78, 870.
[126] van Hooff, J.H.C., J. Catal., *1968*, 11, 277.
[127] Stone, F.S., J. Solid State Chem., *1975*, 12, 271.
[128] Flockhart, B.D.; Leith, I.R.; Pink, R.C., Trans. Faraday Soc., *1969*, 65, 542.
[129] Corio, P.L.; Shih, S., J. Catal., *1970*, 18, 126.

9. LUMINESCENCE SPECTROSCOPY

Harry G. Brittain

Bristol-Myers Squibb Pharmaceutical Research Institute
New Brunswick, NJ 08903

1. Introduction

All substances are capable of absorbing radiant energy, which usually becomes converted to heat after a certain time period. For instance, rocks left out in the sun become hot upon absorption of visible and infrared radiation. A substantial number of naturally occurring mineral substances also exhibit the phenomenon of luminescence, that is the ability to emit light having a wavelength longer than that which was originally absorbed. The property is normally observed by irradiation of the mineral with ultraviolet light (often referred to as "blacklight"), and the subsequent observation of either blue, green, or red light from certain active centers.

The photoluminescence just described should not be confused with thermoluminescence, which is the non-black body radiation emitted by a solid upon its being heated. This latter effect is understood to involve electron/hole centers, and will remain outside the scope of the present review.

Besides providing a spectacular display, mineral luminescence can be used to obtain important information. When it is observed, assignment of the nature of the mineral luminescence is required for full characterization of the material. Specification of two properties are required for this purpose: the wavelength distribution of the emitted light (the luminescence spectrum, measured with a fixed and constant irradiation energy), and the wavelength distribution of the excitation energy (the excitation spectrum, measured with a constant observing emission wavelength).

More detailed studies also consider the time dependence associated with these photophysical processes, yielding a further classification of mineral luminescence properties as being due either to fluorescence or to phosphorescence. No difference can be noted between the two modes if a continuous excitation source is used. The use of pulsed sources permits an empirical distinction: fluorescence will cease immediately as the source is turned off, while phosphorescence will be observed for some time after termination of excitation.

The majority of luminescent minerals contain a metal ion center from which the effect originates. These activator ions can be present in stoichiometric quantities or in trace amounts, but are usually be identified on the basis of their excitation and emission spectra. The most common activator ions are Cr(III), Mn(II), lanthanides, and actinides. The luminescence from metal ion activator sites can be associated either with the mineral surface, or with the bulk interior of the material. The degree of penetration by the excitation energy, or the efficiency with which the luminescence escapes the interior, is governed by the relative transparency of the mineral. The luminescence of clear mineral crystals will largely reflect the bulk property, while the luminescence of opaque minerals is solely a surface phenomenon.

Mineral luminescence can also originate from defect sites existing in a crystal lattice, which are capable of trapping excitation energy for a sufficiently long time to allow the emission of visible light quanta. The luminescence which can be observed with certain diamond samples originates from a defect site where nitrogen had substituted for carbon in the crystal lattice. Other defect sites which can be associated with molecular ions include S_2^-, O_2^-, and F-centers. These defect sites, however, are usually associated with the interior of the mineral and are rarely observed at the surface of a material. For this reason, defect luminescence will be considered outside the scope of the present chapter.

Since all photophysical processes require the adsorption and subsequent emission of light, it follows that in most materials the luminescence can only be obtained over a limited spatial range. As mentioned above, photons do not penetrate deeply into the interior of an opaque mineral, and therefore any observed luminescent phenomena must reflect surface properties. The natural occurrence of highly transparent mineral crystals that are also photoluminescent is not common. The majority of luminescent minerals are opaque, thus permitting study of only the outer material surface. This feature places excitation and emission spectroscopies into the surface science category, and exploitable for study. Quantitative studies of mineral luminescence are extremely useful when combined with the surface techniques described elsewhere in this book.

In the following sections the fundamental nature of luminescence spectroscopy in activator centers will be discussed, and representative examples of luminescent systems used to illustrate aspects of emission phenomena. Since mineral luminescence has not been used extensively to study the surfaces of minerals, it is anticipated that the chapter illustrations will suggest how enterprising workers might use the technology to their advantage.

A significant literature regarding the luminescence of metal ions in solids exists [1-5], although the literature concerning luminescent mineral species is smaller [6,7]. A variety of texts are also available which deal on a more general basis with the topic of luminescent systems [8-12].

2. Principles of Luminescence Spectroscopy

The features associated with luminescence spectroscopy represent transitions among electronic states, which are totally specified by a series of quantum numbers. The important quantum numbers which define a given electronic state are its total orbital angular momentum (L) and the total spin angular momentum (S), quantities which are determined by the vector addition of individual orbital and spin moments. Methods for deducing these state properties are well known [13-15]. It follows that only one combination of orbital and spin angular momentum can have the lowest energy (the "ground" state), and that all other states must be of higher energy ("excited" states). The excitation process is a transition from the ground state to one of the many excited states. One of these excited states must be of lowest energy, and most luminescence is the transition from this lowest excited state down to the ground state.

The photophysical origins of luminescence spectroscopy can be further understood using the energy level diagram representation of Jablonski [16]. A simple version of a Jablonski diagram is shown in Fig. 9-1. Although this representation is most often used to describe the luminescence of molecular species, it is useful to establish the terminology of luminescence phenomena. Each electronic state is further quantized by the vibrational states to which it is coupled, the types and properties of which are defined by the symmetry of the system. For the sake of simplicity, only one vibrational mode is shown for each electronic state in Fig. 9-1 (identified by a vibrational quantum number). In Fig. 9-1, the molecular ground state is characterized by $S = 1$, and is therefore labeled as a singlet state. States having the same spin quantum number as the ground state are also identified as singlet states in the diagram. Other states may exist which have a different spin quantum number, and for the sake of simplicity only triplet states (for which $S = 1$) are shown. The main

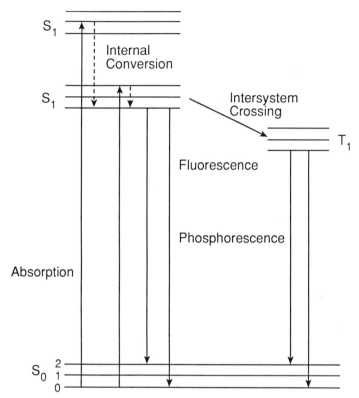

Fig. 9-1. Jablonski-type diagram illustrating the photophysical processes associated with luminescence spectroscopy.

difference between the labels of Fig. 9-1 and actual minerals is that the metal ion energy levels will be characterized by higher values for their spin and orbital angular momenta.

A state identification of any species therefore consists of a specification of the total spin and orbital momentum of the state. The orbital angular momentum of a state is signified in the letter used for its name:

S	=	0	1	2	3	4	5	6
name	=	S	P	D	F	G	H	I

The spin angular momentum of a state is provided as the multiplicity of that state, defined as:

$$\text{multiplicity} \quad = \quad 2\,S\, +\, 1 \qquad\qquad (9\text{-}1)$$

The spin multiplicity is given as a superscript to the left of the state name. The methods for deducing the states of a given molecular species are well established [13-15]. For the simple species depicted in Fig. 9-1, the ground state (having resultant spin and orbital angular momenta equal to zero) would be denoted as ^1S. The presence of partially filled shells of d-electrons or f-electrons in metal ion activators yields states of considerably more complication, where a large number of possibilities exist for the vector addition of individual spin and orbit contributions.

Observation of luminescence will first require efficient promotion of the system from the ground ^1S state to some excited state. One requirement for this process is that the multiplicity of the excited state be the same as that of the ground state. Once the initial excitation has taken place, a sequence of photophysical events may then transpire. If the excitation is into one of the higher excited singlet states internal conversion will take place, and the system will relax nonradiatively to the lowest excited singlet state within 10^{-12} s after the initial excitation. At this point, one of three processes will take place. The system can return to the ground state through radiationless deactivation, and all absorbed energy will be released as heat. The system may return to the ground state with the emission of an appropriate quantum of light, the process identified previously as fluorescence. The lifetime associated with fluorescent processes is approximately 10^{-8} s.

The system may also undergo intersystem crossing, and occupy an excited state characterized by a spin quantum number different from that of the ground state (identified as a triplet state in Fig. 9-1). This state will relax to the ground state either by a radiationless deactivation, or through the emission of a light quantum (now identified as phosphorescence). Since the spin quantum number must change again in this last process, the lifetime of phosphorescent transitions ranges from 10^{-3} to 10^{+3} s. The spin multiplicities associated with the luminescent state of metal ion activators are normally different from that of their corresponding ground states, and consequently the luminescence of mineral systems can usually be considered as being phosphorescence.

The processes just described are suitable for isolated luminescent centers, but complications can arise if the concentration of emissive centers becomes appreciable. Since the activator species remain excited for a comparatively long time period, energy transfer from one activator center to another may take place if these are sufficiently close. The most common result of energy transfer is the enhancement of one metal ion luminescence at the expense of another. This phenomenon is denoted as quenching, and often the quenched species can be rendered totally non-luminescent. In many instances, the excited activator species can be quenched by identical ions in their ground states, yielding a quenching dependent on the concentration of the activator. In many minerals, luminescence of a metal ion center only is observed if the concentration of ions is less than 3%.

Another situation that may result from an energy transfer process is that an ion not luminescent at its concentration level can become emissive in the presence of some other ion. This latter process is termed sensitization, and can be used to pump excitation energy into an activator ion incapable of absorbing energy directly. When the host system is capable of absorbing excitation energy by means of its lattice bands, then an extremely efficient method of excitation can be through the direct UV irradiation of these band systems.

3. Instrumentation for the Detection of Luminescence

The simplest device for measurement of luminescence consists of a source of suitable excitation energy, the mineral sample itself, and the human eye as a detection device. Although this system is useful for qualitative work, quantitative determinations require more sophisticated apparatus.

Fig. 9-2 illustrates a block diagram of a device suitable for the measurement and quantitation of both steady-state and time-resolved

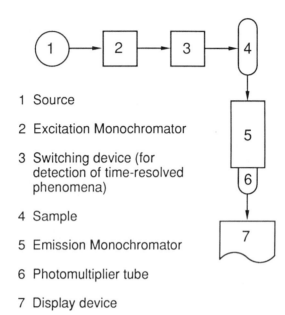

1 Source

2 Excitation Monochromator

3 Switching device (for
detection of time-resolved
phenomena)

4 Sample

5 Emission Monochromator

6 Photomultiplier tube

7 Display device

Fig. 9-2. Block diagram of a spectrometer capable of detecting both steady-state and time-resolved luminescence properties of a mineral system.

luminescence. For the spectrometer to be considered perfect, the source should yield a constant photon output at all wavelengths, the monochromator must be equally efficient at all wavelengths without regard to the polarization state of the light, and the detector must be able to quantitate light of all wavelengths with equal efficiency [12]. These requirements are never met by any instrument, and consequently data adjustments are normally made at each stage of measurement to correct for nonideal behavior.

The most widely used source of excitation energy are high-pressure xenon lamps. These provide continuous output from approximately 200 to beyond 1100 nm, although above 750 nm the lamp output contains a number of sharp line features. High-pressure mercury lamps will provide more intense excitation energies, but here the intensity is concentrated in discrete bands rather than being continuous. If only the emission spectrum of a mineral is required, then the filtered output of a mercury lamp is the preferred source. Suitable filters can be used to isolate any desired band system of a mercury lamp [8].

More use is being made of lasers as excitation sources, although a laser-based spectrometer is considerably more expensive than one using a conventional source. Although they can be extremely powerful, laser sources are extremely limited in the range of their spectral output unless one is working with a laser-pumped tunable dye laser. Lasers can be operated in a continuous or pulsed mode, depending on the design of the system and requirements of the lasing material. Probably the most versatile laser is the argon-ion laser, which provides laser output at discrete wavelengths in the ultraviolet and visible regions.

Most luminescence spectrometers use monochromators to isolate appropriate irradiation wavelengths, and to characterize the emission resulting after the excitation process. All modern devices use diffraction gratings (rather than prisms) for this purpose, although the use of ruled gratings restricts the range for which the spectrometer can be operated with high efficiency. Since mineral work is invariably carried out in the solid state, it is extremely important to note that gratings can yield spurious peaks in excitation and emission spectra due to second-order reflection effects. For instance, a second-order peak at 700 nm (twice the excitation wavelength) will observed in the emission spectrum if an excitation wavelength of 350 nm is used. The detection of second order effects is more pronounced when working in the solid state due to the greater tendency toward reflection of the excitation beam into the emission.

Monochromators are normally calibrated by the output of a low-pressure mercury lamp. The important wavelengths are 253.7, 302.2, 313.2, 365.0, 404.7, 435.8, and 546.1 nm. The measured wavelengths are compared with the known values, and the grating position within the monochromator reset if any difference exists.

Although the human eye is capable of functioning as a multichannel analyzer, quantitative studies of mineral luminescence require the use of mechanical photodetection. Essentially all instruments use a photomultiplier tube for this purpose, and there are literally thousands of designs for specific applications. A photomultiplier tube begins with a photocathode, which ejects electrons when light photons are allowed to impact with its surface. These ejected electrons are amplified by a series of dynodes, where at each stage 10-20 additional electrons are ejected per initial electron. The amplification processes continues until the current pulse arrives at the anode. At this point, a current proportional to the intensity of the initial light beam is available. In many applications the phototube current is simply converted to a voltage signal, and then displayed in a suitable fashion.

Much valuable information can be obtained about a luminescent system through study of the time dependence associated with the photophysical phenomena. Since a finite time is required for a luminescent system to relax completely from its lowest excited state to the ground state, the emission accompanying this transition will also require a finite time. This period is identified as the luminescence lifetime, which is taken as the time required for the luminescence intensity to drop to $1/e$ its original value. As mentioned earlier, fluorescent lifetimes are on the order of 1-300 ns. Phosphorescent lifetimes can be as short as 1 ms and as long as several minutes.

Measurement of a luminescence lifetime requires excitation of the emissive system by a pulsed source, whose duration is negligible when compared to the lifetime of the emission process. The luminescence intensity is then recorded as a function of time, through the use of either transient recorders or boxcar averagers. The luminescence will decay through first-order kinetics, and consequently the decay curve will be in the form of a simple exponential [12]. The analysis of such curves to deduce the luminescence lifetime is relatively easy, as long as the curve consists only of a single exponential.

Since luminescence investigations of mineral systems entail study of the solid state, little freedom concerning sample preparation is possible. The usual practice is to irradiate the material surface and observe the resulting luminescence at right angles to the excitation beam. It is often advantageous to insert a long-pass filter in the emission beam to eliminate any scattered excitation energy, and therefore to reduce the magnitude of second-order grating effects. The sample may be investigated at room temperature, or cooled to cryogenic temperatures so that sharp-line features may be further resolved.

When studying the luminescence of sufficiently large single crystals, it is useful to mount the crystal on a goniometer. For anisotropic systems, the luminescence features can differ along the various crystal axes and these effects can be detected through orientation studies. Study of the polarization properties within the luminescence can also be useful in the characterization process. Excitation of a particular face with linearly polarized light may result in photoselection, where different emission intensities may result when comparing two orthogonal excitation planes. This information can be extremely useful in deducing details regarding the site symmetry of the luminescent species. Since the selection rules governing absorption and emission are derived from group theoretical principles and molecular symmetry [15], the connection between cation site symmetry and spectroscopy can be drawn without much difficulty.

4. Specific Instances of Mineral Luminescence

The most important metal ion activators responsible for luminescence in mineral systems are Cr(III), Mn(II), lanthanide ions, and the uranyl ion. Each of these differs somewhat in the intrinsic photophysical phenomena, and in the mineral systems for each is encountered. These four activator ions will now be discussed in turn.

4.1. Cr(III)

One of the most heavily studied luminescence system is that of Cr(III), since this ion is used as the activator species in the ruby laser [5]. Cr(III) is capable of substituting in a wide variety of metal oxide host systems (e.g., corundum or spinels), imparting bright colors to these otherwise colorless systems. The Cr(III) ion is invariably oxygen-coordinated with six nearest neighbors, and may be in a pure octahedral or a distorted octahedral symmetry site. Upon photoexcitation, luminescence in the red or near-infrared region of the spectrum can be observed. The immediate environment of the Cr(III) ion (denoted as its crystal field) determines the precise wavelength of luminescence [13-15].

Cr(III) exists in a d^3 configuration, and the interelectronic repulsion which results from placing three electrons in the same set of d-orbitals yields several states identified as "free ion" energy levels. These are the states listed on the extreme left of the ordinate in Fig. 9-3. The placement of this ion in a crystal lattice produces a further splitting of the free ion levels, and these are also shown in Figure 3. The exact energy spacing between the energy levels is therefore determined by the strength of the crystal field experienced by the Cr(III) ion. Since different crystal field strengths are anticipated for each mineral host into which Cr(III) can be placed, it may be concluded that the luminescence is a reliable indicator of the mineral structural properties.

The most detailed analysis of the energy levels associated with Cr(III) in the ruby system has been provided by Macfarlane [17]. The important free ion states are the ^4F ground level and excited ^2H level. In an octahedral field, the ^4F level splits into the $^4A_2$ ground state and the excited $^4T_2$ and $^4T_1$ states. The spin allowed transitions that could therefore be used to populate the excited states directly correspond to:

$$525 \text{ nm} \qquad ^4A_2 \rightarrow {}^4T_2 \ \text{(F state)}$$
$$430 \text{ nm} \qquad ^4A_2 \rightarrow {}^4T_1 \ \text{(F state)}$$

The Cr(III) luminescence corresponds to the transition:

$$695 \text{ nm} \qquad ^2E \rightarrow {}^4A_2 \ \text{(F state)}$$

In certain molecular complexes fluorescence from the excited $^4T_2$ level to the ground $^4A_2$ level can take place, but in solid mineral systems the observed luminescence usually corresponds to phosphorescence from the excited ^2E to ground $^4A_2$ states. In the ruby system, the ion site symmetry is lowered from pure octahedral, yielding a splitting of the ^2E excited state. The observed luminescence then consists of a closely spaced doublet. The exact position of Cr(III) phosphorescence

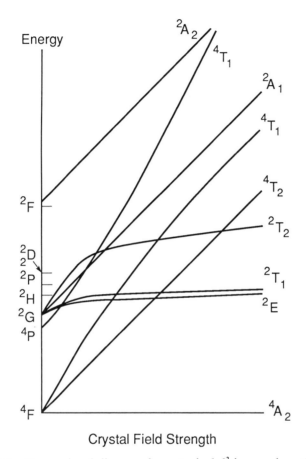

Fig. 9-3. Energy level diagram for a typical d^3 ion, such as Cr(III), in an octahedral environment. The most commonly encountered luminescence in mineral systems corresponds to the $^2E \rightarrow {}^4A_2$ transition.

varies somewhat from host to host, but is normally encountered as sharp line emission between 675 and 750 nm. Examination of Fig. 9-3 reveals that the energy difference between the ground $^4A_2$ and excited ^2E states is not strongly dependent on the magnitude of the crystal field, and hence the spectral range possible for Cr(III) in different hosts is rather small. The bands can contain some asymmetry due to vibrational fine structure, and in extremely high-resolution

Fig. 9-4. Excitation (upper trace) and emission (lower trace) spectra obtained at room temperature for the ruby system, $Al_2O_3:Cr(III)$. The scale of the emission spectrum has been increased by a factor of 2000 to show the extremely weak vibronic bands accompanying the main electronic band at 695 nm.

investigations the natural abundance of three nuclear isotopes has been detected in the vibrational fine structure [18].

An example of the luminescence spectrum of Cr(III) in ruby is shown in Fig. 9-4, together with the excitation spectrum of the system. The particular crystal used for this work was synthetically grown and contained approximately 0.5% Cr(III), and thus no extra peaks due to

Cr/Cr pair luminescence could be observed. The emission peak corresponding to the $^2E \rightarrow {}^4A_2$ transition is observed at 694.5 nm, while weak vibronic bands are also noted at 707 and 714 nm. In the excitation spectrum, the $^4A_2 \rightarrow {}^4T_2$ (525 nm) and $^4A_2 \rightarrow {}^4T_1$ (430 nm) transitions are evident. The presence of vibronic structure on the high-energy side of the $^4A_2 \rightarrow {}^4T_1$ is also evident in the bands observed at 383, 397, and 412 nm. Excitation into any Cr(III) absorption band yields identical luminescence, indicating that all emission takes place out of the ^2E excited state.

Cr(III) luminescence can be observed in any mineral system in which it is present at concentrations less than 3%, and in which it is not quenched by other metal ions. Suitable host systems include corundum (Al_2O_3), spinel ($MgAl_2O_4$), andalusite and kyanite (Al_2OSiO_4), topaz [$Al_2(F,OH)_2SiO_4$], and beryl ($Be_3Al_2Si_6O_{18}$) [19,20]. The luminescence of Cr(III) in the spinels $ZnAl_2O_4$ and $MgAl_2O_4$ has been studied in detail, with crystal field parameters being calculated and full assignment of the vibrational fine structure made [21].

4.2. Mn(II)

Probably the most important metal ion activator in mineral systems is Mn(II), which readily substitutes for Ca(II), Mg(II), or Zn(II) in a wide variety of minerals. The ability of Mn(II) to form luminescent centers of many varied types has led to its widespread use as a lamp phosphor [22]. The luminescence of Mn(II) can be anywhere from green to red, and all possible intermediate colors have been observed. At one time, it was thought that green Mn(II) luminescence indicated substitution at a tetrahedral site and that red Mn(II) luminescence indicated emission from an octahedral site. These variations are now known to be associated with crystal-field stabilization effects, and are not indicative of the metal ion site symmetry.

Mn(II) exhibits a d^5 configuration, and is almost invariably found with all five electron spins unpaired (a situation denoted either as the "weak-field" or "high-spin" condition). The energy level diagram for a high spin d^5 metal ion is shown in Fig. 9-5, where all significant

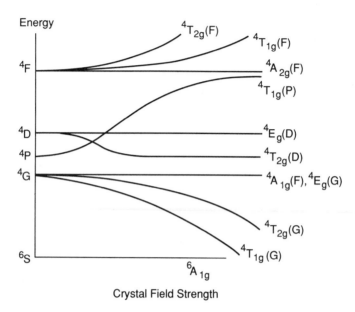

Fig. 9-5. Energy level diagram for the high-spin, weak-field, configuration of Mn(II).

excited states are found to be derived from the ^4G and ^4D free ion states. The ground state represents a unique state ($^6A_1$) whose spin angular momentum is different from all possible excited states. As a result, direct Mn(II) excitation processes are not intrinsically efficient. The highly forbidden nature of absorption transitions characterized by a non zero change in spin multiplicity is a sufficient condition for weak absorption. The Mn(II) luminescence occurs out of the $^4T_1$ excited state, and therefore corresponds to a phosphorescent process (the lifetime of this emission can be as long as several milliseconds). The energy difference between the ground $^6A_1$ and excited $^4T_1$ states is strongly dependent on the magnitude of the crystal field, with the green Mn(II) luminescence being obtained for the weakest crystal field strengths.

As an illustration of Mn(II) spectroscopy, the absorption and luminescence of Mn(II) doped into fluorite (CaF_2) is shown in Fig. 9-6. The peak maximum of the fairly broad Mn(II) luminescence was observed at 610 nm. Although a large number of discrete excitation bands are evident, they are all assigned on the basis of the energy levels diagramed in Fig. 9-5:

515 nm	$^6A_1 \rightarrow {}^4T_1$ (G state)
425 nm	$^6A_1 \rightarrow {}^4T_2$ (G state)
395 nm	$^6A_1 \rightarrow {}^4E , {}^4A_1$ (G state)
365 nm	$^6A_1 \rightarrow {}^4T_2$ (D state)
325 nm	$^6A_1 \rightarrow {}^4E$ (D state)

The Mn(II) luminescence corresponds to:

610 nm	$^4T_1 \rightarrow {}^6A_1$ (S state)

As mentioned earlier, Mn(II) is known to substitute for Ca(II), Mg(II), or Zn(II) in a wide variety of mineral systems. Due to the ubiquitous nature of these ions, it is theoretically possible to observe Mn(II) luminescence almost anywhere. Specific instances include the carbonate class (calcites and aragonite groups), the sulfate class (primarily the barite group), the tungstate and molybdate classes (primarily wolframite and scheelite), and the phosphate class (primarily the apatite group). However, the most important class of minerals known for Mn(II)-activated luminescence is the silicate class, where appropriate groups are members of the epidote, pyroxene, amphibole, feldspar, and zeolite groups. Widely varying luminescence spectral characteristics can be observed depending on the exact nature of the Mn(II) sites in the host system.

4.3.   Lanthanide   Ions

The naturally occurring members of the lanthanide series make up a significant class of activator ions in mineral systems. Although members of the lanthanide series are often identified as the "rare earth" ions, this term is misleading since they are not rare. The most

Fig. 9-6. Excitation and emission spectra obtained at room temperature for Mn(II) doped into synthetic fluorite, CaF_2.

common oxidation state for lanthanide ions is trivalent, but several members are known to exist in the divalent state. The luminescence properties associated with the two oxidation states are quite different and require separate discussion. Lanthanide ions have received extensive use as luminescent centers in phosphor systems, and it can be stated that color video screens would be unsatisfactory without the use of Eu(III) activated phosphors [23].

Unlike the Cr(III) and Mn(II) transition metal ion activators, members of the lanthanide series are characterized by the presence of f-electrons in their valence shells. The angular momenta of these electrons can be arranged with a greater degree of freedom relative to d-electrons, and consequently many more electronic states are available to lanthanide ions. For the trivalent state, these same electrons are not effectively involved in chemical bonding. The nonbonding character therefore precludes appreciable crystal field effects, and the luminescence wavelengths of a given trivalent ion are not found to shift appreciably with the identity of the host system.

An energy level diagram illustrating the energy levels known for the trivalent members of the lanthanide series is shown in Fig. 9-7. The absorption transitions among these levels tend to be weak (due to the spin-forbidden character) and sharp (due to the lack of crystal field interactions). The luminescence bands are also extremely sharp, and characterized by long lifetimes (with the exception of Ce(III), whose lifetime is significantly shorter than the other members of the series). Even though the wavelength maxima of absorption and emission transitions do not shift appreciably with the identity of the host system, the intensities of these same bands are quite sensitive toward environmental details [24-26]. It is a general rule that the most intense luminescence is observed from lanthanide ions around the middle of the series, although in certain host systems all ions are capable of exhibiting strong luminescence intensities. An exhaustive compilation of all measured lanthanide ion energy levels in 26 host systems is available [27].

As an example of lanthanide ion luminescence, the excitation and emission spectra of Eu(III) doped into synthetic xenotime (YPO_4) are shown in Fig. 9-8. A number of generalizations may be drawn regarding lanthanide ion spectroscopy using the Eu(III) data as an example, since the overall trends are followed by all trivalent members of the series. The most striking feature associated with lanthanide spectral data is the sharpness of the transitions. Since crystal field effects are of only minor importance to f-electrons, these are not split appreciably by their environment. Coupling of the electronic bands to vibrational modes is also minimal, and consequently little vibronic broadening is noted. Finally, since the f-electrons are not involved in the chemical bonding, their wavelength positions are essentially independent of the host material.

The strongest luminescence feature responsible for the red color in the Eu(III) emission is assigned as the $^5D_0 \rightarrow {}^7F_2$ transition at 615 nm. Other bands are readily observable, and the total summary is:

580 nm	$^5D_0 \rightarrow {}^7F_0$
595 nm	$^5D_0 \rightarrow {}^7F_1$
615 nm	$^5D_0 \rightarrow {}^7F_2$
650 nm	$^5D_0 \rightarrow {}^7F_3$
685 nm	$^5D_0 \rightarrow {}^7F_4$

Although the wavelength maximum of the $^5D_0 \rightarrow {}^7F_2$ Eu(III) luminescence band does not shift with the identity of the host material, its intensity is strongly influenced by the metal ion site symmetry. In octahedral sites, such as those existing in elpasolite (K_2NaAlF_6), luminescence corresponding to the $^5D_0 \rightarrow {}^7F_2$ transition is totally absent and only the $^5D_0 \rightarrow {}^7F_1$ transition can be observed. With a lowering of the site symmetry from octahedral (such as described above for the xenotime system), the intensity of the $^5D_0 \rightarrow {}^7F_2$ transitions becomes sufficiently large so as to dominate the luminescence spectrum.

The spectroscopy associated with lanthanide ions capable of existing in the divalent state is extremely different from that just described for trivalent ions. Here, the lowest excited states are derived from the 4f/5d combinations rather than being pure f-orbital in character. Since the low-lying d-orbitals are much more sensitive to details of the coordinative environment, the resulting excitation and

Fig. 9-7. (Opposite) Energy level diagram for the members of the lanthanide series. The lowest energy state from which luminescence normally originates are identified by the filled, concave down semicircles (taken from reference [24] with permission of the copyright owner).

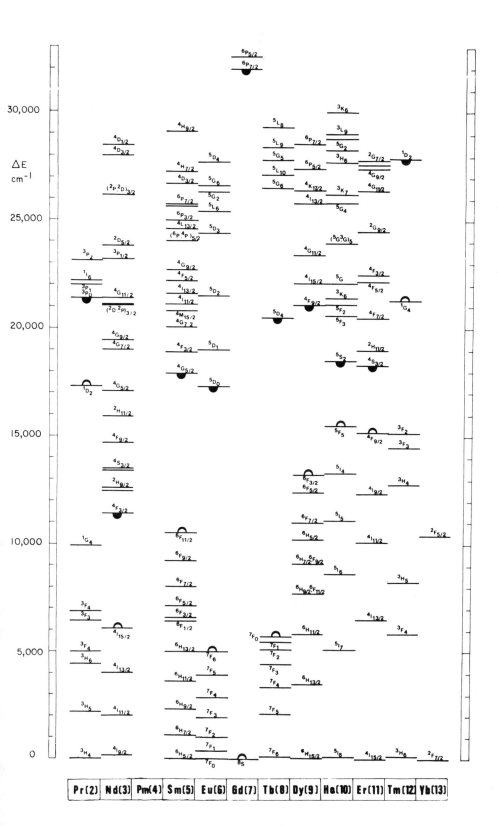

| Pr(2) | Nd(3) | Pm(4) | Sm(5) | Eu(6) | Gd(7) | Tb(8) | Dy(9) | Ho(10) | Er(11) | Tm(12) | Yb(13) |

Fig. 9-8. Excitation (upper trace) and emission (lower trace) spectra
obtained at room temperature for Eu(III) doped into synthetic
xenotime, YPO_4.

emission band systems reflect this sensitivity. Luminescence out of
these levels is broad in character, characterized by much shorter
emission lifetimes, and the observed transition energies vary
considerably with the nature of the host material.

The most important divalent lanthanide ion activator is Eu(II),
which is known to substitute for Ca or Sr in a wide variety of silicates
and aluminosilicates. The efficiency of Eu(II) luminescence in some
of these systems has led to their commercial use as lamp phosphors

[22]. Being isoelectronic with Gd(III), Eu(II) is capable of exhibiting sharp-line, but weak, luminescence corresponding to the $^6P_{7/2} \rightarrow {}^8S_{7/2}$ transition around 310 nm. The strong Eu(II) luminescence corresponds to the transition from an excited $4f^6/5d^1$ state to the ground $^8S_{7/2}$ state, with the emission maxima being strongly affected by the identity of the host material. For example, Isaacs [28] has reported the following Eu(II) band maxima for a series of feldspar materials:

423 nm	$CaAl_2Si_2O_8$
407 nm	$SrAl_2Si_2O_8$
371 nm	$BaAl_2Si_2O_8$

4.4. U(VI)

The science of luminescence spectroscopy actually began with the observations made by Stokes and Becquerel regarding the emission spectra of uranium salts during their studies of radioactive materials [2]. These early observations were extended and systematized by Nichols and Howes [29], but the largest amount of work was performed between 1940 and 1950 during the course of the Manhattan Project. During this time, the fundamental nature of U(VI) luminescence and its associated properties were established.

Although oxidation states of 2+ through 6+ are known for uranium, the only commonly encountered state is U(VI). The U(VI) ion exists primarily as the uranyl ion, UO_2^{2+}, which imparts green, yellow, or orange colors to the mineral systems in which it is found. The same species yields bright green to yellowish-green luminescence upon UV excitation. The luminescence lifetimes are on the short phosphorescence scale, with values between 100 and 500 μs being observed. The uranyl ion is a linear O=U=O unit, which remains intact upon its introduction into a crystal lattice. The two oxygen atoms remain at axial positions, while any ligating groups will enter along the equatorial plane. When oxygen atoms make up the remainder of the coordination environment, it is possible for U(VI) to achieve an octahedral symmetry identified as the uranate ion. This species is generally luminescent only at cryogenic temperatures.

The uranyl ion will bind to a wide variety of species, and is encountered in a large number of minerals such as phosphates, arsenates, vanadates, sulfates, carbonates, and oxides. U(VI) is carried in ground waters as a stable carbonate complex. The uranyl luminescence bands are generally more narrow than those encountered for transition metal ion activators, but not nearly as sharp as bands associated with lanthanide ions. The energy levels of the uranyl ion can only be described with the use of molecular orbital theory, but a complete description of these levels is not yet available.

The excitation and luminescence spectra of uranyl containing minerals are quite similar, and consist of a single pure electronic transition accompanied by a single vibronic progression. The vibrational spacings within these spectra reflect the energies of the totally symmetric uranyl stretching vibrational mode in the ground and excited states. In the ground state this band usually has an energy of 830-860 cm^{-1}, while in the lowest excited state the vibrational frequency decreases to approximately 600-675 cm^{-1}.

As an example of uranyl spectroscopy, the excitation and luminescence spectra obtained for the uranyl ion in autunite, $Ca(UO_2)_2(PO_4)_2 \cdot 10H_2O$, are shown in Fig. 9-9. The excitation spectrum contains a sequence of bands having maxima at 426, 437, 449, 462, 475 and 489 nm. This pattern indicates that the pure electronic absorption transition is located at 489 nm, and that the other bands represent coupling of a 605 cm^{-1} vibration to the electronic state. The luminescence spectrum consists of a sequence of bands having maxima at 490, 512, 535, 561, and 588 nm. This pattern is interpreted as a pure electronic luminescence transition at 490 nm, followed by bands representing coupling of a 860 cm^{-1} vibration to the electronic band.

5.   *Summary*

Luminescence in mineral systems can originate from a variety of factors, but the instances of importance to the surface scientist are associated with the presence of a metal ion activator species. Since

Fig. 9-9. Excitation (upper trace) and emission (lower trace) spectra obtained at room temperature for the uranyl ion in synthetic autunite, $Ca(UO_2)_2(PO_4)_2 \cdot 10H_2O$.

the photophysical processes considered in the present work require the adsorption and subsequent detection of emitted light, it is clear that any observed luminescent phenomena of an opaque mineral must reflect its surface properties. Interpretation of the excitation and emission data can often serve to identify the source of the luminescence, and confirming evidence is available from measurement of the time dependence of the luminescence processes. Quantitative studies of mineral luminescence can therefore be extremely useful when combined with other surface techniques, and multidisciplinary studies clearly represent the best approach toward characterization of mineral surfaces.

References

[1]   Kroger, F.A.   "Aspects of Luminescence of Solids", Elsevier, Amsterdam, 1948.

[2]   Pringsheim, P.   "Fluorescence and Phosphorescence", Interscience, New York, 1949.

[3]   Leverenz, H.W.   "Introduction to Luminescence of Solids", John Wiley, New York, 1950.

[4]   Curie, D. "Luminescence in Crystals", Methuen, London, 1963.

[5]   Goldberg, P., Ed.   "Luminescence of Inorganic Solids", Academic Press, New York, 1966.

[6]   Marfunin, A.S. "Spectroscopy, Luminescence, and Radiation Centers in Minerals", (transl. V.V. Schiffer), Springer-Verlag, Berlin, 1979.

[7]   Walker, G.   In "Chemical Bonding and Spectroscopy in Mineral Chemistry", Berry, F.J., and Vaughan, D.J., Eds., Chapman and Hall, London, 1985, chapter 4.

[8]   Calvert, J.G.; Pitts, J.N.   "Photochemistry", John Wiley, New York, 1966.

[9]   Parker, C.A.   "Photoluminescence of Solutions", Elsevier, Amsterdam, 1968.

[10]   Guibault, G.G. "Practical Fluorescence: Theory, Methods, and Techniques", Marcel Dekker, New York, 1973.

[11]   Schulman, S.G.   "Fluorescence and Phosphorescence Spectroscopy: Physicochemical Principles and Practice", Pergamon Press, Oxford, 1977.

[12]   Lakowicz, J.R.   "Principles of Fluorescence Spectroscopy", Plenum Press, New York, 1983.

[13]   Ballhausen, C.J.   "Introduction to Ligand Field Theory", McGraw Hill, New York, 1962.

[14]   Griffith, J.S.   "The Theory of Transition Metal Ions", Cambridge Press, London, 1964.

[15]   Figgis, B.N. "Introduction to Ligand Fields", Interscience, New York, 1966.

[16]   Jablonski, A. Z. Phys., *1935*, 94, 38.

[17]   Macfarlane, R.M.   J. Chem. Phys., *1963*, 39, 3118.

[18]   Imbusch, G.F.; Yen, W.M.; Schawlow, A.L.; Devlin, D.E.; Remeika, J.P.   Phys. Rev., *1964*,   136A, 481.

[19]   Dana, E.S.; Hurlbut, C.S.   "Manual of Mineralogy", John Wiley, New York, 1941.

[20]   Kraus, E.H.; Hunt, W.F.; Ramsdell, L.S.   "Mineralogy", McGraw-Hill, New York, 1959.

[21]   Geschwind, S.; Kisliuk, P.; Klein, M.P.; Remeika, J.P.; Wood, D.L. Phys. Rev., *1962*,   126, 1685.

[22]   Butler, K.H.   "Fluorescent Lamp Phosphors: Technology and Theory", Pennsylvania State University Press, University Park, PA, 1980.

[23]   Brittain, H.G.   In "Molecular Luminescence Spectroscopy: Methods and Applications - Part II", Schulman, S.J., Ed., John Wiley, New York, 1988, chapter 5.

[24]   Wybourne, B.G.   "Spectroscopic Properties of Rare Earths", Wiley-Interscience, New York, 1965.

[25]   Crosswhite, H.M.; Moos, H.W.   "Optical Properties of Ions in Crystals", Wiley-Interscience, New York, 1967.

[26]   Dieke, G.H.   "Spectra and Energy Levels of Rare Earth Ions in Crystals", Wiley-Interscience, New York, 1968.

[27]   Morrison, C.A.; Leavitt, R.P.   In "Handbook on the Physics and Chemistry of Rare Earths", Gschneider, K.A.; Eyring, L., Eds., Vol. 5; North-Holland Publishing Co., Amsterdam, 1982, chapter 6.

[28]   Isaacs, T.J.   J. Electrochem. Soc., *1971*,   118, 1009.

[29]   Nichols, E.; Howes, H.   "Fluorescence of the Uranyl Salts", Carnegie Inst. Press, Washington, D.C., 1919.

10. SECONDARY ION MASS SPECTROMETRY

James B. Metson
Department of Chemistry
University of Auckland
Auckland, New Zealand

1.Introduction

Secondary ion mass spectrometry (SIMS) is based on the analysis of ions ejected during ion (or other heavy particle) bombardment of a surface. These secondary ions, elemental and molecular, constitute a relative small proportion, typically <1%, of the material ejected during bombardment. Their composition is representative of, although not directly proportional to, the composition of the surface. Mass spectrometric analysis of secondary ions thus gives an analysis of the surface. The method also provides the potential for isotopic resolution, and because of essentially zero backgrounds, provides excellent signal-to-noise characteristics, resulting in a dynamic range extending to seven orders of magnitude or more, and parts per billion sensitivity for favourable elements. The ability to focus the primary ion beam, or use spatially resolved secondary ion optics, allows submicron spatial resolution in imaging and analysis.

Since the appearance of commercial instrumentation in the early 1970s, SIMS has rapidly become indispensable to the semiconductors industry, in particular, and in materials science in general. Contributions in geochemistry, although now substantial, have been less spectacular. The technique, which Lovering [1] suggested would prove the "ultimate geochemical weapon," has as yet fallen well short of this potential. Geologic applications occupy but three pages of the applications section of Benninghoven, Werner, and Rudenauer's 1200-page volume on SIMS [2]. This rather understates the case, in view of the very substantial contributions SIMS has made in areas such as the analysis of lunar samples, identification of isotopic anomalies in meteorites, self- and tracer diffusion studies, and in trace element analysis in general.

The SIMS literature of geologic applications extends to some 160 papers between 1970 and 1987 [3], 110 of these published since 1980 (Fig.10-1). In view of this rapid growth, it is worth considering how the method has developed in geochemical use, relative to early expectations, and where its future lies.

For a technique that has a relatively short commercial history and requires expensive instrumentation, the potential of SIMS in geology was rapidly recognized. The massive sensitivity advantages of SIMS over electron microprobe analysis (EMPA) attracted geochemists, at a time

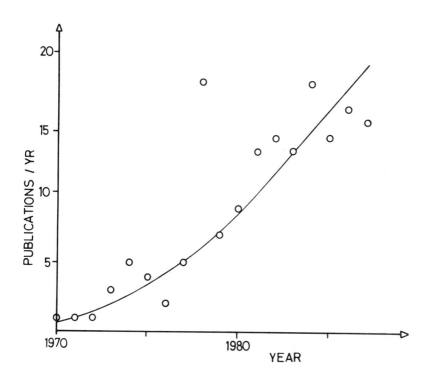

Fig.10-1. SIMS publications in geochemistry since 1970, with the assistance of the compilation of S.J.B.Reed [3].

when the sensitivity limits of the electron probe hindered progress in the solution of a number of geochemical problems, especially where trace element distribution was critical information. The parallels with the development of the electron probe were remarkable, right down to the role of several of the central figures, notably Castaing [4], in both.

The development of SIMS has if anything, suffered through repeated comparison to the evolution of the electron microprobe. There are complexities inherent in the nature of secondary ion production, which make it difficult to devise a universal protocol for matrix correction on a similar basis to that which has evolved in x-ray microanalysis. Asking SIMS to duplicate the role of EMPA with more sensitivity may eventually be routine; currently it is not. It must be recognized that although both are probe-type analytical methods, the two analyses are essentially different.

If we consider the analysed volume in each case, using an ion beam focused to a modest 10-μm diameter, the SIMS analysis volume is conservatively 10^4 times smaller. With a detection limit between 10^2 and 10^4 times smaller (depending on the element), the true power of the method can be appreciated. Of more significance to this work, SIMS is a surface analytical method. An ion can escape the surface only if ejected from the upper two or three atomic layers of the surface.

However, the volume element analyzed by SIMS is progressively removed as analysis proceeds. Thus SIMS is a destructive technique, a feature that complicates bulk chemical analysis but can be profitably exploited in depth profiling element distributions [5]. Furthermore, ions from the primary beam are implanted into (and subsequently sputtered from) the surface. The net effect of this sputtering process is to modify the target matrix until a steady-state condition is reached. Unfortunately ion yields are often acutely sensitive to this matrix composition, making the conversion of observed intensities into surface concentrations difficult.

The complexities in applying the technique quantitatively have distracted attention from the substantial capabilities of SIMS as a surface analysis method, an isotope determination method, and a very high sensitivity trace element analysis/imaging method. Lovering's confidence was well placed, although "ultimate weapon" status is some way from being realized. To understand why, it is necessary to consider present constraints on both instrument performance, and our understanding of the mechanism of secondary ion production, as it applies to mineral surfaces. This constitutes the first section of this chapter, and is followed by a review of the current use of the technique in the surface characterization of geologic materials.

2. Instrumentation and Methods

For a comprehensive history of the evolution of SIMS instrumentation and the range of configurations currently in use, the reader is referred to books and reviews by Benninghoven et al. [2], Liebl [6], and Honig [7]. The essential elements of a SIMS instrument are shown in Fig.10-2:

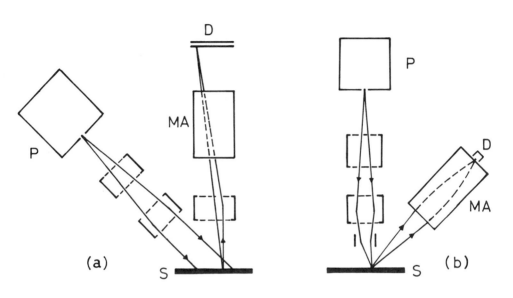

Fig.10-2. Components of (a) an ion-microscope and (b) an ion-microprobe. P primary ion source; S sample; MA mass analyzer D secondary ion detector.

1. Primary Ion Source.   Ion beams are usually generated by electron bombardment, either in an oscillator-type source, an extractor source, or a cold cathode plasma source. The alternative in common usage is the cesium surface ionization source. Most significant of these for geochemical analysis are undoubtedly the plasma-type sources, because of their ability to produce reactive gas, in addition to inert gas, ion beams. A typical duoplasmatron can produce a beam of O_2^+ (uA range), or lower currents of O^- [8,9]. The provision of a negatively charged primary beam is important in the analysis of insulating samples, such as most minerals, because of the capability for internal charge compensation [10,11]. Such charge compensation considerations have virtually prevented the use of Cs^+ sources for mineral analysis.

Liquid metal ion sources, in which a beam of metal ions is extracted from the wetting of a needle tip by a liquid metal, are also widely used, primarily as high brightness sources for probe type imaging [12].

2. <u>Secondary Ion Analysis.</u>   Secondary ions emitted from the sample surface are mass-analyzed, by either a quadrupole mass analyzer, a full magnetic sector spectrometer, or a time of flight (TOF) spectrometer. TOF has been used extensively in polymer analysis, where large structural units are examined [13], but as yet has found little use in geochemistry. Magnetic sector instruments have made the major contribution to geochemical SIMS, the reverse of the situation for organic SIMS, where quadrupoles have been dominant. The reasons for this are discussed in Section 3.3.

Fig.10-3. Layout of the Cameca IMS3f ion microscope/ microprobe.

The Applied Research Laboratories Ion Microprobe Mass Analyzer (IMMA) [14] is an example of a powerful, early SIMS instrument and was the basis of much of Lovering's comment. Of increasing significance in the current literature are instruments based on the ion microscope concepts proposed by Castaing and Slodzian [15] in the early 1960s, and manufactured over several generations by CAMECA. A transfer optical system in the secondary column allows the spatial integrity of the secondary ions to be retained throughout their flight path (Fig.10-3), thus, a magnified, mass resolved image of the sample can be obtained (see Imaging, Section 4.6). The alternative is to operate the SIMS as a scanning ion microprobe, analogous to EMPA [6]. The microprobe mode, particularly when coupled with dynamic emittance matching in the secondary column [16], provides exceptional sensitivity and spatial resolution.

3. The Emission and Analysis of Secondary Ions.

Difficulties in the quantitative interpretation of SIMS spectra arise primarily from the complexity of the emitted spectrum. Ion yields, that is, the ratio of ions to neutrals for a given isotope, in addition to simple intensities, are sensitive to the composition of the lattice from which they are sputtered. This makes application of ionization models rather difficult. To understand the process, two stages must be distinguished in the emission of a secondary ion: the initial primary ion impact, and the subsequent escape of a secondary ion.

3.1. Primary Ion Impact

Primary ions of 1-30 kV are used in SIMS analysis. These span the regime between predominantly scattering processes at low energy and implantation at higher energies. The interaction of such ions with a surface may be modeled as a collision cascade in which the ion is implanted into the surface, transferring energy and momentum to a series of collision partners in the solid [22,23]. During collision and recoil among lattice atoms, sufficient momentum can be transferred to a surface, or near-surface atom, for it to overcome surface binding energies and be ejected into the vacuum. An important consequence of the cascade is that the ejected (secondary) ions usually do not contact the primary ion but emerge as a result of nth order collisions (n = 3 - 7). Secondaries thus have relatively low energies, typically peaking at 5-15 eV. The surface is disordered during the sputtering event [24], making the bulk crystal band structure an unreliable model on which to base interpretation of the ionization process, particularly for insulators. Sigmund's theory [22,23] is largely successful in predicting the total sputtering yield and energy distribution of particles ejected from the surface but makes no prediction of their charge states. Thus the critical factor in quantitation is the ion yield R^+_{ij} :

$$R^+_{ij} \quad = \quad \frac{N^+_{ij}}{N^o_{ij}} \qquad\qquad (10\text{-}1)$$

where N^+_{ij} is the number of positive ions of isotope i of element j, whereas N^o_{ij} is the total sputtered yield of isotope i given by Sigmund's theory. The significance of R_{ij} in quantitative analysis will be discussed later. The lifetime of a single cascade event is 10^{-11} to 10^{-13} s and its dimensions extend 10 nm (this in fact defines the ultimate lateral resolution of the technique). The result of this relatively short lifetime, is that for typical beam densities used in SIMS, cascades are discrete events. It can also be shown [25,26] that a relatively small percentage of incoming trajectories yield the majority of the secondary ions. Thus the intensities observed are the instrument weighted (by collection and detection efficiency) averages of the ions produced in an extended series of discrete events.

Sputtering yield depends on how rapidly the incident particle can dissipate its energy in the sample surface, i.e., the particle range in the lattice. Range is a function of variables such as the mass, energy, and angle of incidence of the incident particle. The effects of channelling for example can be severe for single crystals [27], but are generally not observed for randomly oriented, polycrystalline mineral targets. The ionized portion of this total sputtered yield is then ejected into the vacuum with a characteristic angular distribution [28,29]. The portion collected by the mass spectrometer is the analyzed population and is considered below.

3.2. Secondary Ion Emission

As a result of the collision cascade, some surface atoms gain enough energy to escape from the surface. A small proportion of these, typically 1%, escape as ions and this proportion is extremely sensitive to two variables, the ionization energy (IE), and the chemical environment at the leaving site. Positive ion yields, for example, vary over at least four orders of magnitude through the periodic table and are strongly enhanced (by several orders of magnitude [30]) in the presence of a strongly electronegative element such as oxygen. The inverse is true of negative ion emission, where electropositive Cs is typically used to enhance yields.

Whether oxygen is present through implanting oxygen primaries into the surface, leaked onto the surface via a gas jet [31], or simply present already in the target lattice, is immaterial, provided a local "saturation level", and thus signal stability is reached. Maul and Wittmaack [32] demonstrated similar Si^+ yields from quartz, an anodically grown oxide on Si and an oxygen flooded Si surface. Oxide and silicate minerals and even sulfides, are in a favorable position to achieve this "saturation", particularly if an oxygen primary beam is used. Thus we anticipate and observe, sub-ppm sensitivity in analyzing positive ions of low IP elements sputtered from these minerals, using an oxygen primary beam [33].

A further potential complication is that of preferential sputtering. Numerous studies, both theoretical and experimental [34], have shown that a variety of factors, including both mass and bond strength, affect the rate at which an element is sputtered. This leads to an initial surface enrichment in one or more components in the matrix. The process clearly cannot be maintained, as surface enrichment must lead to higher intensities of the enriched species until a steady state is reached. It is apparent that for accurate quantitative analysis it is important that the steady state has been reached. The problem is not acute in most spheres of mineral analysis, but in mixed organic /inorganic matrices such as coal specimens, preferential sputtering problems would be anticipated [35].

An important point to consider is the ion yield factor Rij and how this is influenced by the matrix from which the secondary particle is ejected. This "matrix effect" is central to the most applications of SIMS in geochemistry, because minerals have an enormous diversity of structure

and composition. The former need not be significant in this context
[36,37], but the latter is often critical. This is reflected in a range of ion
yields for a given element from different matrices. Steele et al. [38]
confronted such variations in Mg and Si yields when attempting the
quantitative analysis of olivines and pyroxenes, (Fig.10-4).

Fig.10-4. Intensity vs. forsterite content for a series of olivines. o = Si, ● = Mg,Δ = Fe.
From Steele et al. [38], with permission, copyright 1981, Mineralogical Society of
America.

Shimizu [39] examined the same problem in secondary ion emission
of Mg, Ca, Al, and Si from silicate glasses. It is noted that both Al and Ca
yields are enhanced by increasing Si content (Fig.10-5).

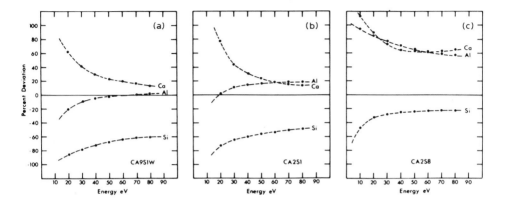

Fig.10-5. Chemical fractionation of secondary ion populations as a function of energy, for low (CA9S1), intermediate (CA2S1), and high (CA2S8) Si glasses. From Shimizu [39(b)], used with permission of the author and Elsevier Science Publishers.

Shimizu considers the collisional reaction:

$$MO + Si^+ \rightleftarrows SiO + M^+ \tag{10-2}$$

Increasing Si levels in the sputtered zone drive the reaction to the right, enhancing the yield of M^+, while the energy distribution of M^+ retains some of the characteristic shape of the MO distribution (a narrow peak shifted toward lower energy Indeed, such a change in energy distributions is observed with yield enhancement [39]. The analysis of high energy secondaries, Shimizu uses 80 ± 10 eV, removes these yield enhancements, presumably due to the decreased residence time of the leaving ion in the near-surface zone where such reactions can take place.

Havette and Slodzian [40,41] developed a more expansive bond breaking model to explain matrix effects and considered positive ion yields as a linear function of all cationic concentrations in the lattice. Thus ionization is viewed as a competitive process where bond strengths and ionicity in the lattice will determine which species escape as ions and which as molecules. Unfortunately such a model is difficult to express in a quantitative form. Havette and Slodzian derived a series of coefficients based on calibration using known standards.

$$Q_\alpha(M_j) = \frac{I(M_j)}{I(M_i)} \times \frac{C_{\alpha i}}{C_{\alpha j}} = \frac{\tau(M_j)}{\tau(M_i)} \qquad (10\text{-}3)$$

where M_i is a reference element (usually Si in silicates). Mj is the cation of interest and has concentration C_{α}j. $\tau(Mj)$ and $\tau(Mi)$ are the respective ionization yields of element j and the reference element. Using such coefficients, Havette and Slodzian have produced tables of relative yields that allow elemental analysis with accuracy of 5% or better.

It must be pointed out that models consistent with a bond breaking approach to secondary ion formation, although satisfactory for analysis of oxygen-containing mineral lattices, still rely on exhaustive calibration. They do not constitute a universal model for secondary ion emission. For example the early local thermal equilibrium (LTE) model of Anderson and Hinthorne [42], which considers the sputtered region as a low temperature plasma in local thermal equilibrium, although conceptually and experimentally flawed when applied to minerals [37], is probably satisfactory for negative ion emission from metals [43]. Yu [44] discusses theories of ion emission, including a comparison of bond breaking and electron tunnelling mechanisms. However a universal model for sputtered ion emission remains elusive [24].

This leaves the immediate problem of adopting a workable approach to quantitative analysis using SIMS while such developments are awaited. Current strategies are discussed in Section 4.5.

3.3. Collection and Analysis

Secondary ions emitted from a surface constitute a minority of the total sputtered particle yield. Furthermore, singly charged monatomic species often constitute a minority of the total ion yield. However in most geochemical applications, there are many elements present and singly charged elemental ions constitute the target population for analysis. Thus it is necessary to resolve elemental ions from isobaric molecular species, dimers, multiply charged ions etc. This can be a severe limitation, quite independent of the complexity of then assessing the ion yield of the particular element/isotope analyzed, and thus converting observed intensity into elemental or isotopic concentration.

Two methods are effective in discriminating against molecular species and both have been widely employed in geochemical application.

1. <u>High Mass Resolution.</u> The least ambiguous method of discrimination is to simply narrow the entrance and exit slits of the mass spectrometer to allow only the exact mass of the target ion to be transmitted. On the basis of their varying mass defects, molecular species are separated from the ion of interest. Pimminger et al. [18] used a mass resolution $M/\Delta M = 6500$ to resolve $^{32}SH^-$ and $^{16}O_2H^-$ from $^{33}S^-$, in a determination of sulfur isotopes in

Fig.10-6. High mass resolution spectrum of the sulfur isotopes in a galena sample. From Pimminger et al. [18], reprinted with permission from Analytical Chemistry <u>56</u>, copyright 1984, American Chemical Society.

galena (Fig.10-6), and there have been many other applications using mass resolution in both isotope (Section 4.4), and elemental analysis. The limited mass defect in many   cases (Fig.10-7a), together with the inevitable loss in sensitivity in narrowing the mass (and energy) window (Fig.10-7b), limits the application of mass resolution. Rare earth element (REE) analysis is a case in point. A mass resolution of >10,000 is necessary to resolve light REE oxides from heavy REEs [45], enough to severely test current instrumental limits (see Section 4.3).

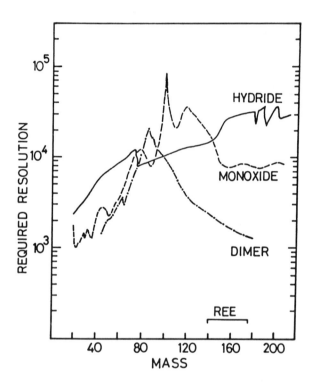

Fig.10-7.(a) Required mass resolution to resolve hydrides, monoxides and dimers. From Shimizu and Hart [45].

2.   Energy Filtering.   The alternative route to a spectrum free of molecular ions is to analyze only ions that leave the surface with a high forward velocity i.e. elevated kinetic energy. MO^+ will have a narrower

velocity (energy) distribution than M^+, with a peak at lower energy. Thus, selecting an energy window away from the low-energy intensity peak effectively discriminates against molecular ions. In a double focusing sector instrument (Fig.10-3), this requires an offset of the secondary accelerating voltage with respect to the pass energy of the mass spectrometer.

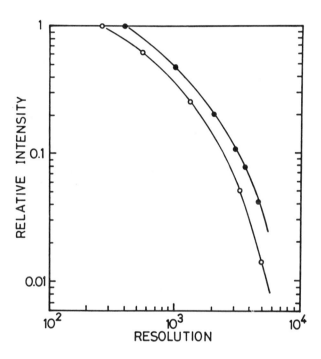

Fig.10-7.(b) Intensity vs. mass resolution, measured on an AEI ion microprobe (o) and a Cameca IMS3f (o). From Shimizu and Hart [45]. Used with permission of the authors and Annual Reviews Inc.

Again energy filtering results in loss of intensity, and thus sensitivity. Typically two to three orders of magnitude are sacrificed in offsetting the energy window by 80-100 eV. This level of offset is not always sufficient to eliminate molecular ions; for example, rare earth oxides are still significant at this offset [46].

Shimizu and Hart [45] and Reed [47] have analyzed the relationship between high mass resolution and energy filtering. At both low atomic number region, and at high atomic number, i.e., Z>140, hydrides are the most difficult to resolve, whereas the monoxides and dimers require highest mass resolution between 70 and 140, (Fig.10-7).

The conclusion from this diagram is that high mass resolution is favoured at low Z, while energy filtering is more effective at mass >70. However, given an instrument designed specifically for high mass resolution, the Z restriction is not a limitation. An example is the super high resolution ion microprobe SHRIMP at Australian National University (ANU). This system was designed to allow mass resolution up to 30,000 while retaining reasonable transmission achieved through use of a large magnetic sector, with a relatively low angle of deflection [17,48]. The instrument has been used to date zircons from a number of locations (see Section 4.5) as well as to identify a variety of isotopic disequilibrium effects which complicate isotopic dating of these samples [49-52].

At the other extreme, a form of energy filtering known as the specimen isolation method has been developed by Metson et al. [19,53]. The principle of the technique is to utilize an extreme energy offset (500-600 eV) to eliminate molecular species, coupled with a broad energy window in order to compensate for the accompanying intensity loss.

If a large positive extraction field is applied between sample and extraction optics, a negatively charged primary beam will cause the surface of an insulating sample to charge negatively with respect to the applied extraction voltage [10]. The extraction voltage is applied to a cover plate, with holes through which the sample is exposed. Charging continues until the potential difference between the sample surface and the surrounding aperture is sufficient to draw secondary electrons to the plate. Charging is therefore stabilized, at a fixed energy offset, determined by the size of the aperture in the plate. Offsets greater than 500 eV are readily obtained by this method, sufficient to remove even the most troublesome of molecular species. Figure 10-8 illustrates the relationship between energy offset and intensity for a typical elemental species and its oxide. The typical level of molecular ion suppression achieved is illustrated in figure 10-9, in a mass spectrum of a titanite specimen obtained under SI conditions. In geochemical use, rare earth element (REE) analysis has been the most successful application of the method [56,57] with detection limits of a few

10s of ppbs. Examples of REE patterns determined by this method are shown later in Fig.10-13. An alternative method of REE analysis has been developed by Zinner and Crozaz [46,58]. A combination of energy resolution (sufficient to remove all but the interfering oxides), together with mathematical deconvolution of peak overlaps, provides almost as much sensitivity, without sacrificing spatial resolution.

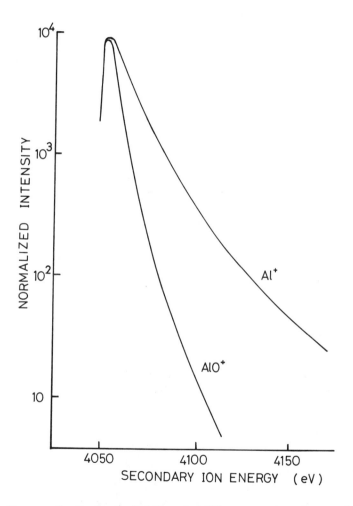

Fig.10-8. Energy distributions of Al^+ and AlO^+ measured under specimen isolated conditions. The peak at 4060eV indicates an offset of 440eV relative to the 4500eV secondary accelerating voltage applied to the sample holder.

It can be seen that a versatile SIMS instrument for geochemical application must be capable of both high mass resolution and energy filtering. The combination of mass resolution in excess of 4000 and high transmission requires a magnetic sector instrument. The ability to energy filter by 100 eV or more is also best achieved by an instrument with a large secondary accelerating field. This is difficult to achieve with a quadrupole instrument because of the very low acceptance energy [21]. Sector instruments offer the geochemist a number of major advantages in both flexibility and sensitivity and in fact dominate the SIMS literature of geochemical applications.

4. Applications

Although there are shortcomings in the theoretical understanding of secondary ion emission, not all applications need such an understanding and the method has been widely and successfully used. Despite expanding at rather less than the exponential rate of total SIMS publications [59], applications in geochemistry are growing rapidly in both number and diversity. In the following section they are arbitrarily divided into those applications where the surface sensitivity of SIMS is paramount and those where the absolute sensitivity of the method is exploited.

Two regimes of operation must be distinguished. Under "static SIMS" conditions, beam densities of up to 10^{-8} A/cm^2 are utilized, giving surface monolayer survival times of some hours. Under "dynamic" conditions, focused beams with densities up to several A/cm^2 are utilized and rapid sample erosion in the beam impact zone results. Because the number of ions detected is loosely proportional to the total sputtered population at steady state, dynamic conditions are much favored for general elemental analysis. Static conditions are more surface sensitive and have been used to analyze mineral surfaces, after cleavage in high vacuum (see Section 4.1).

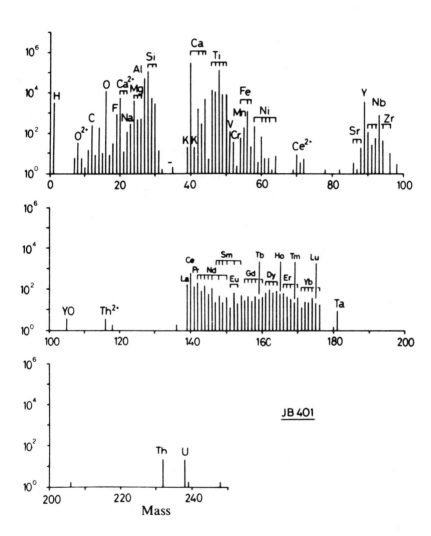

Fig.10-9. Secondary ion mass spectrum of a titanite specimen from Cardiff Mine, Ontario [57].

4.1. Surface analysis

Despite the extreme surface sensitivity of the method, there are relatively few applications where surface sensitivity under static conditions has been exploited in mineral analysis. Dowsett, King, and Parker [60] analyzed a muscovite mica cleaved under UHV conditions using beam densities of $<5 \times 10^{14}$ ions cm^{-2}, i.e., static conditions. The mass spectrum observed was consistent with cleavage in the potassium plane; however minor peaks associated with $^{61}Al(OH)_2^-$ and $^{62}AlOF^-$ were also present. These are attributed to the octahedral layer buried some 0.5 nm below the cleavage plane. The static SIMS spectrum is particularly useful for defining the range of isomorphous replacement for potassium and aluminium, and the presence of Cl^-/F^- substituting for OH^-.

Buhl and Preisinger [61] examined the correlation between surface structure and the SIMS molecular ion spectrum, using single crystal ZnS and CaF_2. The sputtering of structural units is identified, for example, $(Ca_3OF_3)^+$ from the oxidized $\{111\}$ face of CaF_2. Buhl and Preisinger used relatively low-energy primaries (<3 kV) under static conditions to obtain such spectra. The examination of surfaces under static conditions requires considerable experimental care and must usually be combined with fracture under UHV conditions. The method, although useful in identifying structural relationships, provides no information on the depth or spatial distribution of an element; thus profiling under dynamic conditions has been much more widely used in geochemistry.

4.2. Depth Profiling

There are now many examples of SIMS depth profiles in weathering and diffusion studies. A number of experimental and theoretical difficulties are apparent with this type of work, including charging when using high beam currents and changing relative ion yields in inhomogeneous samples. In any depth profile care must be taken to ensure steady-state sputtering conditions and to minimize crater edge effects, i.e., collecting ions only from the base of the sputtered crater [5]. Rough surfaces also introduce primary beam shadowing problems, which tend to "smear out" the true profile.

Beusen and Gijbels [62] examined the effects of hydrothermal treatment on a feldspar, two pyroxenes and a mica, using SIMS depth profiles. Three different primary beams (Ar^+, O_2^+, O^-) were used, with two coated (30-40 nm gold) and one uncoated specimens. Implantation of primary ions modified the matrix of the specimen being sputtered, and thus ion yields also varied (see Section 3.2). The depth profiles obtained reveal remarkable heterogeneity in the feldspar and one of the pyroxenes (bronzite) even before hydrothermal treatment. Bronzite depth profiles, normalized to ^{28}Si as an internal standard, are reproduced in Fig.10-10. Depth profiles after aggressive hydrothermal treatment of an augite phase allowed the behavior of up to eight elements to be monitored. Some were depleted by over two orders of magnitude at the surface. Beusen and Gibjels also exploited the mass spectrometric advantages of the method by using both D_2O and $H_2^{18}O$ to study the role of H and O in surface reactions during treatment. This highlights one of the new dimensions SIMS has added to geochemical analysis: the ability to study H and D directly.

SIMS surface analysis and depth profiling have been somewhat sparingly used in examining the leaching of mineral surfaces, but the information available is extremely significant. Nesbitt and Muir [63] report depth profiles under specimen isolated conditions [19] of naturally weathered plagioclase and observe strong depletion of Si at the surface, leaving an Al-rich outer layer. However Muir, Bancroft, and Nesbitt [64] also report SIMS profiles of labradorite samples leached under laboratory conditions, where Al, Ca, and Na are selectively depleted and a Si-rich layer is observed in the top few hundred angstroms. Both results are confirmed by X-ray photoelectron spectroscopy and highlight the difficulties associated with laboratory simulation of natural weathering processes, in the absence of influences such as the complexing action of humic acids.

In these studies and in several others [65,66] errant behavior of Na is noted. Where any degree of differential charging occurs during ion beam bombardment, Na is quite mobile and care must be taken in interpreting Na depth profiles.

Bancroft et al. [67], Hayward and Checetto [68], and Myhra et al. [69] have examined the leaching behavior of titanites (sphenes) and glasses of titanite composition as potential radioactive waste hosts. SIMS depth profiles provide information on the mechanism of leaching. Leaching in simulated groundwaters is so slow as to be difficult to detect by practically any other method. Titanite crystals appear to resist the progress of leaching by the formation of a Ti-rich external skin formed by selective removal of Ca and Si.

Fig.10-10. Depth profiles showing surface alteration of a bronzite specimen. From Beusen and Gijbels [62] used with permission, copyright Theophrastus Publications SA.

SIMS depth profile measurements are clearly an elegant route to the determination of diffusion constants and this route has been used by several groups. Hofmann et al. [70] examined potassium exchange between a biotite and a ^{41}K enriched hydrothermal alkali chloride solution. $^{39}K/^{41}K$ ratios were measured as a function of depth in the center of an exposed face and locations away from edges of the crystal. As anticipated, much higher (two to four orders of magnitude) diffusion

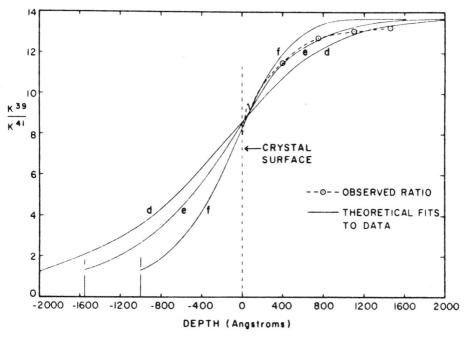

Fig.10-11. $^{39}K/^{41}K$ ratios as a function of depth from the surface of a biotite flake. $^{39}K/^{41}K$ was 1.3 throughout the runs. d, e, and f, correspond to values of 2.17, 1.26, and 0.544 $(x10^{-17}$ cm^2/s) for D. From [70].

exchange rates are measured in the ab plane than down the c axis (Fig.10-11) where the close packed oxygen layers have to be penetrated. A diffusion coefficient of 1 x 10^{-17} cm^2 /s in the c direction, calculated from the SIMS work, agreed well with bulk isotopic transport measurements in the same system carried out previously [71]. Sneeringer et al. [72] carried out a diffusion experiment in the solid state. A synthetic diopside containing no strontium, was pressed into contact with a ppm-level strontium-doped crystal and heated to induce diffusion. The depth profile of strontium in the initially undoped diopside was then used to calculate a diffusion coefficient for diffusion parallel to the c axis (into the undoped crystal). A diffusion coefficient, corrected for knock-in and crater edge effects, of $D_{sr} = 2.6 \times 10^{-15}$ cm^2/s is reported at T = 1250°C.

The major advantage in SIMS measurement of diffusion profiles in such systems is that matrix effects on ion yields are minimal, or more accurately, constant. Careful choice of experimental conditions will minimize crater edge effects, while knock-in effects can be corrected for by profiling a thin film deposited on the surface of the mineral, and observing tailing into the substrate [72]. Knock-in effects arise from ion beam mixing in the initial collision cascade and cannot be eliminated under dynamic sputtering conditions.

SIMS studies of diffusion in silicate melts, critical in elucidating their structure, have been undertaken by Shimizu and Kushiro [73] (self-diffusion of Si and O), Karsten et al. [74] (water), Cunningham et al. [75] (lithium), and Delaney and Karsten [76] (water). Delaney and Karsten quantitatively mapped water content in obsidian samples from a series of hydration experiments, concluding that diffusion coefficients of water in rhyolite melts can vary depending on water content, from 8×10^{-9} cm^2/s at 0.2% H_2O to 1.2×10^{-6} cm^2/s at 3.7% H_2O.

4.3. Trace Element Analysis

Quite distinct from its capabilities as a surface analysis method, SIMS offers the high signal-to-noise advantages of mass spectrometric detection. Thus the SIMS spectrum is characterized by dramatic dynamic range, often $>10^9$ counts per second between the most intense and least intense signal detected. This makes it an extremely sensitive trace element analysis method with the obvious bonus of isotopic resolution.

Trace element and isotopic detection will ultimately be treated separately here, because although they share common demands in both instrumentation and technique, there are significant differences in the role of the matrix from which the analyzed ions are sputtered.

Early trace element applications centered on lunar materials from the Apollo missions. Lunar minerals [77], glasses [78], soils [79], and even the effects of implantation and sputtering by solar winds [80] have been examined. Trace element distribution in meteorites has been another fruitful area of investigation [81,82]. Studies of crystal/melt partitioning are critical in elucidating crystallization processes and have been carried out using SIMS analysis, by a number of groups. High sensitivity is essential in such studies, as it allows trace element partitioning to be studied at natural abundance levels, well within the limits where Henry's law is applicable. In pyroxene crystallization, SIMS analysis [83], showed equilibration was generally reached in ⟨ 4 days at 1275°C, confirmed Henry's law behavior for several trace elements and showed some preference for the Bottinger-Weill melt model [83].

On the other hand, trace element zoning can be used to identify disequilibrium crystallization and a variety of SIMS studies of such zoning have been reported. Meyer et al. [84] describe K, Sr, and Ba zoning in plagioclase from a lunar basalt, while Shimizu [85] examined the effects of such factors as diffusive transport rates on trace element partitioning in a zoned augite phenocryst.

Sulfide mineralization has also received attention, particularly in examining the distribution of trace metals. Dissolved silver and indium were examined by McIntyre et al. [86] (this group has also studied gold distribution in sulfides). Porphyry copper mineralization in the Koloula Igneous Complex was studied by Hendry et al. [87].

A significant advantage of SIMS as a trace and major element method is coverage of virtually the entire periodic table. Thus light elements not analyzable by EMPA are readily accessible. Wilson and Long [88] examined the distribution of Li in Cornish granites. Hutcheon et al. [89] determined its distribution in meteorite inclusions, extended to include Be and B by Phinney et al. [90]. Hervig et al. [91] studied Li and B in topaz, while Dutrow et al. [92] examined Li in staurolite. The ability to determine hydrogen has also been exploited in light element studies of metamorphic amphibole, scapolite, and pyroxene [93] and staurolite [94].

However the rare earths probably provide the most impressive example of SIMS sensitivity in trace element analysis. As described in Section 3.3, rare earth analysis has been successfully demonstrated using both specimen isolated and more conventional energy filtering methods. REE analysis by SIMS has been reported for pyroxenes and plagioclases

[56] Fig 10-12, titanites and other accessory minerals [57,95,19], allanites [96], and chondritic merrillite and apatite [97]. MacRae and Wu [98] compared inductively coupled plasma mass spectrometry (ICP-MS), SIMS, and instrumental neutron activation analysis (INAA), as analytical techniques for REE determination. They conclude that

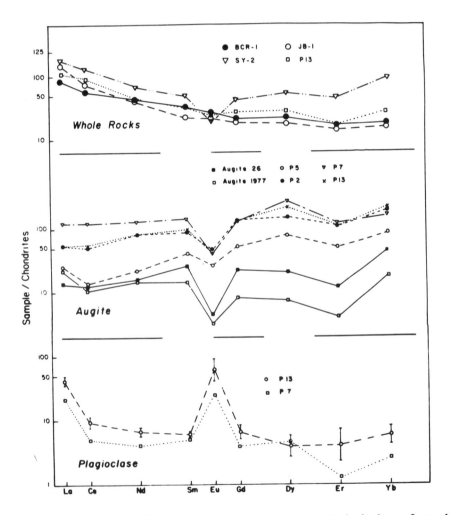

Fig.10-12. REE patterns from whole rocks, pyroxenes and plagioclases from the Pallisade sill measured under specimen isolated conditions. From MacRae et al. [56].

ICP-MS and SIMS are the preferred methods, with comparable sensitivities in the low ppb range. SIMS under SI conditions is somewhat more convenient in terms of the absence of interferences, but ICP-MS offers slightly better precision.

4.4. Isotopic Analysis

From the early studies of ^{26}Mg anomalies in the Allende meteorite to the recording by Black et al. [50] of four zircon ages in one rock, this has been a fertile ground for the ion microprobe. It has taken some time to develop instruments of sufficient precision (for example the Cal Tech instrument [109]); however, important results are now appearing from those who have persevered. Matrix effects do not influence intensities for isotopes of a single element, which leaves only the ionization process and instrumental isotopic fractionation effects in the analysis to be considered. Many authors have noted that isotope ratios in the secondary ion spectrum do not reproduce those in the sample. It is important to understand how these fractionation effects arise. Detector mass discrimination has been noted; however, this should be minimal within the isotopes of a single element - a result confirmed by Shimizu and Hart [100]. Thus fractionation must be accounted for in the sputtering/ionization process occurring at the sample surface.

Slodzian [101], Shimizu and Hart [100], and a number of other authors report that (1) the light isotope is always favored, and (2) isotope fractionation is linear with mass difference. Also of note is the energy dependence of fractionation. Shimizu and Hart [100] suggest this is itself mass dependent. Maximum fractionation is generally observed at lowest secondary ion emission energies (close to 0 eV). In a study of Si and O transport in silicate melts using labeled isotopes, Shimizu and Kushiro [73] used a secondary energy offset of 90 eV specifically to minimize the effect of isotopic fractionation. Hutcheon and Gnaser [99] made a detailed study of time and primary beam fluence effects on isotopic fractionation. They report a strong initial enrichment in lighter isotopes sputtered from mineral surfaces, and a near-exponential dependence on mass difference, attributed to momentum asymmetry in the collision cascade.

The observed isotopic discrimination in the sputtering/ ionization process is usually accounted for as a velocity effect on ionization probability. Selective sputtering - as has been observed in the early stages of bombardment - can only be transitory, until an equilibrium or steady state is reached. Thus isotopic fractionation is attributed to differences in the ionization efficiency of different isotopes. Mass spectrometers analyze a fixed energy bandpass of the ions collected by the immersion lens. Consequently, two isotopes analyzed at the same energy, will have slightly different velocities. If ionization yield then follows velocity, we would anticipate a square root of mass dependence. Shimizu and Hart [100] observe a much greater fractionation, approaching a simple mass ratio dependence.

Slodzian [101] suggests isotopic fractionation is readily explained by electronic exchange following a generalized Landau-Zener model for the dissociation of diatomic molecules. The magnitude of isotopic fraction then follows:

$$\Delta_{ij} = \frac{aK}{V_c} \; x \; \frac{M_{ij}}{M} \qquad (10\text{-}4)$$

where Δij = (observed ratio Rij/standard $R^{\circ}ij$-1); K is a function of the energy of the leaving ion and V_c is the relative velocity of M and the dissociation partner (the anionic group). This type of model is extremely difficult to formulate quantitatively, as pointed out by Williams [102].

Velocity dependence of ion yields is also observed for negative oxygen emission from oxides. Lorin et al. [103] measured a $^{16}O^-/^{18}O^-$ ratio in feldspars 13% higher than the theoretical value. The velocity effect can be minimized by lowering the surface work function - in the case of negative oxygen emission by flooding the surface with cesium or using cesium ions as the primary beam.

Despite the problems described above, considerable progress has been made in applying SIMS to isotopic determination in geochemistry. The measurement of isotopic anomalies in meteorites stands out, as well as the more recent dating work emanating from SHRIMP. Exley [104] evaluated the ion microprobe for examining strontium isotope geochemistry of carbonates. Samples were analyzed without the capability to energy filter signals, so isobaric molecular ions were subtracted by monitoring peaks of the same elemental but different isotopic composition. Although the procedure is questionable, Exley concludes that usable data can be obtained given Sr over 400 ppm, and ~1% precision (2σ) can be achieved for Sr >5000 ppm. $^{87}Sr/^{86}Sr$ was measured in 10-μm spots for calcite in a hydrothermally altered basalt and in individual calcite grains in a veined Iherzolite. Exley and Jones [105] have also published $^{87}Sr/^{86}Sr$ determinations in Kimberlitic carbonates.

4.4.1. Extra Terrestrial Isotopic Anomalies. This area has been reviewed by Armstrong and Hutcheon [106]. It has been an area of conspicuous success for SIMS application because of the ability to analyze small phases "in situ". Anomalies have been observed in meteorite analyses for a variety of elements. ^{26}Mg anomalies in Allende inclusions from ^{26}Al decay have been thoroughly documented from the early papers by Bradley et al. [107] and Hutcheon [108] to "FUN with PANURGE" by Huneke et al. [109]. Interest has broadened considerably to include D/H ratios in interplanetary dusts [110] (D$^-$/H$^-$ was analyzed to avoid interference from H$_2^+$ with the D$^+$ signal). Zinner et al. [111] have also recently reported large isotopic anomalies in both the noble gases and in Si, C, and N in silicon carbide from the Murray meteorite, suggesting the material is pre-solar-system. The ion-probe analysis is the only route to such in situ isotopic analysis and such studies have contributed greatly to understanding of the composition of the "primitive" solar system.

4.5 Quantitative Analysis

As discussed in Section 3.2, secondary ion yields for a given element have been demonstrated to vary as a function of:
- The composition of the matrix from which the ion is ejected
- The angles of incidence of primary ions and extraction of secondaries (and, for single-crystal samples, how these relate to crystallographic axes)

- The presence of strongly electropositive or electronegative elements on the surface.

The use of independent standards for quantitative analysis is thus satisfactory only over a very limited range of sample compositions and instrument conditions. However, given suitable standards and reproducible instrument conditions a "calibration curve" can be generated, usually referencing the element of interest to an internal standard or known reference element in the sample (Si is by far the most useful for silicates). Thus a ratio $I(M)/I(R)$ is measured and the concentration of the element of interest M, in an unknown specimen, is obtained from $[I(M)/I(R)]$unknown = X, where the concentration Cunknown is read directly from the X vs. C calibration plot. Steele et al. [38] demonstrated the limitations of this method when applied to olivine analysis, but Nesbitt et al. [33] successfully used calibration curves, in conjunction with analysis of high-energy secondary ions for quantitative analysis of a series of whole rock samples. Ray and Hart [112] have also demonstrated the quantitative analysis of silicates, using energy filtering to discriminate against molecular ions.

Calibration curves, particularly for trace elements, are often linear at low concentration [33]. Thus the observed secondary ion current is a linear function of concentration.

$$I(M) = I_p S(M) C(M) \qquad (10\text{-}5)$$

where I_p = primary ion current and S = sensitivity factor for element M (equivalent in the simplest case to secondary ion current of species M produced per unit primary ion current). Providing that this sensitivity factor is independent of concentration of M, then

$$\frac{I(M)/C(M)}{I(R)/C(R)} = \frac{S(M)}{S(R)} = S_r(M) \qquad (10\text{-}6)$$

The reference element R is usually silicon for silicate mineral analysis. A table of relative sensitivity factors S_r can thus be defined, which are useful for a given matrix. In the bond breaking approach [40,41], relative ion yields are modified on the basis of the major element composition of the sample:

$$R_{Si}^{M} = \frac{I(M^+)}{I(Si^+)} \times \frac{[Si]}{[M]} \tag{10-7}$$

where $I(M^+)$ and $I(Si^+)$ are the measured secondary ion intensities of M^+ and Si^+ ions, while $[M]$ and $[Si]$ are the respective cationic concentrations of M and Si in the mineral. R_{Si}^{M} depends on the composition of the mineral and is given to a first approximation by

$$R_{Si}^{M} = K_{Si}^{M} [Si] + K_{Al}^{M} [Al] + \ldots \tag{10-8}$$

where K_{Si}^{M} = relative ionization yield of element X in a silicate matrix, etc.

The method is accurate to within 5% but cumbersome in that a series of relative ionization coefficients must be defined by calibration before analysis [40,41].

One method that has shown considerable value in trace element analysis is ion implantation of a known dose of the element being determined [114]. This internal reference avoids the problem of matrix effects, which may significantly alter ion yields between a sample and a standard. The method has been used with considerable success in trace silver determinations in sulfides, where one isotope can be selectively implanted. This avoids the problems associated with tailing in the implant depth profile and provides an extremely accurate measure of silver content at levels of 100 ppb or less.

The process of remote implantation followed by SIMS analysis is cumbersome and expensive, which has led to the suggestion of in situ implants using the SIMS primary ion source [115]. Unfortunately the design of most sources and primary columns, restricts such implants to relatively low energies (<40 kV). This gives shallow and somewhat poorly controlled implants; however, it is clear that some future instruments will be designed with provision for in situ ion implantation as a method of standardization.

In summary, a considerable variety of approaches to quantitative analysis of minerals have been attempted. From the local thermal equilibrium approach, matrix-specific yield factors and empirical calibration, coupled with the analysis of high energy secondary ions, all have had some degree of success. None is entirely satisfactory, which simply reflects that we are still some distance from a convenient and universal model to quantify secondary ion emission.

4.6. Imaging of Mineral Surfaces

One of the most powerful expressions of the sensitivity and spatial resolution of SIMS analysis is the ability to image the distribution of an element across a surface. Two instrument configurations, referred to as the ion microscope and the ion microprobe [6], have been widely used in imaging. The technique proposed and developed by Castaing and Slodzian [15] is a direct imaging or microscope method: The spatial registration of emitted secondary ions is maintained through the mass analyzer and thus a real-time, magnified, and mass-resolved image of the sample surface is obtained at the detector. The alternative is to scan a focused primary beam across the surface and collect or accumulate a scanning ion image at the detector. Because spatial resolution is largely determined by that of the primary beam, the system is referred to as an ion microprobe. The two systems are illustrated in Fig.10-2. Liebl [6] discusses the relative merits of the two techniques.

Because the ion optical elements of the primary column are generally simpler and operate at higher energies than those of the secondary column, ultimate resolution favors the microprobe mode. Both however have been successfully used in imaging mineral surfaces. In their early description of the Applied Research Laboratories IMMA, Andersen and Hinthorne [14] show scanning ion micrographs of the distribution of Ca, Fe, Mg, Zr, Y, P, and K in a lunar basalt.

The Cameca IMS300 was the first commercial ion microscope instrument, and Slodzian demonstrated its capabilities with images (among others) of elements associated with ilmenite exsolution lamellae in a lava sample [116]. Duchesne et al. [117] used a similar instrument in imaging an extraordinary banded Tl distribution in marcasite from the Vedrin mine, Namur. The successors to the IMS-300, the IMS3f and 4f, are powerful imaging instruments and recent developments in charge neutralization methods for insulating samples, particularly in electron guns coaxial with the secondary extraction optics [118], have broadened the scope for geochemical work even further.

Excellent results have also been obtained from sulfides. Dillen and Gijbels [119] imaged cobalt and thallium in pyrites while more recently Chrysoulis and McIntyre [120] have produced high-sensitivity images of precious metals in a sulfide ore concentrate. The high resolution University of Chicago SIMS [121] has been used to study the mode of crystallization of olivine in the Mezo-Madaras chondrite [122]. Images were obtained from Na^+, Mg^+, Al^+, Si^+, Ca^+, Fe^+ and O^- showing the structural relationship of the glass and olivine phases. Some remarkable images have also been obtained from organic matrices, such as coals, containing mineralized material. Figure 10-13 illustrates examples of such images.

5. Future Developments

Future expansions of SIMS applications in geochemistry hinge more on devising suitable experiments and improving access to instruments, than on awaiting developments of the technique itself. With the exception of improvements in instrument stability, and perhaps further exploitation of such areas as the analysis of high-energy secondaries, and multiply charged ions, conventional SIMS is, in the immediate future, unlikely to be any more quantitative than it is now. It is important to emphasize that current approaches as described in Section 4.5, are adequate for many trace element applications, identification of isotopic anomalies, and especially mineral surface analysis and depth profiling.

There is some scope for improvement in the transmission of SIMS instruments, with techniques such as dynamic emittance matching [16] and improved extraction optics. The sensitivity of the method will thus continue to be advanced both in imaging and analysis, but improvements in quantitation and matrix sensitivity will be more hard won. Imaging clearly has a strong future in correlating element distribution with phase relationships. Instrument performance with user friendly specimens is already well within an order of magnitude of the

Fig.10-13. Ion images of surfaces containing both matrix and mineralized material in coal specimens. The images illustrate the use of ion microscopy in examining surface micro-structure, composition, and reactivity. From left to right, top to bottom are shown surfaces of a Pennsylvania fusinite ($^{16}O^-$, $^{24}C_2^-$, and $^{34}S^+$ respectively), and surfaces of a Western Canadian coal after treatment with Tollen's reagent ($^{31}P^-$, $^{40}Ca^+$, and $^{109}Ag^+$ respectively). The images were provided by Professor R.R. Martin, University of Western Ontario.

fundamental limits of the technique (the 10 nm lateral dimension of the collision cascade) [123]. Improvements in source brightness will thus primarily achieve better sensitivities and ease of sample handling at high spatial resolution.

Much has been made of postionization of sputtered neutrals as an alternative to direct secondary ion collection [124-126]. However as yet, none of the methods utilized - such as electron bombardment, thermal or plasma ionization, multiphoton laser ionization - have demonstrated competitive sensitivity or indeed much improvement in quantitative capability over conventional SIMS. Again the efficiency of postionization has been shown to be not simply statistical, but also complicated by a variety of chemical factors.

Accelerator mass spectrometry (AMS), when coupled to a sputtering ion source (Cs^+ bombardment with extraction of negatively charged secondaries), provides an ultrasensitive SIMS analysis method. Postacceleration of secondaries and conversion to unique charge/mass states allows unambiguous identification of targeted ions with extremely high collection efficiency [127]. This method has been used primarily in the analysis of ^{14}C in dating and, although in the early stages of evolution, also in a number of geochemical applications. The composition of carbonaceous materials from Archean gold deposits has been examined [128], also trace levels and isotopic composition of Ag, Pt and Au in sulfides [129]. The demanding nature of the instrumentation and experiments means the application of the method will be confined for some time to areas where extreme sensitivity is required.

Thus, conventional SIMS has spawned a variety of specialized offshoots of considerable potential in geochemical application and there is no reason to suspect such developments will not continue. SIMS applications have only scraped the surface of what is possible within the current capabilities of commercial instrumentation. Although "ultimate weapon" status may overstate the case, SIMS has contributed substantially to the characterisation of mineral surfaces and has a very important future in further geochemical problem solving.

6. References

[1]   Lovering, J.F. NBS Spec.Publ. *1975*, 427,35.

[2]   Benninghoven, A.; Rudenauer, F.G; Werner, H.W. "Secondary Ion Mass Spectrometry." John Wiley, New York, *1987*.

[3]   Compilation of S.J.B. Reed, March *1988*.

[4]   Castaing, R., "Proc. IXCOM 11" ,*1986*,4.

[5]   Zinner, E., Scanning *1980*,3,57.

[6]   Liebl, H., Scanning *1980*,3,79.

[7]   Honig, R.E., Secondary Ion Mass Spectrometry SIMS V Benninghoven, A.; Colton, R.J.; Simons, D.S.; Werner, H.W. Eds. Springer Verlag, Berlin *1986*,2.

[8]   Liebl, H.; Harrison, W.W., Int.J.Mass.Spectrom.Ion Phys. *1976*,22,237.

[9]   Lawrence, G.R.; Beauchamp, R.K.; McKibben, J.C., Nucl.Instr.Methods *1965*,32,357.

[10]   Lau, W.M.; McIntyre, N.S.; Metson, J.B.; Cochrane, D.;Brown J.D., Surf.Interface Anal. *1985*,7(6),275.

[11]   Anderson, C.A.; Roden, H.J.; Robinson, C.F.; Chem.Soc.Comm.*1969*,41,3419.

[12]   Barber, N.; Bordoli, R.S.; Sedgwick, R.D.; Tyler, A.N.J., Chem.Soc.Chem.Comm. *1981*,325.

[13]   Niehuis, E.; Heller, T.; Feld, H.; Benninghoven, A.Secondary Ion Mass Spectrometry SIMS V.A.Benninghoven et al., Eds.Springer Verlag, Berlin *1986*,188.

[14]   Andersen, C.A.; Hinthorne, J.R., Science *1972*,175, 853.

[15]   Castaing, R.; Slodzian, G., J.Microscopie *1962*,1,395.

[16]   Colton, R.J.; Campana, J.E.; Barlak, T.M.; Decorpo, J.J,; Wyatt, J.R., Rev.Sci.Instrum. *1980*,51,1685.

[17]   Clement, S.W.J.; Compston, W., Int.J.Mass Spectrom. Ion Phys.*1972*,10,323.

[18]   Pimminger, M.; Grasserbauer, M.; Schroll, E.; Cerny, I., Anal.Chem. *1984*,56,407.

[19]   Metson, J.B.; Bancroft, G.M.; McIntyre,N.S.; Chauvin, W.J., Surf.Interface Anal. *1983*,5,181.

[20] Shimizu, N.; Semet, M.P.; Allegre, C.J., Geochimica Cosmochim.Acta *1978*,42,1321.

[21] Witmaack, K., Vacuum *1982*,32(2),65.

[22] Sigmund, P., Phys.Review *1969*,184(2),383.

[23] Sigmund, P., Secondary Ion Mass Spectrometry SIMS IV. A. Benninghoven et al., Eds. Springer Verlag, Berlin 1984,2.

[24] Williams, P.; Katz, W., Nucl.Instr.Methods *1980*,168, 373.

[25] Harrison Jr, D.E., Radiat.Effect. *1983*,70,1.

[26] Harrison Jr, D.E.; Mason, W.A.; Webb, R.P. Secondary Ion Mass Spectrometry SIMS IV. A. Benninghoven, et al., Eds. Springer Verlag, Berlin *1984*,24.

[27] Almen, O.; Bruce, G., Nucl.Instrum.Methods *1961*,11, 257.

[28] Garrison, B.J., Surf.Sci. *1986*,167,L225.

[29] Winograd, N., Chemistry and Physics of Solid Surfaces V. *1984*.

[30] Benninghoven, A., Surf.Sci. *1975*,53,596.

[31] Wittmaack, K., Nucl.Instr.Methods *1980*,168,343.

[32] Maul, J.; Wittmaack, K., Surf.Sci. *1980*,47,358.

[33] Nesbitt, H.W.; Metson, J.B.; Bancroft, G.M., Chem. Geol. *1986*,55,139.

[34] Tarng, M.L.; Fisher, D., J.Vac.Sci.Technol. *1978*,15, 50. See also: Hofmann, S., Surf.Interface Anal. *1980*,2,148. Zinner, E., Proc. 13th Ann.Conf. Microbeam Anal.Soc. *1978*,32. Mayer, J.W.;. Poate, J.M., Depth Profiling Techniques. Thin Films Interdiffusion and Reactions. John Wiley and Sons, New York *1978*,119.

[35] Kelly, R., Nucl.Instr.Methods *1978*,149,553. See also Surf.Sci. *1980*,100,85.

[36] Muir, I.J.; Bancroft, G.M.; Metson J.B., Int.J.Mass Spec.Ion Proc. *1987*,75(2),159.

[37] Morgan, A.E.; Werner, H.W., Anal.Chem. *1977*,49(7), 927.

[38] Steele, I.M.; Hervig, R.L.; Hutcheon, I.D.; Smith, J.V., Am.Mineral. *1981*,66,526.

[39] Shimizu, N. Secondary Ion Mass Spectrometry SIMS V. Benninghoven, A. et al., Eds. Springer Verlag, Berlin *1986*,45 see also: Shimizu, N., Int.J.Mass Spectrom.Ion Proc. *1986*,69, 325.

[40] Havette A.; Slodzian, G., J.Physique Lettres *1980*,41, L247.

[41]   Havette A.; Slodzian, G., Secondary Ion Mass Spectrometry
        SIMS III. Benninghoven, A.; Giber, J.; Lazlo, J.; Riedel, M.;
        Werner, H.W. Eds. Springer Verlag, Berlin *1982*,416.
[42]   Anderson, C.A.; Hinthorne, J.R., Anal.Chem. *1973*,45, 1421.
[43]   Bernheim, M; La Bourse., 6th Int. Workshop on Inelastic Ion
        Surface Collisons *1986*.
[44]   Yu M.L., Comments At.Mol.Phys. *1987*,20(2),79.
[45]   Shimizu, N.; Hart, S.R., Ann.Rev.Earth Plant.Sci.
        1982,10,483.
[46]   Zinner, E.; Crozaz, G., Int.J.Mass Spectrom.Ion Phys.
        1986,69,17.
[47]   Reed, S.J.B., Mineral Mag. *1986*,50,3.
[48]   Compston, W.; Williams, I.S.; Meyer, C., Proc. 14th Lunar
        Planet.Sci.Conf., J.Geophys.Res. *1984*,89,B525.
[49]   Compston, W.; Pidgeon, R.T., Nature *1984*,321,766.
[50]   Black, L.P.; Williams, I.S.; Compston, W., Contrib.
        Mineral.Petrol. *1986*,94,427.
[51]   Compston, W.; Kroner, A., Earth Planet.Sci.Lett. *1988*,87,13.
[52]   Compston, W.; Kinny, P.D.; Williams, I.S.; Foster, J.J., Earth
        Planet.Sci.Lett. *1986*,80,71.
[53]   McIntyre, N.S.; Chauvin, W.J.; Metson, J.B.; Bancroft, G.M.,
        J.de Physique *1984*,Colloque C2,C143.
[54]   Metson, J.B.; Bancroft, G.M.; Nesbitt, H.W. Scanning
        Elec.Microscopy *1985*,II,595.
[55]   McIntyre, N.S.; Fichter, D.; Metson, J.B.; Robinson, W.H.;
        Chauvin, W.J., Surf.Interface Anal. *1985*,7(2), 69.
[56]   MacRae, N.D.; Metson, J.B., Chem.Geol. *1985*,53,325.
[57]   Metson, J.B.; Bancroft, G.M.; Nesbitt, H.W.; Jonasson, R.G.,
        Nature *1984*,307(5949),347.
[58]   Crozaz, G.; Zinner, E., Earth Planet.Sci.Lett. *1985*, 73,41.
[59]   See, for example, Honig, R.F. ref.7.
[60]   Dowsett, M.G.; King, R.M.; Parker, E.H.C., J.Vac.Sci.
        Technol. *1977*,14(2),711. See also:
        Dowsett, M.G.; King, R.M.; Parker, E.H.C., Surf.Sci.
        1978,71,541.
[61]   Buhl, R.; Preisinger, A., Surf.Sci. *1975*,47,344.

[62] Beusen, J.M.; Gijbels, R., In "Leaching and Diffusion in rocks and their Weathering Products". Theophrastus Publications S.A. Sept. *1983*,257.

[63] Nesbitt, H.W.; Muir, I.J., Nature *1988*,334(6180),336.

[64] Muir, I.J.; Bancroft, G.M.; Nesbitt, H.W., Geochim.Cosmochim.Acta. *1989*,53,1235.

[65] Havette, A., Scanning Elect.Microscopy *1985*,II,585.

[66] Lovering, J.F., Comments on Earth Sci.Geophys. *1973*, 3,153.

[67] Bancroft, G.M.; Metson, J.B.; Kresovic, R.A. Geochim.Cosmochim.Acta. *1987*,51,911.

[68] Hayward, P.J.; Hocking, W.H.; Doern, F.E.; Cechetto, E.V., Scientific Basis for Nuclear Waste Management V. Lutze, W., Ed.*1982*,319.

[69] Myhra, S.; Savage, D.; Atkinson, A.; Riviere, J.C. Amer.Mineral. *1984*,69,902.

[70] Hofmann, A.W.; Giletti, B.J.; Hinthorne, J.R.; Andersen, C.A.; Comaford, D., Earth Planet.Sci.Lett. *1974*,24,48.

[71] Hofmann, A.W.; Giletti, B.J., Eclogae Geol.Helv. *1970*,63,141.

[72] Sneeringer, M.; Hart, S.R.; Shimizu, N., Geochim. Cosmochim.Acta. *1984*,48,1589.

[73] Shimizu, N.; Kushiro, I., Secondary Ion Mass Spectrometry SIMS IV. Benninghoven, A. et al. Eds. Springer Verlag, Berlin *1984*,469.

[74] Karsten, J.L.; Holloway, J.R.; Delaney, J.R., Earth Planet.Sci.Lett. *1982*,59,420.

[75] Cunningham, G.J.; Henderson, P.; Lowry, R.K.; Nolan, J.; Reed, S.J.B.; Long, J.V.P., Earth Planet.Sci. Lett. *1983*,65,203.

[76] Delaney, J.R.; Karsten, J.L., Earth Planet.Sci.Lett. *1981*,52,191.

[77] For example; Andersen, C.A.; Hinthorne, J.R.; Fredriksson, J., Proc.Apollo 11 Lunar Sci.Conf. *1970*, 159.

[78] Meyer, C., Proc. 9th Lunar Planet.Sci.Conf. *1978*, 1551.

[79] Zinner, E.; Walker, R.M.; Chaumon, J.; Dran, J.C. Proc. 8th Lunar Sci.Conf. *1977*,3859.

[80] Zinner, E.; Walker, R.M.; Chaumond, J.; Dran, J.C. Proc. 7th Lunar Sci.Conf. *1976*,953.

[81]   Fahey, A.; Goswami, J.N.; McKeegan, K.D.; Zinner, E.,
       Astrophys.Journ. *1985*,296,L17.
[82]   Ireland, T.R.; Compston, W.; Heydegger, H.R.
       Geochim.Cosmochim.Acta *1985*,1989.
[83]   Hart, S.R.; Shimizu, N.; Sverjensky, D.A., Econ.Geol.
       1981,76,1873, see also:
       Ray, G.L.; Shimizu, N.; Hart, S.R. Geochim.Cosmochim. Acta
       1983,47,2131.
       Steele, I.M.; Lindstrom, D.J., Geochim.Cosmochim.Acta
       1981,45,2177.
[84]   Meyer, C.Jr.; Anderson, D.H.; Bradley, J.G. Proc. 5th Lunar
       Sci.Conf. *1974*,685.
[85]   Shimizu, N., Nature *1981*,289,575.
[86]   McIntyre, N.S.; Cabri, L.J.; Chauvin, W.J.; Laflamme, J.H.G.,
       Scanning Elec. Micros. *1984*,I,1139.
[87]   Hendry, D.A.F.; Chivas, A.R.; Long, J.V.P.; Reed, S.J.B.,
       Contrib.Mineral.Petrol. *1985*,89,317.
[88]   Wilson, G.C.; Long, J.V.P., Mineral.Mag. *1983*,47,191.
[89]   Hutcheon, I.D.; Steele, I.M.; Solberg, T.N.; Clayton, R.N.; Smith,
       J.V., Meteorite *1977*,12,261.
[90]   Phinney, D.; Whitehead, B.; Anderson, D., Proc. 10th Lunar
       Planet.Sci.Conf. *1979*,885.
[91]   Hervig, R.L.; Kortemeier, W.T.; Burt, D.M., Amer.
       Mineral *1987*,72,392.
[92]   Dutrow, B.L.; Holdaway, M.J.; Hinton, R.W.,
       Contrib.Mineral.Petrol. *1986*,94,496.
[93]   Jones, A.P.; Smith, J.V.; N.Jahrb.Mineral.Mn. *1984*, 228.
[94]   Holdaway, M.J.; Dutrow, B.L.; Borthwick, J.; Shore, P.; Harmon,
       R.S.; Hinton, R.W., Amer.Mineral *1986*,71,1135.
[95]   Muir, I.J.; Bancroft, G.M.; MacRae, N.D.; Metson, J.B.
       Chem.Geol. *1987*,64,269.
[96]   Reed, S.J.B., Chem.Geol. *1985*,48,137.
[97]   Reed, S.J.B.; Smith, D.G.W., Earth Planet.Sci.Lett.*1985*, 238.
[98]   MacRae, N.D.; Wu Tsai-Way. Chem.Geol. (in press).
[99]   Gnaser, H.; Hutcheon, I.D., Surface Science *1988*,195, 499.
[100] Shimizu, N.; Hart, S.R., J.Appl.Phys. *1982*,53(3), 1303.

[101] Slodzian, G., Secondary Ion Mass Spectrometry SIMS III. Benninghoven, A. et al. Eds Springer Verlag, Berlin, *1982,115.*

[102] Williams, P., Surf.Sci. *1979,*90,588.

[103] Lorin, J.C.; Shimizu, N.; Christophe-Michel Levy, M.; Allegre, C.J., Meteoritics *1977,*12,299. See also: Lorin, J.C.;Havette, A., Rapports Isotopiques dans le systeme solaire, CEPAD, France.

[104] Exley, R.A., Earth Planet.Sci.Lett. *1983,*65,303.

[105] Exley, R.A.; Jones, A.P., Contrib.Mineral.Petrol *1983,*83,292.

[106] Armstrong, J.T.; Hutcheon, I.D., Microbeam Analysis - *1985.* San Francisco Press, San Francisco, 273.

[107] Bradley, J.G.; Huneke, J.C.; Wasserburg, G.J., J.Geophys.Res. *1978,*83,244.

[108] Hutcheon, I.D.; Steele, I.M.; Smith, J.V.; Clayton, R.N., Proc. 9th Lunar Sci.Conf. *1978,*1345.

[109] Huneke, J.C.; Armstrong, J.T.; Wasserburg, G.J. Geochim.Cosmochim.Acta *1983,*47,1635.

[110] Zinner, E.; McKeegan, K.D.; Walker, R.M., Nature *1983,*305,119.

[111] Zinner, E.; Ming, T.; Anders, E., Nature *1987,*330, 730.

[112] Ray, G.; Hart, S.R., Int.J.Mass Spectrom.Ion Phys. *1982,*44,231.

[113] Odelius, H.; Lodding, A.R.E.; Werme, L.O.; Clark, D.E. Scanning Elec. Micros. *1985,*III,927.

[114] Leta, D.P.; Morrison, G.H., Anal.Chem. *1980,*52,514.

[115] Smith, H.E.; Bernius, M.T.; Morrison, G.H., Secondary Ion Mass Spectrometry SIMS V. Benninghoven, A. et al. Eds. Springer Verlag, Berlin *1986,*121.

[116] Slodzian, G., Surf.Sci. *1975,*48,161.

[117] Duchesne, J.C.; Rouhart, A.; Schoumacher, C.; Dillen, H. Mineral. Deposita *1983,*18,303.

[118] Slodzian, G.; Chaintreau, M.P.; Dennebouy, R.C., Proc. IXCOM 11 *1986.*

[119] Dillen, H.; Gijbels, R., J.Microsc.Spectrosc.Electron *1981,*6,179.

[120] Chrysoulis, S.; McIntyre, N.S., University of Western Ontario, London, Canada, unpublished results.

[121] Levi-Setti, R.; Wang, Y.L.; Crow, G., Applied Surf. Sci. *1986*,26,249.

[122] Levi-Setti, R,; Wang, Y.L.; Crow, G., J.Physique *1984*,45C9,179.

[123] Bayly, A.R.; Waugh, A.R.; Vohralik, P. Spectrochimica Acta *1985*,40B,717.

[124] Staudenmaier, G.; Hofer, W.O.; Liebl, H., Int.J.Mass Spec.Ion Phys. *1976*,11,103.

[125] Oechsner, H., Topics in Current Physics, Vol.37. Thin Film and Depth Profile Analysis. Oechsner, H. Ed. Springer Verlag, Berlin *1984*,63.

[126] Gruen, D.M.; Pellin, M.J.; Young, C.E.; Calaway, W.F. Appl.Phys.Lett. *1985*,47,521.

[127] Purser, K.H.; Litherland, A.E.; Rucklidge, J.C. Surf. Interface Anal. *1979*,1,12.

[128] Wilson G.C.; Rucklidge, J.C. GSA meeting, Reno, Nevada, Nov 1987.

[129] Kilius, L.R.; Rucklidge, J.C.; Wilson G.C.; Lee, H.W.; Chang, K.H.; Litherland, A.E.; Keiser,W.E.; Beukens, R.P.; Gordon, M.P., Nucl.Instr.Methods in Phys.Res. *1884*,B5,185.

SUBJECT INDEX

Dear Isa,

Hope you enjoy this book!

Thank you for coming to India.

Love

Naomi x

A CLUTCH OF

IND
IAN

MASTERPIECES

Also by David Davidar

The House of Blue Mangoes (2002)
The Solitude of Emperors (2007)
Ithaca (2011)

A CLUTCH OF
IND
IAN
MASTERPIECES

extraordinary short stories
from the 19th century to the present

EDITED BY

DAVID DAVIDAR

ALEPH

ALEPH BOOK COMPANY
An independent publishing firm
promoted by *Rupa Publications India*

First published in India in 2014
by Aleph Book Company
7/16 Ansari Road, Daryaganj
New Delhi 110 002

Published in paperback in 2016

ISBN: 978-93-82277-29-3

7 9 10 8

For sale in the Indian subcontinent only.

Printed in India

To
The memory of K.D. Singh
A truly unforgettable literary icon

And to all my fellow Alephs,
Borgesian marvels in the Indian sun

I divide all literary works into two categories. Those I like and those I don't like. No other criterion exists for me.
—Anton Chekhov,
in a letter to a fellow Russian writer in 1890.

CONTENTS

Our Stories xiii

Our Stories

DAVID DAVIDAR

When I was a boy, I spent summer vacations with my maternal grandparents in a small town near the southern tip of the country. My grandparents lived in a cottage with whitewashed walls, a red-tiled roof, and brown windows. A jasmine creeper, with tiny flowers, bright as stars, climbed across the trellised front verandah, shading the interiors of the house, and ensuring it never got too hot, even when temperatures soared outside. On the verandah was an old-fashioned planter's chair with extendable armrests. I spent most of my days sprawled in this chair, eating fat, crisp banana chips, and reading books that my grandfather procured for me. It was the start of a literary journey that has lasted about five decades now.

My grandfather, or Thatha as I called him, was a strict disciplinarian, and I was genuinely afraid of him—although to my knowledge it was only once that he threatened to thrash me with his walking stick. Tall and gaunt, he would set off every morning in starched white drill trousers, a clean white shirt, cufflinks, black shoes shined to a mirror finish and sola topi, to the school that he was the headmaster of. He believed that little boys were meant only to be seen and not heard, and should only speak when they were spoken to. As he spoke to me only in English, and as my spoken English was extremely poor (I could read and write English fairly proficiently, but until I was four or thereabouts I spoke no language but Tamil; this gradually became a kind of English-Tamil patois until high school, after which English became my first language), there wasn't a great deal of communication between us, even if I'd been able to summon the courage to speak to him. However, my grandfather gave me the gift of literature because every weekend he would bring me a few books from his school library. These tended to be abridged, simplified classics of Western literature—*The Count of Monte Cristo, Lorna Doone, Moby Dick, Tom Sawyer, A Tale of Two Cities* and so on. I found several of them quite absorbing but often mystifying—the names, the customs and the mannerisms of the characters in the books were frequently bizarre. Much as I wanted to, I could hardly discuss these oddities with my Thatha, so I would simply swallow the stories whole.

My grandmother, my Ammamma, a serene and beautiful woman, whom my sister and I doted on, provided an altogether different diet of fiction. Whenever I pestered her for a story, she would tell me tales that were rooted in family lore, or folk tales from the region. I did not need my grandmother's stories to be interpreted for me as they took place in situations and locations that I was familiar with. The stories I liked the best either featured ghosts, or, sometimes, Satan. I found them extremely gripping, even chilling. Cohorts of the Evil One included pigs with backward turned hooves, beautiful women with jasmine flowers in their hair (but whose feet didn't touch the ground), or horrible old women with giant splayed feet who would squat on the windowsills of the dying. I noticed that the only way to detect the presence of Satan was to focus on the feet of those whom he had possessed, so I went through a phase of looking at people's feet before I would look them in the face.

Besides these two streams of stories, my parents (especially my father, who worked for a British company, and was an unabashed Anglophile when he was younger) would give me books by Beatrix Potter and, as I grew older, tales of British public-school life like *Tom Brown's School Days*. This led me to books like Enid Blyton's 'Famous Five' series and the Billy Bunter stories by Frank Richards (the pen name of Charles Hamilton) in my school library, from which I graduated to adult fiction from around the world, thanks to an English teacher who managed to instil in some of her students a genuine love of good books. But there was also a fourth stream of fiction that I immersed myself in—Tamil pulp fiction, as well as tales from the Hindu myths and epics, which my schoolmates would tell me. This was not an unusual experience for Indian children from my socio-economic background.

In his seminal essay, 'Telling Tales', the great poet, linguist, scholar and translator, A. K. Ramanujan, writes about the traditions and sources of Indian stories that were available to the average middle-class Indian child: 'Even in the most anglicized...families or in large cities like Bombay and Calcutta, oral tales are only a grandmother away, a cousin away, a train ride away, and mostly no further away than the kitchen.' He talks about the European stories that he read in books, the Tamil stories that were narrated by a grandmother, an aunt or a cook, and Kannada stories which he heard in friends' houses:

> As we grew up, Sanskrit and English were our father-tongues, and Tamil and Kannada our mother-tongues. The father-tongues distanced us from our mothers, from our own childhoods, and from our villages and many of our neighbours in the cowherd colony next door. And the mother-

tongues united us with them. It now seems quite appropriate that our house had three levels: a downstairs for the Tamil world, an upstairs for the English and the Sanskrit, and a terrace on top that was open to the sky, where our father could show us the stars and tell us their English and Sanskrit names...

We ran up and down all these levels. Sanskrit, English, and Tamil and Kannada (my two childhood languages, literally my mother's tongues, since she too had become bilingual in our childhood) stood for three different interconnected worlds. Sanskrit stood for the Indian past; English for colonial India and the West, which also served as a disruptive creative other that both alienated us from and revealed us (in its terms) to ourselves; and the mother-tongues, the most comfortable and least conscious of all, for the world of women, playmates, children and servants. Ideas, tales, significant alliances, conflicts, elders, and peers were reflected in each of these languages. Each had a literature that was unlike the others'. Each was an other to the others, and it became the business of a lifetime for some of us to keep the dialogues and quarrels alive among these three and to make something of them. Our writers, thinkers, and men of action—say, Gandhi, Tagore, and Bharati—made creative use of these triangulations, these dialogues and quarrels.

A large number of us can draw upon two or three literary traditions, others may have been schooled in more languages or fewer, but one of the reasons Indian literature is so diverse and rich is because of the multiple languages and sources in which it is rooted and created. The country has 24 national languages (including English and Hindi) and the 2011 census recognized 1,635 'mother tongues'. Of these, thirty were spoken by more than a million native speakers, and at least fifteen had long-standing literary traditions. This polyphonic, incredibly complex environment has given rise to some of mankind's most remarkable storytellers.

To this must be added the fact that we've had a lot of practice in the art of telling stories. Our earliest stories were told over 2,000 years ago. Although purists might point to, say, the parables in the Brahmanas as possibly the earliest stories we can lay claim to, even if we started with the Ramayana, the Mahabharata, the Panchatantra and the Jataka tales, all of which were composed within a century or two of each other, ours is an incredibly old literary tradition, pre-dating those of practically every other civilization on the planet, with the exception of the Egyptians and our neighbours in West Asia (the Babylonians of ancient times). The first stories in Sanskrit were followed by tales told in Pali, Tamil, Prakrit, and as we move into the medieval period, Kannada, Telugu, Persian and Urdu.

Some scholars have divided Indian literature into the Great Indian Tradition (pan-Indian and Sanskritic) and the Little Tradition (local literature, folklore and so on) but others reject these classifications. Ramanujan says in his essay, 'Where Mirrors are Windows', that the only way to look at our literary roots and traditions is to see them as 'indissolubly plural and often conflicting but...organized through at least two principles (a) context-sensitivity and (b) reflexivity of various sorts, both of which constantly generate new forms out of old ones'.

This plurality is one of the things that makes the Indian literary tradition unique. Another aspect of our stories that is seen virtually nowhere else in the world is the fact that our oldest tales, dating back a couple of thousand years, are still in circulation, in prose, in verse, in street theatre, television, the movies and in online forums of storytelling.

In the future, as the Indian writing tradition matures and grows in confidence, we will see an ever-decreasing tendency to seek 'approval' from cultural arbiters other than our own peer groups—in other words, we will gradually grow out of the dreadful syndrome known as 'cultural cringe' that so many former colonies have to deal with. All this would seem to project a bright future for Indian literature in the twenty-first century. There are many obstacles that will need to be dealt with—the decline in reading habits, newer and newer forms of entertainment, a lack of resources for writers besides a small group of publishers who are increasingly under siege, and so on—but I'm an optimist when it comes to the power of stories to survive and thrive. Our stories will grow richer, more distinctive, and show us the real India for centuries to come.

As this is not intended to be an essay on Indian literature as a whole, but a brief introduction to the modern Indian short story, I am going to jump ahead to the second half of the nineteenth century when the first modern short stories made an appearance. There was one very specific difference in technique between the older forms of Indian literature and the new forms. Ramanujan explains: 'No Indian [literary] text comes without a context or frame, till the nineteenth century... One might see "modernization" in India as a movement from the context-sensitive to the context-free in all realms.'

II

R. K. Narayan, one of the world's greatest writers, tells an amusing story about creative writing in general, and the short story in particular. He writes: 'Once I was present at a lecture on creative writing. The lecturer began with: "All writing may be divided into two groups—good writing and bad writing. Good books come out of good writing while bad writing produces failures." When touching on the subject of the short story, the lecturer said: "A short story must be short and have a story." At this point I left unobtrusively, sympathizing with the man's predicament.'

The story is amusing but when you come down to it, the short story is devilishly difficult to define if you exclude length as a criterion. Dictionary definitions are banal in the extreme. Here is one example: 'A story with a fully developed theme but significantly shorter and less elaborate than a novel.' If you were taking a creative writing class, your instructor might tell you that your story would need to have the following elements—exposition (the setting up of the story, its backdrop, main characters etc.), conflict, plot, theme, climax and so on. If s/he was of a Chekhovian bent of mind, s/he might tell you to write a 'slice of life' story that was relatively loosely constructed when compared to tightly plotted stories that hinged on events and turning points. There are many other categories that short stories are classified under but these do not need to detain us. Let us instead take a quick look at the origins of the modern short story, and how it spread around the world before speeding ahead to the focus of this introduction—the modern Indian short story.

The short story began to flourish in several parts of the world at about the same time—the nineteenth century. The United States had great practitioners of the form, like Nathaniel Hawthorne, Mark Twain and Edgar Allan Poe (who wrote an essay about short fiction that practically every creative writing course will point you to called 'The Philosophy of Composition'); France had prolific and excellent story writers such as Guy de Maupassant and Alphonse Daudet; in Germany the brothers Grimm published their retold fairy tales; and in England, writers like Thomas Hardy, H. G. Wells and Conan Doyle put out not just literary stories but some of the first modern detective stories and science fiction tales. Modern European and American short fiction followed in the wake of books by writers like Chaucer (*The Canterbury Tales* was published in the fourteenth century) and Boccaccio (*The Decameron*) as well as the great epics of European classical literature like *The Iliad* and *The Odyssey*. The single greatest leap forward in the evolution of the short story in the nineteenth century is attributed to a writer many think of as the father of the modern short story, Anton Chekhov. Chekhov's precursors themselves were among the best modern writers of fiction the world has ever seen, notably Nikolai Gogol, whom the novelist and essayist Vladimir Nabokov considered 'the greatest artist Russia has yet produced'. One of Gogol's contemporaries, the novelist Fyodor Dostoevsky, went further when he famously proclaimed: 'We all come out from Gogol's "Overcoat".'

The British writer William Boyd is effusive in his praise of Chekhov. He writes in *Prospect* magazine:

> Why is Anton Chekhov (1860–1904) routinely and correctly described as the greatest short story writer ever? All answers to this question will seem inadequate but, to put it very simply, the fact is that Chekhov, in his mature stories of the 1890s, revolutionized the short story by transforming narrative. Chekhov saw and understood that life is godless, random and absurd, that all history is the history of unintended consequences... By abandoning the manipulated beginning-middle-and-end plot, by refusing to judge his characters, by not striving for a climax or seeking neat narrative resolution, Chekhov made his stories appear agonisingly, almost unbearably lifelike.

As the nineteenth century bled into the twentieth century, the short story continued to flourish in every corner of the world, driven by increasing rates of literacy, the growth of literary magazines and supplements, especially in the Western world, the packaging and marketing of famous authors as superstars, and so on. It continued to morph into newer and newer forms as the decades went by.

In our country, the modern short story made an appearance almost simultaneously in several languages beginning naturally enough with Bengali. The writer and translator Ranga Rao credits the first modern short story to Poornachandra Chattopadhyay who published 'Madhumati' in 1870 (Poornachandra's older brother Bankim Chandra published *Rajmohan's Wife*, the first Indian novel in English). Rabindranath Tagore soon established himself as one of Bengal's finest short story writers; in Hindi, Munshi Premchand wrote hundreds of stories, many of which appeared in *Hans*, the literary magazine he published; and, in Oriya, the writer Fakir Mohan Senapati published some landmark stories.

Throughout the twentieth century, most of the major literatures in the land threw up great practitioners of the form—Saadat Hasan Manto in Urdu, Kalki in Tamil, Gurzada Appa Rao in Telugu, R. K. Narayan and Raja Rao in English, Thakazhi Sivasankara Pillai and Vaikom Muhammad Basheer in Malayalam, and dozens of notable writers in every corner of the country. Their stories reflected their region, their upbringing, and their (often) cosmopolitan reading. A number of India's pioneering short story writers had a common element in their stories—they were often extremely political in nature. It couldn't have been otherwise in a country trying to free itself from a predatory and oppressive colonial power, while at the same time grappling with a huge variety of hellish social evils. George Orwell writes eloquently about the power of writing that is overtly political: 'I see that it is invariably where I lacked a political purpose that I wrote lifeless books and was betrayed into purple passages, sentences without meaning, decorative adjectives and humbug generally'. Stories without humbug. Stories that are full of life. Many of the stories in this book would fit that description. Others exemplify what William Boyd said of the form: 'Short stories are snapshots of the human condition and of human nature, and when they work well, and work on us, we are given the rare chance to see in them more "than in real life".'

III

Roughly a quarter century ago, a distinguished London publisher asked me to edit an anthology of Indian stories for him. I was flattered, especially as there were no good literary publishing houses in the country at the time, but in the end I turned down the commission. My reason was straightforward— although I had been reading whatever Indian fiction was available in English for a decade or so (since my late teens), I felt I hadn't read enough to be able to attempt an anthology. It would have been easy enough to slap together a few stories, and knit them together, with a seemingly learned introduction, as happens four to five times a year, courtesy publishers who have an eye on the textbook market, but that was not for me.

Earlier this year, after leafing through yet another unsatisfactory anthology, I decided I would finally attempt to put together an anthology of great Indian literary stories. I was aware of the pitfalls that lay in wait for any prospective anthologist. Either the rationale for the anthology hasn't been thought through clearly enough, or the editor has been unable to clear the necessary permissions, or it is too dull and worthy as it tries to please every reader, writer and critic, or it has some sort of agenda that doesn't quite work—for any number of reasons, most anthologies fail to satisfy the reader, either aesthetically or commercially.

I decided to pull this anthology together on the basis of a very simple premise—it would only include stories that I loved, stories that had made their mark on me in the forty years or so that I had been attentively reading serious Indian literature. Like the Chekhov quote I've used as an epigraph, the basic criterion for featuring stories in this book would be whether I liked them or not. I decided to leave out commercial fiction because there would then be no focus to the anthology. There would be no other exclusions. It wouldn't matter to me whether the writer was Dalit or Brahmin, old master or twenty-first-century star, Muslim, Hindu, Christian, Parsi or Sikh, man or woman, straight or gay, Tamil or Kashmiri, Punjabi or Malayali. Nor would it matter if the writers wrote in English (more on this later) or in any of the other Indian languages, whether they lived here and carried the blue passport with the Ashok Chakra, or plied their trade in foreign lands—no, the only thing that would influence the selection of the stories would be whether or not they were, in my subjective view, breathtakingly good Indian stories. What did I mean by Indian? Either the stories would have to be about India or they would have to be written by an Indian or someone of Indian origin. Necessarily, they would need to possess an 'Indian sensibility'. Now, there are learned tomes on 'the Indian sensibility', a sensibility that is rooted in Indian culture, history, society, language, but that is not what I am trying to get at here. No, what I am trying to define is that elusive, ineffable quality to 'serious' poetry and prose that is unmistakably Indian. If you learn this quality from books, or by overflying the subcontinent, all you will be able to produce is a variant of the Inspector Ghote mysteries—entertaining but shallowly rooted writing, and without any great insight into anything of consequence. I do not think an Indian sensibility is to be found only in writers who are Indian by nationality or domicile or language. Rather I think it is inherent to writers who were born here, or have lived here for enough time, for distinctive aspects of this country, this civilization, to have shaped their view of the world, their creative consciousness, and their style. Their writing, whether about India or elsewhere, is informed by this 'Indian sensibility'—when their subject matter is India, they tend not to exoticize, but deepen our understanding of the country.

Having figured out the basic qualifications any story would need to have in order to be included, I refined the criteria for selection. Every story that made the cut would need to be a proper short story. (Vikram Seth's 'The Elephant and the Tragopan' is a short story, although it is told in verse.) This meant I wasn't going to be able to include any extracts from novels, or works in progress. It was also important that none of the stories had dated, or would appear quaint to today's reader. And, finally, every story would need to work perfectly in English, as the anthology was aimed squarely at the reader in English. Other than these qualifications I wouldn't be excluding any writer.

I then made a list of the most memorable Indian stories I had read in the course of my career as a reader. I had read all these stories several times over and they hadn't lost their power every time I re-read them—as good an indicator as any of their brilliance. When I had finished I had about a dozen stories—Rabindranath Tagore's 'The Hunger of Stones', Premchand's 'The Shroud', R. K. Narayan's 'A Horse and Two Goats', Saadat Hasan Manto's 'Toba Tek Singh', Thakazhi Sivasankara Pillai's 'The Flood', Khushwant Singh's 'Portrait of a Lady', Ismat Chughtai's 'Quilt', Mahasweta Devi's 'Draupadi', U. R. Ananthamurthy's 'Mouni', Ruskin Bond's 'The Blue Umbrella', Anita Desai's 'Games at Twilight', and Vikram Seth's 'The Elephant and the Tragopan'. Extensively anthologized, and familiar to anyone with even a nodding acquaintance with Indian literature, these would form the core of the anthology. I then asked my wife and a few friends, all of whom are very well read, for their suggestions on what they thought the anthology should contain. Their picks yielded another bunch of stories for inclusion, which I'd either missed or discarded on account of extremely poor translations. (We commissioned new translations for some of the stories.) The stories in this category are—Anna Bhau Sathe's 'Gold from the Grave', D. B. G. Tilak's 'The Man Who Saw God', Gulzar's 'Crossing the Ravi', Vaikom Muhammad Basheer's 'The Blue Light' and Buddhadeva Bose's 'A Life'. My colleagues and I then sifted through nearly a thousand stories (contained in about fifty anthologies) to add another dozen stories to the longlist. Among the stories from this category are masterpieces like Vijaydan Detha's 'Countless Hitlers', Gopinath Mohanty's 'The Somersault', Amrita Pritam's 'Stench of Kerosene', Harishankar Parsai's 'Inspector Matadeen on the Moon', Nirmal Verma's 'Mirror of Illusion', Sundara Ramaswamy's 'Reflowering', Ambai's 'In a Forest, a Deer', Paul Zacharia's 'Bhaskara Pattelar and My Life', Devanoora Mahadeva's 'Tar Arrives', Irwin Allan Sealy's 'Last in, First Out', Manjula Padmanabhan's 'Feast', and Anjum Hasan's 'Wild Things'. Another category included stories by writers I'd published which had left a deep impression on me. Stories from this category would include—Nisha da Cunha's 'Old Cypress', Vilas Sarang's 'A Revolt of the Gods', and Amrita Narayanan's 'Stolen'. And finally I looked around for a few stories that had never been published before in book form—Githa Hariharan's 'Nursing God's Countries', Cyrus Mistry's

'Proposed for Condemnation', Upamanyu Chatterjee's 'Desolation, Lust', Shahnaz Bashir's 'The Gravestone', or even rarer, had never been published in any form before—'Trying to Discover India' by Shashi Tharoor and Kanishk Tharoor's 'Elephant at Sea'. All these categories put together gave me about fifty stories. I discarded ten for a variety of reasons, and was left with some truly extraordinary fiction. I am sure that there are other Indian stories out there that would have made my list of the greatest stories ever written by Indian writers, but unfortunately these were hidden from me because of the multiple languages in which Indian literature is created—not every language has had its best fiction translated into English. To my great good fortune, with one exception (in which the estate of one of our late great writers did not respond to emails), all the other writers, translators, agents and literary estates were happy to be part of the anthology. And although I determinedly stayed with the original plan of not forcing myself to include writers because of their literary reputations, or regional or other affiliations, I was delighted to find that we had stories from virtually every part of the country with the exception of the Northeastern region (one of the writers in this book, Anjum Hasan, was born there, but there are no stories that reflect the rich and unique literary traditions of the place—I simply couldn't find any that I loved to the extent that they made the shortlist—there were a couple on the longlist). There are also no stories from Gujarati, Sindhi, Konkani, Kashmiri, Maithili and a few other literatures—all of which have accumulated a considerable body of work over time; however, as I have said earlier, stories in some languages were inaccessible to me as there were no good translations to be had in English.

Some of our most admired writers are missing from the anthology. While I was able to include an Amitav Ghosh translation of Tagore in the book, he has written no short fiction that I know about. Similarly, other writers of fiction I admire like Aravind Adiga, Kiran Desai, Shrilal Shukla, Purnachandra Tejasvi, Arundhati Roy, Shivarama Karanth, M. T. Vasudevan Nair, Sunil Gangopadhyay, and Qurratulain Hyder (to pick just a few of the more prominent ones) do not feature for one of three reasons—they haven't written short stories; or, if they have, I prefer their novels to their stories; or, in the case of writers who needed to be translated, some of their best work didn't travel particularly well into English.

Once the selection was done I had to figure out how the stories would be organized—by century, region, language? In the end I decided to keep it simple and arrange the stories in chronological order—based on the author's date of birth, not the date on which the story was published because with the older stories it was not always possible to unearth the date of first publication. This arrangement yielded some very nice pairings and juxtapositions.

Time now to address two contentious issues that have simmered for decades. Let me deal with the easier issue—the authenticity or otherwise of Indian writing in English. I think it is fatuous to consider Indian writing in English unauthentic for two reasons: (1) Those of us who write in English do so because that is the language we are most comfortable with ('our father-tongue') and it makes us no less Indian, nor our reality any less Indian and (2) English has been an Indian language for many hundreds of years now, and is as rooted in this soil as any of the other 'Indian' languages that arrived from beyond our borders. Vikram Chandra, one of our best story-tellers, built up a fine head of steam on the issue a few years ago. Here's a brief excerpt of what he had to say:

> Indians have lived in many languages simultaneously for thousands of years. Did the great Sanskrit playwright Kalidasa speak Sanskrit at home? Maybe he did, and maybe he spoke a Prakrit. We'll never know for sure. But we do know for certain that the Bombay poet Kalidas Gupta, whose takhallus or nom-de-plume is 'Raza', was born in Jullunder, Punjab, in a Punjabi-speaking household. Raza first wrote in Farsi, then in Urdu and English...
>
> I was born into a household that on a census form would undoubtedly be tagged as 'Mother Tongue: Hindi'. But I called my mother 'Mummy' and my father 'Daddy'. They spoke to me in Hindi sprinkled with English. Sitting on my mother's lap, I read newspapers in English. English was everywhere in the world I grew up in, and continues to be an inextricable thread in the texture of every day I live in Bombay and in India. English is spoken on the playgrounds, and we tell folk tales in it, we riddle each other and joke with each other in it, and we make up nonsense verse and nursery rhymes and films in it. Along with many other languages, it is spoken in the slums, on the buses and in the post offices and the police stations and the courtrooms. English has been spoken and written on the Indian subcontinent for a few hundred years now, certainly longer than the official and literary Hindi that is our incompletely national language today... If Hindi is my mother-tongue, then English has been my father-tongue. I write in English, and I have forgotten nothing, and I have given up nothing.

Just as absurd as the notion that Indian writing in English isn't a major strand of Indian literature is the idea that there isn't anything of consequence taking place in Indian languages other than English. Every major language in this country has writers who have created indelible masterpieces. Every major language has had its fair share of innovators and writers who have pushed creative boundaries as far as they can go. Many of these masters find a place in this anthology.

The major problem that will persist for the foreseeable future where Indian literature is concerned is this—great literature created in one language is often inaccessible to readers in other languages and there doesn't seem to be any practical way to deal with the problem. I will talk here only of the difficulties of translating literary work from the other Indian languages into English (as I have no real knowledge of the problems faced by publishers, readers and writers who are trying to translate works into their own languages). Let's start with the fundamental problem—the impossibility of making an exact translation that reproduces every nuance and wrinkle of the story in the original language, but that is at the same time a smooth read without awkward, clumsy passages in the language that the book or story has been translated into. This is something the world's greatest translators, to varying degrees—whether it is Gregory Rabassa in Spanish or A. K. Ramanujan in Tamil and Kannada—are agreed on. The best they can do is what Rabassa (whose translation of *One Hundred Years of Solitude* won Gabriel Garcia Marquez's unqualified admiration) puts in the following way: 'There's a musicality that underlies a book, and I think that if you can move that into English, you can catch it and you've got it.'

In this country, compounding the problem is the fact that very few translators who are capable of making outstanding translations actually ever do so—it is simply not worth their while, either in terms of recognition or monetary compensation. Unfortunately, for as long as sales of translations remain small, as has been the case for decades now, it is hard to see this situation becoming better. Maybe if a large-hearted philanthropist—without an agenda and a real interest in Indian culture—decided to fund world-class translators to work with writers in the various Indian languages, to sculpt great translations of our finest stories into other languages, things might improve, but until that happens, we'll just have to muddle along as best we can. What this means, in real terms, is that we will need immensely talented writers and translators to voluntarily undertake to translate books or stories. And we will need more organizations like Katha, which has done extraordinary work in the field of translation, to spring up. It would also help if government organizations tasked with publishing translations imposed more quality control on the books they put out.

For this anthology I was extremely fortunate to have found the work of some of our most accomplished translators. Amitav Ghosh, Khushwant Singh, Ranga Rao, Gita Krishnankutty, Rakhshanda Jalil, Geeta Kapur, O.V. Usha, Sitakant Mahapatra, Lakshmi Holmström, S. Krishnan, H.Y. Sharada Prasad, C. M. Naim, Arshia Sattar, A. K. Ramanujan, Manu Shetty, Arunava Sinha, Vernon Gonsalves and Gayatri Chakravorty Spivak (most of them writers of stature themselves) have all produced great translations—the stories they have translated work brilliantly for the reader in English.

Just one last comment about the selection before I wind down. What was it that I was looking for in the stories? Why did I like these ones, and not others? At first glance practically every form of short fiction is represented in this selection—humorous sketches, carefully plotted stories, domestic dramas, Chekhovian slice-of-life stories, stories that revolve around a single unforgettable character, ghost stories, vampire stories, erotic stories, fables, satire, adventure stories, stories that can be read by both young adults and adults, science fiction, fantasy, political stories, stories within stories... But if you look closer you will find that the majority of the stories have a vivid sense of place and an exceptionally strong voice. Many of them are rooted in classical Indian forms of story-telling or unselfconsciously use Indian myth and legend in their narratives. In addition, they possess in great abundance the Indian sensibility I was referring to earlier. There is nothing ersatz about these stories. In them, you will encounter an India that is sharper, clearer and imprints itself more deeply on your consciousness than anything you will find in real life. This is what serious literature is meant to do, and that is the hallmark of these stories across genres.

For the past few months I have been building a memory palace. This is a mnemonic device invented by the Romans and Greeks centuries ago to help people retain masses of information. If you're a fan of Benedict Cumberbatch as Sherlock, you have probably seen him construct a memory palace to organize clues to help him track down a killer. The concept is simplicity itself. All you will need to do is imagine a structure in your mind to which you attach whatever information it is that you want to remember. My memory palace is modelled on an actual palace—the opulent former residence of the Maharaja of Mysore that I remember from a visit when I was in college. I have a hazy recollection that it sprawled over acres of land and had a lot of white marble in every room; I didn't much care for the wastepaper baskets made of elephant feet and the numerous shikar trophies on the walls, and on the floors. My memory palace, unlike the maharaja's palace, has onyx wastepaper baskets and modern Indian masters on the walls (cost is no object, naturally). Also, unlike the actual maharaja's palace, this one has an unlimited number of rooms. Some of the rooms are dimly lit, and the others are dark (which indicates that there is no activity going on within them). When I enter any of the rooms that are faintly illuminated, the entire room blazes with light. At the same time the room begins to hum with activity, with various characters engaged in all manner of actions. Each of the lighted rooms represents a favourite Indian book, story or poem—collected on a long journey that began with the absorbing and mystifying books I read on my grandfather's jasmine-wreathed verandah. It is the place I call home.

A CLUTCH OF

IND IAN

MASTERPIECES

THE HUNGER OF STONES
RABINDRANATH TAGORE

Translated from the Bengali by Amitav Ghosh

We met him on the train, my cousin and I, on our way back to Calcutta after a trip around the country during the Puja holidays. At first we took him for a north Indian Muslim, because of the way he was dressed. As for his conversation, it left us utterly baffled. He held forth on every conceivable subject, and with such confidence that you would think the Creator himself never moved a finger except on his advice. We'd had no idea that there were so many unheard-of goings-on in the world: that the Russians had advanced so far, that the British had so many hidden designs, that there was so much trouble brewing amongst our own rajas and maharajas—we had been entirely at peace with the world till then, not having known anything about all this. But then, as our newfound friend said with a tight little smile, 'There happen more things in heaven and earth, Horatio, than are reported in your newspapers.' We were real innocents: this was the first time we had been away from home. He held us spellbound: on the slightest of pretexts, he would switch from lecturing us on science to expounding the Vedas or reciting Persian poetry—and since we knew nothing about science or the Vedas or Persian poetry, our awe of him increased with every word he uttered. My cousin, who was a theosophist, was even convinced that he had some sort of supernatural power—some magnetism or divine force, or an astral body, or something of the kind. He hung upon the lightest word from this unusual man with the rapt attention of a devotee, even jotting down notes in secret. I felt the great man too knew what was going on and, although he didn't let on, he was not in the least bit displeased.

When we got to the station where we were to change trains, we went off to the waiting room together. It was half past ten. We learnt there was a long wait ahead: our train had been delayed for some reason or the other. I made up a bed for myself on one of the tables, hoping to catch some sleep. But just then the great man launched upon a story.

There was not to be any sleep for me that night.

When I went to work for the Nizam's government in Hyderabad, having quit my job in Junagadh state because of certain disagreements over administrative matters, my new employers chose to send me, because of my youth and good health, to the outlying town of Barich to handle the collection of cotton revenues.

Barich stands upon a very romantic site. Beneath a range of lonely mountains, the fast-flowing river Shusta (from the Sanskrit 'svachchhatoa', limpid) runs like a nimble dancer through towering forests along its winding, rippling course. Right on its banks, looming in solitude at the foot of the mountains, at the top of a flight of one hundred and fifty stone steps, rises a palace built of white marble. There is no habitation anywhere nearby. The village and the cotton market of Barich are a good distance away.

Some two hundred and fifty years ago, Shah Mahmud II had built this palace here, upon this remote and lonely site, as his house of pleasure. At that time, secluded deep within the mansion's cool moist interior, rosewater flowed from fountains in the bathing-chambers; and young Persian women, their hair loosened before their bath, sat on smooth wet stone seats with their bare feet in the clear water of the pool, strumming sitars and singing ghazals of the vineyard.

The fountains are silent now; there are no songs, and no fair footsteps resound on the white marble. Today the palace serves as an enormous empty residence for lonely womanless revenue collectors like myself. An elderly clerk in my office, Karim Khan, warned me repeatedly against living in that palace. 'Go there during the day if you must,' he said, 'but don't on any account spend the night there.' I laughed. My servants said they would work there during the day but would not spend the night. I agreed. The palace's reputation was such that even thieves would not go there at night.

At first the emptiness of that abandoned marble mansion bore down on me like a crushing weight. I spent as much time away as I could, working through the day without a pause; and when I went back at night, I would fall asleep at once in exhaustion.

But before a week had passed, the house began to assail me like a strange addiction. It is hard to describe the state I was in, and just as hard to make it sound credible. The palace was like a living thing, slowly ingesting me in its entrails.

Perhaps the process had started the moment I set foot in the place; but I still remember, perfectly clearly, the day when I first became aware of the way it was working on me.

The hot weather was just beginning to set in, so the market was slow and I didn't have much to do. It was a little before sunset; I was sitting by the river, at the bottom of the great flight of steps, relaxing in a long-armed chair. The river was running low then: a broad stretch of sand had appeared on the far shore and was glowing with the colours of the sunset sky; on this side, close at hand, pebbles glistened on the steps that lay beneath the clear shallow water. It was very still that day. The air was heavy with the thick scent of wild basil, mint and fennel wafting down from the nearby mountains.

The moment the sun slipped below the mountain-tops, a low dark curtain descended upon the stage of the day. Because the hilly terrain shut out the last light of the sun, twilight was of short duration here. Toying with the idea of going for a ride, I had half-risen from my chair when I heard a footstep on the steps behind me. I turned to look: there was no one there.

My hearing had played a trick on me, I decided; but no sooner had I sat down than I heard a sound again, a number of footsteps this time, running quickly down the stairs. I was transfixed, seized by an excited delight not unmixed with a tinge of fear. There was nothing in front of me, no discernible form; yet I knew, as well as if I could see them, that a group of high-spirited young women had just stepped into the river to bathe at the end of the hot summer's day. Even though there was not so much as a whisper anywhere that evening, neither in the mountains, the river nor the house, I still heard the bathers perfectly clearly as they swept past me one after another, laughing and chattering like a playful mountain stream. They did not notice me: I seemed to be just as invisible to them as they were to me. The river was as calm as ever, but I still felt clearly that the waters were being stirred by many braceleted arms, as the women laughed and splashed water on each other, and their feet sent the water drops arcing through the air like fistfuls of pearls.

A great ferment of excitement was stirring within me now; whether it grew out of fear or joy or curiosity I cannot say. I began to wish I could see everything properly, but there was nothing to see; I thought, if only I listened carefully, I would be able to hear everything they were saying—but no matter how intently I listened, all I could hear was crickets chirping in the forest around me. I thought: swinging in front of me is a dark curtain, two and a half centuries old. Let me pick up a corner of it and cast an apprehensive glance inside: I would see a great gathering of people. But there was nothing visible in that inky darkness.

Suddenly the pall of stillness was swept aside by a gust of wind. The Shusta's surface was quickly teased into wavelets, like the braids of a heavenly nymph, and the shadowy evening forest gave a deep sigh, as if waking from a bad dream. Dream or reality, the unseen mirage from two hundred and fifty years ago that had presented itself before me vanished in the twinkling of an eye. The magical creatures that had brushed past me with swift disembodied steps and shrill silent laughter to plunge into the Shusta did not make their way back, wringing the water from their dripping clothes. They vanished in that single breath of spring, as a whiff of perfume is lost upon the breeze.

Then, all of a sudden, I had a fright. I began to wonder whether the muse of poetry had chosen to descend on me, finding me off my guard in that desolate place—perhaps I, who had to sweat for my living by collecting cotton revenues, had suddenly been doomed to the curse of poesy. I thought to myself, I must make sure to eat properly: it's when your stomach's empty that obstinate sicknesses of all kinds crowd in on you. I called my cook and ordered a truly Mughal dinner, rich and spicy.

Next morning the episode began to seem ridiculous. I clapped a sola topi on my head like an Englishman, and went rattling off to work in the best of spirits, driving the trap myself. I expected to be home late that day, as I had my quarterly report to write. But no sooner did evening come than I began to feel that the house was summoning me back. Who summoned me, I cannot say; but I kept thinking, it won't do to stay any longer, everybody will be waiting. I left my report unfinished, put on my sola topi, and drove back to that great silent palace below the mountains, startling the lonely twilit tree-lined path with the rumbling of my wheels.

At the top of the palace's main stairway was an enormous room. Three rows of monumental columns held up the ceiling on intricately carved arches.

The very emptiness of that vast room sent its echoes resounding through the house all day and all night. It was still early in the evening, so no lamps had been lit yet. The moment I pushed the door open and entered the great room, I had the distinct feeling that I had caused an uproar—that a large gathering had suddenly broken up, and people were tumbling through the windows and out of the doors, fleeing where they could in every direction, down the terraces and corridors. In my astonishment, seeing nothing anywhere, I stayed exactly where I was. A rapture stole over my body; the mild scent of vanished perfumes and pomades, relics from another epoch, wafted past my nostrils. Standing there in that great dark empty room, amongst those long rows of ancient columns, I heard the gurgling of fountains upon the marble floor and the sound of sitars playing an unknown tune. Somewhere, a copper gong was striking the hour; from somewhere else came the ringing sound of anklets and gold jewellery; musical instruments were playing far away; crystal chandeliers tinkled in the breeze; bulbuls sang in cages on the terrace, and the palace cranes called in the garden, the whole weaving a ghostly music around me.

I was in such a trance that I began to imagine that this ineffable, unattainable, unreal setting was the only reality on earth, that everything else was a mirage. That I was the person I was—So-and-so, eldest son of the late So-and-so, who earned a salary of four hundred and fifty rupees collecting cotton-revenues, who went to his office every morning in a trap wearing a sola topi and a short jacket—all this seemed such an absurd, unfounded lie that I began to shout with laughter, standing in the middle of that great silent room.

At that moment, my Muslim servant entered the room with a lighted kerosene lamp. Whether he thought me mad or not I do not know, but as for myself I recalled at once who I was, that I was indeed the worthy Mr So-and-so, son and heir of the late So-and-so. I also bethought myself that only seers and poets can tell whether, in this world or beyond it, water can really spout endlessly from invisible fountains, and unending melodies sound on illusory sitars; but it was certain that I earned four hundred and fifty rupees a month collecting taxes in the cotton markets of Barich. The thought of my recent trance began to seem ridiculous; sitting down at my lamp-lit camp-table, with a newspaper in my hands, I soon succumbed to laughter.

After a meal of rich Mughlai food, having read my newspaper, I took myself off to the small corner room that served as my chamber, put out the lamp and lay down on my bed.

A brilliant star shone through an open window: perched high above the dark forested Aravalli mountains, from its exalted place in the sky millions of leagues away, it fixed its gaze upon the humble Mr Tax Collector on his humble camp-cot. Diverted by this odd conceit, I soon drifted off to sleep: for how long I cannot say, but suddenly I felt myself shiver, and I was awake again. It was not as though there had been any sound in the room, or that anyone had entered it. But by that time, the dim glow of the waning moon was creeping in diffidently through the open window, while the star that had gazed so fixedly upon me had dipped beneath the gloomy mountains.

I could not see anybody in the room. But I had a clear sense that someone was nudging me, ever so gently. The moment I sat up, five beringed fingers beckoned and, without a word being said, gestured to me to follow cautiously behind.

I rose stealthily to my feet. I knew there was no living soul but me in that immense hundred-chambered palace, filled with a great emptiness, with sleeping sounds and waking echoes; yet with every step I took along those deserted, echoing corridors, I was stricken with fear, terrified that somebody would awaken suddenly. Most of the rooms in the palace were kept shut, and I had never ventured to enter any of them. I cannot clearly tell where I went that night, following that invisible, beckoning figure with silent step and hushed breath. I could not begin to count the dark narrow passages, the broad corridors, the still sombre council hall and small hidden airless chambers through which we made our way.

Even though I had not set eyes on my guide, my mind was not entirely ignorant of her appearance. She was from Arabia: her firm rounded arm, looking as if it were carved from marble, showed below her broad sleeve, a fine veil hung down from her cap across her face, and a curved dagger glinted in her waistband.

It seemed to me that a night from the Thousand and One Nights had transported itself here from the realms of fiction; that I was stealing through the narrow unlit alleyways of the sleeping city of Baghdad on a dark night, on my way to some perilous assignation.

All of a sudden, my guide came to a halt and gestured towards the bottom of an indigo curtain. There was nothing there, but the blood froze in my veins. I sensed that at the foot of the curtain, swathed in a robe of silk brocade, lay the drowsing form of a gigantic Kafir eunuch, his legs outstretched, a drawn sword resting on his lap. My guide stepped lightly over his legs and lifted up a corner of the curtain.

A part of a room was visible inside, its floor covered with a Persian rug. There was a seated figure too, but all that could be seen of her was the lower part of a pair of loose saffron-coloured leggings and two beautifully shaped feet in gold-worked slippers, resting on a cushion of pink velvet. On a table beside her, arranged on a blue crystal platter, were apples, pears, oranges and bunches of grapes; and beside them, as though in expectation of a visitor, two wine glasses and a gold-encrusted decanter. An intoxicating scent of incense drifted from the room and overcame me. Heart pounding, I raised a foot to climb over the eunuch's outstretched legs when he suddenly awoke, his naked sword falling to the floor with a clatter. I started at the sound of a piercing shout, and found myself sitting on my camp-cot, drenched in sweat. The waning moon had turned pale in the first light of dawn, like a sick man after a sleepless night; and crazy old Meher Ali was marching down the empty road as was his custom, shouting, 'Stay away, stay away.'

Thus untimely ended the first of my Arabian nights—with a thousand left, yet remaining.

A strange feud now arose between my days and my nights. By day, I would take my weary body off to work, heaping curses upon my beguiling nights full of empty dreams. But once evening had set in, it was my workaday daytime existence that seemed trivial, false and absurd.

Once evening came, I would feel myself caught in a web of rapture. I would become a different being, a character in an unrecorded history of centuries ago. My short English jacket and my tight pantaloons would begin to seem oddly incongruous; with the greatest care, I would put on a red velvet fez, loose leggings, a flowered shirt and a long silk achkan, with a coloured attar-scented handkerchief. Then, putting away my cigarettes, I would light a great hubble-bubble filled with rosewater, and sink into a high upholstered sofa. And thus I would sit, as though I were waiting in the most eager suspense for some extraordinary night-time tryst.

As the darkness gathered around me, strange things would happen that are impossible to describe. It was as though the pages of some extravagant romance were blowing through the strange rooms of that vast palace on sudden gusts of summer breeze—episodes that could be followed only to a certain point and no further. Setting out in pursuit of those swirling fragments, I would wander from room to room all through the night.

Whirling through those disjointed dreams—gusts of wind moist with scented waters, whiffs of henna and snatches of sitar music—I would sometimes catch fleeting glimpses of a woman, like flashes of lightning. Hers were the saffron leggings, the soft pink feet shod in upturned gold-worked sandals. Her breasts were tightly bound in a flowered bodice with gold braid, and a fringe of gold hung from her red cap to veil her forehead and cheeks.

I was besotted with her: it was to meet her that I would roam every night among the alleyways of that labyrinthine dream world, in the subterranean realm of sleep.

Sometimes, standing before my wide candle-flanked mirror, changing into my princely night-time attire, I would catch sight of her, that Persian woman, reflected back at me beside my own mirror image, bending her neck, glancing passionately, painfully, sensually out of her great dark eyes, hinting with full red lips at some unspoken utterance, pirouetting with her slim youthful figure in a light graceful dance—and then, in a trice, melting into the glass in a shower of incandescence from her pain and desire and rapture, laughter and sidelong glances and shimmering jewellery. Then a strong wind, redolent of all the scents of the forest, would blow out my candles; I would abandon my elaborate costume, close my eyes, and stretch out my ecstatic body on the bed in the corner of my dressing-room. In the breezes that blew over me, in all the mingled scents borne from the Aravalli Hills, I would discover many kisses, many caresses, many a touch of a soft hand floating in that solitary darkness. I would hear murmuring voices, feel the warmth of perfumed breath upon my forehead and the gentle scented touch of a woman's kerchief blown across my cheek, again and again. It was as though a bewitching she-serpent was binding me in her intoxicating coils: with deep sighs, my benumbed body would sink into a heavy sleep.

One evening, I thought of going out for a ride. I have no idea who it was that kept dissuading me, but I paid no heed. My sola topi and short jacket were hanging from a wooden rack. Just as I was about to change into them, a sudden whirlwind swept down, carrying the sand of the Shusta and dead leaves from the Aravallis like a pennant, and bore away my jacket and my hat. They went cartwheeling through the air: a sweet chorus of laughter swirled along with them, rising through several octaves, sounding every note on the scale of derision, until finally it dissolved into the sunset.

I did not go riding that day; and never again did I wear my absurd little English jacket and sola topi.

That very midnight, sitting up suddenly on my bed, I heard the sound of weeping—racking, broken-hearted sobs. It was as though beneath my bed, beneath the floor, among the foundations of this vast mass of stone, a voice was calling out from a dark and musty grave, crying: 'Take me away, give me my deliverance; break down the doors of this rooted illusion, this deep sleep, this futile dream. Put me on your horse, take me in your embrace—carry me away through the forest, over the mountains, across the river into your own sunlit room. Give me my deliverance!'

I! But who was I? How could I save her? What lovely creature of desire was I to draw out of the flowing, whirling torrent of dreams in which she was immersed? Where did you live and when, you other-worldly beauty? Where were you born, in which palm-fringed oasis, by which desert stream? What desert-dwelling nomad woman brought you into this world? What Bedouin raider plucked you from your mother's arms like a tender flower from its parent creeper, and carried you off on his lightning-swift horse across the searing sands to the slave market before some royal palace? What servant of the emperor counted out his gold upon seeing your first bashful bloom of youth, transported you across the seas, and then carried you on a litter of gold as a gift for his master's harem? And what then was your history? The sound of the sarangi, the tinkling of ankle-bells, the golden wine of Shiraz— and interspersed among them, the glint of a dagger, the sting of poison, a wounding glance. Limitless wealth, perpetual imprisonment.

Diamonds glitter on the bracelets of slave-girls as they wave their fans on either side; the Shah-en-shah Badshah lies beneath those gleaming feet, beside those pearl-embroidered slippers; a gigantic Ethiopian stands at the door, drawn sword in hand, like a messenger of death in angel's dress. And then that rich, envy-spuming, blood-soaked, intrigue-ridden torrent sweeps over you, fearsomely bright—to what abyss of cruel death, my fragile sprig of the desert, or still more cruel shore of noble living?

Then Meher Ali began to shout: 'Stay away, stay away. It's a lie, all of it's a lie.' Opening my eyes, I saw that it was morning: a messenger from my office handed me the day's mail, and the cook came to ask what meals he should prepare that day.

I told myself, 'It won't do to live here any longer.' That very day I packed my things and moved into my office. The old clerk, Karim Khan, smiled wryly upon seeing me. His smile annoyed me. Without responding, I began on the day's work.

As the day wore on, my mind began to wander. It was as though I had an appointment to keep. The cotton revenue accounts lost their urgency, indeed the affairs of the Nizam's entire estate dwindled into insignificance. Everything that was actual and current, everything happening around me—people coming and going, eating and working—seemed utterly mean, trivial, devoid of sense.

Flinging my pen aside, I shut my enormous account book and leapt into my trap. It seemed to halt of its own accord at the gates of that great marble palace at the very point of dusk. I raced up the staircase to the vast room at the top. Everything was quiet today, as though the chamber had taken offence and was sulking. I was stricken with remorse—but to whom could I express it, of whom could I ask pardon? I wandered vacantly from room to room, wishing that I had an instrument in my hands so that I could sing to one absent: 'O flame, the moth that tried to fly away from you has come back to die. Give it its absolution, set its wings alight and turn it to ashes.'

Then, suddenly, I felt a couple of teardrops falling on my cheek. The sky above the Aravallis had been heavy with rain-filled clouds that day. The darkness-shrouded forest and the Shusta's inky waters were still with fearful expectation. Suddenly, the water, the earth and the sky quivered, all at once; and a howling storm burst forth out of the distant forest like a madman that had burst his chains, baring its teeth in lightning-flashes. The palace's empty cavernous rooms began to shriek in torment, beating their doors wildly in the wind.

All my servants were at the office that day, and there was nobody at hand to light the lamps. Thus it happened that there, in the impenetrable darkness of that room, on that overclouded, moonless night, I sensed, with perfect clarity—a woman, lying face down on a rug beside the bed, tugging at her unbound hair with clenched fists, blood pouring from her ivory forehead: sometimes bursting into fierce arid laughter, sometimes into heart-rending sobs, tearing away her dress and beating upon her bared breasts with both hands—and all the while, driven by the roaring wind, sheets of rain poured in through an open window, drenching her entire body.

Neither the storm nor the weeping stopped that night. I spent those hours wandering in the dark from room to room, grieving helplessly. There was no one anywhere; no one to whom I could offer solace. Whose was this dreadful sorrow? What was it that lay behind such a perturbation?

Then there came the lunatic's cry: 'Stay away! Stay away! It's a lie, all of it's a lie.'

I saw that it was dawn. Meher Ali was making his rounds and shouting out his cry as usual, even on this unpropitious morning. It occurred to me then that perhaps Meher Ali too had once lived like me in this palace. Emerging deranged by the experience, he returned each morning even now to wander around the palace, still held in thrall by the great stone monster.

Right then, in the pouring rain, I ran up to the crazy old fellow and asked: 'Meher Ali, what is it that's a lie?'

He thrust me aside without an answer and went on his way, circling around and around in a mesmeric trance, like a bird caught in the hypnotic spell of a python's gaze. But over and over again as though his life depended on it, he kept warning himself: 'Stay away! Stay away! It's a lie, all of it's a lie.'

I left for my office at once, careering like a madman through the storm. I summoned Karim Khan and demanded: 'Tell me clearly, what does all this mean?'

The gist of what the old man said is this. There was a time once when many flames of unfulfilled desire and demented lust had teemed and flared inside that palace. Every block of stone within it is still hungry, still athirst, from the curse of that anguished and frustrated longing. Whenever they find a living human being within their grasp, they seek to devour him like ravening demons. Of all the people who had spent three nights in that place, Meher Ali was the only one who had emerged alive, although he too had lost his reason. No one else has ever been able to elude its grasp.

I asked: 'Is there no way to save me?'

The old man said: 'Only one, and a very difficult way it is too. I'll tell you what it is—but to do that I must first tell you the old story of the Persian slave-girl in that rose garden. In all the world there has never been another tale so strange or so affecting...'

Just then the porters came to inform us that our train was pulling into the station. So soon? By the time we had repacked our bedrolls, the train had arrived. An Englishman, just up from sleep, thrust his head out of a first-class compartment to read the name of the station. Spotting our fellow-traveller, he cried out 'Hullo' and invited him into his compartment. Ours on the other hand was a second-class carriage. So we never learnt who the gentleman was, nor did we get to hear the end of his story.

I said, 'The man took us for fools and had a good laugh at our expense. The story was all made up from beginning to end.'

The argument that followed led to a lifelong rupture between me and my theosophist cousin.

two

~

THE SHROUD

MUNSHI PREMCHAND

Translated from the Hindi by Arshia Sattar

Outside the hut, father and son sat before the dying embers in silence. Inside, the son's young wife, Budhiya, was thrashing about in labour. Every now and then, a blood-curdling shriek emerged from her mouth and they felt their hearts stop. It was a winter night, the earth was sunk in silence and the whole village had dissolved into the darkness.

Ghisu said, 'Looks like she's not going to make it. She's been like this all day. Go take a look.'

Madhav replied irritably, 'If she's going to die, why doesn't she do it quickly? What's the point of taking a look?'

'You're pretty harsh. You've had a good time with her all year, and now? Such callousness?'

'Well, I can't stand to see her suffer and thrash about like this.'

This clan of cobblers was notorious in the village. If Ghisu worked a day, he would rest for three. Madhav was such a shirker that if he worked for half an hour, he'd smoke dope for one. Which was why they were never hired. If there was even a fistful of grain in the house, they took it to mean they didn't have to work. When they'd been starving for a few days, Ghisu would climb a tree and break off some branches and Madhav would sell them in the bazaar. As long as the money lasted, they'd loaf around here and there. And when starvation hit them they would break off more branches or look for work. There was no shortage of work in the village, it was a village of farmers and there were at least fifty jobs for a hard-working man. But these two were called in only when you had to be satisfied with two men doing the work of one.

Had they been renunciants, they would have had no need to exercise control or practice discipline in order to experience contentment and fortitude. Theirs was an unusual existence—apart from a few mud pots, there were no material possessions in their house. They went on with their lives, covering their nakedness with rags, free of worldly cares, burdened with debt. They'd suffer abuse, they'd suffer blows, but they had not a care in the world. They were so wretched that even though there was no hope of being repaid, people always loaned them something. During the potato harvest, they'd pull up peas or potatoes from other people's fields, cook them in some fashion and eat them. Or, they'd uproot a few stalks of sugarcane and suck on them at night. Ghisu had lived out sixty years with such supreme detachment and now Madhav, his worthy son, walked in his father's footsteps, determined to become even more illustrious.

At this moment, too, they were roasting potatoes, which they had dug up from someone else's field, in the embers. Ghisu's wife had died many years ago. Madhav had married only the previous year. After the woman had come, she had laid the foundations for some kind of discipline in the household and managed to fill those shameless stomachs. And since she'd arrived, the two had become even more inclined to relax and had even started acting pricey. If someone called them in to work, they'd ask for double wages without batting an eyelid. Today, that woman was dying in childbirth and it was quite likely the pair was waiting for her to die so that they could get a good night's sleep.

Ghisu pulled out a potato and, peeling it, he said, 'Go and see what's happening to her. There'll be the business of a witch, you can bet on it.'

Madhav was afraid that, if he went into the hut, Ghisu would grab a larger share of the potatoes. He said, 'I'm scared to go in there.'

'What's there to be scared of? I'm right here.'

'So why don't you go and see, then?'

'When my wife was dying, I didn't move from her side for three days. This one, she'll be embarrassed in front of me, won't she? I've never even seen her face. Now to look at her uncovered body! She'll be uncomfortable. If she sees me, she won't be able to throw her arms and legs around so freely.'

'I'm wondering what will happen if there's a child—ginger, jaggery, oil—

there's nothing in the house.'

'Everything will come when god is good and ready. This lot, who aren't giving us any money now, these same people will call us tomorrow and give us cash. I've had nine sons, there was never anything in the house, but god got us through the mess somehow.'

In a society where people who toil day and night are not much better off than these two, and instead of farmers it's those who exploit them that grow rich, it's no surprise that attitudes like this develop. Let's say that Ghisu was cleverer than the farmers, that instead of joining those simple-minded peasants, he'd joined the company of conmen. Of course, he did not have the capacity to follow that company's rules and regulations, which was why others of his ilk had become chiefs and headmen in the village while he remained the one at whom fingers were pointed. Still, he had the consolation that, however badly off he was, he didn't have to work as achingly hard as the others and that people could not take undue advantage of his simplicity and helplessness.

The two of them pulled out the potatoes and devoured them, hot as they were. They had eaten nothing since the previous day. They didn't even have the patience to let them cool so, every now and then, they scalded their tongues. When it was peeled, the outer part of the potato did not seem that hot but, as soon as it was bitten into, the inner part burned the tongue, the throat and the palate. Instead of holding that burning coal in one's mouth, it seemed wiser to send it down as soon as possible so it could cool down in the stomach. That's why they were swallowing so quickly, although the effort made their eyes water.

Ghisu thought back to a landlord's wedding feast that he had been to twenty years ago. The contentment he had felt at that feast was worth remembering for a lifetime and, even today, the memory was fresh. He said, 'I'll never forget that meal. I've never eaten that kind of food—or that much of it—ever again. The girl's family fed everyone as many puris as they could eat. Everyone. The rich, the poor—everyone ate those puris. And they were made with pure ghee, mind you. Chutney, raita, three kinds of greens, one curried vegetable, curds—I can't tell you how delicious that food was. There was no holding back. Ask for whatever your heart desired; eat as much as you want. People ate and ate, so much that they couldn't even drink water. But those who were serving, they kept putting freshly cooked, perfectly round fragrant kachoris on to our plates. We refused, we covered our plates with our hands, but they just kept serving! And when we were done, we even got paan and cardamom. I was in no shape to take the paan, I could barely stand. I went off immediately and wrapped myself in my blanket and lay down. That's how big-hearted he was, that landlord. Like an ocean!'

Madhav savoured those delicacies in his mind and said, 'No one gives us a meal like that now.'

'There's no one to feed us like that any more. That was a different time. Now everyone's counting pennies—don't spend on weddings, don't spend on religious festivals. Ask them, where will they stash all the money they take from the poor? There's no problem stashing the money, but when it comes to spending, then they think of thrift.'

'You must have eaten about twenty puris, no?'

'I ate more than twenty.'

'I would have eaten fifty.'

'I ate no less than fifty. I was pretty sturdy those days. You're not half of what I used to be.'

They ate the potatoes, drank some water, curled up, covered themselves with their dhotis and fell asleep right there by the embers, like two enormous pythons that had eaten their fill.

And still, Budhiya moaned.

In the morning, when Madhav looked inside the hut, his wife lay there, stone cold, flies buzzing around her face, her expressionless eyes rolled upwards. Her body was covered with dust, the child had died in her womb. Madhav ran to Ghisu. They started to wail loudly and beat their chests. The neighbours heard the weeping and wailing and came and, as was customary, began to console the two unfortunates. But this was not the time for full-throated lament, the shroud and the wood had to be considered. Money disappeared from that house like a piece of meat in a kite's nest. Father and son went, wailing, to the village landlord who could not stand the sight of them. He'd beaten them himself often enough, for stealing, for not showing up for work after they had promised to. He asked, 'What is it, Ghisua, why are you crying? I don't see you around much these days, seems like you don't want to live in this village any more.'

His eyes filled with tears, Ghisu touched his head to the ground and said, 'Master, I am ruined. Madhav's wife died last night. She suffered all night, master. The two of us sat by her side half the night, we gave her all the medicines we could. But she has abandoned us. And now there's no one to give us even a piece of bread, master. We've been destroyed, our home has been uprooted. I am your slave! There's no one but you—who will organize her funeral? Who else can I turn to except you?'

The landlord was a compassionate man but having pity on Ghisu was like trying to dye a black blanket. In his heart, he felt like saying, 'Get away from here! You don't come when you are called and now, when you need me, you come here and flatter me! Bastard! Rascal!' But this was not the moment for anger or for retribution. He tossed a reluctant two rupees at him but not a single word of consolation escaped his lips. He did not even look at Ghisu, as if he'd rid himself of a burden.

Once the landlord had given two rupees, how could the village merchants and traders have the courage to refuse? And Ghisu knew how to use the landlord's name to his advantage. Some gave two annas, others gave four. Within an hour, Ghisu had collected the healthy sum of five rupees. Grain came from one place, wood from another. In the afternoon, Ghisu and Madhav went off to the bazaar to buy the shroud. People began to cut bamboo poles and the soft-hearted women of the village would come and stare at the corpse and shed a few tears at Budhiya's misfortune.

What a sad custom, that the woman who didn't even have rags to cover her body while she was alive now needed a shroud. After all, the shroud burned with the body. And then what's left? If the same five rupees had come earlier, there might have been some medicine. Ghisu and Madhav were trying to gauge each other's thoughts. They wandered around the bazaar, from this cloth shop to the next. They looked at all kinds of fabric, from silk to cotton, but nothing seemed right. Eventually, it became evening. And who knows by what divine inspiration the pair landed up in front of a bar and, as if they'd planned it earlier, sauntered in. They stood around uncertainly for a while. Then Ghisu went up to the counter and said, 'Mister, give us a bottle.' Soon, snacks arrived and then some fried fish, and the two of them sat on the porch, drinking calmly. After knocking back a few rather quickly, their spirits rose.

Ghisu said, 'What's the point of the shroud? It only gets burned, it's not as if it goes with her.'

Madhav looked at the sky, as if calling the gods to witness his innocence, and said, 'It's the way of the world. Otherwise, why would people spend thousands feeding Brahmins? Who knows whether you benefit in the other world? Rich people have money, let them blow it. What do we have to waste? But we're still answerable to others. They're sure to ask, "Where's the shroud?"'

Ghisu laughed. 'Let's say I dropped the money. That we looked and looked but could not find it anywhere. They won't believe a word, but the same lot will give again.'

Madhav also laughed at this unexpected good luck. He said, 'She was a good woman, poor thing. She's dead, but she's given us food and drink.'

More than half the bottle was gone. Ghisu ordered two rounds of puris and chutneys and pickles and liver. There was an eating place just in front of the bar. Madhav leapt across and brought all the food back on two leaf plates. Another one and a half rupees well spent. There was only a little change left. The two of them sat eating their puris, as grandly as if they were lions hunting in the jungle. They were not afraid of being responsible to anyone, nor did they worry about their reputations. They had conquered those virtues long ago.

Ghisu said philosophically, 'We're feeling good. She'll get some credit for that, won't she?'

Madhav bowed his head piously and said, 'Of course. Definitely. Lord, you are present in each of us, let her go to the highest of heavens. We're both blessing her from the bottom of our hearts. The meal we've had today! We've never eaten like this in our lives.'

A moment later, a tiny doubt rose in Madhav's mind. He said, 'We'll go there one day, too, won't we, Father?'

Ghisu ignored the naïve question. He wasn't going to ruin the pleasure of the moment with thoughts of the world beyond.

'She's there. If she asks us why we didn't provide her with a shroud, what are we going to say?'

'We'll say, go to hell!'

'She's sure to ask.'

'And you're sure that she's not going to have a shroud? You think I'm an ass? You think I've spent sixty years on earth just digging up grass? She'll have a shroud. And a finer one than this.'

Madhav was still doubtful. He said, 'Who's going to give it? You've spent all the money. And I'm the one she'll ask, I'm the one that married her.'

Hotly, Ghisu said, 'I'm telling you, she'll have a shroud. Why don't you trust me?'

'Why don't you tell me who's going to give it?'

'The same people who gave us this one! Though, this time, we're not going to see the cash.'

As the darkness deepened and the light of the stars grew brighter so the bar grew more radiant. Some sang, some prattled, some embraced their companions, some pressed a cup to their friends' lips. The atmosphere was heady, the air intoxicating. So many came here and got high on a single sip. More than the drink, it was the air that got them drunk. They came, drawn there by the drudgery of living and, for a short while, they forgot whether they were alive or dead. Or neither.

Meanwhile, father and son tippled away happily. Everyone was staring at them. How favoured by fortune they were—they had an entire bottle between them. Having eaten his fill, Madhav gave the leftover puris to a beggar who had been watching them with hungry eyes. And, for the first time in his life, he felt the pride, the pleasure, the exultation of giving.

Ghisu said, 'Here. Eat it all. And bless us! The one who earned this is dead. But your blessings will surely reach her. Bless her with every part of your being, this is hard-earned money.'

Madhav gazed up at the sky again and said, 'She'll go to heaven, Father. She'll be the queen of heaven!'

Ghisu stood up and, swimming as he was through waves of joy, said, 'Son, she's going to heaven. She never bothered anyone, never hassled anyone. She's fulfilled the biggest wish of our lives by dying. If she doesn't go to heaven, you think those fat cats will, those fellows who loot the poor with both hands and then bathe in the Ganga and make offerings of holy water in temples to wash off their sins?'

But this rush of piety soon passed, for impermanence is the essence of intoxication. Sadness and despair crept in. Madhav said, 'But, Father, she suffered a lot in her life. She endured so much before she died.' He covered his eyes with his hands and began to weep, shrieking more and more loudly.

Ghisu reassured him, 'Why are you crying, son? Be glad for her—she's been freed from this world of illusion, she's been released from the cage. She's the lucky one, she's already broken the bonds that tie us to the world.'

And then, they stood there, both of them, and started to sing loudly, *'Liar! Why do you lower your eyes, you liar?'* They began to dance. They leaped, they jumped, they wriggled their hips, they even fell. They expressed emotion with their eyes, they acted out feelings and, finally, they surrendered to their drunkenness and slumped down in a heap.

three

~

A HORSE AND TWO GOATS
R. K. NARAYAN

Of the seven hundred thousand villages dotting the map of India, in which the majority of India's five hundred million live, flourish, and die, Kritam was probably the tiniest, indicated on the district survey map by a microscopic dot, the map being meant more for the revenue official out to collect tax than for the guidance of the motorist, who in any case could not hope to reach it since it sprawled far from the highway at the end of a rough track furrowed up by the iron-hooped wheels of bullock carts. But its size did not prevent its giving itself the grandiose name Kritam, which meant in Tamil 'coronet' or 'crown' on the brow of this subcontinent. The village consisted of less than thirty houses, only one of them built with brick and cement. Painted a brilliant yellow and blue all over with gorgeous carvings of gods and gargoyles on its balustrade, it was known as the Big House. The other houses, distributed in four streets, were generally of bamboo thatch, straw, mud, and other unspecified material. Muni's was the last house in the fourth street, beyond which stretched the fields. In his prosperous days Muni had owned a flock of forty sheep and goats and sallied forth every morning, driving the flock to the highway a couple of miles away. There he would sit on the pedestal of a clay statue of a horse while his cattle grazed around. He carried a crook at the end of a bamboo pole and snapped foliage from the avenue trees to feed his flock; he also gathered faggots and dry sticks, bundled them, and carried them home for fuel at sunset.

His wife lit the domestic fire at dawn, boiled water in a mud pot, threw into it a handful of millet flour, added salt, and gave him his first nourishment for the day. When he started out, she would put in his hand a packed lunch, once again the same millet cooked into a little ball, which he could swallow with a raw onion at midday. She was old, but he was older and needed all the attention she could give him in order to be kept alive.

His fortunes had declined gradually, unnoticed. From a flock of forty, which he drove into a pen at night, his stock had now come down to two goats which were not worth the rent of a half-rupee a month the Big House charged for the use of the pen in their backyard. And so the two goats were tethered to the trunk of a drumstick tree which grew in front of his hut and from which occasionally Muni could shake down drumsticks. This morning he got six. He carried them in with a sense of triumph. Although no one could say precisely who owned the tree, it was his because he lived in its shadow.

She said, 'If you were content with the drumstick leaves alone, I could boil and salt some for you.'

'Oh, I am tired of eating those leaves. I have a craving to chew the drumstick out of sauce, I tell you.'

'You have only four teeth in your jaw, but your craving is for big things. All right, get the stuff for the sauce, and I will prepare it for you. After all, next year you may not be alive to ask for anything. But first get me all the stuff, including a measure of rice or millet, and I will satisfy your unholy craving. Our store is empty today. Dal, chilli, curry leaves, mustard, coriander, gingelly oil and one large potato. Go out and get all this.' He repeated the list after her in order not to miss any item and walked off to the shop in the third street.

He sat on an upturned packing case below the platform of the shop. The shopman paid no attention to him. Muni kept clearing his throat, coughing and sneezing until the shopman could not stand it any more and demanded, 'What ails you? You will fly off that seat into the gutter if you sneeze so hard, young man.' Muni laughed inordinately, in order to please the shopman, at being called 'young man'. The shopman softened and said, 'You have enough of the imp inside to keep a second wife busy, but for the fact the old lady is still alive.' Muni laughed appropriately again at this joke. It completely won the shopman over; he liked his sense of humour to be appreciated. Muni engaged his attention in local gossip for a few minutes, which always ended with a reference to the postman's wife who had eloped to the city some months before.

The shopman felt most pleased to hear the worst of the postman, who had cheated him. Being an itinerant postman, he returned home to Kritam only once in ten days and every time managed to slip away again without passing the shop in the third street. By thus humouring the shopman, Muni could always ask for one or two items of food, promising repayment later. Some days the shopman was in a good mood and gave in, and sometimes he would lose his temper suddenly and bark at Muni for daring to ask for credit. This was such a day, and Muni could not progress beyond two items listed as essential components. The shopman was also displaying a remarkable memory for old facts and figures and took out an oblong ledger to support his observations. Muni felt impelled to rise and flee but his self-respect kept him in his seat and made him listen to the worst things about himself. The shopman concluded, 'If you could find five rupees and a quarter, you would pay off an ancient debt and then could apply for admission to swarga. How much have you got now?'

'I will pay you everything on the first of the next month.'

'As always, and whom do you expect to rob by then?'

Muni felt caught and mumbled, 'My daughter has sent word that she will be sending me money.'

'Have you a daughter?' sneered the shopman. 'And she is sending you money! For what purpose, may I know?'

'Birthday, fiftieth birthday,' said Muni quietly.

'Birthday! How old are you?'

Muni repeated weakly, not being sure of it himself, 'Fifty'. He always calculated his age from the time of the great famine when he stood as high as the parapet around the village well, but who could calculate such things accurately nowadays with so many famines occurring? The shopman felt encouraged when other customers stood around to watch and comment. Muni thought helplessly, my poverty is exposed to everybody. But what can I do?

'More likely you are seventy,' said the shopman. 'You also forget that you mentioned a birthday five weeks ago when you wanted castor oil for your holy bath.'

'Bath! Who can dream of a bath when you have to scratch the tank-bed for a bowl of water? We would all be parched and dead but for the Big House, where they let us take a pot of water from their well.' After saying this Muni unobtrusively rose and moved off.

He told his wife, 'That scoundrel would not give me anything.'

'So go out and sell the drumsticks for what they are worth.'

He flung himself down in a corner to recoup from the fatigue of his visit to the shop. His wife said, 'You are getting no sauce today, nor anything else. I can't find anything to give you to eat. Fast till the evening, it'll do you good. Take the goats and be gone now,' she cried and added, 'Don't come back before the sun is down.' He knew that if he obeyed her she would somehow conjure up some food for him in the evening. Only he must be careful not to argue and irritate her. Her temper was undependable in the morning but improved by evening time. She was sure to go out and work—grind corn in the Big House, sweep or scrub somewhere, and earn enough to buy foodstuff and keep a dinner ready for him in the evening.

Unleashing the goats from the drumstick tree, Muni started out, driving them ahead and uttering weird cries from time to time in order to urge them on. He passed through the village with his head bowed in thought. He did not want to look at anyone or be accosted. A couple of his cronies lounging in the temple corridor hailed him, but he ignored their call. They had known him in the days of affluence when he lorded over a flock of fleecy sheep, not the miserable gawky goats that he had today. Of course he also used to have a few goats for those who fancied them, but real wealth lay in sheep; they bred fast and people came and bought the fleece in the shearing season; and then that famous butcher from the town came over on the weekly market days bringing him betel leaves, tobacco, and often enough some bhang, which they smoked in a hut in the coconut grove, undisturbed by wives and well-wishers. After a smoke one felt light and elated and inclined to forgive everyone including that brother-in-law of his who had once tried to set fire to his home. But all this seemed like the memoirs of a previous birth. Some pestilence afflicted his cattle (he could of course guess who had laid his animals under a curse) and even the friendly butcher would not touch one at half the price…and now here he was left with the two scraggy creatures. He wished someone would rid him of their company too. The shopman had said that he was seventy. At seventy, one only waited to be summoned by God. When he was dead what would his wife do? They had lived in each other's company since they were children. He was told on the day of their wedding that he was ten years old and she was eight. During the wedding ceremony they had had to recite their respective ages and names. He had thrashed her only a few times in their marriage, and later she had the upper hand. Progeny, none. Perhaps numerous progeny would have brought him the blessing of the gods.

Fertility brought merit. People with fourteen sons were always so prosperous and at peace with the world and themselves. He recollected the thrill he had felt when he mentioned a daughter to that shopman; although it was not believed, what if he did not have a daughter?

His cousin in the next village had many daughters, and any one of them was as good as his; he was fond of them all and would buy them sweets if he could afford it. Still, everyone in the village whispered behind their backs that Muni and his wife were a barren couple. He avoided looking at anyone; they all professed to be so high up, and everyone else in the village had more money than he. 'I am the poorest fellow in our caste and no wonder that they spurn me, but I won't look at them either,' and so he passed on with his eyes downcast along the edge of the street, and people left him also very much alone, commenting only to the extent, 'Ah, there he goes with his two great goats; if he slits their throats, he may have more peace of mind.' 'What has he to worry about anyway? They live on nothing and have nobody to worry about.' Thus people commented when he passed through the village. Only on the outskirts did he lift his head and look up. He urged and bullied the goats until they meandered along to the foot of the horse statue on the edge of the village. He sat on its pedestal for the rest of the day. The advantage of this was that he could watch the highway and see the lorries and buses pass through to the hills, and it gave him a sense of belonging to a larger world. The pedestal of the statue was broad enough for him to move around as the sun travelled up and westwards; or he could also crouch under the belly of the horse, for shade.

The horse was nearly life-size, moulded out of clay, baked, burnt and brightly coloured, and reared its head proudly, prancing with its forelegs in the air and flourishing its tail in a loop. Beside the horse stood a warrior with scythe-like mustachios, bulging eyes and aquiline nose. The old image-makers believed in indicating a man of strength by bulging out his eyes and sharpening his moustache tips. They had also decorated the man's chest with beads which looked today like blobs of mud through the ravages of sun and wind and rain (when it came), but Muni would insist that he had known the beads to sparkle like the nine gems at one time in his life. The horse itself was said to have been as white as a dhobi-washed sheet, and had had on its back a cover of pure brocade of red-and-black lace, matching the multi-coloured sash around the waist of the warrior. But none in the village remembered the splendour as no one noticed its existence. Even Muni, who spent all his waking hours at its foot, never bothered to look up. It was untouched by the young vandals of the village who gashed tree trunks with knives and tried to topple off milestones and inscribed lewd designs on all the walls. This statue had been closer to the population of the village at one time, when this spot bordered the village; but when the highway was laid through (or perhaps when the tank and wells dried up completely here) the village moved a couple of miles inland.

Muni sat at the foot of the statue, watching his two goats graze in the arid soil among the cactus and lantana bushes. He looked at the sun; it had tilted westwards no doubt, but it was not the time yet to go back home; if he went too early, his wife would have no food for him. Also he must give her time to cool off her temper and feel sympathetic, and then she would scrounge and manage to get some food. He watched the mountain road for a time signal. When the green bus appeared around the bend he could leave, and his wife would feel pleased that he had let the goats feed long enough. He noticed now a new sort of vehicle coming down at full speed.

It looked both like a motor car and a bus. He used to be intrigued by the novelty of such spectacles, but of late work was going on at the source of the river on the mountain and an assortment of people and traffic went past him, and he took it all casually and described to his wife, later in the day, not everything as he once did, but only some things, only if he noticed anything special. Today, while he observed the yellow vehicle coming down, he was wondering how to describe it later when it sputtered and stopped in front of him. A red-faced foreigner who had been driving it got down and went round it, stooping, looking, and poking under the vehicle; then he straightened himself up, looked at the dashboard, stared in Muni's direction, and approached him. 'Excuse me, is there a gas station nearby, or do I have to wait until another car comes—' He suddenly looked up at the clay horse and cried, 'Marvellous!' without completing his sentence. Muni felt he should get up and run away, and cursed his age. He could not readily put his limbs into action; some years ago he could outrun a cheetah, as happened once when he went to the forest to cut fuel and it was then that two of his sheep were mauled—a sign that bad times were coming. Though he tried, he could not easily extricate himself from his seat, and then there was also the problem of the goats. He could not leave them behind.

The red-faced man wore khaki clothes—evidently a policeman or a soldier. Muni said to himself, 'He will chase or shoot if I start running. Sometimes dogs chase only those who run—O Shiva, protect me. I don't know why this man should be after me.' Meanwhile the foreigner cried, 'Marvellous!' again, nodding his head. He paced around the statue with his eyes fixed on it. Muni sat frozen for a while, and then fidgeted and tried to edge away. Now the other man suddenly pressed his palms together in a salute, smiled, and said, 'Namaste! How do you do?'

At which Muni spoke the only English expressions he had learnt, 'Yes, no.' Having exhausted his English vocabulary, he started in Tamil: 'My name is Muni. These two goats are mine, and no one can gainsay it—though our village is full of slanderers these days who will not hesitate to say that what belongs to a man doesn't belong to him.' He rolled his eyes and shuddered at the thought of evil-minded men and women peopling his village.

The foreigner faithfully looked in the direction indicated by Muni's fingers, gazed for a while at the two goats and the rocks, and with a puzzled expression took out his silver cigarette-case and lit a cigarette. Suddenly remembering the courtesies of the season, he asked, 'Do you smoke?' Muni answered, 'Yes, no.' Whereupon the red-faced man took a cigarette and gave it to Muni, who received it with surprise, having had no offer of a smoke from anyone for years now. Those days when he smoked bhang were gone with his sheep and the large-hearted butcher. Nowadays he was not able to find even matches, let alone bhang. (His wife went across and borrowed a fire at dawn from a neighbour.) He had always wanted to smoke a cigarette; only once had the shopman given him one on credit, and he remembered how good it had tasted. The other flicked the lighter open and offered a light to Muni. Muni felt so confused about how to act that he blew on it and put it out. The other, puzzled but undaunted, flourished his lighter, presented it again, and lit Muni's cigarette. Muni drew a deep puff and started coughing; it was racking, no doubt, but extremely pleasant. When his cough subsided he wiped his eyes and took stock of the situation, understanding that the other man was not an inquisitor of any kind. Yet, in order to make sure, he remained wary. No need to run away from a man who gave him such a potent smoke. His head was reeling from the effect of one of those strong American cigarettes made with roasted tobacco. The man said, 'I come from New York,' took out a wallet from his hip pocket, and presented his card.

Muni shrank away from the card. Perhaps he was trying to present a warrant and arrest him. Beware of khaki, one part of his mind warned. Take all the cigarettes or bhang or whatever is offered, but don't get caught. Beware of khaki. He wished he weren't seventy as the shopman had said. At seventy one didn't run, but surrendered to whatever came. He could only ward off trouble by talk. So he went on, all in the chaste Tamil for which Kritam was famous. (Even the worst detractors could not deny that the famous poetess Avvaiyar was born in this area, although no one could say whether it was in Kritam or Kuppam, the adjoining village.) Out of this heritage the Tamil language gushed through Muni in an unimpeded flow. He said, 'Before God, sir, Bhagwan, who sees everything, I tell you, sir, that we know nothing of the case. If the murder was committed, whoever did it will not escape. Bhagwan is all-seeing. Don't ask me about it. I know nothing.' A body had been found mutilated and thrown under a tamarind tree at the border between Kritam and Kuppam a few weeks before, giving rise to much gossip and speculation. Muni added an explanation, 'Anything is possible there. People over there will stop at nothing.' The foreigner nodded his head and listened courteously though he understood nothing.

'I am sure you know when this horse was made,' said the red man and smiled ingratiatingly.

Muni reacted to the relaxed atmosphere by smiling himself, and pleaded, 'Please go away, sir, I know nothing. I promise we will hold him for you if we see any bad character around, and we will bury him up to his neck in a coconut pit if he tries to escape; but our village has always had a clean record. Must definitely be the other village.'

Now the red man implored, 'Please, please, I will speak slowly, please try to understand me. Can't you understand even a simple word of English? Everyone in this country seems to know English. I have got along with English everywhere in this country, but you don't speak it. Have you any religious or spiritual scruples for avoiding the English speech?'

Muni made some indistinct sounds in his throat and shook his head. Encouraged, the other went on to explain at length, uttering each syllable with care and deliberation. Presently he sidled over and took a seat beside the old man, explaining, 'You see, last August, we probably had the hottest summer in history, and I was working in shirt-sleeves in my office on the fortieth floor of the Empire State Building. You must have heard of the power failure, and there I was stuck for four hours, no elevator, no air conditioning. All the way in the train I kept thinking, and the minute I reached home in Connecticut, I told my wife Ruth, "We will visit India this winter, it's time to look at other civilizations." Next day she called the travel agent first thing and told him to fix it, and so here I am. Ruth came with me but is staying back at Srinagar, and I am the one doing the rounds and joining her later.'

Muni looked reflective at the end of this long peroration and said, rather feebly, 'Yes, no,' as a concession to the other's language, and went on in Tamil,

'When I was this high,' he indicated a foot high, 'I heard my uncle say...'

No one can tell what he was planning to say as the other interrupted him at this stage to ask, 'Boy, what is the secret of your teeth? How old are you?'

The old man forgot what he had started to say and remarked, 'Sometimes we too lose our cattle. Jackals or cheetahs may carry them off, but sometimes it is just theft from over in the next village, and then we will know who has done it. Our priest at the temple can see in the camphor flame the face of the thief, and when he is caught...' He gestured with his hands a perfect mincing of meat.

The American watched his hands intently and said, 'I know what you mean. Chop something? Maybe I am holding you up and you want to chop wood? Where is your axe? Hand it to me and show me what to chop. I do enjoy it, you know, just a hobby. We get a lot of driftwood along the backwater near my house, and on Sundays I do nothing but chop wood for the fireplace. I really feel different when I watch the fire in the fireplace, although it may take all the sections of the Sunday *New York Times* to get a fire started,' and he smiled at this reference.

Muni felt totally confused but decided the best thing would be to make an attempt to get away from this place. He tried to edge out, saying, 'Must go home,' and turned to go. The other seized his shoulder and said desperately, 'Is there no one, absolutely no one here, to translate for me?' He looked up and down the road, which was deserted in this hot afternoon; a sudden gust of wind churned up the dust and dead leaves on the roadside into a ghostly column and propelled it towards the mountain road. The stranger almost pinioned Muni's back to the statue and asked, 'Isn't this statue yours? Why don't you sell it to me?'

The old man now understood the reference to the horse, thought for a second, and said in his own language, 'I was an urchin this high when I heard my grandfather explain this horse and warrior, and my grandfather himself was this high when he heard his grandfather, whose grandfather...'

The other man interrupted him with, 'I don't want to seem to have stopped here for nothing. I will offer you a good price for this,' he said, indicating the horse. He had concluded without the least doubt that Muni owned this mud horse. Perhaps he guessed by the way he sat at its pedestal, like other souvenir-sellers in this country presiding over their wares.

Muni followed the man's eyes and pointing fingers and dimly understood the subject matter and, feeling relieved that the theme of the mutilated body had been abandoned at least for the time being, said again, enthusiastically, 'I was this high when my grandfather told me about this horse and the warrior, and my grandfather was this high when he himself...' and he was getting into a deeper bog of reminiscence each time he tried to indicate the antiquity of the statue.

The Tamil that Muni spoke was stimulating even as pure sound, and the foreigner listened with fascination. 'I wish I had my tape recorder here,' he said, assuming the pleasantest expression. 'Your language sounds wonderful. I get a kick out of every word you utter, here'—he indicated his ears—'but you don't have to waste your breath in sales talk. I appreciate the article. You don't have to explain its points.'

'I never went to a school, in those days only Brahmins went to schools, but we had to go out and work in the fields morning till night, from sowing to harvest time...and when Pongal came and we had cut the harvest, my father allowed me to go out and play with others at the tank, and so I don't know the Parangi language you speak, even little fellows in your country probably speak the Parangi language, but here only learned men and officers know it. We had a postman in our village who could speak to you boldly in your language, but his wife ran away with someone and he does not speak to anyone at all nowadays. Who would if a wife did what she did? Women must be watched; otherwise they will sell themselves and the home,' and he laughed at his own quip.

The foreigner laughed heartily, took out another cigarette, and offered it to Muni, who now smoked with ease, deciding to stay on if the fellow was going to be so good as to keep up his cigarette supply. The American now stood up on the pedestal in the attitude of a demonstrative lecturer and said, running his finger along some of the carved decorations around the horse's neck, speaking slowly and uttering his words syllable by syllable, 'I could give a sales talk for this better than anyone else... This is a marvellous combination of yellow and indigo, though faded now... How do you people of this country achieve these flaming colours?'

Muni, now assured that the subject was still the horse and not the dead body, said, 'This is our guardian, it means death to our adversaries. At the end of Kali Yuga, this world and all other worlds will be destroyed, and the Redeemer will come in the shape of a horse called Kalki; this horse will come to life and gallop and trample down all bad men.' As he spoke of bad men the figures of his shopman and his brother-in-law assumed concrete forms in his mind, and he revelled for a moment in the predicament of the fellow under the horse's hoof: served him right for trying to set fire to his home...

While he was brooding on this pleasant vision, the foreigner utilized the pause to say, 'I assure you that this will have the best home in the USA. I'll push away the bookcase, you know I love books and am a member of five book clubs, and the choice and bonus volumes really mount up to a pile in our living room, as high as this horse itself. But they'll have to go. Ruth may disapprove, but I will convince her. The TV may have to be shifted too. We can't have everything in the living room. Ruth will probably say, "What about when we have a party?" I'm going to keep him right in the middle of the room. I don't see how that can interfere with the party—we'll stand around him and have our drinks.'

Muni continued his description of the end of the world. 'Our pandit discoursed at the temple once how the oceans are going to close over the earth in a huge wave and swallow us—this horse will grow bigger than the biggest wave and carry on its back only the good people and kick into the floods the evil ones—plenty of them about,' he said reflectively. 'Do you know when it is going to happen?' he asked.

The foreigner now understood by the tone of the other that a question was being asked and said, 'How am I transporting it? I can push the seat back and make room in the rear. That van can take in an elephant'—waving precisely at the back of the seat.

Muni was still hovering on visions of avatars and said again, 'I never missed our pandit's discourses at the temple in those days during every bright half of the month, although he'd go on all night, and he told us that Vishnu is the highest god. Whenever evil men trouble us, he comes down to save us. He has come many times. The first time he incarnated as a great fish, and lifted the scriptures on his back when the floods and sea waves...'

'I am not a millionaire, but a modest businessman. My trade is coffee.'

Amidst all this wilderness of obscure sound Muni caught the word 'coffee' and said, 'If you want to drink "kapi", drive further up, in the next town, they have Friday market, and there they open "kapi-otels"—so I learn from passers-by. Don't think I wander about. I go nowhere and look for nothing.' His thoughts went back to the avatars. 'The first avatar was in the shape of a little fish in a bowl of water, but every hour it grew bigger and bigger and became in the end a huge whale which the seas could not contain, and on the back of the whale the holy books were supported, saved, and carried.' Having launched on the first avatar it was inevitable that he should go on to the next, a wild boar on whose tusk the earth was lifted when a vicious conqueror of the earth carried it off and hid it at the bottom of the sea. After describing this avatar Muni concluded, 'God will always save us whenever we are troubled by evil beings. When we were young we staged at full moon the story of the avatars. That's how I know the stories; we played them all night until the sun rose, and sometimes the European collector would come to watch, bringing his own chair. I had a good voice and so they always taught me songs and gave me the women's roles. I was always Goddess Lakshmi, and they dressed me in a brocade sari, loaned from the Big House...'

The foreigner said, 'I repeat, I am not a millionaire. Ours is a modest business; after all, we can't afford to buy more than sixty minutes' TV time in a month, which works out to two minutes a day, that's all, although in the course of time we'll maybe sponsor a one-hour show regularly if our sales graph continues to go up...'

Muni was intoxicated by the memory of his theatrical days and was about to explain how he had painted his face and worn a wig and diamond earrings when the visitor, feeling that he had spent too much time already, said, 'Tell me, will you accept a hundred rupees or not for the horse? I'd love to take the whiskered soldier also but I've no space for him this year. I'll have to cancel my air ticket and take a boat home, I suppose. Ruth can go by air if she likes, but I will go with the horse and keep him in my cabin all the way if necessary,' and he smiled at the picture of himself voyaging across the seas hugging this horse. He added, 'I will have to pad it with straw so that it doesn't break...'

'When we played Ramayana, they dressed me as Sita,' added Muni. 'A teacher came and taught us the songs for the drama and we gave him fifty rupees. He incarnated himself as Rama, and he alone could destroy Ravana, the demon with ten heads who shook all the worlds; do you know the story of Ramayana?'

'I have my station wagon as you see. I can push the seat back and take the horse in if you will just lend me a hand with it.'

'Do you know Mahabharata? Krishna was the eighth avatar of Vishnu, incarnated to help the Five Brothers regain their kingdom. When Krishna was a baby he danced on the thousand-hooded giant serpent and trampled it to death; and then he suckled the breasts of the demoness and left them flat as a disc though when she came to him her bosoms were large, like mounds of earth on the banks of a dug-up canal.' He indicated two mounds with his hands. The stranger was completely mystified by the gesture. For the first time he said, 'I really wonder what you are saying because your answer is crucial. We have come to the point when we should be ready to talk business.'

'When the tenth avatar comes, do you know where you and I will be?' asked the old man.

'Lend me a hand and I can lift off the horse from its pedestal after picking out the cement at the joints. We can do anything if we have a basis of understanding.'

At this stage the mutual mystification was complete, and there was no need even to carry on a guessing game at the meaning of words. The old man chattered away in a spirit of balancing off the credits and debits of conversational exchange, and said in order to be on the credit side, 'O honourable one, I hope God has blessed you with numerous progeny. I say this because you seem to be a good man, willing to stay beside an old man and talk to him, while all day I have none to talk to except when somebody stops by to ask for a piece of tobacco. But I seldom have it, tobacco is not what it used to be at one time, and I have given up chewing. I cannot afford it nowadays.' Noting the other's interest in his speech, Muni felt encouraged to ask, 'How many children have you?' with appropriate gestures with his hands. Realizing that a question was being asked, the red man replied, 'I said a hundred,' which encouraged Muni to go into details, 'How many of your children are boys and how many girls? Where are they? Is your daughter married? Is it difficult to find a son-in-law in your country also?'

In answer to these questions the red man dashed his hand into his pocket and brought forth his wallet in order to take immediate advantage of the bearish trend in the market. He flourished a hundred-rupee currency note and asked, 'Well, this is what I meant.'

The old man now realized that some financial element was entering their talk. He peered closely at the currency note, the like of which he had never seen in his life; he knew the five and ten by their colours although always in other people's hands, while his own earning at any time was in coppers and nickels. What was this man flourishing the note for? Perhaps asking for change. He laughed to himself at the notion of anyone coming to him for changing a thousand or ten-thousand-rupee note. He said with a grin, 'Ask our village headman, who is also a moneylender; he can change even a lakh of rupees in gold sovereigns if you prefer it that way; he thinks nobody knows, but dig the floor of his puja room and your head will reel at the sight of the hoard. The man disguises himself in rags just to mislead the public. Talk to the headman yourself because he goes mad at the sight of me. Someone took away his pumpkins with the creeper and he, for some reason, thinks it was me and my goats...that's why I never let my goats be seen anywhere near the farms.' His eyes travelled to his goats nosing about, attempting to wrest nutrition from minute greenery peeping out of rock and dry earth.

The foreigner followed his look and decided that it would be a sound policy to show an interest in the old man's pets. He went up casually to them and stroked their backs with every show of courteous attention. Now the truth dawned on the old man. His dream of a lifetime was about to be realized. He understood that the red man was actually making an offer for the goats. He had reared them up in the hope of selling them some day and, with the capital, opening a small shop on this very spot. Sitting here, watching the hills, he had often dreamt how he would put up a thatched roof here, spread a gunny sack out on the ground, and display on it fried nuts, coloured sweets, and green coconut for the thirsty and famished wayfarers on the highway, which was sometimes very busy. The animals were not prize ones for a cattle show, but he had spent his occasional savings to provide them some fancy diet now and then, and they did not look too bad. While he was reflecting thus, the red man shook his hand and left on his palm one hundred rupees in tens now. 'It is all for you or you may share it if you have a partner.'

The old man pointed at the station wagon and asked, 'Are you carrying them off in that?'

'Yes, of course,' said the other, understanding the transportation part of it.

The old man said, 'This will be their first ride in a motor car. Carry them off after I get out of sight, otherwise they will never follow you, but only me even if I am travelling on the path to Yama Loka.' He laughed at his own joke, brought his palms together in a salute, turned round and went off, and was soon out of sight beyond a clump of thicket.

The red man looked at the goats grazing peacefully. Perched on the pedestal of the horse, as the westerly sun touched the ancient faded colours of the statue with a fresh splendour, he ruminated, 'He must be gone to fetch some help, I suppose!' and settled down to wait. When a truck came downhill, he stopped it and got the help of a couple of men to detach the horse from its pedestal and place it in his station wagon. He gave them five rupees each, and for a further payment they siphoned off gas from the truck and helped him to start his engine.

Muni hurried homewards with the cash securely tucked away at his waist in his dhoti. He shut the street door and stole up softly to his wife as she squatted before the lit oven wondering if by a miracle food would drop from the sky. Muni displayed his fortune for the day. She snatched the notes from him, counted them by the glow of the fire, and cried, 'One hundred rupees!

How did you come by it? Have you been stealing?'

'I have sold our goats to a red-faced man. He was absolutely crazy to have them, gave me all this money and carried them off in his motor car!'

Hardly had these words left his lips when they heard bleating outside. She opened the door and saw the two goats at her door. 'Here they are!' she said. 'What's the meaning of all this?'

He muttered a great curse and seized one of the goats by its ears and shouted, 'Where is that man? Don't you know you are his? Why did you come back?' The goat only wriggled in his grip. He asked the same question of the other too. The goat shook itself off. His wife glared at him and declared, 'If you have thieved, the police will come tonight and break your bones. Don't involve me. I will go away to my parents...'

four

~

A LIFE

BUDDHADEVA BOSE

Translated from the Bengali by Arunava Sinha

Gurudas Bhattacharya, Vachaspati, the seniormost teacher of
Sanskrit at Khulna's Jagattarini School, was brought up short
by a word while teaching Bengali literature to Class IX students.

'Amaar projagawn amaar cheye tahare bawro kori mane...' The
pandit found a word in the sentence confusing. Cheye? Did that refer
to the Bengali word for glance? Or for desire? After some thought,
he explained the sentence as, 'The king says his subjects want him,
they desire his sanctuary, but they respect the King of Kaushal more.
Grammar has been distorted a little here.' The boys on the first bench
exchanged glances. Then one of them stood up to say, 'It's fine, sir.
The word "cheye" in this case is used for comparison, in the sense
of "than". My subjects consider him more noble than me. See, it
says a little later, "Are you so bold as to imagine you can be more
pious than me?"'

'If only I were an Arab Bedouin rather than this,' the boy next
to him recited.

Gurudas did not respond. Accepting the correction made by his
students, he continued teaching the poem. The bell rang.

It was the last period. Collecting their umbrellas and books, the
other teachers left for their homes, while Gurudas made his way to

the school library. The library was nothing but three cupboards full of books in one corner of the staff room—most of them textbooks obtained as free samples. Among the more valuable volumes were several hardbound sets of the Bengali literary magazine *Probashi*, a Philips atlas of twenty years' vintage, a Chambers Dictionary, and three Bengali and English-to-Bengali dictionaries used by students. Clearing his throat, Gurudas said, 'Can you unlock this cupboard, Nabakeshto?'

Not even the servants at school paid much attention to the Sanskrit teacher. And Nabakeshto donned the mantle of bearer, doorman and gardener single-handedly. 'The library is closed, sir,' he answered with a touch of insolence.

'Never mind, just unlock it. I need some books.'

'But I have to leave for Rasoolpur right away—my daughter's in-laws have invited us...'

'That's all right, you can go. Leave the keys with me.'

'All right then. Don't forget to give them back to me before eleven tomorrow. You know how strict the new headmaster is. And lock the door of the room before you go...here's the padlock, see?'

Unlocking the cupboard, Gurudas planted himself in front of it; with a glance at his back, Nabakeshto gathered his bundle, wrapped in a gamchha, from its place beneath the table—he was taking a bunch of grapefruits from the tree in the schoolyard as a gift for his son-in-law.

No one was allowed to take the dictionaries home; Gurudas spent a good deal of time leafing through the two Bengali dictionaries. The light grew dim, the silence of a provincial evening thickened inside the room. He forgot to sit down, forgot his hunger, his internal senses seemed to soak up the rows of letters. Today's incident had wounded him—he had not been able to capture the meaning of a word which millions of adults and children used every single day without a thought. How could he—he was a teacher of Sanskrit. He had learnt Sanskrit, but not Bengali. But he was a Bengali—that was the language he spoke. He seemed to realize for the first time that the Bengali language was not Sanskrit, not even a corrupt form—it was a complete, living, changing, evolving, independent language, the spoken language of seventy million people, their mother-tongue. 'A living language, the mother-tongue'—he repeated the words in his head several times. But prowess in one's mother-tongue was not automatic, it needed nurture.

Gurudas noticed that none of the dictionaries included the word he had tripped over that morning. He was reminded of other words used in similar fashion—'thekey', the Bengali word used for 'from' or 'since', or 'dyakha', used for 'seeing' or 'meeting' or 'looking after'. This was how the Bengali language perfumed the Sanskrit verb ending. None of this was in the dictionary. There were mistakes—mistaken explanations, even mistaken spelling. How were the students to learn? And I—how am I to learn?

It was late evening by the time Gurudas returned home. His wife Harimohini asked, 'So late?' Gurudas did not answer. He ate his dinner in silence. 'Are you ill? You aren't eating.' 'I am not ill.' He went to bed early that night.

Jagattarini School began at eleven in the morning, and the district school, at ten-thirty. Gurudas went to the district school around a quarter past eleven the next morning, spending half an hour in the library before breathlessly entering his own class in the nick of time. It was Saturday the next day—from the school he went to the only college in Khulna. He had a nodding acquaintance with the Sanskrit teacher there (here, too, it was he who taught Bengali). They conversed for some time, and he flipped through three or four books in the library—but his restlessness did not leave him.

No, he had not found it—he had not found what he was looking for anywhere. Could there not be a complete Bengali dictionary, which had room for every single word (both Sanskrit and vernacular) in the language, which included every combination, every application, every colloquial usage, which would enable the Bengali language to be learnt, its nature to be understood, its unique creative spirit to be appreciated? The college professor had said there was not a single such book. There were a few good ones among those he examined, but in a workmanlike way—where was the dictionary that one could use for real scholarship?

The biggest bookshop in town was Victoria Library. In the evening Gurudas asked to see a major Bengali dictionary and the Oxford English Dictionary. Having leafed through them for a few minutes, he said softly, 'There's something I want to discuss, Rebati babu.'

In a small town, everyone knew everyone else. The owner looked at Gurudas over the rims of his glasses.

'It's Saturday—may I take these two books home? I'll return them to you first thing Monday morning.'

'Take them home?'

'I'll handle them very carefully—won't soil them, won't crease them—I'll look after them. I need them urgently, you see.'

'Someone's already ordered those books, Pandit mashai.'

'I see.' Gurudas's fair, lean face reddened. A little later he said, 'Then I'd better buy them.' He had to wage a terrible war against himself, but…he had spoken, he couldn't take his words back now.

'Pack these books for Pandit mashai…' Rebati babu made no further reference to the books having been ordered.

'But I can only pay next month.'

'Hmm…' Gurudas sent up a silent prayer, 'Let him not agree, O Lord.' But Rebati babu's mouth softened.

'Very well. But on the first of the month, don't forget. We run a very small business, you know…sign here, please.'

He had got them at a discount by virtue of being a teacher. Thirteen rupees and fourteen annas—nearly a third of his salary.

Gurudas browsed through the two books late into that night by the glow of a lantern. His grasp of English was poor, but he had no difficulty in realizing the difference in the presentation of the two books. And yet this was just a condensed version, he had heard that Oxford had a giant dictionary too.

Before going to sleep he mused over Panini, considered the sheer extent of the Sanskrit dictionary *Shabdakalpadruma*, and recollected Vidyasagar. An extraordinary talent for grammar, unbeatable enthusiasm for analysis, bottomless vocabulary. He used to have them all. What had happened to them?

⌁

Harimohini had planted flowers in a fenced-in corner of the small yard. She was watering them with her daughter on Sunday morning when Gurudas came up to them, smiling.

'Shibani, go check if Nidhu's ma has brought the milk.'

'Later. Listen to me first.'

Harimohini paused and looked at him.

'I'm about to start something new.'

A ray of hope flashed across Harimohini's face. Had the match for Shibani been finalized with the Chatterjees of Nimtala then? Their elder daughter Bhabani had been married into a high-born family—this was the other daughter. She had turned fifteen, if she didn't get married now, then when?

'Have they sent word?'

'Who?'

'The Nimtala Chatterjees.'

'No, that's not it. I am going to write a dictionary of the Bengali language. I made up my mind last night.'

There was no flicker of expression on Harimohini's face.

'You know what a dictionary is, don't you? A collection of words. The meaning of words in the Bengali language, similar words, usage of words, and so on. There isn't a book like this at the moment.'

'Not a single one? You're going to write it?' Harimohini felt a burst of pride. 'Will it say anything about gods and goddesses?'

'Everything.'

Yes, everything. Unknown to Gurudas, a smile spread across his face. He had fallen asleep the previous night as soon as he had come to this decision—a deep sleep. And when he awoke this morning, he discovered his mind was calm, his heart, cheerful, and his body, healed and rested, while support for his endeavour radiated from the branching rays of the sun in the sky. As though nature had been waiting these last few days only for his resolve to do this: as soon as he accepted it, satisfaction spread across the heavens, and the movements in his body acquired an easy rhythm. Gods and goddesses—of course he would have to include them. But all the gods? All their names? He would have to determine which of them belonged to an encyclopaedia and which to a dictionary. Which of the Sanskrit words could be considered Bengali? What to do with Brajabuli? What were the indications that a word was part of the Bengali language? Would he have to add words which were not in circulation but might be required? There was so much to think about. So much to think—but even Harimohini's flowering plants were urging him to start at once.

Gurudas had been to Puri once as a student, he was reminded now of his visit. He could see just such an ocean stretching ahead of him—a succession of waves, hollows, whirpools, effort…the horizon in the distance. On this ocean his raft would have to float, this was the sea he would have to cross. For a moment Gurudas felt his skin prickle.

After lunch he brought the subject up again with his wife.

'I was thinking of the dictionary.'

'Yes, what?'

'The thing is, I need some material. Books and things.'

'Very well.'

'Very expensive books. I was thinking, Chakraborty mashai had made an offer for that acre of land back home…'

'You'll sell it?' A shadow fell across Harimohini's face. 'We have nothing else, and the girl's growing up too.'

'We can survive on what we have.' Gurudas could not inject too much confidence into this assertion, so he tried to compensate with a gentle smile.

'That is to say, I will survive, and once your son's grown up you'll have nothing to worry about.'

'The things you say! I think only about myself all the time, don't I? But I shan't let Nobu be a teacher like you. You know Netai, my nephew? He's passed his matriculation examination and joined the Railways. Sixty rupees a month already—and extra earnings on top of that.'

Gurudas did not approve of the final statement, but swallowing his criticism, he returned to the original subject.

'Just the Railways? Nobu might even become a deputy magistrate like my brother,' said Gurudas, throwing a sidelong look at his wife. It was a calculated ploy—he was fully aware of Harimohini's reverence for his stepbrother's status as a deputy magistrate.

'Do you suppose I could ever be so lucky? But then, everything is possible if the gods smile on us, isn't that right? That reminds me, I'd sent you sweets after Lakshmi Puja the other day, but Shibani said you didn't eat them.'

'I touched my forehead with them—that's better than eating them. Listen, I'm giving the land to Chakraborty mashai then, all right?'

'Giving it? We hardly have anything anyway—and there's not only the girl who has to be married off but also the boy whom we must leave something to.'

'Everything will be done. But I cannot turn back now.'

'Cannot turn back—what do you mean?'

'Wealth is by nature temporary, but...' The pandit groped for the right word, and then turned helplessly to emotional appeal. 'I have made up my mind—are you going to stop me now?'

The land they owned was in Nandigram, about an hour away by steamer. Gurudas paid a visit during the Janmashtami holidays. A house, fruit-bearing trees, a small pond, some farmland. Some? It was about seventy acres in his grandfather's time. After being divided up, about eight acres had come to Gurudas. He had had to sell nearly two acres for his elder daughter's wedding, and now another acre. Never mind, at least he was getting a hundred and fifty rupees. Rummaging through the books at home, he even found an old Sanskrit dictionary printed in Bombay—it had belonged to his father—and, how fortunate, the Sanskrit grammar that he had borrowed from a schoolmate and forgotten to return. The first thing he did on returning to Khulna was to buy two reams of the cheapest paper, which Shibani laced into a notebook.

On the first day of the Puja holidays at school, Gurudas travelled to Calcutta, where he had to put up for three days at a boarding house in Sealdah. Two more Bengali dictionaries; Suniti Chatterji's book on linguistics; an ancient (but excellent) Sanskrit-to-Bengali dictionary found after scouring the pavements of College Street and Chitpur; a Bengali grammar written by an Englishman; Tekchand's novel, and *Hutom Pyanchar Naksha* published by Basumati; and Kaliprasanna Sinha's Mahabharata published by Hitabadi. He didn't dare ignore Rabindranath Tagore's 'The Theory of Words' when it caught his eye—poets were the creators of language, might as well find out what he had to say. All this accounted for nearly fifty rupees. Then there were the new clothes for Durga Puja, a pair of shankhas each for Harimohini and Bhabani, a dhoti for his son-in-law, a pair of rubber slippers for Nobu costing a rupee and thirteen annas. He had to spend eight annas on his way back on a porter to carry all the books—that really pinched.

They had a wonderful time back home that year during Durga Puja. Harimohini stayed back with the children, while Gurudas returned to Khulna the day after Lakshmi Puja. He cooked his own meals, and read all day. He found the English difficult, but managed to make sense of it, and it grew easier the more he read. On the day before Kali Puja, in one of the notebooks made by Shibani, he wrote the first letter of the Bengali alphabet, 'Aw', in a bold hand. Fifty words were written that day. The school opened three days later, the family returned, and his leisure hours shrank.

Gurudas set himself a routine. He woke up at five in the morning to write for two hours, and then drank his share of milk, went out to take private tuition, bought the day's provisions, and returned. This gave him a little time before his bath. He had to take private classes in the evening too—the exams were approaching—but he didn't go to bed until he had put in a couple of hours of writing. Gurudas was making smooth progress.

Winter came. There was no light before six in the morning, and this was when the pressure of checking annual exam papers intensified. But the Christmas holidays were approaching.

He had to visit Calcutta again during the Christmas vacations. The subject was like Draupadi's sari—unfolding constantly, an unending mystery, one whose depths you kept sinking into. How would he prove equal to this task—he, a mere Gurudas Bhattacharya, a minor Sanskrit scholar? He did not know his way on this road, had no clear idea of where he would find the bricks and cement needed to build this structure. In Calcutta he laid siege to the Imperial Library: the days passed navigating his way through the dense jungle of comparative linguistics. Many of the books were written in German, with an abundance of Greek letters and a thick growth of Latin, Gothic and Persian references, as though the immense vegetation of the Aryan languages had stretched up to the sky, spreading its branches far and wide. Sanskrit alone had never given him this feeling of kinship with the West, with the entire world. For the first time he set eyes on the Monier Williams Sanskrit dictionary, he discovered Skeat's etymological dictionary too. Ten days passed cramming his notebooks with jottings.

When he was about to set off for Calcutta again during the summer holidays, Harimohini could not keep herself from objecting mildly.

'Why must you go to Calcutta again?'

'Do you need me here?'

'I was thinking of the expenses. The boarding house costs money too.'

Gurudas had thought about this as well. The examination season was in the past, and not many studied Sanskrit these days, he had no private tuitions. Thanks to a supply of food from the land back home they managed to survive on forty-five rupees—but barely. They could afford coarse rice and dal, and their clothes—anything more was virtually impossible. But...he simply had to go.

'Doesn't your mother's brother live in Calcutta?' said Harimohini. 'You could always...'

'Of course not, how can I stay a month at someone else's house? And he's only my mother's cousin—I haven't met him in years... it's impossible. But I'll manage—don't worry.'

'It's all very well for you to say that, but I spend sleepless nights.'

'But why?'

'Have you decided that Shibani will remain a spinster?'

That was true. He had to accept that his daughter was showing signs of womanhood. It was time for her to be married. But...how?

'Why so anxious? She's not even fifteen yet. Many people don't even think of marriage till eighteen these days.'

'You, of all people, are saying this? Your family, the Brahmin pandits of Nandigram didn't allow their daughters to pass the age of ten.'

'Why shouldn't I? Didn't Rammohan Roy speak up against idol worship? Didn't Vidyasagar introduce widow remarriage? They were Brahmin pandits too.'

'Those who get their daughters married at eighteen also give them the chance to go to school and college, right? They don't let them rot at home and turn into liabilities. Do you have it in you to educate her?'

Gurudas's lean, fair face grew pale. She was right. He had no response. He must try to arrange a match.

From the matchmaker he learnt that Rameshwar Banerjee of Hatkhola in Calcutta was looking for a bride for his third son. Rameshwar had been a professor at Sanskrit College during the single year that Gurudas had read there. He decided to plead with Rameshwar in Calcutta to provide a safe passage for his daughter.

In Calcutta Gurudas rented a 'seat' in the cheapest room in a boarding-house he was familiar with. His meals were at a 'pice hotel' (which he had discovered on his previous visit; for four paise you could eat so much that you didn't need a second meal). His days were spent at the Imperial Library, at the university library, wandering among second-hand bookshops, and seeking audiences with renowned professors. He had sensed a new requirement: instructions, advice, discussions—he had brought along all the pages in his notebook, in case anyone had any constructive comments to offer. It wasn't easy to meet professors—some had gone to Darjeeling, others were busy. Only two deigned to meet him. Leafing through the notebooks apprehensively, both of them said, 'Excellent, it's coming along very well, you must complete it.' When he enquired whether a detailed discussion was possible, he learnt that both were engaged as chief examiners for the BA exams, and did not have the leisure even to die at present.

One day he overheard a young man at a bookshop on College Street. The buyer was looking for a book on the history of Bengali literature; turning over the pages of two or three, the words he uttered were clearly weighed down with nausea. 'Dead! All dead! Rotting and infested with worms which this swarm of professors is picking out to eat. They collapse when they see living literature. Rabindranath was born in vain.' The young man disappeared, his sandals flapping.

Chuckling, the shopkeeper said, 'Subrata Sen speaks as forcefully as he writes.'

'Who was that?' Gurudas stepped forward. 'What did you say his name is?'

'Subrata Sen. You haven't read him? Very powerful.'

At the boarding house he normally drank a large glass of water and went to bed—his exhaustion taking him beyond the hot weather, the stench, and his hunger in an instant. But sleep eluded him that night, the young man's statement ringing in his ears constantly. And you, Gurudas Bhattacharya, engaged in composing a dictionary of the Bengali language—what do you know of Bengali literature? Ishwar Gupta, Bankim, Michael—and that was it. The young man had named Rabindranath—some people said he had injected new life into the Bengali language, but you know nothing about him, you haven't read him at all. And these new writers—take Subrata Sen, for example—language lived through transformation in every era. It would die if it were to lose this power. And if a dictionary could not provide a portrait of this evolution, what use was it?

He had to think of the whole thing afresh. A dictionary was not a compendium of explanations for students, not a list or collection, not an immovable, static, ponderous object. Its essence lay in the flow, in movement, in showing the path to the future—to move ahead it had to gather sustenance from the creative work that writers were engaged in constantly. It would have to be replete with hints, allusions, advice, even imagination—just like a flowing waterfall glinting under the light. He would have to read literature—living works, current, changing literature—all that was being written, read, said, heard in the Bengali language—all these were his ingredients.

He came back home bathed in a new glow. Within five minutes of his return Harimohini asked, 'Did you meet Rameshwar Banerjee?'

'I did.'

'What did he say?'

'In a minute.' Gurudas sat down on a mat, leaning back against a post. 'They have many demands. They're well-off, you see.'

'Who'll marry your daughter on the strength of her appearance alone?'

'A thousand rupees in cash. Twenty-five bhori gold. All expenses. Provided they like the girl. But…can we afford all this? I'd better make some more enquiries…'

Sighing, Harimohini went away. Evening fell.

This time Gurudas had brought a ream of foolscap paper from Calcutta. It was cheaper there, and available at even lower prices if bought by the ream. He had nearly exhausted his older notebooks. He had to scribble copiously—scratch out bits, make changes, there was new information every day. And yet he wasn't even done with the first letter, 'Aw'.

Gurudas got down to work calmly. Some of it involved reading. He had avoided reading the newspapers all this time, but now he had to scan a couple of Bengali dailies every evening at the public library. And he left no Bengali book he could get hold of untouched. Happening to read Rabindranath's *Ghare Baire*, he was astounded. Could the Bengali language actually be used in this way? This was not Hutom, this was Kalidasa. Not even Kalidasa, something else altogether.

His notebook and pencil were always in his pocket. He took voluminous notes. Most of them would not prove useful, but who could predict what would?

The Bengalis' forms of self-expression became the subject of his discoveries, something he paid careful attention to. He listened closely when his wife, son or daughter spoke; with so much interest that he often did not grasp the content, and forgot to answer when they asked him a question. What he wanted to know was not what they were saying but how they were saying it. When the younger students raised an uproar during the lunch-break at school, he lurked unobserved behind them. At the market he kept his ears peeled for rural dialects. When he went home on holidays, he sought out Muslim peasants and made unnecessary conversation with them—they had a special way of speaking, which he wanted to grasp.

And he had to go to Calcutta during the longer vacations. He learnt the Greek alphabet, took help from a priest at St Xavier's School to understand the rules of Latin grammar, even had to visit madrasas for Arabic and Persian. Hardly any books were available in the provinces—for this too he had to visit Calcutta.

How did he afford all this? Cheap boarding houses and pice hotels, but still. Gurudas had made arrangements, getting rid of another acre of land, this time without telling his wife. He didn't know anyone in Calcutta particularly well, feeling beleaguered if he had to speak in English. Nor did his soiled clothes evoke respect from anyone. He had to discover everything he needed all by himself, with the help of that eternal quality, effort, the capital that God endowed every human being with. Effort, endeavour, waiting, patience. It took him four hours to do an hour's work—he was lighting rows of fireflies and pushing through the darkness. But there were lights at every street corner—like signals for trains in the blackness of night.

Summer holidays once more, the monsoon once more. The rains were torrential that year. Earthworms burst through the kitchen floor in July. Leeches in the front yard. Snakes here and there. On some nights water streamed through gaps in the tin roof—having found dry spots for the children to sleep, the parents stayed up all night. After seven days of incessant rain, Gurudas opened his safe one day to get the shock of his life. Instead of his best books, what he saw were millions of termites wriggling about. Fifty pages of Suniti Chatterji's book were missing, the third volume of the Mahabharata was in shreds, the Sanskrit dictionary from his father's time crumbled in his hands when he picked it up. The day passed battling the termites—he poured in four annas worth of kerosene.

Immediately after this accident there was a bit of hope; Shibani's marriage suddenly seemed a possibility. The groom was from Barishal, recently posted here at the Khulna steamer station. The groom's family approved of the bride, and made no demand for dowry—only the cost of the wedding, and shankha and sindoor for the bride. This was no cause for concern—Harimohini still had some ornaments.

The wedding would not take place before March, but Bhabani was overcome with joy when she heard. At long last she would be able to visit her mother. She lived in a large family, surrounded by her in-laws, at Madaripur—she didn't even have the chance to visit her own family during Durga Puja.

Shibani came down with a fever after the rains. When the fever didn't go down even after a week, Gurudas sent for the kaviraj. He prescribed plenty of red and black pills—but to no avail.

On the twenty-first day the official assistant surgeon turned up. His fees were four rupees, and he stomped about in boots. Typhoid, he said, after examining the patient. Give her nothing but glucose. Pour water over her head morning and evening. Here are the medicines. Note down the temperature at four-hour intervals. Inform me after three days.

The medicines were bought with borrowed money. The doctor came once a week—paying his fees was a near-impossible task. Milk and fish were stopped; Harimohini's deity was given a quarter of her regular rations.

Shibani lost weight, the fat disappeared from her cheeks, her discoloured teeth grew bigger and uglier. Then came the day when her hair had to be cut on the doctor's orders. Her scalp needed water, the more the better. Harimohini poured water over her daughter's head every hour, but Shibani was delirious.

At the time of her death, her limbs had withered away to resemble sticks, her chest was like that of a seven-year-old boy's. And this girl was sixteen, and had been healthy, full of grace. The ornaments put aside to pay for the wedding were used to clear the debt to the doctor.

Gurudas returned home at ten at night after the cremation. It was the end of February, winter was on its way out. He felt rather cold—wrapping a shawl around himself, he sat down next to his wife, who was slumped on the floor. The night passed in the same position.

A long night, but the sun rose finally. Harimohini had fallen asleep, while Nobu was curled on the floor in the cold. Covering his son with the shawl, Gurudas carefully slipped a pillow beneath Harimohini's head. Then he went out, spread a mat and sat down with his notebook. This last one had also been made by Shibani. For a moment, all the letters blurred. Wiping his eyes on the end of his dhoti, he set down more letters next to the blurred ones.

Five more years passed, the dictionary was in its seventh year. He was done with twenty-four letters, up to 'Thaw'.

The words no longer flowed. What had started as an extraordinary, thrilling joy had now turned into work. Work, duty, responsibility, compulsion.

The madness of discovery was gone, the excitement of gathering material had dissipated. He had an enormous quantity of information at his disposal now, the roads were familiar. It was time to work, it was time for nothing but work. Daily work, weekly work, monthly, annual, continuous. No likes, no dislikes, no reluctance either. This was an immaculately conceptualized world, where individuality was dead.

That year saw the fruition of a long drawn-out effort of the Jagattarini School—the government finally approved grants. Teachers' salaries were increased; Gurudas's monthly earning leapt to fifty-five rupees—it could even get to seventy or seventy-five eventually. In that same year Nobu, or Nabendu, vaulted over the hurdle of the matriculation examination. Not just that, he got a job almost immediately. A job with the Railways, as his mother had hoped.

A few months later there was tragic news: Bhabani had become a widow. And within two months she appeared in her father's yard with close-cropped hair, dressed in a widow's garb and holding three children by the hand. Her late husband's parents were no longer willing to shoulder the burden of their daughter-in-law, without whom they couldn't survive a moment at one time. 'They are not as well-off as before, my brothers-in-law have several children, and he didn't leave anything for us, Baba.'

Her father said, 'Don't worry. Nobu has a job now. I'll look after all of you.'

Gurudas went to Calcutta during the summer holidays that year—after a gap of two years. He couldn't postpone things any more, it was time to find a publisher.

In his canvas shoes, holding a dusty umbrella, he scoured the summer pavements from Goldighi to Hedua with his manuscript stuffed into a tin trunk. Finally he came across Bharat Press in a lane off Sukia Street. They published old Sanskrit and Bengali books, and were inclined towards dictionaries. But the proprietor Bipin babu said, 'We cannot judge how good your dictionary is. If you can get a recommendation from someone worthwhile, we'll think about it.'

'Such as? Whose...' Gurudas was too embarrassed to utter the word recommendation.

Bipin babu mentioned three or four people. The very first one was that of the vice-chancellor of the university.

Gurudas arrived at this gentleman's house the next day. About a dozen people were waiting in a small room. As the day progressed, a crowd of people waiting for an audience filled the open space in front of the house. Dhotis, Western suits, Madrasis, Punjabis, even saffron. Some paced up and down, some leaned against the railing, some peeped over the swing door before ducking behind it. Young men, old men, women, helpless faces, grave expressions—but all of them similarly afflicted by the need for recommendations. The clacking of typewriters, the ringing of telephones, the bustle of orderlies and clerks—it was impossible to tell who had got an audience and who was waiting in despair. From seven the clock moved on to eleven—there was no hope of a meeting today.

Gurudas slipped while getting off the tram on the way back, hurting himself. Putting tincture of iodine on his bruises, he rested on a plank in the boarding house all day. When he awoke the next morning, his hips were aching. But still he got into the second-class coach of the tram with his trunk.

No luck that day either—four hours passed, alternately sitting and standing. Four successive days went by this way.

On the fifth day he arrived even earlier, in case he could get in before anyone else. He discovered there were only two people already there. A man of dignified appearance walking across the yard stopped suddenly on seeing him.

'What's the matter? Here again?'

'I had to come again, because…'

'You haven't met him yet? Haven't I been seeing you every day? Well, what do you need?'

'I have composed a dictionary of Bengali. It's about this dictionary…'

'Oh, a dictionary? Of Bengali?' The man surveyed Gurudas from head to toe, not omitting his tin trunk. 'You've actually brought your manuscript?'

'Just…in case he wants a look…if he has the time.'

'Very well, sit down. Go straight in as soon as he arrives. Through this door here—there's nothing to be afraid of.'

He really did get an audience, along with a slip of paper with the words, 'I endorse this book for publication' along with a signature.

Five hundred copies each of several slim volumes would be published, each costing one rupee. The books would not be bound. Half of whatever was left over after paying for costs would go to the author, but if expenses were not recovered within a year, the writer would recompense the publisher.

These were the terms of the contract. Bipin babu kept the manuscripts for the first four letters, 'Aw' through 'Dirgho-ee', and Gurudas received the proofs within a week of returning to Khulna.

Six volumes were published in a year; the vowels were done. But Bipin babu welcomed him sombrely the next summer. 'The books aren't selling at all. There they are—see for yourself. An entire dictionary is available at ten rupees, who's going to pay six for just the vowels? And who cares for so many details? I couldn't cover my costs, but I know you cannot recompense me. I can absorb this loss, but if you want to publish further you'll have to pay half the costs. If the books sell, I'll recover my costs first, plus 30 per cent commission. The rest will be yours.'

'Half the costs? How much?'

'It takes between two hundred and two-fifty to print each volume. You'll get bills.'

Gurudas left another six volumes of his manuscript with the printer. For each volume that was printed, he sold half an acre of land. Eventually nothing but the homestead was left, and then that was sold too.

By then ten more years had passed. Gurudas was almost through with 'Baw'; all the letters up to 'Dontyo-naw' had been published. Meanwhile his hair had greyed, he wore thick lenses in nickel-framed glasses—but despite the spectacles everything seemed blurred at night. Harimohini was suffering from arthritis, she couldn't do the household tasks any more. The entire family was under the care of the lean, indefatigable Bhabani. She paid a little extra attention to her father, offering him whatever she could—a little milk or fruit, or some juice. When she had a few moments to spare, she leafed through his dictionary. Gurudas had taught her, the first child of his youth, a little Sanskrit and Bengali. She knew her grammar, and had even picked up proofreading skills. There were times—perhaps on the morning of a holiday—when Gurudas sat outside the house, writing, while Bhabani sat at his side, turning over the pages of books, not talking. They never spoke—but they were happy, both of them.

Nabendu now had a salary of seventy-five rupees. He lived in Calcutta, his job was to check tickets on trains leaving from Sealdah Station. His days passed travelling on trains, but he rushed home whenever he could, and he handed over a decent sum of money to Gurudas every month. It was thanks to him that they survived—even with three growing children. Gurudas could afford to go to Calcutta from time to time, and Harimohini did not come to know that they didn't own any land any more, that they actually had to buy all their provisions now.

Harimohini busied herself in finding a match for her twenty-seven-year-old son. Nabendu wasn't willing, he said he was trying to get the post of stationmaster—it would be better to marry after he had settled down. Actually, it was the state of the family that had made him reluctant to add to his financial burden. But Harimohini insisted, and he was married in May.

Along with new quilts and sheets, a painted box of toiletries, and the fragrance of Vaseline and scent, the new bride brought a wave of joy into the house. A beautiful girl of fifteen. A little pain was unavoidable too; reminded of Shibani, Harimohini wiped her eyes covertly.

Nine months after his wedding, Nabendu slipped while trying to climb into a moving train and fell on the tracks. By the time he was pulled out, his heart was still beating in his mangled body, but not long enough to make it to the hospital.

His wife was seven months pregnant at the time. She fell unconscious when informed, and delivered a premature, dead baby four hours later. She never succeeded in getting back on her feet; overcome by 'childbed' fever, suffering for six months, she finally vanished into the shadows like an insubstantial shadow herself.

Gurudas received one thousand five hundred rupees from Nabendu's provident fund, and another two thousand rupees as 'compensation'. And a few months later, just before Durga Puja, the war between Germany and England broke out.

From 'panchambahini'—fifth column—to 'anubidaran'—splitting the atom—Gurudas collected many new words during the six years of the war. These would have to be added to the appendix. But his work didn't progress significantly during this period, he only got as far as the Bengali letter 'Law'. Nor could he publish beyond the Bengali letter 'Raw'; printing had become four times as expensive, and paper was hard to come by. Meanwhile, the landlord suddenly demanded seventeen rupees as rent for the house for which Gurudas had been paying seven and a half rupees all this time; the price of rice vaulted from four rupees per maund to forty; kerosene became too expensive for lanterns. And his eyes began to trouble him. The doctor said he had developed a cataract in one of them, and that surgery was necessary. This meant a trip to Calcutta and a cost of about a hundred and fifty rupees. He dismissed the proposition as soon as he heard it—it was more important to remain alive, even if on only one square meal a day.

They survived on Nabendu's three and a half thousand rupees. He dipped into it to pay for Bhabani's daughter's wedding, which cost about five hundred. Despite controlling his expenditure strictly, the rest melted during the war years like ice put out in the sun. He had returned his daughter-in-law's jewellery to her father.

It was during the war that Harimohini learnt that they no longer owned a home of their own. But she was not perturbed—she had lost that ability. She had turned inert after her son's death—indeed, she was not quite right in the head. She seldom spoke, just ate her meals and stayed in bed most of the time, suffering from arthritis. Her teeth had fallen out, she was an old woman now.

Bhabani stood like a pillar, resilient. Her sons Amal and Bimal were in school. The elder one passed his matriculation examination and joined Khulna College, where Gurudas interceded with the principal to ensure that he would not have to pay any fees. Bimal gave up his studies suddenly and, applying his own judgement, got a job at the ration shop, where he learnt to pilfer. When the sixteen-year-old's mother found out, she used a piece of wood to take the skin off his back.

Gurudas was penniless when the war ended. His salary and allowance at the school amounted to sixty-three rupees, but because of his age the authorities were pleading with him to retire. After much begging, he secured an extension of two years—he would have to leave after that.

But suddenly the problem of employment became a trivial one. Rivers of blood began to flow as the country's independence approached. After Independence, Khulna was allotted to Pakistan. After waiting and watching developments for a while, Gurudas decided to go to India with his family.

It's best not to talk about how the journey was made. Partly on foot, partly by train, occasionally on a boat across a river. Their belongings (such as they were) were left behind; they took only absolutely essential clothes, a few utensils, and his case of books. The published copies, handwritten notes, and...and virtually nothing else. All those books he had collected with so much effort since childhood had to be left behind.

Although they were unencumbered, the journey was not an easy one. He had grown old, his vision was dimmed. His wife hobbled. Amal and Bimal actually had to carry their grandmother at times—but how far can you walk bearing the weight of a heavy old woman? They had to pause for rest beneath trees, while Harimohini shrieked with arthritic pain. Rain. Sun. Dust. Droppings. Flies. And hordes of helpless people. Two babies were crushed to death by the crowd at Ranaghat Station.

It took ten days to get to Calcutta. They passed a week at Sealdah Station, eating nothing but muri, and were then transferred on a lorry to a camp at Bongaon, where they were served a lump of rice and dal at two every afternoon. Gurudas recovered a little on this diet, but there was no respite from Harimohini's cries of pain.

Finally the Lord took pity on her. Cholera broke out in the camp, and her heart gave way after she had emptied her stomach out several times. They could not cremate her themselves; government officials gathered bodies wholesale and took them away in a black vehicle.

Two months later they were given shelter at a refugee colony near Kanchrapara. Rows of one-room bamboo shanties with a little space to cook in. A pond nearby, a tube well for fresh water at a slight distance. Still, Bhabani set up a household despite the limitations. Amal got a job at a nearby mill, which helped them survive somehow. Bimal began to lead a dissolute life, spending all his time outside the house, smoking and watching films, though no one knew how he got the money for it.

Gurudas pulled out his notebooks again. One eye was clouded over with cataract, the other had dimmed too. Every moment of daylight was priceless. He went outside as soon as the sun rose, while Bhabani brought him a cup of tea and a little muri. Bhabani had to have her tea with her father—he insisted on it. Gurudas had discovered tea towards the end of the war. It really provided energy, and suppressed hunger too. Starting with the first light of day, he worked till the last rays of the sun faded. He sat cross-legged, his notebooks on a small stool, and just two or three books open around him—whatever he had been able to salvage from Khulna. When his back ached, he placed a book beneath the small of his back and lay down for a few minutes. It brought relief.

The next month Bhabani made him a bolster. And that same day he wrote a postcard to Bipin babu at Bharat Press.

The reply came two days later. Bipin babu had asked after him, expressing pleasure at hearing from him after such a long time. Demand had picked up for his dictionary recently, the previous editions had almost sold out. It was necessary to publish the subsequent volumes now. The money realized from the sales of the earlier volumes would be enough to publish the new ones— Gurudas would not have to pay any more money. Bipin babu wrote that he would be obliged if Gurudas could inform him when the new manuscripts would be available.

After a few more letters had been exchanged, Bipin babu agreed to provide a monthly 'assistance' of fifteen rupees. Gurudas saved some of it to get some new books all over again. Several volumes were published in succession over the next two years; he got as far as the letter 'Dontyo-shaw' meanwhile.

The following year Gurudas finished his dictionary; it took another two years to publish all the volumes. He had to read everything in print once more: the corrigenda, the appendices, everything. The *Great Bengal Dictionary* was completed in fifty-two volumes. It had taken him thirty years. He had been a young man of forty when he began—now the hair on his head was white, his back was bent, his cheeks were like crevices, his veins protruded on his skin. He was blind in one eye, and had marginal vision in the other.

Gurudas took to his bed a few days later. The task for which he had conserved the last drops of his energy had been completed, he no longer needed it. He recalled Shibani, Nobu, Nobu's wife. He recalled his wife. 'Don't perform my last rites, Bhabani,' he told his daughter. 'I don't believe in any of it.'

But he simply suffered in bed. Death wasn't at his beck and call.

Meanwhile, there were murmurs in Calcutta about his dictionary. One Gurudas Bhattacharya had apparently composed a dictionary—an outstanding achievement. It was talked about in the university, at literary gatherings and newspaper offices. Those who bought the dictionary praised it, those who didn't praised it even more.

Eventually a young journalist appeared in a jeep one day, accompanied by Bipin babu from Bharat Press. Gurudas did not speak much—he had no strength. Covering her face, Bhabani answered all their questions in a soft tone. A sensational report appeared in the next day's paper, peppered with magnificent words like sacrifice, dedication and devotion.

And so Gurudas became famous.

It was the fifth year after Independence. The government had announced literary awards. Someone on the committee proposed Gurudas for an award. Gurudas Bhattacharya? Oh, the dictionary. Well…well, one has to admit he has accomplished a mammoth task, written thousands of pages. And, we hear he's in financial difficulties, eking out an existence in a refugee colony—it would be a splendid gesture. Something to capture the popular imagination with. You've seen how *Swadeshi Bazaar* has praised him, haven't you?

Gurudas was chosen to receive the award.

In reply to the official communication, Bhabani wrote that her father was ill and unable to visit Calcutta in any circumstances.

One of the younger ministers said, 'Very well, let us go to him. People will approve.'

As a result, an enormous car drew up at the Kanchrapara refugee colony at ten o' clock one morning, escorted by a jeep. A minister of the independent state emerged from the car, accompanied by two high officials, and two orderlies in shining red uniforms. The young journalist who had written the report, a government clerk, and a photographer with a camera jumped out of the jeep. The car could not come all the way to the door. As children and women stared, the visitors walked along the narrow path between rows of shanties to Gurudas's hut. The tiny space was suddenly filled with people.

There was no room to sit—the ceremonies were conducted with everyone standing. The minister said a few words. A silk shawl, a bouquet of flowers, and one hundred rupee notes tied with a silk ribbon, amounting to five thousand rupees, were placed on Gurudas's bed. The cameras clicked, Gurudas's weak eyes blinked as the flash-bulbs went off.

He lay still on his back, his hands gathered at his chest. His expression did not betray whether he was aware of what was going on. But when the guests had moved away from his bed, their demeanour suggesting they wanted to depart but were staying back only because they were embarrassed to be leaving so quickly, Gurudas spoke clearly but faintly, 'Turn me on my side, Bhabani. This is very funny, but if I laugh I will be insulting all these people. Make me face the other way.' The eye with cataracts was still, but laughter flashed in the other eye for an instant. Bhabani turned him over on his side carefully.

He died the same afternoon. His grandsons and the young men from the neighbourhood took him to the crematorium draped in the same silk shawl and covered with the same flowers.

He had made a single statement before dying. 'Keep the money, Bhabani, it'll prove useful to you.'

five

~

TOBA TEK SINGH

SAADAT HASAN MANTO

Translated from the Urdu by Khushwant Singh

Acouple of years or so after the partition of the subcontinent, the governments of Pakistan and India felt that just as they had exchanged their hardened criminals, they should exchange their lunatics. In other words, Muslims in the lunatic asylums of India should be sent across to Pakistan; and mad Hindus and Sikhs in Pakistan asylums be handed over to India.

Whether or not this was a sane decision, we will never know. But people in knowledgeable circles say that there were many conferences at the highest levels between bureaucrats of the two countries before the final agreement was signed and a date fixed for the exchange.

The news of the impending exchange created a novel situation in the Lahore lunatic asylum. A Muslim patient who was a regular reader of the *Zamindar* was asked by a friend, 'Maulvi sahib, what is this thing they call Pakistan?' After much thought he replied, 'It's a place in India where they manufacture razor blades.' A Sikh lunatic asked another, 'Sardarji, why are we being sent to India? We cannot speak their language.' The Sardarji smiled and replied, 'I know the lingo of the Hindustanis.' He illustrated his linguistic prowess by reciting a doggerel.

Hindustanis are full of shaitani
They strut about like bantam cocks.

One morning a mad Mussalman yelled the slogan 'Pakistan Zindabad' with such vigour that he slipped on the floor and knocked himself senseless.

Some inmates of the asylum were not really insane. They were murderers whose relatives had been able to have them certified and thus saved from the hangman's noose. These people had vague notions of why India had been divided and what Pakistan was. But even they knew very little of the complete truth. The papers were not very informative and the guards were so stupid that it was difficult to make any sense of what they said. All one could gather from their talk was that there was a man called Muhammad Ali Jinnah who was also known as the Quaid-i-Azam. And that this Muhammad Ali Jinnah alias Quaid-i-Azam had made a separate country for the Mussalmans which he called Pakistan.

No one knew where this Pakistan was or how far it extended. This was the chief reason why inmates who were not totally insane were in a worse dilemma than those who were utterly mad: they did not know whether they were in India or Pakistan. If they were in India, where exactly was Pakistan? And if they were in Pakistan how was it that the very same place had, till recently, been known as India?

A poor Muslim inmate got so baffled with the talk about India and Pakistan, Pakistan and India, that he got madder than before. One day while he was sweeping the floor he was suddenly overcome by an insane impulse. He threw away his broom and clambered up a tree. And for two hours he orated from the branch of this tree on Indo-Pakistan problems. When the guards tried to get him down, he climbed up still higher. When they threatened him he replied, 'I do not wish to live either in India or Pakistan; I want to stay where I am, on top of this tree.'

After a while the fit of lunacy abated and the man was persuaded to come down. As soon as he was on the ground he began to embrace his Hindu and Sikh friends and shed bitter tears. He was overcome by the thought that they would leave him and go away to India.

Another Muslim inmate had a Master of Science degree in radio engineering and considered himself a cut above the others. He used to spend his days strolling in a secluded corner of the garden. Suddenly a change came over him. He took off all his clothes and handed them over to the head constable. He resumed his peregrinations without a stitch of clothing on his person.

And there was yet another lunatic, a fat Mussalman who had been a leader of the Muslim League in Chiniot. He was given to bathing fifteen to sixteen times during the day. He suddenly gave it up altogether.

The name of this fat Mussalman was Muhammad Ali. But one day he proclaimed from his cell that he was Muhammad Ali Jinnah. Not to be outdone, his cellmate who was Sikh proclaimed himself to be Master Tara Singh. The two began to abuse each other. They were declared 'dangerous' and put in separate cells.

There was a young Hindu lawyer from Lahore. He was said to have become unhinged when his lady-love jilted him. When he heard that Amritsar had gone to India, he was very depressed: his sweetheart lived in Amritsar. Although the girl had spurned his affections, he did not forget her even in his lunacy. He spent his time cursing all leaders, Hindu as well as Muslim, because they had split India into two and made his beloved an Indian and him a Pakistani.

When the talk of exchanging lunatics was in the air, other inmates consoled the Hindu lawyer with the hope that he would soon be sent to India—the country where his sweetheart lived. But the lawyer refused to be reassured. He did not want to leave Lahore because he was convinced that he would not be able to set up a legal practice in Amritsar.

There were a couple of Anglo-Indians in the European ward. They were very saddened to learn that the English had liberated India and returned home. They met secretly to deliberate on problems of their future status in the asylum: would the asylum continue to have a separate ward for Europeans? Would they be served breakfast as before? Would they be deprived of toast and be forced to eat chapatis?

Then there was a Sikh who had been in the asylum for fifteen years. And in the fifteen years he said little besides the following sentence: 'Uper di gurh gurh di annexe di, bedhyana di, moong di dal of di laltain.'

The Sikh never slept either at night or in the day. The warders said that they had not known him to blink his eyes in fifteen years. He did not as much as lie down. Only on rare occasions did he lean against the wall to rest. His legs were swollen down to the ankles.

Whenever there was talk of India and Pakistan, or the exchange of lunatics, this Sikh would become very attentive. If anyone invited him to express his views, he would answer with great solemnity: 'Uper di gurh gurh di annexe di, bedhyana di, moong di dal of the Pakistan government.'

Sometime later he changed the end of his litany from 'of the Pakistan government' to 'of the Toba Tek Singh government'.

He began to question his fellow inmates whether the village of Toba Tek Singh was in India or Pakistan. No one knew the answer. Those who tried got confused when they tried explaining how Sialkot was at first in India and was now in Pakistan. How could one guarantee that a similar fate would not befall Lahore—from being Pakistani today, who could say it would not become Indian tomorrow? For that matter, how could one be sure that the whole of India would not become a part of Pakistan? All said and done who could put his hand on his heart and say with conviction that there was no danger of both India and Pakistan vanishing from the face of the globe one day!

The Sikh had lost most of his long hair. Since he seldom took a bath, the hair of his head had matted and joined with his beard. This gave the Sikh a very fierce look. But he was a harmless fellow. In the fifteen years he had been in the asylum, he had never been known to argue or quarrel with anyone. All that the older inmates knew about him was that he owned land in village Toba Tek Singh and was a prosperous farmer. When he lost his mind, his relatives had brought him to the asylum in iron fetters. Once a month, some relatives came to Lahore to find out how he was faring. With the eruption of the Indo-Pakistan troubles their visits had ceased.

The Sikh's name was Bishen Singh but everyone called him Toba Tek Singh. Bishen Singh had no concept of time—neither of days, nor weeks, nor of months. He had no idea how long he had been in the lunatic asylum. But when his relatives and friends came to see him, he knew that a month must have gone by. He would inform the head warder that 'Miss Interview' was due to visit him. He would wash himself with great care; he would soap his body and oil his hair and beard before combing them. He would dress up before he went to meet his visitors. If they asked him any questions, he either remained silent or answered, 'Uper di, annexe di, bedhyana di, moong di dal of di laltain.'

Bishen Singh had a daughter who had grown into a full-bosomed young woman of fifteen. But he showed no comprehension about his child. The girl wept bitterly whenever she met her father.

When talk of India and Pakistan came up, Bishen Singh began to question other lunatics about the location of Toba Tek Singh. No one could give him a satisfactory answer. His irritation mounted day by day. And now even 'Miss Interview' did not come to see him. There was a time when something would tell him that his relatives were due. Now that inner voice had been silenced. And he was more anxious than ever to meet his relatives and find out whether Toba Tek Singh was in India or Pakistan. But no relatives came. Bishen Singh turned to other sources of information.

There was a lunatic in the asylum who believed he was God. Bishen Singh asked him whether Toba Tek Singh was in India or Pakistan. As was his wont God adopted a grave mien and replied, 'We have not yet issued our orders on the subject.'

Bishen Singh got the same answer many times. He pleaded with 'God' to issue instructions so that the matter could be settled once and for all. His pleadings were in vain; 'God' had many pressing matters awaiting 'His' orders. Bishen Singh's patience ran out and one day he let 'God' have a bit of his mind: 'Uper di gurh gurh di annexe di, bedhyana di, moong di dal of waheguru ji ka khalsa and waheguru di fateh! Jo Bole So Nihal, Sat Sri Akal!'

This was meant to put 'God' in his place as the God only of the Mussalmans. Surely if he had been God of the Sikhs, he would have heard the pleadings of a Sikh!

A few days before the day fixed for the exchange of lunatics, a Muslim from Toba Tek Singh came to visit Bishen Singh. This man had never been to the asylum before. When Bishen Singh saw him he turned away. The warders stopped him: 'He's come to see you; he's your friend, Fazal Din,' they said.

Bishen Singh gazed at Fazal Din and began to mumble. Fazal Din put his hand on Bishen Singh's shoulder. 'I have been intending to see you for the last many days but could never find the time. All your family have safely crossed over to India. I did the best I could for them. Your daughter, Roop Kaur...'

Fazal Din continued somewhat haltingly, 'Yes... she too is well. She went along with the rest.'

Bishen Singh stood where he was without saying a word. Fazal Din started again. 'They asked me to keep in touch with you. I am told that you are to leave for India. Convey my salaams to brother Balbir Singh and to brother Wadhawa Singh...and also to sister Amrit Kaur... Tell brother Balbir Singh that Fazal Din is well and happy. Both the grey buffaloes that they left behind have calved—one is a male, the other a female...the female died six days later. And if there is anything I can do for them, I am always willing. I have brought you a little sweet corn.'

Bishen Singh took the bag of sweet corn and handed it over to a warder. He asked Fazal Din, 'Where is Toba Tek Singh?'

Fazal Din looked somewhat puzzled and replied, 'Where could it be? It's in the same place it always was.'

Bishen Singh asked again: 'In Pakistan or India?'

'No, not in India; it's in Pakistan,' replied Fazal Din.

Bishen Singh turned away mumbling: 'Uper di gurh gurh di annexe di, bedhyana di, moong di dal of the Pakistan and Hindustan of dur phittey moonh.'

Arrangements for the exchange of lunatics were completed. Lists with names of lunatics of either side had been exchanged and information sent to the people concerned. The date was fixed.

It was a bitterly cold morning. Busloads of Sikh and Hindu lunatics left the Lahore asylum under heavy police escort. At the border at Wagah, the superintendents of the two countries met and settled the details of the operation.

Getting the lunatics out of the buses and handing over custody to officers of the other side proved to be a very difficult task. Some refused to come off the bus; those that came out were difficult to control; a few broke loose and had to be recaptured. Those that were naked had to be clothed. No sooner were the clothes put on them than they tore them off their bodies. Some came out with vile abuse, others began to sing at the top of their voices. Some squabbled; others cried or roared with laughter. They created such a racket that one could not hear a word. The female lunatics added to the noise. And all this in the bitterest of cold when people's teeth chattered like the scales of rattlesnakes.

Most of the lunatics resisted the exchange because they could not understand why they were being uprooted from one place and flung into another. Those of a gloomier disposition were yelling slogans: 'Long Live Pakistan' or 'Death to Pakistan'. Some lost their tempers and were prevented from coming to blows in the very nick of time.

At last came the turn of Bishen Singh. The Indian officer began to enter his name in the register. Bishen Singh asked him, 'Where is Toba Tek Singh? In India or Pakistan?'

'In Pakistan.'

That was all that Bishen Singh wanted to know. He turned and ran back to Pakistan. Pakistani soldiers apprehended him and tried to push him back towards India. Bishen Singh refused to budge. 'Toba Tek Singh is on this side,' he cried, and began to yell at the top of his voice, 'Uper di, gurh gurh di, annexe di, bedhyana di, moong di dal of Toba Tek Singh and Pakistan.' They did their best to soothe him, to explain to him that Toba Tek Singh must have left for India; and that if anyone of that name was found in Pakistan he would be dispatched to India at once. Bishen Singh refused to be persuaded. They tried to use force. Bishen Singh planted himself on the dividing line and dug his swollen feet into the ground with such firmness that no one could move him.

They let him be. He was soft in the head. There was no point using force; he would come round on his own—yes. They left him standing where he was and resumed the exchange of other lunatics.

Shortly before sunrise, a weird cry rose from Bishen Singh's throat. The man who had spent all the nights and days of the last fifteen years standing on his feet, now sprawled on the ground, face down. The barbed-wire fence on one side marked the territory of India; another fence marked the territory of Pakistan. In the No Man's Land between the two barbed-wire fences lay the body of Bishen Singh of village Toba Tek Singh.

THE FLOOD

THAKAZHI SIVASANKARA PILLAI

Translated from the Malayalam by O. V. Usha

The temple stood on a rise, the highest ground for miles around. Despite this, its deity was submerged in water up to its neck. There was water everywhere you looked. In the three-roomed uppermost storey of the building, sixty-seven children, three hundred and fifty-seven adults and assorted dogs, cats, goats and fowl had taken refuge from the floodwaters.

Most of the locals had fled their homes, and made their way to dry land. If the family owned a boat, one of them would remain behind to guard their house and their possessions.

Chennan, the pariah, had been standing in water an entire night and day. It had been three days since his master had escaped to safety. Chennan did not own a row boat so inside his hut he had built an elevated platform out of coconut fronds and twigs that jutted above the level of the floodwaters. The family spent two days in this primitive loft, hoping the floodwaters would recede soon. Chennan was concerned that if they left the hut, his five banana trees that were heavy with fruit and his hayrick would be stolen. The waters continued to rise. Chennan was now knee-deep in water as he stood on his platform. A couple of rows of thatch on the roof slipped under water. Chennan shouted for help, but who was there to hear him? He was convinced that he and his dependents—his pregnant wife, four children, a cat and a dog—would die within the next twelve hours, as, by then, the waters would have risen above the roof. The downpour, which had started three days ago, showed no signs of abating.

Undoing a portion of the thatch from inside the hut, Chennan somehow scrambled on to the roof and looked around. He saw a big boat moving north. He yelled to the men in the boat for help. They heard him, and began rowing towards his hut. Chennan pulled out his children, his wife, the dog and the cat through the opening he had made in the roof. When the boat came alongside the hut, the children began climbing into it. Just then they heard a voice.

'Chennacho… Poohay…!'

It was Matiyathara Kunheippan, calling out from the roof of his house.

Chennan hastily helped his wife into the boat. The cat also jumped in. No one remembered the dog. It was sniffing at something on the western edge of the roof when the boat began to move away from the house.

$$\Longleftarrow$$

When the dog eventually returned to the spot on the roof from which the family had made its escape, the boat was far away. The animal began to run around frantically on the roof, sniffing here and there, whining all the while. A frog which had been sitting on the roof was alarmed by the commotion the dog was creating and jumped into the flood as the dog drew close. The movement frightened the dog; it reared back, then stared intently at the ripples caused by the frog when it had dived into the water.

The dog was hungry now, and began sniffing around for food. It disturbed another frog, which urinated on its nose, before jumping into the water. The dog sneezed and snorted, shook its head violently, began cleaning its face with its forepaws.

The rain which had stopped for a while, began lashing down again, and the dog crouched down miserably under its onslaught. Meanwhile, its master and family had got to safety in Ambalappuzha.

Night fell. An enormous crocodile, half submerged in the water, drifted by the hut. The dog put its tail between its legs in terror, and howled, but the crocodile paid no attention to it and was soon out of sight.

$$\Longleftarrow$$

Starving, terrified, tired, and without any protection from the driving rain, the dog crouched on the roof of the hut. It howled piteously and the gusting wind carried its cries of distress across great distances. The few people who remained in that flooded wilderness guarding their huts probably felt sorry for the dog when they heard its anguished howls.

Its master was probably eating his dinner. He was probably rolling a ball of rice for the dog between his fingers, as was his practice at the end of a meal. The dog's howls began to weaken, then died away. It pricked up its ears when it heard a man reciting the Ramayana. It remained silent for a while, almost as if it was listening to the chanting of the verse, and then began barking again as loudly as it could manage. The melodious chanting of the Ramayana rose clear and strong into the still night. The rain had stopped. The dog listened intently to the sound of that human voice. The voice died away into the cold air. Now there was no sound except that of the gusting of the wind, and the lapping of waves against the hut.

The dog lay down on the roof. It snuffled miserably. A fish splashed in the dark water. A frog jumped. The dog barked once or twice, then fell quiet.

When the sun came up, the dog began a low howling, almost as if it were keening. Rows of frogs, lined up on the roof of the house, would eye the dog, and if it made a move, would jump into the water. The dog looked at the thatched roofs poking out of the floodwater in the vicinity, hoping that they would contain food; but there was no food to be had. It began nuzzling at its fur, feeding on the fleas it found there; it scratched the flea bites with its hind legs, and then it briefly fell asleep in the feeble warmth of the sun.

Dark clouds enveloped the sun. The wind rose, and with it the waves. The carcasses of drowned animals floated past the hut. The dog looked longingly at them, and barked once or twice.

Every now and again small row boats would come into view and the dog would look at them hopefully. One of them went past in the distance and disappeared behind a thick clump of ketaki shrubs. It began to drizzle again. Sitting on the roof, the dog presented a picture of utter helplessness.

The rain stopped. A small boat appeared in the distance and steadily drew closer. It moored under a coconut tree close to the hut. The dog wagged its tail, stretched, yawned, growled under its breath. The man in the boat climbed the tree, lopped off a coconut, sliced it open, drank its contents and then rowed away.

A crow flew down and landed on the huge rotting carcass of a dead buffalo that was floating by, and began to rip and tear at the flesh. Chennan's dog barked longingly at the prospect of a meal. The crow ate its fill and flew away. A green bird landed on a banana trunk and chirped. An ant's nest that had become dislodged from somewhere upriver snagged on the eaves of the hut. Thinking that it was something it could eat, the dog nosed at the nest, and got bitten for its pains. It jumped away, snorting and sneezing, its tender nose flushed and swollen.

That afternoon, a small row boat drifted by, with two men in it. The dog began barking joyfully, jumping up and down in its happiness. Its demeanour, the way it was trying to express itself, seemed almost human. It went down to the water's edge, ready to jump into the boat.

'Look, a dog,' one of the men said. The dog began to whimper in a peculiar key—it was almost as though it was responding to the compassion in the man's voice.

'Let it be,' said the second man.

The dog whimpered, opened and closed its mouth, it seemed as though it was praying to the men to rescue it. It tried twice to jump into the boat, and was rebuffed both times. The boat began to move away from the house. The dog howled. It was heart-rending. One of the men looked back.

'Ayyooo...' The sound wasn't from either of the men but the dog. 'Ayyooo...' It was an exhausted, pitiful sound. It was almost as if the animal was bidding goodbye to the world.

The men did not look back again. The boat drew steadily away, and the dog remained where it was on the roof, watching it go. Its moaning and the expression on its face seemed to say that it would love human beings no more. The boat disappeared from sight.

The dog lapped up some water, looked at some birds flying overhead. A water snake was borne along by the flood at great speed towards the hut. Startled, the dog retreated along the roof. The snake slipped into the opening in the thatch through which Chennan and his family had made their escape. The dog peered into the opening and began to bark furiously.

The barking gradually tapered away into a whimpering. It was a sound of lament, full of the hunger the animal hadn't been able to sate, its despair at the hopelessness of its situation, and its fear that it might not survive. What it was trying to express would have been clear to anyone—even to a man from Mars.

With nightfall, a storm picked up. The roof of the hut began to shake as waves battered against it. The dog nearly fell off—twice. A long, sinister head rose out of the dark water. The dog began to bark frantically when it saw the crocodile. From somewhere close by came the sound of chickens squawking.

Another boat came into view, heavily loaded with coconuts and bunches of bananas. It stopped beside a banana tree close to the hut.

'Where is that barking coming from? Hasn't the hut been vacated?' one of the men in the boat asked.

The dog turned towards the sound of the men's voices and began barking. As one of the men began to climb the banana tree, the dog's barking grew in fury and menace.

'Better watch out, that dog sounds really angry, it might bite you,' the man who had remained in the boat warned his companion who had climbed the tree. No sooner had he spoken than the dog leapt at the thief who was trying to steal Chennan's bananas. The man tumbled down into the water, the dog with him. The thief's companion pulled him into the boat, while the dog swam back to its rooftop. It clambered back on, and shook the water off its coat, barking furiously all the while. The thieves lopped all the bananas off the trees, and loaded them on to the boat.

The men goaded the angry animal on to even greater rage. 'Just you wait, we've got something to show you,' they said grimly.

They transferred as much hay as they could from Chennan's hayrick on to their boat. Then one of the men got on to the roof and advanced threateningly towards the dog. It attacked him immediately, and sank its teeth into his leg. Howling in agony, the man jumped back into the boat, while his companion took hold of an oar and brought it crashing down on the dog's head. The animal yelped, and retreated.

The man who had been bitten lay in the bottom of the boat tossing and turning in agony. As the boat receded into the distance, the dog sent it on its way with a volley of barks. It was close to midnight by then. An enormous bloated carcass of a cow drifted up to the house, and snagged on the roof. The dog kept an eye on the carcass. After a while, pushed by the current, the carcass began to move away slowly. Galvanized into activity, its tail wagging, grunting, the dog rushed down the roof to where the carcass was still within reach, and began tearing at the flesh, sating its terrible hunger. Just then, with a loud cracking noise, the entire carcass was dislodged from the roof, and the dog, which was eating from it, went with it. The carcass rolled over in the current, and the dog disappeared from view. Now one could only hear the sound of the gushing wind, the cacophony of frogs, and the sibilant whisper of the waves. There was no other sound.

Those who still stood guard in some of the flooded homes could no longer hear the dog's piteous whining, or heart-rending moaning. Every so often, another bloated carcass of some drowned animal would float by on the flood. Crows fed on some of them. Everywhere you looked there was nothing but desolation. And the only people who thrived in this wilderness of water and destruction were thieves and looters.

The hut collapsed after a while, and sank out of sight. And the loyal animal, which had stood guard over its master's property for as long as it could, was gone as well. The crocodiles had it now. It was all over, the waters covered everything.

~

The flood receded slowly. One day, Chennan swam across to the spot where his hut had once stood, looking for his dog. Under one of the coconut trees, he saw the corpse of a dog, gently rocking in the eddies of the shallow water. Chennan turned it over with his foot to check if it was his dog, he thought it might be. One of the animal's ears was missing. You couldn't even tell what colour the dog was, for its skin had rotted and sloughed away.

seven

~

THE BLUE LIGHT

VAIKOM MUHAMMAD BASHEER

Translated from the Malayalam by O. V. Usha

This story is about an extraordinary, no, miraculous event in my life. That I call it miraculous…well, what else can I say?

This is what happened.

No need for the date, month or year.

I was on the lookout for a house. Nothing new about that. Those days I was always looking for a house. I never came across a house or even a room to my liking. As far as the place, which called itself a hotel, where I stayed, a hundred flaws, more, had come to light there. But to whom to complain? Leave? And go where?

And so it went, hating where I stayed, constantly looking for a place and finding only those that I hated even more. I lost count of the number of houses I looked at and disliked. And, as is the way with rented houses, someone who liked them took them.

Still, it was a time when there was a dearth of houses for rent. What could be had at one time for ten rupees was not available now for even sixty.

So I spent my days, roaming the town daily, until, suddenly, one forenoon, there it stood—a house, no, my house! It was a small two-storeyed building. 'Bhargavinilayam'. Far from the bustle and noise of the town. Yet, close to the municipality border. A decrepit board at the gate said 'For rent'.

I liked the place instantly. It was an old house. There was something strange about it even at first glance. Didn't matter; it was perfect for me. Two rooms and an open verandah upstairs. Four rooms downstairs. Also, there was a bathroom, a kitchen and a pipe with running water. The only thing missing was electricity.

In front of the kitchen there was an ancient well with a stone wall round it. A small way off, in a corner of the walled compound around the house, stood a toilet. Trees in abundance. The public road ran right before the property.

I was surprised and delighted! Why had no one snapped up this house yet? A lady of great beauty she was. Ah! She shouldn't be seen by anyone. Should hide her behind purdah! Such unusual thoughts and feelings that old house evoked in me. I was in a happy daze; I was entranced.

I quickly made whatever arrangements needed to be made, borrowed money from various friends, paid the two months' advance rent and got the keys in my hand. Wasting not a moment, I moved into the upstairs of the old house. I bought a hurricane lamp and kerosene. I swept the rooms myself, upstairs and downstairs, kitchen, bathroom and all. I ritually cleansed the place, sprinkling water. There was a good deal of accumulated waste; there was plenty of dust. After the sweeping, I thoroughly washed all the rooms until they were spotless. I saw one room downstairs that was locked. I did not try to open it.

I had a bath, and felt a great sense of relief and well-being. I came out and perched myself on the wall round the antique well. Such bliss! One could sit and daydream here, or walk or run or whatever in complete privacy inside the walled compound around the house.

There should be a garden in the front yard, I thought, mainly of beds of roses. Jasmine too. A cook? No, that would be a bother. I could fill the thermos with tea from the tea shop when I went for breakfast after my morning bath. Lunch could be arranged from the old hotel. Hopefully, they would send me dinner as well.

The postman had to be informed about the change of address, and tipped not to reveal my whereabouts to anyone. Nights of lovely solitude; days of lovely solitude: I would be able to write a lot! Borne along by these and other happy thoughts, I peered down into the well. The tangle of plants growing across its mouth and from its walls made it impossible to see if it had any water. I threw a pebble into the well.

'Bhllllum!' it rang after a moment. There was water.

All this by eleven in the morning.

The previous night I hadn't slept a wink. I settled my account with the hotel. I met the houseowner. I folded the canvas cot and secured it. Gramophone, records—all neatly packed. Boxes, papers, armchair, shelf, everything—all my assets had been set in order. At the crack of dawn, I set out with my luggage loaded on to a pair of pushcarts. The men who brought my luggage left it outside the gate…as if they were frightened to come into the place.

Arriving at my new residence, I brought my possessions in myself, then, elated, locked the house, walked humming to the gate, fastened that and went out on to the road with a fine new sense of pride, walking on air.

As I went along, I wondered: with whose song should the new house be inaugurated? I had more than a hundred records in my possession. English, Arabic, Hindi, Urdu, Tamil, Bengali. Nothing in Malayalam. There are talented singers. They have cut records. But the music direction and finish of all of them is poor. But good directors and singers are coming up. Yes, I must buy some Malayalam records.

Whose song should I play first today, I wondered. Pankaj Mullick, Dilipkumar Roy, Saigal, Bing Crosby, Paul Robeson, Abdul Karim Khan, Kanan Devi, Kumari Manju Dasgupta, Khurshid, Juthika Roy, M. S. Subbulakshmi… some twenty names passed through my mind. Finally, I decided: there is a song which says that someone from a distant land is here. It begins with '*Door desh ka rehne wala aaya*'. Whose voice was it? Male or female? I could not remember. I would find it when I got back home.

Thus, I went on my way, happily. First of all I met the postman. Talked to him. When I told him which house I had moved into, he looked unnerved.

'Ayyo, sir! There was an unnatural death in that house. That's why it's been lying empty for so long. No one will stay there.'

An unnatural death? I was a little flustered. I asked, 'What unnatural death?'

'Did you see the well in the backyard? Someone jumped into it and died. After that there is no peace in that house. Many tenants came but none stayed. At night doors would bang shut on their own. Taps would open by themselves and water would gush out.'

Doors banging shut on their own. Taps opening by themselves. Startling indeed. I'd seen that both taps were secured with locks. Travellers scaled the compound wall and bathed at the taps, that was why they were locked—that was the houseowner's version. But what was the need for a lock on the bathroom tap inside the house—it did not occur to me to ask.

The postman continued, 'Some invisible thing or person would try to choke anyone sleeping inside that house! Did no one tell you, sir?'

I thought, oh, good! And I've paid two months' rent in advance. Putting a brave face on it, I said, 'All that doesn't matter. The whole thing will just take a mantra to settle. Whatever it is, bring my letters and other mail there.'

Though I spoke evenly, I did feel a ripple of anxiety, as any other man might. Call me a coward, if you will. My pace on the road slackened. What could I do? I do not usually court eerie experiences. But if they happen anyway, what does one do? But what could really happen?

I went to the hotel and drank some tea. I did not feel hungry. It was as if a strange fire had been lit in my gut, driving out my appetite. I spoke to the man in the hotel about sending regular meals to my house.

When I told him where I was living, he said, 'I can arrange to send you food during the day. At night none of the boys will come there. A woman killed herself by jumping into the old well in the yard. She still hangs around the place. Aren't you afraid of ghosts, sir?'

It was a woman; half my fear melted away.

I said, 'Oh, nothing to all that. Moreover, there's the mantra.'

I had no idea what the mantra was. But it was a woman...some softness there. I walked into the bank nearby. I had a few friends among the clerks there. I talked to them. They were furious with me: 'How foolish could you be?! It's a haunted house and it's mostly men that are harmed.'

She hated men! That was nice to know.

One of my friends said, 'Couldn't you have told us you were planning to take Bhargavinilayam on rent?'

I said, 'Who knew it was like this? Let me ask you a question—why did that woman jump into the well?'

'Love!' said another friend. 'Her name was Bhargavi. She was twenty-one. She passed her BA. Before that she was in love with someone. Big love. Then the fellow married another woman. And Bhargavi jumped into the well and died.'

Most of my fear was gone. So this is why she hated men.

I said, 'Bhargavi will not harm me.'

'And why is that?'

I smiled, 'Mantra! Mantra!'

'We'll see about that. You're going to howl in your bed at night.'

I did not say anything more.

I went back to the house. Opened doors and windows. Went downstairs and walked to the well.

'Bhargavi kutty!' I called softly. 'We don't know each other. I have come to stay in this house. I believe that I am a good man. I've been celibate all my life, you know? I have already heard many rumours about you Bhargavi kutty. That you do not allow anyone to live here. That you open the taps at night. That you bang the doors shut. That you strangle men, especially, with invisible hands. Yes, I've heard a lot. What do I do now?

'I've already paid two months' rent as advance, and I have very little money. Also I like this house, this house that belongs to you, Bhargavi kutty, I like it so much! It's named for you, isn't it? Bhargavinilayam!

'I need to stay here and work. I must write some stories. Bhargavi kutty, do you like stories? If you do, I'll read everything that I write to you. I have absolutely no quarrel with you, Bhargavi kutty. No cause for that either, is there? A while ago I did drop a pebble into the well. I did it without thinking. Nothing like that from me in future, I promise.

'Listen Bhargavi kutty, I have a first-rate gramophone. Some two hundred excellent songs too. Are you fond of music?'

I fell quiet after saying that much. Who was I talking to? To the well, ready with its mouth open to swallow anything? To whom—the trees, the house, the air, the universe? Was it to the disturbance in my mind? I was talking to an idea, an abstraction—Bhargavi! I had not seen her. A young woman of twenty-one. Deeply in love with a man. She dreamt of a life as his wife, his life partner. But that dream…yes, it passed as a dream. Despair gripped her, and shame, disgrace and betrayal too.

'Bhargavi kutty!' I said, 'you needn't have done what you did. Don't think I am blaming you. The man you were fond of did not love you enough. He loved another woman more.

'Life turned bitter for you. That much is true. However, all of life is not full of bitterness. Forget about it! And, for you, now history will not repeat itself.

'Bhargavi kutty, don't think for a moment I am blaming you. Was it really for love that you died? Love is only the golden dawn of eternal life. Silly girl! You didn't know a thing about it. That's what your hatred of men proves. Consider that you knew just one man. Also let's accept, for argument's sake, that he wronged you grievously. But is it right to view all men through the tinted glass of that one experience? Had you lived instead of killing yourself, you would have known through experience that your extreme reaction was not right. There would have been someone to love you, to adore you, to call you "my goddess!" Now, didn't I say, for you history will not repeat itself? Bhargavi kutty, how can I get to know your whole life story?

'Whatever it is, don't harm me, I beg you. No one will question you if you were to choke me to death tonight. Not that it's anyone's business to question you. All I mean is that there is no one to do it. You know why? I have no one.

'Now you know how things stand. We, you and I, are living here. That is, I mean to live here. Having paid rent, legally, the house, well and everything else is mine. But let that be. You use the well and the four rooms downstairs. We share the kitchen and the bathroom equally. What do you say? Do you like my idea?'

Night fell. I came back with a tea-filled thermos after having eaten dinner at the hotel. By the light of my electric torch, I lit the hurricane lamp. The room was flooded in yellow light.

I went downstairs with the torch. I stood still in pitch darkness for some time. My intention was to turn the taps off tightly. I opened all the windows wide. I approached the kitchen, near the well. Suddenly, I had the strong feeling that I must not tighten the taps.

I closed the doors, climbed the stairs, and had some tea. After that I lit a bidi and sat in my armchair. I was just beginning to write, when I had the feeling that Bhargavi was behind my chair.

I said uneasily, 'I don't like anyone watching while I write.'

I turned my head; no one there.

I'm not sure why, but I didn't want to write any more. I drew up another chair in front of mine.

'Sit down in this, Bhargavi kutty.'

The chair stayed empty. I got up and began strolling through the two rooms. No breeze. The leaves of the trees outside didn't stir. I looked down through the window: a light!

Blue, red or yellow—I couldn't tell. It was gone in a second.

Just my imagination, I told myself. I can't swear that I saw that light. Still, how could one think one sees without actually seeing? Was it a glow-worm perhaps?

I spent a long time pacing the rooms. I spent a long time standing at the windows. There was nothing unusual. I tried to read something but couldn't concentrate. The chair remained empty.

Let me go to sleep early, I thought. I made my bed and turned off the lamp. Then I wanted to play a song! I lit the lamp again. Opened the gramophone. Fitted a new needle into the playing arm, and then I wound up the machine.

Whose song should I play? The world was still and silent. Then, an uncanny whooshing filled both my ears! Like the wind, like the sea. And then absolute quiet again. Fear seized me. I must shatter this ominous silence into a thousand pieces. Whose song would best serve that purpose?

I looked through my collection, and chose a record of the black American singer Paul Robeson. The gramophone began playing. A man's voice rich and majestic: *Joshua fit the battle of Jericho…* That song ended. Pankaj Mullick was next:

Tu dar naa zaraa bhi…
You don't be afraid at all!

The next was a sweet, soft and alluring voice:

Kaatrinile varum geetam…
The song that arrives on the breeze…

And in time, M. S. Subbulakshmi finished that song.

Somehow, after these three songs, I was at peace. I sat quietly, for quite a while. Finally, I invited Saigal in. He sang in his languid voice, full of melancholy and sweetness:

So jaa raajkumaari, so jaa…
Go to sleep, princess; may beautiful dreams visit your sleep.

That also ended.

'That's all for now, the rest tomorrow,' I said aloud and, lowering the lid of the gramophone, lit a bidi, turned off the lamp and went to bed. Next to me I had the torch, a watch and a dagger. Also the empty chair.

I closed the doors to the verandah before getting into bed. It must have turned ten now. I lay alert, my ears strained to the night.

There was no sound other than the tik-tik of the watch. Minutes passed, and hours. There was no fear in my mind. What there was…was a cold wakefulness. This was not new to me. I'd had some unusual experiences in my life over the years, in many lands, many places, during my lonely life of twenty years, none of which I could find the meaning of. Because of this, my attention constantly shifted between the past and the present.

Would there be a knock at the door? Would there be the sound of water flowing from the taps? Would I be throttled in my sleep? I stayed awake and vigilant till three in the morning.

I heard nothing, felt nothing. Finally, I fell into a deep and tranquil sleep. I did not have a single dream. I woke up at nine the next morning.

Nothing had happened.

'Good morning, Bhargavi kutty, thank you so much! I have understood one thing for certain—people are spreading false rumours about you. Let them talk, no? Who cares?'

My days and nights passed in this manner. I would think of Bhargavi kutty...her mother, father, siblings, there must be so many unknown stories.

I wrote most nights. When I got tired, I played records. I would announce the singer, the content, the mood and so on, before the song began. I would say, 'Here, Pankaj Mullick, the great Bengali singer. The song is filled with sadness, and invokes memories. These are bygone times, you know. Listen carefully.'

Guzar gayaa voh zamaana kaisa, kaisa...
How wonderful were those days that are now gone.

Or else I would say: 'And now Bing Crosby. *In the moonlight*—it means... Oh, I beg your pardon, you must know, you're a graduate!'

I would talk like this, to myself. Two and a half months passed. I planted a garden. When flowers appeared, I said: 'All the flowers are for Bhargavi kutty!'

In the meantime, I finished writing a short novel. Many of my friends visited me. At times, some stayed overnight. Before going to bed, I would go downstairs without their knowing it. And I would look into the darkness and whisper, 'Bhargavi kutty, listen! Some friends of mine are here. Don't strangle them. If you do, the police will arrest me. Be careful. Good night!'

Usually before going out I would say, 'Bhargavi kutty, take care of the house. If any thieves get in, you can strangle them. Don't leave the bodies here though. Dump them some miles away, or we will be in trouble.'

When I returned from the cinema after a late show, I would announce, 'It's me!'

All this I said in the newness of the first days. As time passed, I forgot Bhargavi. Well, my long chats to her stopped. I would still remember her occasionally, but that was all.

I will tell you what sort of remembering that was. Millions have died since human life began in this world...haven't countless men and women died? All of them have mingled with this earth as smoke or dust. We know this. Bhargavi was reduced to a memory of that kind.

Then, as life went by without any great event, something happened. That's what I am going to tell you about now.

One night. It was ten o' clock. From around nine I had been writing a story. It was emotionally charged and I was developing it briskly. I became aware that the light was getting dim.

I lifted the lamp and shook it gently to check the oil. There was no kerosene. Still, I thought, let me write one more page. I was entirely absorbed by the story I was writing, and in the meanwhile the light had faded. What could one do at a time like that? Check if there was oil in the lamp. That was what I did. I raised the wick a little; I continued writing. After a while, the light faded further. I raised the wick again, resumed my writing. The light dimmed again; again, I raised the wick. Eventually, the wick turned into a smoking red ember four inches long.

I turned on my torchlight and lowered the wick completely. The lamp was snuffed out.

I said to myself, 'How can I get some light?'

I needed kerosene. I could borrow some kerosene from the bank clerks' stove.

I locked the door and went out with my torch and kerosene bottle. I went down and out of the house, locking the front door behind me. I went down the path and out of the gate, latching it. I walked along the deserted road. A faint moon lit my way. There were heavy clouds above, too, so I strode along quickly.

I came to the bank building and, from the street, called out the name of one of the bank clerks. I called twice or thrice before one of them came down and opened the side gate. We went up using the staircase behind the building. I found that the three of them were playing a game of cards.

When I talked about kerosene, one of them asked with a laugh, 'Couldn't you have asked your girlfriend to get you kerosene? Have you finished writing her life story?'

I didn't say anything. Must write, I thought. As one of them was pouring kerosene into the bottle, it began to pour outside.

I said, 'Give me an umbrella also.'

They said, 'One? There is not even a quarter of one. Let's play cards for a while. You can go when the rain lets up.'

So we played cards. My partner and I did three rounds of salaam. Mostly because my mind was on my story and I was distracted while playing. The rain stopped at about one. I got up and picked up the torch and kerosene bottle. By now, my friends were also ready to sleep. After I went down to the street they switched off their light.

There was neither movement nor light in the street as I walked back home. When I took the turning towards my house the world lay immersed in mist and hazy moonlight. I didn't remember what thoughts ran through my mind then. Perhaps I was not thinking at all. I continued walking along the dark road, empty and silent, flashing my torch. No living thing stirred anywhere.

I reached my house, opened the gate, shut it, and then opened the front door. I entered the house and latched the door from inside. I had no cause to think that anything unusual would have happened upstairs. But something had.

For no reason whatever, a vast sorrow abruptly filled my heart and I felt like crying. I can laugh easily but never shed a tear. Tears just don't come to me. When sorrow overwhelms me, a divine exaltation takes hold of me, as it did now—a great compassion!

In that state of mind, I climbed the stairs. And what I saw when I arrived on the upper floor was extraordinary, no, it was miraculous.

When I locked my room and left, you remember, there was no oil left in the lamp; the flame had gone out and I had withdrawn the wick into the base of the lamp. Darkness had filled the room. Later, while I was out, it had rained and some hours had passed. Now, I saw light shining clear through the chink below my door. Well, my eyes saw the light but my mind did not yet register or believe what I saw.

Out of sheer habit, I took out my key and flashed the torchlight on the lock. The lock shone like silver; why, it seemed to smile.

I opened the door and went in. And now, what I saw before me struck me with full force. An indescribable sensation surged through every atom of my being. No, I did not quake with fright. Rather, as I stood transfixed, my mind was flooded by a tide of tenderness.

Blue light!

White walls, the whole room, drenched in ethereal blue light. That light came from the lamp, in which a blue flame blazed.

Who had lit this blue light in Bhargavinilayam?

eight

~

THE SOMERSAULT

GOPINATH MOHANTY

Translated from the Oriya by Sitakant Mahapatra

The day Jaga Palei of Sagadiasahi defeated Ramlawan Pande of Darbhanga to enter the finals of the All-India Wrestling Competition that was being held in the Barabati Stadium the sky was shredded by the roars of approval from thousands of spectators. It was not so much the victory of Jaga Palei they were celebrating as the fact that this was Orissa's victory. Orissa had won. This was the feeling everywhere.

At that moment, Jaga Palei became a symbol, the symbol of the glory, and the fulfilment of the hopes and aspirations of the Oriya people. A sea of humanity surged forward to greet him, to meet this hitherto unknown, unheard-of wrestler. The waves broke on each other, there was a stampede. At least twenty-one persons had to be removed to hospital. The situation became so riotous and uncontrollable that the police had to be called out.

Among the crowd that returned home that evening there were some who'd had their shirts torn; others had bruised bodies or had lost watches and fountain pens. It did not matter. Each of them carried in his heart the pride of being Oriya. And something which might be termed the intoxication of heroism. As if each one of them was a Jaga Palei! Newspapers flashed photographs of the wrestling match.

All the Oriya papers raved about Jaga Palei. 'Jaga Palei—Orissa's glory'; 'Jaga Palei—Orissa's honour'; 'Jaga Palei, the unparalleled Oriya wrestler'; 'Never-heard-of-before wrestling at Cuttack!'; 'Jaga Palei, Emperor of Athletics'; 'the Newest Success of the Unbeaten Wrestling Artist' and so on.

Excitement spread rapidly to the rural areas as soon as the newspapers published the news. Many cursed their bad luck that they hadn't been able to see such an epoch-making event.

The week that followed could legitimately be called 'Jaga Palei Week'. In buses and in trains, in hotels and in the village Bhagabat-tungi, the talk was only about Jaga Palei's feat. This news completely overshadowed all the other news whether it was 'Rocket to Mars', 'Man's Flight in Space', 'Death of Lumumba' and the subsequent political developments in the Congo, 'Success and Failure in Panchayat Samiti and Zila Parishad Elections', and other such exciting events. Since there were no auspicious marriage dates to be had in the coming year, hundreds of marriages took place in the fortnight following the event; during these festivities a frequent subject of discussion was Jaga Palei's wrestling.

'Did you go to see the wrestling match?'

'Did you like it?'

Even if one had not been present, one would have to answer, 'Oh, yes, of course; it was simply wonderful.' It was almost as if to say otherwise was to do worse than confessing to a hidden guilt.

During Jaga Palei Week, a small, five-page booklet went on sale in crowded places. The poet: Abid. The price: Ten paise. Hawkers could be seen hawking the songbook to the accompaniment of harmonium music in front of the cutchery, railway station, bus stand and market squares. Framed photographs of the wrestling match went up on the walls of photographers' studios, and also at sweetmeat-and-tea stalls and paan shops in town. Alando Mahila Mandali, Olangsha Yubak Mandal, Gababasta Grama Samaj, Bamphisahi Truckers Club, Ganganagar Sanskritika Sangha, Uttarward Kuchinda Minamandali, Ghusuri Abasor Binodan Samaj and many other institutions passed resolutions congratulating Jaga Palei and sent them to the press.

Even though his name had become widely known, Jaga Palei of Sagadiasahi still continued his traditional profession of carrying gunny bags in the maal godown. He had done this job since he was fifteen—from the day his father Uddhab Palei had returned the bullock carts of the moneylender, come home, gone to sleep on the spread-out end of his dhoti and had never woken up again. Uddhab Palei had died of pneumonia. When he had fallen ill, the Chhota Mian of Mohamaddia Bazaar tried to exorcise the evil spirit. Govinda Ghadei of Janakasahi, who kept different tablets in his shop inside the cutchery premises for curing different diseases, had administered four different kinds, bitter, kasa, raga, and saline respectively. For this he had charged one rupee seventy-five paise. Karuna Gosain, the monk of Tinigheria, had prescribed that he should feed eighteen bundles of straw to stray cattle on a Wednesday and then prostrate himself on the street. Uddhab had followed this prescription as well but nothing had helped. He died without discovering whether mankind had discovered a cure for pneumonia.

It was thus that Jaga Palei was left fatherless in the big city, with no job, no savings, no help and the greedy eyes of the well-to-do on his two-roomed thatched house and three gunths of land. He was responsible now for his widowed mother and two minor siblings—Khaga and a twelve-year-old sister, Sara. The well-wishers arrived and proffered advice to the family: 'Sell the plot of land, build a small house elsewhere and with the balance start a business.' On the face of it, this advice was reasonable. The ancestral plot of land may have been in a congested locality of town but a little way away was the main road where a gunth sold for seven hundred rupees. With two rooms on it, it could fetch three thousand rupees. Wouldn't it be so much cheaper to purchase land and build a house in Tulsipur Bidanasi, Uttampur or around the Dairy Farm?

This is what the plot of land looked like: at its back ran a dirty, dark drain; on the right a tank whose foul water threw up bubbles constantly; on the left a washerman's house and a sprawling basti; in the front, a lane hardly six to seven cubits wide and the back of the boundary wall of the double-storeyed building belonging to the moneylender, Garib Das. Black water seeped out of the chinks in the boundary wall, and gave off a putrefying stench.

But Uddhab Palei had never sold that small plot of his ancestors, so neither did his wife and son sell it. The advice of the well-wishers remained unheeded.

Another bit of advice from the same well-wishers was that members of the household should take up service as domestic servants. Or else how would they be able to look after themselves? At fifteen Jaga looked quite a man. He received various offers: to become an apprentice bullock cart driver or to become a machine operator in a saw mill or become a shop assistant. A babu suggested he do some domestic service with the chance to become a peon after a while. Another person came and told him that Jaga was very fortunate as his sahib wanted him to be his personal valet. No work—Jaga only had to accompany him wherever he went, run the occasional errand and there would be no end to the good food, tips and a salary to cap it all! Jaga was shown the dream of riding in cars and planes, sleeping on thick mattresses, wearing expensive clothes and eating good food. Many would come seeking little favours through him, flattering him in diverse ways. It would be for him to make or break them. He would become an important man. What hadn't the babu been able to do for people who depended on him? After all, for him money was just like mud and stones!

Jaga Palei listened to everything in silence. Somebody seemed to whisper inside him: 'Do not listen, Jaga, close your eyes, say "No". No, you will not take to servanthood. However ferocious a dog with a thick blanket of fur, thick tail, huge body and large teeth—a dog remains a dog at his master's call. It can only lick his boots and lie chained to a post. A dog seen looking out from behind the windows of his master's car may excite the onlooker's admiration. Nobody, however, can ever forget that he remains only a dog.'

To fifteen-year-old Jaga Palei such thoughts came naturally; for in his blood was the tradition of endless ancestors—people who tilled the soil and preserved an unbending tensile dignity, which three generations of town life had not corroded.

⌐⌐

Jaga turned his back on all the offers and chose the life of a daily wage earner, carrying loads every day. His mother did not object. With the help of her daughter she opened a small snack shop in the front room of the house. His mother knew how to prepare good and tasty food. So sales were brisk. Khaga went around hawking groundnut and bara bhaja. After a while, he took a job rolling bidis in a factory. Somehow, the family of four members lived on; nobody died, the house was not sold. From the outside everything looked the same. Four persons became of one mind, suffered hardships and privation. Nobody came to know anything of this.

Jaga had one obsession in life. Physical fitness. Early inspiration for this came from his father's godfather, the old khalipha of Sagadiasahi. Jaga remembered his mango complexion, the body of a young man, the flowing beard, the look of a child in his small blue eyes, and the green turban. Once he had tugged at Jaga's shoulder and asked him why he did not attend his akhara. He had asked Uddhab to hand Jaga over to him so that he could make a wrestler out of him. Uddhab had smiled and agreed. That was the beginning.

A couple of small rooms in an old building near a tamarind tree with a compound wall. That was the khalipha's house. There was no wife or children to be seen; nobody knew if he'd ever had any. He had just a single pleasure in life—the akhara inside his compound. Early in the morning, before the darkness lifted, Jaga would go to the akhara and do various types of gymnastic exercises; he would practise with the club and the lathi, and wrestle. Many joined the akhara; many also dropped out. But there wasn't a sunrise when Jaga Palei would not come out from the akhara after his exercise.

When wrestlers from elsewhere came to town the khalipha would organize matches. Jaga was unbeaten in these competitions. Those who cared for wrestling soon came to know about him. They would praise his iron-hard body, the lightning speed of his reflexes and the marvellous tricks he had learnt from the khalipha. But rarely were these people from among the higher circles of society.

Mostly they were shopkeepers, tailors, butchers, drivers, carpenters and so on. This lack of fame in all sections of society was in part due to the khalipha's regulations. No showing off, no publicity. Only during Dussehra and Muharram would he allow his team to go round town demonstrating their skills.

As his reputation as a wrestler grew, more job offers came his way. One was a watchman's job—guarding somebody's house with a rifle or a lathi. Good pay. The other proposal was even more astonishing. Plenty of food, a monthly salary, and a special bonus for special kinds of work. The work would be of the age-old, time-honoured variety: he was expected to act as a Kichaka, or what was known in modern terminology as a goonda. King Virata had been defended by Kichaka. Now new empires had opened up in business, trade and industry. And empires always needed Kichakas. It was for the master to point a finger at his enemies. Then there would be work of all kinds: staring hard at somebody; breaking someone's legs; breaking someone else's neck; imprisoning someone; throwing stones at somebody's house at night; accosting and roughing up somebody and so on. If dragged to court, the Kichaka would be defended by his master's lawyers; the master himself would always remain invisible. Jaga Palei was not interested.

There was another proposal. Somebody wanted him for a son-in-law. He would be expected to do nothing except lead a life of leisure and enjoy his father-in-law's property. Jaga Palei turned this down as well. What remained was the old work—carrying grey bags of cement from one place to another and getting paid per bag.

After the big wrestling match he found strangers crowding round him. Lights flooded him from many directions and photos were taken. Then came the questions. Questions—and more questions even before he could begin answering any of them. 'How long have you been wrestling? Who is your guru? Ah, Omar khalipha! Who are the wrestlers you have defeated earlier? Please give us a list. What prizes have you won? What is your diet? How much do you eat? Are you married? How many children? What do you consider necessary for health and a long life?'

Somebody from the crowd shouted, 'Do you agree that vegetable ghee is good for health? Ah, you have never taken that!'

More questions. 'How many cups of tea do you drink every day? What tea? You never drink tea? Please tell us the truth, sir. What bidi do you prefer? Which gurakhu do you use? Which party do you support? What do you think of the recent changes in the country? Oh, when can you grant an interview? We would like to publish your photograph along with your signature and your views on our commodities: flash it in cinema slides... Please, your autograph please.' As the questions were being hurled at him there was much jostling, which often saved him the trouble of answering because it was difficult for his questioners to stand their ground. Jaga Palei felt suffocated. He stood in grim silence throughout with folded hands. That too was photographed. Then he turned and ran as fast as he could through the crowd. He was afraid that they might follow him but he managed to get away.

First, he went to his guru and fell at his feet. The khalipha embraced him, his flowing beard touching his chest and said: 'You're a good boy; you have upheld my good name.' This was what mattered to Jaga Palei; he hardly noticed the praise from other quarters. Another reason he wasn't unduly moved by all the praise that had come his way was because he was a realist. He knew that there was always somebody who would win and somebody who would lose. Just as in this contest he had won and the other man was defeated.

From the khalipha he went to the temple and listened for a time to soothing music sung to the accompaniment of the tambourine. On the way back from the temple he heard news about the match blaring from radios. A little later, newspaper vendors carrying bundles of papers, were shouting the same news. His head was reeling. Instead of going home, he went to the Kathjuri embankment. When he eventually got home, late at night, he found an elaborate meal waiting for him: rice, dal, mashed potato, fried brinjal, fish curry. His family members embraced him and praised him in their own way. Excepting a few neighbours, no one else came to gawk at him. He was relieved.

Before dawn the next morning he was back at his exercise and then the daily carrying of bags. He did not say a word to anybody about his profession and his private life. However, the newspapers revealed the fact that he was a labourer. He was unaware of how exactly news about him had spread, but news of his achievements reached the maal godown where he earned his daily wage; people would stop him to congratulate him and ask about his wrestling.

They would tell him about the high standing he had in the world of Indian wrestling and how he had raised the prestige of Orissa. They said he would have a great future if he won the last round. That would bring him even greater prestige and status than he enjoyed at the moment and would take him to wrestling matches outside Orissa and even outside India. He would go to Ceylon, Singapore, Mongolia, Peking, Japan, Russia, Germany, America, Africa, and so on. Along with prestige he would also earn a lot. For all this, he had only to win the last round of the All-India Wrestling Competition.

And there was also a lot of useful advice! He should take greater care of his diet and health, his workouts would need to be carefully regulated; he must eat plenty of fruit, mutton, milk, vitamins; in sum, he would need to take good care of himself. After all, he had to uphold the prestige of Orissa and, later, of India.

The flood of advice wearied him. His reality was that the only mutton he saw was in the shops and eating places he passed as he walked down the tired streets of the town. Milk was a dream. And as for fruits, all he could afford were bananas and coconuts. All that he aspired to was a seer of chura per day but his domestic budget was tight and rarely permitted more than half a seer.

A few days later a large number of unemployed labourers came to town from down south. They camped in the open under a tree. All that they wanted was to earn some wages and somehow make a living. The wage rate went down. To his utter misfortune, his younger brother, Khaga, met with an accident while returning from the bidi factory. He had fractures and multiple injuries and was carried to the hospital. This added to the woes of the family and Jaga's daily worries.

A newcomer opened a small hotel at the end of the street and started selling various types of delicacies and sweets and cakes and tea. Benches and chairs were provided and food was served on sparkling clean plates with a fan overhead and music from the radio. Customers started dwindling at the shop run by Jaga's mother and sister. The family began to be crushed by want and deprivation.

And yet he persisted with his wrestling. His diet came down from half a seer to a quarter seer of chura and fried rice worth only four annas a day; he could only afford a coconut every three days. He would fill his stomach with some rice and whatever green leafy vegetables were available. It was never enough, and hunger would burn fiercely in his stomach. On days when he had no work, Jaga could be seen sitting silent and forlorn, lost in thought. He would think about how alone and friendless he was, how he had been forsaken by everybody just a few days after his great victory.

Three months passed. Then came the fateful day of the final test: Dilip Singh of Punjab versus Jaga Palei of Orissa. When it was over, the newspapers were full of reports and analyses of the match. All were agreed that the wrestling, the artistry and skill which Jaga applied against the heavily-built, massive Dilip Singh were exemplary but that the odds had been heavily against him. It appeared that Dilip Singh would fall flat, but ultimately he won. Profiles of Dilip Singh appeared in the papers. All the great men in the wrestling world were his patrons. There was also information about the variety and quality of his diet, how his weight was taken every day and many other facts about him. Jaga Palei faded back into anonymity. Fresh discussions started in trains and buses and wherever people gathered. Some were angry with the man who had soiled Orissa's name; many others were unhappy and crestfallen. But everyone forgot quickly.

And the day after the wrestling match, like any other day, Jaga Palei quietly went back to his exercise and the carrying of bags.

PORTRAIT OF A LADY
KHUSHWANT SINGH

My grandmother, like everybody's grandmother, was an old woman. She had been old and wrinkled for the twenty years that I had known her. People said that she had once been young and pretty and had even had a husband, but that was hard to believe. My grandfather's portrait hung above the mantelpiece in the drawing room. He wore a big turban and loose-fitting clothes. His long white beard covered the best part of his chest and he looked at least a hundred years old. He did not look the sort of person who would have a wife or children. He looked as if he could only have lots and lots of grandchildren. As for my grandmother being young and pretty, the thought was almost revolting. She often told us of the games she used to play as a child. That seemed quite absurd and undignified on her part and we treated them like the fables of the prophets she used to tell us.

She had always been short and fat and slightly bent. Her face was a criss-cross of wrinkles running from everywhere to everywhere. No, we were certain she had always been as we had known her. Old, so terribly old that she could not have grown older, and had stayed at the same age for twenty years. She could never have been pretty; but she was always beautiful. She hobbled about the house in spotless white, with one hand resting on her waist to balance her stoop and the other telling the beads of her rosary. Her silver locks were scattered untidily over her pale, puckered face, and her lips constantly moved in inaudible prayer. Yes, she was beautiful. She was like the winter landscape in the mountains, an expanse of pure white serenity breathing peace and contentment.

My grandmother and I were good friends. My parents left me with her when they went to live in the city and we were constantly together. She used to wake me up in the morning and get me ready for school. She said her morning prayer in a monotonous sing-song while she bathed and dressed me in the hope that I would listen and get to know it by heart. I listened because I loved her voice but never bothered to learn it. Then she would fetch my wooden slate which she had already washed and plastered with yellow chalk, a tiny earthen ink pot and a reed pen, tie them all in a bundle and hand it to me. After a breakfast of a thick, stale chapati with a little butter and sugar spread on it, we went to school. She carried several stale chapatis with her for the village dogs.

My grandmother always went to school with me because the school was attached to the temple. The priest taught us the alphabet and the morning prayer. While the children sat in rows on either side of the verandah singing the alphabet or the prayer in a chorus, my grandmother sat inside reading the scriptures. When we had both finished, we would walk back together. This time the village dogs would meet us at the temple door. They followed us to our home, growling and fighting each other for the chapatis we threw to them.

When my parents were comfortably settled in the city, they sent for us. That was a turning point in our friendship. Although we shared the same room, my grandmother no longer came to school with me. I used to go to an English school in a motor bus. There were no dogs in the streets and she took to feeding sparrows in the courtyard of our city house.

As the years rolled by we saw less of each other. For some time she continued to wake me up and get me ready for school. When I came back she would ask me what the teacher had taught me. I would tell her English words and little things of Western science and learning, the law of gravity, Archimedes' principle, the world being round, etc. This made her unhappy. She could not help me with my lessons. She did not believe in the things they taught at the English school and was distressed that there was no teaching about God and the scriptures. One day I announced that we were being given music lessons. She was very disturbed. To her music had lewd associations. It was the monopoly of harlots and beggars and not meant for gentlefolk. She rarely talked to me after that.

When I went up to university, I was given a room of my own. The common link of friendship was snapped. My grandmother accepted her seclusion with resignation. She rarely left her spinning wheel to talk to anyone. From sunrise to sunset she sat by her wheel, spinning and reciting prayers. Only in the afternoon she relaxed for a while to feed the sparrows. While she sat in the verandah breaking the bread into little bits, hundreds of little birds collected round her, creating a veritable bedlam of chirrupings. Some came and perched on her legs, others on her shoulders. Some even sat on her head. She smiled but never shooed them away. It used to be the happiest half-hour of the day for her.

When I decided to go abroad for further studies, I was sure my grandmother would be upset. I would be away for five years, and at her age one could never tell. But my grandmother could. She was not even sentimental. She came to leave me at the railway station but did not talk or show any emotion. Her lips moved in prayer, her mind was lost in prayer. Her fingers were busy telling the beads of her rosary. Silently she kissed my forehead, and when I left I cherished the moist imprint as perhaps the last sign of physical contact between us.

But that was not so. After five years I came back home and was met by her at the station. She did not look a day older. She still had no time for words, and while she clasped me in her arms I could hear her reciting her prayer. Even on the first day of my arrival, her happiest moments were with her sparrows, whom she fed longer and with frivolous rebukes.

In the evening a change came over her. She did not pray. She collected the women of the neighbourhood, got an old drum and started to sing. For several hours she thumped the sagging skins of the dilapidated drum and sang of the homecoming of warriors. We had to persuade her to stop to avoid overstraining. That was the first time since I had known her that she did not pray.

The next morning she was taken ill. It was a mild fever and the doctor told us that it would go. But my grandmother thought differently. She told us that her end was near. She said that, since only a few hours before the close of the last chapter of her life she had omitted to pray, she was not going to waste any more time talking to us.

We protested. But she ignored our protests. She lay peacefully in bed, praying and telling her beads. Even before we could suspect, her lips stopped moving and the rosary fell from her lifeless fingers. A peaceful pallor spread on her face and we knew that she was dead.

We lifted her off the bed and, as is customary, laid her on the ground and covered her with a red shroud. After a few hours of mourning we left her alone to make arrangements for her funeral.

In the evening we went to her room with a crude stretcher to take her to be cremated. The sun was setting and had lit her room and verandah with a blaze of golden light. We stopped halfway in the courtyard. All over the verandah and in her room right up to where she lay dead and stiff, wrapped in the red shroud, thousands of sparrows sat scattered on the floor. There was no chirping. We felt sorry for the birds and my mother fetched some bread for them. She broke it into little crumbs, the way my grandmother used to, and threw it to them. The sparrows took no notice of the bread. When we carried my grandmother's corpse off, they flew away quietly. Next morning the sweeper swept the breadcrumbs into the dustbin.

ten

~

QUILT

ISMAT CHUGHTAI

Translated from the Urdu by Rakhshanda Jalil

In the winters, when I pull a quilt over me, its shadow on the
wall looms like an elephant and, immediately, my mind races back
to gambol in the pages of a world I have left far behind. And I am
reminded of all sorts of things.

Excuse me, but I am not about to launch into a romanticized
account of my own quilt. In any case, I can't associate any great
romance with a quilt. I do feel that while a blanket may be less
comfortable than a quilt, its shadow can never be as terrifying as
the swaying images conjured upon a wall by a quilt. I am talking
of the days when I was a little girl and would spend the entire day
fighting and playing with my brothers and their friends.

Sometimes I wonder why I was such a quarrelsome little wretch!
Especially since every girl my age was busy collecting admirers and
I was hell-bent on picking quarrels and getting into scrapes with
every boy and every girl I knew.

Perhaps, that was why, when Amma was about to leave for Agra
for a week, she left me with a lady whom she considered as close
as a sister. Amma knew well enough that there wasn't even a baby
mouse in that house and so there was no one I could get into scrapes
with. What a fitting punishment for me! So, Amma went off leaving

me with Begum Jan...the same Begum Jan whose quilt remains imprinted on my memory like the mark of a hot iron. It was the same Begum Jan whose impoverished parents had agreed to make Nawab Sahib their son-in-law for, though ripe in age, he was essentially a decent man who stayed away from the company of whores and dancing girls. Nawab Sahib had himself been on the Haj and sent many others on pilgrimage too.

But the Nawab Sahib had one very strange pursuit. Others are known to keep pigeons, or gamble on fighting quails or indulge in cockfights; he held all such vulgar pastimes in utter disdain. He only had students staying over at his home—fair, young boys with slender waists—whose expenses were borne entirely by Nawab Sahib.

Nawab Sahib had married Begum Jan, brought her home and kept her among his other household goods and promptly forgotten all about her. And she, frail and delicate as she was, began to melt away in the sorrow of her loneliness.

Who can say where her life began: when she had made the mistake of being born? Or when she arrived in this house as the Nawab's bride and began her life on the ornate four-poster bed? Or, when the stream of boys entering Nawab Sahib's household increased steadily? Platters of rich and delicious foods would be sent for the boys and Begum Jan would peep through the cracks in the sitting room doors and see the nimble-waisted boys with their taut calves and sheer, scented, soft muslin kurtas, and feel as though she were rolling on live embers.

Or, did her life start when she stopped all her prayers and entreaties and admitted defeat, when the nightlong vigils ended and all the charms and talismans she used failed to move the Nawab? There's no point applying leeches to a rock. Has anyone ever heard of a leech sticking to a rock? Nawab Sahib did not budge an inch and, eventually, Begum Jan gave up. She turned her attention towards knowledge, but here too she found nothing. Romantic novels and emotional poetry heightened her overwhelming sense of defeat. A restful night's sleep deserted her altogether and Begum Jan abandoned all semblance of life and living and turned into a bundle of despair and longing.

She lost all interest in clothes and jewellery. After all, one dresses to impress someone; the Nawab Sahib could not spare the time to leave the sheer kurtas and come to her nor would he permit her to go out and about. Ever since she had married him and come here, all sorts of relatives would come and stay with her for months while she remained confined in the house like a prisoner.

The sight of the relatives enraged her all the more; they would descend in droves to enjoy the good life, eat well and stock up on their winter beddings and quilts whereas she, despite her quilt of freshly-carded cotton, would spend the entire winter stiff with cold. Every time she turned and twisted on her bed, the quilt would cast shadows on the wall beside her and create new images. And not one of those images was enough reason for her to live. Why, then, should she live? But Begum Jan had to stay alive for as long as she was destined to live. And so she decided to live her life, and how!

Rabbo saved her as she began to decline. Within no time, Begum Jan's frail sickly body began to fill up. Her cheeks shone and her beauty burst forth. Life rippled through Begum Jan as some strange sort of oil was rubbed into her. Excuse me, but the secret ingredients of that massage oil cannot be found in the finest of journals.

Begum Jan must have been forty or forty-two years old when I first saw her. How elegantly she lay propped up against the cushions while Rabbo sat beside her, rubbing her back. A purple shawl was draped over her legs and she looked as regal as a queen. I loved to look at her face. I wanted to gaze at her face from up close for hours on end. Her complexion was extremely pale with not the slightest trace of reddishness. Her hair was black and drenched in oil and I had never seen the parting of her hair look crooked. God forbid that a single strand of hair be out of place! Her eyes were black and, since her brows had been plucked of stray hairs, they appeared to be arched. Her eyes seemed somewhat taut, their lids heavy and slightly swollen, and lined with thick lashes. But the most remarkable and the most riveting thing by far about her face were her lips. They were mostly painted red. The upper lip was shadowed with a faint moustache and long strands of hair hung from her temples. Sometimes, if one gazed at her face long enough, it began to seem strange, almost like an adolescent boy's.

The skin on her body was pale and smooth; it looked as though someone had stretched it taut and tight. Often, when she bared her calves to be massaged, I looked stealthily at their gleaming skin. She was quite tall and, because she had added some flesh to her bones, she appeared to be a big woman. But her body was proportionate and well formed. Her hands were large, white and smooth, and her waist slender. And Rabbo would sit beside her and scratch her back...I mean scratch her back for hours on end. Getting her back rubbed or tickled or scratched was among the necessities of Begum Jan's life, or perhaps even greater than the necessities of life.

Rabbo had no other household chores; all she did was sit on the four-poster bed beside Begum Jan and press her head or legs or some other part of her body. Sometimes, the sight of Rabbo annoyed me: she was forever pressing or massaging something or the other. I don't know about anyone else but, speaking for myself, my body would have disintegrated if someone were to touch me all the time.

As though the daily massages were not enough, God save us on the day Begum Jan took her bath. Two hours before the bath, her body would be rubbed with fragrant unguents and lotions, and rubbed so much and for so long that the very thought was enough to break my heart. The doors to her room would be closed, coal braziers would be lit and a succession of massages would start. More often than not, only Rabbo would remain in the room. The other maid servants would grumble and hand over anything that might be required at the doorstep.

Begum Jan was said to be suffering from an itchy disease. The poor thing suffered from such a terrible itch that thousands of different oils and potions were rubbed into her body but the itch remained. Doctors and hakims would say, 'There is nothing wrong with her; her body is smooth and clean. Perhaps there is some problem beneath the surface.'

'No, no, these doctors are fools. May your enemies fall sick! It is nothing but the heat in your blood,' Rabbo would say with a smile and stare at Begum Jan with narrowed eyes. And Rabbo...she was as dark as Begum Jan was fair, and as ruddy as Begum Jan was pale. She had faint pockmarks on her gleaming black face. Her body was sturdy and agile, her hands small and nimble, her belly taut and small, and her large and full lips seemed perpetually moist. She gave off a strange odour. And her small, rounded hands were swift as they darted from the waist to the buttocks or ran over the thighs and then raced towards the ankles. Whenever I went to sit beside Begum Jan, all I wanted to see was where Rabbo's hands were and what they were doing.

Winter and summer, Begum Jan wore Hyderabadi lace kurtas. She wore dark pyjamas with her snowy white kurtas. And in summer, when the fans would be working, she always covered her body with a light coverlet. She loved winters, and I too liked being in her home in winter. She seldom went out and preferred to lie on the carpet, having her back scratched as she munched on dry fruits.

The other maid servants disliked Rabbo, the witch, because she ate with Begum Jan and stayed with her and, God help them, even slept with her. Begum Jan and Rabbo were the subject of salacious gossip everywhere; whenever their names cropped up in genteel conversation, respectable ladies erupted into peals of laughter. People cracked all sorts of jokes about them. But the two of them did not meet anybody; they existed in their world, and their itch.

As I said, I was quite young then and completely infatuated with Begum Jan. She, too, was very fond of me. As luck would have it, Amma had to go to Agra; she knew, left on my own, I would get into fights with my brothers and go loafing and get into trouble. And that is why she left me with Begum Jan for a week. I was happy, and so was Begum Jan.

Now the question arose: where should I sleep? Naturally, it could only be in her room. And so a small bed was placed next to her four-poster bed. Till eleven or so, Begum Jan and I chatted and played cards; then I went to sleep on my bed. And as I drifted off to sleep, Rabbo was sitting there scratching her back. 'What a bhangan she is!' I said to myself.

Sometime in the night I woke feeling a strange kind of fear. The room was plunged in darkness and in that darkness Begum Jan's quilt was swaying as though an elephant was trapped inside it. 'Begum Jan!' I called out fearfully. The elephant stopped moving. The quilt fell flat.

'What is it? Go to sleep,' said Begum Jan's disembodied voice.

'I am scared,' I sounded like a mouse.

'Go to sleep...what is there to be scared about? Recite the Ayatul Kursi.'

'All right,' I said and tried to hurriedly recite the Ayatul Kursi but, each time, got stuck at 'Ya'lemu ma baina' even though I knew the entire verse by heart.

'Shall I come to you, Begum Jan?'

'No, my dear, go to sleep,' she said in a stern voice.

And then the sound of two people whispering to each other could be heard. Oh dear, who was this second person? I was even more scared.

'Begum Jan...do you think thieves have broken in?'

'Go back to sleep, child...what thieves?' I heard Rabbo's voice and quickly ducked my head inside the quilt and went to sleep.

In the morning, I had no memory of the night's terrifying sight. But then I have always been like that...all through my childhood I would get scared at night, wake up and mumble or run around. In fact, people would say that I was haunted by ghosts! Therefore, I didn't give the previous night's occurrence a second thought. The quilt looked quite innocent in the morning.

The next night when I woke up I found a fight being resolved in great silence between Rabbo and Begum Jan on the four-poster bed, and for the life of me I could not figure out what it was all about, and whether it was

ever resolved. Rabbo cried great hiccupping sobs and then the slurping sounds
of a cat licking a bowl could be heard. Terrified, I went back to sleep.

Rabbo had to go and meet her son, who was a great one for picking
fights. Begum Jan had done a great deal for him, got him a shop, sent him
to a village, but nothing seemed to have worked. He even stayed at Nawab
Sahib's house for some time. Fancy clothes were stitched for him but, God
knows why, he ran away and never ever came back, not even to meet Rabbo.
And so Rabbo had to go to meet him at some relative's home. Begum Jan
did not want Rabbo to go but what could Rabbo do; she was helpless.

All day long, Begum Jan waited anxiously. Every bone in her body ached,
yet she couldn't bear anyone to touch her. She did not eat a morsel and spent
the entire day lying about listlessly.

'Shall I scratch your back, Begum Jan?' I asked eagerly as I dealt the cards.
Begum Jan looked closely at me.

'Shall I scratch your back...really, I mean it,' I said as I put the cards down.

For some time, I kept scratching her back and Begum Jan lay quietly.

The next day, Rabbo was expected back but there was no sign of her.
Begum Jan's temper grew steadily worse. She drank so much tea that she gave
herself a headache.

Once again, I began to scratch her back...her back that was smooth as
a table top. Gently, slowly, I kept scratching. It gave me such pleasure to do
this small chore for her.

'Scratch a bit harder...undo the clasp,' Begum Jan said. 'Here...ai hai...here,
here, just beneath the shoulder...wah...that is wonderful...wah...ah...ah...' She
sighed with pleasure as I continued my ministrations.

'A little more to this side...' she would say, even though she could well
have reached the spot with her own hand. But she wanted me to attend to her
and it gave me a lot of satisfaction to do so. 'Yes, here...ouch...you are tickling
me...wah...' she laughed. I kept talking as I scratched her under her clothes.

'I will send you to the bazaar tomorrow...what will you buy? That doll
that sleeps and wakes as it opens and closes its eyes?'

'No, Begum Jan, I don't buy dolls... Do you think I am a child?'

'So, if not a child, are you an old woman?' she laughed. 'All right, buy a
boy doll then...let me put on my clothes... I will get you lots of clothes, all
right?' she said and turned on her side.

'All right,' I answered.

'Here,' she took hold of my hand and guided it to the spot that was
itching her. Wherever she felt the itch, she would take my hand and guide

it there and I, lost in my thoughts of the boy doll, kept scratching her like a mindless machine. All the while, she kept talking to me ceaselessly.

'Listen, you don't have enough frocks... I will send for the tailor tomorrow. I will ask him to stitch frocks from the cloth your mother has left with me.'

'I don't want that red cloth...it looks like the material worn by the chamars!' I was so busy talking nonsense that I scarcely noticed where my hand had wandered to. Begum Jan was lying flat on her back... Arre... I quickly snatched my hand back.

'Uii, girl...can't you see where you are scratching...you are poking my ribs out,' Begum Jan smiled mischievously and I grew flustered.

'Come here and lie down beside me,' she said and she made me lie down with my head pillowed on her arm.

'Aii hai, you have become so thin...your ribs are sticking out.' She began to count my ribs.

'Owooo...' I bleated.

'Am I going to eat you... What a tight sweater someone has knitted! And you are not even wearing a warm woollen vest...' I began to fidget.

'How many ribs do we have?' she asked, changing the subject.

'Nine on one side and ten on the other,' I answered, trying to recall what I had learned in the Hygiene class in school.

'Move your hand... Yes...one...two...three...'

I wanted to run away. But she held me tightly.

'Owww...' I shivered. Begum Jan began to laugh loudly. Even today when I recall her face as it looked then, I feel a strange disquiet. Her eyelids looked heavier and the shadow on her upper lip appeared darker. Despite the cold, tiny droplets of sweat gleamed on her lips and nose. Her hands were cold, but also soft and smooth as though the skin had been peeled off. She had taken off her shawl and her body, visible through her thin lacy kurta, shone white like freshly kneaded dough. Heavy, bejewelled gold buttons swung on one side of her open kurta. Evening had fallen and the room was darkening. An unknown fear held me in its grip. Begum Jan's deep, dark eyes looked at me. I began to cry inwardly. She was holding me as though I was a toy made of clay. The heat of her body was making me nervous. But she was like a woman possessed and I, I was in such a state that I could neither scream nor cry.

After some time, she fell back exhausted and spent. Her face became pale and wan and she began to draw long deep breaths. I thought she was about to die any minute, and so I quickly got up and ran away as fast as I could.

Thank God Rabbo returned that night. I quickly got under my quilt and pretended to sleep. But I couldn't fall asleep; I just lay there quietly for hours.

My mother was showing no signs of coming back. I had begun to fear Begum Jan to such an extent that I would spend the entire day sitting with the maids. I was terrified of going into her room, but who could I share my fear with? In any case, what could I say? That I was scared of Begum Jan? The same Begum Jan who adored me?

Today, Begum Jan and Rabbo had another fight. You can call it my bad luck, or whatever else you'd like to call it, but as a result of their quarrel Begum Jan suddenly seemed to realize that I was roaming around in the cold outside and that I might die of pneumonia.

'Girl, will you cause me no end of shame? If something were to happen to you, I'd get into trouble.' She made me sit down beside her. She was rinsing her mouth and hands in a basin. There was tea on a low stool beside her.

'Pour the tea...give me a cup too,' she said, as she wiped her face on a towel. 'I'm going to get changed.'

I sipped my tea as she changed her clothes. Usually, when she sent for me to run some errand for her as the barber's wife soaped her back, I would enter her bathroom with my head averted so I could avoid looking at her. And I would run out as soon as I could. Now, I felt queasy as she changed her clothes in front of me. I kept sipping my tea and turned my head away from her.

'Hai, Amma!' my heart cried out with a sniffle. 'After all, how much can you punish me for fighting with my brothers...' My mother had always disliked my habit of playing with boys, as though boys were lions or cheetahs who would gobble up her darling daughter. And which boys, after all? My own brothers and a few of their wretched little friends! But no, my mother believed in keeping women locked up behind seven locks. And here I was... more terrified of Begum Jan than I would have been of all the rogues in the world. If I'd had my way, I would have run out on the street and not stayed in this place a minute longer. But I felt abandoned. I sat there helplessly with my heart feeling as heavy as a stone.

She changed her clothes and embarked on an elaborate toilette with all sorts of warm unguents rubbed into her skin. Hot and glowing, she turned to shower her attention on me.

'I want to go home...' I said in response to all her overtures, and began to cry.

'Come, come near me... I will take you to the bazaar... Listen...listen to me...'

But I was past caring. I didn't want her toys and sweets. All I wanted to do was go home.

'Your brothers will beat you up...you witch!' she said, and slapped me affectionately.

'Let my brothers beat me...' I said to myself and kept sitting there rigid and angry.

'Unripe mangoes are tart, Begum Jan,' the resentful Rabbo said caustically. Begum Jan had a fit. The golden necklace that she was cajoling me to wear just a short while ago, broke into pieces. Her dupatta, made of the finest lace, was torn to shreds. And her hair, which had always been immaculate and perfectly coiffed, became dishevelled and wild.

'Ow, ow, ow...' she began to shout. I dashed out of her room.

It took a long time for Begum Jan to grow calm. When it was time for me to sleep, I tiptoed into her room and saw Rabbo sitting beside her and rubbing her body.

'Take off your shoes...' she said as she scratched Begum Jan's ribcage and I ducked under my quilt like a little mouse.

Surr surr phutt kuchh... Begum Jan's quilt was again swaying in the dark like an elephant.

'Allah! Aa...' I let out a faint squeal. The elephant under the quilt reared up and sat down. I too fell silent. But the elephant went on the rampage again. Every pore in my body quivered. Tonight, I had resolved to gather my courage and switch on the lamp beside my bed. The elephant was throwing itself around in a frenzy and seemed to be trying to sit down. The slurping sound of someone eating could be heard...as though someone was licking lip-smacking chutney. Aha! Now I understood... Begum Jan had not eaten a thing all day. And that wretched Rabbo was forever greedy for food. Surely she was eating some delicacy under the quilt. I flared my nostrils and sniffed the air, trying to get a sense of what it was. But I could smell nothing except the warm scent of sandal oil and henna.

The quilt began to swell once again. I tried my best to lie quietly. But that quilt began to make the strangest of shapes and I trembled with terror. It seemed as though a gigantic toad was making the oddest of sounds as it puffed out its chest; any minute now, it would leap on top of me.

I gathered up my courage and tried to say something. 'Aaaa... Amma...' But it had no effect. The quilt seemed to have entered my brain and was swelling in there. Terrified though I was, I put my feet down on the other side of my bed and, feeling my way in the dark, switched on the light. The elephant under the quilt turned a cartwheel and subsided. But as it flipped over, it raised one edge of the quilt by a foot or so. Allah! I ducked into my bed without another word.

STENCH OF KEROSENE

AMRITA PRITAM

Translated from the Punjabi by Khushwant Singh

Outside, a mare neighed. Guleri recognized the neighing and ran out of the house. The mare was from her parents' village. She put her head against its neck as if it were the door of her father's house.

Guleri's parents lived in Chamba. A few miles from her husband's village, which was on high ground, the road curved and descended steeply downhill. From this point one could see Chamba lying a long way away at one's feet. Whenever Guleri was homesick, she would take her husband Manak and go up to this point. She would see the homes of Chamba twinkling in the sunlight and would come back with her heart glowing with pride.

Once every year, after the harvest had been gathered in, Guleri was allowed to spend a few days with her parents. They sent a man to Lakarmandi to bring her back to Chamba. Two of her friends, who were also married to boys outside Chamba, came home at the same time of the year. The girls looked forward to this annual meeting when they spent many hours every day talking about their experiences, their joys and sorrows. They went about the streets together. Then there was the harvest festival. The girls would have new dresses made for the occasion. They would have their dupattas dyed, starched and sprinkled with mica. They would buy glass bangles and silver earrings.

Guleri always counted the days to the harvest. When autumn breezes cleared the skies of the monsoon clouds she thought of little besides her home in Chamba. She went about her daily chores—fed the cattle, cooked food for her husband's parents and then sat back to work out how long it would be before someone would come for her from her parents' village.

And now, once again, it was time for her annual visit. She caressed the mare joyfully, greeted her father's servant, Natu, and made ready to leave the next day.

Guleri did not have to put her excitement into words: the expression on her face was enough. Her husband, Manak, pulled at his chillum and closed his eyes. It seemed either as if he did not like the tobacco, or that he could not bear to face his wife.

'You will come to the fair at Chamba, won't you? Come even if it is only for the day,' she pleaded.

Manak put aside his chillum but did not reply.

'Why don't you answer me?' asked Guleri in a temper. 'Shall I tell you something?'

'I know what you are going to say: "I only go to my parents once in the year!" Well, you have never been stopped before.'

'Then why do you want to stop me this time?' she demanded.

'Just this time,' pleaded Manak.

'Your mother has not said anything. Why do you stand in my way?' Guleri was childishly stubborn.

'My mother...' Manak did not finish his sentence.

On the long awaited morning, Guleri was ready long before dawn. She had no children and therefore no problem of either having to leave them with her husband's parents or taking them with her. Natu saddled the mare as she took leave of Manak's parents. They patted her head and blessed her.

'I will come with you a part of the way,' said Manak.

Guleri was happy as they set out. Under her dupatta she hid Manak's flute.

After the village of Khajiar, the road descended steeply to Chamba. There Guleri took out the flute from beneath her dupatta and gave it to Manak. She took Manak's hand in hers and said, 'Come now, play your flute!' But Manak, lost in his thoughts, paid no heed. 'Why don't you play your flute?' asked Guleri coaxingly. Manak looked at her sadly. Then, putting the flute to his lips, he blew a strange anguished wail of sound.

'Guleri, do not go away,' he begged her. 'I ask you again, do not go this time.' He handed her back the flute, unable to continue.

'But why?' she asked. 'You come over on the day of the fair and we will return together. I promise you, I will not stay behind.'

Manak did not ask again.

They stopped by the roadside. Natu took the mare a few paces ahead to leave the couple alone. It crossed Manak's mind that it was this time of year, seven years ago, that he and his friends had come on this very road to go to the harvest festival in Chamba. And it was at this fair that Manak had first seen Guleri and they had bartered their hearts to each other. Later, managing to meet alone, Manak remembered taking her hand and telling her, 'You are like unripe corn—full of milk.'

'Cattle go for unripe corn,' Guleri had replied, freeing her hand with a jerk. 'Human beings like it better roasted. If you want me, go and ask for my hand from my father.'

Amongst Manak's kinsmen it was customary to settle the bride price before the wedding. Manak was nervous because he did not know the price Guleri's father would demand from him. But Guleri's father was prosperous and had lived in cities. He had sworn that he would not take money for his daughter, but would give her to a worthy young man of a good family. Manak, he had decided, answered these requirements and very soon after, Guleri and Manak were married. Deep in his memories, Manak was roused by Guleri's hand on his shoulder.

'What are you dreaming of?' she teased him.

Manak did not answer. The mare neighed impatiently and Guleri, thinking of the journey ahead of her, rose to leave. 'Do you know the bluebell wood a couple of miles from here?' she asked. 'It is said that anyone who goes through it becomes deaf.'

'Yes.'

'It seems to me that you have passed through the bluebell wood; you do not hear anything that I say.'

'You are right, Guleri. I cannot hear anything that you are saying to me,' replied Manak with a deep sigh.

Both of them looked at each other. Neither understood the other's thoughts.

'I will go now. You had better return home. You have come a long way,' said Guleri gently.

'You have walked all this distance. Better get on the mare,' replied Manak. 'Here, take your flute.'

'You take it with you.'

'Will you come and play it on the day of the fair?' asked Guleri with a smile. The sun shone in her eyes. Manak turned his face away. Guleri, perplexed, shrugged her shoulders and took the road to Chamba. Manak returned to his home.

Entering the house, he slumped listless on his charpai. 'You have been away a long time,' exclaimed his mother. 'Did you go all the way to Chamba?'

'Not all the way; only to the top of the hill,' Manak's voice was heavy.

'Why do you croak like an old woman?' asked his mother severely. 'Be a man.'

Manak wanted to retort, 'You are a woman; why don't you cry like one for a change!' But he remained silent.

Manak and Guleri had been married seven years, but she had never borne a child and Manak's mother had made a secret resolve: 'I will not let it go beyond the eighth year.'

This year, true to her decision, she had paid five hundred rupees to get him a second wife and now, she had waited, as Manak knew, for the time when Guleri went to her parents' to bring in the new bride.

Obedient to his mother and to custom, Manak's body responded to the new woman. But his heart was dead within him.

Early one morning he was smoking his chillum when an old friend happened to pass by. 'Ho Bhavani, where are you going so early in the morning?'

Bhavani stopped. He had a small bundle on his shoulder: 'Nowhere in particular,' he replied evasively.

'You must be on your way to some place or the other,' exclaimed Manak. 'What about a smoke?'

Bhavani sat down on his haunches and took the chillum from Manak's hands. 'I am going to Chamba for the fair,' he replied at last.

Bhavani's words pierced through Manak's heart like a needle.

'Is the fair today?'

'It is the same day every year,' replied Bhavani drily.

'Don't you remember, we were in the same party seven years ago?' Bhavani did not say any more but Manak was conscious of the other man's rebuke and he felt uneasy. Bhavani put down the chillum and picked up his bundle. His flute was sticking out of the bundle. Bidding Manak farewell, he walked away. Manak's eyes remained on the flute till Bhavani disappeared from view.

Next afternoon when Manak was in his fields he saw Bhavani coming back but deliberately he looked the other way. He did not want to talk to Bhavani or hear anything about the fair. But Bhavani came round the other side and sat down in front of Manak. His face was sad, lightless as a cinder.

'Guleri is dead,' said Bhavani in a flat voice.

'What?'

'When she heard of your second marriage, she soaked her clothes in kerosene and set fire to them.'

Manak, mute with pain, could only stare and feel his own life burning out.

The days went by, Manak resumed his work in the fields and ate his meals when they were given to him. But he was like a man dead, his face quite blank, his eyes empty.

'I am not his spouse,' complained his second wife. 'I am just someone he happened to marry.'

But quite soon she was pregnant and Manak's mother was well pleased with her new daughter-in-law. She told Manak about his wife's condition, but he looked as if he did not understand, and his eyes were still empty.

His mother encouraged her daughter-in-law to bear with her husband's moods for a few days. As soon as the child was born and placed in his father's lap, she said, Manak would change.

A son was duly born to Manak's wife; and his mother, rejoicing, bathed the boy, dressed him in fine clothes and put him in Manak's lap. Manak stared at the newborn baby in his lap. He stared a long time, uncomprehending, his face, as usual, expressionless. Then suddenly the blank eyes filled with horror, and Manak began to scream. 'Take him away!' he shrieked hysterically. 'Take him away! He stinks of kerosene.'

twelve

~

GOLD FROM THE GRAVE
ANNA BHAU SATHE

Translated from the Marathi by Vernon Gonsalves

Hearing that a powerful moneylender had died in a nearby village, Bhima sprang to his feet. He was exhilarated. His joy wouldn't subside. Looking in the direction of the village, he suddenly turned to glare at the sun in the sky.

The sun was setting. Rain clouds crowded the sky. They had the rough, battered look of freshly ploughed land. The retreating light, filtering through those nasty looking clouds, streamed down over Mumbai.

There was a gentle breeze. The fifty or so huts in this suburb in the jungle began to creak in the breeze. The huts were made of old tin sheets, mats, planks and sacks. And those houses contained people. Cast-off things sheltering a cast-off people. Burnt out after the day's fight for food, they now rested. The kitchen fires were alight. White smoke loitered through the green trees. Children were playing.

Bhima sat lost in thought beneath a massive tamarind tree. He was terribly agitated. Drawn relentlessly towards that dead moneylender, his spirit was racing back and forth between that village's cemetery and the tamarind tree. He repeatedly glanced at the sun and then at that village. He needed the dark, so he was getting all fidgety. His beloved daughter Narbada was playing close by and his wife was in the house, patting bhakars into shape. Bhima looked awe-inspiring. His Satara outfit comprised a long red turban, a yellow dhoti and a shirt of thick coarse cloth. He looked a proper wrestler. His massive forehead, thick neck, dark eyebrows, flamboyant moustache, and broad yet fiery features had struck fear into many a ruffian.

Bhima's village was a long way off, on the banks of the Warna. However, seeing that even his bull-like strength could not fill his stomach there, he had moved to Mumbai. He had combed the entire city in search of work. But he hadn't found any. As his many dreams of getting a job, becoming a worker, bringing home a pay packet, making his wife a coin-necklace were shattered, Bhima had lost hope and had moved to this suburb in the jungle. Mumbai had everything, except work and shelter. So he had got upset with Mumbai. However, just after shifting to the suburb, he had found work in the quarry on a nearby hill.

On finding work and shelter, Bhima was happy. As he put all his bull-like might to work, he seemed to almost challenge the hill. He lifted his pickaxe and the hill would recoil. As his sledgehammer rose, the dark rock face would flinch. The contractor was happy with him. Bhima too was contented, as he was getting a wage.

But, within a space of just six months, the quarry closed down. When he got to work one day, he learned that the quarry had shut. Hearing that he had lost his job, Bhima was thrown into a daze. Hunger danced before his eyes. Anxiety and indecision gripped him. What was he going to do, he wondered despondently.

Clothes under his arm, Bhima turned back from the quarry. On the way, he stopped at a stream. He bathed there and prepared to make his way home, devastated beyond belief. It was then that his eyes fell on a mound of ashes. They were the ashes of a dead body. As he looked at the charred human bones Bhima grew even more despondent. Must be some jobless wretch; poor chap must have given up on life. I'll also die like this! Starvation will start in a couple of days, then Narbada will sit crying. My wife will fall into a deep depression and there'll be nothing I can do about it…

Suddenly he saw something sparkling in the heap of ashes. When he looked closely he discovered that the sparkle came from a gold ring of about a tola. Overjoyed, he grabbed hold of the ring. One tola of gold and that too from a corpse's ashes! He was delighted by his discovery—there was gold to be found in the ashes of a corpse. He had found a new means by which to live.

From the next day onwards Bhima began visiting crematoriums, and cemeteries on the banks of rivers and streams. He would sift through the ashes of bodies and pick up a fragment of gold here, an ornament there—earrings, nose rings, a gold coin, bracelet or anklet; he would find something of value every day.

Bhima's new venture began to flourish. He discovered that gold ornaments which were left on bodies that were being cremated would melt with the heat of the fire and enter the bones. So he would crush burnt bones and remove the gold. He'd break skulls. He'd crush wrists. But he'd get the gold.

In the evening he would go to Kurla, sell the gold and collect cash. On the way home he would get dates for Narbada. Business was steady.

Bhima lived by sifting through the ashes of corpses. He soon lost sight of the difference between life and death. What he understood was that if there was gold in the ash it was the ash of the rich and if there was no gold it was that of the poor. Sometimes he would rave to whoever was within earshot—so it is the rich who should die and the rich who should live; the poor should never die. Continuing with his rant, he would loudly proclaim that the lowly lot had absolutely no right to live or to die. Happy was the man who died with a gold tola in his molars, is what he believed.

The brutal reality of unemployment had made him brutal. Night and day he hovered around cremation grounds and graveyards. Corpses had become his means of existence. His life had become one with the dead.

Before long, people began to notice that bizarre things were happening in those parts. Buried bodies were rising from their graves. The corpse of the young daughter-in-law of a moneylender had moved mysteriously from the burial ground to the river bank. People living in the area were terrified by all this. Suspecting that somebody was digging up the bodies, the police began keeping watch. But keeping watch on corpses is no easy task.

The sun had set. Darkness covered the land. As his wife served his meal, Bhima ate in grim silence. When she realized that he was preparing to go out, she said softly, 'You're going somewhere, aren't you? I don't think what you are doing is right. You should find some other way to make a living. Corpses, corpses' ashes, gold, this existence, it's all wrong. People brand—'

Bhima was upset by what his wife was saying. 'Be quiet,' he said irritably. 'How does it matter what I do? If my home fires go cold who's going to come and light them up?'

'It's not like that...' she said quietly, noticing her husband's angry face, 'It's not good to roam around like some ghoul or ghost. I'm saying whatever it is I'm saying because I'm afraid.'

'Who told you that there are ghosts in the graveyards? Listen, this Mumbai is a ghosts' bazaar. The real ghosts stay in houses and the dead ones rot in those graveyards. Ghosts take birth in the village—not in the wild,' raved Bhima.

In the face of his anger she kept her mouth shut as he made preparations to leave. He growled, 'I didn't get work even after going to Mumbai. But sifting through corpses' ashes, I've got gold. When I broke hills they gave me two rupees. But now that ash easily gives me even ten rupees...' Saying this he left the house. It was quite late by then. It was quiet and peaceful outside.

Bhima had tied a muffler round his head. Over that he had put on a hooded cloak-like covering made of sackcloth, which he cinched at the waist. Carrying a pointed crowbar, he was walking with big strides. It was pitch-dark but he felt no fear. A sari, one petticoat and a blouse, dates in the morning was all he had on his mind. He was in a wild mood today.

There seemed to be a certain amount of tension in the air and it was getting tenser by the moment. A pack of jackals ran past him. A snake crossed the path and slithered away. An owl hooted in the distance adding to the eerie atmosphere. Nothing moved in the desolate jungle. Straining to catch every sound, Bhima drew near the village in which the moneylender had died. He sat down and surveyed the surroundings. All was silent in the village. Occasionally someone would cough. A lamp winked in a hut. When he saw that there was nothing to be worried about he slipped swiftly into the cemetery and started searching for the new grave of the moneylender. Pushing aside shattered pots and battered biers he jumped from this grave to that. He advanced from row to row looking carefully for the moneylender's grave. Clouds filled the sky. They deepened the darkness. Then abruptly lightning shone, dancing in the nooks and crannies of the clouds. It looked like it would rain. That made Bhima panic. Worried that he'd not be able to find the new grave if it rained, his search grew frenzied. He began to sweat, he felt he was losing his mind. By midnight he had searched the whole burial ground. From one end he reached the other end and slumped to the ground, distraught and confused. The wind was building up. It rattled the old poles of broken biers. It was almost as if someone were gnashing his teeth. Then a fearsome snarling erupted out of the night. Something was snarling, sobbing and scraping at the mud. Fearfully, he moved towards the sound. It died out at once. But almost immediately he felt as if someone was dusting his hands and feet, and he got startled. He stopped abruptly. Fear ran like an electric current through his body and struck him right inside his head. For the first time in his life he was afraid.

In the next instant he got a grip on himself when he realized what was actually happening. He felt somewhat ashamed that he had been so scared. The new grave was close by. Ten to fifteen jackals were busy digging it up as they had scented the dead body. As stones had been placed on top of the grave, they had started tunnelling into it from all sides. As they scraped away the earth, they snarled and snapped at each other; each one was desperate to be the first to get at the corpse. Bhima was enraged by the sight of the jackals. He took a giant leap and landed right on top of the grave. He began lifting the stones on the grave and hurled them at the pack of jackals. In the face of this sudden onslaught the jackals retreated. Determined to dig up the grave before the jackals renewed their assault on it, Bhima began scraping away the mud. The jackals, only momentarily deterred, attacked him. In a fit of madness one of the jackals pounced on Bhima. It bit him and leaped back. Excitement and anger surged through Bhima's body. He had wrapped his hand in the sacking he had brought with him. He removed the sacking and grabbed hold of the crowbar. When the jackal that had bitten him returned to the attack, he brought the crowbar down on its body with all his tremendous strength. The jackal yelped and died. Bhima began digging up the grave once more. The remaining jackals attacked in a solid, snapping mass. A desperate fight broke out. Bhima lunged at the snarling animals with his crowbar. The jackals were coming at him from every side, and he was getting bitten all over his body. But his flailing crowbar was finding its mark as well and he was wounding the jackals every time he connected.

And so battle was joined between this modern Bhima, heir to the legacy of Kunti's son Bhima, and the jackals. He struggled with all his strength—if he was going to have a meal tomorrow he needed to get to that corpse. Nature was asleep. Mumbai was resting. The village was quiet. And in that burial ground the clash over gold and corpse was reaching its climax. Bhima was attacking and felling the jackals, who yelped in agony every time they were hit. Even as some were wounded, others dodged the blows and bit Bhima, who moaned and cursed every time a bite was taken out of his flesh. Curses, growls, screams, the sound of the crowbar making contact with the jackals—all this mayhem sent tremors through the cemetery.

After a really long time, the jackals stopped attacking and retreated into the darkness. Taking advantage of this, Bhima removed the remaining mud from the grave. He wiped the sweat from his face. He jumped into the grave. At that the jackals again charged him. He began furiously lashing out at them once more, and finally managed to drive them off. As the last of the animals scurried away, Bhima got hold of the body, shoved his hands under its armpits and scooped it out of the straw in which it had been wrapped. As the man had been dead for a while, his body was stiff and unyielding. He leaned it upright in the grave, and began examining it for loot. He found a ring on its finger and pulled it off. It had an earring in one ear, which he clawed off. There would definitely be some gold in the mouth. He tried to insert his fingers into the corpse's mouth but its jaws were clenched tight. Quickly, Bhima took the crowbar and pried open the dead man's mouth. Propping it open with the crowbar he inserted his fingers into the open jaws. Just then, the jackals, which had been skulking unseen in the darkness, began howling in unison. Their wailing and howling woke up the village dogs, which began barking, and running around. The commotion woke up the villagers. Someone yelled: 'The jackals are eating the body, come on...'

Afraid that the villagers would find him robbing the corpse, Bhima hurriedly put the ring he had stolen inside his pocket and began rooting around inside the corpse's mouth. He yanked out the crowbar that he had used to prop open the mouth, without remembering to extract his fingers first. The dead man's teeth clamped down on his fingers like a nutcracker on a betel nut. Bhima writhed in pain.

He could see the men from the village approaching with lanterns. Desperately he tried to extricate his fingers. When they wouldn't move, he became furious with the corpse. He swung the crowbar at the dead man's jaw. The blow only succeeded in jamming his hand deeper into the dead man's mouth. He felt the corpse's teeth cut into his fingers. He froze, thinking: This is a real ghost, today it will catch me and hand me over to the people who will kill me because I am desecrating this body. Or they will hand me over to the police. As all this went through his head, Bhima lost control and began to savagely attack the corpse. 'Pimp, let me go...' he began to yell, before he realized any noise would give away his position; he struggled on in silence. The villagers were drawing close. Bhima forced himself to calm down, to think. He realized what he must do. He pushed the crowbar back into the jaws of the corpse, levered them apart and slowly pulled out his fingers. They had been almost bitten through. Cradling his wounded hand with the other one, he leaped out of the grave and ran into the night.

When he reached home he had a high fever. His wife and daughter wept when they saw the state he was in. A doctor was summoned and he amputated two of Bhima's fingers. The same day, news arrived that the quarry was resuming operations. Hearing this, the elephant-like Bhima started sobbing like a small child. He had lost two of his 'hill-breaker fingers' for the sake of gold from the grave.

THE MAN WHO SAW GOD

D. B. G. TILAK

Translated from the Telugu by Ranga Rao

I

The news that Gavarayya's wife had eloped was spread gleefully through the entire village. That only the day before China had violated the borders of India and that the war was still on suddenly became unimportant. Some even forgot about it. It seemed that in the entire Telugu country only the people of this village recognized or appreciated the special significance of a woman eloping!

At street junctions, in the coffee shack (there was only one in that village), on the low ridges of agricultural fields and near the panchayat building, the men; near the compound walls of their homes, at the wells, and on the steps of the public pond, the women: everyone was talking about this most wonderful development. A simple elopement may not have created such a sensation; specifically, it was the fact that the woman had eloped with a useless tailor working on a sewing machine on a pyol opposite the house that made the incident so unusual and so worthy of gossip. It would have been less damaging to the husband's reputation, in the opinion of experienced village elders, if she had eloped with someone else, a gentleman. Many young men felt hurt that Gavarayya's wife had unnecessarily insulted and wronged them.

'What did she see in that man to elope with him?' asked a young woman of her mother-in-law for the third time, unable to curb her curiosity.

'Mind your own business. Why didn't you yourself elope with him, you would have found out,' said the mother-in-law, irritably.

The consensus among the villagers was that Gavarayya got what he deserved.

～

Everyone in the village hated Gavarayya. Physically, he was dark and ungainly. Face pitted with pockmarks. Lips thick and coarse. Brows bushy, as though caterpillars had been stuck there. Gavarayya lived on the outskirts of the village in an old-fashioned house with rooms opening off a central courtyard.

'I knew something like this was going to happen,' said Avadhani, with his eyes half-closed. Avadhani was the trustee of the village temple.

Chalapati, the munsif, who was in charge of law and order, and Narasimham, the karanam or revenue officer, nodded their heads. Seeing he had their attention, the temple trustee annotated the Upanishad-like statement he had just made.

'Will the Lord Venugopala overlook such lapses? The first spouse had died anyway. Now this second wife has perpetrated something worse than death. Yet, has he done one charitable act? Has he accepted one word of good counsel?'

The munsif, Chalapati, tapped his walking stick on the floor and said, 'Has he allowed anyone to get close to him? A wicked man, this Gavarayya, sir. Too much ego, too much arrogance...'

'Sinful wealth, sinfully earned, sir! Won't the consequences follow? The other day all the young fellows went and asked him to donate one rupee for the bhajan, just one rupee, and he chased them away, it seems...' said the karanam.

～

Twenty years ago, at the age of twenty, Gavarayya had arrived in the village. A paternal aunt lived here. He had neither a father nor mother.

When he landed up he had some cash with him. There was a town about two miles from the village. Every day he would go to the town early in the morning and return after sunset. When the villagers came to know that he was engaged in the hides business everyone uttered God's name, 'Hari, Hari', with their hands over their ears. They made it clear to the paternal aunt that Gavarayya's business was utterly sinful; no good could come of it. But the aunt didn't do anything about it. Some even took the liberty of trying to persuade Gavarayya that trading in the hides of animals was not ethical.

'Why only the hides of animals, I would even sell human skins, only they are too thin and useless,' was apparently what Gavarayya said in response. He was a stubborn, headstrong fellow. Wouldn't talk to anyone; wouldn't socialize with anyone.

Why, no one had even seen Gavarayya laughing. Worse, it appears Gavarayya did not distinguish between 'Sin' and 'Virtue'.

It was generally believed that during the last twenty years he had made more than a lakh of rupees. The elders of the village—because they were generous people and righteous—forgave his sinful ways of making money and in the interest of his welfare, with a desire to ensure his pleasant tenure in the next world, advised him—patiently and ceaselessly—to make donations and endow charities, construct the temple pavilion, pick up the bill for the construction of the school building, perform week-long religious ceremonies. Narrowing his round, deep-set eyes, pressing his cheroot tightly between his lips, Gavarayya would say sharply, 'I won't give a single copper. Go and complain to whoever you want.' He has no sense of decency or proportion, they all said behind his back. Because he had a cash balance of one lakh rupees they did not say it openly to him, for by birth they were wise.

A few days after he arrived at his aunt's house, Gavarayya brought his first wife across. He wouldn't let her go anywhere. She wasn't even allowed to go to the neighbours' houses. The paternal aunt, the wife and Gavarayya—the three of them were like three ghosts. A rumour circulated in the village that Gavarayya's wife was possessed from time to time by some spirit, and that without anyone's knowledge, at midnight, while everyone slept, an exorcist would come and perform mysterious rites and offer fumigations to try and cure her.

How they survived in such isolation no one could understand. A little while later, a bangle seller visited the village to sell his wares. He claimed to have come from Gavarayya's native place. He supplied more details about Gavarayya's past.

Gavarayya's father had been a very wicked man, it appeared. He gambled and drank. His wife, a sick woman, was perpetually confined to her bed. As a result Gavarayya's father had a mistress in the neighbouring town. As a child Gavarayya did not know anything of a mother's care and attention. Because of his father's evil reputation, no one made friends with Gavarayya. He was not even allowed to join the village school. The founder of the school was a renowned landlord of the area. Gavarayya's father fought endlessly with that man. The landlord was a man of considerable standing in the community; every year he would organize the religious celebrations of Lord Subbarayudu, and feed the public. Ganging up with a bunch of rowdies, Gavarayya's father would create trouble at these celebrations. Gavarayya's father alleged that once, during a life-and-death crisis, the landlord had made him sign a mortgage for twice the amount that he had actually lent him and then unjustly seized the little land he owned. But of course the elders in the village were not so foolish as to believe the words of a rowdy. Besides they themselves were clandestinely engaged in usury on the same principles as the landlord's. As for the general public, not even in its dreams would it believe that such a gentleman and a great devotee of the Lord like the landlord would commit such an atrocity.

Distressed by Gavarayya's loneliness and isolation, his father brought him a puppy and two kittens and told him to play with them. 'These are better than humans,' he would remark to his sick wife. Gavarayya spent his entire childhood playing with dogs, trees and walls. After some time his sick mother passed away. With her death, the wrongdoings of Gavarayya's father increased greatly. Gavarayya's father had loved his wife very much. He had spent a lot of money on her treatment. Now, when in her last days, he could not find the money for her medical treatment, he blamed it all on the landlord's perfidy and went crazy. One evening when a labourer of the landlord was returning from the fields, someone broke his head with an iron rod. The landlord said that his suspicion centred entirely on Gavarayya's father. They arrested Gavarayya's father and charge-sheeted him. All the villagers gave irrefutable evidence. Though they may or may not have witnessed the murder, they honoured the orders of the pious landlord as divine commands. Gavarayya's father was sentenced to life. He pointedly told Gavarayya: 'Don't trust anyone. Stand on your own feet. All humans are traitorous wretches, venomous serpents.' He was then taken away by the police.

Without a mother or father, Gavarayya survived in that house in a state of pure terror. Pining for them, without food or water for two whole days, sitting in a corner of that house, the fourteen-year-old boy wept disconsolately.

No one came anywhere near him, didn't even greet him. All those good souls in the village probably thought that the father being a rowdy and a murderer, the child had inherited his traits and that the farther they kept away from him the better it would be for them. Gavarayya fell ill. No question of medical treatment or anything remotely like it. The fever turned out to be smallpox. The entire village could hear pitiful, terror-stricken cries coming from that house: 'Amma, I am dying.'

The moment they came to know that it was smallpox, people stopped going anywhere near the house. 'If the son too were to die, it would be good riddance for the village,' some said. One day in the evening a jutka stopped before Gavarayya's house. A forty-year-old woman climbed out of the horse-drawn carriage. She was covered in jewellery from top to toe. She was tall, well-built. Not even the hair at the temples had greyed. The whole street came out and stood gawking at her. She didn't glance this way or that, but went straight into the house. A servant followed with boxes and luggage and the door shut behind him.

Through the servant it transpired that she had had a liaison with Gavarayya's father. She had been on pilgrimage for the past six months to various religious centres, and had only returned two days earlier. Having come to know that Gavarayya's father had been sentenced to life imprisonment and that this had effectively made the son an orphan, she had left at once for the village. That marked a turning point. Though Gavarayya looked uglier than ever after his recovery from his illness, she did not flinch from him. One or two women took the liberty of trying to convince her of the folly of taking over the responsibility of bringing up the ugly son simply because she had had a liaison with the father. But as was her nature she didn't bother about anyone's opinion. After some time she found a bride for Gavarayya and married him off. Just before she died, when Gavarayya was twenty, she passed on to him her jewellery and cash of ten thousand rupees.

Everyone in the village was familiar with the next few phases of Gavarayya's life: his arrival at his paternal aunt's house in this village, his starting the hides business, and after seven or eight years, the loss of his wife. However, now that they knew about his origins courtesy the bangle seller, their hatred of Gavarayya began to seem entirely justified. His father was a wicked man, a murderer; this fellow, playing since his childhood with cats and dogs, had himself acquired animal traits. Moreover he had been brought up by a fallen woman. That such an inhuman fellow had landed in their village of all places made them feel deeply anguished.

But then elders like Avadhani did not despair at first. True, Gavarayya was a wicked man. A picture of ugliness—that too was true; a miser who wouldn't part with a copper—agreed; mulish fellow, uncultured—everyone knew that. All the same, a rich man. If his riches could be made to support good acts he, and his elders, would find salvation. With that altruistic intention they approached Gavarayya frequently, counselled him on the numerous ways in which he could become a virtuous man. Support the celebrations of the Lord Venugopala's festival, they said. Construct a pavilion in the temple, they said. Must repair the temple compound wall which had tilted, they said. Gavarayya did not bite once at any of the suggestions being made. At least come to the temple, have the Lord's darshan and take His prasadam, they said. They hoped that because it wouldn't cost him anything he might consider at least this suggestion sympathetically, and that over time his visits to the temple would lead to the seed of piety sprouting in his stony heart. But not once did Gavarayya go anywhere near the temple.

Unable to stand the sight of a fellow creature turning an atheist and a sinner, prompted by their own good nature and devotion to duty, the village elders employed the weapon of boycott. But this didn't work either. Because Gavarayya had boycotted the village from the beginning. And besides, the washerman, the trader and the barber kept serving him clandestinely. As he paid their dues promptly, not one of them wanted to lose him as a customer.

When Gavarayya's first wife slipped and fell into the well and drowned, Avadhani and the others called on him and condoled with him. All this was the result of his karma, he should wake up now at least, they exhorted him. But Gavarayya did not care. Not just that, within a year he married a girl who was fifteen years younger than him and brought her home. Though she wasn't much of a beauty, she decked herself out fashionably. Gavarayya took care of her with great love and affection, people used to say. But she too, for her part, never stepped outside her house. If out of curiosity anyone called on her, she chatted with them pleasantly. If anyone made fun or spoke sarcastically of either Gavarayya's looks or his age, she would say, 'He is a very good man.' Often she went by cart to the neighbouring town to watch a movie. That Gavarayya relaxed all his rules for his wife came as a surprise to everyone. But then even that relaxation was within certain limits: the residents of the house had to stay away from the villagers and from the happenings in the village of course.

How she had developed enough intimacy with the sewing machine man on the pyol of the opposite house as to elope with him, nobody knew. But this mishap gave the munsif, the karanam and others great joy. In the absolute darkness around them they now saw a ray of hope. Take advantage of the situation, they said to themselves.

All four of them together called on Gavarayya one day. It was a very big house. An old-style house, large and spacious, with rooms on all four sides opening out to a central courtyard: a manduva house. At the rear, the haystack, the cattle-shed, trees and the like. And beyond there were no more houses; it was all fields. Gavarayya had steadily expanded the small dwelling of his paternal aunt till it had reached these proportions.

This house would never be like any of the others. The buzz and stir of life was missing and a terrifying loneliness had accumulated within it, it appeared. As if at night ghosts with reversed feet roamed all over the house and then at daybreak climbed up and hid in the attic or in the corners of the eaves. Even those four, enjoying authority, status and an inborn courage, felt, it is true, a little dread as they stepped into that house.

His eyes closed, Gavarayya sat leaning against the wall. Through the banian he wore, his powerful chest and muscles, pock-marked all over, looked like dented babul timbers. Hearing their steps, he opened his eyes and looked at them. With streaks of red, his eyes looked like the eyes of a drunk. Avadhani, the munsif, the karanam and another village elder sat down on the bench next to Gavarayya. Gavarayya gave them a questioning look.

The karanam began to speak: 'You didn't deserve this calamity, Gavarayya. You are quiet and peaceable by nature, you never bother others or interfere in their lives. When the news came, believe me, we thought, "O, what a pity!" Our minds were convulsed with compassion.'

Gavarayya did not react. He just kept looking at them.

The munsif said: 'Whether it be in the matter of payment of taxes or of salaries to the field hands—he always punctually fulfilled his duty. He never kept with him money due to others. To that fellow, that field hand Subbayya, even before the year was over Gavarayya measured out his grain. That fellow has been singing his praises, "Our landlord is an extremely large-hearted man..."'

Gavarayya still did not respond. Whether he had heard them or not, they had no clue. He just kept looking at them.

'But then there is one small matter, Gavarayya babu,' said Avadhani. Greatly skilled at speech, he said whatever it was that he was expressing softly, smoothly, solemnly, blending it all with a special pity and tenderness. As the temple trustee, his words had even greater authority. 'Take it for granted that no one can cross this worldly ocean happily without support from God. You are an excellent man, a straightforward man—even if, as far as your business is concerned, it is totally against the laws; this is unambiguously asserted by our shastras. But then in the age of Kali, certain exceptions have been made by men. Therefore, it can be overlooked. Even then they insisted on one thing—whatever it is, if you don't stop thinking of God, all your hardships will evaporate like the mist. "With fruit, flower, leaf or even water"—ancestors said—"God is easy to please...'"

Gradually, a smile of triumph appeared on the faces of all four men. The normally snarling and growling Gavarayya was keeping quiet today. What everyone believed was that an impregnable fort wall was now showing a crack. They felt that Gavarayya could be reformed slowly. They rose, took their leave. Gavarayya still said nothing, just kept staring at them.

II

A year passed. Off and on, the karanam, the munsif and others would call on Gavarayya at his house. They would invite him to have a darshan of the deity, or for a harikatha recital, or for something else, and would personally escort him to the event in question. Now, even the villagers would raise their hands in namaskaram when they passed him on the street. But Gavarayya would not speak a word. Like a man enchanted he would sit in the temple or through the harikatha. Sometimes he would get up abruptly and walk away. The elders would shake their heads. 'Rakshasa, son of a whore! Is he going to change easily?' they would say.

Avadhani would respond, 'Does he have a choice in the matter? Defying society and dharmam where can this fellow go?'

The compound wall of Lord Venugopala's temple was on the verge of collapse. They couldn't put off repairs any longer, the elders decided. Not less than twenty-five thousand it would cost. Nobody could come up with such a huge sum except Gavarayya, they said. During the year Gavarayya had made huge profits. Even if you ignored the hides business, he had bought a three-quarters share in the oil mill in the town. In wonder, the villagers all told each other that in just this one oil business he had made a lakh of rupees. They thought that the reason for this prosperity was the newly sprung piety in him, but, lest he forgot, made it a point to tell him this again and again. On an auspicious day, the munsif, the karanam, the trustee of the temple and the wealthy people in the village together called on Gavarayya and explained the need to repair the temple compound wall. They were unanimous in their opinion that he alone should take it up.

'Your name will become immortal, Gavarayya. In your name we will have puja and religious rites performed,' they added.

'Why does God need a temple? And where is the need for a compound wall around the temple?' asked Gavarayya, chewing the end of the cheroot in his mouth.

They were all dumbstruck. 'Blasphemy! Blasphemy!' said the karanam and slapped his own cheeks in a gesture of expiation.

The temple trustee winked at the karanam.

'The question which Gavarayya has posed is not an ordinary one. Even philosophers and maharishis have been given no end of trouble by this knotty question. But then, even without our knowledge, inside Gavarayya there is saadhana at work, a spiritual process. If we can be patient for a few more days, a great man like Gavarayya will himself find an answer to his question. The favour of the Lord will not come directly and fully at one go. Comes in gradual stages, let me tell you. When that day of Grace comes, our Gavarayya himself will, I assure you, say, "Here, Avadhani garu, take this ten thousand—get the compound wall constructed." Won't he?'

Gavarayya, who had never done it before, now raised his hands and made a namaskaram to Avadhani. 'I have to go now, so please show yourselves out. I have a job to do in town.' Saying this, he strode out. Avadhani was taken aback. They looked at each other.

'Avadhani, you have conquered. He bowed his head and made a namaskaram to you! Curse him! He is changing. Surely changing,' said the karanam with glee.

'By the time the rainy season is over—take my word for it—he will drop the twenty-five thousand before us. Along with the compound wall we will build parapet walls for our houses,' said the munsif, chuckling into his moustache.

The rainy season came. The rain poured down heavily, angrily, powerfully. With thick clouds, darkened sky, lightning and thunder, nature was all hustle and bustle like a wedding pandal. Though the streets were slushy, womenfolk, among them newly married girls, held the pleats of their saris above their ankles, and with much gaiety and merriment took part in auspicious ceremonies at various houses. The farming community was busy in the fields. Though the munsif's wife suffered from arthritis and Avadhani's daughter retched from the nausea of pregnancy, these problems did not in any way affect the collective happiness of the village. Though the karanam's widowed sister kept staring through the window at the young man who had come to visit his relatives, and kept waving her hand to him ceaselessly, because of his short-sightedness and the shadows cast by the clouds he missed her signals. The coolies and others such continued to live on the outskirts of the village; besides the village's inhabitants, the gutters, the gulleys, the diseases—everything was in proper order. In the library building, constructed by the panchayat, the card players continued their games round the clock.

All was right with the world, until an unexpected development shattered the village's sense of peace and equilibrium. The munsif and a couple of others were sitting on the munsif's pyol, covered in shawls, puffing happily at their cheroots, absorbed in a spiritual discussion. Paanakaalu came running up to them. He looked distraught. From his appearance it appeared that right behind him followed some catastrophe—an earthquake or a deluge.

'Hey you. What's happened?' asked the munsif.

'She has come sir. She has returned, sir,' said Paanakaalu.

Two months earlier, the munsif's bull calf had slipped its tether and run away. The munsif said with a smile, 'Then there's nothing to be worried about? I shall come and see. Tie it up in the cattleshed.'

'Not the young bull, babu! It's Gavarayya's wife!' said Paanakaalu.

His announcement made them all straighten up. 'What, what, what did you say?' they all said simultaneously. In a hushed voice Paanakaalu gave them his version of what had happened. The night before, very late, after watching the late-night movie in the town, as Narasimham from the cloth shop approached the village, he noticed someone moving in the dark of the trees by the roadside. It struck him that it might be a ghost and he was scared, his heart beat fast. He ran and hurriedly woke up Paanakaalu sleeping in a hut a little distance away. Both returned to investigate. They found a young woman, in dirty clothes, her hair in disarray, plodding along laboriously. She was carrying a small bundle. They went up to her to see who she was, but failed to recognize her. They accosted her; she did not respond. She quickened her pace. The two followed her at a distance. She cut across the fields and entered the thatched shed that lay between Gavarayya's backyard and the fields. It was overflowing with firewood, old tin cans and odds and ends. For a moment they thought of waking up Gavarayya and informing him of his uninvited visitor but were afraid he might react violently and beat them up.

In the morning Paanakaalu remembered the incident and went to look inside the shed. The visitor of the night was none other than Gavarayya's wife! Her belly bulged considerably; she must be due any time now. She lay on the straw, groaning.

'Do you think Gavarayya knows about this...?' asked the munsif.

'No, babu. He leaves for town early in the morning, as you know,' said

Paanakaalu.

The munsif rose quickly. Put on his chappals and started for Avadhani's house. On the way everyone greeted him and enquired of him, 'What, munsif, it seems Gavarayya's wife has returned. What is happening to this village and the people, babu? Right and Wrong, Merit and Sin, all things are trampled into the dust! If those who elope and conceive are allowed to come back and are accepted by their husbands, won't the young in the village too go out of control! What has happened to this village, babu, and to these elders!' said an old woman past sixty seated on a pyol. Even before he reached Avadhani's house the munsif realized that the whole village was discussing Gavarayya's wife. Avadhani looked testy, angry.

'Do you know?' asked the munsif, climbing the steps.

'Every child and adult knows! Only you, supposed to be performing the munsif's duties, don't know. With what audacity has this sinful female returned to this village? Doesn't she have any respect for dharmam? Doesn't she fear God?'

In no time at all the karanam, the munsif and the elders of the village, were seated in conference. They said they needed to ascertain unambiguously if Gavarayya was going to drive away the fallen woman immediately or leave the village himself. That very morning someone in the untouchable quarter had contracted cholera and the reason for that was the sinful woman's stepping into the village, said one elder. Avadhani's elder sister pushed the door slightly ajar and said, 'We are all dying of shame. If these violators of our laws remain in the village what will be the fate of householders, annayya? Please think about this. Please tell us if you want us to remain in the village or jump into the pond.'

'Gavarayya is not at home. As soon as he returns, you elders go and tell him what you think of all this. How can he not respect your views? Whatever you may say, Avadhani garu, this man is not the old Gavarayya. He has now learned to discriminate between Sin and Virtue. He has developed respect for God, and more than that, respect for you,' said Seshagiri, the biggest landowner of the village.

'How many hopes we pinned on Gavarayya!' said the munsif with a sigh, remembering the compound wall of the temple, and the twenty-five thousand.

'Doesn't Gavarayya's aunt still live in that house? What is she doing? Keeping quiet?' asked the karanam.

'What aunt, babu? She has fallen ill and is now bedridden, poor woman. She has grown really old,' said someone.

At dusk Gavarayya got off his cycle. The moment he stepped into his house, he found Avadhani, the karanam and the munsif waiting for him. When he raised his eyebrows quizzically at them, Avadhani told him the whole story. Gavarayya's eyes became red. His thick lips shook. He rushed to pick up a crowbar from the corner of the room. 'I shall kill that whore,' he shouted.

Avadhani hurriedly said, 'No, don't! No need for such a drastic act. Gavarayya babu, your devotion to morality is well known, why get into a murder case now? It will create trouble for you, and create trouble for us as well…throw her out of the house, that's all.'

Gavarayya did not listen to him. He rushed out of the house. His visitors, realizing that in his present state there was no telling what he might do to them, slipped quickly out of the house and returned to their own homes. Gavarayya went into the thatched shed. It was all cluttered and gloomy inside.

Gavarayya strained his eyes to see. He saw his wife lying supine on the ground in the corner. He went over to her and raised the crowbar. Then he paused. Unsure as to whether she was dead or alive, he knelt down and peered at her face. It was pale and emaciated. Her hair was dirty and knotted. Pitiful, paralysed by fear of death, her face looked awful. He called her name, 'Chitti.' Shook her by the shoulders. Chitti opened her eyes. Recognized Gavarayya. Tears trickled from her eyes. She lifted her shaking hands and made a namaskaram to him. 'Don't do anything to me, I shall go away,' she said in a feeble voice.

Gavarayya recalled that his mother had, just before her death, done exactly this—raised her hands and made a namaskaram to his father and looked as pitiful. Chitti's hands slid down and lay helplessly at her sides.

He sat there for a while, lost in thought. Then, he went in and brought back water in a chembu and a bowl full of rice. 'Chitti,' he said softly. Chitti opened her eyes. 'Get up, eat this rice and go away at first light. Don't show me your face again, is that clear?' he said. Chitti nodded her head. Gavarayya stood up and said, 'If you are seen in this shed in the morning, you won't escape death.' He returned to the house.

That night Gavarayya had no appetite. The food was tasteless. He felt as though fires had been lit underneath his feet. Outside in the sky thick clouds gathered. The wind screamed through the leaves of the trees, crashed into the eaves of the roof and ricocheted off with a terrible moaning sound. After tossing and turning for a while, Gavarayya fell into a restless sleep. In his sleep, he had a horrifying nightmare. His father and his mother had been tied to a stake and fires had been started all around them. The people who had started the fire appeared to be Avadhani, the karanam and the munsif. Both his father and his mother were screaming in agony and terror, 'Babu-i, babu-i.' In a while they were reduced to red and black ashes. Out of the ashes rose somebody. He quickly strode towards the thatched hut in Gavarayya's backyard. That person had a conch in one hand, and the divine discus in the other, and Vaishnava marks on his forehead. He looked like the image of Lord Venugopala that he had seen in the temple. Tears were flowing out of the eyes of Lord Venugopala. God was crying. He took the infant from Chitti's bed into his arms and began consoling Chitti, 'Don't be afraid. I am with you.' Meanwhile, armed with burning torches, a mob of furious villagers came and set the thatched shed on fire. The blaze enveloped the shed. The flames were reaching the sky. God was burning. Chitti was burning. The innocent little infant was burning.

Gavarayya woke up with a start. His forehead was dripping with perspiration. The casement shutters were flapping in the wind. The wind roared, thunder and lightning crashed—it was terrifying. Gavarayya wiped his face. He gulped down a glass of water and looked out into the darkness. The dark outside blended with the rain, like black poison sliding down from the heavens, black sin surging like a rising sea. In that terrifying darkness, these men, houses and homes, fears, desires, everything appeared unreal. Out of the furious tandava of Nature, the awe-inspiring dance of the elements, a singular truth emerged, and struck him like a whiplash. There was a huge flash of lightning. The tip of the gopuram of Lord Venugopala's temple was starkly outlined by the light. Something flashed in the depths of Gavarayya's mind. Gavarayya coughed. Lantern in hand he went into the backyard. It was a huge yard. Shrill, loud moans came from the thatched shed. He went in. Chitti was writhing in labour. Unable to watch the sight, he left the lantern there and came out and stood in the cold, wind and rain. He didn't know how long he stood there. Suddenly, there was a flash of lightning, followed by a dreadful sound.

It seemed like lightning had struck somewhere close by. Chitti screamed aloud, 'Babu-i.' Gavarayya froze for a minute. Presently, he heard an infant puling—'kayre, kayre'.

A smile spread across Gavarayya's hard-bitten, ugly face. In that downpour, in that darkness, finding his way by the lighting flashes, Gavarayya ran to fetch the village midwife.

It was eight in the morning. Avadhani, the karanam, Seshagiri, the munsif, were all seated in the verandah of Gavarayya's house. The storm had cleared away and everything was washed by soft sunlight. In that sunlight everything—the trees, leaves, grass and the black wings of the crow on Gavarayya's house—shimmered. Only the faces of Avadhani, the munsif and the others didn't shine. Anger swirled in their frustrated eyes. Outside the house, the village servant and a couple of servants who had come with their masters stood as if ready to carry out their slightest bidding, their arms deferentially crossed over their chests, embodiments of obedience and resolve.

After a while Gavarayya came out of his house. He looked like an iron girder on the move. Narrowing his little eyes, he looked directly, disdainfully at them. Avadhani said fiercely, 'I didn't expect you would do this, Gavarayya! Thought you were more of a gentleman than this. Would you let a fallen woman live in your house? Would you bring up a baby boy fathered by some stranger? You have not only lost both shame and self-respect, but committed a sin, a horrible sin. Do you think that God will forgive and tolerate you?'

Gavarayya sat relaxed, leaning against the wall. From the waist-folds of his dhoti he took out a leaf of tobacco and rolling it, said, 'I have been advised by God himself.'

'What did He tell you?' asked Avadhani.

'I am going to adopt that boy. That girl will stay put in my house,' said Gavarayya.

'You will play husband to a woman who eloped?' blustered Seshagiri, unable to believe his own ears.

'If I marry a girl half my age what else will she do? If I throw her out, where will she go? Give her food and clothes and she will stay here and take care of the boy,' said Gavarayya.

Avadhani guffawed contemptuously.

'Did you say God had told you? Which God, I ask? The God of adultery? Those who have engaged in great penance have not been rewarded, but...'

'God appeared before me. I have seen God,' said Gavarayya stubbornly. His eyes shone when he said this.

'You had better send her and her child packing. We are not going to accept such a blatant act of immorality,' said the karanam.

'We won't allow you to come anywhere near the temple,' said Avadhani.

'The village as a whole will unite against you, Gavarayya! We are telling you calmly and respectfully what your next course of action should be, but you are not accepting our advice. We are not going to sit idle, I warn you,' said the munsif.

Suddenly Gavarayya rose, picked up the crowbar lying by the wall and thundered, 'Who dares challenge me, let me see? Who can challenge me? Have I stolen someone else's money? Did I ask any of you to come here? Who are you to come and preach to me? Whoever interferes in my affairs I shall cut in half with a single blow—let whoever wishes to come, come forward.' He was shaking with rage. He was like a volcano erupting. His appearance, his fury terrified the village servant who stood gaping in fear. Avadhani stood up. The others too stood up.

'Then you say you will not abandon her,' said Avadhani.

'I won't leave her,' said Gavarayya.

'Then we are ostracizing you. The washerman and the barber won't come to you. What's more, I'd like to see which labourer will dare to work in your fields!' said Avadhani.

The munsif looked at the village servant and said, 'Make sure the whole village knows what has been decided. Announce it everywhere on your tabor.'

Gavarayya gave them a thunderous look, and said, 'Get away from me, you so-called village elders! I wouldn't live in this godforsaken village even if you asked me to. I won't allow my child to be brought up in the midst of such wretched people, thoo!' He hawked and spat, and glared at the departing Avadhani and the rest of the delegation. Before seven days, a week passed, Gavarayya's house was locked up. It transpired that Gavarayya had bought a house in the town and had put his house and land in the village up for sale.

INSPECTOR MATADEEN ON THE MOON
HARISHANKAR PARSAI

Translated from the Hindi by C. M. Naim

Scientists say there is no life on the moon. But the senior inspector, Matadeen, says, 'The scientists lie. There are men, just like us, on the other side of the moon.'

Science has always lost out to Inspector Matadeen. Let experts argue till they are hoarse that the prints on the dagger do not match the fingerprints of the accused, Inspector Matadeen will still manage to put his man in prison.

Matadeen says, 'These scientists, they never investigate a case thoroughly. Just because they can see only the bright side of the moon they've declared there's no life on it. I've been to the dark side. There are men living there.'

That has to be true. When it comes to dark sides, Inspector Matadeen is the recognized expert...

But, you might ask, why did he go to the moon? As a tourist? To catch a fugitive?

No. He went under the Cultural Exchange Scheme, to represent India. The Government of Moon wrote to the Government of India, 'We are an advanced civilization, but our police force is still not good enough. They often fail to catch or punish criminals. We understand you have established Ram Rajya in your country.

Please send one of your police officers to give our men proper training.'

The home minister told the home secretary, 'Send an IG.'

He replied, 'Sir, we cannot send an inspector general. It's a matter of protocol. Moon is only a small satellite of Earth. We cannot send someone of too high a rank there. Let me depute some senior inspector.'

And so they chose Inspector Matadeen, the investigating officer of a thousand and one cases, and the Moon Government was asked to send an earth-ship to fetch him.

Meanwhile, the home minister sent for Inspector Matadeen. 'You're going there,' he said, 'to represent the glorious traditions of the Indian Police. Make sure you do a good job. Make the universe applaud our department, so that even the prime minister hears about us.'

On the appointed day, an earth-ship arrived from the moon. Bidding everyone goodbye, Inspector Matadeen started walking towards the ship. He was chanting a chaupai under his breath: 'Pravisi nagara kijai sab kajaa, hridaya rakhi kausalpur raja...'

On reaching the ship, Inspector Matadeen suddenly called out to his clerk Munshi Abdul Ghafoor. 'Munshi!'

Abdul Ghafoor clicked his heels, saluted, and said, 'Yes, Pectsa.'

'Did you remember to pack some FIR forms?'

'Yes, Pectsa.'

'And a blank copy of the Daily Record Register?'

'Yes, Pectsa.'

Inspector Matadeen then sent for Havaldar Balbhaddar and said to him, 'When it's time for my wife to deliver, send your wife to lend a hand.'

Balbhaddar replied, 'Yes, Pectsa.'

'You needn't worry, Pectsa,' Abdul Ghafoor added, 'I'll send my wife too.'

Inspector Matadeen then turned to the pilot. 'You have your driver's licence?'

'Yes, sir.'

'And your headlights work?'

'Yes, sir.'

'They'd better,' growled Inspector Matadeen to his men, 'otherwise I'll challan the bastard mid-space.'

The pilot overheard him and said, 'In our country, we don't talk to people in this manner.'

'I know, I know,' Inspector Matadeen sneered, 'no wonder your police is so weak-kneed. But I'll kick them into shape soon enough.'

He had placed one foot inside the earth-ship's door when Havaldar Ram Sanjivan came running. 'Pectsa,' he said, 'SP Saab's wife would like you to bring her a heel-polishing stone from the moon.'

Inspector Matadeen was delighted. 'Tell Bhai Saab I'll definitely get her one.'

Finally he climbed in and took his seat and the earth-ship took off. It had barely crossed the earth's atmosphere when Inspector Matadeen shouted to the pilot, 'Abbe, why aren't you honking?'

'There's nothing for millions of miles!' the pilot replied.

'But a rule is a rule,' Inspector Matadeen snarled. 'Keep your thumb down on the horn.'

The pilot pressed his thumb down on the horn, and kept it that way until they arrived on the moon.

Senior officers of the Moon Police had come to receive Inspector Matadeen. He swaggered out of the earth-ship and ran an eye over their shoulder-patches. None had a star on it, or even a ribbon. Inspector Matadeen decided it wasn't necessary to click his heels or salute. He also thought: After all, I'm now a Special Advisor, not just an inspector.

The welcome party took him to the local Police Lines and put him up in a fine bungalow.

After a day's rest, Inspector Matadeen decided to begin his work. First, he went out to inspect the Police Lines. In the evening he expressed his surprise to the host inspector general. 'There's no Hanuman temple in your Police Lines! In our Ram Rajya, every Police Lines has its Hanumanji.'

The IG asked, 'Who is Hanuman? We've never heard of him.'

Inspector Matadeen explained. 'Every policeman must have a daily darshan of Hanumanji. You see, Hanumanji was in the Special Branch in Sugriv's administration. It was he who discovered where Ma Sita was being held forcibly. It was a case of abduction—Section 362 IPC, you know. Hanumanji punished Ravan right on the spot—set fire to his entire property. The police must have that kind of right. They should be able to punish a criminal as soon as they catch him. No need to get bogged down in courts. But, sad to say, we are yet to achieve that in our Ram Rajya.

'Anyway, Bhagwan Ram was highly pleased with Hanumanji. He took him to Ayodhya and assigned him the city beat. That same Hanumanji is our patron god. Here, I brought his photograph along. Use this to get some figures cast, then have them set up in all the Police Lines.'

A few days later, an idol of Hanumanji was enshrined in each and every

Police Lines on the moon.

In the meantime, Inspector Matadeen began to study how the local police worked. It seemed to him that the Moon Police was careless and lacked in enthusiasm, that it showed little concern for crime. But the reason for this attitude was not apparent.

Suddenly, a thought occurred to Inspector Matadeen. He sent for the salary register. One glance at it and everything was clear. Now he knew why the Moon Police behaved the way it did.

That evening he reported to the police minister. 'Now I know why your men are so lackadaisical. You pay them large salaries, that's why. Five hundred to a constable, seven hundred to a havaldar, and a thousand to a thanedar! What kind of foolishness is this? Why should your police try to catch any criminal? In our country, we give the constables just one hundred, and the inspectors two. That's why you see them running around catching criminals. You must immediately reduce the salaries.'

'But that would be highly unfair,' the police minister protested. 'Why would they work at all if they are not given good salaries?'

Inspector Matadeen replied, 'There's nothing unfair about it. In fact, as soon as the first reduced pay cheques are sent out, you'll see a revolutionary change in your men's attitude.'

The police minister ordered a cut in the salaries. Sure enough, in a couple of months, a drastic change was evident. The policemen suddenly became extremely zealous in their performance. There was panic in the criminal world. When the police minister sent for the records kept at the police stations, he was amazed to see that the number of registered cases was several times higher than before. He said to Inspector Matadeen, 'I must praise your keen insight. You have brought about a revolution! But do tell me how it works.'

'It's very simple,' Inspector Matadeen explained. 'If you pay an employee little money, he won't be able to live on it. No constable can support a family on just one hundred rupees a month, nor can an inspector live with dignity on two hundred. Each will have to make some extra money. And he can do that only if he starts catching criminals. Immediately, he becomes concerned about crime, and turns into an alert and dutiful policeman. That's why we have a most efficient police system in our Ram Rajya.'

The news of this miracle spread all over the moon. People began to come to look at the man who could reduce salaries and yet create efficiency. The policemen were the most happy. They said to Inspector Matadeen, 'Guru, if you hadn't come we'd have continued living on our salaries alone.' The Moon

Government was also delighted, for it could now have a surplus budget.

Half the problem was taken care of in this way. The police had started catching criminals. Now only the investigative process remained to be reformed—how to get a criminal sentenced after one had caught him. Inspector Matadeen decided to wait for some major incident so that he could use it as a model to display his special methods.

One day, some people quarrelled and one of them was killed. When Inspector Matadeen heard about this, he marched to the police station, sat down at a desk, and declared, 'I shall investigate this case to show you how it's done. All of you just watch and learn. This is a murder case. And in a murder case one must have rock solid evidence against the accused.'

The station officer said, 'Before we start collecting evidence against anyone, shouldn't we first try to discover who did the killing?'

Inspector Matadeen replied, 'No, why work backwards? First, make sure of your evidence. Did you find any blood? On someone's clothes or elsewhere?'

One of the inspectors said, 'The assailants ran away as the victim lay dying on the road. A man who lives near the spot picked him up and brought him to the hospital. His clothes did have some blood on them.'

'Arrest the man immediately.'

'But, sir,' the station officer remonstrated, 'he only tried to help the dying man!'

'That may well be true,' explained Inspector Matadeen, 'but where else are you going find blood spots? You must grab the evidence which is readily available.'

The man was arrested and brought to the police station. He protested, 'But I carried the dying man to the hospital! Is that a crime?'

The local officers were visibly moved, but not Inspector Matadeen. Everyone waited to see how he would respond.

'But why did you go to where the fight occurred?' Inspector Matadeen asked the man.

'I didn't go there,' he replied, 'I happen to live there. The fight took place right in front of my house.'

It was clearly a test of Inspector Matadeen's genius. He quietly responded, 'True, your house is there, but why go where a fight is taking place?'

There could be no answer to that question. The man could only repeat and go on repeating, 'I didn't go there. I live there.'

And each time Inspector Matadeen responded, 'That's true, but why go where a fight is taking place?'

This line of questioning greatly impressed the local officers. Inspector Matadeen settled back and explained his investigative principles. 'Look,' he said, 'a man's been killed. This means someone definitely killed him. Someone is the murderer. Someone has to be convicted and punished. You might ask, who is guilty? But, for the police, that's not so important. What is important is who can be proven guilty or, better still, who should be proven guilty?

'A murder has occurred. Eventually, someone will be convicted. It's not for us to worry if it is the actual killer or someone innocent. All human beings are equal. In each of them is present a bit of the same god. We don't discriminate. We're humanists.

'So the question actually is who ought to be proven guilty? That depends on two things. One, has the man been a nuisance to the police, and two, will his conviction please the men at the top?'

Inspector Matadeen was told that though the arrested man was otherwise a decent person, he was given to criticizing the police whenever they made a mistake. As for the question of pleasing the men at the top, the man belonged to the Opposition party.

'This is a first-rate situation,' Inspector Matadeen declared, thumping the table. 'Rock solid evidence, plus support from the top!'

One inspector tried to protest, 'But we can't let a decent man be convicted of a crime he didn't commit!'

Inspector Matadeen explained patiently, 'Look, I've already told you that the same God resides in all of us. Whether you convict this man or the actual killer, it is *God* who will hang. Further, in this instance, you're getting blood-spattered clothes. Now where would you find bloodstains if you let him go? Go ahead, file the FIR as I tell you.'

Inspector Matadeen dictated the First Information Report leaving a few spaces blank for future needs.

Next day, the station officer came to Inspector Matadeen and said, 'Gurudev, we're in deep trouble. Numerous citizens are demanding why we are trying to frame that poor innocent man? It has never been done before. What should we say? We feel so ashamed...'

'Don't worry,' Inspector Matadeen said consolingly. 'In this job, one always feels a bit of compunction in the beginning. But later you'll feel ashamed for letting innocent people go free. Now, you should understand that every question has an answer. The next time someone comes to you to question this decision, tell him, we know the man is innocent, but what can we do? Those at the top want it so.'

'In that case they'll go to the SR.'

'Let him say, "Those at the top want it so."'

'Then they'll complain to the IG.'

'He too should say, "It's the men at the top who want it so."'

'They'll then go to the police minister.'

'So what? He should say the same thing, "Friends, what can I do? Those at the top want it so."'

'But the people won't give up. They'll go to the PM.'

'The PM should respond in the same way, "I know he's innocent, but those at the top want it so."'

'Then...'

'Then what?' Matadeen stopped him short. 'Who can they go to next? To God? But has anyone ever come back after going to God?'

The station officer remained silent. Such brilliant logic left him dumbfounded.

Inspector Matadeen continued, 'That one sentence—"Those at the top want it so"—has always come to the rescue of our government in the last twenty-five years. You too should learn it well.'

They began to get the case ready for trial. Matadeen ordered, 'Bring me a few eyewitnesses.'

'How can we do that?' the station officer asked. 'How can there be eyewitnesses when no one saw him kill that man?

Matadeen smacked his head in despair. 'God, what fools I have to deal with! They don't even know the ABC of this business.' Then he added angrily, 'Do you know who an eyewitness is? An eyewitness is not someone who actually sees an event take place, he's one who claims that he saw.'

'But why would someone make such a claim?' the station officer protested.

'Why not?' thundered Inspector Matadeen. 'I can't see how you people manage to run your department at all. Arre, the police must always have a ready list of eyewitnesses. When one is needed, you just pick a name from that list and present the person in court. In our country we have people who "eyewitness" hundreds of cases every year. Our courts have recognized that these men possess some divine power that lets them foresee the place where some incident is going to happen, allowing them to get there beforehand.

'I'll get you eyewitnesses. Bring me some bad characters. You know the kind—petty thieves, gamblers, goondas, bootleggers.'

Next day, half a dozen fine specimens showed up at the police station. Inspector Matadeen was delighted. It had been a long while since he had last seen such men. He had missed them. His voice melting with affection, he asked them, 'You saw that man assault the deceased, didn't you?'

They replied, 'No, sir, we didn't see a thing. We weren't even there.'

Inspector Matadeen knew it was the first time for them. He patiently continued, 'I know you weren't there. But you saw him attack with a lathi, didn't you?'

The men decided they were dealing with a lunatic. Who else would talk such nonsense? They began to laugh.

'Don't laugh!' said Inspector Matadeen sternly. 'Answer my question.'

They again replied, 'How can we say we saw it when we weren't even there?'

Inspector Matadeen lost his temper. 'I'll tell you how,' he snarled. 'I have here detailed reports on what you fellows have been up to. I can have each one of you locked up for at least ten years. Now, tell me, do you wish to stay in business or would you rather go to jail?'

Terrified, the men said, 'No, sir, we don't want to go to jail.'

'In that case, you saw that fellow beat the victim with a lathi, didn't you?'

'Yes, sir, we did. We saw him come out of his house and start hitting the man with a lathi until the poor fellow fell to the ground.'

'Good. In future too, you'll see more such incidents, won't you?' Matadeen said firmly.

'Yes, sir. We'll see what you tell us to.'

The station officer was overwhelmed by this miracle. He couldn't move for a few minutes. Then, getting up from his chair, he threw himself at Inspector Matadeen's feet.

'Here now, let go. Let me do my work,' Inspector Matadeen remonstrated, but the station officer clung to him and kept repeating, 'I want to spend the rest of my days at your feet.'

In due course, Inspector Matadeen put together the entire dossier and, in the process, taught the local police everything he knew—how to substitute FIRs, how to leave some pages blank for future use, how to change entries in the Daily Record, how to win over hostile witnesses... The man he had arrested was sentenced to twenty years.

The Moon Police was now fully trained. Case after case was brought before the courts and, in every instance, a conviction was won. The Moon

Government was delighted. The Moon Parliament passed a resolution to thank the Government of India. It noted the remarkable efficiency the Moon Police had achieved under Inspector Matadeen's guidance. Inspector Matadeen was given a civic reception. Covered with garlands, he was taken around in a procession in an open jeep. Thousands of people lined the road and shouted his praises. Inspector Matadeen responded in the style of his home minister—with folded hands, lowered eyes—full of humility. But this was his first time and he felt somewhat ill at ease. He had never even dreamt, when he had entered the service some twenty-six years ago, that one day he would be so honoured on Moon. He wished he had remembered to bring along a dhoti-kurta and a Gandhi cap.

On Earth, the Indian home minister watched the proceedings on television. 'This may be the time for me to make a goodwill visit,' he mused.

A few more months passed.

Then, suddenly one day, the Moon Parliament met in an emergency session. It was a stormy but secret meeting, and so its report was not made public. We can only offer what was faintly heard by people outside the chamber. The members seemed enraged and could be heard shouting:

'No one takes care of sick parents!'

'No one tries to rescue a drowning child!'

'No one helps if a house catches fire!'

'Men have become worse than animals!'

'The government should resign immediately!'

'Resign! Resign!'

Next day the Prime Minister of Moon sent for Inspector Matadeen. Inspector Matadeen could see that the Prime Minister had visibly aged, that he seemed not to have slept for a few nights. He looked quite disconsolate as he said, 'Matadeenji, we are extremely grateful to you and to the Government of India but you should go back tomorrow.'

'No, sir,' Matadeen replied, 'I'll return only after I've finished my term here.'

'We'll give you your full term's salary,' the Prime Minister said. 'Double the amount...triple, if you wish.'

Inspector Matadeen was polite but firm. 'No, sir, I'm a man of principles. My work is more dear to me than money.'

In the end, the Prime Minister of Moon sent a confidential letter to the Prime Minister of India.

Four days later, Inspector Matadeen received orders from his own IG to return immediately. Picking up a heel-polishing stone for the wife of his SP, Inspector Matadeen climbed aboard the earth-ship and bade farewell to the moon. The entire Moon Police burst into tears as the earth-ship lifted off.

What happened on the moon that he had to leave so suddenly? What did the Prime Minister of Moon write to the Prime Minister of India? These questions remained unanswered for a long time.

Then someone got hold of that confidential letter and made parts of it public.

Thank you for lending us the services of Inspector Matadeen, but now you must recall him immediately. We had thought India was our friend, but only an enemy could have done what you did to us. We were innocent and trusting, and you deceived us. Ever since Inspector Matadeen has trained our police, things have come to a terrible pass. No one comes to the help of an assault victim for fear he might himself be accused. Sons abandon their sick parents, lest they be charged with murder. Houses catch fire and burn down, but neighbours don't help for fear they might be accused of arson. Children drown before people's eyes but no one comes to their rescue lest they be accused of drowning them. All human relations are breaking down. Your man has destroyed almost half of our civilized life. If he stays around longer he'll destroy the remaining half. Please call him back immediately to your own Ram Rajya...

fifteen

~

DRAUPADI

MAHASWETA DEVI

*Translated from the Bengali by Gayatri Chakravorty Spivak**

I

Name: Dopdi Mejhen, age 27, husband Dulna Majhi (deceased), domicile Cherakhan, Bankrajharh, information whether dead or alive and/or assistance in arrest, one hundred rupees...

An exchange between two medallioned *uniforms.*

FIRST MEDALLION. What's this, a tribal called Dopdi? The list of names I brought has nothing like it! How can anyone have an unlisted name?

SECOND MEDALLION. Draupadi Mejhen. Born the year her mother threshed rice at Surja Sahu (killed)'s at Bakuli. Surja Sahu's wife gave her the name.

FIRST. These officers like nothing better than to write as much as they can in English. What's all this stuff about her?

SECOND. *Most notorious* female. *Long wanted in many...*

Dossier: Dulna and Dopdi worked at harvests, *rotating* between Birbhum, Burdwan, Murshidabad and Bankura. In 1971, in the famous *Operation* Bakuli, when three villages were *cordoned off* and

*English words in the original Bengali story have been italicized in this translation.

machine gunned, they too lay on the ground, faking dead. In fact, they were the *main* culprits. Murdering Surja Sahu and his son, occupying *upper-caste* wells and tube wells during the *drought*, not *surrendering* those three young men to the police. In all this they were the chief instigators. In the morning, at the time of the body count, the couple could not be found. The *blood sugar* level of Captain Arjan Singh, the *architect* of Bakuli, rose at once and proved yet again that diabetes can be a result of anxiety and depression. Diabetes has twelve husbands—among them *anxiety*.

Dulna and Dopdi went underground for a long time in a *Neanderthal* darkness. The Special Forces, attempting to pierce that dark by an armed search, compelled quite a few Santhals in the various districts of West Bengal to meet their Maker against their will. By the Indian Constitution, all human beings, regardless of caste or creed, are sacred. Still, accidents like this do happen. Two sorts of reasons: (1) the underground couple's skill in self-concealment; (2) not merely the Santhals but all tribals of the Austro-Asiatic Munda tribes appear the same to the Special Forces.

In fact, all around the ill-famed forest of Jharkhani, which is *under* the jurisdiction of the police station at Bankrajharh (in this India of ours, even a worm is under a certain police station), even in the southeast and southwest corners, one comes across hair-raising details in the eyewitness records put together on the people who are suspected of attacking police stations, stealing guns (since the snatchers are invariably not well educated, they sometimes say 'give up your *chambers*' rather than give up your gun), killing grain brokers, landlords, moneylenders, law officers and bureaucrats. A black-skinned couple ululated like police *sirens* before the episode. They sang jubilantly in a savage tongue, incomprehensible even to the Santhals. Such as:

Samaray hijulenako mar goekope
and,
Hendre rambra keche keche
Pundi rambra keche keche

This proves conclusively that they are the cause of Captain Arjan Singh's diabetes.

Government procedure being as incomprehensible as the Male Principle in Sankhya philosophy or Antonioni's early films, it was Arjan Singh who was sent once again on *Operation Forest* Jharkhani. Learning from Intelligence that the above-mentioned ululating and dancing couple were the escaped corpses, Arjan Singh fell for a bit into a *zombie*-like state and finally acquired so irrational a dread of black-skinned people that whenever he saw a black person in a ball–bag, he swooned, saying 'they're killing me', and drank and passed a lot of water. Neither uniform nor Scriptures could relieve that depression. At long last, under the shadow *of a premature* and *forced retirement,* it was possible to present him at the desk of Mr Senanayak, the elderly Bengali *specialist* in combat and extreme-Left politics.

Senanayak knows the activities and capacities of the opposition better than they themselves do. First, therefore, he presents an encomium on the military genius of the Sikhs. Then he explains further: is it only the opposition that should find power at the end of the barrel of a gun? Arjan Singh's power also explodes out of the *male organ* of a gun. Without a gun even the 'five Ks' come to nothing in this day and age. These speeches he delivers to all and sundry. As a result, the fighting forces regain their confidence in the *Army Handbook*. It is not a book for everyone. It says that the most despicable and repulsive style of fighting is guerrilla warfare with primitive weapons. Annihilation at sight of any and all practitioners of such warfare is the sacred duty of every soldier. Dopdi and Dulna belong to the *category* of such fighters, for they too kill by means of hatchet and scythe, bow and arrow, etc. In fact, their fighting power is greater than the gentlemen's. Not all gentlemen become experts in the explosion of *chambers*; they think the power will come out on its own if the gun is held. But since Dulna and Dopdi are illiterate, their kind have practised the use of weapons generation after generation.

I should mention here that, although the other side make little of him, Senanayak is not to be trifled with. Whatever his *practice,* in *theory* he respects the opposition. Respects them because they could be neither understood nor demolished if they were treated with the attitude, 'it's nothing but a bit of impertinent game-playing with guns'. *In order to destroy the enemy, become one.* Thus he understood them by *(theoretically)* becoming one of them. He hopes to write on all this in the future. He has also decided that in his written work he will *demolish* the gentlemen and *highlight* the message of the harvest workers. These mental processes might seem complicated, but actually he is a simple man and is as pleased as his third great-uncle after a meal of turtle meat. In fact, he knows that, as in the old popular song, turn by turn the world will change. And in every world he must have the credentials to survive with honour. If necessary he will show the future to what extent he alone understands the matter in its proper perspective. He knows very well that what he is doing today the future will forget, but he also knows that if he can change colour from world to world, he can represent the particular world in question. Today he is getting rid of the young by means of '*apprehension and elimination*', but he knows people will soon forget the memory and lesson of blood. And at the same time, he, like Shakespeare, believes in delivering the world's *legacy* into youth's hands. He is Prospero as well.

At any rate, information is received that many young men and women, *batch by batch* and on jeeps, have attacked police station after police station, terrified and elated the region, and disappeared into the forest of Jharkhani. Since after escaping from Bakuli, Dopdi and Dulna have worked at the house of virtually every landowner, they can efficiently inform the killers about their targets and announce proudly that they too are soldiers, *rank and file*. Finally the impenetrable forest of Jharkhani is surrounded by real soldiers, the *army* enters and splits the battlefield. Soldiers in hiding guard the falls and springs that are the only source of drinking water; they are still guarding, still looking. On one such search, army informant Dukhiram Gharari saw a young Santhal man lying on his stomach on a flat stone, dipping his face to drink water. The soldiers shot him as he lay. As the .303 threw him off spread-eagled and brought a bloody foam to his mouth, he roared 'Ma-ho' and then went limp. They realized later that it was the redoubtable Dulna Majhi.

What does 'Ma-ho' mean? Is this a violent *slogan* in the tribal language? Even after much thought, the Department of Defence could not be sure. Two tribal-specialist types are flown in from Calcutta, and they sweat over the dictionaries put together by worthies such as Hoffman–Jeffer and Golden–Palmer. Finally the omniscient Senanayak summons Chamru, the water carrier of the *camp*. He giggles when he sees the two specialists, scratches his ear with his bidi, and says, The Santhals of Maldah did say that when they began fighting at the time of King Gandhi! It's a battle cry. Who said 'Ma-ho' here? Did someone come from Maldah?

The problem is thus solved. Then, leaving Dulna's body on the stone, the soldiers climb the trees in green camouflage. They embrace the leafy boughs like so many great god Pans and wait as the large red ants bite their private parts. To see if anyone comes to take away the body. This is the hunter's way, not the soldier's. But Senanayak knows that these brutes cannot be dispatched by the approved method. So he asks his men to draw the prey with a corpse as bait. All will come clear, he says. I have almost deciphered Dopdi's song.

The soldiers get going at his command. But no one comes to claim Dulna's corpse. At night the soldiers shoot at a scuffle and, descending, discover that they have killed two hedgehogs copulating on dry leaves. Improvidently enough, the soldiers' jungle scout Dukhiram gets a knife in the neck before he can claim the reward for Dulna's capture. Bearing Dulna's corpse, the soldiers suffer shooting pains as the ants, interrupted in their feast, begin to bite them. When Senanayak hears that no one has come to take the corpse, he slaps his *anti-Fascist paperback* copy of *The Deputy* and shouts, '*What?*' Immediately one of the tribal specialists runs in with a joy as naked and transparent as Archimedes's and says, 'Get up, *sir*! I have discovered the meaning of that "hende rambra" stuff. It's Mundari *language*.'

Thus the search for Dopdi continues. In the forest *belt* of Jharkhani, the *Operation* continues—will continue. It is a carbuncle on the government's backside. Not to be cured by the tested ointment, not to burst with the appropriate herb. In the first phase the fugitives, ignorant of the forest's *topography*, are caught easily, and by the law of confrontation they are shot at the taxpayer's expense. By the law of confrontation, their eyeballs, intestines, stomachs, hearts, genitals, and so on become the food of fox, vulture, hyena, wildcat, ant and worm, and the untouchables go off happily to sell their bare skeletons.

They do not allow themselves to be captured in open combat in the next phase. Now it seems that they have found a trustworthy *courier*. Ten to one it's Dopdi. Dopdi loved Dulna more than her blood. No doubt it is she who is saving the fugitives now.

'They' is also a *hypothesis*.

Why?

How many went *originally*?

The answer is silence. About that there are many tales, many books in press. Best not to believe everything.

How many killed in six years' confrontation?

The answer is silence.

Why after confrontations are the skeletons discovered with arms broken or severed? Could armless men have fought? Why do the collarbones shake, why are legs and ribs crushed?

Two kinds of answer. Silence. Hurt rebuke in the eyes. Shame on you! Why bring this up? What will be will be...

How many left in the forest? The answer is silence.

A *legion*? Is it *justifiable* to maintain a large battalion in that wild area at the taxpayers' expense?

Answer: *Objection*. 'Wild area' is incorrect. The battalion is provided with supervised nutrition, arrangements to worship according to religion, opportunity to listen to 'Bibidha Bharati' and to see Sanjeev Kumar and the Lord Krishna face-to-face in the movie *This is Life*.

No. The area is not wild.

How many are left?

The answer is silence.

How many are left? Is there anyone *at all?*

The answer is long.

Item: *Well, action* still goes on. Moneylenders, landlords, grain brokers, anonymous brothel keepers, ex-informants are still terrified. The hungry and naked are still defiant and irrepressible. In some *pockets* the harvest workers are getting a *better wage.* Villages sympathetic to the fugitives are still silent and hostile. These events cause one to think...

Where in this picture does Dopdi Mejhen fit?

She must have connections with the fugitives. The cause for fear is elsewhere. The ones who remain have lived a long time in the primitive world of the forest. They keep company with the poor harvest workers and the tribals. They must have forgotten book-learning. Perhaps they are *orienting* their book-learning to the soil they live on and learning new combat and survival techniques. One can shoot and get rid of the ones whose only recourse is extrinsic book-learning and sincere intrinsic enthusiasm. Those who are working practically will not be exterminated so easily.

Therefore *Operation* Jharkhani *Forest* cannot stop.

Reason: the words of warning in the *Army Handbook.*

II

Catch Dopdi Mejhen. She will lead us to the others.

Dopdi was proceeding slowly, with some rice knotted into her belt. Mushai Tudu's wife had cooked her some. She does so occasionally. When the rice is cold, Dopdi knots it into her waistcloth and walks slowly. As she walked, she picked out and killed the lice in her hair. If she had some *kerosene,* she'd rub it into her scalp and get rid of her lice. Then she could wash her hair with baking soda. But the bastards put traps at every bend of the falls. If they smell *kerosene* in the water, they will follow the scent.

Dopdi!

She doesn't respond. She never responds when she hears her own name. She has seen in the Panchayat office just today the notice for the reward in her name. Mushai Tudu's wife had said, What are you looking at? Who is Dopdi Mejhen! Money if you give her up!

How much?

Two—hundred!

Oh God!

Mushai's wife said outside the office: A lot of preparation this time. A–ll new policemen.

Hm.

Don't come again.

Why?

Mushai's wife looked down. Tudu says that sahib has come again. If they catch you, the village, our huts...

They'll burn again.

Yes. And about Dukhiram.

The sahib knows?

Shomai and Budhna betrayed us.

Where are they?

Ran away by train.

Dopdi thought of something. Then said, Go home. I don't know what will happen, if they catch me, don't know me.

Can't you run away?

No. Tell me, how many times can I run away? What will they do if they catch me? They will *kounter* me. Let them.

Mushai's wife said, We have nowhere else to go.

Dopdi said softly, I won't tell anyone's name.

Dopdi knows, has learned by hearing so often and so long, how one can come to terms with torture. If mind and body give way under torture, Dopdi will bite off her tongue. That boy did it. They *kountered him.* When they *kounter* you, your hands are tied behind you. All your bones are crushed, your sex is a terrible wound. *Killed by police in an encounter... unknown male...age 22...* As she walked thinking these thoughts, Dopdi heard someone calling, Dopdi! She didn't respond. She doesn't respond if called by her own name. Here her name is Upi Mejhen. But who calls?

Spines of suspicion are always furled in her mind. Hearing 'Dopdi' they stiffen like a hedgehog's. Walking, she *unrolls the film* of known faces in her mind. Who? Not Shomra, Shomra is on the run. Shomai and Budhna are also on the run, for other reasons. Not Golok, he is in Bakuli. Is it someone from Bakuli? After Bakuli, her and Dulna's names were Upi Mejhen, Matang Majhi. Here no one but Mushai and his wife know their real names. Among the young gentlemen, not all of the previous *batches* knew.

That was a troubled time. Dopdi is confused when she thinks about it. *Operation* Bakuli in Bakuli. Surja Sahu arranged with Biddi babu to dig two tube wells and three wells within the compound of his two houses. No water anywhere, drought in Birbhum. Unlimited water at Surja Sahu's house, as clear as a crow's eye.

Get your water with *canal* tax, everything is burning.

What's my profit in increasing cultivation with tax money?

Everything's on fire.

Get out of here. I don't accept your Panchayat nonsense. Increase cultivation with water. You want half the paddy for sharecropping. Everyone is happy with free paddy. Then give me paddy at home, give me money, I've learned my lesson trying to do you good.

What good did you do?

Have I not given water to the village?

You've given it to your kin Bhagunal.

Don't you get water?

No. The untouchables don't get water.

The quarrel began there. In the drought, human patience catches easily. Satish and Jugal from the village and that young gentleman (was Rana his name?) said a landowning moneylender won't give a thing, put him down.

Surja Sahu's house was surrounded at night. Surja Sahu had brought out his gun. Surja was tied up with cow rope. His whitish eyeballs turned and turned, he was incontinent again and again. Dulna had said, I'll have the first blow, brothers. My great-grandfather took a bit of paddy from him, and I still give him free labour to repay that debt.

Dopdi had said, His mouth watered when he looked at me. I'll put out his eyes.

Surja Sahu. Then a *telegraphic message* from Shiuri. *Special train. Army.* The *jeep* didn't come up to Bakuli. *March-march-march.* The *crunch-crunch-crunch* of gravel under hobnailed boots. *Cordon up.* Commands on the *mike.* Jugal Mandal, Satish Mandal, Rana *alias* Prabir *alias* Dipak, Dulna Majhi-Dopdi Mejhen *surrender surrender surrender. No surrender surrender. Mow-mow-mow down the village.* Putt-putt-putt-putt—*cordite* in the air—putt-putt—*round the clock*—putt-putt. *Flame thrower.* Bakuli is burning. *More men and women, children…fire…fire. Close canal approach. Over-over-over by nightfall.* Dopdi and Dulna had crawled on their stomachs to safety.

They could not have reached Paltakuri after Bakuli. Bhupati and Tapa took them. Then it was decided that Dopdi and Dulna would work around the Jharkhani *belt*. Dulna had explained to Dopdi, Dear this is best! We won't get family and children this way. But who knows? Landowners and moneylenders and policemen might one day be wiped out!

Who called from the back today?

Dopdi kept walking. Villages and fields, bush and rock—*Public Works Department* markers—sound of running steps at the back. Only one person running. Jharkhani forest still about two miles away. Now she thinks of nothing but entering the forest. She must let them know that the *police* have set up *notices* for her again. Must tell them that that bastard sahib has appeared again. Must change *hideouts*. Also, the *plan* to do to Lakkhi Bera and Naran Bera what they did to Surja Sahu on account of the trouble over paying the field hands in Sandara must be cancelled. Shomai and Budhna knew everything. There was the *urgency* of great danger under Dopdi's ribs. Now she thought there was no shame as a Santhal in Shomai and Budhna's treachery. Dopdi's blood was the pure unadulterated black blood of Champabhumi. From Champa to Bakuli the rise and set of a million moons. The blood could have been contaminated; Dopdi felt proud of her forefathers. They stood guard over their women's blood in black armour. Shomai and Budhna are half-breeds. The fruits of war. Contributions to Radhabhumi by the American soldiers stationed at Shian-dange. Otherwise crow would eat crow's flesh before Santhal would betray Santhal.

Footsteps at her back. The steps keep a distance. Rice in her belt, tobacco leaves tucked at her waist. Arijit, Malini, Shamu, Mantu—none of them smokes or even drinks tea. Tobacco leaves and limestone powder. Best medicine for scorpion bite. Nothing must be given away.

Dopdi turned left. This way is the *camp*. Two miles. This is not the way to the forest. But Dopdi will not enter the forest with a cop at her back.

I swear by my life. By my life Dulna, by my life. Nothing must be told.

The footsteps turn left. Dopdi touches her waist. In her palm the comfort of a half-moon. A baby scythe. The smiths at Jharkhani are fine artisans. Such an edge we'll put on it Upi, a hundred Dukhirams—thank god Dopdi is not a gentleman. Actually, perhaps they have understood scythe, hatchet and knife best. They do their work in silence. The lights of the *camp* at a distance. Why is Dopdi going this way? Stop a bit, it turns again. Huh! I can tell where I am if I wander all night with my eyes shut. I won't go in the forest, I won't lose him that way. I won't outrun him. You fucking jackal of a cop, deadly afraid of death, you can't run around in the forest. I'd run you out of breath, throw you in a ditch, and finish you off.

Not a word must be said. Dopdi has seen the new *camp,* she has sat in the *bus station,* passed the time of day, smoked a bidi and found out how many *police convoys* had arrived, how many *radio vans.* Squash four, onions seven, peppers fifty, a straightforward account. This information cannot now be passed on. They will understand Dopdi Mejhen has been kountered. Then they'll run. Arijit's voice. If anyone is caught, the others must catch the *timing* and *change* their *hideout.* If *Comrade* Dopdi arrives late, we will not remain. There will be a sign of where we've gone. No *comrade* will let the others be destroyed for her own sake.

Arijit's voice. The gurgle of water. The direction of the next *hideout* will be indicated by the tip of the wooden arrowhead under the stone.

Dopdi likes and understands this. Dulna died, but, let me tell you, he didn't lose anyone else's life. Because this was not in our heads to begin with, one was kountered for the other's trouble. Now a much harsher rule, easy and clear. Dopdi returns—good; doesn't return—bad. *Change hideout.* The clue will be such that the *opposition* won't see it, won't understand even if they do.

Footsteps at her back. Dopdi turns again. These three and a half miles of land and rocky ground are the best way to enter the forest. Dopdi has left that way behind. A little level ground ahead. Then rocks again. The *army* could not have struck *camp* on such rocky terrain. This area is quiet enough. It's like a maze, every hump looks like every other. That's fine. Dopdi will lead the cop to the burning 'ghat'. Patitpaban of Saranda had been sacrificed in the name of Kali of the Burning Ghats.

Apprehend!

A lump of rock stands up. Another. Yet another. The elder Senanayak was at once triumphant and despondent. *If you want to destroy the enemy, become one.* He had done so. As long as six years ago he could anticipate their every move. He still can. Therefore he is elated. Since he has kept up with the literature, he has read *First Blood* and seen approval of his thought and work.

Dopdi couldn't trick him, he is unhappy about that. Two sorts of reasons. Six years ago he published an article about information storage in brain cells. He demonstrated in that piece that he supported this struggle from the point of view of the field hands. Dopdi is a field hand. *Veteran fighter. Search and destroy* Dopdi Mejhen is about to be *apprehended.* Will be *destroyed.* Regret.

Halt!

Dopdi stops short. The steps behind come around to the front. Under Dopdi's ribs the *canal* dam breaks. No hope. Surja Sahu's brother Rotoni Sahu.

The two lumps of rock come forward. Shomai and Budhna. They had not escaped by train.

Arijit's voice. Just as you must know when you've won, you must also acknowledge defeat and start the activities of the next *stage*.

Now Dopdi spreads her arms, raises her face to the sky, turns towards the forest, and ululates with the force of her entire being. Once, twice, three times. At the third burst the birds in the trees at the outskirts of the forest awake and flap their wings. The echo of the call travels far.

III

Draupadi Mejhen was *apprehended* at 6.53 p.m. It took an hour to get her to *camp*. Questioning took another hour exactly. No one touched her, and she was allowed to sit on a canvas camp stool. At 8.57 Senanayak's dinner hour approached, and saying, Make her. *Do the needful,* he disappeared.

Then a billion moons pass. A billion lunar years. Opening her eyes after a million light years, Draupadi, strangely enough, sees sky and moon. Slowly the bloodied nailheads shift from her brain. Trying to move, she feels her arms and legs still tied to four posts. Something sticky under her ass and waist. Her own blood. Only the gag has been removed. Incredible thirst. In case she says 'water' she catches her lower lip in her teeth. She senses that her vagina is bleeding. How many came to make her?

Shaming her, a tear trickles out of the corner of her eye. In the muddy moonlight she lowers her lightless eye, sees her breasts, and understands that, indeed, she's made up right. Her breasts are bitten raw, the nipples torn. How many? Four-five-six-seven—then Draupadi had passed out.

She turns her eyes and sees something white. Her own cloth. Nothing else. Suddenly she hopes against hope. Perhaps they have abandoned her. For the foxes to devour. But she hears the scrape of feet. She turns her head, the guard leans on his bayonet and leers at her. Draupadi closes her eyes. She doesn't have to wait long. Again the process of making her begins. Goes on. The moon vomits a bit of light and goes to sleep. Only the dark remains. A compelled spread-eagled still body. Active *pistons* of flesh rise and fall, rise and fall over it.

Then morning comes.

Then Draupadi Mejhen is brought to the tent and thrown on the straw. Her piece of cloth is thrown over her body.

Then, after *breakfast*, after reading the newspaper and sending the radio message 'Draupadi Mejhen apprehended,' etc., Draupadi Mejhen is ordered to be brought in.

Suddenly there is trouble.

Draupadi sits up as soon as she hears 'Move!' and asks, Where do you want me to go?

To the Burra sahib's tent.

Where is the tent?

Over there.

Draupadi fixes her red eyes on the tent. Says, Come, I'll go.

The guard pushes the water pot forward.

Draupadi stands up. She pours the water down on the ground. Tears her piece of cloth with her teeth. Seeing such strange behaviour, the guard says, She's gone crazy, and runs for orders. He can lead the prisoner out but doesn't know what to do if the prisoner behaves incomprehensibly. So he goes to ask his superior.

The commotion is as if the alarm had sounded in a prison. Senanayak walks out surprised and sees Draupadi, naked, walking towards him in the bright sunlight with her head high. The nervous guards trail behind.

What is this? He is about to cry, but stops.

Draupadi stands before him, naked. Thigh and pubic hair matted with dry blood. Two breasts, two wounds.

What is this? He is about to bark.

Draupadi comes closer. Stands with her hand on her hip, laughs and says, The object of your search, Dopdi Mejhen. You asked them to make me up, don't you want to see how they made me?

Where are her clothes?

Won't put them on, *Sir*. Tearing them.

Draupadi's black body comes even closer. Draupadi shakes with an indomitable laughter that Senanayak simply cannot understand. Her ravaged lips bleed as she begins laughing. Draupadi wipes the blood on her palm and says in a voice that is as terrifying, sky splitting and sharp as her ululation, What's the use of clothes? You can strip me, but how can you clothe me again? Are you a man?

She looks around and chooses the front of Senanayak's white bush shirt to spit a bloody gob at and says, There isn't a man here that I should be ashamed. I will not let you put my cloth on me. What more can you do? Come on, *kounter* me—come on, *kounter* me—?

Draupadi pushes Senanayak with her two mangled breasts, and for the first time Senanayak is afraid to stand before an unarmed target, terribly afraid.

COUNTLESS HITLERS

VIJAYDAN DETHA

Translated from the Rajasthani by Christi A. Merrill and Kailash Kabir

The five were only men. Some younger, some older, all between
thirty and fifty. The eldest was beginning to grey here and there,
but the others had heads of hair as black as bumblebees. They looked
like men: eyes where eyes should be, noses where noses should be,
teeth where teeth should be. Arms and legs where arms and legs
should be. Copper-coloured complexions. White turbans, some old,
some new. Cholas of white muslin, like their dhotis. Knotted gold
earrings in their ears. Gold pendants around their necks hung from
black cords. Each man spoke like a man. Each man walked like a man.

All were farmers. They worked the land and reaped the yields.
The dry womb of the earth turned green with their wheat and
fennel, mustard, cumin and fenugreek. After Independence, these
mighty farmers had done well. They cast seeds in the dirt with their
eyes closed, and then gathered up the fruits. The five looked as if
they had been born not of woman's flesh but from the earth's own
womb. As if they had grown up and blossomed among the kareel,
aak, khejari and acacia trees. As if the grass, the trees, the shrubs, the
flowers were their kin.

The five were brothers, cousins of near about the same stock.
They were going to Jodhpur to buy a tractor. Each had bundles of

rupee notes stashed in the undershirt pocket at his breast. The heat of it made their faces glow. The roots of wealth may lie deep in the heart, but the sheen of such invisible fruits shines clear for all to see.

They stepped off the bus with their hands in their pockets and headed off, their strides long and brisk, towards the tractor showroom as arranged. If it were in their power, they wouldn't have let their feet even touch that pavement black as rot. Once they reached the showroom, they recognized the owner through the window. As soon as their eyes fell on his shiny bald pate they cried, 'We're in luck! Om ji himself is here today.'

A blast of ice-cold air rushed over them as soon as they pulled open the door. They walked into the shop, and one sighed, 'Here he's enjoying heaven, while we toil like beasts of burden.'

Om ji smiled a thin smile and said in a delighted voice, 'If you want to exchange your farm for my shop, I wouldn't object.'

'Hah! You'd regret it!'

'That remains to be seen.'

The eldest cousin scolded them, 'We've only just walked in the door and already you're talking about regrets. Each person must follow his own fate and do the work that suits him best.'

Sitting on those cushiony chairs felt like sitting on nothing. They poked and prodded the soft cushions two, three times to make sure the seats would hold their weight. Satisfied, they settled into the chairs, elbows on the armrests. After the perfunctory duas and salaams one of the cousins began, 'Somehow or the other our number has finally come. We need to have the tractor today. We started out this morning at an auspicious hour. We need to return to our village before the day is done. We would consider it a favour if you could arrange for it somehow.'

'Every customer I meet makes the same demand. You have waited more than two years and now you cannot even wait two more days?'

The youngest cousin said, 'Two days would be too long. At this point we cannot wait another two hours. Our women have been standing at the doors ever since we left this morning watching for our return to bless the tractor. Charge a little extra if you have to, but you must deliver it today!'

Om ji smiled at their impatience, then said, 'I know how you rustics are. I made sure the tractor was ready yesterday. Take it whenever you wish.'

Their joy knew no bounds. It was as if they had suddenly been handed the whole world to rule! The middle cousin looked at Om ji's head shining like the moon and said, 'How could a man with such a lucky brow ever shirk his work? May you live long.'

The cousins were familiar with Om ji. One or the other would visit him from time to time to check their number on the waiting list. He became as friendly with them as business demanded. His manner was easy, his words pleasant. Every bit of him looked like it had been manufactured in a factory, like the parts of the tractor. There was a bald spot where a bald spot should be, fringed on three sides with thinning hair. A neck where a neck should be. A smile as the occasion required.

He scanned the five faces before him and said, 'You must be relieved. You've spent your whole day bouncing up and down inside a bus. Now sit back and relax, have some cold water,' and he reached for his buzzer as he continued to make polite conversation. A man came in at once. Om ji asked him to bring some lassis. When the man disappeared, he began apologizing, 'I will not be able to offer anything to rival what you have in your village. The milk here is water-thin. The curds will turn your stomach. All you get in cities is cooled air, icy water, soft cushions and bright lights. The grandiosity of the adulterated and the ostentation of the fake. You cannot find good grain and spices at any price. I am ashamed to offer you anything at all.'

One of the cousins laughed and said, 'If you really mean to offer, there are plenty of luxuries to be had around here. The envy of the gods above. Otherwise, we'll just have to cool down with a lassi instead.'

The hint was clear enough. Om ji laughed loudly and said, 'No, we cannot have any of that here in the store. But if you can wait till evening, I will be able to offer you real hospitality at my home.'

'Your invitation alone is enough, Om ji! Where's our tractor? Let's just take a quick peek.'

'First, have your lassis and then we'll go down and have a look.'

'The lassis aren't going to run away, are they? The sight of the tractor will cool us down. Then the lassis will taste sweeter.'

Om ji went with them himself. The tractor stood ready in the workshop. A blood-red Massey Ferguson, vivid as a mound of birbahuti bugs. The sight of it made them flush in their hearts. They patted the tractor and inspected it closely. Then they all went back to the office. Their glasses of lassi were sitting on the table, carefully covered.

Om ji eased himself back into his chair and began musing, 'How times have changed! There used to be just one thakur who ruled over the area. But now you big peasants have become the new thakurs. You are the ones who have really taken advantage of Independence. Where before people used to dream of having buttermilk, now they order all the luxuries as if they were water. In the old days people couldn't even afford a plough and a spade but now no one even gives a second thought to spending thousands of rupees on a tractor. Yaar, enjoy this independence, have as much fun as you can, don't even think twice.'

The fourth cousin interrupted him. 'I wouldn't call this khak fun! Nothing to eat but grain and you barely fill your belly. We've suffered for a thousand generations. Now the one-eyed lady puts on make-up and you begrudge her airs? Thanks to Gandhi baba we actually live like human beings now. How else would our villages' have got all those motors, tractors and radios?'

'And soon we'll have to fill our stomachs with paper notes. Before too long we won't even be able to buy grain.'

'You just keep giving us tractors and we'll keep giving you grain. Draw up a contract if you like.'

The eldest cousin spoke up. 'No one gives anything to anyone just like that. The water buffalo grazes only to fill its own belly. Everyone everywhere wracks his brain just to find a way to meet his own needs. One does it by selling a tractor, and another by buying it.' When his words reached his own ears the eldest cousin realized his talk had gone down the wrong path and he tried to steer the conversation back to better terrain by adding, 'Still, what you say is true. Due to Gandhi baba's grace, we're better off since Independence. Heaps of grain in every home, milk and curds flowing freely…'

Om ji began shaking his bald head and cut in, 'No, not in every home, that's not true. It's only a small number of you big farmers who have all you could want.'

The youngest cousin had been to college. He said, 'What do you mean *all* we could want? The best you can say is that the jaws of misery's grip have loosened a little. Just enough to give us room to breathe. But happiness is still as distant as the moon.'

Wanting to put an end to all this nonsense the middle cousin said, 'What's the use of wishing for the moon? Let's get down to business. Take the money out of your pockets to give it to Om ji so we can get our goods and return. We're wasting time talking.'

Suddenly they remembered why they had come. A moment later their hands were in their pockets, pulling out rupee notes, piling them on the table. A 50 horsepower foreign-built tractor with trolley, harrow and plough. A sixty-thousand-rupee transaction.

Om ji got busy counting the money and putting it away in his drawer while the five cousins all stood up at the same time and went down to the garage for their merchandise. The eldest cousin sent the youngest off to the bazaar for garlands, mounds of gur, rum, and bright red gulal powder. The four cousins helped to load the plough and harrow on to the trailer. They had just caught their breath when the youngest returned. They celebrated by passing around the gur and festooning the tractor's hood with marigold garlands. Then they painted a gleaming red swastika on the front of the hood in gulal. The youngest three were able drivers.

The day had passed quickly. The sun was just about to slip behind its western veil. From the Ajmer-Jodhpur toll gate the road looked clear, smooth and wide. The garlands fluttered in the breeze to the rhythm of the engine's roar. Sitting atop the tractor the five cousins felt as if heaven itself were gliding beneath their wheels. And the earth curving towards the horizon before them seemed punier than a coconut shell. As if the sinking sun had paused in the sky just to gaze at them. As if the thrumming wind were trying to sweep away any inauspiciousness. All the happiness in the world tossed inside their hearts. Even the long journey of the setting sun's rays seemed to be made worthwhile at the touch of the goddess sparkling in their pendants. The tractor's clanging sent birds hidden in roadside thickets and trees flying in all directions. But to the cousins, it was their own happiness taking wing.

Suddenly a shrill cry broke into their reverie. They looked around, startled. A hawk was swooping down, wings spread wide, on a baby hare it had spotted hiding in the brush. It seized the trembling body in its talons and soared upwards, back into the sky. The cousins smiled and looked at one another. The eldest observed, 'One's fate can never be postponed. It was destined that his death should take place in this very bush, by this very hawk, at this very moment.' They gazed into the sky until the hawk faded away. The tractor continued to roar along the road. They were approaching a small overpass. The fourth cousin urged the driver on, 'As much as we're hurrying, we're still running late. So far everything has been auspicious—there were good omens when we left the village.'

A steep slope lay just ahead. As they came over the crest they noticed a cyclist riding on the road, a few furlongs ahead. The cyclist heard the roar of the engine and turned to look behind him. A tractor coming. He turned back and began pedalling furiously. The men sitting in the tractor noticed him speed up, and watched as the gap between them widened. The youngest cousin was at the wheel. He muttered, 'Fool! Pedal as fast as you like, you'll never beat a tractor!' He gave the throttle a little tug, and it roared even louder.

The engine's roar rattled louder in the cyclist's ears. He pedalled faster, and the gap widened again. The driver couldn't stand to see the distance between them. He accelerated even more, saying, 'Little mother-lover! He'll tire out in the end, let him enjoy his little triumph while he can.' The middle cousin added, 'You never know what's going on inside the skulls of those bareheaded punks.'

The tractor was racing along by now. The garlands began flapping even more wildly. The eldest cousin agreed, 'Of course he'll wear out. Why bother speeding up? A poor cycle can't compete with a tractor!'

A piercing shriek struck their ears as a hawk swooped down from the sky and pounced on a mouse scurrying desperately to get to his hole underground. A moment later, the shrieks faded away. The sun was half-sunken. Now the sun would also disappear for the night. Scarlet light radiated from the setting sun, red as gulal, as if reflecting the tractor's red gleam. The brothers turned from the setting sun and looked at the road ahead. Arre! He was even further ahead! The same thought pinched everyone inside: a two-hundred-rupee cycle against a sixty-thousand-rupee tractor. No match! Does a mouse dare to wrestle an elephant?

The second cousin spat out, 'If he pumps those pedals till his lungs burst, it's his family he'll be leaving behind.' The fourth cousin said, 'Ram only knows when he'll leave his family behind; all I can see is that he's leaving our tractor in the dirt.' The youngest cousin eased out the throttle a little more. The tractor was brand-new. It wasn't good to race along at full throttle.

The cyclist looked back. He had quite a lead now. And his exhilaration made him pedal even faster. His feet were spinning round and round like reels. The cycle slipped down the road as easy as water down a mountainside. As if the cyclist had turned into a whirlwind, or even that he were riding a whirlwind.

All the eyes on the tractor were riveted on the cyclist. Quite a gap lay between them now. And it was only growing wider. A foreign tractor. Worth sixty thousand rupees. Festooned with marigold malas. And a two-paisa cycle! A college punk. Head bare. Wearing shorts.

A sharp gust of wind snapped one of the garland threads. The garland began to flap around. Doubling up, unfurling straight. Another garland snapped. The tractor driver felt every thump of the marigold garland on the hood like a thorny cane beating against his breast. He ground his teeth together and pulled the throttle out to the limit. The tractor catapulted forward like a shot from a cannon. The sound of the revving engine echoed in the air. The sky that moments ago seemed to be falling beneath their wheels now seemed to be rising higher and higher over them.

The gap began to close. Even more. Ah, now they were really close.

The world seemed as small as a coconut, reduced to two little dots. The tractor. The bicycle. A sixty-thousand-rupee machine. And a two-paisa piece of junk.

As it happened, two army trucks came bumping down the road from the other direction just at that moment and the tractor was forced to slow down. The cyclist saw his chance and clipped ahead.

The middle cousin said, 'These city punks are worthless! Taking advantage of a chance like that!' The eldest cousin said, 'If the poor fellow wants to show off for now, then let him. How long can he carry on like this? He's bound to run out of breath. Pagla, squandering his energies like this. Once his internal piping starts sagging, he won't even be able to do it with his woman. Were such drives meant to be spent on a cycle?'

Now that the road was clear the youngest cousin opened up the throttle. Like gunpowder suddenly touched with a spark. The tractor was like a dust storm trying to catch the wind. And gradually the gap began to diminish.

The cyclist heard the tractor just behind him and looked around. He snapped his head forward in a fury. And his feet began to spin like reels. They became speed itself, speed and nothing else.

Now he had begun to sweat. He was the fastest cyclist in all of Rajasthan. And, yes, he was also a man. Arms where arms should be, legs where legs should be. Breath where breath should be. Dreams where dreams should be. He had been working out on his bicycle, sixty or seventy miles a day for the past two months. If he came first in the All India Bicycle Championship next month, then he might get to go to Paris. He felt confident enough after two months of dedicated training. But today's little contest would prove it for certain. He clenched his teeth and poured all his strength into spinning the pedals.

He went to college with a young woman who had fallen in love with him the first time she saw him race and proposed to him. But he had not been able to reply with a forthright 'yes' or 'no'. They kept meeting and talking and spending time together, and once they had begun to know each other in their souls, it became clear what they had to do. He had promised to marry her as soon as the All India Championship was over. He had been raised in tight circumstances. And she had grown up in a house of plenty. But they lived only for one another. They ate as if with the same mouth. And on their priceless wedding night the moon would smile on their bridal bed.

Suddenly her face appeared before his eyes. As if she had turned into the breeze to watch the race. His vigour increased tenfold. As if his feet had grown wings. What power did that lifeless tractor have compared to the shimmery image of his beloved? The cyclist pulled further and further ahead. Before long, the distance between them had doubled.

Now the tractor was at full throttle. They could do no more. Their insides started writhing. The whistling wind was being swallowed up by the roar of the engine. Their reign over the whole world had been grabbed from their hands in a dash. The new tractor shot down the road like a cannonball. It looked as if a whirlwind had taken over that bareheaded boy's feet. His beloved's face shone before his eyes. The distance grew and grew. His lungs didn't quaver, and his breath didn't break.

Half of the marigold garlands had snapped and fallen. But what could the cousins do?

No one can see what the ephemeral future holds. Suddenly the feet fast as a whirlwind were spinning emptily. The chain had come off. Still the boy didn't worry. He figured his feet could match the tractor's speed. Images of his beloved's face surrounded him. There could be no greater power than this in the world. He stopped the cycle and quickly dismounted. He leaned the bicycle on the kickstand and patiently began putting the chain back on.

Slowly the distance was decreasing. The air could not contain the tractor's roar, nor the five cousins' happiness. Well, who knows when luck will smile on you? It didn't matter how, but this sixty-thousand-rupee matter of honour was saved. If people want to deceive themselves into believing in fraudulent victories, then who would stop them?

The tractor's roar sounded closer. It was taking much too long to get the chain back on in the flurry. Before long the tractor was right there. And still he had confidence in his strength, and the power of his beloved's face before him.

The tractor roared past. All five cousins shouted out words typically human as they sped by. A flock of crows began cawing overhead as if in one voice. The voices of the humans couldn't be heard over the cawing of the crows and the roar of the engine.

The tractor was already one or two farm-lengths ahead when the cyclist got the chain back in place and started off again. Four of the cousins turned back to watch him. They thought to themselves, the bastard was just pretending his chain came off! Maybe the race was too much for him.

But the chain was back on and he had turned into a tornado again. The distance between them slowly began to decrease as he came closer and closer.

The scenery was beginning to merge with the darkness. The four cousins were straining to see the boy behind them. He was gaining ground!

Now it was an all-out race. The tractor couldn't go any faster. They gnashed their teeth. The red of the tractor began to dissolve in the fading light. The youngest cousin asked, 'Where is that haraami now?'

The fourth cousin said through clenched teeth, 'Looks like he's going to pull ahead.'

'Hah! Even his father wouldn't have dreamed of it!' As he said this, the youngest cousin started to hear first the hawk's shrieks, then the mouse's squeals, echoing in his ears in turns. After a moment the shrieks were in one ear, the squeals in the other, and wouldn't stop. It seemed as if the entire universe were about to rip apart. The tractor's roar got swallowed up in that echo.

A whole different world was glittering in the eyes of the cyclist. Everywhere he looked, images of his beloved's face were twinkling—in the soft scattering of stars, in the trees and shrubs, in the sand dunes, in the tractor's trolley up ahead. Today would be the test. If he could get ahead of the tractor, then he would get married as soon as possible. Tomorrow, if she agreed. If not, then the day after. Or the day after that. Whenever she wanted. Why wait to pass them? All the world was in the palm of his hand. The warp and woof of golden dreams was being woven in front of his eyes.

Meanwhile, the hawk's shrieks and the mouse's squeals were smothering every particle of air. The four cousins shouted through clenched teeth, 'That bare-headed fellow is making us lick the dirt off our turbans!'

Then they came up with a new plan. 'Make the tractor swerve as soon as he gets close. What will the little haraami have to say to that…' The hawk's shrieks and the mouse's squeals had now found human voices.

And meanwhile the images of his beloved's face began growing brighter and brighter. Each image became more and more distinct.

Now he had moved up, beside the trolley. The shrieks and squeals hid themselves away in the driver's head and assumed a posture of silence.

The next moment the speeding cyclist crashed into the tractor. Lightning flashed before his eyes and the lights of his beloved's faces extinguished one by one. The tractor's rear tyre passed over his bare head, mashing it into chutney. The rest of the faces were snuffed out.

A human voice hissed once more in the wind, 'Mother-lover, he had nerve trying to overtake a tractor!'

The youngest cousin had been to college. He pulled the tractor over, grabbed a bottle out of a sack and said, 'Let's give the poor guy some rum!'

Then he went over to him, walking on two legs like a man. Opened the bottle above the cyclist. Emptied half the bottle of rum into the boy's mouth. Then he broke the bottle near the boy's head and ran back to the tractor. The tractor roared as he took off. The women must be standing in the doorway waiting for them. How happy they would be to see them return!

Human laughter echoed in the wind.

A picture was left behind them on the road, waiting for expert appraisal. Brain-white smudges on a blood-red background. Shards of broken glass. A man's dead body. White shorts. Bloodied sky-blue undershirt. Mashed dreams. Streams of love. The painting wasn't bad!

But…paintings of the two World Wars, pictures of Hiroshima and Nagasaki, of Vietnam, of Bangladesh…those are the true masterpieces. Compared to this one, those are so much more refined, so much more complex and nuanced. This one doesn't compare. Still, considering it was done by a band of rustics, it wasn't so bad.

Yes, the five were only men. Each man spoke like a man. Each man walked like a man.

seventeen

~

MIRROR OF ILLUSION

NIRMAL VERMA

Translated from the Hindi by Geeta Kapur

Layers of burning sand had settled on the tin roof. When the wind rose, a bright curtain of sand flapped around the house. The wartime barracks were being demolished, there were mounds of rubble everywhere and it looked like the dust road had developed lumpy warts.

One could see everything from the window. Coloured shadows slid across the hillocks all day long. In the distance one heard, continuously, the sound of the stone-breaking machine. Like a growling giant— grr, grr, grr.

Noises intruded upon the brittle edges of her afternoon sleep and Taran woke up with a start. She touched her forehead. Strands of hair were sticking to her brow, the powder of her bindi had trickled down to the bridge of her nose. I have been awake all this while, she thought: it seemed to her she had been thinking the same thing while she was asleep…it's always like that in the afternoon—you drift in and out of sleep.

She washed her eyes, wiped off her bindi, worked the pump and splashed her face with water. From the bathroom window she could look out over the fields where they were pulling down the barracks. Half demolished, they stood around like broken-down skeletons. The sand was ablaze. Taran could feel it crackling between her teeth.

'See if Babu is awake, Taran. If he is, take the hookah to him,' Bua called from the room adjoining the kitchen. Even at her age Bua remembered everything. It seemed that while she did her chores, or even sat dozing by the door, her attention was tugged about by Babu's every need.

As soon as evening fell the Diwan Sahib waited for his guests impatiently. Even if he went out to the small train station for a stroll he returned hurriedly in the hope of some unexpected guest and immediately asked Bua if someone had visited them in his absence. She never said more than 'yes' or 'no'. After so many years she still felt an obscure dread in his presence. Even as a child she had kept her head bowed before him. When she became a widow the Diwan had provided her with a small pension. Now, in her dotage, she lived in his house. When Taran's mother died the girl was left all alone. Had it not been for Bua she could never have endured a moment in this dismal place.

With sunset there was a regular little crowd on the Diwan's verandah. The government supervisor, Mr Das, the wealthy contractor, Meharchand, they all came after their day's work and sat together for a while. What was there to do, after all, in this wilderness; where could one enjoy an evening out? A few tribal mud-huts, a couple of shacks selling paan-bidi, some wayside eateries, and over there on the hill the temple of Kali Devi. All said and done there was just Diwan Sahib's house and these men from different parts flocked to it eagerly.

'Taran, will you fill the chillum for Das babu,' the Diwan called, turning his face to the door. Now his veneer of indifference vanished. Das babu had arrived, the others would be coming along soon. 'How is it you got so late today, the siren went off long ago,' he said.

Das babu's round, flabby body slumped into the chair. When he spoke his two rows of yellow teeth rattled faintly. 'I'd gone across the canal to see some land. On my way back I stopped at the new petrol pump. Now there won't be any petrol problem here Diwan Sahib.'

When Taran brought the chillum to him he shrank into himself. He was getting to be over fifty but he still felt awkward around a woman. Taran turned away and he relaxed, cleared his throat. Still, when he spoke his voice was strained: 'Why don't you make a trip to Haridwar, Rishikesh for a few days, Diwan Sahib? If nothing else it will be pleasant for your bitia. She's alone all day long, doesn't she get bored?'

Das babu would not call Taran by her name. Had she been a little younger he would not have felt this embarrassment. Had she been older the name could have been pronounced naturally. Between these two boundaries there was a certain awkwardness. Her youth seemed to have got stuck in the marsh somewhere…how was he supposed to regard her?

Diwan Sahib sat there, saying nothing in response. He enjoyed talking to his friends, to be sure, but in his heart he always held himself aloof from them. There was a line of privacy no one was allowed to cross.

Taran stopped short on hearing Das Babu's comment, then she went back inside. Bua was sitting there mending clothes. She avoided her and went into her own room. She shut the door and stood leaning against it for some time. The thick silence of the house pressed against her, the voices from the verandah were left floating out there, detached and unfamiliar.

From one of her windows she could see the verandah. If on an evening his friends did not turn up, Babu sat there all alone with his eyes half-closed. How terribly remote he looked in his stillness! She had often thought—I should go and sit with him, talk to him about this and that. After all there are only the two of us left; we can at least share each other's memories. But she had never taken the step. She just watched him from the window.

There was a strong wind those days. Gusts of hot dust came swirling up to the house, rattled the doors and then scattered in the courtyards. Far out in the distance they were dynamiting the boulders. Every now and then there was an explosion, then a vast echo. The earth vibrated underfoot. And one could see flags fluttering a warning all over the landscape.

Taran sat dozing by the window. Then she suddenly got up, startled, as though someone had come and touched her. The twilight glow had spread across the fields and crept quietly up to her.

Bua came into the room and, seeing her sitting in the dark, admonished her: 'How many times have I told you not to sit indoors in the evening? Why don't you go out for a bit, you can take Shambhu along.'

But just then someone climbed up the stairs. Taran looked out excitedly. Engineer babu had arrived. How strange he is, she thought, the way he comes up thumping his feet and shaking the whole house.

He had come to the township four or five months ago as a government architect, but everyone always called him Engineer babu. He had simple, easy ways, and he talked in such a friendly manner—one would have thought he had lived in the place for years. He was not one of the regular visitors to the house. It was difficult to say if he could be called Babu's friend at all. Babu was twice his age and felt a little awkward in his company.

Taran went to the mirror and quickly tidied her hair, combing it, waving it here and there. She powdered her face and her eyes fluttered. As she raised her hand to put on a bindi, below the parting of her hair, right in the middle of her forehead, she stopped and asked herself if this wasn't some delusion. No, she did not delude herself any more. She was not attractive: this hardly even depressed her now. Years ago if someone looked her up and down in the street her whole body would start to flush and quiver. She would run back to her room and gaze at herself in the mirror for hours. What is it people notice in me…this question would tantalize her and she would think up all sorts of answers, her heart racing. Now if anyone glanced at her she only felt a wry sort of surprise; and stepping out of herself she would regard her own body with careless inquisitiveness.

'You are here, all by yourself, at this hour?'

Taran's train of thought was interrupted. Engineer babu stood at the door.

'I was just coming out to the verandah, have you had some tea?'

'I'll have tea another time. Some day when you agree to sit with us out on the verandah. I have to hurry off now.'

But Taran insisted that he stay for a bit. 'Wait, you must have something to eat. You've just arrived,' and she started off for the kitchen.

'No, really, don't bother, please…I'm just on my way home from town and I've swallowed so much dust my stomach is all filled up.'

When he laughed, Taran always felt a little wistful. Everyone in town addressed him respectfully as Engineer babu, yet he looked so young, even younger than herself. The first time she saw him sitting with her father's friends she was astonished. He looked like a college student among all those venerable men.

'We haven't seen you in our neighbourhood recently. Montu often asks after you.' Montu was his little servant. Every time Taran went for a walk along the railway line he would come and greet her.

'I'll come one of these days, will you be there?'

'Come next week, I have a lot of work to get through in the next four or five days.'

Before leaving he stopped, took out his handkerchief and wiped his glasses. Taran raised her eyes and looked at him timorously, and her gaze hung on the spot even when he had moved away. What a man this Engineer babu is, she whispered to herself. When he pounds up the stairs the whole house trembles!

The dying sun had left a blood-red trail across the sky. It was beginning to get dark. She could just make out the labourers' shacks huddled in between the boulders, and further away, on top of the hill, the temple of Kali, wrapped in a cloud of smoke.

She left the window and came to her bed. A half-written letter lay under her pillow. Every time she began writing to her brother his face emerged and hovered above the grey mass of time. Not the face he had worn when he left home after clashing with Father. That was a distorted face, and even now her heart winced as she recalled it. No, not that face, but the one that had looked on, so young and so forlorn, when their mother died... She had looked up to him for consolation, not daring to cry before Babu, and he had whispered to her, 'Don't you see, Taran, how lonely Babu is now? We must stand by him. In a few days everything will be normal again.'

And did it ever become 'normal' again? She was too young then, she had not understood why her brother had left home, why Babu had not been able to stop him. Now she knew. Mother had been the only link between Babu and the rest of them. When she died they fell apart, drifted away.

Bua came to call her for dinner. She looked at the sheet of paper in Taran's hand and asked, 'Why, was there a letter?'

'I was writing to Bhai. His letter came yesterday.'

'He's asked you to come, hasn't he?'

'Yes, he's been asking me to come and spend some time with him. Bua, tell me, should I go?'

Bua looked at her in amazement—all the way to Assam? She could not imagine Taran going so far away, and all alone. 'Ah, but if he feels so lovingly towards his sister why hasn't he come and met her in all these years? He has a quarrel with the father, but should he cut you off as well?'

Bua suffered from asthma, she could not speak for too long. She brought up a few words and the others drowned in her rising breath. Taran looked at the tears in her eyes and couldn't tell whether they were for her estranged brother or were squeezed out by the fit of coughing.

'You go along, Bua, I'm just coming,' she said casually.

A dull silence filled her room. She could hear her father's footsteps going up and down the verandah. Outside there was pale moonlight and the mounds of rubble threw long, thin shadows across the empty fields.

A hazy image formed within her. Tea plantations...lush green sprawled all over, up and down the sloping valleys... Somewhere there hidden inside a grove of trees is my brother's house, she thought. They say you have to board a steamer to get there…wonder how it will feel to travel on a steamer!

Taran stood and looked at the green signal beside the railway station. The wheels of the train rumbled close by the house—the hillocks all around and the distant boulders reverberated with the sound. For a moment the frightful sound of the stone-breaking machine was drowned. The beam from the engine's headlight lit up dry shrubs along the tracks. Then again the same dense, heavy silence shrouded everything.

That night Bua came and sat in Taran's room. She kept chopping betel nut for herself. Every few minutes she would look at Taran and let out a long sigh, 'Are you asleep, Taran?' she would ask in a doubting tone.

'No, Bua, not yet.' Taran knew she was leading up to something. She lay there with her eyes closed, waiting.

'I went into your mother's room today,' Bua started slowly. 'I was so amazed, Taran. How much she had collected and preserved over the years. There was her wedding sari too, carefully folded and put away in the big trunk.'

Taran's heart stirred with curiosity. Mother too had been a bride once. How difficult it was to imagine that.

'At that time your Babu had just been appointed Diwan… The wedding was such a grand affair. In the Sikh court he was the only Hindu Diwan who had unbridled access to the king.'

An old irretrievable dream floated into Bua's eyes, the betel nut chopper lay still in her hand. 'One day the English Resident of the State came to meet him. Everyone in the locality came out and looked on in awe. But your Babu was so particular about his caste, as soon as the Englishman left, he bathed and got the servants to throw away all the dishes.'

Taran sat up in bed. How many times she had heard this story from Bua! Yet every time she heard it, she felt a vague thrill, as if each time she was being led anew into a strange, illusory world.

'Bua, you've known Babu in those days, were you always so afraid of him?'

'Ah, but who wasn't afraid of your father then? It's the same fear that continues to this day. Your mother was even more timid, she would just keep gazing up at him. The day he went to court, I peeped at him through a chink in the curtains. He used to wear a dazzling white churidar, a silk achkan and a superb pink turban…we couldn't take our eyes off him.'

Bua's hands were on the betel nut chopper, but her gaze was fixed far away on the past.

'At times I wonder if your Babu's insistence on pedigree and a noble family for your match is justified. After all it's no longer the same. What is our status now that we should demand these things? But who will ever tell him this?'

Her face clouded over with bafflement, she couldn't work out how something that was long past and altogether finished could still cling to them like a leech. There were no honours attached to the family now, their properties had been sold and all their possessions of value lost a long time ago. The house was the last remaining piece of their ancestral heritage. And of course, the tattered, dust-laden title of Diwan. You could wrap it around you, flap it about, but it was of no consequence. No one was likely to defer to it.

There was a lump in Bua's throat but a smile stuck to her face, as though she had forgotten to wipe it away. Taran did not like this part of Bua's ruminations. She liked to remember how her mother teased her, showing her what she was to take as her dowry. That used to fill her with a lovely sensation...not the thought of her wedding, not even the jewellery...it was just an unknown happiness that welled up in her—she would float in it all alone. Taran lay down again. She looked out of her window and at the signal, far out in the dark. Across the canal were the barracks. Night had sifted down upon them. Somewhere out there, she thought, in a narrow little room, Engineer babu lives. She closed her eyes.

For a moment she became oblivious of herself. She did not know whether Bua was still talking. A sweet weariness overcame her, a long-forgotten feeling welled up inside her. She was lying in a tub full of water, her body stretched out before her, soft and naked. Nothing had happened between then and now, whatever had been gained and lost with time was all present, drifting above the water's surface.

'Have you gone to sleep, Taran?' Bua asked. Taran was startled by the question, she couldn't tell if she was on this or the other side of sleep... shadows flitted over the water and she lay underneath in limpid silence.

Then there were days when the Diwan would not emerge from his room. The verandah would be deserted. The chairs would stand around, unused and covered with dust. Bua would go to Babu's door and come back. He would ask for his food to be sent in to him. If he came out and encountered Taran, he would refuse to look at her, or if he looked at her it would be as if he had trouble recognizing her. And he would turn away abruptly.

Taran had begun to understand why he avoided her for no reason, why there was tension in the house. At first this had baffled and enraged her. But that time was past. Now a dry apathy filled her heart...why doesn't he get rid of me, she wondered. Bua had urged him many times to approach people. Letters were written, matters seemed to be taking shape. Her photograph and horoscope were sent off. But somehow things always stopped midway. Taran couldn't understand why that happened.

Even now when Taran remembered that night her body shuddered...

Bua had come to her room just before midnight. She was awake, listening to Bua's footsteps coming towards her. She held her breath and kept lying on her bed.

Bua's voice was trembling. 'Did you hear anything?'

Taran had sat up in bed. 'What is it?'

'I can't stay in this house any longer.'

'What happened, Bua?'

'Is there anything left to say, Taran?' Her voice had choked.

Numbly, Taran had stared at the shadow in the dark that was Bua. 'Did he talk to you?'

'I was sitting in my room, he came himself... I ask you, if he has something to say why doesn't he say it direct to you? You aren't a baby anymore, why on earth does he drag *me* into it?'

'But what did he say?'

'I can't figure him out. He started off saying it would have been best if it had all happened in your mother's lifetime. I used the opportunity and said—who goes by the family's name these days? It's more than enough if one can find a decent fellow. As soon as he heard this he stiffened. Then he strode out of the room and slammed the door. After a while when he came back I could hardly recognize him. His hair was dishevelled and his eyes were bloodshot. That's how he looked when your mother died, as if he wouldn't ever forgive her... He had a bundle under his arm. He came and threw it in front of me and said—Here's her mother's jewellery, she can take it all and go wherever she pleases. When her brother left home I didn't perish, did I? If she leaves nothing will happen either... I was quite bewildered. Taran, how can he talk like this about his own daughter?'

That night Bua's question hovered about her in the dark. She didn't know what her father wanted of her. He had become suspicious of her—that she would run away from him one day. Like her brother. She had never considered the possibility, but that night her father's dread became her own. She felt restless and guilty. Did she really want to live on in this house? She asked herself the question again and again, and she realized that Babu was right. She had a horror of the house, the empty rooms. She had tried deluding herself but the question wouldn't stop nagging her—do you want to go on living here?

Such a simple question and yet she wouldn't dare answer it. She would curl up all night in bed, her face hidden in the pillow, quailing before her own dreams.

Then, one day, she suddenly decided she would visit her brother for a few days. But she could not bring herself to tell Babu. She went across to his room several times but always turned back.

Finally, she asked Bua to go and tell him. Bua gazed at her doubtfully but later when she had thought it over it seemed like a good idea to let her go away for a while.

That day Babu called her to his room. She went and stood hesitantly at his door. Her breath caught in her throat, she felt stifled.

'Come in and sit here,' Babu said in a dull voice. He sat resting on a pillow against the wall, very quiet and very still. A thought flashed in her mind—there is still time, I should turn and go back to my room, quietly, just as I came in, and it will all pass away... But her feet were suddenly glued to the floor.

'I believe you want to go away for a while?' Taran sat there, alert in her silence. She thought—he does not want to mention my brother, he never refers to him. Every time a letter came from him he would send it on to her without opening it.

'You don't feel happy here, Taran?' There was bleak curiosity in his voice, but also an innocence, as though he had thought of this possibility only for the first time.

Taran looked at him and her heart began to pound with the tension she felt. Maybe Babu will ask me to stay, she thought feverishly, maybe he too feels lonely without me. She began to tremble inwardly. If only he would ask me to stay, she thought, I would immediately drop the idea of going... just let him ask me...

But Babu didn't say a word. Taran lowered her eyes.

'It's all right, if you want to go you can. Don't worry about me,' he said finally. His voice was poised, unfeeling.

Going out of the room, she faltered a little. She had hoped he would call her back, say something else. But the room was filled with a terrible silence.

That afternoon she lay in her room, willing herself to sleep. She would have liked to be able to draw sleep to herself any moment of the day, like a shroud. This compulsion to keep awake, to look around with wide-open eyes, it seemed unnecessary and false. She had begun to wonder if all her waking moments all these years had not been some delusion...they had hovered about at sleep's edge and left her untouched.

In the evening Taran came out of her room. Her body dragged because of the day's lethargy. The wind had blown all day. The sky was lined with thick yellow layers of sand. Far off, the sun glittered on the rocks. Taran noticed Babu had not come out to the verandah, his door was still shut.

Bua sat in her own room, coughing and mumbling something to herself. Every time the breeze swept through the house all the doors rattled. Taran went back to her room quickly and put on her sandals. She told Bua she was going out for a walk. But she couldn't say if Bua had heard her at all; she could hear her gurgling cough as she descended the steps.

The sand fields stretched out to the far horizon. The sun had left a mottled layer of gold on the ground. Beside the new road, rubble heaps stood about like pyramids. Taran walked along these and reached the water tank.

She was very familiar with this barren landscape, had been witness to its changing contours ever since she was a child. During wartime, when the barracks were being erected, a whole lot of military trucks would drive around, raising clouds of dust. She was there then…now, when those giant boulders were being cut down to make a new road and the barracks stood half demolished, she was still around, watching everything from her window, all day long.

But soon she was going to be rid of all this, she thought with a little tremor. And when she saw Engineer babu coming down the road she broke into a smile. He came and stopped by the water tank. His shirtsleeves were rolled up and she could see his hairy forearms were coated with dust. There was sweat on his neck, all along the open shirt collar. His eyes were always alert behind his glasses, restless and yet extremely serious.

'How come you're standing here?'

'Oh, I'd come out for a stroll. It's so humid and still in the house, I couldn't bear it… I saw you from a distance, Engineer babu, though I could hardly recognize you in that sola topi.'

He laughed and Taran remembered the first time she had heard that laugh. She had hidden herself behind the door of the verandah where Babu entertained his friends and she had watched him, saying to herself—he is even younger than I am…

'Have you been to the town, Engineer babu?'

'No, how could I?' He recounted his problems with such style, one would have thought he quite relished them. 'How could I? The lorry has not been out for three days. I can't get there on my own, nor can Montu.'

'The lorry hasn't been? But who brings the food then? There isn't even a decent restaurant here.'

'Ah, but you don't know Montu,' he laughed boyishly. 'We are both lazy about going into town to shop, so he's discovered a nice dhaba here. He gets food for us both from there.'

Taran gazed at him in wonder. What a man this Engineer babu is, she thought. He's come all the way here, leaving behind his home and family. And there is no one here in this wilderness he can call his own, except his servant boy.

'Come, you'd set out for a stroll, let's go on.' They started walking on the rough, uneven path.

Every time the wind blew they had to swallow a mouthful of dust. In the Kali temple right above the labourers' hovels the evening lamp had been lit. A white curtain of sand fluttered around it, screening it from the last rays of the sun.

Engineer babu stopped suddenly. 'Do you see the boulders behind those huts?' His gaze rested on them. Taran looked at him curiously. 'After the road is made all those will be razed to the ground. The land around the railway line will be prepared for cultivation. On the far side of the canal factories will come up. You'll see everything change before your very eyes.'

His voice had grown eager. The glass of his spectacles glinted in the evening sun. He always related such exciting stories. And Taran gazed at him fascinated, thinking to herself—he looks just like a college boy, yet how much he knows! But one detail always made her laugh. He spoke to her so forcefully about his plans, almost as if he was afraid she was going to thwart them!

They stopped at the gate beside the railway line. He had fallen silent now, as if the gathering dusk had hushed him and left him suddenly forlorn. A slender moon lit the sky. The jagged cliffs which looked harsh in the daylight seemed to soften in the evening glow. They lost their aggressive bulk and huddled closer to each other.

'Engineer babu, have you ever been to Assam?'

'Assam, no, why, what's there?'

'Nothing, I just thought of it. My brother lives there. He's about your age.'

'Oh,' he said, a little uninterested, and she felt embarrassed. The wind had fallen, it was almost night. She had to turn back.

'Perhaps you should return home now. It's getting late. Shall I send Montu with you?'

'No, I'll manage, it's hardly any distance.'

He crossed the rail tracks and started walking across the vast bare fields. Taran turned round and kept looking after his receding figure. Then she started walking back and an inexpressible happiness rose inside her... All the worries that had been nagging at her seemed irrelevant, she could not explain to herself why she had been afraid for so long, what, after all, had made her so afraid? Look at Engineer babu, she said to herself, here he is, so far from home, how must he feel? Walking along the dirt road she felt as if all the staleness of the past years had been washed away from her. No, now I will never return to this house, she repeated to herself, I shall live my own life, I will not let anyone drag me back into this wasteland.

Climbing the steps of her house Taran noticed the verandah was deserted; the whole house stood in absolute silence. There was light only in the kitchen and it stretched out in a faded patch up to Babu's door.

The door was slightly ajar. Taran began to feel uneasy—was Babu sitting there all by himself in the dark? As she came up to the room, she pushed open the door a little further. Now, her hands were trembling. At first she couldn't see anything in the dark. Then, suddenly, she shrank back. Yes, Babu was there, walking about in the room as if in a delirium. She watched him. He stopped suddenly in the middle of the room, as if he was trying to capture a forgotten memory.

Then, as abruptly, he turned round and walked up to a photograph that had stood for years in the little alcove. She saw him pick it up and wipe off the dust with unsure hands... An image gleamed through the faded curtain of memories. The photograph had been taken at the silver jubilee celebrations of George V. Babu sat there, amidst other state officials, flanking the English Resident... He stood there with the image in his hand, hypnotized to see himself there, sitting with them in such glory.

Taran came in and stood by the door, petrified. This was the first time she had seen old age so nakedly. Babu's brittle grey hair, his bony hands with the blue veins sticking out, those innumerable wrinkles which made his face so pitiful...he had grown so very old before her eyes.

'Babu,' Taran's lips quivered. She approached him in the darkness. This was perhaps the first time she had dared to go so close to him.

He raised his head and looked at her. His eyes had glazed over. 'What do you want?' he asked tremulously.

She turned and left the room, then stood for a long time on the verandah outside. A certain fear began to crawl around her mind... Babu will never let me go, she thought, I will live here alone, bound to his shadow... Even when he is gone his wretchedness will cling to me all my life.

The moment of courage that had come and touched her that evening had fled already. Perhaps it had never belonged to her...no, it would not come to her again.

All night she could hear Bua coughing in her room. Sometime in the middle of the night Taran went to Babu's room and stood there, leaning against the door. She felt as if her mother had died again that night and that she was weeping the tears that had remained choked within her all those years.

Then she came back to her room and stood by the open window. There was moonlight over the barren fields. A few lights glimmered beyond the railway line. Engineer babu lives somewhere there, she thought. And suddenly she remembered what he had said to her: that in a few years everything would change. Was he right? She smiled now, a dry, wan smile.

She came away from the window and lay down on the bed. Her eyes were heavy but she could not sleep for a long time. Once she was swept by a light wave of sleep and dreamt her brother stood before her. That dear, familiar face with its lonely eyes... It was so long ago that she had seen him, would she know him now if she ever chanced upon him all of a sudden?

And then it all came floating before her eyes... Somewhere, very far away, hidden in the shade of a plantation, was her brother's house. They say you have to board a steamer to get there...wonder how it feels to travel on a steamer!

~

REFLOWERING

SUNDARA RAMASWAMY

Translated from the Tamil by S. Krishnan

Amma was lying on the cot and I was curled up on the floor. We were late risers and had earned that privilege after fighting for it for years as ours is a family that believes in rising early—for generations now, we've all bathed before sunrise. But then, Amma and I were invalids. Amma had asthma and I suffered from joint pain, both of which could play up early in the morning.

Outside, there were sounds of a horse shaking its mane, its bells jangling, which meant the horse buggy was ready. Appa would have picked up the bunch of keys for his shop. The clock would be inching towards eight thirty. Appa would now put on his slippers. Kweech. Kweech. Then, we would hear the abrupt sound of the umbrella opening, closing. The daily umbrella-health-test, that.

The door would open slightly and a thin streak of sunlight would prance into the room, a shifting glass pipe of light, dust swirling inside it.

Appa!

I see him in profile—one eye, spectacles, half a forehead streaked with vibhuti and a dot of chandanam paste, golden-yellow, topped by a vivid spot of red kumkumam.

'Ambi! Get up!' Appa said.

I closed my eyes. I did not move a limb and pretended to be fast asleep. 'Ai! Get up. You good-for-nothing fellow,' Amma said. 'Appa's calling.'

On the sly I looked at Appa. He looked affectionate, even gentle. I opened my eyes slowly and laboriously—as though I was waking up from deep sleep.

'Get ready, Ambi. Eat and then go to Aanaipaalam,' said Appa. 'Go and bring Rowther to the shop straightaway. I'll send the buggy back for you.'

I looked at Appa, then at Amma. I had told her about the squabble between Appa and Rowther in the shop the previous day.

'Can you or can you not manage without him?' asked Amma. 'This farce has gone on far too long,' she said. 'Making up one day and parting the next!'

Appa's face reddened. If it grew any redder, it seemed as though blood might start dribbling from the tip of his nose.

'Onam is round the corner. *You* can come to the shop and make the bills,' he shouted. Anger twisted his lips, slurred and flattened out the words.

'Is Rowther the only person in this whole world who knows how to make bills?' asked Amma.

'Shut your mouth!' yelled Appa. Abruptly he turned to me. 'Get up, you!' he ordered.

I sprang up from my bed and stood taut as a strung bow.

'Go. Do what I told you to,' he growled.

I got out of the room hastily. I heard the buggy leave.

I got ready in a hurry. I put on a veshti over my shorts, and a full-sleeved shirt, all in the hope that they would give me some confidence when dealing with Appa. I wasn't angry with him as I often was. Nor did I feel sad. I even felt a little affectionate. Poor man! He had got himself into a fix. On an impulse, he'd spoken harshly to Rowther. He could have been more calm. Now, if a person is merely short-tempered, one can talk of calmness. But if he is *anger personified?*

Impressed with my insight I went to Amma and said, 'If Appa is anger personified where is the question of calmness?' Amma laughed. Then she made her face stern and added, 'Smart, aren't you?' 'Now, if you are a clever boy, you'll take Rowther to the shop.' Placing her right hand over her heart, she said, 'Tell him whatever *he* may have said, I apologize for it.'

~

As I climbed into the waiting buggy I thought that we could not manage the Onam festival sales without Rowther. Who could do sums like him? His mental arithmetic was lightning-quick. Five people sitting in a row with paper and pencil could not equal one Rowther and his brain. Remarkable. Even regular buyers who flocked to him to have their bills tallied were amazed. 'Is this a mere human brain?' many wondered aloud. 'If the man can be this fast just by listening to the figures, what would he not do if he'd been granted sight?' And to think that Rowther had only studied up to the third class. That was two grades less than Gomathi who worked in the shop, fetching and cleaning.

The dispute between Appa and Rowther had started mildly enough the previous evening. 'Look here, Rowther, what are you going to do if you let your debts keep mounting like this?' Appa had asked. Appa's question had been provoked because Rowther had chosen clothes for his entire family, and piled them up by his side before thinking of asking Appa for credit. It was quite clear that Appa did not like this.

'What can I do, ayya? My house is full of women. My sons are useless. My sons-in-law are useless. Four sons, four daughters-in-law, eight granddaughters, eight grandsons. How many is that? If I gave them just one item of clothing apiece, even then the cost is high.'

You could see by the way Appa was staring at Rowther that he thought things were getting out of hand. Rowther would need to be sorted out immediately.

'Kolappan, wrap up the clothes and give me the bill,' said Rowther.

How dare he take the things before permission had been granted? Appa's face reddened. 'It is not possible for me to give you credit this time,' he said.

'So, you're saying you don't want our relationship to continue, no, ayya? All right. Girl, take me home.'

Rowther stood up. Gomathi took his right arm and placed it on her left shoulder. They went down the steps. When the shop closed in the evening, Rowther would usually look in the direction of my father and take permission to leave. This particular evening he did not take permission.

I thought I would first pick up Gomathi and take her with me to Rowther's house. That would perhaps lessen his hurt. But Gomathi was not at home. 'Rowther had sent word that he was not coming. She's just left for the shop,' her mother said.

I took a shortcut through the grove, and reached Rowther's house through a narrow lane. A tiled house with a low roof. In the front yard, there was a well on the right, its parapet wall stark, unpainted, broken. Velvet moss sprang around it in bright patches. Stone steps led to the house. A strip of gunny sack curtained the entrance.

'It's me, Ambi!' I said loudly.

A little girl came out followed by another who was obviously her twin.

'Who is it, child?' I heard Rowther yell from inside the house.

'It's me. Ambi,' I said.

'Come! Come,' said Rowther. He sounded happy.

I pushed aside the sack curtain and went inside. The floor had been swabbed smooth with cow-dung. Rowther was sitting cross-legged, like a lord. His arms reached out for me. 'Come, come,' his mouth kept saying.

I went and knelt in front of him. He put his arms around me. His eyes stared and stared, as if trying to recapture the vision they had lost long ago. He dragged me down beside him. His emotions seemed to overwhelm him.

'Ah! You seem to be wearing a veshti today!' he said.

'Just felt like it.'

'What's the border like?'

'Five-striped.'

'Just like ayya, huh? The boys in the shop tell me that you look just like your father, too. It is my misfortune that I can't see you.'

He ran his fingers over my face, my nose, my mouth, my neck, my eyes, my ears, my forehead. 'Everything in place, thank the Lord.' He laughed.

I thought that this was the right moment to tell him why I had come. But the words stuck in my throat, as if held there by an unseen hand.

'Amma...' I started to say, making a tentative start.

Rowther interrupted me. 'How is amma's health now?'

'As usual.'

'I have thuthuvalai kandankattri leghiyam. No better medicine for asthma. Only, ayya likes to see English labels on his medicine bottles. I don't have English here. Only medicine,' he said, enjoying his own joke hugely.

This was the right moment to tackle him.

'Amma wants me to take you to the shop. She wants me to tell you that *she* is very sorry if Appa has said anything to hurt you. You are not to misunderstand him. She says please don't turn down her request.'

Rowther's face brightened visibly. He raised his hands in salute. 'Amma, you are a great woman,' he called out. 'Get up, let's go to the shop at once,' he said.

That year the sales during Onam were very good. Rowther was in his element. With great élan he supervised the shop boys who scurried around him. He looked like Abhimanyu in the Mahabharata fighting a whole battalion single-handed. He would state the price of a piece of cloth as soon as the cost and quantity of the material were mentioned to him. Only the good Lord knew what spark fired his brain, what genius made him spit out an answer instantaneously. A brain that could multiply and add up the cost of sixteen different items in a trice to announce, 'Items sixteen. Grand total—one thousand four hundred fourteen rupees twenty-five paise.' How could that be called an average brain? Even if the whole thing were written down on a blackboard, I would have easily taken half an hour to work it out. But for him, answers slipped out like lightning. He had never until now made a single mistake. Amma told me that in the early years of their association, Appa would sit up half the night, checking Rowther's calculations. It seems he'd say, 'That man is getting carried away with his own prowess. I must find at least an error or two.' But he never could. He just lost a good night's sleep.

One day, a cart drawn by a single bullock, heavily curtained on both sides, stopped in front of the shop. From inside came the wailing of women and children.

'Sounds like the women from my household,' Rowther said.

Rowther's house had come up for public auction! Apparently the amina was taking all the household things and flinging them on to the street.

Rowther started crying like a child and called on God to help him out. Even as he wept, Kolappan came with a bill saying, 'Forty-five metres and seventy centimetres at thirteen rupees and forty-five paise.' Rowther stopped his keening for a moment and said to him, 'Write this down, six hundred fourteen rupees and sixty-six paise.' He turned to my father who was at the cash counter and sobbed. 'Ayya, I have to pay the court the loan and the interest on it, more than five thousand rupees. Where will I go for the money?'

Appa took Rowther in the buggy to see a lawyer.

Rowther did not show up for work the next day. The shop assistant Kolappan said he had, with his own eyes, seen Rowther reciting the bills in the Chettiar's cloth shop.

'What injustice! I have just come back after paying the court the entire amount for his debts. He's let me down, the ungrateful wretch!' Appa shouted.

Kolappan also whipped himself into a fury. 'He knows how to calculate, but he's an idiot. Wait, I'll go this minute and drag him here by his hair,' he said, and jumped on to his bicycle.

Appa sat down on the floor, devastated. He started to mumble. 'This is a wicked world,' he said. 'These days you can't even trust your own mother.'

In a little while, Kolappan returned. Rowther was sitting behind him, on the carrier. He marched a stone-like Rowther to the cash counter.

'I lost my head, ayya,' said Rowther as he stood before Appa, his hands folded in supplication.

'A time will come when you will be cut down to size,' said Appa.

'Please don't say such things, ayya,' pleaded Rowther, "Come work for me and I'll pay your debts," the Chettiar said. And I lost my head.'

Appa only repeated, 'The time will come when you will be cut down to size.'

To everyone's surprise, events began to unfold that made it look as if Appa was going to be right after all. When he returned from Bombay that year after seeing his wholesalers, he brought back a small machine and showed it to Amma. 'This can do calculations,' he said.

'A machine?'

'It can.'

Amma made up a sum. Appa pressed a few keys. The machine gave the answer.

I quickly worked it out on a piece of paper. 'The answer is correct, Amma!' I shouted.

'Have they transformed Rowther's brain into a machine?' Amma asked.

I kept trying out the calculator all day. That night, I kept it by my side when I slept. I gave it the most difficult sums I could think of. It answered correctly every time. I remembered something Gomathi had once told me. 'Thatha! How can you do sums in a nimit?' she had asked Rowther, mixing up, as she always did, the Tamil and the common English word. It seems Rowther had said, 'Child, I have three extra nerves in my brain.'

Now, how had those extra nerves got inside this machine? I couldn't control my excitement. I showed the calculator to Gomathi. She also worked out many many sums.

'Even I am getting it all right,' she said, 'This machine is more cunning than Thatha!'

One evening Rowther was totalling up the accounts for the day, Gomathi was sitting there, the calculator balanced on her lap, checking his calculations. At one point, very impulsively she said, 'You are correct, Thatha.'

'Are you telling me I am right?' asked Rowther.

'I have worked it out,' said Gomathi.

'Hmm,' said Rowther. 'I'll give you a sum. Answer.'

Rowther gave her a sum. Gomathi gave the right answer. He tried sum after sum on her. She had the correct answer each time. Rowther turned pale. 'Dear God. I am so dumb I cannot understand anything,' he muttered.

'I'm not doing the sums, Thatha,' said Gomathi. 'It's the machine.'

She stuffed the calculator into his hands.

Rowther's hands shook as he took the calculator. His fingers trembled. He touched the front of the calculator, then the back.

'Is *this* doing the sums?' he asked again.

'Yes,' said Gomathi.

'You keep it,' he said and thrust the calculator back at her.

After this, Rowther grew very quiet. He went into a sort of trance, leaning against the wall, and would not say anything. That day, Gomathi and I took care of all the billing. After a long time, Gomathi dug her finger into his thigh and asked, 'Thatha, why don't you say something, Thatha?' He did not respond.

He continued to come to the shop regularly but after that looked and acted like a walking corpse. It seemed as if all the laughter, happiness, repartee, teasing, sarcasm, had been drained from him. His voice was slow, hesitant. Even his body looked thinner.

Appa had stopped asking him to do the bills.

One afternoon, it was a busy time in the shop. Murugan had a pile of cut pieces with him. I was working out the cost. Suddenly, Rowther interrupted him, 'What did you say was the price of poplin?'

Murugan stopped calling out and looked at Rowther's face, 'Fifteen rupees and ten paise per metre.'

'Wrong. Get the material out and look—it is sixteen rupees and ten paise per metre.'

Appa came over to Rowther.

Murugan's face fell as he checked the price. 'You are right,' he mumbled.

'You have sold ten metres. You could have lost ten rupees. Are you here to give away ayya's money to everyone who comes in from the street?'

'So, you know the price?' Appa asked Rowther.

'Only a memory, ayya.'

'Do you remember all the prices?'

'It is God's will,' said Rowther.

'What is the price of the smallest towel then?' asked Appa.

'Four rupees and ten paise.'

'And the biggest one?'

'Thirty-six rupees and forty paise.'

Appa kept asking. The answers kept coming.

Appa looked amazed. He said, 'From now onwards, when the bills are being made, please check the prices.'

'I will do my best, ayya,' said Rowther. Then he looked up and said, 'Oh, by the way, have you paid your electricity bill, ayya? Today is the last date for payment.'

'Oh, no!' said Appa, calling out to Kolappan.

Rowther said, 'He hasn't come today, ayya.'

'How do you know?' asked Appa.

'Everybody has a voice, a smell. Today I missed Kolappan's voice, his smell,' said Rowther, and then he complained about Murugan.

'Yesterday, he told a customer that we had no double veshtis. Please reprimand him.'

'I don't understand,' said Appa.

'Ayya, you put out ten double veshtis for sale. Weren't only seven sold? There should be three left.'

Appa asked for the veshtis to be brought.

Sure enough, there were three unsold.

A sardonic smile played on Rowther's face. He said to Murugan, 'Oh Lord Muruga, you merrily send customers away by telling them we don't have what we do actually have. Are we here for business or for charity?'

That evening Rowther moved away from the bill-making section and went and sat down beside Appa.

'If I am by your side I will be more helpful, ayya,' he said and, without missing a beat, 'and if you increase the speed of the fan a little, I will also get some breeze.'

Appa gave the appropriate order.

'It is time to pay your advance income tax, ayya. Shouldn't you see your auditor?' asked Rowther.

'Yes, I must go see him,' said Appa.

It was time to close the shop.

'Ayya, you had wanted to get some medicine for amma. Have you bought it, yet?'

'I'll buy it.'

Appa was tugging at the locks to check if they had been locked properly.

'Ayya, you were saying that your mother's tithi was due soon. Why not ask Murugan to notify the priest on his way home?'

'Good idea,' said Appa.

The employees left one by one.

Gomathi took Rowther's hand, placed it on her shoulder and started moving.

'Won't you be doing the bills any more, Thatha?'

'Ibrahim Hassan Rowther is no longer a mere adding machine. He is now the manager. It is God's will,' Rowther replied.

MOUNI

U. R. ANANTHAMURTHY

Translated from the Kannada by H. Y. Sharada Prasad

Bhavikere Kuppanna Bhatta and Sebinakere Appanna Bhatta are like the cobra and the mongoose. Theirs is an old enmity, its origin lost to memory. Their houses stand on two hillocks half a mile apart. A fence in the valley separates their estates. Both of them are tenants of Narasimha, the deity of the Sri Matha. Years ago both of them had come here to work on the areca plantations, with just a copper tumbler each and not a coin between them. Now both have accounts in Gopala Kamti's grocery shop six miles away.

Kuppanna Bhatta is the older of the two. His face looks like a sour dried mango, his fifty-odd years etched on it. There was a time, in fact till just two years ago, when he was known for his sharp tongue—one lash from it was enough to split anything. His short, hairy frame has been shrinking with each passing day, but the way he carries himself has not changed at all. The same shining bald head, the same pudgy nose, that curved jaw heightening the resemblance to a mango. But the small eyes, fiery once, are now milky, and look as if they are covered with a layer of ash. The banana stains on his dhoti and the dhotra thrown over his shoulder, like his debts, have stayed with him through the years. He still keeps a wad of tobacco tucked into one cheek. Earlier, when angry, he would wait with drawn lips until his adversary stopped speaking, stride to a corner of the courtyard, spit out the tobacco, and return to utter something which was like a tight slap. The other man would be left speechless. And Kuppanna Bhatta would mix another little pellet of tobacco with lime in the palm of his hand and, transferring it into his mouth, lapse into a long silence. Nowadays, his face is like a locked door. If anyone tries to pick a quarrel with him, the closed expression on his face seems to say, 'You and I have nothing to do with each other.'

Appanna Bhatta, on the other hand, has always been popular. He is a man who gets along with everyone. Words slide off his tongue as smoothly as a strand of hair drawn through freshly-churned butter. Even the Brahmins of Buklapura, the agrahara two miles away, are his friends. More so because Kuppanna Bhatta has antagonized them. Appanna Bhatta keeps a good house and his estate is pleasing to the eye. There are chrysanthemums at his doorstep, a canopy of coconut fronds shades his courtyard. Coloured mats are spread out for visitors and no one leaves his house without being offered a cup of coffee. A far cry from Kuppanna Bhatta's courtyard. Cowdung here, children's shit there, garbage and flies everywhere, and leeches which latch on to the soft flesh between the toes. Appanna Bhatta often says, 'Even the hut of Koraga the sweeper is better kept than this.'

Kuppanna Bhatta's arrears with Gopala Kamti have mounted from year to year. He must owe Kamti more than a thousand rupees. Everyone knows that a messenger from Gopala Kamti goes to Kuppanna Bhatta every few days to demand payment. Appanna Bhatta is very different. At the end of each year he clears his account to the last rupee. He is equally prompt in paying his dues to the matha from which both Kuppanna Bhatta and he have leased their areca estates. Not a single nut does he withhold. Not only that, the agent of the matha who comes to collect the dues is shown the utmost respect, and served a special meal with payasam.

As for Kuppanna Bhatta, he owes the matha some five thousand rupees. Every year he is short ten or twelve maunds of areca nut.

The matha authorities have warned him to either pay up or surrender his tenancy. The agent is also worried about Kuppanna Bhatta's neglect of the plantation. He has often said that the only way to save the areca palms is to hand the land over to Appanna Bhatta.

Before the Swamiji of the Sri Matha went to Kashi, Appanna Bhatta had invited him over to his house. Humbly, he had made an offering of fifty silver rupee coins on a silver plate and prostrated himself at the holy feet.

Even his own brother-in-law believed that Kuppanna Bhatta was to blame for his unpopularity. Little wonder then that the Brahmins of Buklapura nodded when Appanna Bhatta said, 'I can understand his resenting my shadow. But what wrong have you people done to him? Could he not have performed the upanayana of his oldest son here and invited you all to the feast? Did he have to go all the way to Agumbe by bus and spend twice as much money in the temple there?' Then he continued in a soft voice, winking, 'One should not stretch one's feet beyond one's bed, our elders have said. But, believe my words, this man has got some concealed wealth. Otherwise, how could he have afforded such a lavish show?'

—

Why does he sit so listlessly, wondered Kuppanna Bhatta's wife, Gowramma, who is confined by asthma to a corner of the hall. Even after the evening birds have fallen silent, the cows have come home and been milked, and his daughter Bhagirathi has lit the lamp in the puja room, he has to be reminded three or four times to sit down to his evening prayers. In the past he had never allowed the children to light the kerosene lantern for fear that they would break the glass. Now the older son wipes the glass, lights the wick and places the lantern on the jagali.

Gowramma remembers...two years ago, her brother Subrahmanya had come to visit her. Hardly anyone comes to their house now. But never mind. At least one of her brothers is now a prosperous farmer, earning enough to live a decent life. That makes her happy. But why wasn't her husband able to do this? Appanna Bhatta too had come here in search of his fortune. He had done well and was now even able to invite the Swamiji of the matha to his house! After the meal was over, Swamiji had sent a boy to fetch Kuppanna Bhatta. But how much should her husband tolerate? How could he step into his adversary's house, over a threshold he had not crossed for thirty years? How would it have been possible for him to stand there and be humiliated by Swamiji in front of Appanna Bhatta? So he had done something he had never done before in his life. He had taken refuge in the kitchen and said to Bhagirathi, 'Tell the boy I am not at home.' Naturally, Appanna Bhatta added his own twist to the story when he narrated this to Swamiji.

It is all one's fate. Otherwise, why would he be tied down to someone like her, an invalid, racked by asthma twenty days out of thirty? Across the fence, Appanna Bhatta's palms are laden with bunches of areca nut. Here, in her husband's plot, the nuts fall off the diseased trees. No labourer stays with him. They all go to the other estate. He has antagonized everyone. There is not a single two-legged creature in the entire region he has not quarrelled with. But he has never taken advantage of people. He does not know what deceit is and has never told a lie to save his skin. But what a temper! It is all in his stars. Just as it is in mine to suffer before I die. Fate. That's all.

What was I thinking about? Ah, Subrahmanya's visit. He had not been patronizing. In fact he had shown great deference. 'Bhavayya,' he had said, 'you have no friends here. I hear the matha people too are asking you to hand over the land. Bhagirathi has come of age. How long can you keep a girl who has shot up like a banana plant in the house? When you approached those Tirthahalli people, Appanna Bhatta came in the way. He is always sharpening his knife... I have no right to advise an elder. But frankly, this plantation does not seem to suit your horoscope. Give it up and come away with me. You know that I have some five hundred areca palms in Halasoor. You can take care of as many as you wish to. I shall look after the rest.'

No matter how humble and hesitant Subrahmanya had been when he had said all this, Kuppanna Bhatta, like Parasurama, had flown into a towering rage. 'Wait till I am dead. After that you can look after your sister and her children. Until then, mind your business,' he had said harshly, dismissing Subrahmanya.

That was two years ago. Where was all that anger now? Since the bill collector saab from Kamti's shop and that rude man from the matha had begun to pester him, he had shrunk to half his size.

Then one day, he who had never asked for anything, came to his wife and said, 'Will you give me your ornaments? I will return them to you in six months.' The note of supplication in his voice drove a dagger through her heart. All she thought of was, let my husband and my taali, the symbol of my marriage, be safe. She opened her box, and took out the jewels which had come to her at her wedding and which she had saved for Bhagirathi's marriage—earrings, a four-strand necklace, a chain, a gold belt, an ornament for the hair. She gave them to him, taking care not to let him see the tears in her eyes. She made only one plea to God: 'Protect my husband, and let Bhagirathi get married into a good home.' She kept telling herself, 'What do I need these ornaments for? His self-respect is more important.' But who would look at a girl who had no gold? She had kept the ornaments for only one reason. Her own disease-racked frame had no use for them. Tormented by the change in her husband's manner she waited anxiously for Subrahmanya to visit them again. She could confide in him and lighten her grief.

Bhagirathi was in the kitchen trying to coax some flames from the damp firewood. Her mother could hear her coughing from the smoke in her lungs.

A few minutes later, Bhagirathi came out, wiping the tears streaming down her cheeks, her hair tousled. 'Amma, there is nothing to make hooli with,' she said. Gowramma told her, 'Then make cucumber-seed saaru.'

Why couldn't her husband go to Halasoor? How long could he carry on here? Appanna Bhatta has been pursuing him like a Yamadootha. If he can build a bund to prevent his plantation from flooding, can't he build a fence to keep other people's cattle from straying in? Or at least chase them away? Why did he have them sent to the pound four miles away? Her husband had waited and waited for the cattle to come home, and when they didn't, he trudged all the way to the pound, paid the fine and had them released. Appanna Bhatta surely has his eye on their plantation.

Ganapa, who was five, and whose belly was swollen from disease, went weeping to his mother. 'Give this child some of the water you have washed the rice in,' she told Bhagirathi.

No one in the neighbourhood cares whether we are alive or dead. Not once has anybody given my daughter a string of flowers or asked her how she was. Suddenly, the memory of the jewellery brought forth all the tears she had held back when she had given it to her husband. A fit of coughing followed. Wheezing, gasping for every breath, she stared at the darkness in the house around her, too weak to move from her corner.

Kuppanna Bhatta had decided that if his wife demurred even a little he would not touch the jewels. But she had not said a word. She had just opened the box and given him the small bundle wrapped in silk and he was overwhelmed. 'This is my last chance to get rid of my debts,' he had thought as he pledged them in the bank for two thousand and five hundred rupees. He vowed to redeem the jewels in six months and restore them to his wife.

Then Kuppanna Bhatta sought out the smaller cultivators, those who had only a hundred or two hundred areca palms, and advanced them twenty-five rupees a maund, with the promise that he would pay another twenty-five after the sale. Even Appanna Bhatta had never gone beyond forty-seven or forty-eight rupees a maund, and had never paid more than fifteen in advance. So the smaller growers agreed to go along with Kuppanna Bhatta, who was then able to transport a hundred maunds to the wholesale mandi at Shimoga. 'When the price goes up a little, I will come and conclude the sale. Let me know how the market is,' he told the traders.

All this infuriated Appanna Bhatta. His regular clients had been snatched away from him. He told the whole town that here was clear proof that his rival had come into money. Kuppanna Bhatta had failed to pay the rent, but had enough for speculating on the market.

Meanwhile Kuppanna Bhatta made a careful calculation of his liabilities. He scrutinized all the figures he had noted in the margins of the panchanga. He owed five thousand rupees to the matha, one thousand to Gopala Kamti, and three hundred for the medicines for his wife's treatment. There was the interest on the sum borrowed from the bank. And the money he needed for Bhagirathi's marriage. The first thing he had to do was to get the jewels back. He could pledge them again the following year.

Areca nut should sell at sixty this year. That would give him a clear thousand by way of profit. He would pay two hundred rupees to Kamti, fifty to the doctor, two hundred and fifty to the matha agent. He would get the gold back with the remaining amount and use it as his reserve capital. A few more such careful transactions and he would be a free man. He took the cowries from his bag, laid them out and looked for the omens.

A fly flew up from the manure-pit and sat on his nose. Mango-gnats and mosquitoes hovered around his face. As it was summer, the courtyard was dry. Every whiff of breeze brought with it the smells of the cowshed. A withered jackfruit tree stood in one corner. The child with the malarial belly was defecating a few feet away, a black cur near his feet. A grey dog dozed on the ash-heap.

Kuppanna Bhatta rose, scratching the spot on his foot where a mosquito had bitten him. He went out and came back with a twig of tube cactus which he tied to a beam in the roof to attract mango-gnats. Just then Bhagirathi came out with a pot of water and asked Ganapa, 'Have you finished?' The black cur and the grey mongrel rushed to the spot vacated by Ganapa and Kuppanna Bhatta, disturbed by their barking, shouted, 'Hacha, hacha,' to chase them off.

From a corner of the room came the sound of wood being sawed. It was Gowramma's wheezing. 'Give her some medicine, Bhagirathi,' Kuppanna Bhatta told his daughter and began pacing up and down the jagali.

The tiles will have to be re-laid before the monsoon. Or the damp walls will crawl with hairy caterpillars. There are a hundred holes in the floor where the water has dripped from the roof. It is difficult to find a dry spot for the children to sleep. And there is mildew everywhere—on the mat, in the rice container, even under the box which contains the family deities. It does not grow on people's buttocks, yet. That's all.

The flies covered the octopus-like branches of the cactus, making its green turn black in no time. Kuppanna Bhatta stood there, looking out in front of him. There was nothing in the landscape to hold his attention. Paddy fields

in front of the house and then the jungle. A hill in the distance. A cart-track
cut across the jungle from left to right. Someone seemed to be walking along it.

'Who is it?' shouted Kuppanna Bhatta.

'It is me, Manja,' said the man who was carrying a large load of dried
branches, without slackening his pace.

'Wait a moment. When will you come to our place to split the firewood?'

'After I have finished the work for Sebinakere ayya,' replied Manja, moving
away.

Feeling utterly helpless, Kuppanna Bhatta picked up a knife and went into
the garden to bring in some banana leaves. By the time he returned, the three
older boys had come back from school, which was four miles away. Bhagirathi
was waiting for the cattle. The sound of wood being sawn continued to be
heard from Gowramma's corner. The medicine had had no effect. Ganapa
sat in the corner, stuffing his mouth with dry avalakki and pieces of jaggery.

Areca nut that season went up to fifty-one, fifty-two, and, in a month,
to fifty-five rupees a maund. Kuppanna Bhatta waited for it to touch sixty.

'Your money has gone down the drain,' Appanna Bhatta mocked those who
had sold their stock to his adversary. They in turn gave Kuppanna Bhatta a
hard time. Every day, ten to twelve people came to his doorstep and demanded
immediate payment. Making excuses and holding them off exhausted Kuppanna
Bhatta. With abject humility, he pleaded with them to wait for a month more.
But they would not listen. So he was compelled to go to Shimoga and raise
a loan of two thousand and five hundred rupees at 15 per cent interest on
his stock. He paid his suppliers ten rupees per maund and promised to pay
the remaining fifteen in a month's time.

Seeing all these transactions being made by his rival, Appanna Bhatta
instigated Gopala Kamti and the agent of the matha to demand a settlement
of their dues. 'This is the right time for it,' he whispered to them.

Gopala Kamti's bill collector turned up one day at Kuppanna Bhatta's house
and planted himself on the jagali. No amount of pleading could move even
a hair on his moustache. He shouted at the top of his voice that he would
not stir until he had been paid at least five hundred rupees.

If Kuppanna Bhatta had been the sort of person to weep he would have
cried. As the bill collector's voice rose, Gowramma's breathing became more
agonized. Like a man in the grip of a nightmare, Kuppanna Bhatta seemed
incapable of any movement. Whenever someone passed by the estate, the bill
collector's voice became louder.

At last, Kuppanna Bhatta went in, opened his box and brought out four one-hundred rupee notes. The bill collector took them, mounted his bicycle, and, still muttering, rode off. On the way back, he met Appanna Bhatta, reported to him what had happened and was treated to a tumbler of coffee.

Hardly had the bill collector left when the doctor's compounder presented himself. His peace was purchased with a hundred rupees. There were still a thousand rupees left. The loan would have to be repaid now from out of the profit. It was all right, Kuppanna Bhatta reassured himself.

But early next morning, the agent of the matha turned up. Kuppanna Bhatta laid out a seat for him and asked him if he would have some coffee. The agent refused. 'I have come to you even though there is a swelling on my leg. Anybody else would have treated you roughly. You do not show your face at the matha on your own. Even when Swamiji sent for you from Appanna Bhatta's house, you did not come. He who deceives the Sri Matha Narasimha, will not prosper, Bhatt-re. You had better settle your debts,' continued the agent, as he sat there fanning himself. He had come to know that Gopala Kamti had received four hundred rupees.

The more Kuppanna Bhatta pleaded, the angrier the agent became. 'I will bring a search warrant and a notice of attachment tomorrow,' he said, preparing to leave.

Kuppanna Bhatta resigned himself to the inevitable. He counted out eight hundred rupees and handed them over to the agent. 'You must pay the remaining amount by the next date or else you will have to give up the plantation,' he warned before he left. And he hurried off to Appanna Bhatta's house for his afternoon meal.

The next day some of those who had sold areca nut to Kuppanna Bhatta besieged him. 'We are in financial difficulties, Bhatt-re. The rainy season is coming. We need to buy things for the house,' they said. The whole world seemed to know that Kuppanna Bhatta had some money!

The price of areca nut did not rise beyond fifty-five. In fact it tumbled to fifty. In dismay Kuppanna Bhatta hastened to Shimoga by bus and sold off the stock. He paid back the short-term loan to the last rupee, including the interest. He also paid his suppliers in full. Only the jewels remained in the bank.

Just before the rains set in, Subrahmanya turned up in his bullock cart and implored Kuppanna Bhatta to go with him. 'Come with me, Bhavayya,' he urged. 'I will find a groom for Bhagirathi. Gowramma's health has deteriorated and Ganapa is as thin as a stick.'

Turning to Gowramma, he said, 'Akka, my wife is pregnant. If you and Bhagirathi come and stay with us for a few months, you will do me a favour.

There is no one to take care of her. Bhagirathi will be a great help. I will look for a boy for her. Don't listen to your husband. Come with me to Halasoor... Why did you give him all your gold, Akka? He is a selfish, stubborn man. He cares only about his pride, his self-respect. His wife and family mean nothing to him. Why has he not done well like the other areca growers, tell me? I just don't understand any of this.'

Gowramma replied, 'How can I leave him in this condition...'

'I am asking you to come only for a few months, Akka. Till the confinement,' Subrahmanya persisted.

Kuppanna Bhatta did not say yes, nor did he say no. 'It's her wish,' he told Subrahmanya and left it at that. He refused to go with them. 'There is no one to look after the plantation,' he said.

Gowramma was torn between anxiety for her husband and the pull of the home of her birth. But Bhagirathi's marriage had to be thought of. And without the gold to give as dowry, finding a husband for her had become more difficult.

Kuppanna Bhatta spent the whole of the monsoon alone, cooking his own kanji. Without his wife and her wheezing, the house seemed to be swallowed up by silence. As the rain poured down, Kuppanna Bhatta spent many nights awake, his eyes wide open. And he came to a decision. He would get the gold back from the bank. He did not want to be obliged to anyone.

~~

Where Kuppanna Bhatta had expected a hundred maunds of nuts from his own plantation, the yield that year was only fifty. Swamiji was away in Kashi at that time. Kuppanna Bhatta wrote a letter, and with great humility, asked for leniency just this once. He undertook to hand back the land the following year.

He waited for fifteen days. There was no reply. If he waited any longer, Kamti's bill collector and the agent would descend on him and carry away the crop. The ornaments in the bank would remain unredeemed.

The areca nut had to be moved to the market before the debt collectors could lay their hands on it. If he allowed the ornaments to remain in the bank, the interest would mount and he would not see them again in this lifetime.

He went to the town to inquire whether there was any lorry bound for Shimoga. There was only one, it belonged to Gopala Kamti's relation. He struck a deal, thinking, whatever has to happen will happen.

The very next day, Kamti's bill collector presented himself outside Kuppanna Bhatta's house. His heart sank. Whatever he tried to say was drowned out by the bill collector's ranting. The man insisted that Kamti had ordered him to come back with not a coin less than six hundred rupees, plus fifty rupees more by way of interest.

The argument was conducted at such a pitch that one of Appanna Bhatta's servants heard it and carried the news to his master. Appanna Bhatta yoked his bullocks to the cart and promptly proceeded to the matha.

By the afternoon, Kamti's man, who had gone back to report to his master, returned with some gunny bags and scales. 'Sell us fifteen maunds of areca nut at forty-five rupees a maund, then we will have to pay you twenty-five rupees,' he said, taking the money from his pocket.

The selling price in Shimoga was fifty rupees. Besides, Kamti's weighing scales were adjusted, so he bought at one weight and sold at another. 'Not possible,' said Kuppanna Bhatta gruffly.

The bill collector saab made bold to step into the house. For Kuppanna Bhatta, a Muslim crossing the threshold was the limit. He sprang towards the room in which the areca nut was stored and shouted, 'You will see me dead before you touch this door.'

The man was taken aback at the outburst. Wiping his face with a towel, he sat down, the scales in front of him. 'What can you do when grown men behave like children?' he muttered. Kuppanna Bhatta made no reply. He shook with emotion. His lips trembled. But he continued to sit outside the door, his arm stretched across it. The bill collector just stared at him.

Fifteen to twenty minutes had passed, when the sound of the bells of a bullock cart were heard. Kuppanna Bhatta recognized it as the cart from the matha. For a moment, his body went limp and he shut his eyes. But he quickly steadied himself. The agent of the matha walked straight in, accompanied by the shanbhog—the village accountant—the matha storekeeper, and a servant with another set of weighing scales. 'What is going on here?' asked the agent loudly. Kamti's bill collector told him.

'He is our tenant. God's dues will be paid first, the annual rent as well as arrears. The grocer's claim can only come later. Go tell your master that,' the agent said to the bill collector. But the saab was not so easily cowed. He said Kuppanna Bhatta had promised that Kamti's debt would be cleared first. This was not true. He also described how he had foiled Kuppanna Bhatta's plan to move the areca nuts secretly to Shimoga.

'So this is what you have fallen to, Kuppanna Bhatta? You think you can swallow and digest what belongs to Narasimha?' asked the agent contemptuously.

With a meekness he had never shown in all his life, Kuppanna Bhatta told him about the petition he had addressed to Swamiji at Kashi. He took his janewara between his thumb and forefinger and vowed that if he was given a reprieve this time, he would clear the dues the following year.

'Clear the dues, he says! With what face can you say it? You have money to speculate on the market, but you have no money to pay what you owe to God. You were arrogant enough to defy Swamiji and not come when he sent for you. You are eating his food, God's food! You don't bother to take care of the plantation. Look at the way you have kept this place. Thath! Anyone would think that it belonged to an outcast! Anyway, this is Swamiji's order.' He took a piece of paper from his pocket and read it out. It said that the estate should be taken away from Kuppanna Bhatta and entrusted to Appanna Bhatta for cultivation.

The shanbhog put on his reading glasses and added up Kuppanna Bhatta's dues—unpaid rent, loan and interest—and pronounced, 'Six thousand.' The agent told the storekeeper, 'Weigh the areca nuts kept in the room.'

The storekeeper had recently been employed by the matha, after having lost all his money through trading in buffaloes. He was a dark, hefty man, built like Bhima. As soon as he heard the agent's words, he scrambled up to impress him. 'Move. Move...' he bellowed.

'Hoi, fifteen maunds of it should go to Kamti.' The bill collector was persistent.

'Will you clear out of the way, or should we set you right?' the agent threatened the bill collector, looking as if he was ready to swallow him up. The bill collector hurriedly climbed onto his bicycle and left unceremoniously, wondering what excuse to give when his master scolded him.

Once again the storekeeper shouted, 'Move out of the way,' to Kuppanna Bhatta, who sat there, barring the door.

'Before you touch one areca nut in this room, you will have to go over my dead body,' said Kuppanna Bhatta, loudly and firmly.

The shanbhog and the storekeeper were speechless.

'What did you say?' the agent challenged.

'That you will see me dead before you get to the areca nuts,' Kuppanna Bhatta replied, his whole body burning with the fire of his resolve.

'You, born a Brahmin and you are descending to this, Bhatt-re!' said the agent. The shanbhog put in his bit of advice. But Kuppanna Bhatta was unmoved.

'Very well, you will see what you are in for,' the agent said, as he made ready to leave with his entourage. On the way he went to Appanna Bhatta's house, instructed him to keep an eye on Kuppanna Bhatta in case he tried to move the areca nuts at night, and went off without having any coffee.

Kuppanna Bhatta sat for a long time where he was, until his legs went dead under him. Late in the afternoon, he got up and stretched his limbs to get the blood flowing in them again. He latched the front door, bathed, and boiled some kanji for himself. When the cows came back a couple of hours later, he tied them up in the shed, gave them bran-water and milked them. He lit the lantern and sat down in the courtyard. He did not feel like cooking, so he warmed some milk, drank it and lay down right there. But sleep eluded him. After a long time he opened the door and went out. From the clearing in front of the house, he looked up at the sky. Through the still silence of the trees, he could see the silver of pre-dawn spread across the sky. He calculated the hour, then went back into the house, bolting the door behind him. His eyelids drooped with weariness.

It was only when the cowherd called his name that he woke up. After the cows were led away, he drew water from the well and poured it over his head. He put a vessel of water on to boil.

Soon he heard someone pounding on the door.

'Bhatt-re, it is me, the ameen. Open the door.'

Kuppanna Bhatta made no reply to the bailiff. He went and sat at the threshold of the room where the areca nuts were stored. After a while, there was the sound of a bicycle bell. Then came the voice of Kamti, 'Open the door, Bhatt-re. I shall pay you forty-seven per maund. You can clear my account.' Kuppanna Bhatta did not move.

Kamti's bill collector shouted, 'Will you open the door or not?'

An altercation between the ameen and Kamti ensued. The ameen hurried away, threatening to summon the agent of the matha.

Again Kamti said, 'Bhatt-re, open the door. I will pay you fifty rupees. Settle my account.'

When there was no answer even to this, he began abusing him.

'If I had known what kind of a shameless man you are, I would never have let you set foot in my shop!' he said.

Then he said to his bill collector in a loud voice, 'Budan, go to this man's brother-in-law in Halasoor and demand a settlement of our accounts. Do not move from there until you are paid the money. He cannot be as shameless as this man is.'

Just then the agent arrived. He and Kamti began to quarrel. They sounded like the raucous crows that gather around the rice balls offered to them during a shraddha.

The agent used his ultimate weapon. He told the town crier, 'Go and announce to the whole of Buklapura that Kuppanna Bhatta's movable property is going to be publicly sold.'

The man began his job right there. Kamti's protests were drowned in that noise. Gradually, the drumbeats moved farther away.

'I will go to court,' threatened Kamti.

'Why don't you?' replied the agent.

The sound of the town crier's drum faded into the forest. Silence descended on Kuppanna Bhatta's house. Outside, the noonday sun blazed.

'You have to take the next step now,' the agent told the ameen.

'Do whatever is necessary. The storekeeper will help you. If Kuppanna Bhatta gives any trouble, send for me. I shall be at Appanna Bhatta's place,' he added, before he left.

'Bhatt-re, for the last time, will you open your door, or do you want us to break it?' The storekeeper called out.

There was no answer. He picked up a crowbar and rammed the door with it. The bolt rattled.

'Open the door now at least.'

There was still no response. One more blow and the door flew open. Like a Yamadootha, the ameen stood in front of Kuppanna Bhatta who looked as if he was deep in meditation. The ameen signalled to him to get up but Kuppanna Bhatta remained sitting, his eyes closed.

The storekeeper picked up Kuppanna Bhatta's shrunken body as if it were a handful of sticks and put him down outside, in the hot sun. One by one, he piled up the areca nut sacks in another corner of the courtyard. He was in high spirits. 'Are the vessels and other things to be sold too?'

The ameen pulled out a piece of paper from his pocket and read out the order, 'Everything movable and immovable is to be sold.'

The storekeeper darted into the kitchen. He brought the milk vessel into the courtyard and poured the milk on the ground. Then he brought all the goods out. A copper handi, a set of rice vessels, kadais, a set of bronze thalis, a silver jug, a copper teertha vessel (the same one that Kuppanna Bhatta had brought with him from the plains), a sankha and a brass thali used for the puja, a number of tubs for storing water at feasts, a cradle, a collection of ladles, a torn mat and a patched blanket, a wooden bench, a dozen wooden planks, a framed silver picture of Krishna, a silk dhoti, a pair of brooms—everything except the holy box containing the family deity, the saligrama. Kuppanna Bhatta sat amidst his worldly possessions, his hand to his forehead, his eyes lowered.

'Aren't the cattle part of movable property?' the storekeeper asked.

'Of course,' said the ameen.

The town crier had finished his job and come back. 'Go, bring home the cattle grazing on the slopes of the hillock,' the ameen told him.

Then he went inside and fetched a gunny bag for the assortment of vessels. As he shook the gunny bag, four or five baby mice, newborn and still pink, fell out of it. They writhed and wriggled in the hot sun. 'Eesh,' the ameen exclaimed in disgust. 'Thath!' said the storekeeper, as a crow hopped in stealthily and, lunging suddenly, carried off one of the baby mice.

A number of people from Buklapura stood behind the trees around the house. They had ropes and sickles in their hands and pretended to be there in the course of their work. No one dared to step into the courtyard. Not one person made a bid for the goods. 'Why should we commit the sin of coveting another's belongings?' some thought. 'Cheh! Poor man,' cried several others, feeling sorry for him. One by one, they moved away from there. Only a few, who could not conquer their curiosity, stayed back to see what would happen.

The agent had the areca nut sacks loaded on to the cart. The ameen made a detailed list of all the articles they had seized. Everything except the brooms, the silver picture, and a couple of broken vessels was placed on the cart. Without looking at him, the agent admonished Kuppanna Bhatta. 'No one who misappropriates that which belongs to Narasimha can survive, let alone prosper. Understand, Bhatt-re? From today, neither the land nor the house is yours. The tenancy has been made over to Appanna Bhatta.'

As he heaved himself up into the cart, he fired a last salvo, 'Anyway, why should you worry? You have already sent your hoarded wealth with your wife to her parents' place!'

There was an old ajji in Buklapura whose name was Sitakka. No one knew how old she was. She seemed to have been living there from the beginning of time. She addressed everyone in the singular, showing no respect for anyone. Married and widowed when she was still a child, she lived all by herself in a dilapidated little hut. Sitakka had no one to call her own, not even someone to offer a ball of rice at her shraddha when she died. Every day, she begged for a handful of rice to make some kanji for herself, her only meal of the day.

Sitakka had a sharp tongue, and everyone in the agrahara was terrified of her. If the shadow of a lower caste person fell across her path, if children accidentally touched her while they were playing, if she found anyone retreating into their homes when they saw her (because she was a widow and hence a bad omen), out would pour a stream of imprecations. It was almost as if she could not sleep if she did not spend a couple of hours a day raving and ranting at someone. Perhaps she thought that unless she picked a quarrel and invoked their forefathers for several generations and cursed these evil times, people would take her for granted. Afraid of her abuse, people gave her rice even before she asked for it! After her meal, she spent her time sitting either on the stone steps leading to the river or on the platform outside the temple, twisting cotton into wicks.

At the sound of the town crier's drum that day, Sitakka's ears perked up. Quickly, she took the mandatory three dips in the river, gulped her kanji down and went to find out what had happened. The agrahara seemed to have been emptied of all men. None of the women would tell her anything. Finally she got the information from a farm labourer. Sitakka marched into Kuppanna Bhatta's courtyard. There she saw a couple of misshapen vessels and a few brooms strewn around. Kuppanna Bhatta sat among them, his bald head exposed to the fierce sun.

Sitakka's harsh outburst shattered the oppressive silence. 'Which son of a whore has ruined your home? May his house crumble. Let his cattle be seized by tigers. But what is the matter with you? Why are you sitting like this? Have you too begun to believe that a poor man who loses his temper loses his teeth? Where is your arrogance now? Your insolence? You never cared to talk to me, to ask whether I was dead or alive.' Kuppanna Bhatta made no reply. Sitakka was puzzled.

'Let the fire swallow your enemy...' Sitakka began again but stopped when she noticed someone standing under a nearby tree. 'Who is that lurking there like a ghost?' she asked, crossing over to him. It was a young man called Narasimha Bhatta from Buklapura. But she did not recognize him. 'Who are you? Come here, will you,' she said. The man came forward timidly.

Together the two of them took hold of Kuppanna Bhatta by the arms, picked him up and seated him in the shade of the jagali. Quietly, the young man sneaked away.

Cursing Appanna Bhatta, the agent, Kuppanna Bhatta, the other Brahmins of Buklapura and anyone else she could think of, Sitakka picked up the pots and pans scattered around and placed them on the jagali. She saw the lock on the door of the storeroom and heartily cursed the hand that had put it there.

'May a snake bite it,' she muttered. Then she hastened to her own home, but before that, she stood in front of Appanna Bhatta's house and showered curses on his heartlessness. She got together some avalakki, asked someone for some curds to mix it with, crushed a green chili into it, added a pinch of salt and took it back to Kuppanna Bhatta. She wrapped it in a banana leaf, which she pushed in front of him and ordered, 'Eat this. Don't starve yourself like an idiot.' Then she went home, had a bath and lay down on the bare floor.

After the sun had set, Appanna Bhatta called one of his labourers and told him, 'Go and see what Kuppanna Bhatta is doing—whether he is there or has slunk away to his wife's place. Take care that you're not seen.' The servant did as he was told. In the light of the moon, he saw the still, motionless figure on the jagali, and was frightened. He rushed back to report what he had seen.

'See what a schemer he is!' Appanna Bhatta said to his wife. 'This is another one of his tricks. Do you know what a drama he put on for the benefit of the agent? I tell you, if he does not creep to his wife's home by the first light of the day, my name is not Appanna Bhatta.'

The next day dawned. The man who brought in the firewood announced, 'He is still on the jagali.' Curious to see what was happening Appanna Bhatta took the key the agent had given him and set off. As he neared Kuppanna Bhatta's house, his heart started thudding. For a while he lingered behind a large tree, pretending to cut down something.

Kuppanna Bhatta was sitting on the jagali, like a stone, his dark body leaning against the wall, his dhotra thrown over his head. Appanna Bhatta's heart missed a beat. As he went nearer, he thought to himself, 'If Kuppanna Bhatta picks a quarrel, I will give him a fitting reply. I will ask him about the wealth that he has already dispatched to his in-laws' place.' He felt reassured and stepped into the courtyard. A grey dog lay in front of Kuppanna Bhatta, gazing intently at him. A plantain leaf with a small heap of dried-up avalakki on it lay untouched. Before he realized it, Appanna Bhatta had called out, 'Bhatt-re!' There was no reply. With a shaking finger, he jabbed at the figure who continued to sit with his eyes closed.

All the blood had drained out of Kuppanna Bhatta's aged face, leaving it wan and pale. Appanna Bhatta shook him by the shoulder.

Kuppanna Bhatta opened his eyes.

'You need not have been so rough with the agent,' Appanna Bhatta stated.

There was no reaction, but Kuppanna Bhatta's eyes were open. Appanna Bhatta felt emboldened.

'What do you lack, Bhatt-re? You have carried your hatred of me this long. But I don't hold it against you. After all, you are older than me. As an elder, you should wish me well. That's what I want. You should not leave with hard feelings against me. So I have come to speak to you. I shall send you to your brother-in-law's house in my own cart. Come and eat a meal in my house before you go. Let us forget this enmity of thirty years.'

Appanna Bhatta had not meant to go so far. One word had led to another without his realizing it. Kuppanna Bhatta's continued silence had made him raise his own voice. But it appeared as though his words did not penetrate the other man's eardrums. Not a muscle on Kuppanna Bhatta's face moved. His stillness hit Appanna Bhatta with the force of a blow. And Appanna Bhatta was speechless. After a while, he recovered. As he paced up and down the courtyard, he continued, more softly this time, 'I have not done anything to ruin you, Bhatt-re. For thirty years I have endured your resentment. But did I ever try to compete with you? It is true that I informed the agent that Narasimha's share of the produce was being diverted to Gopala Kamti's account. If I am not telling the truth, let my tongue split into two. Call the elders of Buklapura and let them adjudicate. If they find that I am in the wrong, I shall fall at your feet.'

Appanna Bhatta's voice became shrill. 'Last year you took away all those growers who used to supply areca nut to me. When you had money to trade, was it right to default on your payments of rent? Wouldn't anyone get angry at that? Aren't there some rules in such matters? Like you, I am a man with a family. Like you, I have a daughter to marry off. I also have my troubles. For the past year, I have been suffering from some ailment of the stomach. I am in constant pain. Last month, my buffalo, which used to give a bucketful of milk, suddenly died. It is all our fate, Bhatt-re. Come on. Get up and come with me. Have a meal at my place. I will get the bullock cart ready and you can start for your brother-in-law's place in the cool hours of the evening.'

The grey mongrel looked at one face, then at the other, and then at the avalakki on the plantain leaf. Kuppanna Bhatta's eyes stared vacantly past Appanna Bhatta and into the forest beyond. Appanna Bhatta felt his throat go dry in the oppressive silence. He kept repeating, 'Bhatt-re, Bhatt-re...' as if he were reciting the names of his ancestors during a shraddha. The dog rose, stretched its limbs, and lay down again. Mango-gnats hovered around. The sun blazed. Kuppanna Bhatta sat on his haunches, resting his head on his palm and his elbow on his knees, a towel over his head. Appanna Bhatta looked at him again, as if seeing him for the first time. There was a churning feeling in his stomach. The avalakki on the banana leaf was dry, though it had once been soaked in curds. Flies sat on it. And flies sat on Kuppanna Bhatta's nose. Every few seconds, he blinked.

'Bhatt-re, I don't want your property. Stay on in your house. I will let you cultivate this land as long as you live. Just pay me the rent every year. I am not a butcher. Like you I came from the coast with an empty copper vessel in my hand. All I want is that I should not drop dead on a jungle path as my father did. There was no one even to pour a few drops of water down his throat as he was dying. When I found him, lying there in the sun, the crows were pecking at him. I pray that kind of death does not come to me, that's all, Bhatt-re. Here, take your keys. Keep them,' Appanna Bhatta cried. Overcome with emotion, he blew his nose.

The man who sat there on his haunches, as if waiting for his last breath, did not react. This still, shrouded figure seemed to burn into Appanna Bhatta's mind, and his tongue, never lacking in fluency, grew silent too. Appanna Bhatta put his towel over his head and he too squatted on the jagali, opposite the silent one.

OLD CYPRESS

NISHA DA CUNHA

The Visit

It was the garden really that decided everything, or most things. And we said, yes this is it. And again, oh yes no question, no question at all. Mark said, 'Don't you want to think about it a bit more, I mean, you do have at least till tomorrow to decide—why not sleep on it, don't you think?' I liked that. I mean, I realized that he ended most sentences with 'don't you think?' I liked it because it helped my own feeling of indecision, or panic, at such a big step. Also, it wasn't really indecision so much as helping one to work out things rhetorically, a help in every sense.

Except for nodding from time to time and gazing at the view, my husband hadn't actually said very much. But then that is very much his way, I mean, he wasn't really going to turn cartwheels on the old lawn out of sheer happiness—he wasn't that sort of man and I should know. But the old tennis court at the back of the bungalow, I think, really swung things for him. I do tend to dither. Especially if it's very important. Also, if the decision is not just for me. My husband said, 'Perhaps we could come back in the evening, if that's convenient? We do have to leave tomorrow—early.' That was the perfect solution. In every way. I realized that, because apart from everything else, we would look at things in the evening light, see if it was a kindly light as the morning had been friendly. There's a lot to light, I thought, yes that's what we must do—if the light is friendly, we will know, because we will see what happens to the garden in the evening. So we will know. And the house? Oh, but the garden dominates this house, so if the light is friendly in the evening then the house had to be perfect. The garden is the house. Mark and my husband had strolled on and I realized I was talking to myself. I'm glad because anyway, after a certain point, nobody listens to anybody else—as long as they listen to bits and pieces it's a lot to be grateful for. There was a bit of an old wall. I sat on it and dangled my legs. I could look at the house from here. I realized I hadn't been quite fair, I don't think I could have meant that the garden dominated the house, no. As I sit here I see the house from the garden and I feel both garden and house—and I know that that is what I felt when we were inside the house, that the garden came right in. That's it. It's a real unit—it's not separate at all. So really they complement each other not dominate—also it's odd even though we have looked at so many houses, and quite liked one or two, this one is quite different. This one feels quite other. As if I knew what it might feel like when one was actually living in it. Something about it, even now as I sit on this old wall, I feel as though I've done it before. A sun-warmed old wall—as though I have looked at the house from here before, and then been called out to, then jumped down from the wall and crossed this bit of lawn with long shadows on it, and gone straight through the side verandah and into the study. But it's odd. How can I know when I have never done that before? It shouldn't feel familiar, surely? It must be a good sign, a friendly sign. This is the house that will look after us on our new journey far from everything we have known before—at an age when we are not old certainly, but not young any more. That's behind us. Being young—whatever we may feel inside of us. Now I must find the others and tell them this is our house, our home. I've had a sign. My husband is a cautious man, so he said, 'Yes, well, but we'll come back in the evening anyway,' and Mark said, 'Yes, that would be best,' and I said, 'Please don't let anyone else look at it before we come back,' and Mark said, 'You must be joking! Anyone else look at this house before you come back this evening. Dear lady, nobody has shown the least bit of interest in Old Cypress except you—and I've been here for the last fifteen years. I think word's got round about the cypress.' I said, 'Cypress, you mean the cypress near the old tennis court? But it's magnificent! I've never seen such a cypress before in a private garden, it must be at least a hundred years old. I mean thereabouts,' and Mark said, 'Well, not quite a hundred but certainly very old, but it's a bit gloomy at certain times of the day or season, and of course it spreads.' I said, 'How do you mean, spreads? You mean the tree, but, it has to,' and Mark said, 'No well, not quite, I mean that too, its roots are spreading and causing problems to some of the paving stones near the old tennis court but that's to be expected, also its branches, but they can always be pruned, but no, what I really meant was that word about the cypress spreads, that it's rather gloomy and dark and the sun never goes near it so it has a kind of reputation that hasn't been exactly kind—oh you know the kind of thing—so of course the house has been neglected too.' Good, I thought, I'm glad, I thought, nobody will come near it, and that's why we will live in it because other people haven't loved it the minute they saw it, like we did. Old Cypress is part of the house and garden, they all go together, the person who built the house planted the tree, laid out the garden and that old wall I sat on, and the tennis court, everything. It's one whole and it's all evolved together.

'We'll come back this evening.' And that was that. A mist came curving in from somewhere as we walked the same road that evening. I wonder if it's much thicker in the winter. It will soon enclose us all. I'm glad we are all close together. It's strange. It's thick. I can hardly see two feet in front of me. I'm glad Mark is with us. He'll know where the gates of the house are. Where they ought to be. The ground feels gravelly, quite suddenly. So we must have turned into the gates. Rain, that must be why there is this mist. It's raining somewhere. Soon it will rain here; we don't need rain just now—it's our evening for deciding and I can't see a thing. Perhaps I said some of these things out aloud without realizing because Mark said reassuringly, 'It won't rain, it's not going to. Wrong season, you'll soon get used to this mist, it goes as it comes, I should have explained it, of course it does give one a bit of a shock creeping up like that but not to worry, no rain. Just put it down to these hills. Perhaps it's good for the tea. The sun will come back for you, to your bungalow, ma'am, don't worry.' Dear Mark, I must say if I hadn't taken to him already I did right then for saying not 'the bungalow' or 'it', but specifically 'your bungalow'. Also his calling me 'ma'am' made me realize how much I missed that. There had been a long time in my life when it was the most familiar sound—'But surely, ma'am, Yeats couldn't have meant that!' or 'Do you think, ma'am, that we could have a free lecture?'—yes, that was a time. And Mark had just called

me ma'am. 'Of course some people,' he was saying, 'don't get used to it too
easily. Young Anand hasn't, it still depresses him on a lovely clear sunny late
afternoon to see the mist suddenly appear from nowhere, but then he hasn't
really been here all that long, it's bound to pass, don't you think? I mean his
depression. Anand is my young assistant, lives in a bungalow further down this
road. He'd be your closest neighbour. We might go and see him after we've
finished here.'

By now we had reached the end of the long driveway and we could
just begin to pick out the bungalow—the sloping roof, a bit of wall, a pillar,
steps from the garden, 'See what I meant about the mist? It goes as it came.'

And it was true. He was quite right—the light brought the shape of the
bungalow back first and then gradually all the details, the marvellous windows,
the deep bow window almost touching the lawn, and the wooden pillars of
the verandah, and the quaint covered bit leading off to the store rooms and the
servant's room, the stone flagging, a chimney, bits of creeper—indiscriminate
details shone through the mist.

Colour returned.

I think at this point a word or two of explanation might be in order.
Why we were here in the first place. Also, how Old Cypress happened at all.
Mark and so on. To start at the start, a dear gay friend of ours, mine really,
had said, 'When you're there among those hills why not try those tea estates—
sometimes an old cottage or an old bungalow left over, you know the sort of
thing? Not quite dilapidated, but well on the way, sort of left over, you know,
darling—weedy now but once wonderful: planned garden, bird-bath-in-corner,
low wall, creepers, low sloping roof, old trees, an old Aga in the kitchen, a
bit of parquet flooring—I know I'm getting a bit carried away, darling, but
I think your search might well be over this time—you can thank me later if
you find it—but somehow this time I feel, yes, try the old tea places—after
all they knew how to build exactly what they wanted, and I know what you
want. Everything romantically gone to seed, you know the kind of thing?
You might just find something to suit your rather outdated tastes (darling, I
do mean this absolutely affectionately) and there's sure to be a spare room
and I might come and stay—for a bit—once you're all organized.'

'Good idea,' I said and thought—just what we need, lots of good ideas—
even inspired ones, and we would accept because time, I felt, was running
out. This business of time running out was very real but also needs some
explaining because it's part of early retirement and takes every individual
rather differently. There is a kind of urgency about it for some because it's

not merely a question of a decision but it's also a kind of psychological landmark and, to mix metaphors, it's a kind of plunge, a high dive the first time you try, scared as hell even though it's water down below. And if you wait too long looking down from that great height it stops being water, it stops being friendly. It's just fear, it's just panic then. And so it is with early retirement—you have to move fast once you decide or the faint heart part of one takes over. Now to move fast you have to have a roof over your head and a prospect. Not just dreams. And this is where the real problem lies. It can't just be any old roof, nor can it be any old prospect. Because as my other good friend said, 'This time it's for keeps, baby! Who needs it? You do—I can see that—so if you're sure you're sure, then there's no need to get your knickers in a twist now, is there?' Sweet reasonableness is the hallmark of that friend—as you can see. I suppose he's right anyway; if you can begin to extricate yourself from his rather vivid speech. I mean, once you move and you put everything you have into that move there'll be no more changing of minds—it's the last move and you have got to be 109 per cent sure. So it's tricky, to say the least. Mind you, at fifty plus you do feel certain of what you like and what you dislike—also of what you love and what you absolutely loathe. But the problems don't get less for being certain, they seem to multiply. Then I suppose my colourful friend would say, 'No sweat, baby! Just chicken out. No crime. Go back on the whole idea. Early retirement! Why not just go on as before, everybody you know is going on as before. Why not live as if life goes on and on? So why? For what?' Now this is the real danger zone. This is when you have to get out, get that roof over your head and the prospect that pleases and then you're fine, home and dry. More or less.

Really that friend who mentioned tea estates and leftover bungalows did us more than a good turn, actually perhaps helped us find it. Old Cypress—I like what it's called, the bungalow whose gates we have turned into, with the mist cleared and the place feeling like home already and Mark saying, 'Just down these steps and to the tennis court and one last look at the old tree and then you can be quite sure.' And so we did that. The steps were old, bits of flagging broken at the top and a garden urn and then a kind of graveyard darkness because of the tree. There it was, very dark and looming over the uneven, weed-covered tennis court. People in white had actually flashed about calling out, 'Beautiful shot! My serve, I think.'

'What do you think?' said Mark.

'I think that this is going to be our home.'

'Good, that's decided then,' my husband said.

A sudden strangeness in the stomach.

'Now,' said Mark, 'let us pay a visit. Your closest neighbour, Anand. I mentioned him earlier, remember? My new assistant and his feeling about the mist? He'll feel reassured, I mean, you can tell him how the mist disappeared so quickly and that's how you decided about your new home, don't you think?' 'Yes,' I said, 'that's a nice idea, but won't he mind us just barging in like that?' 'Nonsense, rubbish, it will jolly him up no end, and we'll take him back for tea. Velu will be really pleased, he usually makes more scones than I can eat and I hate to disappoint him.' 'Velu and scones?' I said and Mark said, 'Velu is my chap, I inherited him from the manager here before me. I think he goes with the territory, he really does make the most incredible scones, they are his best thing after his soups—crumpets too, but not quite.' But that was in another land. By now we were walking down a winding lane and I thought, Mark really is full of surprises, there seems no end to the surprises of Mark, tea gardens and scones and sudden mist, an old cypress, what more. Truly my cup runneth over.

Later that night, undressing for bed, I said, 'I love that bungalow.'

'I know.'

'Is that all?'

'That's a great deal—good, I'm glad.'

'Really? Are you really glad? Not just for me?'

'Look if it's all right with you, Radha, I'm tired, really bushed. I must sleep—we have a very long day tomorrow. All that travelling. Start so early.'

'I wish we could talk for a while about today, there's so much, I'm so excited—'

'I know there's so much, but no, we can't talk about it tonight, please I need to sleep. Try and sleep. Goodnight, Radha.'

I tried and tried to sleep and became more and more awake.

I saw the gate and the garden and the old wall and the creepers and the sloping roof and the verandah and the wooden pillars and the urn and then the broken steps down to the tennis court and then the great tree spreading dark, almost black, spreading and—it's useless. 'Why don't you take something? A Calmpose—and don't try so hard.'

She thought, I wish two people could want to talk about things at the same time, I wish excitement for two people could mean the same thing, I wish sleep could come to two people at the same time, I wish that he didn't feel so bad about leaving Bombay, he loves Bombay he's always loved it, no matter what has happened to Bombay he still loves it. If trees are cut down, if garbage heaps get higher and higher, if neighbours are rude, if neighbours chuck sanitary towels out of windows, he still loves Bombay. If taxi drivers are rude he thinks it's me—he meets nice polite ones! I suppose while he's working he doesn't think about it all as I do, but when he's not working any more he'll see how terrible Bombay has become. Bombay is unfriendly when you're growing old. It's different and that difference is sad. We have a choice, we're lucky, so many people don't have a choice.

She had watched the streets with no pavements left and huge trenches left uncovered and the pushing and shoving and she'd watched how the old were treated. Old people in Bombay. Nobody had time for them. She had watched living in flats for old people—the staircase, the lifts, the crossing of roads for old people, how they cowered and pretended to be brave, how they were pushed aside, brushed aside, how they could never get into buses any more and had to pretend they preferred to walk. She'd watched the crowds—old people shopping, old people in queues, the rudeness of people when you were old. No city for old people. And soon we will be old. Oh, we've been through all this before. That's why we are here. But he loves Bombay. He hates leaving and all that it implies. It's a watershed. Well! What about me? I love Bombay, in my fashion. It's what we've known together for thirty years. I hate leaving. It's every bit as difficult for me as it is for him. So he could try. Perhaps it's different for him. Perhaps. He's lucky though, he can sleep. She saw old people staring out of upper balconies at kites. At nothing.

Packing

She hugged her delight in the place—it helped her in all the weeks of packing and wrapping and chucking out of stuff and mostly keeping stuff that she knew ought to be thrown out—especially old letters and postcards that went back years and years—surely she'd need to look at them! Going away didn't mean throwing away one's past, and the letters and postcards were like all the old photographs and all the books—they were one's life, proof of everything or most things that had happened in one's life—so she looked at them all and read them all and then pushed them all back into a spare suitcase and felt she'd need them always. Punting on the river Cam—David—Ruth—Suzanne—Joan—Christopher—Kim—Lizabeth—Colina—impossible to throw anything away, anyway it was all very neat and she felt extra tired that day and when Rohan asked her that evening she said, 'Oh, today I've been dealing with huge sections of my life and yours, so I'm very exhausted.' 'What can you mean, what have you packed, specially?' She said, 'Letters, postcards, bibs and bobs.'

Rohan said, 'Oh God, Radha, are you carting all that around? I'd hoped at last you'd chuck it all out, just collects dust, has done all these years, don't tell me you're taking all those old half-used bottles of turp and linseed oil and filthy, squeezed-up tubes of Flake White?' 'Of course,' she said, 'we might feel like painting again, and we'll have the time, anyway all that's done already. Today's packing was quite different, it made me curiously sad—shall we have a drink?—it all somehow made me realize how something would never happen again, how I'd never be young again, never sit on an elephant in Jaipur with my mother and father again, do you know what I mean?'

'I'll get us a drink and, yes, of course I know what you mean. I wish you wouldn't go on packing day after day—you're very tired. I told you we could get the packers in—for everything—'

'Not for most things, they can't deal with one's past—packers, I mean.'

'They wouldn't be here to deal with your past, your blessed, aching youth and happy-sad-times, they'd just pack impersonally and finish a jolly sight faster than you and without being on the verge of another nervous breakdown.'

'Surely that was a bit unnecessary, cruel. Have you had a very bad day again at your work?'

'No, nothing unusual. But I'm fed up of coming home evening after evening to all this muck all over the place and you looking a mess.'

'What on earth can you mean? Muck and mess? Where do you expect all the packing to be done—it's not muck, it's our life. As for my looking a mess, this isn't a movie where people look nice even when they're cooking a four-course meal.'

'I'm sorry, really I am. Let's have a drink. Let's talk about something else. All right, Radha, I said I was sorry. Okay and speaking of a "four-course meal", I hope there's something reasonable to eat...?'

'Of course, bits and pieces, leftovers, when have you ever had a square meal in this house? Let's get sloshed and not think any more.'

'Good idea, probably means anyway that you hadn't thought about what we'd eat, right?'

'Naturally. Rohan, do you think about Old Cypress a lot? I mean do you keep remembering things about it, I mean quite suddenly?'

'No, I don't, as it happens. I've got far too much on my plate at the moment. I suppose you do. Do you?'

'Yes. I do, it helps me to get through all this business of winding up here. You know, I think of that garden and the old wall and how the sun will warm it and then I think of that front room with the deep bow windows almost touching the grass. We'll have long, old-fashioned breakfasts and—'

'We might start by getting a toaster, we both love toast but for thirty years we've never had a toaster.'

'Right, we'll get a toaster. Tomorrow I'll go and buy one. Oh, I think we are going to be very happy, I feel that. Do you?'

'I feel very hungry—let's eat.'

'Oh you are so, so somehow without excitement. I thought you might begin to feel, when are you going to say anything about Old Cypress, I mean, I think about it so much of the time and I want to talk about it to you—the old tennis court we saw together, we are going to live there together quite soon, isn't there anything about it that you feel?'

'No. If you must know I feel nothing at all. There. You made me say that. I wasn't going to say anything—I just hoped you'd realize on your own—you always want things said. Always. Now let's eat and not talk. I'm tired.'

'You are so awful. I do hate you so often, I don't understand why we have lasted so long. Do you?'

'There you go again, and since you ask me I'd say no, really I don't know how we have lasted so long, as you put it. Perhaps we haven't lasted at all!'

'You always try to spoil everything. Why do you have to say such things? You are a hurtful man. But I won't listen to you. I'll pretend you never said it. That you never meant it.'

'But I did mean it. You live in a world of your own—you have for a long while. You don't see anything that's really happening. Oh, let's just leave it.'

'There you've done it again, you say dreadful things and then retreat, how can you say the things you say and then say "leave it"? I can't leave it. So tell me.'

'No, I said no. Not now. Soon. Some other time. Not now. I'm going out for a bit, I'll get a bite to eat. You carry on, don't wait up.'

'Don't go. Please don't go, not now, not just now.'

'I must. I'll not be long. But I must.'

≈

One evening he'd come back from work to find her an island in a sea of open tea chests and mountains of old newspaper; she was tired but dogged—Judas 2 lying in an exhausted, drugged heap near her. She said cheerfully,

'Another few days and I'll be done, I think. Just the breakables—like this!'
And held out a bowl, a white bowl with blue flowers at the rim. It was
seeing that in her outstretched hand that made him say, 'I can't go with you,
I mean I'm not coming with you.' She said, not understanding, 'You mean
I'm to go ahead, before you, you'll follow after, is that what you mean? Is
something up at work?'

'No,' he said, 'nothing's up at work. I'm not leaving here. I'm not going
to Old Cypress.'

'You mean not just now, or never?'

'Never.'

She said, 'But it's ours, our new home, Old Cypress belongs to us now.
It's all arranged, what can you mean? Are you feeling all right? I mean, are
you ill? Are you joking? Because if you are I don't feel much like joking, I
feel tired like death, I feel like a jolly old drink. What's the matter? Why are
you looking like that—looking like this—at me?'

'I'm not coming to Old Cypress. It's like a full stop, that old bungalow.
Here I'll keep going. With you and Old Cypress I'll grow older quicker, it's
like the last stop with nothing further. Death the next stop. It's too quiet
there. And that mist every evening! No, I'm not leaving here—here it's noisy,
it's dirty, it's crowds and it's traffic but I'll always be with people.'

'Oh, you've just got cold feet, I get like that sometimes thinking about
it. Sometimes it feels so final, so unfamiliar. But it won't be like that once
we're there. Soon you'll love it. You're tired. You're anxious and worried.
Don't be. Please. You didn't really mean those awful things you said—I hope.
Not all of them, anyway.'

'I do mean them. And worse. I hate that place. Miles and miles of forest
and then miles and miles of tea bushes and shade trees and absolute quiet. A
graveyard would be more lively. And then that old tree—as I said, a graveyard's
the proper place for that tree. It looms and glooms over that old house—no
wonder nobody wants to live there.'

'But I didn't do all this on my own, in fact if you hadn't thought about
retiring early none of this would even have started happening, don't you
remember? And about Old Cypress, we saw the sale notice about it together
at the club and we both followed it up and then we met Mark who took
us there—good God man! Have you forgotten everything? And you said you
liked it—we both liked it. We saw it twice—all morning and it was sunny, and
then we went back in the evening—you never once said you didn't like it, in
fact, you didn't say anything about graveyards or anything distressing about the

place, why else did we follow it up? Why else did we buy it? Why else am I packing and packing as if my life depended on it? You said that Mark had told you early autumn would be a good time to aim for so that we could be settled, more or less, before winter set in—'

'Oh, hang Mark and anything he said, I'm not going. And about liking it, well—you loved it so immediately and I thought I might get to like it—given time. But, I don't have the time. And I'm not coming with you. I'm staying here. I'm not stopping you. You must go.'

'Do you mean it's to be a holiday place—closed most of the year and then opened up for a month every year? Have you gone quite mad? That wasn't the idea at all—we are going to live there always. You're retiring in less than a month and we're leaving. It's here that we might visit from time to time. Oh, what is the matter? Please tell me?'

'No, I mean, I'm not coming at all. I'm involved with someone—it's become very important to me—so I—'

'What are you talking about—who are you involved with? What has become so important to you—what?'

'I'm, well, very close to a woman, and I'm not leaving her. That's about it.'

'What's about it? You're not that kind of person! You mean you love some woman? I mean, someone has become that important to you and you've never told me? Never given me so much as a hint? It's not possible. It's the one thing that just is not possible. It's not you. I'm not listening.'

Early next morning she woke him up with, 'How old is she—tell me? Did last night happen? Did you say those things?'

'What's the time?—are you mad—I have to sleep.'

'You can sleep later, dammit—how old is she? You must know. I want to know. You've slept quite a lot anyway—how can you sleep?'

'Must you know? Must we go on and on—'

'Yes, and I don't remember going on and on—'

'Half the night—'

'I don't remember, anyway, yes, I want to know. How old is she?'

'She has a name—just in case you plan long conversations about her.'

'How old is she?'

'She's younger, younger than—'

'You mean younger than I am or much younger than you are?'

'Younger than both of us. Actually she's—'

'Why can't you get on with it, why can't you just say it? If it was all right carrying on with her for so long, why do you find it difficult to tell me how old she is? Has the enormity of it all just struck you? She's not a minor, is she? She can vote, can't she? Why is it worse for you to name it?'

'She's twenty-six.'

'Bravo! There, now that's said. I'm sure you feel a lot better. Lucky you. I can't say I feel a whole lot better, but then that's not why I asked you, is it? To use an old, hackneyed phrase of attack—or abuse—isn't it cradle-snatching? Our son is—rather will be—twenty-seven this year. I suppose that's neither here nor there?'

'Yes, well, you did ask.'

'Yes, I did ask, as you put it. Asking for trouble, I am. Is that what you mean? With all my questions. As to your sensitive observation about her having a name, do you and she talk of me using my name? Do you—I mean it is quite simple really—why do I find it difficult?'

'We don't talk about you—I mean it's possible. We just don't.'

'Yes. I can see that. Anyway she and you have a head start—I've been in the dark. What did you tell her every time we went looking for a place to live in? Surely she's a wee bit curious sometimes? How did you pretend so well, how could you disguise so well, how could you make me believe—or rather her believe, and me just assume or imagine—all was well? Or do you mean I should have guessed? Say something, anything. What about trust, loyalty, fidelity? What about friendship?'

'I don't know, I wasn't sure—I—'

'Bastard!' she said, 'you weren't sure—you mean it was always either, or? And me not to know, me get the raw end of the deal either way—be left or bereft, if you came, or if you didn't? We weren't a unit any more anyway for you to worry about? I was not to know—and, oh horrors—I would never know if things hadn't worked out for you. How could you?'

At last he said, 'Look, I didn't mean to hurt you—you have to believe that. Let's leave it at that, no need to try and—'

'Try and what? Try and understand? No need, you say? Every need, there is every need. I need to know. I have to. So that I never ever again—oh hell, is it men? Is it true, then, that men are so different? What has so often been said but that I thought couldn't be really true. Have you and I been so different all these years. Us. Perhaps you are right, there is no need to try and understand because there is nothing to be gained. Nothing. I suppose deception, betrayal happen stealthily, not brazenly, openly. But. Not you, not you. Not us. Because you are my friend. And I thought that I was yours.'

'But you are, you are—I really have no other friend like you—nor will

I ever. But there was something I needed—'

'Yes, so it would seem. But, as you say, friend, what can you mean not telling me, not warning me, lying to me, being untrue to me? Think if it had been me? Can you think of that—and there were times—I did feel a great deal for someone else—I really did. We were, you and I, having what was more than a bad patch. I was very lonely, work was going badly, I was not all that well and yes, there really was someone who cared a great deal for me—really cared and made me feel loved and desired and special. But there was always you. God knows why I worried so much—why I didn't allow anything to happen. I wish I had. It's no use now. I wish women were more like men; I suppose some are. Anyway, so you have a "chick"? What do you feel when you are with her? Remember how you and I used to joke about a chick on the side?'

'I feel younger, she makes me feel younger.'

'Younger? Yes well—'

'Yes younger, altogether, I mean as though, well, like I used to feel…'

(God, she thought, I asked for this, this I really did ask for.)

'Look,' he said, 'I know this hurts you, is hurtful, but I didn't want to hurt you, didn't mean—didn't ever mean—'

'Oh shut up, how dare you talk of didn't want to hurt, you don't begin to understand what you have done—undone—over twenty-nine years together. More than that really. Something built over more years than most people alive. And what about the years before, before the twenty-nine years? All those years?'

'I thought it would pass, I mean, I thought it might not last,' he said. 'I didn't realize how much it meant to me,' he said.

She hated him so much that she could scarcely breathe. And she thought, the other times I've hated him, wanted to leave him, get away, go away from him, from my life with him—and never done it—never done it because there was something good we had at solid rock bottom. She thought, what I felt was true must be so for both of us. Apparently not. More fool she. Me. Me miserable which way I fly—myself am—

Another day she said, 'Then why did you come to the hills, why did you plan with me, why did we do all that, why did you go ahead with all that? I mean, if all the time…if all that time…' she said.

He said, 'I wasn't certain it would work out.'

She said, 'Work out? You weren't certain it would work out with me or with her? Can't you be more precise, even now? What is it with you, anyway? Who is she? I really do need to know. Also I want to know.'

'Why, why must you?'

'Why must I? At best, why not? At worst, curiosity—I mean apart from being young, other things. Is she tall, is she small, is she thin or is she fat, does she laugh a lot or not at all, does she read, has she a dimple? I'd like to know the kind of person you fell for to put it in crude or youthful terms.'

'Stop it! Just stop it.'

'No, no, I'll not stop it. How can it possibly upset you? And you're surely not worried about upsetting me, are you? I mean, having done the worst, and having told me the worst, this is all small cheese. So. What is she like? What does she look like? Who is she?'

'Well if you must—'

'Yes, I must.'

'Well it's Aditi…'

'You mean the Aditi in your office? That one? Not possible.'

'Yes, that one. And yes, more than possible. You asked—remember that.'

'Yes, I did ask. And I do remember that. The one I had taught. The one who'd joined you soon after her first child was born. Good Lord, we went to her home for a ceremony for the child—I didn't know where to hide my peda and you said wrap it up in your handkerchief, no one would notice. But then—oh, it's not possible, it just doesn't bear thinking about.'

'Then don't. Just leave it. Leave it alone. It's bad enough.'

'Bad enough, is it? What do you mean? Bad enough for you, or her? Or me? Or just the whole thing? And I won't leave it, I just can't believe it. Give me time. After all, you've had a great deal of time, a real head start! But how can you stand her? When she's not gushing, she's coy—when she's not coy—she's embarrassing.'

'You don't know her; you don't know her at all. You haven't the right to discuss her.'

'Of course I do, I have every right. We've often discussed her, you and I. I have every right. My husband is about to marry someone. Are you? My husband has had a long and fruitful affair with someone he is going to marry, my husband is throwing up—giving up—a long relationship contained in a marriage that has lasted twenty-nine years and seven months. I have every right to discuss her—to talk about her, to tell you what I think about her, and to tell you what I think about you and her. I've never fancied myself. It's you that fancied me. Remember? And now. Oh well, I must say I'm disappointed in you. I thought it would be someone spectacular. In every sense. After all, you are giving up quite a lot. You are, you know. And I don't mean me. I mean our life, our ways together. Our quarrels, our journeys, our illnesses, our child. Our child when he was little, and now when he is not. My God, does she know you snore? Oops sorry, of course she does, it hasn't been all platonic has it, for so long? How on earth did you manage? I mean her husband, your wife, her child, your child? How did you juggle it all, keep all the balls in the air and not drop any one of them? How? And when? And the secrecy, the hypocrisy, the others in the office, and me at home? And tennis right through it all! I do wish—I wish—that I had known or guessed. It all seems such a waste. That I worked quite hard at it all because it mattered so much to me. A waste. All, all of it a waste and all the beastly, unbearable hurt to other people because you mattered. I can't bear it. I don't think I can bear it. How did you? How can you bear it? Do you, did you, just stop thinking? Or didn't you think at all? Or is it really that different being a man? And guilt, was there none of that either? No sleepless nights, days, anxiety, off your food-drink? Good Lord we went cottage hunting together. Friends. You and I. Friendship. You and I. Always you and I. What a fool I've been. I assumed, I took for granted, our life. You pretended so well. I didn't think you loved me. But I felt you liked me more than anyone else. Just as I do, did. That's so much after so many years. You and I have talked about that too—rather often—about how lucky we were, are. We compared ourselves to other people, we—oh let's stop this anyway. I want to finish the packing. The crates will have to leave quite soon. You didn't help much, did you? I mean, in the way of packing. I suppose I'll have to unpack some of the stuff. I mean, it's all been packed as though we were both going. I take it you will want some of it. Not just clothes, some of our shared things.'

'No, not to worry.'

'What does that mean—not to worry? We'll have to separate our things, specifically—books, paintings, furniture, oh for God's sake almost everything. How are we going to do it? When are we going to do it? Why didn't you make up your selfish, beastly mind sooner—before I'd started packing? You're a coward, you knew all the time. And it wasn't kindness to me, it was actually weakness. You didn't know how you'd put it—say it—or some such finicky detail. Having done the worst damage you could ever do, you suddenly found it difficult to tell me. Coward, worse than coward. There is no word for what you are, for what you have done. I still can't somehow believe all this has happened, is happening. Is it true? Did you not so long ago tell me that you were leaving me? Did you tell me that you were leaving me for someone you'd loved for a long while and now were planning to marry? Were these things said—done? Have they happened? Somewhere between wrapping blue and white bowls bought in Batot in a little tea shop—it was so cold and twelve white bowls with blue flowers on the rim lay on a dusty shelf and we bought them for a rupee each. We both liked them very much, they were the first things, almost, that we bought. I wanted eight and you said six would be fine. We used them for soup, one broke when I fell down the stairs in our first home, one when the child was two—that left four. That's when they became bowls for nuts. Judas knocked two of them off a low table with his great tail. Now there are just the two, semi-wrapped in newspaper. You take one. I'll pack the other.'

'I don't want it. Please take them both, please don't cry, don't cry.'

'Don't touch me, leave me alone. And what do you mean, you don't want it? You don't want a white bowl with blue flowers on the rim? Why not? How can you not? Don't you want it to remember, don't you want it to forget? We'll divide everything up. A division of the spoils! That's it! We'll divide it all up—whatever we can. Till we come to Judas 2 and the child! That way it will be fair. That way you'll feel pain, begin to feel some of this. Why should you be spared? Even bastards feel pain, don't they? Don't come near me—I'm not a child, I'm not a pet, don't come near me with your warm familiar ways of bringing comfort to me. All the times—when my mother died, all the things I couldn't bear, thinking I'd die of grieving—you brought me back. But not this time. This time you've brought death, betrayal, everything, so don't dare come near me—cold comfort, dissembling, pretending. And now you take one of these white bowls with blue on it and think back on a very cold morning high in the mountains, you think of us two young people. Two graves with white and purple iris. A tea shop. Very hot, sweet, too sweet tea but so welcome and opposite a dusty counter and twelve white bowls of thick ceramic with a blue pattern on them. Can you see them? And I said, "Why don't we buy those bowls," and you said, "By all means," and I thought we'll buy all twelve or at least eight so if any of them broke we'd still have plenty and you said, "Six will be plenty." Mean, but sensible. And the man at the counter smiled and said they were Chinese and that's why they would cost a rupee each and wrapped them up in old newspaper and there were six small rings of dust on that counter. Can you forget? It was only twenty-eight years ago! Too long? You always did say, "Travel light, travel light," that's why I suppose you can be up and off like the young these days, everything they need stuffed into a small rucksack, and a "peach on the pillion". I'm very tired. I'm going to bed.

I wish I had my own room. I've always wanted a room of my own. You'll put out the lights. Come on, Judas 2. I wonder if he's guessed—he usually knows when something is up.'

At the Station

'Thank you for coming to the station, you dislike...'

'Don't be silly, you must know...'

'It's a Saturday, where will you go from here? Have a good read at the Will? Ignore me, I wish the train would leave now...'

'You're lucky to get this coupé—I mean, what with Judas 2...'

'Yes. I hope he's good at stations—I mean performing—he'll miss you a lot...'

Judas 2 straining at the leash, trying to get to Rohan and yet frightened and sitting very close to her.

'We never took him on a journey before—to get used to this sort of thing...'

'No, we didn't.'

'But, finally, Rohan why? Why has all this happened?'

'I can't answer that, I wish I could. Okay then, take care...'

'Yes, you too.'

Dog straining, now seriously worried. 'Stop it.'

'I think it's time.'

'Yes.' I can't believe it's me, I can't believe it's him.

As the train started pulling out she said, 'I'll miss you—us—you know?'

'Yes,' he said. 'We always had a lot of fun...'

'Yes'—I wish he'd say some more—'What? I can't hear you—what?'

'I said not just a lot of fun—I mean a lot of things—together—also I really do love you...'

'Yes, but you love her more.'

'No, it's not like that, it's more complicated.'

'No, it's as horribly simple as that. Leave it now.'

'Look if you need anything, I mean...'

'Need what? Like what? It's done now. You must know I wouldn't ask you or...look, let's just leave it.'

She sat down and pulled Judas 2 very close, turned her back to the window; this time the train really did leave. She was glad they were alone. Now she could cry—that's what stations were for. That's what trains were for.

The Bad Times

One life to live—lead. Enjoy—say the young rather quaintly, and quite right too!

In the middle distance—two horses cropping non-existent grass. Tails flicking flies. One very dark black-brown and the other white with a large brown patch. Lovely and lonely in the sun those two. Also in the shade. A bell tinkles as they move. You can't really be thinking about death on a day like this—with one almost painted green bird poised at the very end of a twig deciding—shall I or shall I not? There was a rhyming game when they were all young—girls played it when they were skipping. Not boys. It was sissy, they said, to skip.

Two little dicky birds sat upon a wall—
One named Peter, the other named Paul.
Fly away Peter, fly way Paul—
Come back Peter, come back Paul.

Silly when you said it just like that but it was nice if you skipped well—sometimes you skipped together with a special friend, facing her—oh yes it was fun—such a long-ago time that. When?

I contain that time in me. Now. Both then and now.

Always went backwards—I mean I did—and I still do. Long ago and then again not so long ago. Rohan would say, 'If only you could try and not let your mind dwell on the past—let it go—just let it go...' and Radha would say, 'How—how? It's with us all the time—except sometimes more than at other times—surely, surely—the pastness of the past—the past is not chronological—it isn't, you can grieve today for that flower that died, very now, and twenty years ago a death—you can feel the sharp pain, and then the dull, on-and-on pain, a death twenty years ago it doesn't leave you, it's like today—Rohan you know exactly what I mean though I've put it so badly. The pain doesn't pass—it's there all the time—in every way. In sounds, in smell, in touch. The smell of a sari worn by someone. The smell of mogra in the hair—the long-ago person is there. Is here.

'The feel of the person, the sound, and it's not memory, and it's not just nostalgia. It's the here and the now. It's real in the present. Longing to hold, hug, talk, walk. The "thereness" of a person doesn't go merely because it was a long-ago time. Present, past, it's the same thing, it's just been made convenient to separate it. The past lives securely in the present because it's the same thing. When people say—"Forget it, put it aside, put it away, don't relive it, don't think of it, blot it out, think of something else, don't dwell on it"—that's it. That's the point. There is no way you can do those things. It's unnatural. But I suppose there would be no way to carry on living. So. We try to force ourselves to do something that is unnatural. Or we pretend.'

And Rohan with great patience, 'But try—that's all I said. Try. Or, Radha, you'll always be grieving for everything that goes—let it go, naturally, like the leaves on a tree...'

The now Radha wanted to say, 'Have you done it—do you remember, Rohan? Nothing, anything of our days and ways?' She wanted to say, she wanted to shout, wanted to cry out, 'Look—remember that time—cats in Venice—the Dali retrospective, wine and cheese and olive sharp on the tongue as if this minute shared—how can you forget, how can you say, laugh it off, forget, the past is done, it's over?'

How are you managing, I wonder—I suppose, very well. But damn you, damn him. I miss him. I miss. I miss him and that's why I can't—don't—get on with this unpacking and living each day as I should. Try. Start now. Start. I will tomorrow. Yes, I really will tomorrow. Just now I'll go to bed. I don't want to try and sleep. I want to sleep easily each night, as head touches pillow. Not think and think about—betrayal. I am so burdened with sorrow. I don't need it, I don't want it. Tomorrow. Come on, Judas 2. I wonder what your thoughts are? Do you think of Rohan? Do you miss him? Do you have long thoughts—as you lie there—about betrayal? Judas 2—if he came up the path just now, would you recognize his footsteps, his smell, his touch? Would your tail wag and wag till it nearly fell off? Would you love him as though nothing had happened? Wouldn't you take sides? Would you leave me? Judas 2, I hope you realize he left you too and just remember that. One day he thought about us both, and he said, "Can I do without them?" He thought about us both—separately, and then together. After a long while (hopefully) or a short while, he said, "Of course—off with their heads—nothing to it—who needs it—travel light!" So you see, Judas 2, he moved on. We are his past. And he has let us go naturally, as the leaves on a tree.

'He bid me take love easy, as the leaves that grow on the tree; But I, being young and foolish, with him would not agree...'

Anyway, let's try and sleep, old Judas 2—and I'm sorry trying to influence you like this—they say divorced parents try and do that to their children, I'll not do it to you, Mr Judas 2—you needn't take sides if you don't want to—and you can go on forgiving Rohan or anyone you like. And tomorrow you and I have serious work to do.

And you and I are going to listen to our music again. Telemann and

Vivaldi and then the *Well-Tempered Clavier.*

Mark can't go on spoiling us, she thought, as she lay in the dark—I have to do something and soon. That's all there is to it. She thought of the first evening, arriving at Old Cypress and finding a note from Mark—'Wish I could be here to welcome you, I hope the long journey didn't feel too long. I've had a fire lit in the living room and the master bedroom and had all the rooms aired. Velu has left you candles in the kitchen cupboard and something to eat—tomorrow will feel better—when the sun comes out and you've had a night's rest. Welcome to your new home.'

For many days the house stayed aloof, very cold, hostile. Endless bending over trunks, books so familiar, now seeming not hers at all. Cupboards not closing, shelves odd, out of place. The old Aga smoking and smoking without really cooking anything not even heating anything—but one day it felt different. Distinctly different. Curtain up. Rocking chair near the fireplace, rugs beginning to belong to Judas 2.

Perhaps she had come home.

But not the nights. Not the dead of night. She was afraid, heard voices. Sounds.

Saw shadows move. Wished she had learnt to pray—she hugged Judas 2. We have followed too much the devices and desires of our own hearts—I wish morning would come.

The Bad Times (Contd.)

One day as they walked Mark said, 'Is it all right if I ask you some questions, just a very few?'

'Yes, a few, of course.'

'Is your husband coming? I mean, when is he coming—it's now you need a lot of help so I wondered—not prying, just asking.'

'No, he's not coming. He's not following after or anything like that. I'm here. I've come to stay. About help—well, I'll unpack and uncrate slowly, perhaps you know of someone who might help with the heavier stuff? I don't mean to impose. Truly. Perhaps your Velu might know of someone who might help with the heavier stuff, do you think?'

'Yes—I think. And if it's all right with you, after work I could drop by and help. I'm quite good with nails and hammer even an old Aga that needs starting with more than a kick, that sort of thing.'

'Yes, soon. I'll ask you. You are very kind—'

'Will you, rather are you, always so formal? Don't be. It's often such a

waste of time. Anyway, I'll wait for you to ask me. Don't be too long, soon
winter will come and before that you should be properly set up.'

'Yes, before winter comes, and thank you.'

Another day, on his way into the nearest town, Mark dropped by to ask
if she needed anything and when she said a hot-water bottle he said, are you
feeling the cold already and she said, yes, and he asked if she was sleeping all
right, and she said, more or less, but sometimes she woke at odd hours and
heard things. He said, what kind of things, you mean the wood creaking sort
of sounds? And she said, not-in-the-house sounds but outside-sounds. Like
people playing-tennis sounds and Old Cypress creaking. Silly, of course, but
it had happened several nights. Even Judas 2 heard it—heard them. Mark said,
why don't you come into town with me now, we could talk and you might
see other things that you need. Bring Judas 2, he'll like the drive, and so she
had gone with him. Judas 2 had his head stuck out of the window—looking
amazed.

Soon she said, 'Judas 2 likes it here, he's always had a raw deal. For a big
dog he's never had any space—where we lived he couldn't ever roam free
or run, his walks were sedate up-and-down-the-lane kind of walks—dodging
garbage and wild cats and filthy crows and the children would call out, "Here
comes Judas's mummy!"—They loved him and always wanted to hold his leash
for a bit—but he never liked them much, he really likes it here, he has the
garden, unbelievable freedom.'

'Do you miss it then? I mean what you have left—the lane and the
children calling you Judas 2's mum?'

'Yes, yes, I do, it was so many years, no other reason. Anyway, the children
in the lane were beginning to grow up—some had just started karate and
some were learning ballet so anyway soon they wouldn't have had the time
for Judas 2 and me. I shall buy a toaster as well. I've just decided. We—I—love
toast and I've never got round to buying one—amazing how lazy one can
get. I mean to be lazy like that, selective lazy.'

'Wait till you see our "everything" shop, it's a real "Ye old shoppe" with
marvellous wood floors that creak and counters with the most amazing things—
home-made jams and chocolate fudge and hot-water bottles and socks with
clocks and toasters and books by Mrs Gaskell—'

'No! You're just making it up as you go along, aren't you?'

'Well, just wait till you get there and you'll see it's true, and I'm glad
you're going to get a toaster. Also they have strange things in large old glass
jars—we're nearly there. About those sounds you hear some nights, voices, are

you frightened? Or do you manage to get back to sleep? And Judas 2—you said he heard them—what did you mean?'

'Exactly that—Judas 2 hears those sounds and those voices, they are playing a game of tennis and they call out—'

'Here we are then—Judas 2 is allowed in with us—I told you it was a remarkable old shop—now for your toaster and the hot-water bottle. Why is he called Judas 2?'

'We had a dog called Judas—he was bitten by a Russell's viper—many years ago.'

'I am glad you have Judas 2.'

The Old Tennis Court

One day Mark said, 'Why don't we try and deal with the old tennis court, I've been thinking about it, it seems a shame not to try. I mean it's there, isn't it? And if it's unusable tennis-wise, you can still put it to some use because we'll have cleared it of years and years of nettles and brambles and goodness knows what. I notice Judas 2 has become quite fond of that part of the garden.'

'Yes, he has and always comes back really filthy when he's been rooting about there—but about the tennis court, I never thought to touch it somehow—I mean when Rohan decided not to come, that is.'

'Don't you play—?'

'Ye-es, I do and I don't. It was really a passion of Rohan's.'

'But if the court proved usable would you play? Are you fond of the game?'

'Yes, yes, I am, do you play? I didn't think to ask. You do so much else, but do you?'

'Yes, I used to play quite a lot in college, but here all these years it's been too far, somehow, to go for a game, so I gave it up. But I'd start again—if you would. And then there's Anand—it would be good for him—let's give it a try, anyway.'

'But won't it be a fearful lot of work? I mean, work on the court—won't it?'

'Well, we'll soon find out. It has a wonderful natural boundary—the old wall and then the tree. We have a heavy roller down in one of the factory sheds and there's any amount of gravel—it will be a red gravel court. Really, it seems a shame not to use it, I mean, there it is, waiting. Let's give it a try. Of course, you'd tell me if you hated the idea because it reminded you too much of things you want to forget, or pained you too much, you would, wouldn't you? I don't want to push you too hard...'

'Yes, I would. Tell you, I mean. And, thank you, Mark, for thinking as you do...'

'Anytime—Radha. For you. Anytime.'

'I suppose partly you're trying to rid me of night fears—sounds from the tennis court...?'

'Yes—partly.'

'You know, Mark, the person who really will be pleased about this is Ram Swaroop, he's been itching to clear all the tangled stuff and the bushes, he hates a jungle—he doesn't think much of my idea of a garden.'

'Don't I know it,' Mark said, 'he loves clearing, he loves clipping, chopping and mowing, he loves very neat flower beds with edges clearly demarcated, you know the kind of thing, if he had his way he'd make a bird sculpture of your Old Cypress! Be firm with him—he's a very knowledgeable mali— but don't let him bully you or touch the way you want your garden...'

'I want a clearing-in-a-forest sort of garden.'

'Right. You tell him to help you achieve exactly that, all right? About tomorrow, you haven't forgotten, it's the Saturday I'm taking you and Anand to our famous game sanctuary—we'll leave very early and come back by nightfall. I feel certain you will like it very much, and if we're lucky we might see a tiger or a leopard—apart from all the other game.'

'Yes,' she said, already excited. 'I've only been once to a game sanctuary when I was very young. I remember being frightened of the silence—what do we take—I mean, what shall I—?'

'Nothing, nothing, Anand and I have arranged everything. Anand has never been, you know—I suspect he's coming because you are!'

Very early the next morning they arrived to fetch her. It was still dark and she was worried about leaving Judas 2.

It felt mysterious in the jeep, warm between Mark and Anand, enveloped in a kind of cocoon—all round the light darkly mauve and the headlights on. Gradually she felt like a child again, the day not quite begun, neither night still nor day yet. What delight hid, unknown, behind each bend in the road? The jeep wound round bends and slowly, very gradually, light touched trees and bushes that had been there all the time; it was magical. Radha began to hum and Mark completed the old tune. 'Oh,' she said surprised, 'it's an old favourite. I thought not sung any more, before your time!'

Anand half-asleep woke with a startled jerk, 'What? Where are we? Are we there? I don't want—'

'Go back to sleep, Anand. What's it called then? That old tune—the one you were humming?'

'I don't know but it has such a quaint story to tell, I love it—about a boy called Willy who goes a-courting, every day he is asked about his beloved's skills and Willy is so proud about her spinning and her weaving and her baking and then in the last verse, which goes:

> *How old is she now, my boy Willy?*
> *How old is she now, Willy won't you tell me now?*
> *Twice six, twice seven, twice twenty and eleven—*
> *But she is too young to be taken from her Mamie.*

'That's the final verse. I hadn't thought of it for a very long time, maybe it was this setting off on an adventure so early that reminded me—made me remember. When our son was little we often went down to a village near the sea and we'd sing this song on the way. He loved the words, he's twenty-seven now. That's where Judas was bitten—by the snake.'

The sun was high in the sky now, the hillscape very different. They stopped for coffee and sandwiches near a break in the hills from where they could see a deep gorge far down, water cascading down, deep green, deep green silence—all round in the near and far distance, violet-blue hills.

'We're nearly there now, it's our lucky day. We are going to see a lot of wildlife,' said Mark.

'How do you know? Why do you think we'll be lucky?'

'Just because. Also I saw a marvellous blue kingfisher just now—it flashed past Anand's left ear. That's how I know! Ready everyone?'

Later, much later, Anand was to tell the doctor, 'I was frightened—it was a lonely, dark journey and when we stopped for breakfast a dark bird flew close by me, a death bird, it chose me, not Radha, not Mark, it chose me. And then we drove down to death valley—it was waiting for me. Everything was dead—near a small stream there was a dead skin, or pelt, of an animal—the others didn't see it. Only I saw it, it was waiting for me. Like the yew hedge in my garden—Ram Swaroop was not frightened of it, only I could see what it was, it was death all the time, hovering, waiting at every turn—even on the way to the sanctuary there was a greasy, greeny-black sort of hut shaped like a wigwam—doctor, you do understand me, don't you? Radha was interested because she had never seen such a strange shape and when Mark told her we could stop and take a look she was intrigued. But I knew immediately that it was a bad thing, again a death thing, an evil thing.

I was afraid and stayed in the jeep. I tried not to look, though I knew what lived inside that evil, smoking tent. But I couldn't stop them—how could I—if I told them they would guess how afraid I was, they would guess everything and write to my father. I was so afraid because I knew it was part of the black yew hedge in my garden—it was part of the message that came every evening in the mist. But I couldn't warn them—what would I have said?'

When they got back into the jeep Radha said she'd never have thought that the strange hut could have been put together to extract eucalyptus oil, was that the way it was usually done? And Mark said, it was one of the more primitive ways, yes. And simple, really quite ingenious—to build a tent wigwam sort of shape entirely of eucalyptus leaves and a slow burning fire inside and huge bowls or vats to collect the pure oil.

'Interesting, wasn't it? I wish you'd come with us, Anand, didn't you want to see? You are lazy.'

'So you see, doctor, that's what they thought because that's what they saw, but I knew different. It was a death hut, it was waiting for me, if I had gone with them I would never have been able to come out again. It was the same colour as the yew hedge—and that smoke coming out of it was another kind of mist, waiting for me, for me. And then we went on to the sanctuary—I couldn't change anything I couldn't make them turn back. I was chosen—marked—the only one. I was so afraid.'

And Radha had thought, what more can this marvellous day hold for us, it's been just as Mark had said it would be when he saw that blue kingfisher flash past Anand's ear, oh the very air seemed green, and there was a wonderful stillness, sometimes only our footfalls and the sudden clear call of a bird; there was a clear stream with large, rounded pebbles and we paddled our feet and it was so clear and so cold we cooled our beer bottles in it—small fish darted past the brown, shiny, wet bottles, the most ordinary things had turned into new images of happiness. We found a silvery skin in the water—a water snake perhaps. Mark laid it out in the sun to dry and said I must take it back to remember this day. It's odd but I think I saw, or half saw, Anand push it back into the water—he said it was an accident. I don't think it was. He didn't want me to have that beautiful thing. In the late afternoon we went for an elephant ride deep into the jungle. Our elephant was called Durga.

And Anand told the doctor, 'When the sun was nearly gone we went for an elephant ride, it was to be the grand finale to the day according to Mark because he said now we would see some big game. Our elephant was called Durga and our mahout was a cruel looking man, I didn't like him and he guessed who I was, he took us to the deepest part of the jungle, it was not a jungle at all it was death, it was a kind of lunar landscape—dead burnt-out trees in a desert, trees charred and burnt out, skeletons hung everywhere—and these dead white-yellow shapes that were once trees. And the deathly silence, doctor, it was so quiet—only two sounds in that long, long ride the susurration of dead long grass when the elephant's huge, padding feet brushed past it and the loud terrible thwack and again thwack when that evil mahout beat the elephant on the head, it seemed between his eyes, so close to where I was sitting—and that poor Durga just went on and on and we never saw any wildlife—how could we—it was all dead like the trees, like the grass, like the birds—there was only Durga and the mahout and us—I don't know why Mark or Radha didn't say anything or do anything—we went on and on in that deadness.'

On the journey home, as the jeep took one of the hairpin bends at the crest of the mountains, their headlights picked up a shape. Mark turned off the engine and they watched a beautiful large cat move across the road then turn—great eyes in the headlights, great beautiful cat's eyes—and spring into the jungle. Radha said, 'Oh Mark—like you promised—we did see some wildlife! How beautiful she looked...'

But Anand told the doctor, 'The terror never stopped even when we left the sanctuary. On the way home death stalked me—death looked at me at a turn in that dark road. It had waited all day and found and looked at me—pretending to be a leopard.'

About Mark

Mark was good at most things. At the beginning she realized he was good at all things useful. Shelves that wouldn't hang straight, crooked nails, uneven stone flagging, drainpipes clogged with last autumn's leaves, the Aga in the kitchen—most things. He worked very quietly and unlike most people who are good at useful things he didn't irritate her, didn't ever make her feel she was bad at most things.

She missed him the evenings he couldn't come.

One day she said, 'Tell me about you.' He was oiling a stubborn cupboard door.

'Very little to tell—'

'Tell anyway—when did you come to these hills, to this tea estate?'

'Oh, I came here a long time ago, I was Anand's age—in fact I lived in his cottage—in our family, to use a cliché, we've always been in tea.'

'Your family—and where are they now?'

'There's really only my mother now. I like it here very much. Always have, right from the beginning, that's why I'm sometimes unable to understand Anand. In fact, I wouldn't be anywhere but here. My mother feels I'm lost now.'

'Why lost? Doesn't she like it here? Or doesn't she like it for you?'

'No, not quite that, she's always been used to this sort of life but then there was a family. She'd like me to start a family—she thinks it's easy. I think it's very difficult. In the early years she'd bring friends, you know with daughters—or photographs—sometimes even lists of names. She thought I might get to like someone, she thought I'd be unhappy otherwise. She likes people to marry. Then she's happy. She doesn't believe I'm settled yet. I'll never be more settled, I tell her—it's over fifteen years that I've been here. No other life to live, I tell her. She can't and won't understand that.'

'And now…?'

'Oh, she still comes on her annual visit, she still feels anxious—but things are better now, she has begun to come just to be here with me. I like that. I like her enormously.'

'That's nice.'

'Yes, I like her visits very much, because there's a lot she likes to do—in fact, apart from this worrying about seeing me settled, she's an unusual mother. I think. She loves walking—I mean really walking—she reads a great deal—so, in fact, there's never a worry when she comes. In fact, I have often wondered why it has taken her so long to realize how and why I like the solitude that comes with the territory, so to speak—I obviously take after her…'

'What does she say?'

'She says that's different because she had known the—"other". And now has this—the quiet. The extreme solitude.'

'Yes, I can understand that.'

'About Anand, I mean, she's met him only once. She told me he'd not last. I laughed at her because he had only just come, not settled down at all. It was wrong to decide from that. But now I think about it, about what she said, and it does trouble me. Not a great deal, but I do watch him. I do think about him quite a lot.'

'Let's go to his cottage and take him for a walk, bring him back here, do you think we might?'

'Why not? He might even come. Seeing it's you—I think this door is done.

It won't stick—I'm sure it won't. Also that silly squeak has gone. Wrap up.'

It was cold out—but very nice. They walked quickly. Anand's cottage was in darkness. They unlatched the gate and went in. It made an odd sound in the darkness.

'Anand, Anand,' they called out just in case. After a while they heard his voice. Disembodied.

'Here, I'm here, what is it? Who is it?'

He was at the far end of his garden, in the dark, near the hedge.

He seemed pleased to see them—only he also seemed odd, standing there in the dark, his cottage in darkness. 'Would you like to come, a short walk and back to my place, we'll have something by the fire? Come—do.'

'Yes, I'd like that, only I have to watch this hedge—it's about now that it happens—I can't leave now.'

'What happens? Is it prowlers? Jackals? What happens to the hedge, Anand?'

'Something—I'm not sure yet. That's why I have to watch. I have to wait.'

About Anand

One day she told Mark, 'There's a bird that comes every day and picks away at that perfect Easter lily—every day, ever since it opened completely. I've watched for it—it leaves a single small feather in return. Judging from the feather it's a very small, very plain bird. I feel so angry because it's destroying something so perfect but I can't do anything about it—because I really do understand why the bird is so destructive, after all there's a whole garden of things to choose from, why this most perfect flower—so white with those slashes of scarlet just painted on and with such a beautiful name, amaryllis. The bird doesn't really need to eat it, does it? The bird can't become the lily. Jealousy. It's jealous—I suppose pecking away at it makes it look so much less perfect. Anyway, where do you think we ought to put the bird bath? Mark, are you listening? Do you think near the old copper beech or don't you think the birds will come there? Do you know I saw one morning, very early, a jackal. Really, it was a jackal—and to come so near—I was worried for Judas 2. What could a jackal do to Judas 2? Or is it not a worry? I know so little. I must make a list of things for you—all the things I mean to ask you and tell you about—there's so much—I'm sorry, do I exhaust you with all my questions? Is it all a most frightful bore? But you can take your time, I mean go through the list as slowly as you like, you know, I'll keep adding things. All right? What would I do without you?'

'Oh, you'd manage. You'd manage very well. I can tell.'

'But I'm so glad you're here.'

'I'm glad too.'

'Is Anand all right? I dreamt about him last night. Sometimes—quite often—I worry about him, I'm not sure why.'

'I'm going to see him, would you come with me, let's look him up, all right? Let's go now. I made him take a week off. He looked tired. He looked terrible.'

'Yes, let's go now—very now.'

With a strange sense of urgency they left Old Cypress and walked quickly down the road to Anand's cottage. Even from the gates they could see Anand at the far end of his garden staring at something.

'Oh, Anand, what is it? What are you staring at? Tell me or tell Mark—what is it—what?'

'I can't see what's behind this hedge. It's a yew hedge and it's so dark and heavy it shouldn't be here, it belongs in the graveyard, I can't see beyond it and I have to if I'm to manage—'

'But, Anand, does it really trouble you, because then—'

'Yes, yes, of course, I think about it all the time, I don't like it, it's so tall—dense—it makes a darkness. I don't like that—the sun never comes through and it traps that terrible evening mist—and it keeps coming at me—all the time. It waits for me to get back from work and then it comes for me, the hedge comes and the mist comes. You don't believe me, come inside the house. It's there as well.'

They went with him into the little house—on the round dining table they found a place laid just for one—a meal uneaten—and in the bedroom a bed unslept in, and all the lights on.

'Look, there it is behind that cupboard and there under the bed and behind the bathroom door! Can you see it, can you smell it?'

Yes, yes, they said, and now we are here and so lie down and try to rest, we'll bring something to help...

'I don't want anything except to kill that hedge before it kills me, Ram Swaroop refuses to do it because he says it's old and beautiful, he refuses to cut it or cut it down, so I shall do it because I can't bear it, it's taken over my whole life and I can't see anything. Even in the game sanctuary it was there.'

It's all right, they said, it's all right.

Mark went to fetch the doctor saying, 'Stay with Anand. Don't leave him alone, I'll be back as soon as I can, poor bastard.'

She made him lie down and sat with him and he talked and talked about that thing and loneliness and not being able to see the sun and now there was no sun it had killed the sun and was waiting to kill him but he would kill it first and would she help him, would she, would she and when, because they would have to hurry and had she got an axe they could start soon before the hedge killed him. He was sorry—he was sorry to be so much trouble to her and to Mark but he hadn't told anyone how awful things were—and it was terrible to be like this to feel like this—he didn't dare sleep in case it came for him while he was asleep, the hedge, that dark thing, would she please look out of the window and see if it had moved and thank you for staying with me—that time when she had come and they'd had tea with Mark—that long-ago time he had been so happy so pleased that she was coming to live here and so near him in Old Cypress but she'd taken so long to come—why had she taken so long? The hedge had grown and grown taller and taller and now she mustn't go away again, was she going? His mother didn't know— nobody knew—about his fear. He hadn't told his mother because she would have told his father he was a failure. His father would think he was a failure. She wouldn't, would she, tell his father? He wanted to know if she thought he was a failure, no good at anything? But he was afraid, really, up here in the silence and the loneliness he was afraid. He heard that hedge in the night and sometimes even in the day, could she hear it?

Try to rest, perturbed child, try, try.

You won't go away will you?

No. No.

He was afraid to close his eyes. It might come.

Didn't you like that little tea-party that long-ago time at Mark's?

So long ago—the sun never comes.

But it wasn't so long ago, perhaps it was. In soul time, in despair time it was a long, long time. Why, she had been happy—she and Rohan. Then. Odd. After seeing Old Cypress they had gone to Anand's cottage and then gone back to Mark's for tea. A warm room with a fire in the hearth and the curtains drawn and Velu as amazing as he had sounded when Mark had spoken about his prowess at scones. It had been a magical evening, the friendliness and the warmth and the scones and the smell of wood-smoke and the wonder that they had found Old Cypress and now it was theirs. But. There had been a hidden and dark thing. Not just that fear that Anand had—but. The fact that she hadn't known about Rohan's woman. And Rohan was hiding his dark secret, and Anand was hiding the terrible fears he had—hiding them from Mark and his mother and his father, so there they all were. Nobody knowing about anyone else. Nobody with intuition. How could she blame Mark for not realizing about Anand's deep, terrible fear? Each of them had had something to hide, something to pretend about. She herself, who had always prided herself on her awareness—her sensitive awareness about things—she had been proved the biggest fool of all. She had not been aware of the slightest hint of anything between Rohan and his 'chick'. They'd all been together in the firelight, friendly and warm, laughing and talking and together—or at any rate pretending so well—to be so happy together. So much for intuition, so much for good simulation. What worlds away—what worlds undone—and all less than six months ago. Not possible. Don't think. Don't think. Don't think about yourself. Think only about Anand. Hold his hand.

After what seemed hours she heard a car door slam. She heard voices. Anand said, 'Don't go, don't leave me. I wish I was like Mark, he's not afraid of anything. He's alone and he's not afraid. Ever.'

'Don't think like that, it's just that you're not well. Soon you'll be fine. Now that the doctor has come. Everything will be fine. You'll see. I promise.'

'Are you ever frightened?' Yes, oh yes, often I am. But now we won't let you be alone, all right? All right. We'll all be together. Hang together. Please God, let Anand get well again. Make him not be frightened. Let him find some light not darkness all the time. Let him make it. Please.

Flora and the Bird bath

Come on dog, old faithful, best friend, grow old along with me—not yet though—why must you keep sitting near that crate, why not one of the other ones? She came and sat down close to the crate and watched Dog crouching even closer—what is it? What? What are you watching, waiting for—ears very keen, hair bristling, what? And then she saw what it was. A squirrel in the crate—a very small face emerged through an opening, expression watchful, disappeared again. Dog, even more watchful, sprang. Two paws atop the crate quivering with the insult, barking. On the crate in black paint she saw—'This Side Up—Handle With Care—Flora, Bird bath'. That time with the dog sitting next to her she'd packed them with great care and not expertise in old newspaper and soft old clothes. Well, they'd have to come out now. Clever squirrel—I'll do it today—now. Why on earth did I wait so long—because my heart with rue was laden.

Once, long ago, Rohan and the child on a Saturday morning disappeared not taking her—she had thought for a haircut or some such thing. Appeared again having combed Chor Bazaar and found these two. Dusty, cobwebby and perfect. 'Flora' they had instantly named the first. A marvellous woman in marble with one great wing and one most delicate hand holding a lily—broken lily but clearly a lily—this most perfect woman set in an old heavy frame out of which a lizard came startling them all—'It's for you, Mom, for your birthday, it's for when we have a garden, Mom.' Time—God, that was years ago, the child now twenty-seven, then what? Still little, still able to shout with excitement, still hug, still hold one's hand. So what did one feel then, feel now? Rohan ever so slightly embarrassed, shuffling his Saturday-chappalled feet, so pleased at having found such perfect presents, knowing how pleased I was, feeling so loving enabling me to express what I felt. That was Flora. And then with a flourish—that golden, raining-gold Saturday—emerged out of the boot of the car—'Close your eyes, Mom, don't look don't look till we tell you'—Bird bath—perfection, pure genius for them to have found such treasure and now she had a garden. Find the pliers, get on with things. Start.

Good old squirrel for making her realize it was time to start unpacking. My birthday came and went. All our birthdays came and went. A family. Find the pliers. When Rohan had finally told her about his 'chick' she'd said, 'Dog is mine he comes with me and Flora and the bird bath.' 'Naturally—they're yours, after all...' He could have meant anything—tone-wise he'd never given much away. She had persisted, 'Would you like to keep Flora and I'll take the bird bath?'

'Yes,' he had said abstractedly.

'Yes what—? Will you keep Flora?'

'No, no, I don't want her, for God's sake I don't want anything—'

'Except my life—except my life,' bitterly she thought now, and hoped she had said then. And now Flora and the bird bath with its two perched, ready-to-fly birds. She would put them in the garden, near the far wall, and real birds would come to drink—at last. When did time pass? When did time happen? When?

Letter from Radha to Her Son

It's been a very long time since I wrote to you. But then I can't really remember when last I had word from you. When you were much younger you used to get rather embarrassed when your father and I quarrelled—we always did, quarrel I mean, but those were important quarrels. Necessary. I suppose what has happened now between us has made you feel you can't communicate with either of us—I do hope not for too long. And don't feel divided. I suppose what I really mean is, don't put it off for too long. I miss you. Especially now. I don't think sons should die before their parents. I mean children. A young boy, man, died in our hills and it has just happened and it has distressed me greatly, enough to make me write to you. I miss you and if you felt like visiting—it would not put too great a strain on either of us! Do you think? I know how busy you are, always. But. Permit me to use that old cliché. Life is short. I suppose a lot shorter than we think.

The boy was called Anand. I mean, the boy who died. He had a father, unlike yours, who pushed him into a way of life that was really impossible for him. He got lonelier and more and more lost. Then he became very ill. Now he is lost. He is lost to us. Us is Mark and me.

You might like to see Judas 2. Also what Flora looks like and the bird bath—in a garden.

You don't have to come—but seeing as we were once quite good friends— why not?

Why not indeed! Ay, there's the rub. I shouldn't think your father would mind, in fact, knowing him, he'll be rather relieved somebody is checking on how I'm getting on—as long as it's not him!

Do come—if the spirit moves you—

There are some good walks—the weather's very nice when it's nice.

Very quiet.

Also there is in my back garden, near the tennis court, a very old cypress.

Love, of course

Mama

Letter from a Dear Friend Who Wrote

This morning a letter—how could a friend know, exactly know, however dear, dear, dearest friend know enough to write: 'Loss, I think is an opaque pane of glass in a window, vista lost, but also, in direct contradiction, it is a winding pathway into the past—think you will, of course you will, but try, try my friend to stay buoyant—all the days, early days till now, where memories and Rohan whom you loved—love—cling like burrs. Just keep going, live ordinary days, slipping, sliding, ordinary days and you'll work it out. Bleak days for a long, long time—but remember why you are there in those hills you chose, and live. Live all you can. Soon.'

That's how the letter went and she read it several times and then folded it very carefully and tucked it into the pocket of her shirt. She went down the steps and into the garden and to the swing that Mark had set up for her—this was her own swing. She had told him that once on a journey to another hill station they had stopped at a small hotel just before the ghats began. While Rohan was busy filling up petrol and checking the water she had seen a little, bright, fussy playground. There was a slide with sand to land in, a see-saw, a miniature merry-go-round and, her best love, a swing. There were no children around and she had gone and sat on the swing with the lovely sense of how she would swing really high and then come rushing down with a familiar feel of fear in her stomach, when a man in a red and white uniform came up to her saying rudely and rather loudly, 'Perhaps you cannot read—see that notice over there.' She had left the swing feeling silly and sad and angry with the odd notice that hung on a railing. The notice said, *'Children Below The Age of 12 Years and Adults Above The Age Of 12—Strictly Not Allowed to Play.'* Spoilers of fun she had thought and anyway what on earth did they mean? She'd gone back to the car and told Rohan about it, spoilers of fun she had told him, and he had said, 'For God's sake, act your age.' After a bit she'd stopped being angry with Rohan and the notice but she hadn't forgotten. That long-ago time. Standing in the garden one day she had told Mark about it and he'd said, 'Everyone needs to, wants to, has to, once in a lifetime have a swing of their own—and I know a perfect place for yours. Come and see if you agree.' And it was perfect, of course, between two of the trees quite close to the old wall in her front garden—so she had sun and she had shade and Mark had said she could read there or do nothing there because that's what good swings were for. It was a very good swing. She sat down and began to swing, first very gently kicking her feet but feeling the grass and then a little higher and higher and then really high with that wonderful feeling of fear and exhilaration in the pit of her stomach. 'And I ride'—yes I ride, she thought. Oh and I'm lucky, I'm really lucky, and I've just realized it, I've been stupid and selfish and a misery. Here I am swinging, sometimes so high I can see over the wall. And what do most people have as they grow older—what do they have for the vistas stretching ahead—I mean, as they grow older and older, not really needed by anyone in that very real sense any more, what? Living with their grown-up children—more than a little unwanted—facing a smaller and smaller space, trying to be unobtrusive, making oneself invisible, finally not even a room of their own, a part of a room—sharing and not wanted, babysitting in a flat, gazing out of a window or a bit of a balcony—one dusty potted plant and a teeny bit of sky—upwards a kite, a crow, a cloud and downwards people hurrying—people running-off-to-work, or back-from-work sounds. Apart. Not wanted—a bit of a bore if truth were told. And then the next stage and the next. I'm lucky, she almost shouted. I am. I have. A home. A garden. Trees. This swing. The sun. My limbs and the use of them. Sky. Old Cypress. And later, much much later, if it's a cane or a wheelchair, I can be everywhere still, not a bother to anyone. Worrying about help at lifts and stairs and at traffic lights. I'll be me and so near there'll be Mark (because he's going to live forever) he'll look in from time to time and see that all is well. Truly, this is the life. Freedom, and I don't and won't depend on anyone to be sad or happy. Me. And I have friends. And I am a friend to some. On my own terms. They like me. I like them. There'll always be books, there'll always be music. So it will all work out. The letter was really saying 'Hang in there' and I will, I am. Doing just that. So it doesn't matter about the old tennis court, nobody will play there any more perhaps. But they did play there once. So it's a legacy. And if I hear the sounds of people playing in the dead hours of the night I'll not be afraid. I should be happy—they are happy sounds—people happy, calling out to each other, friendly, playing-tennis sort of sounds. I hear them. So they are calling out to me. Happily. They lived here. Now I live here. They are as important as I am. They don't resent me. And Old Cypress protects me, protects us all. Living, I must go now and tell Mark about this. He'll be busy but he'll understand that this is important. So I'll tell him. And he'll know. He'll know exactly what has happened.

Reply to a Letter from a Dear Friend Who Wrote

Dear friend you thought to write—thank you. You must know, of course, that I mean much more than thank you. Also, I could not write for so long, though I meant to and now I am. Writing, I mean. You wrote to me of loss—you wrote of memories that cling like burrs—of trying to remain buoyant. How did you know, how can you know so well? Thank you for knowing and for writing as you did. It is summer now. But then it was still autumn and a long bleak winter. It is over now, I mean the very cold and most of what went with it. Unpacking was bleak and lonely but my home has begun to seem lived-in and Judas 2 has begun to have his favourite places in the sun and in the shade and close to the fire when I light it. Do I make it seem idyllic? It's because I leave out what's bad or sad or very bleak. Already we have a death. What's worst is the nights. No one to talk to when I can't sleep or wake in fear. No familiar person next to me in bed. Familiar hand, familiar comfort, silly old jokes, old habits. No one to share things with. Also there were strange sounds in the night from the tennis court at the back and the tree that names my home never looked too friendly at night. But. It's summer now and the sun is so friendly and bits and pieces of the garden look ready for you to see. Why not come? If the spirit moves you. I mean it. The garden isn't quite what it will be soon but in a strange sort of way it has begun to take shape, soon I shall feel a healthy, instinctive anger if anyone should steal a white hibiscus just opened (my side of the wall) because it's mine. I enjoyed planting it, so the rewards are mine. Also, I cannot bear to mow the lawn in one miraculous corner near the old wall because it's thick with daisies—I don't want to see their lopped, chopped heads flying—even if Ram Swaroop disapproves (he's my one-hour-a-day mali shared with Mark). So whatever the rules about manicured lawns—that corner is snow-like grass. I think my garden at best will be like a disciplined jungle, no, not jungle, a clearing in a forest. To rest, to breathe, to be at peace. Sometimes a white pigeon comes and drinks from the bird bath—why not come?

Now I will learn the name of ferns and wild flowers and weeds. Now I will learn to recognize birds. Also their calls.

Now I will do many things.

I might bottle fruit. Bake bread. Make jams. I might. I might think about Anand. I might try to grieve less. Think about death less. Listen to music and not be sad.

It's odd though, it's not all gloom now as it used to be. Of course, sometimes, some days, it feels a bit bleak. The prospect before and after—but that might happen anyway. Sometimes I'm quite peaceful—peaceable—happy, like now. Quite content to look at the sun creeping over this wall and the difference it makes to the moss growing on it out of its old crevices. It glistens. Of course, sometimes when birds come to the bird bath I do hear an excited voice calling out, 'It's for our garden, Mom, Pop found it, I—I helped him find it, it's for your birthday, Mom.' Well, that's natural too. Nothing really lasts. But it did. For a long time. So. That was the nature of our days.

Now this is the nature of my days.

'Sometimes it gets a wee bit lonely in these hills especially towards evening— but keep your eyes firmly fixed upwards—on the trees and the hills—you'll be all right then.'

I remember the woman who got into the train when it started to climb— that's what she said. She knew.

She said it like a warning—she said it like a benediction.

I try hard—sometimes I fail. Most times.
Because, well, because.
The poets say it best, remember that Carver fragment?

Fear of anxiety.
Fear of having to identify the
body of a dead friend.
Fear of running out of money.
Fear of confusion.
Fear of waking up to find you
gone.
Fear of death.
Fear of living too long.
Fear of death.
I've said that.

THE BLUE UMBRELLA

RUSKIN BOND

I

'Neelu! Neelu!' cried Binya.

She scrambled barefoot over the rocks, ran over the short summer grass, up and over the brow of the hill, all the time calling 'Neelu, Neelu!' Neelu—Blue—was the name of the blue-grey cow. The other cow, which was white, was called Gori, meaning Fair One. They were fond of wandering off on their own, down to the stream or into the pine forest, and sometimes they came back by themselves and sometimes they stayed away—almost deliberately, it seemed to Binya.

If the cows didn't come home at the right time, Binya would be sent to fetch them. Sometimes her brother, Bijju, went with her, but these days he was busy preparing for his exams and didn't have time to help with the cows.

Binya liked being on her own, and sometimes she allowed the cows to lead her into some distant valley, and then they would all be late coming home. The cows preferred having Binya with them, because she let them wander. Bijju pulled them by their tails if they went too far.

Binya belonged to the mountains, to this part of the Himalayas known as Garhwal. Dark forests and lonely hilltops held no terrors for her. It was only when she was in the market town, jostled by the crowds in the bazaar, that she felt rather nervous and lost. The town, five miles from the village, was also a pleasure resort for tourists from all over India.

Binya was probably ten. She may have been nine or even eleven, she couldn't be sure because no one in the village kept birthdays; but her mother told her she'd been born during a winter when the snow had come up to the windows, and that was just over ten years ago, wasn't it? Two years later, her father had died, but his passing had made no difference to their way of life. They had three tiny terraced fields on the side of the mountain, and they grew potatoes, onions, ginger, beans, mustard and maize: not enough to sell in the town, but enough to live on.

Like most mountain girls, Binya was quite sturdy, fair of skin, with pink cheeks and dark eyes and her black hair tied in a pigtail. She wore pretty glass bangles on her wrists, and a necklace of glass beads. From the necklace hung a leopard's claw. It was a lucky charm, and Binya always wore it. Bijju had one, too, only his was attached to a string.

Binya's full name was Binyadevi, and Bijju's real name was Vijay, but everyone called them Binya and Bijju. Binya was two years younger than her brother.

She had stopped calling for Neelu; she had heard the cowbells tinkling, and knew the cows hadn't gone far. Singing to herself, she walked over fallen pine needles into the forest glade on the spur of the hill. She heard voices, laughter, the clatter of plates and cups and, stepping through the trees, she came upon a party of picnickers.

They were holidaymakers from the plains. The women were dressed in bright saris, the men wore light summer shirts, and the children had pretty new clothes. Binya, standing in the shadows between the trees, went unnoticed; for some time she watched the picnickers, admiring their clothes, listening to their unfamiliar accents, and gazing rather hungrily at the sight of all their food. And then her gaze came to rest on a bright blue umbrella, a frilly thing for women, which lay open on the grass beside its owner.

Now Binya had seen umbrellas before, and her mother had a big black umbrella which nobody used any more because the field rats had eaten holes in it, but this was the first time Binya had seen such a small, dainty, colourful umbrella and she fell in love with it. The umbrella was like a flower, a great blue flower that had sprung up on the dry brown hillside.

She moved forward a few paces so that she could see the umbrella better. As she came out of the shadows into the sunlight, the picnickers saw her.

'Hello, look who's here!' exclaimed the older of the two women. 'A little village girl!'

'Isn't she pretty?' remarked the other. 'But how torn and dirty her clothes are!' It did not seem to bother them that Binya could hear and understand everything they said about her.

'They're very poor in the hills,' said one of the men.

'Then let's give her something to eat.' And the older woman beckoned to Binya to come closer.

Hesitantly, nervously, Binya approached the group.

Normally she would have turned and fled, but the attraction was the pretty blue umbrella. It had cast a spell over her, drawing her forward almost against her will.

'What's that on her neck?' asked the younger woman.

'A necklace of sorts.'

'It's a pendant—see, there's a claw hanging from it!'

'It's a tiger's claw,' said the man beside her. (He had never seen a tiger's claw.) 'A lucky charm. These people wear them to keep away evil spirits.' He looked to Binya for confirmation, but Binya said nothing.

'Oh, I want one too!' said the woman, who was obviously his wife.

'You can't get them in shops.'

'Buy hers, then. Give her two or three rupees, she's sure to need the money.'

The man, looking slightly embarrassed but anxious to please his young wife, produced a two-rupee note and offered it to Binya, indicating that he wanted the pendant in exchange. Binya put her hand to the necklace, half afraid that the excited woman would snatch it away from her. Solemnly she shook her head.

The man then showed her a five-rupee note, but again Binya shook her head.

'How silly she is!' exclaimed the young woman.

'It may not be hers to sell,' said the man. 'But I'll try again. How much do you want—what can we give you?' And he waved his hand towards the picnic things scattered about on the grass.

Without any hesitation Binya pointed to the umbrella.

'My umbrella!' exclaimed the young woman. 'She wants my umbrella. What cheek!'

'Well, you want her pendant, don't you?'

'That's different.'

'Is it?'

The man and his wife were beginning to quarrel with each other.

'I'll ask her to go away,' said the older woman.

'We're making such fools of ourselves.'

'But I want the pendant!' cried the other, petulantly.

And then, on an impulse, she picked up the umbrella and held it out to Binya.

'Here, take the umbrella!'

Binya removed her necklace and held it out to the young woman, who immediately placed it around her own neck. Then Binya took the umbrella and held it up. It did not look so small in her hands; in fact, it was just the right size.

She had forgotten about the picnickers, who were busy examining the pendant. She turned the blue umbrella this way and that, looked through the bright blue silk at the pulsating sun, and then, still keeping it open, turned and disappeared into the forest glade.

II

Binya seldom closed the blue umbrella. Even when she had it in the house, she left it lying open in a corner of the room. Sometimes Bijju snapped it shut, complaining that it got in the way. She would open it again a little later. It wasn't beautiful when it was closed.

Whenever Binya went out—whether it was to graze the cows, or fetch water from the spring, or carry milk to the little tea shop on the Tehri road— she took the umbrella with her. That patch of sky-blue silk could always be seen on the hillside.

Old Ram Bharosa (Ram the Trustworthy) kept the tea shop on the Tehri road. It was a dusty, un-metalled road. Once a day, the Tehri bus stopped near his shop and passengers got down to sip hot tea or drink a glass of curd. He kept a few bottles of Coca-Cola too, but as there was no ice, the bottles got hot in the sun and so were seldom opened. He also kept sweets and toffees, and when Binya or Bijju had a few coins to spare, they would spend them at the shop. It was only a mile from the village.

Ram Bharosa was astonished to see Binya's blue umbrella.

'What have you there, Binya?' he asked.

Binya gave the umbrella a twirl and smiled at Ram Bharosa. She was always ready with her smile, and would willingly have lent it to anyone who was feeling unhappy.

'That's a lady's umbrella,' said Ram Bharosa. 'That's only for memsahibs. Where did you get it?'

'Someone gave it to me—for my necklace.'

'You exchanged it for your lucky claw!'

Binya nodded.

'But what do you need it for? The sun isn't hot enough, and it isn't meant for the rain. It's just a pretty thing for rich ladies to play with!'

Binya nodded and smiled again. Ram Bharosa was quite right; it was just a beautiful plaything. And that was exactly why she had fallen in love with it.

'I have an idea,' said the shopkeeper. 'It's no use to you, that umbrella. Why not sell it to me? I'll give you five rupees for it.'

'It's worth fifteen,' said Binya.

'Well, then, I'll give you ten.'

Binya laughed and shook her head.

'Twelve rupees?' said Ram Bharosa, but without much hope.

Binya placed a five-paise coin on the counter.

'I came for a toffee,' she said.

Ram Bharosa pulled at his drooping whiskers, gave Binya a wry look, and placed a toffee in the palm of her hand. He watched Binya as she walked away along the dusty road. The blue umbrella held him fascinated, and he stared after it until it was out of sight.

The villagers used this road to go to the market town. Some used the bus, a few rode on mules and most people walked. Today, everyone on the road turned their heads to stare at the girl with the bright blue umbrella.

Binya sat down in the shade of a pine tree. The umbrella, still open, lay beside her. She cradled her head in her arms, and presently she dozed off. It was that kind of day, sleepily warm and summery.

And while she slept, a wind sprang up.

It came quietly, swishing gently through the trees, humming softly. Then it was joined by other random gusts, bustling over the tops of the mountains. The trees shook their heads and came to life. The wind fanned Binya's cheeks. The umbrella stirred on the grass.

The wind grew stronger, picking up dead leaves and sending them spinning and swirling through the air. It got into the umbrella and began to drag it over the grass. Suddenly it lifted the umbrella and carried it about six feet from the sleeping girl. The sound woke Binya.

She was on her feet immediately, and then she was leaping down the steep slope. But just as she was within reach of the umbrella, the wind picked it up again and carried it further downhill.

Binya set off in pursuit. The wind was in a wicked, playful mood. It would leave the umbrella alone for a few moments but as soon as Binya came near, it would pick up the umbrella again and send it bouncing, floating, dancing away from her.

The hill grew steeper. Binya knew that after twenty yards it would fall away in a precipice. She ran faster. And the wind ran with her, ahead of her, and the blue umbrella stayed up with the wind.

A fresh gust picked it up and carried it to the very edge of the cliff. There it balanced for a few seconds, before toppling over, out of sight.

Binya ran to the edge of the cliff. Going down on her hands and knees, she peered down the cliff face. About a hundred feet below, a small stream rushed between great boulders. Hardly anything grew on the cliff face—just a few stunted bushes, and, halfway down, a wild cherry tree growing crookedly out of the rocks and hanging across the chasm. The umbrella had stuck in the cherry tree.

Binya didn't hesitate. She may have been timid with strangers, but she was at home on a hillside. She stuck her bare leg over the edge of the cliff and began climbing down. She kept her face to the hillside, feeling her way with her feet, only changing her handhold when she knew her feet were secure. Sometimes she held on to the thorny bilberry bushes, but she did not trust the other plants, which came away very easily.

Loose stones rattled down the cliff. Once on their way, the stones did not stop until they reached the bottom of the hill; and they took other stones with them, so that there was soon a cascade of stones, and Binya had to be very careful not to start a landslide.

As agile as a mountain goat, she did not take more than five minutes to reach the crooked cherry tree. But the most difficult task remained—she had to crawl along the trunk of the tree, which stood out at right angles from the cliff. Only by doing this could she reach the trapped umbrella.

Binya felt no fear when climbing trees. She was proud of the fact that she could climb them as well as Bijju. Gripping the rough cherry bark with her toes, and using her knees as leverage, she crawled along the trunk of the projecting tree until she was almost within reach of the umbrella. She noticed with dismay that the blue cloth was torn in a couple of places.

She looked down, and it was only then that she felt afraid. She was right over the chasm, balanced precariously about eighty feet above the boulder-strewn stream. Looking down, she felt quite dizzy. Her hands shook, and the tree shook too. If she slipped now, there was only one direction in which she could fall—down, down, into the depths of that dark and shadowy ravine.

There was only one thing to do; concentrate on the patch of blue just a couple of feet away from her. She did not look down or up, but straight ahead, and willing herself forward, she managed to reach the umbrella.

She could not crawl back with it in her hands. So, after dislodging it from the forked branch in which it had stuck, she let it fall, still open, into the ravine below.

Cushioned by the wind, the umbrella floated serenely downwards, landing in a thicket of nettles.

Binya crawled back along the trunk of the cherry tree. Twenty minutes later, she emerged from the nettle clump, her precious umbrella held aloft. She had nettle stings all over her legs, but she was hardly aware of the smarting. She was as immune to nettles as Bijju was to bees.

III

About four years previously, Bijju had knocked a hive out of an oak tree, and had been badly stung on the face and legs. It had been a painful experience. But now, if a bee stung him, he felt nothing at all: he had been immunized for life!

He was on his way home from school. It was two o'clock and he hadn't eaten since six in the morning. Fortunately, the kingora bushes—the bilberries—were in fruit, and already Bijju's lips were stained purple with the juice of the wild, sour fruit.

He didn't have any money to spend at Ram Bharosa's shop, but he stopped there anyway to look at the sweets in their glass jars.

'And what will you have today?' asked Ram Bharosa.

'No money,' said Bijju.

'You can pay me later.'

Bijju shook his head. Some of his friends had taken sweets on credit, and at the end of the month they had found they'd eaten more sweets than they could possibly pay for! As a result, they'd had to hand over to Ram Bharosa some of their most treasured possessions—such as a curved knife for cutting grass, or a small hand-axe, or a jar for pickles, or a pair of earrings—and these had become the shopkeeper's possessions and were kept by him or sold in his shop.

Ram Bharosa had set his heart on having Binya's blue umbrella, and so naturally he was anxious to give credit to either of the children, but so far neither had fallen into the trap.

Bijju moved on, his mouth full of kingora berries. Halfway home, he saw Binya with the cows. It was late evening, and the sun had gone down, but Binya still had the umbrella open. The two small rents had been stitched up by her mother.

Bijju gave his sister a handful of berries. She handed him the umbrella while she ate the berries.

'You can have the umbrella until we get home,' she said. It was her way of rewarding Bijju for bringing her the wild fruit.

Calling 'Neelu! Gori!' Binya and Bijju set out for home, followed at some distance by the cows.

It was dark before they reached the village, but Bijju still had the umbrella open.

—

Most of the people in the village were a little envious of Binya's blue umbrella. No one else had ever possessed one like it. The schoolmaster's wife thought it was quite wrong for a poor cultivator's daughter to have such a fine umbrella while she, a second-class BA, had to make do with an ordinary black one. Her husband offered to have their old umbrella dyed blue; she gave him a scornful look, and loved him a little less than before. The pujari, who looked after the temple, announced that he would buy a multi-coloured umbrella the next time he was in the town. A few days later he returned looking annoyed and grumbling that they weren't available except in Delhi. Most people consoled themselves by saying that Binya's pretty umbrella wouldn't keep out the rain, if it rained heavily; that it would shrivel in the sun, if the sun was fierce; that it would collapse in a wind, if the wind was strong; that it would attract lightning, if lightning fell near it; and that it would prove unlucky, if there was any ill luck going about. Secretly, everyone admired it.

Unlike the adults, the children didn't have to pretend. They were full of praise for the umbrella. It was so light, so pretty, so bright a blue! And it was just the right size for Binya. They knew that if they said nice things about the umbrella, Binya would smile and give it to them to hold for a little while—just a very little while!

Soon it was the time of the monsoon. Big black clouds kept piling up, and thunder rolled over the hills.

Binya sat on the hillside all afternoon, waiting for the rain. As soon as the first big drop of rain came down, she raised the umbrella over her head. More drops, big ones, came pattering down. She could see them through the umbrella silk as they broke against the cloth.

And then there was a cloudburst, and it was like standing under a waterfall. The umbrella wasn't really a rain umbrella, but it held up bravely. Only Binya's feet got wet. Rods of rain fell around her in a curtain of shivered glass.

Everywhere on the hillside people were scurrying for shelter. Some made for a charcoal burner's hut, others for a mule-shed, or Ram Bharosa's shop. Binya was the only one who didn't run. This was what she'd been waiting for—rain on her umbrella—and she wasn't in a hurry to go home. She didn't mind getting her feet wet. The cows didn't mind getting wet either.

Presently she found Bijju sheltering in a cave. He would have enjoyed getting wet, but he had his schoolbooks with him and he couldn't afford to let them get spoilt. When he saw Binya, he came out of the cave and shared the umbrella. He was a head taller than his sister, so he had to hold the umbrella for her, while she held his books.

The cows had been left far behind.

'Neelu, Neelu!' called Binya.

'Gori!' called Bijju.

When their mother saw them sauntering home through the driving rain, she called out: 'Binya! Bijju! Hurry up, and bring the cows in! What are you doing out there in the rain?'

'Just testing the umbrella,' said Bijju.

IV

The rains set in, and the sun only made brief appearances. The hills turned a lush green. Ferns sprang up on walls and tree trunks. Giant lilies reared up like leopards from the tall grass. A white mist coiled and uncoiled as it floated up from the valley. It was a beautiful season, except for the leeches.

Every day, Binya came home with a couple of leeches fastened to the flesh of her bare legs. They fell off by themselves just as soon as they'd had their thimbleful of blood, but you didn't know they were on you until they fell off, and then, later, the skin became very sore and itchy. Some of the older people still believed that to be bled by leeches was a remedy for various ailments. Whenever Ram Bharosa had a headache, he applied a leech to his throbbing temple.

Three days of incessant rain had flooded out a number of small animals who lived in holes in the ground. Binya's mother suddenly found the roof full of field rats. She had to drive them out; they ate too much of her stored-up wheat flour and rice. Bijju liked lifting up large rocks to disturb the scorpions who were sleeping beneath. And snakes came out to bask in the sun.

Binya had just crossed the small stream at the bottom of the hill when she saw something gliding out of the bushes and coming towards her. It was a long black snake. A clatter of loose stones frightened it. Seeing the girl in its way, it rose up, hissing, prepared to strike. The forked tongue darted out, the venomous head lunged at Binya.

Binya's umbrella was open as usual. She thrust it forward, between herself and the snake, and the snake's hard snout thudded twice against the strong silk of the umbrella. The reptile then turned and slithered away over the wet rocks, disappearing into a clump of ferns.

Binya forgot about the cows and ran all the way home to tell her mother how she had been saved by the umbrella. Bijju had to put away his books and go out to fetch the cows. He carried a stout stick, in case he met with any snakes.

First the summer sun, and now the endless rain, meant that the umbrella was beginning to fade a little: From a bright blue it had changed to a light blue. But it was still a pretty thing, and tougher than it looked, and Ram Bharosa still desired it. He did not want to sell it; he wanted to own it. He was probably the richest man in the area—so why shouldn't he have a blue umbrella? Not a day passed without his getting a glimpse of Binya and the umbrella; and the more he saw the umbrella, the more he wanted it.

The schools closed during the monsoon, but this didn't mean that Bijju could sit at home doing nothing. Neelu and Gori were providing more milk than was required at home, so Binya's mother was able to sell a kilo of milk every day: half a kilo to the schoolmaster, and half a kilo (at reduced rate) to the temple pujari. Bijju had to deliver the milk every morning.

Ram Bharosa had asked Bijju to work in his shop during the holidays, but Bijju didn't have time—he had to help his mother with the ploughing and the transplanting of the rice seedlings. So Ram Bharosa employed a boy from the next village, a boy called Rajaram. He did all the washing-up, and ran various errands. He went to the same school as Bijju, but the two boys were not friends.

One day, as Binya passed the shop, twirling her blue umbrella, Rajaram noticed that his employer gave a deep sigh and began muttering to himself.

'What's the matter, babuji?' asked the boy.

'Oh, nothing,' said Ram Bharosa. 'It's just a sickness that has come upon me. And it's all due to that girl Binya and her wretched umbrella.'

'Why, what has she done to you?'

'Refused to sell me her umbrella! There's pride for you. And I offered her ten rupees.'

'Perhaps, if you gave her twelve...'

'But it isn't new any longer. It isn't worth eight rupees now. All the same, I'd like to have it.'

'You wouldn't make a profit on it,' said Rajaram.

'It's not the profit I'm after, wretch! It's the thing itself. It's the beauty of it!'

'And what would you do with it, babuji? You don't visit anyone—you're seldom out of your shop. Of what use would it be to you?'

'Of what use is a poppy in a cornfield? Of what use is a rainbow? Of what use are you, numbskull? Wretch! I, too, have a soul. I want the umbrella, because—because I want its beauty to be mine!'

Rajaram put the kettle on to boil, began dusting the counter, all the time muttering: 'I'm as useful as an umbrella,' and then, after a short period of intense thought, said: 'What will you give me, babuji, if I get the umbrella for you?'

'What do you mean?' asked the old man.

'You know what I mean. What will you give me?'

'You mean to steal it, don't you, you wretch? What a delightful child you are! I'm glad you're not my son or my enemy. But look, everyone will know it has been stolen, and then how will I be able to show off with it?'

'You will have to gaze upon it in secret,' said Rajaram with a chuckle. 'Or take it into Tehri, and have it coloured red! That's your problem. But tell me, babuji, do you want it badly enough to pay me three rupees for stealing it without being seen?'

Ram Bharosa gave the boy a long, sad look. 'You're a sharp boy,' he said. 'You'll come to a bad end. I'll give you two rupees.'

'Three,' said the boy.

'Two,' said the old man.

'You don't really want it, I can see that,' said the boy.

'Wretch!' said the old man. 'Evil one! Darkener of my doorstep! Fetch me the umbrella, and I'll give you three rupees.'

V

Binya was in the forest glade where she had first seen the umbrella. No one came there for picnics during the monsoon. The grass was always wet and the pine needles were slippery underfoot. The tall trees shut out the light, and poisonous-looking mushrooms, orange and purple, sprang up everywhere. But it was a good place for porcupines, who seemed to like the mushrooms, and Binya was searching for porcupine quills.

The hill people didn't think much of porcupine quills, but far away in southern India, the quills were valued as charms and sold at a rupee each. So Ram Bharosa paid a tenth of a rupee for each quill brought to him, and he in turn sold the quills at a profit to a trader from the plains.

Binya had already found five quills, and she knew there'd be more in the long grass. For once, she'd put her umbrella down. She had to put it aside if she was to search the ground thoroughly.

It was Rajaram's chance.

He'd been following Binya for some time, concealing himself behind trees and rocks, creeping closer whenever she became absorbed in her search. He was anxious that she should not see him and be able to recognize him later.

He waited until Binya had wandered some distance from the umbrella. Then, running forward at a crouch, he seized the open umbrella and dashed off with it.

But Rajaram had very big feet. Binya heard his heavy footsteps and turned just in time to see him as he disappeared between the trees. She cried out, dropped the porcupine quills, and gave chase.

Binya was swift and sure-footed, but Rajaram had a long stride. All the same, he made the mistake of running downhill. A long-legged person is much faster going uphill than down. Binya reached the edge of the forest glade in time to see the thief scrambling down the path to the stream. He had closed the umbrella so that it would not hinder his flight.

Binya was beginning to gain on the boy. He kept to the path, while she simply slid and leapt down the steep hillside. Near the bottom of the hill the path began to straighten out, and it was here that the long-legged boy began to forge ahead again.

Bijju was coming home from another direction. He had a bundle of sticks which he'd collected for the kitchen fire. As he reached the path, he saw Binya rushing down the hill as though all the mountain spirits in Garhwal were after her.

'What's wrong?' he called. 'Why are you running?'

Binya paused only to point at the fleeing Rajaram.

'My umbrella!' she cried. 'He has stolen it!'

Bijju dropped his bundle of sticks, and ran after his sister. When he reached her side, he said, 'I'll soon catch him!' and went sprinting away over the lush green grass. He was fresh, and he was soon well ahead of Binya and gaining on the thief.

Rajaram was crossing the shallow stream when Bijju caught up with him. Rajaram was the taller boy, but Bijju was much stronger. He flung himself at the thief, caught him by the legs, and brought him down in the water. Rajaram got to his feet and tried to drag himself away, but Bijju still had him by a leg. Rajaram overbalanced and came down with a great splash. He had let the umbrella fall. It began to float away on the current. Just then Binya arrived, flushed and breathless, and went dashing into the stream after the umbrella.

Meanwhile, a tremendous fight was taking place. Locked in fierce combat, the two boys swayed together on a rock, tumbled on to the sand, rolled over and over the pebbled bank until they were again thrashing about in the shallows of the stream. The magpies, bulbuls and other birds were disturbed, and flew away with cries of alarm.

Covered with mud, gasping and spluttering, the boys groped for each other in the water. After five minutes of frenzied struggle, Bijju emerged victorious.

Rajaram lay flat on his back on the sand, exhausted, while Bijju sat astride him, pinning him down with his arms and legs.

'Let me get up!' gasped Rajaram. 'Let me go—I don't want your useless umbrella!'

'Then why did you take it?' demanded Bijju. 'Come on—tell me why!'

'It was that skinflint Ram Bharosa,' said Rajaram.

'He told me to get it for him. He said if I didn't fetch it, I'd lose my job.'

VI

By early October, the rains were coming to an end. The leeches disappeared. The ferns turned yellow, and the sunlight on the green hills was mellow and golden, like the limes on the small tree in front of Binya's home. Bijju's days were happy ones as he came home from school, munching on roasted corn. Binya's umbrella had turned a pale milky blue, and was patched in several places, but it was still the prettiest umbrella in the village, and she still carried it with her wherever she went.

The cold, cruel winter wasn't far off, but somehow October seems longer than other months, because it is a kind month: the grass is good to be upon, the breeze is warm and gentle and pine-scented. That October, everyone seemed contented—everyone, that is, except Ram Bharosa.

The old man had by now given up all hope of ever possessing Binya's umbrella. He wished he had never set eyes on it. Because of the umbrella, he had suffered the tortures of greed, the despair of loneliness. Because of the umbrella, people had stopped coming to his shop!

Ever since it had become known that Ram Bharosa had tried to have the umbrella stolen, the village people had turned against him. They stopped trusting the old man, instead of buying their soap and tea and matches from his shop, they preferred to walk an extra mile to the shops near the Tehri bus stand. Who would have dealings with a man who had sold his soul for an umbrella? The children taunted him, twisted his name around. From 'Ram the Trustworthy' he became 'Trusty Umbrella Thief'.

The old man sat alone in his empty shop, listening to the eternal hissing of his kettle and wondering if anyone would ever again step in for a glass of tea. Ram Bharosa had lost his own appetite, and ate and drank very little. There was no money coming in. He had his savings in a bank in Tehri, but it was a terrible thing to have to dip into them! To save money, he had dismissed the blundering Rajaram. So he was left without any company. The roof leaked and the wind got in through the corrugated tin sheets, but Ram Bharosa didn't care.

Bijju and Binya passed his shop almost every day. Bijju went by with a loud but tuneless whistle. He was one of the world's whistlers; cares rested lightly on his shoulders. But, strangely enough, Binya crept quietly past the shop, looking the other way, almost as though she was in some way responsible for the misery of Ram Bharosa.

She kept reasoning with herself, telling herself that the umbrella was her very own, and that she couldn't help it if others were jealous of it. But had she loved the umbrella too much? Had it mattered more to her than people mattered? She couldn't help feeling that, in a small way, she was the cause of the sad look on Ram Bharosa's face ('His face is a yard long,' said Bijju) and the ruinous condition of his shop. It was all due to his own greed, no doubt, but she didn't want him to feel too bad about what he'd done, because it made her feel bad about herself; and so she closed the umbrella whenever she came near the shop, opening it again only when she was out of sight.

One day towards the end of October, when she had ten paise in her pocket, she entered the shop and asked the old man for a toffee.

She was Ram Bharosa's first customer in almost two weeks. He looked suspiciously at the girl. Had she come to taunt him, to flaunt the umbrella in his face? She had placed her coin on the counter. Perhaps it was a bad coin. Ram Bharosa picked it up and bit it; he held it up to the light; he rang it on the ground. It was a good coin. He gave Binya the toffee.

Binya had already left the shop when Ram Bharosa saw the closed umbrella lying on his counter. There it was, the blue umbrella he had always wanted, within his grasp at last! He had only to hide it at the back of his shop, and no one would know that he had it, no one could prove that Binya had left it behind.

He stretched out his trembling, bony hand, and took the umbrella by the handle. He pressed it open. He stood beneath it, in the dark shadows of his shop, where no sun or rain could ever touch it.

'But I'm never in the sun or in the rain,' he said aloud. 'Of what use is an umbrella to me?'

And he hurried outside and ran after Binya.

'Binya, Binya!' he shouted. 'Binya, you've left your umbrella behind!'

He wasn't used to running, but he caught up with her, held out the umbrella, saying, 'You forgot it—the umbrella!'

In that moment it belonged to both of them.

But Binya didn't take the umbrella. She shook her head and said, 'You keep it. I don't need it any more.'

'But it's such a pretty umbrella!' protested Ram Bharosa. 'It's the best umbrella in the village.'

'I know,' said Binya. 'But an umbrella isn't everything.'

And she left the old man holding the umbrella, and went tripping down the road, and there was nothing between her and the bright blue sky.

VII

Well, now that Ram Bharosa has the blue umbrella—a gift from Binya, as he tells everyone—he is sometimes persuaded to go out into the sun or the rain, and as a result he looks much healthier. Sometimes he uses the umbrella to chase away pigs or goats. It is always left open outside the shop, and anyone who wants to borrow it may do so; and so in a way it has become everyone's umbrella. It is faded and patchy, but it is still the best umbrella in the village.

People are visiting Ram Bharosa's shop again. Whenever Bijju or Binya stop for a cup of tea, he gives them a little extra milk or sugar. They like their tea sweet and milky.

A few nights ago, a bear visited Ram Bharosa's shop. There had been snow on the higher ranges of the Himalayas, and the bear had been finding it difficult to obtain food; so it had come lower down, to see what it could pick up near the village. That night it scrambled on to the tin roof of Ram Bharosa's shop, and made off with a huge pumpkin which had been ripening on the roof. But in climbing off the roof, the bear had lost a claw.

Next morning Ram Bharosa found the claw just outside the door of his shop. He picked it up and put it in his pocket. A bear's claw was a lucky find.

A day later, when he went into the market town, he took the claw with him, and left it with a silversmith, giving the craftsman certain instructions. The silversmith made a locket for the claw, then he gave it a thin silver chain. When Ram Bharosa came again, he paid the silversmith ten rupees for his work.

The days were growing shorter, and Binya had to be home a little earlier every evening. There was a hungry leopard at large, and she couldn't leave the cows out after dark.

She was hurrying past Ram Bharosa's shop when the old man called out to her.

'Binya, spare a minute! I want to show you something.'

Binya stepped into the shop.

'What do you think of it?' asked Ram Bharosa, showing her the silver pendant with the claw.

'It's so beautiful,' said Binya, just touching the claw and the silver chain.

'It's a bear's claw,' said Ram Bharosa. 'That's even luckier than a leopard's claw. Would you like to have it?'

'I have no money,' said Binya.

'That doesn't matter. You gave me the umbrella, I give you the claw! Come, let's see what it looks like on you.'

He placed the pendant on Binya, and indeed it looked very beautiful on her.

Ram Bharosa says he will never forget the smile she gave him when she left the shop.

She was halfway home when she realized she had left the cows behind.

'Neelu, Neelu!' she called. 'Oh, Gori!'

There was a faint tinkle of bells as the cows came slowly down the mountain path.

In the distance she could hear her mother and Bijju calling for her.

She began to sing. They heard her singing, and knew she was safe and near.

She walked home through the darkening glade, singing of the stars, and the trees stood still and listened to her, and the mountains were glad.

twenty-two

~

CROSSING THE RAVI

GULZAR

Translated from the Hindi by Rakhshanda Jalil

It is a wonder that Darshan Singh did not go mad. His father had died, he'd had to abandon his mother in what remained of the gurdwara and his wife had given birth to two babies at the same time. Twins...both boys! Darshan Singh did not know if he should laugh, or cry. Fate had dealt him a strange deal...taken away with one hand what She had given with the other.

~

It was rumoured that freedom had come or was coming—though it was hard to tell when it would reach Todarmalpur. Hindus and Sikhs were surreptitiously making their way to the safe haven of the gurdwara. Shahni had been moaning in pain for the past few days and nights. These were the last days of her pregnancy; it was the first time she was giving birth.

Every day, Darshan Singh would tell the family a new story about the riots. Every day, his father would comfort him.

'Nothing will happen, son; nothing at all. Has any Hindu or Sikh home been attacked so far?'

'But the gurdwara was attacked, Bhapaji. It has been set on fire twice.'

'And yet you want to go there?'

Darshan Singh fell silent. But all around him, people were leaving their homes and making their way to the gurdwara.

'It is comforting to gather in one place, Bhapaji. There is not a single Hindu or Sikh left in our alley. We are the only ones here.'

One night, ten or fifteen days later, the sound of Bhapaji falling down in the courtyard woke everyone up. Cries of 'Jo Bole So Nihal...' could be heard from the gurdwara in the distance. Bhapaji had been woken up by the noise, and had gone up to the terrace to investigate. On his way down, he had slipped and tumbled down the stairs into the courtyard where his head had smashed into an axe that was lying there.

Somehow or the other, they had managed to perform Bhapaji's last rites. Then they stuffed whatever little they possessed into a pillowcase and the three of them made their way to the gurdwara. There was no dearth of frightened souls inside the gurdwara. Nevertheless, Darshan Singh felt a little safer. He said, 'After all, we are not alone; if nothing else we are close to our Blessed Guru.'

Groups of armed young volunteers patrolled outside constantly. People had collected flour, dal and ghee from their homes and the community kitchen churned out meals throughout the day and night. But for how long could their provisions last...it was the unvoiced question everyone had. However, they tried to be hopeful... Surely the government would send some help.

'But which government?' someone would ask. 'The English have left.'

'Pakistan has been created but there is no Pakistani government in place yet.'

'I have heard that the military is everywhere; they are taking caravans of sharanarthis up to the border under their care.'

'Who are the sharanarthis?'

'Refugees.'

'We have never heard these words before.'

A group comprising two or three families, who could no longer tolerate the burden of anxiety, decided to set off for the border.

'We are leaving. We have heard that there are trains leaving from the railway station. How long can we sit about waiting?'

'After all, brother, one has to find the courage. The Blessed Guru will not carry us on His shoulders.'

Another quoted from the Sacred Book to strengthen his argument, 'Nanak is the ship; he who boards the ship will cross safely.'

No sooner had the people left than a bubble of emptiness would form in their absence. But as soon as a fresh set of people came with news from the world outside, the bubble would burst.

'Do you know...a huge camp has been set up at the railway station!'

'People are dying of hunger and because they have been eating rotten food. Diseases are spreading.'

'Five days ago, a train passed this way. There wasn't even room for a seed of sesame in it; people were even packed on the roof.'

The next morning was Sankrant. Verses from the Holy Book were being recited throughout the day and night. Shahni gave birth to her twins at a very auspicious moment. Of the two, one baby was extremely weak; it had little chance of survival but Shahni kept it alive through sheer force of will.

That night someone said, 'A special train has come to get the refugees; let us go.'

A large mass of people set out from the gurdwara. It included Darshan Singh and Shahni who, though extremely weak, decided to leave for the sake of her sons. The mother refused to budge.

'I will come, son; I will leave with the next lot. You take my daughter-in-law and my grandsons and go.'

Darshan Singh tried to dissuade her from staying but the priest intervened. The volunteers in the gurdwara too tried to bolster his courage.

'Leave now, sardarji. One by one, we shall all go across the border. We will bring your mother with us.'

And so Darshan Singh set out. He put his two babies in a roll-top wicker basket as though he was a street hawker who had set out with his family on top of his head.

$\Longrightarrow$

The train was at the railway station but there was no space inside the train. People seemed to be growing out of its roof like grass.

People saw the weak and sickly mother and her newly-born babies, pulled her on to the roof, and made some space for her.

After about ten hours the train shuddered and began to move. The evening was red, as though bloodied, its face livid and ablaze.

Shahni's breasts were suckled dry by her babies. She would lift one baby away, then put the other to her breast to suckle. The two infants, wrapped in filthy rags, looked like something that had been picked up from a garbage heap.

After some time, as the train slowly pushed its way through the night, Darshan Singh noticed that while one baby moved its arms and legs, and even let out the occasional cry, the other was absolutely still. He thrust his hand inside the bundle and found it was stone cold.

When Darshan Singh burst into tears, the people sitting around him realized what had happened. They tried to take the baby away from Shahni but she seemed to have turned to stone too. She sat there with the wicker basket clasped to her chest.

'No, one doesn't drink without his brother.'

Despite everyone's efforts, Shahni refused to let go of the basket.

The train stopped ten times, and moved ten times.

People kept speculating about the progress they were making, although no one could see anything in the dark.

'I am sure we have just crossed Khairabad.'

'Surely this is Gujranwala.'

'Just an hour more...as soon as we reach Lahore, it will be as though we have reached Hindustan.'

In their nervous exhilaration, they began to chant religious slogans.

'Har Har Mahadev!"

'Jo Bole So Nihal!'

A wave seemed to ripple through the crowd as the train climbed a bridge.

'It is the Ravi river.'

'It is the Ravi! That means we have reached Lahore!'

In that clamour, someone whispered in Darshan Singh's ear.

'Sardarji, throw the dead baby into the Ravi. That will be the end of its journey. What good will it do to take it across the border?'

Gently, Darshan Singh extricated the basket from his wife's grasp. Then, in one swift move, he plucked one of the bundles of rags out and flung it into the Ravi with a loud cry of 'Waheguru'.

The faint cry of a baby was heard in the darkness. Terrified, Darshan Singh turned to look at Shahni. A dead baby was clinging to her breast.

A wave of noise erupted all around him...

'Wagah! Wagah!'

'Long Live India!'

GAMES AT TWILIGHT

ANITA DESAI

It was still too hot to play outdoors. They had had their tea, they had been washed and had their hair brushed, and after the long day of confinement in the house that was not cool but at least a protection from the sun, the children strained to get out. Their faces were red and bloated with the effort, but their mother would not open the door, everything was still curtained and shuttered in a way that stifled the children, made them feel that their lungs were stuffed with cotton wool and their noses with dust and if they didn't burst out into the light and see the sun and feel the air, they would choke.

'Please, Ma, please,' they begged. 'We'll play in the verandah and porch—we won't go a step out of the porch.'

'You will, I know you will, and then—'

'No—we won't, we won't,' they wailed so horrendously that she actually let down the bolt of the front door so that they burst out like seeds from a crackling, overripe pod into the verandah, with such wild, maniacal yells that she retreated to her bath and the shower of talcum powder and the fresh sari that were to help her face the summer evening.

They faced the afternoon. It was too hot. Too bright. The white walls of the verandah glared stridently in the sun. The bougainvillea hung about it, purple and magenta, in livid balloons. The garden outside was like a tray made of beaten brass, flattened out on the red gravel and the stony soil in all shades of metal—aluminium, tin, copper, and brass. No life stirred at this arid time of day—the birds still drooped, like dead fruit, in the papery tents of the trees; some squirrels lay limp on the wet earth under the garden tap. The outdoor dog lay stretched as if dead on the verandah mat, his paws and ears and tail all reaching out like dying travellers in search of water. He rolled his eyes at the children—two white marbles rolling in the purple sockets, begging for sympathy—and attempted to lift his tail in a wag but could not. It only twitched and lay still.

Then, perhaps roused by the shrieks of the children, a band of parrots suddenly fell out of the eucalyptus tree, tumbled frantically in the still, sizzling air, then sorted themselves out into battle formation and streaked away across the white sky.

The children, too, felt released. They too began tumbling, shoving, pushing against each other, frantic to start. Start what? Start their business. The business of the children's day which is—play.

'Let's play hide-and-seek.'

'Who'll be It?'

'You be It.'

'Why should I? You be—'

'You're the eldest—'

'That doesn't mean—'

The shoves became harder. Some kicked out. The motherly Mira intervened. She pulled the boys roughly apart. There was a tearing sound of cloth, but it was lost in the heavy panting and angry grumbling, and no one paid attention to the small sleeve hanging loosely off a shoulder.

'Make a circle, make a circle!' she shouted, firmly pulling and pushing till a kind of vague circle was formed. 'Now clap!' she roared, and, clapping, they all chanted in melancholy unison: 'Dip, dip, dip—my blue ship—' and every now and then one or the other saw he was safe by the way his hands fell at the crucial moment—palm on palm, or back of hand on palm—and dropped out of the circle with a yell and a jump of relief and jubilation.

Raghu was It. He started to protest, to cry 'You cheated—Mira cheated— Anu cheated—' but it was too late, the others had all already streaked away. There was no one to hear when he called out, 'Only in the verandah—the porch—Ma said—Ma *said* to stay in the porch!' No one had stopped to listen, all he saw were their brown legs flashing through the dusty shrubs, scrambling up brick walls, leaping over compost heaps and hedges, and then the porch stood empty in the purple shade of the bougainvillea, and the garden was as empty as before; even the limp squirrels had whisked away, leaving everything gleaming, brassy, and bare.

Only small Manu suddenly reappeared, as if he had dropped out of an invisible cloud or from a bird's claws, and stood for a moment in the centre of the yellow lawn, chewing his finger and near to tears as he heard Raghu shouting, with his head pressed against the verandah wall, 'Eighty-three, eighty-five, eighty-nine, ninety...' and then made off in a panic, half of him wanting to fly north, the other half counselling south. Raghu turned just in time to see the flash of his white shorts and the uncertain skittering of his red sandals, and charged after him with such a bloodcurdling yell that Manu stumbled over the hosepipe, fell into its rubber coils, and lay there weeping, 'I won't be It—you have to find them all...all... All!'

'I know I have to, idiot,' Raghu said, superciliously kicking him with his toe. 'You're dead,' he said with satisfaction, licking the beads of perspiration off his upper lip, and then stalked off in search of worthier prey, whistling spiritedly so that the hiders should hear and tremble.

Ravi heard the whistling and picked his nose in a panic, trying to find comfort by burrowing the finger deep-deep into that soft tunnel. He felt himself too exposed, sitting on an upturned flowerpot behind the garage. Where could he burrow? He could run around the garage if he heard Raghu come—around and around and around—but he hadn't much faith in his short legs when matched against Raghu's long, hefty, hairy footballer legs. Ravi had a frightening glimpse of them as Raghu combed the hedge of crotons and hibiscus, trampling delicate ferns underfoot as he did so. Ravi looked about him desperately, swallowing a small ball of snot in his fear.

The garage was locked with a great heavy lock to which the driver had the key in his room, hanging from a nail on the wall under his work-shirt. Ravi had peeped in and seen him still sprawling on his string cot in his vest and striped underpants, the hair on his chest and the hair in his nose shaking with the vibrations of his phlegm-obstructed snores. Ravi had wished he were tall enough, big enough to reach the key on the nail, but it was impossible, beyond his reach for years to come. He had sidled away and sat dejectedly on the flowerpot. That at least was cut to his own size.

But next to the garage was another shed with a big green door. Also locked. No one even knew who had the key to the lock. That shed wasn't opened more than once a year, when Ma turned out all the old broken bits of furniture and rolls of matting and leaking buckets, and the white anthills were broken and swept away and Flit sprayed into the spider webs and rat holes so that the whole operation was like the looting of a poor, ruined, and conquered city. The green leaves of the door sagged. They were nearly off their rusty hinges. The hinges were large and made a small gap between the door and the walls—only just large enough for rats, dogs, and, possibly, Ravi to slip through.

Ravi had never cared to enter such a dark and depressing mortuary of defunct household goods seething with such unspeakable and alarming animal life but, as Raghu's whistling grew angrier and sharper and his crashing and storming in the hedge wilder, Ravi suddenly slipped off the flower pot and through the crack and was gone. He chuckled aloud with astonishment at his own temerity so that Raghu came out of the hedge, stood silent with his hands on his hips, listening, and finally shouted, 'I heard you! I'm coming! *Got* you...'and came charging round the garage only to find the upturned flowerpot, the yellow dust, the crawling of white ants in a mud hill against the closed shed door—nothing. Snarling, he bent to pick up a stick and went off, whacking it against the garage and shed walls as if to beat out his prey.

Ravi shook, then shivered with delight, with self-congratulation. Also with fear. It was dark, spooky in the shed. It had a muffled smell, as of graves. Ravi had once got locked into the linen cupboard and sat there weeping for half an hour before he was rescued. But at least that had been a familiar place, and even smelled pleasantly of starch, laundry, and, reassuringly, of his mother. But the shed smelled of rats, anthills, dust, and spider webs. Also of less definable, less recognizable horrors. And it was dark. Except for the white-hot cracks along the door, there was no light. The roof was very low. Although Ravi was small, he felt as if he could reach up and touch it with his fingertips. But he didn't stretch. He hunched himself into a ball so as not to bump into anything, touch or feel anything. What might there not be to touch him and feel him as he stood there, trying to see in the dark? Something cold, or slimy—like a snake. Snakes! He leapt up as Raghu whacked the wall with his stick—then, quickly realizing what it was, felt almost relieved to hear Raghu, hear his stick. It made him feel protected.

But Raghu soon moved away. There wasn't a sound once his footsteps had gone around the garage and disappeared. Ravi stood frozen inside the shed. Then he shivered all over. Something had tickled the back of his neck. It took him a while to pick up the courage to lift his hand and explore. It was an insect—perhaps a spider—exploring *him*. He squashed it and wondered how many more creatures were watching him, waiting to reach out and touch him, the stranger.

There was nothing now. After standing in that position—his hand still on his neck, feeling the wet splodge of the squashed spider gradually dry—for minutes, hours, his legs began to tremble with the effort, the inaction. By now he could see enough in the dark to make out the large solid shapes of old wardrobes, broken buckets, and bedsteads piled on top of each other around him. He recognized an old bathtub—patches of enamel glimmered at him, and at last he lowered himself on to its edge.

He contemplated slipping out of the shed and into the fray. He wondered if it would not be better to be captured by Raghu and be returned to the milling crowd as long as he could be in the sun, the light, the free spaces of the garden, and the familiarity of his brothers, sisters and cousins. It would be evening soon. Their games would become legitimate. The parents would sit out on the lawn on cane basket chairs and watch them as they tore around the garden or gathered in knots to share a loot of mulberries or black, teeth-splitting jamun from the garden trees. The gardener would fix the hosepipe to the water tap, and water would fall lavishly through the air to the ground, soaking the dry yellow grass and the red gravel and arousing the sweet, the intoxicating scent of water on dry earth—that loveliest scent in the world. Ravi sniffed for a whiff of it. He half-rose from the bathtub, then heard the despairing scream of one of the girls as Raghu bore down upon her. There was the sound of a crash, and of rolling about in the bushes, the shrubs, then screams and accusing sobs of 'I touched the den—' 'You did not—' 'I did—' 'You liar, you did *not*' and then a fading away and silence again.

Ravi sat back on the harsh edge of the tub, deciding to hold out a bit longer. What fun if they were all found and caught—he alone left unconquered! He had never known that sensation. Nothing more wonderful had ever happened to him than being taken out by an uncle and bought a whole slab of chocolate all to himself, or being flung into the soda man's pony cart and driven up to the gate by the friendly driver with the red beard and pointed ears. To defeat Raghu—that hirsute, hoarse-voiced football champion—and to be the winner in a circle of older, bigger, luckier children—that would be thrilling beyond imagination. He hugged his knees together and smiled to himself almost shyly at the thought of so much victory, such laurels.

There he sat smiling, knocking his heels against the bathtub, now and then getting up and going to the door to put his ear to the broad crack and listening for sounds of the game, the pursuer and the pursued, and then returning to his seat with the dogged determination of the true winner, a breaker of records, a champion.

It grew darker in the shed as the light at the door grew softer, fuzzier, turned to a kind of crumbling yellow pollen that turned to yellow fur, blue fur, grey fur. Evening. Twilight. The sound of water gushing, falling. The scent of earth receiving water, slaking its thirst in great gulps and releasing that green scent of freshness, coolness. Through the crack Ravi saw the long purple shadows of the shed and the garage lying still across the yard. Beyond that, the white walls of the house. The bougainvillea had lost its lividity, hung in dark bundles that quaked and twittered and seethed with masses of homing sparrows. The lawn was shut off from his view. Could he hear the children's voices? It seemed to him that he could. It seemed to him that he could hear them chanting, singing, laughing. But what about the game? What had happened? Could it be over? How could it when he was still not found?

It then occurred to him that he could have slipped out long ago, dashed across the yard to the verandah, and touched the 'den'. It was necessary to do that to win. He had forgotten. He had only remembered the part of hiding and trying to elude the seeker. He had done that so successfully, his success had occupied him so wholly, that he had quite forgotten that success had to be clinched by that final dash to victory and the ringing cry of 'Den!'

With a whimper he burst through the crack, fell on his knees, got up, and stumbled on stiff, benumbed legs across the shadowy yard, crying heartily by the time he reached the verandah so that when he flung himself at the white pillar and bawled, 'Den! Den! Den!' his voice broke with rage and pity at the disgrace of it all, and he felt himself flooded with tears and misery.

Out on the lawn, the children stopped chanting. They all turned to stare at him in amazement. Their faces were pale and triangular in the dusk. The trees and bushes around them stood inky and sepulchral, spilling long shadows across them. They stared, wondering at his reappearance, his passion, his wild animal howling. Their mother rose from her basket chair and came towards him, worried, annoyed, saying, 'Stop it, stop it, Ravi. Don't be a baby. Have you hurt yourself?' Seeing him attended to, the children went back to clasping their hands and chanting, 'The grass is green, the rose is red...'

But Ravi would not let them. He tore himself out of his mother's grasp and pounded across the lawn into their midst, charging at them with his head lowered so that they scattered in surprise. 'I won, I won, I won,' he bawled, shaking his head so that the big tears flew. 'Raghu didn't find me. I won, I won—'

It took them a minute to grasp what he was saying, even who he was. They had quite forgotten him. Raghu had found all the others long ago. There had been a fight about who was to be It next. It had been so fierce that their mother had emerged from her bath and made them change to another game. Then they had played another and another. Broken mulberries from the tree and eaten them. Helped the driver wash the car when their father returned from work. Helped the gardener water the beds till he roared at them and swore he would complain to their parents. The parents had come out, taken up their positions on the cane chairs. They had begun to play again, sing and chant. All this time no one had remembered Ravi. Having disappeared from the scene, he had disappeared from their minds. Clean.

'Don't be a fool,' Raghu said roughly, pushing him aside, and even Mira said, 'Stop howling, Ravi. If you want to play, you can stand at the end of the line,' and she put him there very firmly.

The game proceeded. Two pairs of arms reached up and met in an arc. The children trooped under it again and again in a lugubrious circle, ducking their heads and intoning

> The grass is green,
> The rose is red;
> Remember me
> When I am dead, dead, dead, dead...

And the arc of thin arms trembled in the twilight, and the heads were bowed so sadly, and their feet tramped to that melancholy refrain so mournfully, so helplessly, that Ravi could not bear it. He would not follow them, he would not be included in this funereal game. He had wanted victory and triumph— not a funeral. But he had been forgotten, left out, and he would not join them now. The ignominy of being forgotten—how could he face it? He felt his heart go heavy and ache inside him unbearably. He lay down full length on the damp grass, crushing his face into it, no longer crying, silenced by a terrible sense of his insignificance.

A REVOLT OF THE GODS
VILAS SARANG

Translated from the Marathi by the author

I

It was the time of the Ganesh festival. In the evening I had to go out to photograph one of the numerous festival shows that were being staged all over Bombay. It took me longer than I expected, and the next morning I rose late. Dalvi was already at the studio when I arrived and was busy developing negatives in the darkroom. We had been getting a lot of extra business during the festival, having been called in to photograph several shows. Amateur singers and actors were particularly eager to be photographed on stage. While Dalvi handled such jobs expertly, I was much happier doing portraits in the studio. During the festival, however, there were often two assignments on the same evening, and then I had to go out too. Dalvi enjoyed these stints; I hated making my way among the noisy crowds, toting a camera and other paraphernalia, and moving from one end of the stage to the other in search of a better angle, often crouching awkwardly in order to catch someone in a good pose. It's different in the studio: an elegant curtain in the background, lights placed at strategic points, and in their midst the customer seated silently and meekly, if somewhat uncomfortably. This was the last day of the Ganesh festival, and I was relieved that there would be no more spells of working late. There was still a lot to do in the studio of course—making prints, enlargements, and so on.

I set about touching up some passport photos. Touching up—that's my forte. It's mostly Dalvi who looks after developing and printing; he's better at it than I am. We've set up our studio in partnership, relying largely upon Dalvi's technical skill; I take care of things like touching up, cutting prints, and keeping the accounts.

A young man who looked like someone in search of a job came in to have some passport photos taken. I asked him if he wanted to wear a tie, and he said yes. Inspecting the tie I handed him, he remarked that it seemed old-fashioned and asked if we had another one he could use. We only had one tie, I said curtly. I was annoyed. If he was so particular, he should've come with his own tie, I thought. He couldn't even manage a decent knot, and had probably seldom worn a tie before. I had to tie it for him.

I found it difficult to keep working until one o'clock, and my eyes were heavy from lack of sleep. Just as we were about to close down, a blind man walked up, tap-tapping his cane and asked if his pictures were ready. Four days ago he had come in, feeling each step with his cane before climbing it, had stood at the counter for a few moments taking in my presence, and then asked, 'Is this the Gajanan Photo Studio?' Apparently some institution was sending him to the US to study music, and he needed pictures for a passport. He was talkative, as all blind persons seem to be, and was excited about his prospective journey. He told me that the blind had a difficult time in India, and how a blind person could hope to develop his abilities in Western countries. I listened absent-mindedly as I led him to the seat. It occurred to me that I was taking a blind man's picture for the first time. I switched on the lights; he stared with eyes wide open. Usually, people blink in the glare, or turn their heads away. The blind man sat motionless, with his sightless eyes shining defiantly under the violent lights.

I had told him that he could pick up the pictures today; I searched about but couldn't find them. They must be lying somewhere amidst the heaps of festival pictures, I thought. 'I'm sorry, could you come back in two days?' I said. He went away, tap-tapping his stick. I locked the studio and went home. I was glad that the studio was closed for the day. That evening Dalvi had to go to witness the immersion of his household Ganesh.

I slept for an hour after lunch before being awakened by the noise of kids in the courtyard below. The members of the festival committee for our building were getting things ready for the Ganesh immersion. Like countless other buildings in Bombay, ours too had its Ganesh, with shows every evening for eleven days, the whole affair financed by donations from each tenant. I got out of bed and made myself a cup of tea. Refreshed, I went out for a stroll. I thought I'd catch up with the procession bearing our building's Ganesh somewhere along the way.

The evening shadows were lengthening. I walked, looking at the shadows of buildings, of electric poles, and of men and women. I like observing shadows more than watching things and human beings. Buildings and poles and men remain the same for years, or change very little, very slowly. On the other hand, the shapes and angles of shadows change minute by minute. They are sometimes clearly etched, and at other times like badly focused photographs. Men, stolid and heavy, merely dream of transfigurations, while shadows change endlessly. A man is nothing compared to his shadow. Without shadows, life would have been far duller.

The main street was crowded. Small groups of people went by with their household clay figures of Ganesh. One man carried upon his head a small figure on a wooden platform, and the rest walked behind him singing hymns. The large civic figures were brought down later in the evening. I sat in an Irani restaurant on the way to the seashore. After some time I observed people from our building approaching with the clay image. I got up and merged into the tail end of the procession.

As always at the yearly immersion, the beach was swarming with people. Amid the sea of human beings milling about one could spot here and there the colourful figure of the elephant-headed god. The civic images, perched comfortably upon handcarts and trucks, were in greater evidence. The figure from our building was medium-sized. Members of the festival committee had complained that the collection was unusually small this year. I was sure, though, that they had managed to pocket their usual share of the money. Anyway, I didn't take much interest in their dealings.

The roads had become so congested that all the processions had more or less ground to a halt. The sea was only a short distance away, but it would take a long while to reach the water's edge.

Then I noticed a disturbance at the head of our procession. Being at the tail end, I didn't know what was wrong at first. I craned my neck, but couldn't see our Ganesh figure and wondered if it had fallen off the handcart. Then someone said that the Ganesh had jumped off. Naturally, I thought he was joking. But the commotion seemed to be spreading. Indeed the figures in the different processions had begun to disappear, apparently deserting their seats on the wooden platforms and handcarts and trucks. In the melee I couldn't quite see what was happening, but I did observe a huge civic figure jumping down from a truck. I saw a man fall to the ground at a blow from its trunk.

Here and there one glimpsed clay images that had come to life, as it were, and were sprinting away. But scarcely anyone possessed the composure to watch this miraculous spectacle. Confused and terrified, everyone ran back towards the city. Even the traffic cops had abandoned their positions and were running with the rest. Knocked about in that disorderly retreat, I lost one of my slippers. For a while I ran with one slipper, but then realized that there was no point in returning that way. I cast it off too and ran barefoot.

When I reached the building, I noticed that almost everyone else was already there. People gathered in the courtyard and discussed the event in excited tones. Since everyone was talking at the same time it was hard to understand what was being said. The women, many of whom chanted prayers breathlessly, looked particularly frightened. Children cried.

After some time the clamour became more intelligible. An attempt was made to discover the significance of the supernatural event. At first everyone seemed to be convinced that it was a manifestation of divine wrath.

Mr Kini—who was an accountant in a government office, and was commonly known as Accountant Kini—appeared even more excited than the rest. 'Look,' he said in a cracked voice. 'For a long time I've had the feeling that sooner or later something of this sort was going to happen. True, Lord Ganesh is called lambodara—pot-bellied—but how many sins can you expect him to swallow? I bet now he won't return to heaven without giving a good lesson to all the black marketeers, smugglers, food adulterators, and politicians. They're the ones who've brought this on.'

'But all these images that have come to life—just what are they going to do now?' Akshikar, who lived on the third floor, asked. 'How will they punish the guilty?'

It was something to think about.

'Well, who can tell what the gods will do?' someone said. 'We just have to face up to whatever happens next without complaining.'

Professor Matkari had a different view. He said, 'I see no reason for being so frightened. You must have seen that the gods just jumped off their seats and went away. It didn't seem as if they were lashing about at people, or wanted to hurt anyone.'

'That's right,' Subodh, who was studying physics in college, said. He was on the festival committee in our building. 'I see no point in talking about sin and the like. The gods have not arisen to dispense punishment. They could've done that without resorting to miracles. They're gods, after all. I think they were simply outraged by the whole spectacle—the sweaty crowds, the hurly-burly, and the disorder. Why should the gods put up with it? When they couldn't stand it any longer they just got up and walked away.'

The discussion continued. I slipped away and went upstairs to my apartment. I stretched out in bed. The excitement and the exertion had tired me.

II

The next day there was a great stir in the city. People wondered if the absconding gods had vanished completely, or were still prowling about. At first it seemed that they had disappeared, but then fleeting glimpses of them were caught here and there. The elephant-headed god's trunk would appear around a corner, or, climbing a flight of stairs, one would suddenly become aware of a shadowy figure disappearing ahead. A lonely passer-by might receive a blow on the head from a god's trunk. No one was hurt seriously, but some people were sent into a state of shock for a few hours. They would behave strangely for a while, say unintelligible things, and then gradually return to normal life. That was all the gods did. They didn't unleash any terrifying forces, or throw the daily life of the city out of gear. In a few days people calmed down and went about their business. And of course the people of Bombay have a reputation for not being thrown off balance or surprised by anything. Still, people tended to return home earlier in the evening, and avoided walking through isolated areas. There was no telling when one might receive a blow on the head.

The day after the revolt of the gods I didn't go to the studio. Like everyone else, I thought I'd see how things turned out first. Then we went back to business as usual. For a week or so, there were very few customers. But we had plenty of work at the studio. Dalvi spent most of his time in the darkroom. I had taken pictures at the show in the Makarand Housing Society, and they had all come out poorly. The entire roll of film had been exposed at the wrong speed. Dalvi wasn't happy about it. I told him not to worry; I'd touch up all the pictures and set them straight.

One day the blind man returned. I had forgotten about his pictures in the meantime. I searched for them again without success. I didn't know what to do. I didn't feel I should make a blind man come back time and again. For a moment I thought of handing him someone else's pictures: after all, he couldn't see them. I said, 'Please come back in four days; I'll definitely have them ready.' He seemed upset and went away, tap-tapping his cane noisily.

Another day Accountant Kini came to my room with a roll of film. Photography is his hobby and that has made us friends. He comes in often to talk over technical matters. Some of his pictures of flowers and the like have appeared in the Sunday newspapers, so he feels he is an artist.

While leaving Kini said, 'For the last two or three days we've been bothered by mice. We've never had them in our apartment before. These must have come in recently. Last night I set a trap; in the morning I found two. Then it suddenly occurred to me—when the Ganeshas arose and left, the mice that had invaded my house were the same ones on whose backs the gods are supposed to ride.'

'What did you do then?' I asked.

'Well, I didn't know what to do,' Kini said. 'I was appalled at the idea that the mice I was going to kill might belong to the gods.'

'You could've taken the cage with the trapped mice to the sea and drowned them,' I said. 'As it is, don't we immerse the mice along with the Ganeshas?'

'Oh, no,' Kini said. 'It's one thing to carry Lord Ganesh and his mouse reverently to the sea for immersion, and quite another to drown mice.'

I saw no difference, but didn't have enough interest in the matter to argue about it.

Touching up the pictures I had taken at the Makarand Housing Society show turned out to be an exhausting job. It wasn't the same as touching up a studio portrait. There were often a number of persons in a single picture, with faces much smaller than in a portrait. Touching up each little face was tedious. Then one day their man came and picked up the pictures. The next morning he was back at the studio. 'Are these our pictures, or someone else's?' he asked. 'We can hardly recognize a single face.' I shrugged. 'What's so special about a face?' I said. 'We won't come back next year,' he said.

Then I spent an entire morning looking for the blind man's pictures, again without success. I was exasperated. I decided that I'd take his picture again, free of charge of course. He was supposed to come in the next day. The next morning I waited for the sound of his cane, but he didn't turn up.

As I was getting ready to go to bed that evening there was a knock at the door. My first thought was of the blind man. Then I laughed, realizing how absurd the idea was. He didn't know where I lived; besides, photographers aren't called at night like doctors. I wondered who it could be. Opening the door, I saw Accountant Kini. I thought he had come to ask about the roll of film he had given me the other day. I said I hadn't been able to develop it, what with all the work on the festival pictures. But he stayed on, making small talk. I saw that his mind wasn't on the conversation, but on something else he wasn't mentioning. He kept fidgeting.

Then the conversation faltered; I deliberately remained silent. Kini looked around aimlessly at the calendar on the wall, at my books, at the curtains. Then he looked at me and said, 'There's something I wanted to talk to you about.'

'Go ahead,' I said.

'I've never mentioned this to anyone until now. My wife is the only person who knows about it,' Kini said. 'In the last few years I've developed a strange habit—that of cursing the gods.'

Kini paused and stared at the foot of my bed. I remained silent.

'In our family we worship Lord Ganesh,' Kini continued without raising his head. 'I've known the hymn to Ganesh by heart since childhood. Yet, instead of hymns and prayers, what comes into my mind are curses. Unspeakable, abominable curses. I was brought up well as a matter of fact. Never learned bad words as a child. Even today I wouldn't say something like "damn". And yet when I think of God my mind literally spews out curses. At first I was shocked to discover that I knew so many.'

There was a knock at the door. 'Papa, papa,' little Sushma shouted. 'What a nuisance,' Kini said, and rose to open the door. 'What do you want?' he asked his daughter.

'Mummy's calling you,' Sushma said. Then she seemed to want to get in. But Kini pushed her away. 'I'll be back in a few minutes,' he told her.

'And don't you come knocking again.'

He bolted the door and returned to his seat.

'So that's my story. Usually, the cursing goes on silently in my mind. Often, when I'm at home, I mutter to myself. On a few occasions, when my wife and Sushma have gone out, I've shut the doors and windows and, standing in front of the images of the gods, shouted curses until I was hoarse, I found it nauseating to my own ears, but my mouth wouldn't stop.'

Kini paused. He suddenly looked tired.

'At first I tried to control my mind, but without success. Then I tried different kinds of charms and amulets. I made offerings to many deities. Nothing worked.'

'Perhaps it's due to something nagging at your mind,' I said. 'Deep in your mind there may be resentment at some injustice, or some disappointment. Perhaps it comes to the surface in this way.'

'Oh, no,' Kini said. 'I'm as happy in life as anyone could be. My only son—a bright boy, and well-behaved—is studying to be an engineer. We're all in good health and there's no great worry on my mind. Oh yes, I've been happy. And yet, there's this horrible thing that won't go away.'

'Have you been to a psychiatrist, then?' I asked.

'I don't believe in that sort of thing. This isn't a matter of mere chemicals, or of nerves. Scientific mumbo-jumbo would be of no use. It's only God who can save me.'

From Kini's apartment came the sound of his wife shouting, and then of Sushma crying. Kini listened for a moment and then ignored it.

'Then this thing happened the other day,' he said. 'The gods became angry. Since that day, I've been thinking—did that happen because of my dreadful habit? I had a feeling that something of this sort would happen one day. Sins must bear fruit, sooner or later.'

For a few minutes I didn't speak. Then I said, 'Look, Mr Kini, what you've said doesn't make much sense. Think about it—hundreds of Ganesh images in Bombay got up and went away. That couldn't be the fruit of one person's sins. I mean, there'd be too large a gap between cause and effect. God doesn't watch you alone. His eye is on everyone, you know.'

I lit a cigarette and continued, 'But I don't find it strange that such an idea came into your head. When a shattering public event like this occurs, everyone begins to see his own life in its light. I'm sure thousands of persons in Bombay at this moment have similar ideas. On the other hand, your petty sins may not count for much in God's account books.'

Kini remained silent. My words seemed to have made some sense to him, but I doubted that he would change his mind.

Lying in bed, I thought for a long while about what Kini had told me. It was past midnight when I fell asleep, but I woke up almost immediately. I was wondering what had awakened me, when I heard small noises coming from the kitchen. I listened. Mice, I thought. I hadn't had mice in my apartment before. Then I remembered that Kini, too, had spoken about mice the other day. Mice must be invading the whole building, probably because of the squalor in the new shanty town growing up behind it.

The noise of the mice made it hard to get back to sleep. Sounds in the dark are particularly irritating. As long as things are visible, noise recedes into the background. But once things have become invisible, noises—be they loud or small—begin to rule. I got out of bed, went into the kitchen and knocked a knife-handle upon the kitchen table. I went back to bed. Things were quiet for a few minutes. Then the noises began again. Suddenly I remembered Kini's notion: these were Lord Ganesha's mice. Sacred mice. I smiled.

Listening to the sound of mice in the dark, it seemed to me as though someone were scratching words that I couldn't decipher upon a black slate with a rough slate-pencil. It occurred to me that I had some rat poison in the house, and I thought of scattering some on the kitchen floor. But instead of getting out of bed I lay in the dark with my eyes wide open. Perhaps someone was photographing me under powerful lights, although I was unable to see them. Then the photographer, with a black cloth still over his head, began to help me knot a tie around my neck. I pulled away the tie with both hands and jumped out of bed. I switched on the light and splashed cold water on my face. Then I dressed and went out.

III

The streets were deserted. I felt like roaming about, and thought of going to the beach. I often went to the beach at night. In the evenings the beach is crowded. It's nice at night, with few people around and the sea babbling in sleep. Once in a while a plane passes overhead, blinking red and green lights.

It was different now, though. It wasn't safe to wander around alone at night any more. You didn't know when you might be struck by an elephant's trunk. I thought it over for a minute or two and decided to go to the beach all the same. I didn't care. Perhaps somewhere inside I wanted that blow on the head.

I arrived at the beach and sat on the huge pipe that pours sewage water into the sea. I could see a few forms wrapped in blankets sleeping by the beach wall. The sea shimmered under the moonlight. It was midway between high and low tide, and a cool breeze was blowing in over the calm water towards the beach.

After an hour or so I saw two or three figures walking towards the beach. Some sleepless wanderers like me, I thought. The figures were short, and I wondered if they were children. What were children doing out by themselves at this hour? Then I saw something waving like a tail, and thought they were dogs. The figures came forward, and walked out on to the sand. Then I realized in amazement that they were Ganesh images.

More figures began to arrive. The ones that had come first reached the water and went straight in. I watched them disappear under the surface, legs first, then belly and shoulders, and then the elephant head. One after the other the images came down to the beach and entered the sea. Their sleek bodies shone in the moonlight. They walked with such an air of freedom, with such grace! As each entered the sea, the water seemed to make way for the god, compliantly breaking into ripples. Some of the gods, as they entered the water, playfully waved their trunks in the air. All was quiet, and the gods were disappearing noiselessly one by one into the sea.

Spellbound, I watched the unearthly sight. The gods were quietly going home, spurning all rituals. They had come down the beach in a wave of pure joy; their clay figures melted into the sea like lumps of sugar. I thought their melting figures would make the salt sea sweet.

Somewhere in my mind questions were fluttering. Why were the gods leaving so suddenly? Had they accomplished their mission? Why had they risen up? Were they returning because they had failed to achieve some objective? Perhaps they had decided to leave the affairs of men to men themselves. Perhaps they had undertaken this experiment to undo some divine entanglement of their own which had nothing to do with the world of men. Or they had staged the uprising for fun, and now, having tired of the game, they were departing. I remembered the tale of the Pied Piper and the rats. I thought that a Pied Piper who can lure away the gods must be very powerful indeed.

But such questions and ideas didn't linger long in my mind. They appeared pointless, entranced as I was by the sight before me. I watched for a long time. Then the number of figures coming down to the sea grew smaller. In the end a very small god came skipping down and threw himself into the water with abandon. Then the beach was deserted; the water shimmered quiescently. I fell asleep without realizing it.

The sun was up when I awoke. Slum-dwellers were coming down to the edge of the sea to defecate. The sea was low. I felt stiff. I stretched myself and started for home.

I pondered over the previous night's spectacle as I walked. I wondered if it had really happened, or whether it was just a dream. Who was I to be blessed among all men with the sight of such a vision? I thought I could find out the truth when I got back to the city. If the gods had truly left, people couldn't have failed to have noticed it. Also, on other beaches of the city a few people must have seen the gods leaving. But then I didn't really care if what I had seen was real or the play of my imagination. To have seen it was enough.

I recalled an incident from my childhood. In the village where I grew up there lived an old astrologer who was something of a crank. One day he made some calculations and declared that on a certain day there would be great floods and storms and everything would be washed away. He began to warn the villagers, telling them to leave their huts and go to the top of a mountain on the appointed day. Everyone laughed at him.

The astrologer became increasingly frantic, running about to save the villagers from what he believed was their folly. On the day before the world was to end, he packed a few of his belongings and, warning the people for one last time, climbed up to the top of the mountain behind our village. It was a hard climb, and the old man must have been half dead with exhaustion when he reached the summit. That night there arose by chance a great storm; sheets of rain poured down for hours, and trees fell. In the morning, when the sky had cleared, a small band of villagers climbed the mountain. They found the astrologer's stiffened body.

When I was small I hadn't realized it, but many years later it struck me: braving the violent winds and the rain, the old man must have died in the conviction that he was indeed witnessing the end of the world. That was the truth as he saw it; and, in the final analysis, what other truth is there?

I reached the building where I lived. Climbing up to my room, I stopped and knocked on Accountant Kini's door. Kini opened the door, and looked at me with sleepy eyes.

'Mr Kini,' I shouted. 'They've gone. The gods have left.'

'What, the gods have left?' Kini muttered, rubbing his eyes 'How? And when? Why did they leave?' And, with half-shut eyes, he began to mumble curses as though he were reciting prayers.

twenty-five

~

IN A FOREST, A DEER

AMBAI

Translated from the Tamil
by C. T. Indra

Those nights are unforgettable, nights when we listened to stories. It was Thangam Athai who told them. They were not the usual ones about the crow and the fox, the hare and the tortoise. These stories were her own. Some like snatches of poetry, some like never ending songs, they were without beginnings or ends, unfolding in a myriad enticing ways. She would conjure up all kinds of images in our minds. Even asuras and devas assumed new shapes in her stories. Surpanakai, Tatakai and other such demons were transformed into beings with feelings and emotions. Her Mantarai would bring tears to our eyes. Characters hidden in the dark corners of the epics would be drawn out by Athai's words, much like a bird with broken wings being revived by caressing hands. Whether it was the magic of night, the expansive central hall of that ancient house, or the proximity of cousins lying next to each other, I do not know, but those stories still reverberate in my mind like the constant hum of a bee.

~

I see Thangam Athai now in my mind's eye in that house with its old pillars
and its vast central hall. Leaning against the huge wooden door. Carrying
the lighted earthen lamp, its flame protected with the loose end of her sari,
and setting it gently in its niche. Serving food to her husband, Ekambaram.
Drawing water from the well, one leg firmly placed against its parapet wall.
Manuring the plants.

Thangam Athai was dark and beautiful—her face was smooth, with not
a wrinkle, her hair silvered over. In Athai's house, there was an old-fashioned
harmonium with foot-operated bellows. Only Athai used it. Playing it, she
would softly sing a range of songs—from the devotional Thevaram to the
love song Vathaname Chandra Bimbamo to the playful Vannan Vanthane—her
long, tapering fingers flitting over the reeds of the harmonium like dusky
butterflies.

A sense of mystery surrounded Thangam Athai. There was pity in the
sympathetic looks others directed at her, the way in which they would tenderly
stroke her, in their moist eyes. Ekambaram Mama had another wife besides
Athai. But he treated Thangam Athai as he would a flower. No one ever
heard him call her using the all too familiar 'di'. He called her Thangamma.
Even so, Athai seemed distant, as if she existed behind a screen of smoke. It
was Muthu Mama's daughter, Valli, who solved this mystery—not that what
she said made much sense to us. According to Valli's mother, Thangam Athai
had never 'flowered'.

'What does that mean?' many of us asked.

'It means she never came of age.' Valli was the one who had graduated
to wearing half-saris.

'But her hair is all grey?'

'That's different.'

After that we began to look closely at Athai's body. We tried to discover
what a body which had not flowered would be like. It was not at all clear.
When she came out of the bath, wrapped in wet clothes, she appeared no
different from any other woman. Nor when she stood in her green sari, the
red blouse held by a firm knot under her bosom, her hair gathered into a
knot. Valli's mother had once told Valli, 'Her's is just a hollow body.'

Where was the hollow? Could it be that, like the broken wing of a bird,
it was not visible?

One evening, they cut down a huge withered tree in the garden. The
final stroke of the axe brought it crashing down, with a rush of rustling leaves.
When cut open, we saw that its inside was hollow. Valli dug her elbow into

my waist and whispered, 'There, that's hollowness.' This cloven tree, exposing itself so fully that everyone could see the emptiness inside, how could we compare it to Athai's glistening dark body?

⌐⌐

What secret did her body keep hidden? How was her body different? Summer afternoons, she would lie down to rest in the storeroom. She would place her arm gently round us when we lay down next to her, our heads nestling against her bosom that had been freed from its taut blouse. Enjoying the total security afforded by her bosom, her hips, her arms, we would wonder what was hollow inside her. Her body was warm to the touch. It seemed to be brimming with energy, like a fruit laden with juice, and its regenerating vitality would wash over us like a surging river—when she touched us, caressed us or, with a firm hand, massaged oil into our bodies. The cow yielded milk only at her touch. Seeds she sowed invariably sprouted.

Amma said she had a lucky touch. When my little sister was born, Athai had come down to stay with us. 'Akka, stay close to me, keep holding me. Only then will I not feel the pain,' Amma moaned as we were being sent out of the room. When we inched back to the door and looked in, Thangam Athai was gently stroking Amma's bulging stomach, saying in a soft voice, 'Don't panic. Nothing will go wrong.'

'Oh Akka! If only you too...' Amma sobbed.

'Come on, what is it that I lack? I live like a queen, my house is full of children,' said Athai. Ekambaram Mama's second wife had seven children.

'Your body cannot...' Amma choked back another sob, unable to complete her sentence.

'Why, what's wrong with my body?' Athai asked. 'Isn't my appetite normal? Don't I sleep well? This body is like all other bodies. When hit, it feels the pain, its blood clots. When a wound festers, pus forms. Food gets digested. What else does one want?'

Amma took Athai's hand and pressed it to her cheek. 'Oh! How they wounded and gored your body...' Amma wailed, holding on to Athai's hand.

Valli's mother had told her that there was no medicine that had not been tried on Athai's body. She was sure to receive the formulations of every vaidyan visiting the town. Athai was given English medicines too. Sometimes, after taking certain medicines, she would fall into a bottomless sleep. They even tried ritual worship, complete with neem leaves and the accompaniment of the udukkai drum. In the hope that a sudden fright would help, they had a dark, shrouded figure jump on Athai when she went into the backyard late one evening. Athai fell down shrieking, and knocked her head against the granite stone used for washing clothes. She still has the scar on the corner of her forehead. When the next vaidyan arrived, Athai apparently cried out, 'Leave me alone, please leave me alone.' On the night that they went to arrange a second wife for Ekambaram Mama, Athai ground the seeds of the arali flower into a potion and drank it. They somehow managed to save her by giving her an antidote. And Mama said, his eyes brimming with tears, 'I don't want anything that would hurt you.' After that, Athai herself selected a girl for him. That was how Sangamalam came into the family. All this, of course, was information gathered by Valli.

Without releasing her hand from Amma's hold, Athai was stroking her head with the other, murmuring, 'Forget, forget everything. Why do you rake up my story when you are about to deliver?'

That night, my little sister was born.

It was later, during one of my visits home, that Athai narrated this story. It was the rainy season. One night, a jamakkaalam had been spread at one end of the hall and a few pillows were scattered on it. Some of the pillowcases were stained with hair oil. Some of the pillows were without cases, made of thick darkish cloth and stuffed with cotton that had grown lumpy. They were not pillows that were in daily use. These were given to us children when there were guests. Children who had played all day long, eaten their fill and who fell asleep the moment their heads touched the pillows—would such children even notice minor discomforts?

We heard the noise of the kitchen being cleaned. Then the metallic clang of the brass vessel, the screech of a door opening, the thud of a sodden broom being placed behind the door. Then, the tin of kolam powder grated against cement—the earthen stove must now be getting a kolam. Closing the kitchen door, Athai should soon be coming through the hall.

None of us was asleep. We waited.

When Athai arrived, it was Somu who spoke.

'Athai, tell us a story...please, Athai.'

'So, you are still awake, all of you?'

She stood for a moment, looking at us, then came over and sat down. Kamatchi and Somu crept towards her and, settling their heads on her lap, looked up expectantly. The others leaned forward on elbows rested firmly upon the pillows.

Athai was tired. Sweat glistened on her forehead. She closed her eyes and thought for a while.

'It was a big forest...' she began.

'All the animals lived happily in that forest. There were many fruit bearing trees there. A stream ran through it. When thirsty, the animals would drink from the stream. Everything the animals needed was available right there. There was no fear of hunters in that forest. The beasts roamed about without the least fear of an unexpected arrow piercing them, threatening their lives. Not that there were no forest fires there, or strangers coming to fell trees, pluck fruits, shoot birds, or hunt the fleeing pig. Yet, it was a place the beasts and birds dwelling there were familiar with. They knew on which tree the owl would sit, and how it would hoot when the forest lay wrapped in the silence of night, on which stone the frog would squat and when it would suddenly give a gurgle-throated croak, and when and where the peacock would dance.

'One day a herd of deer went to the stream for a drink of water. As they were returning, one young deer lost its way. Suddenly, it found itself in a different forest. A forest with no paths. Where all the trees bore the marks of arrows. There was a waterfall there too, one that tumbled and roared. The forest looked desolate to the little deer, as if it was a deserted, friendless place. The deer trembled. It ran here and there in a panic, crying out in fear all the while. Night set in. The deer's terror did not subside. The sound of the cataract frightened it. At a distance, a hunter had made a fire, and was roasting the game he had killed. The deer noticed the fire and moved away stealthily. It wandered round and round the forest till, exhausted, it lay down.

'Many days went by. One full moon night, the deer saw that the cataract had gilded itself with moonlight and seemed transformed. Its appearance was no longer terrifying. The moonlight fell softly on everything and, as if touched by a magic wand, the deer shed its fear. It liked the forest. Every nook and cranny of the forest became clear to it. Though a different forest altogether, this too has everything, the deer thought. There's the waterfall. And there are the trees, the plants. Gradually the deer was able to see the animals and the birds, the beehive hanging from a tree, the grass shot through with green. This new forest no longer held any secrets for the deer, and it roamed everywhere. With all fear gone, the deer became calm.'

Athai finished the story. Only the part of the hall we were in was lit. The rest was in darkness. As we had listened to the story, we had imagined the dark spaces to be the forest. We had made friends with the deer and in the end we too were at peace. Everyone fell asleep, hugging pillows. Drowsy, resting on my coarse blue, yellow and black pillow, I looked up. And there, sitting in our midst, arms crossed over her bosom, palms clasping her shoulders and head resting on her knees, was Thangam Athai.

twenty-six

~

BHASKARA PATTELAR AND MY LIFE
PAUL ZACHARIA

Translated from the Malayalam by Gita Krishnankutty

I first saw Bhaskara Pattelar one evening in Udina Bazaar. Squatting on the edge of a shop verandah, I was brooding upon my sorrows. The sound my frayed mundu made as it gave way little by little kept pace with my thoughts. Suddenly I heard someone call out to me.

'Hey!'

I looked up to see who it was.

The man who had hailed me sat on a chair in the verandah of a shop across the road. He must have been about thirty-five years old. He was as tall as he was large. He wore a silk jubba and was fair-skinned. His eyes and hair had a coppery tint. He had a big moustache which curved downwards and his lips were stained red with betel juice. His big body barely fitted into the chair. Half-a-dozen people stood around him respectfully.

'Come here, you whore's son!' Pattelar called out in Kannada. I sprang up, my palms joined together. Who could this be? What did such a great man want of me?

As I jumped down from the verandah, my mundu tore along its entire length. With one hand, I held the torn ends of the garment together behind me, covered my mouth respectfully with the other, and crossed the road.

The giant got up from the chair and came to the edge of the verandah. 'What is that in your hand, rascal?'

'Nothing, master,' I said.

'Why then, dog, are you holding one hand behind you? Do you want to shit? Don't you know how to show respect to those who deserve respect? Come closer, you rascal.'

I quickly took away the hand that held the mundu together, joined my palms, bowed deeply, and moved closer. My torn mundu fell apart like the halves of a stage curtain. Laughing loudly, Pattelar gave me a kick. I fell backwards, naked and staring skywards. Pattelar spat betel juice at me. 'Thoo!'

I rolled on my stomach, pressed my face into the sand, joined my palms, and begged. 'Show me mercy, yejamanare!'

It was then that I saw the gun leaning against Pattelar's chair. It was as large and terrifying as Pattelar himself. The people around him said something in Malayalam and laughed. What! They were Malayalis!

'Get up, dog,' said Pattelar.

I got up and tied one of the pieces of the mundu around my waist.

'Where are you from, rascal?'

'From Kochi, master.'

'Where do you live?' He switched to Malayalam.

'In Ichilampadi, master.'

'How many acres have you encroached upon?'

'Only five, master!'

'Are you married?'

'Yes, master.'

'How old is your wife?'

'Twenty-one, master.'

'Good,' said Pattelar. 'Is she pretty?'

I said nothing.

'Speak, son of a dog, is she a beauty?'

'No,' I managed to say.

'Ha, ha, ha,' laughed Pattelar. His men laughed with him. I could hear people in the other shops laugh as well.

'You liar!' Pattelar took his gun and aimed it at me in jest.

He said, 'I'll have a look and see if she's pretty, and give her a certificate.' The betel-juice-stained teeth smiled through the stubble on his pink face.

'Run!' he said.

I began to run in the direction of Ichilampadi.

'Not that way! This way.' Pattelar pointed with his gun in the opposite direction. I gathered my mundu with one hand and ran along the road in the direction he was pointing. The sun was setting. I looked back once.

'Don't look back, rascal!' Behind me, I heard the loud report of the gun. I sobbed aloud and kept running towards a bend in the road some distance away. When I had rounded it, I stopped and leaned against a tree, gasping. Then I clambered down the slope to the river bank and lay on a big, flat rock. I saw a star fall from the sky. When I got my breath back, I rolled down into the water, shivering, and washed away the betel-flecked spittle from my body. I wrung out my torn mundu and wiped myself dry with it. Then I crossed the river and ran to Ichilampadi. Even as I neared my hut, I heard Omana sobbing. On the other side of the hut, torch beams went down to the river and crossed it.

Pattelar sent for me the next day. He bought me a mundu and Omana a sari from the jauli shop. He said to me, 'From today you can work in the toddy shop as a server—I've spoken to Varkey chettan. Omana is pretty. You lied to me. Will you tell lies again?'

'No, yejamanare,' I said.

Everyone knew that it was the Malayali hangers-on who had corrupted Pattelar. He had been just another proud janmi and a man of pleasure. When he sat on the shop verandah, the hangers-on would sometimes point to a woman and say, 'Yejamanare, look at that one, she's so beautiful! Why don't we have a word with her?'

Pattelar would say in Kannada, 'Why bother? It's not worth it.'

They would then say, 'Maja madana, let's have fun, Pattelare! We are with you!'

'All right.' Pattelar would get up and go after the woman.

If someone walked through the bazaar unmindful of Pattelar and his group, they would whisper. 'Pattelar, look at him, he's behaving as if you're not here. What cheek! Should we give him one?'

Pattelar would gulp down his toddy and say, 'No. Let's not have trouble. Galatta beda.'

'We are with you! Give him one, Pattelare. Teach him to show respect.'

'All right!' Pattelar would then get up.

In time, Pattelar came to believe in all this and to go along with it. When I think of it, I feel sad for him. What a pity that such a good man should have come to this! Whose sins was he paying the price for, by sinning over and over again? And what bond from a previous birth was it that brought me to him all the way from the land of my birth?

And so I became Pattelar's servant. As soon as he took his seat on the shop verandah, I would bring him a pot of toddy and then wait in the yard, bowing respectfully, hands folded across my chest. Whenever he went woman-hunting, he would ask me to go with him. In a narrow deserted gully somewhere, he would cover some whimpering Gowda girl's mouth with one hand and grip her hands with the other, before pushing her down on the dry leaves. Once in a while, he would ask me, 'Want her, rascal?'

'No, yejamanare,' I would say.

Without letting go of the girl, Pattelar would aim a friendly kick in my direction.

'Go away, scoundrel; go where I can't see you.'

I would move away and while waiting for Pattelar to finish, look up at the sky and at the mountains in the distance. If someone came by, I would signal to them, 'Pattelar is here!' At once they would take another path. Indeed, people understood what was going on as soon as they saw me. Whenever I heard a girl cry out, I would think of Omana, of that first day. Omana had stopped crying as time went by. That was a great relief to me. Omana is such a simple soul. So am I. How good it is not to have to cry.

One day, at dusk, Pattelar was seated in his armchair on the shop verandah in Udina Bazaar, drinking toddy tapped that evening. I stood in the yard and filled his glass from the pot. Every now and then I ran across the street to the tea shop and brought him vadas and bolis. The usual group stood around Pattelar, some on the verandah and some down in the yard. I brought glasses for them too from the toddy shop and poured them drinks.

It was then that the last bus from Dharmasthala to Hassan groaned and slid to a stop in the bazaar. One or two people got off. Spreading a reddish glow behind it, the bus disappeared into the darkness to make its way up the Shiradi hills, past Sakaleshpuram determined to reach Hassan somehow before midnight. One of the passengers who had alighted crossed the road and approached us. I recognized him at once. He was a dhani—a rich merchant from Arshanamukki. He had areca nut groves and, with his younger brother, owned a lorry. He looked very worried. I ran up to him, joined my palms respectfully, and said, 'Ayyo, dhani! Why have you come by bus today and why are you looking so worried?'

He said, 'My brother went to Yenijira early this morning with the lorry and is not back yet. He should have returned at noon. Have any of you seen him? Did you hear of an accident anywhere?'

Suddenly Pattelar pointed with his foot at the newcomer from his armchair and asked me, 'Who is this, rascal? Why are you grovelling? Is he your father, dog?'

I stood between them on the roadside, in the huge shadows cast by the Petromax lamp, beginning to perspire and aware that my joined palms were growing limp with fear. I said, 'Yejamanare, this is Yousappicha from Arshanamukki. He's a dhani. His younger brother is missing.'

Pattelar flung the toddy from the glass in his hand on Yousappicha's face. He said, 'Do we keep your brother here? Look at the bloody dhani!'

The shops were closing for the day and everyone was going home. Only a few people lingered in the tea shop and the toddy shop. They were listening silently. No one came out.

Pattelar jumped down into the yard, swaying a little. My voice trembled as I begged him to spare Yousappicha. 'Please forgive me, yejamanare! Yousappicha is one of our own.'

Pattelar gave me a terrible blow across my right cheek and ear. I slumped to the ground; my ears filled with the noise of a world splitting apart. When I raised my head, Pattelar had Yousappicha on the ground and was kicking him. He kicked him in the head, stomach and spine.

'This dog of a dhani comes to look for his brother here! So late in the evening!' shouted Pattelar. I was too scared to get up from where I cowered. My God, I thought, Pattelar is beating Yousappicha to a pulp. A man he is seeing for the first time! Why hit a man who had come to look for his brother? Yousappicha lay still on the roadside, in the shadows of Udina Bazaar, like a muddy bundle of rags. Not moving and hardly daring to breathe, I stared at the sprawled figure.

Pattelar climbed back on to the verandah, sat down on his chair and summoned me. My head buzzed violently as I jumped up and rushed to him. In the dark I stumbled over one of Yousappicha's outstretched legs and fell. He knew nothing of all this. I scrambled up again, ran and poured toddy for Pattelar from the pot. Pattelar said, 'A dhani, is he? And he comes at dusk, like a bad omen. You be careful too! Next time you invite people like this, you'll get the same treatment.'

'Ayyo, yejamanare.' I wailed, 'I didn't invite Yousappicha. His younger brother—'

'Shut your mouth, dog!' shouted Pattelar.

I shut my mouth.

⌇

It was late when Pattelar finally got up to go home. He said to us, 'Put him in the back of the jeep.'

We carried Yousappicha and put him in the jeep. It was I who held his hands. The pulse is beating, I said to myself.

Pattelar's wife, Sarojakka, did not like noisy gatherings in the house at night. So Pattelar usually went to play cards in the watchman's hut in the yard where areca nuts are dried. Tonight, there were six or seven people with him. Yousappicha was dumped in a corner. Blood oozed from his mouth in a long thread and vanished into the dusty floor. I sat near him and watched the card game, dozing off occasionally. Now and then, when I woke up and saw the way Yousappicha lay, I felt sorry. Once, I almost whispered, 'Yejamanare, may I give Yousappicha a drop of water?' But my voice stayed trapped in my throat. Sometime near dawn, Yousappicha groaned once. Then I fell asleep. When I woke up, it was daylight and only Yousappicha and I were in the hut. He was cold and dead. 'Ayyo, my Yousappicha, that you had to die this way,' I said sadly, looking at his corpse. Suddenly I thought of Sarojakka. Ayyo, Yousappicha must be moved before she finds out.

Shivering in the morning cold, I ran, slipping and sliding on the areca nuts, to the window of Pattelar's room. I knocked on it softly.

It was Sarojakka who opened the window.

'What is it?' she asked.

'Nothing, nothing, yejamanathi,' I stammered.

Then Pattelar got up, came to the window and asked, 'What is it?'

I said, 'Nothing, master.'

Pattelar looked searchingly at me and said, 'I see. Go back now; I'll come.'

I waited near Yousappicha, trembling in the cold. Pattelar came, wearing a shawl, his ears covered with a monkey cap, and said, 'Get two or three people, tie him up in a sack and bury him in Arabi Majal before Sarojakka comes out to milk the cow. I'll take care of the police.'

I felt very relieved.

⌇

One day, Pattelar said to me. 'Listen, I'm going to kill Sarojakka. I need you to help me.'

I shook with fear. Sarojakka was such a good person. She gave me something to eat or drink every day. She was aware of Pattelar's wayward ways but treated him with affection. She gave him good advice and never quarrelled with him. I felt very sad. I said. 'Why do you want to do that, yejamanare? She is such a good akka. Please don't...'

Pattelar said, 'I shall make it look like an accident. You will be the witness. You will call her out to the verandah. I will sit there pretending to load my gun. It will look as if the gun went off by mistake. She must not know that I shot her. I do not want her to die with that sorrow. There's no need for you to feel sad either. Just think of it as an accident, that's all. You'll have to give proper evidence. Her brothers are nasty fellows.'

I thought sadly: Pattelar wants to kill Sarojakka to get hold of her share of the property.

'Yejamanare,' I said, 'we don't need Sarojakka's share.'

He gave me a long look.

That afternoon, Pattelar sat on his chair on the verandah, put the gun on his lap, opened the bag of gunpowder and pretended to load his gun. I squatted on the edge of the verandah some distance away. The gun turned slowly in my direction. I told myself, Pattelar is just loading his gun, that's all he's doing. There's nothing to worry about.

When Pattelar signalled with his eyes, I looked towards the kitchen and called, 'Sarojakka!'

She answered.

I said, 'Please give me something to drink.'

'Shall I give you some kanji?' she asked.

'All right.' I said. Poor Sarojakka, I thought to myself. She will bring me kanji to satisfy a hunger I do not feel. And she will die.

With sudden fear, I called out again, 'Sarojakka, please don't add salt!' I did not look at Pattelar.

Sarojakka came out. She had the bowl of kanji in one hand and a glass of buttermilk in the other. She gave Pattelar the buttermilk and said. 'Drink this. You did not eat anything this morning.'

Pattelar flinched, but took the glass from her.

Sarojakka walked towards me with the bowl of kanji. My eyes were on Pattelar. He tried hurriedly to put down the glass of buttermilk in one hand. At the same time, he tried to point the gun with the other at Sarojakka. I saw him pull the trigger. The glass fell down and broke. At the same moment, I heard the shot... I was staring at Sarojakka, wanting to cry for her, when suddenly I felt a searing pain in my stomach, as if someone had stabbed me with a red-hot stick. It felt like a fire beginning to blaze inside me. I could not bear the pain and jumped up. Blood spurted from my stomach! Ayyo! Is it I who've been shot! Pattelar was staring at me wide-eyed. Sarojakka screamed.

I shouted, 'Yejamanare, save me. It's I who have been shot!'

'Shut your mouth, scoundrel,' shouted Pattelar, 'or I'll finish you off right now!'

Sarojakka went out into the courtyard and called for help. People came running. I fainted then. When I regained consciousness, I was lying on the floor of the grain-shed, with a cloth tied around my stomach.

I heard Pattelar say to the others. 'Don't waste time. He's finished. Carry him to Arabi Majal. I'll bring my gun. We can bury him there.'

I whimpered, 'Yejamanare, Omana will have no one. Don't kill me, please.'

Pattelar looked at me in astonishment. While he watched me uncertainly, Sarojakka suddenly came in and said, 'What are you saying? I'll not allow it. Take him to the hospital!'

⌒

I lay unconscious in the hospital for a week. I had many dreams about Sarojakka and Omana. It was while I dreamt that I was lying with my head on Sarojakka's lap that I woke up. My head swam in a wave of happiness. The warmth and softness of Sarojakka's lap clung to me for a long time. I laid my hand on my wounded stomach, feeling very happy that Sarojakka was alive.

After I was discharged from the hospital, I said to Omana, 'Do you know, I had a dream that I lay with my head on Sarojakka's lap?' Omana did not believe me.

⌒

Pattelar came to my house one night, woke me up and said, 'Come. We're going to catch the fish in the temple ghat at Arshanamukki.'

'Ayyo! My yejamanare!' I said, trembling with fear.

'Why are you scared?' asked Pattelar. 'I am here.' He held up a cloth bag which contained dynamite sticks, and shook it. 'You and I will do it. No one else has the courage. Cowardly dogs!'

I looked out fearfully. Moonlight filtered faintly through the Makara mist. This was the time the yakshi of the Arshanamukki temple came out to stroll around. The fish that lived in the ghat belonged to the yakshi. The Udina River flowed wide in front of the temple, looking like a lake. Even in summer, the

current was strong enough to make your legs totter. In the rainy season the
roaring floodwaters resembled a stormy sea. But the five thousand fish in the
bathing ghat by the temple always lay there undisturbed. They were huge, sleek,
gleaming fish, with knowing eyes. They did not fear men, but allowed no one
to touch them. They splashed and writhed in the water, clustered together as
if they were in a fish basket. If you threw them puffed rice, you could not
see the water for the thrashing of the giant fish as they fought to get to the
food. When they jumped and fell back, the sound was like coconuts falling.
I have often felt frightened as I watched them. What if I were to fall among
them? Wouldn't they eat me up in a flash? I would edge backwards a couple
of steps before throwing them more puffed rice. At other times, my mouth
would water as I looked at the fish. If Omana could get just one of them, it
would last her a whole week—to make curry, deep-fry, make roast chutney
and also pickle some. Just one fish would be enough. But they belonged to
the yakshi. The yakshi caught and ate anyone who caught and ate her fish.
She would tear him open, bring her fish back to life from where it lay in
his greedy intestines, give it a loving kiss on its pouting lips and, her laughter
filling the night like the tinkle of crystal balls, shoot it back into the river
like a silver arrow flashing through the darkness. Then, smacking her lips, she
would bend down to savour slowly the one who had eaten her fish.

'Ayyo, yejamanare,' I wailed. 'Ayyo! Ayyo!'

Pattelar took the headlamp out of his shoulder bag and strapped it on to
his forehead. He gave me the cloth bag of dynamite sticks and said, 'Carry it
carefully.' Then he picked up his gun. As we went down into the courtyard,
Pattelar said, 'Bring two or three baskets with you.'

All along the way, in the meadows and valleys filled with moonlight,
I sensed a restless wariness. I heard the tinkling of anklets. The yakshi was
pursuing us to protect her fish. Carefully placing my feet in Pattelar's footprints,
and trembling with fear, I walked under the eye of the moon, to the river by
the Arshanamukki temple.

Once, I mustered enough courage to say, 'Yejamanare, the yakshi...' Pattelar
brandished the gun and said, without turning around, 'Fool, the yakshi is afraid
of guns. Don't you know a gun is made of iron? She's afraid of you as well.
You're from another religion and because of that she doesn't like your blood!'
In the darkness, Pattelar shook with laughter.

When we had gone quite a way, I whispered, 'Yejamanare, if anything
happens to me... Omana...'

Pattelar laughed again. 'Don't I look after Omana even when you are here?

If you're not here, you don't have to worry at all. Ha, ha, ha!'

We reached the river even as I was asking myself whether Omana would grieve if something happened to me. There was only the soft sound of the summer waters. And the moonlight. And the mist drifting over the water. And the stillness of the temple. Shivering with terror, I looked up at the sky, standing on the steps of the bathing ghat. The moon was only a faint glow in the foggy sky. Veils of mist and cloud floated in its shadowy light. In the waters of the river, all was quiet. Where were all the fish, I wondered. Were they asleep? Do fish sleep? Were we going to kill the fish when they were sleeping? Ayyo! That would be so sad. Ayyo, my fish, I said to myself, you doze and dream in still corners under the moving sheet of water. But now you will not be able to complete your dreams. The thunder of these dynamite sticks will, like a hammer, shatter your dreams and your hearts. Then I will jump into the water with my basket. I will stack your gleaming bodies in it till it is full. The moonlight will fill your unclosing eyes. The mist will kiss your gasping mouths. I will weave through the water like an otter, lift up your corpses and stack my basket high, over and over again. The shreds of your broken dreams will cling to me like your white scales. And then, while the yakshi is devouring me, you will fly out of these baskets with throbbing hearts and return to the river through the mist. I will drip from the yakshi's mouth as blood and marrow, dribble down her thighs and become manure for the earth.

I said to Pattelar, 'Yejamanare, I am afraid.'

I looked once towards the temple, took a stick of dynamite and held it out to Pattelar. He said, 'Shitting coward, throw it yourself!' He pressed the switch on the battery-box tied to his waist and bent his head so that his headlamp shone directly into the water. I lit the stick of dynamite and threw it into the centre of the patch of light. It sank and disappeared. Both of us waited for the explosion that would resound from the envelope of water. One! Two! Three: I counted. I counted to fifty and said, 'The stick did not explode, yejamanare!'

'Throw the next one, rascal,' said Pattelar.

I lit the next stick and threw it. When I had counted to fifty, Pattelar said, 'The next one!' I counted to fifty again.

Only the light from the lamp on Pattelar's forehead floated on the water.

I took the next stick in my hand and looked at Pattelar. 'Yejamanare,' I stammered, 'you throw this one, yejamanare.'

Pattelar looked at me and muttered something. The beam from his forehead wove its way through the night. I was terrified. Pattelar was growling obscenities.

His face was distorted as he hurled dirty words at the yakshi. Then he grabbed the stick of dynamite from me and, spilling light from his headlamp over the steps, he ran up to the temple yard.

'Baddimagale! You whore's daughter!' he screamed, lit the stick of dynamite and threw it with all his might at the closed door of the temple. I covered my ears with both hands and crouched down. Look at yejamanare, he's wrecking the temple. The explosion comes now! I closed my eyes tight, dug my fingers into my ears, and sat there, frozen. But there was only silence everywhere. When light filtered through my closed eyes, I opened them and looked up. Pattelar was bending down and peering at me. I thought, Pattelar is mad with rage and is going to kill me. Ayyo! My Omana, I shall not see you again. My God, here I come. Was Pattelar picking up his gun? I stared into the blinding light. Suddenly, he switched off the lamp. He stood there without moving for some time. When the ghost-light of the headlamp had faded from my eyes, I stood up. I picked up the cloth bag in which I had brought the dynamite. I put the baskets on my head.

'Shall we go, yejamanare?' I asked.

Pattelar said nothing. He walked up to the temple door and picked up the unexploded stick of dynamite lying there. Then, with a grunt, he threw it far into the river, as if aiming at the other bank. I heard it fall with a plop somewhere in the middle of the river, beyond where the fish lived. As we were going out through the temple yard, a sound came from the river, 'Bhum!'

The sound of the exploding dynamite pursued us like something awful the river had uttered, making me feel as if a world had ended.

'Bhum!' the river had said.

As we walked back under the whirling stars of the midnight sky, I heard footsteps around us and bells tinkling from the shadows, and I shivered in fear. Pattelar moved silently, like a sleepwalker, with the gun on his shoulder. Shadowy forms of bats circled in the sky above us. For a moment I thought they were flying fish.

━━━

People regularly brought Pattelar gifts so that he would not harass them. They would prod me and say, 'Pattelar might forget who brought this. You must remind him.' I used to put away the balls of jaggery, the dried and salted chunks of deer meat, the rough boards of rosewood, ripe jackfruit, fowl and big packets of pickled fish wrapped in spathes of the areca nut palm. In the evening, I would usually give Pattelar the details of who had brought what.

Pattelar would clear his throat loudly, spit and say, 'The spineless scoundrels!' I would carry all the gifts on my head to Pattelar's house, taking care not to leave anything behind.

Sarojakka would say, 'Do people give these things out of love? They give them because they're afraid. I don't like to keep such things at home.'

'Ayyo!' I would exclaim, looking towards the front verandah, where Pattelar was seated. Sarojakka always secretly gave me a portion of all the gifts. And I would take them in a basket to Omana. 'See, Sarojakka gave me these!'

One day, Pattelar said to me, 'You rascal, the fish pickle I ate at your house with the toddy yesterday tasted exactly like the pickle that Kuttapparai brought to my house. How did this happen? Does Kuttapparai also have an account in your house? Tell me the truth!'

I said, 'Ayyo, yejamanare. I don't know.'

Pattelar burst out laughing. 'So you need my help to finish the pickle your Sarojakka gives you with such love, do you?' He shook with laughter.

Early one morning, I was lying with Omana, hugging her close. She was still enveloped in the fragrance of Pattelar's perfume. I took a deep breath of the scent which I loved and pressed Omana to me with great pleasure. I said to myself, although she smells of Pattelar's perfume, she belongs only to me. Some day, I'll buy her this perfume. It was then that I heard a knock. I jumped up, startled, opened the door a little and peered out. There were four or five people standing in front of the house. I made out Kuttapparai, Ahmed—the panchayat president's younger brother—and a couple of Malayalis. I began to tremble uncontrollably. They had come to kill me. Perhaps they had killed Pattelar before they came here. I stood there, unable to move. When Omana peered over my shoulder, her breasts brushed my back. I thought, I will never feel her breasts rub against me again. Kuttapparai said from the courtyard, 'Come out. We have come to ask you something.'

I closed the door without replying, hugged Omana and said, 'If these people kill me, Omana, you must commit suicide. I think Pattelar is already dead.' Then I opened the door and went out.

They grabbed hold of me by the shoulder, took me under the elanji tree in the compound, and said, 'Look, don't be scared! Pattelar has to be killed. What is the best way to do it?'

I stared at them. Kill my yejamanare! Ayyo, who would I have then to turn to?

They said, 'Listen, everyone is fed up with Pattelar. You know that no man or woman can walk safely in the village because of him. Pattelar is useful to you, we know. But you're a good fellow, and a harmless one. If you help us, we'll help you. We'll give you a hundred rupees. And the job of a peon in the panchayat office.'

Sunrise had touched the sky. I thought, if Pattelar dies, the village and the villagers will benefit. I'll have a good job as well. But Omana, won't she feel sad? Or would she be happy? I had never asked her this question. Suddenly, I was afraid. If Pattelar died, would Omana then belong to me? And only me? I was both thrilled and frightened by this thought.

My mind in a whirl, I asked Ahmed, 'What sort of help do you want, muthalali?' My words tumbled into my ears as if somebody else had spoken them. Ahmed said, 'It's not easy to kill Pattelar, you know. His companions are always with him. All we want you to do is to get Pattelar to sit for a while on your verandah this evening. We will shoot him from a hiding place. It will happen right in front of his men, so no one will suspect you.'

I stared at them, and said, 'Don't let them shoot me or Omana, please.'

They laughed soundlessly. 'We don't shoot as badly as your Pattelar, you fool.'

At noon, scratching my head, I asked Pattelar, 'I wonder if you're going to Ichilampadi this evening, yejamanare?'

Pattelar said, 'How does it matter to you whether I go or not?'

I said, 'Omana is preparing a meal specially for yejamanar. Yejamanar must come.'

'What is she making?'

'Mushrooms fried with coconut, roasted deer meat, fried fish and arrack. She's frying eggs as well, yejamanare.'

'All right,' said Pattelar.

When Pattelar arrived at my house with his men that evening, the lamp had been lit. The men squatted here and there in the courtyard. Pattelar told me, 'I don't feel like drinking arrack. Go and bring me two bottles of fresh toddy.'

I heard suppressed laughter from the courtyard. It doesn't matter, I whispered, you won't laugh after today. Although I'm scared, Omana will soon be mine alone.

I was anxious about the people hiding somewhere in the darkness with their guns. What if Pattelar ate and left before I returned with the toddy? Who would get him to sit on the verandah?

I looked helplessly into the darkness, then left my house and started running down the road to the toddy shop. I then ran back again, a bottle of toddy in each hand. As I set foot in the courtyard, Pattelar opened the door and came out on to the verandah. 'Ah, you've come,' he said. 'I've just finished eating.'

His men were dozing in the shadows of the courtyard. I did not have the courage to look farther. What were the assassins going to do now? Would they do something in haste? As I climbed on to the verandah, I shrank into myself. Where would they shoot from? My God, don't let them shoot Omana. I kept trying to think of a way to make Pattelar sit down. But he stood on the upper verandah, waiting to wash his hands. I knew that the thatch of the hut hung so low that the killers would not be able to see him from where they were hiding.

Omana came to the verandah with a lamp in her hand. I put down the toddy and said, 'I'll draw water for you to wash your hands, yejamanare.'

The well was next to the lower verandah. It was a deep well; they had had to dig right down to the nether world to find water. Pattelar stepped down to the lower verandah and walked to the well. I lowered the bucket into the well, drew water and straightened up. As Pattelar bent and extended his hands to wash them, I heard the shot. I wept silently, and looked at Pattelar. Then I looked at Omana. She was running into the house, crying. Pattelar covered his right ear and cheek with one hand, and tried to jump to safety. The second shot came then, and dislodged earth and stones from the wall of my hut. I dropped the bucket and rope. Pattelar had fallen on the well's rim and was beginning to slip in. He was silently clawing at the earth, desperately trying to hold on. In a moment he would be gone. 'My yejamanare!' I shouted. I leaped forward, caught hold of both his hands, stretched myself out on the ground for support and pulled him from the mouth of the well. Pattelar lay on the wet earth and breathed through his open mouth like a fish that has been thrown on land. I saw Pattelar's men crashing through the shrubbery, flashing their torches. Shots rang out and there were shouts. I helped Pattelar get up, took him into my hut and made him sit down on a mat spread on the floor. There was a wound on his ear. His right cheek looked black and blue and he could not open his right eye properly. Omana leaned against the wall, weeping. I poured a glass of toddy from the bottle I had brought from the shop and said, 'Drink this, yejamanare.'

I stood looking at Pattelar and thought, why did I save this man? A small push with these hands of mine would have been enough. I stared at Omana. Was this really my Omana? Who was I? Who was this wounded man, sitting on my torn mat? In the shadows thrown by the flame of the kerosene lamp that was flaring in the breeze, it seemed to me that Pattelar and Omana were turning into shapeless, writhing forms. My head reeled. I fainted.

As I lay in a swoon, I remember seeing, like a picture in a calendar, Sarojakka and Omana flying somewhere through the clouds, looking like white cranes.

A few days after the failed attempt to kill Pattelar, I was squatting on the verandah of the toddy shop on a rainy evening. The shop was closed because the toddy had not yet arrived. Suddenly, Pattelar came by in his jeep and asked me to get in. I got into the back and he drove off like a mad man. Now and then, the jeep grazed against culverts. At one of the bends, it skidded and nearly overturned. I leaned forward, full of fear, and forcing my voice against the screaming wind, I cried, 'Yejamanare, please go slow; otherwise we'll be killed!' Pattelar did not answer. The jeep roared on, soaking the forests on either side of the road in the blaze of its headlights. Eventually it rattled over the stones of the rough pathway to his house, and stopped. Pattelar turned to me and said, 'I've killed Saroja. We have to make it look like suicide. You have to help me.'

I stared at Pattelar, unable to take my eyes off his face. 'Ayyo, my yejamanare!' I said. 'My Sarojakka! My Sarojakka!' Crying silently in my heart, I stumbled out of the jeep and started running. Pattelar got into the jeep and gave chase through the peringalam groves and the thickets of weeds. I heard the jeep growling behind me, its eyes glittering in the dark. The headlights threw enormous shadows that writhed through the bushes and pursued me. I was sure Pattelar was going to run me over and kill me too. Sarojakka, I cried to myself, here I come! Omana, what will you do? Suddenly I was blocked by an embankment. I put my back against it and looked at the shining eyes of the jeep racing towards me. I did not want to die! I screamed, 'My yejamanare, please don't kill me!'

Pattelar stopped the jeep, got down, took me by the hand, pushed me inside the vehicle and said, 'Don't run away, you wretched rascal. What are you afraid of? I need you. Be a little brave.'

Sarojakka lay on her back on a cot. She had been strangled.

Her tongue was hanging out. Her neck was black and blue. Her face was distorted. I put a trembling hand to my chin and gazed at Sarojakka. Sarojakka, who had fed me sweets and kanji, given me old clothes. Sarojakka, whom I had once betrayed, but whom I loved. Sarojakka, who had refused to let me be killed. And now she was dead. At the moment of death, what thoughts had gone through her head as she looked at Pattelar's face. Had she pleaded for her life? No, she would have looked at Pattelar with eyes full of astonishment and sorrow.

Then Pattelar said, 'She didn't know I killed her. I covered my face with a towel, came in through the window in the darkness and caught hold of her. But she gripped my hands. That's what worries me. Would she have known they were my hands when she touched them?'

I did not say anything.

Pattelar was quiet for a while and stood looking at Sarojakka.

Then he said to me, 'I've run through all my money. So I had to kill her. Don't you know that I have nothing against Sarojakka?'

He then bent down and gazed at Sarojakka's face. He said, 'Saroja, I didn't kill you.'

I drew back, afraid. I thought Sarojakka was going to get up to answer Pattelar with her tongue hanging out.

Then Pattelar turned to me and said, 'Never mind. Come here and take hold of her legs.'

I caught hold of her ankles in both my hands. It was the first time I had touched Sarojakka. Holding her slim, pale, soft ankles tightly in my hands, I looked at Sarojakka's face. That distorted face was not my Sarojakka's. I looked once at Pattelar, and then I bent down, pressed my face into the soles of her feet and kissed them.

Pattelar said softly, 'You liked Sarojakka.'

I climbed on the bed without saying anything, hoisted Sarojakka on to my shoulders, then put my arms around her and lifted her up as one would a child. Pattelar tightened a rope around her neck and pulled her up. I squatted on the bed, pressed my face to her cold feet and sobbed aloud. Pattelar did not scold me. He stood quietly, looking out of the window.

I looked back once, as we left the room. My Sarojakka hung from the rafters like a broken reed.

Pattelar said again, 'She wouldn't have known, would she?' I followed him without replying. All I wanted to do was to get back to Omana. I wanted to embrace her feet and weep all night, to overcome my sin and sorrow.

As we walked to the jeep. Pattelar said, 'I'm going to the police and to her brothers now to tell them that she killed herself. Remember, you are the witness.'

He stopped near the jeep. I waited in front of him in the dark. Suddenly Pattelar flashed his torch into my face. With dazzled eyes, I searched confusedly for his face behind the torch. Pattelar shone the light into my blinking eyes and asked in a small voice. 'Tell me, look at my face, and tell me, can one recognize a person by just touching his hands?'

I said, 'Who knows, yejamanare?'

Pattelar spat out an obscenity and threw the torch on the ground. Its light came rolling towards my feet. I bent, picked it up and switched it off. I heard Pattelar gasp. In the darkness, stretching both his hands towards my blinded eyes, Pattelar said. 'Touch my hands.'

My fingers brushed his outstretched hands.

'Is this me?' he whispered.

'Yes, yejamanare,' I said and drew my hands back. Pattelar growled in a voice that came from deep within him. 'I made a mistake. I should have asked you to kill her.'

The next day I heard that Sarojakka's brothers had got some people together and attacked Pattelar's house. I ran there, crept up behind the crowd and peered over their shoulders. The house had been set on fire, but the fire had been put out. Shattered household items were strewn all over the courtyard. The jeep looked battered. I did not wait to see more but ran back home before anyone could recognize me. When I got home, I shut the door and said to Omana, 'It's all over. Pattelar's house has been attacked and burned. I don't know where Pattelar has gone. They are sure to come looking for me now.'

'What will you do?' Omana asked, holding on to my shoulder and weeping.

No one came looking for me. All that day and the next I sat at home not knowing what to do. When I was certain that no one was interested in me, I asked a passer-by, 'Do you have any news of Pattelar?' He gave me a hard stare, then said, 'Yes. They say he's gone into hiding. But why are you asking me, it is you who should know.' Worried, I sat on the verandah and thought, what do I do now?

At midnight, as soon as I heard a knock on the door, I said to Omana, 'That's Pattelar.' I got up and opened the door. Pattelar stood on the verandah. The rain had drenched him. He did not have a torch. He had his gun and bag of gunpowder. All he wore was a mundu. There were bruises on his body.

'My yejamanare!' I said. I brought him into the hut and shut the door.

Pattelar squatted on the ground. I felt very sad, looking at him. I sat next to him with folded hands and said, 'My yejamanare, to think that this should have happened to you! Tell me what I must do.'

Omana brought some hot kanji for Pattelar. As I stood by the well in the dark, pouring water for Pattelar to wash his hands and feet, I remembered many things. Pattelar had his kanji, then he asked me, 'Will you go with me?'

'Where to, yejamanare?'

Pattelar said, 'We'll go to Kodagu. I can hide in my nephew's house.'

He took some money out of the fold of his mundu, put it on the mat and said to Omana, 'This is for your expenses until this fellow comes back.' Then he raised his head and looked at Omana. Her face was in the shadows thrown by the lantern.

Would Omana feel sad, I wondered. It seemed to me that the fragrance of Pattelar's perfume filled the room. Omana came up to me, held my hands and began to cry. She put her face between my palms and sobbed. Pattelar sat staring at the ground. Omana's crying made me feel suffocated. Trying to lift her face, I said, 'Omana, I am there for yejamanar.'

Slinking through the night, we reached the main road, hailed a night lorry passing by and arrived in Madikere. As soon as we entered the courtyard of Pattelar's nephew's house, he hurried us into the shed used for distilling lemon grass oil, closed the door, and said, 'It's dangerous for you here as well. I've just heard that Aunt's brothers are in town with gunmen; you had better go into the forest, Uncle.'

Pattelar's face fell. 'Who told you?'

'People who saw them in town. The story has spread to these parts, Uncle.' His face showed pity and fear.

We stayed in the shed till nightfall. Pattelar spoke to his nephew about the possibility of coming to an understanding with Sarojakka's relatives.

The nephew replied, 'What I heard is that they have already taken over Aunt's share. In which case, they'll hardly be interested in making the peace.' Pity touched his face once more. I could not bear it. I squatted in my corner and shut my eyes tight.

At night, we packed up the roast meat, salt, chillies and coconuts that Pattelar's nephew gave us. I put the bundle on my head. Pattelar took the gun, the bag of gunpowder and a torch. We entered the forest. It was peaceful and quiet. My fears vanished as we moved like two ants beneath the shade of the sky-high trees. Walking briskly and enthusiastically, I said, 'Yejamanare, this is like our old hunting trips.' Pattelar didn't reply. He just turned and looked at me.

That night and the following day we ate our dried meat and coconut chutney, rested now and then, and walked. I followed Pattelar, with the load on my head, thinking in wonder of the paths my life had taken.

The hiding place that Pattelar had in mind was on the far side of a river. We heard the murmur of the river from quite a distance. We were then walking under a wild champakam tree full of flowers. I paused beneath the tree. 'Yejamanare,' I called, 'the fragrance of your perfume!'

Pattelar turned. He called me by my name, 'Thommi.'

'Yejamanare,' I answered.

Pattelar stood on the fallen wild champakam blossoms, gun in hand. The roar of the river filled my ears. Pattelar said, 'You must never again make Omana cry.'

'My yejamanare,' I said, 'Omana is my life. Will I ever make her cry?'

Once again, there was only the sound of the rushing river. Pattelar leaned against the champakam tree. He said softly, not looking at me, 'Saroja would not have known, would she?' For a moment, the sunlight drifting through the leaves cast a net over Pattelar's face. I looked into his eyes, which were caught in the meshes of the net and said, 'No! No! No, yejamanare.' We walked to the river.

The river hissed amongst the rocks. From the hills on the other side, a brook sprang down a rocky slope, descending as a waterfall. Pattelar began crossing the river, gun and bag held high over his head. I followed with the bundle on my head, through the chest-deep foaming water. Pattelar was more than halfway across the river when I suddenly saw something strike a protruding rock, making splinters fly. A shot! I had not heard the sound of the shot above the roar of the water. I counted six gunmen running down the hill on the other bank. Pattelar was thrashing through the water towards the same bank. More shots hit the river, flinging up water and bits of rock. 'My God,' I said, 'they are using big rifles!' I turned and waded back like a mad man. I stumbled, fell, went under and, flailing about, somehow reached the river bank we had started out from. Crouching behind the thickets at water's edge, I looked back. Pattelar had reached the far shore and was now running, bent double, under cover of the bushes. Two of the gunmen were just behind him. I saw two others running through the trees to cut him off in front. Suddenly, Pattelar stopped. He held the gun above his head in both hands and stood still. I cried, 'My yejamanare!' Shots came one by one, then together. I saw blood gush from Pattelar's chest, forehead and stomach. He fell over and lay still. I then proclaimed my great sorrow—my voice carrying across the white expanse of the river, 'My yejamanare! Ayyo! My yejamanare!'

Someone shouted, 'Don't let him go!' A few gunmen jumped across the rocks, waded through the water and rushed to the bank where I was hiding. I left the shelter of the thickets and ran for my life. 'Ayyo! I don't want to die!' Panting like a dog, I raced headlong, the fear of death in me. I heard the gunmen somewhere behind me.

Beyond the wild champakam, there was a steep slope. I rolled down the slope into a grove of enormous trees. I clambered up one, clinging to it like a chameleon. The dusk-like darkness beneath the branches wrapped itself around me. I stopped climbing only when I felt as if my arms and legs were being torn off. I looked down. Where was I? My head reeled. I closed my eyes and wound my arms tight around the tree. The gentle warmth of urine flowed down my legs. Pressing my face into the moss-covered bark, I became yet another shadow amongst the dim branches. I stayed there like a lizard that had forgotten to move. Even when everything had become silent and quiet below, I did not move. After a long time I grabbed hold of a branch and straddled it. In the evening, it rained. Even after the rain stopped, the leaves dripped on me all night long.

In the morning I limped, shivering, to the river. I saw no one. Only Pattelar lay there. I crossed the river, walked up to his body and looked at his face wonderingly: it was Bhaskara Pattelar who lay on the grass by the river! The body had begun to smell. I held my nose with one hand and asked, 'Yejamanare, is this really you?' Pattelar lay cold on the grass with his eyes closed. Bhaskara Pattelar was dead! Bhaskara Pattelar, who had bought me mundus, who had found me a job, who had loved my wife, was dead! I squatted there and wept. Forgive me, master, for having once betrayed you! I looked at his swollen, discoloured face and continued to weep.

I opened his dead fingers and took the gun from his hand. For a moment, I held his wrist. I said, 'Get up, yejamanare. Sarojakka might be waiting for us.'

I covered Pattelar with grass and twigs. Then I took his gun, walked along the bank of the river, climbed on to a rock and threw it into the rainbows that fluttered in the waterfall beneath me. I felt a kind of relief. And also, a kind of courage. Crossing the river, I ran to Ichilampadi to tell Omana that Pattelar was dead.

twenty-seven

~

TAR ARRIVES

DEVANOORA MAHADEVA

Translated from the Kannada by Manu Shetty and A. K. Ramanujan

An Overview of the Village

The dirt track, that's just good enough for a bullock cart to amble along, starts like an alley from the village and winds on for three miles before joining the main road on which the buses ply. The same route back from the main road dips down to the village, hedged in on both sides by cacti, meanders through the banyan grove in which the spirits reside, divides into three, and then runs into the village. On either side of the alleys are houses so close together that at first sight they seem to be choking each other.

It is not a village that's known for anything. If one counts, there would at the most be some eighty households in it. There's not even a little hotel there, like everywhere else. You can judge the place from that. In the village live four young men who have studied up to high school, either in neighbouring Nanjangud or Mysore. Their names are Lakuma, Rajappa, Madu and Shambu. The village patel and these characters don't quite get along. After all, how could he suffer these twits, born before his own eyes, strutting around with their heads held high? It wasn't like this in his father's time. These young fellows of upper caste or lower caste, run around with that untouchable Lakuma, as if they were born out of the same womb. They had been warned not to do this by their fathers. You'd think they'd listen to their fathers. Everyone is waiting for those boys to be caught molesting some girl. That will be an opportunity to drag them to the village hall, strip them down to their underpants, tie them to a pole and flog them. People feel that the boys are certain to oblige sooner or later.

A Road is Ordered

It was during such times that the order sanctioning a road for the village came from above. The patel's body puffed up even more. The news spread rapidly through the village. It may be recalled in this connection, that some seven or eight years ago, when the minister had visited the village, the patel had personally garlanded him and given him lemons, and had said that for a village to be considered a village it should at least have a road. He had pleaded with the honourable minister to be magnanimous and sanction one.

The evening that news about the order reached the village, the patel, who had just returned from Nanjangud, announced that he himself would be the road contractor and that, from now on, for months together, no one would need to look around for work. With the surplus government money the village temple would be renovated. Every house in the village was electrified by the news.

In Front of the Village

It's been many days now since all this has happened. Now the village is a regular battlefield. The banyan trees that once stood spread out against the sky, as high as the eye could see, have all been cut down to the ground. To those who once saw the village hidden behind the dense mat of banyan trees, it must now look like a desert. There's not even a trace left to show that the trees were once there. Where they once stood, there are now machines shuttling back and forth. The sight of these machines, with their stone wheels that stand as tall as a man, moving around and belching smoke, fascinates not just the children but also the men and women of the village. The clamour that rises from the machines crosses the village and goes beyond. To anyone observing the village from a distance, it will look as though there is an upheaval going on there. But if you come closer, it is clear its people are still the same, and belong to the same village. If you talk to them you will know them to be the same people. But they have changed beyond recognition—with the black tar that is spread on the ground smeared all over them, nobody can any longer recognize the way they look, the way they move!

Off to one side, ten or fifteen young women sit and crush gravel, the pallus of their saris tied around their heads. One among them raises her thin voice and sings, '*Come, come soon, Chenna Basavayya.*' Others prod her on, joining her in a chorus, which is heard by even those working at a distance. With the workers engrossed in their work, the whole place, for as far as a furlong, has a festive air about it. Measuring and digging, digging and levelling, levelling and sprinkling water, sprinkling water and bringing the blend of gravel and tar in a barrow and filling it in, filling it in and spreading it—all this goes on without pause. There is a machine that goes backwards and forwards. A dhug-dhug machine that levels. A fire-spitting machine that mixes the gravel, sand and tar together. For everything a machine. How many of them! Each one a black-black weird form. Where the measuring is going on the patel is standing with his hands clasped in front of him.

Hands and Mouths Tarred

Children, everywhere, are the same. They gather around the machines. When the elders scold them they retreat; when they've turned away, the children are back again. On the whole, all the children spend their evenings where the tar is melted. Even after night has fallen they do not return to their houses. The elders have to come there, spank them, and take them back with them. But then, even the elders who come to take the children away can't help spending at least a few moments watching the miracle of road-making before they leave. And if, while going home, they hold the hands of the children, the tar from the hands of the children will smear theirs. That's what happened when Rangappa came to collect his son. Angered, he spanked the boy, saying: 'Is this what your master taught you in school?' The boy started to cry and didn't stop. When Rangappa, losing his patience, hit the child on the mouth, he cried even louder and put his hands into his mouth. The tar on his hands stuck to his mouth, his mouth was sealed and the crying stopped. Many similar things happened every evening. It became a habit with the women to bring home a ball of tar when returning from work. Apparently, some woman tried to plug a leak in a pot with the tar, and it worked like magic. Since then, in whichever corner of the village there is a leaking pot, the tar finds its way there. It has also become a practice to borrow tar from the households that maintain a ready stock of it.

A Letter to the Editor

There is this skinny man who drives the big machine. He commutes every day from Nanjangud. One day a rumour spread that the paper he brought along with him carried news of the village and everyone, without getting down to work, crowded around the skinny driver to find out whether it was true. The patel was fuming. The news the paper carried about the village read as follows:

> *Sir,*
>
> *This is with reference to the road that the Government in its magnanimity has sanctioned to our village and the construction that is presently going on. We have learnt that the road contractor, who is also the village patel, plans to utilize government funds for the renovation of the temple. This is misuse of public funds. It is our prayer to the concerned authorities that they should ensure that no such thing takes place.*
>
> *Aggrieved*

The Village Council is Called

One cannot say with certainty whether the roadwork went on that day, or whether it did not. Shall we say it went on in name only? What had started off as one thing had begun to turn into something entirely different. The fear as to what would be the outcome if all this spread through the streets of the village was a very real one. As evening approached, the plot grew ever more complicated. The patel had the drummer announce that every household should send one person to the village hall for a meeting of the Village Council. Outside it was so dark that people couldn't see each other's faces. The dim light from within the houses fell out of the windows and was lost in the darkness.

The Village Hall

In the village hall the lantern was spewing abundant smoke and light. Already a few men had spread themselves out as if they had nothing to do with what was going on. People started dropping in one by one. As the crowd grew, the bustle grew with it. After a little while, the din that rose from the village hall began to drown out the noise of the village. The patel was seated in the centre of the gathering. The light from the lantern was falling straight on to his face. You could see his face sweating lightly and his forehead rhythmically wrinkling, unwrinkling and wrinkling again. You could see who was sitting in which far corner when their faces lit up whenever they struck a match to light a bidi. There was such a mob of people there.

The first man: 'Why don't you start?'

The second man: 'Everyone knows it. What is there to start, except the dressing down?'

Rajappa: 'What does that mean, brother?'

The third man: 'You are the ones who know about meanings and things like that. Have we gone to school, like you?'

The shanbhog: 'Be quiet for a while. Patel, why don't you speak?'

The patel: 'What is there to say, tell me?'

(*The shanbhog looks towards Lakuma, Rajappa, Shambu and Madu.*)

The shanbhog: 'Everyone knows that you are the ones who wrote the letter to disgrace the village. Are you people inclined to accept that?'

(*They nod their heads as if to say, yes. There is an immediate uproar.*)

The patel: 'What was wrong with what we were doing?'

(*The patel's words crackle through the gathering. No one opens his mouth.*)

Then Rajappa says: 'If you put the money that's meant for something into something else, what do you think will happen?'

(*The clamour rises again. The shanbhog waves his hand. The noise dies out gradually.*)

The patel: 'Did I swallow the money for my children's sake? Or did I do it for the sake of the village and for God's work?'

Madu: 'You might do it for anything, sir. You didn't use it for its intended purpose.'

(*The clamour rises again, this time louder than before. The shanbhog tries to restore order as usual. This time one or two men stand up.*)

The first man: 'Don't you fellows respect gods, elders and things like that?'

The second man: 'If they did, would they get up to their tricks?'

The third man: 'Enough, keep quiet. Why does everyone have to say something?'

The fourth man: 'Flog them. That'll set them right. They will talk straight.'

Rajappa, Shambu and Shambu's father: 'Who's that man? Come beat us then. We'll handle that. Your writ runs only till the tip of your nose.'

The patel (*raising his voice*): 'Yes. What you did was the right thing to do, I say.'

Shambu: 'And you think what you did is right? Why talk of it?'

The patel: 'What did you say? You have the audacity to talk back to me like that! Have I become so feeble that I have to listen to such nonsense from you?'

So saying, the patel clenched his fist and got up in a rage. His head hit the lantern, and broke the glass. Shards of glass fell tinkling to the ground. The flame flickered and flickered until it finally died out. The place filled with shouts, yelling and scolding. Everyone struggled to get out of the village hall. It went on like this for a while, and then everything grew becalmed. Everyone closed the doors of their houses and turned in for the night.

News from Hosur

It could be said the village was still sleeping; dreaming perhaps, of what had happened the previous night. As if to spoil the dream, a man from neighbouring Hosur passed by in the early hours of the day. The words he left behind began to creep along the alleys of the village:

Sir, I am from Hosur. In our village there was a seductress. It had been her practice to seduce all the teenage boys of the village. How long could one try to talk her out of it? She went about things her own way. They tried scolding her, but she carried on as usual. Finally, they did what had to be done. They called a meeting of the Village Council and dragged her in front of it. They stripped her naked and flogged her with a tamarind branch. Yet, she continued to be arrogant and displayed no remorse about her behaviour.

And she went even further. She went to the police station and told the police everything in detail: they did this to me, they did that to me. The police van arrived and rounded up all the landlords. Nobody knows what happened after that.

The news from Hosur blended with what was already brewing in the village, and the ferment rose higher and higher.

The patel: 'Hmm.'

The shanbhog: 'Ha.'

The foursome: 'My sons! They were lording it over. Now they've been shown their place.'

The first man: 'Ayyo, Siva!'

Tar in the Pit, in the Pitch of the Night

Things didn't stop there. Remember the tar drums that had been heated and kept in front of the village? The next morning not one was to be found. If you followed their trail, it led you outside the village to a pit around which, their mouths wide open, the drums were disgorging tar in thick lumps. As the sun rose higher and higher the tar continued pouring into the pit, stretching gently from the mouths of the drums. Those who had gathered round to see stood watching. Those who hadn't seen it yet, flocked to the pit. After they had watched and watched, and the blackness of it had filled their eyes, they would slowly move away. Not one had a word to say.

Even the patel didn't have a word to say. One man said to the patel: 'You should have just said yes, we would have beaten them last night.' Even at that, the patel only gritted his teeth. Didn't say a word. He went straight to the police station from there. It was evening when he came back. It became known that the police would arrive to investigate early the next morning.

The One Who Went to the Hamlet Hasn't Returned

The situation grew ever more complicated. Towards nightfall a further wrinkle was added. By now it was quite dark, all the houses had shut their doors, and nobody was to be found on the streets. But Rangappa's boy still hadn't come back home. Apparently he had headed home after lunch. They questioned the schoolmaster who said he had seen the boy walk homewards; he was the kind of dumb kid who headed straight to wherever he was going. Lantern in hand, his concerned parents knocked at every house in the village. They asked his playmates. Everyone said that they had not seen him. Not a single street, not a single alley was left unsearched. It went on like this for a long while. Gradually his mother's crying was the only sound in the silent village; she wailed on throughout the night.

Investigations

At dawn the police van came bumping along the rutted dirt path and stopped in front of the village, close to where the tar had flowed out of the drums into the pit. The inspector asked the man who was at the police vehicle to fetch the patel. The patel hurried over when he received the summons as did everyone else in the village.

The tar in the pit had grasped Rangappa's boy by the feet. The tar drums clasped both his hands. His hands, his body and face were all covered with tar. And, if you took a closer look, you could see that in that child's body life was still throbbing.

~

LAST IN, FIRST OUT
IRWIN ALLAN SEALY

A wise man would have gone home when he heard the tube light smash, but my wife calls me an unwise man and I must be, since I smoke as well as drive an autorickshaw on Delhi roads, and I butted in.

For that matter, a wise man would have finished his BCom and gone into marketing, but I thought: No office for me, no boss for Baba Ganoush. And this looked like the life back then, not that I'm saying it isn't still, some days, maybe even many days. But autorickshawry has its own traps and it's always tempting to get that last fare, just one more, and that's the one that takes you out of your way—when it doesn't land you in shit.

God knows there's shit enough by day on Delhi roads. And three wheels aren't the steadiest undercarriage when the going gets rough. Better than two is all you can say, and probably not all the time either. You see some sights on the road that you'd like to forget, and when it comes to the crunch, the guy with the least steel is the loser. I've seen some two-wheeler accidents where the helmet didn't help much more than the severed head. *Bastard Blue Line buses*, people rant, me too, but might is right in the jungle.

Keep well in, I tell my passengers, and they do. (As if it would make a whole lot of difference when the bus rams you.) But a wraparound shield is better than nothing—even if the dents are starting to join up on my Bhavra. The Bee is what I named her in the good old black-and-yellow days before this greenie shift.

You could say I own the buzzer. I've paid back most of the deposit on her to the Punjab National Bank, and I can usually go home by nine, maybe ten. Mornings I start early with schoolkids, twelve monsters packed in with a little removable wooden bench, schoolbags outside. And I don't always work late. I've saved a bit of money in term deposits at the PNB. If I overdraw on the current account, they automatically take it out of the next deposit: last in, first out.

Most days I wear a clean white polyester safari to work. Impractical, I know, and the wife never fails to remind me, though secretly she likes me in it. No pen in my pocket, no comb either. Good Agra sandals, size eleven, and I don't tuck one foot under me as I drive. It's hard enough having to double over just to get into the driver's seat. No holy pictures along the top of the windscreen, just the Shah Rukh poster at the back on the one side and Deepika on the other. I have noticed men sit right up against my life-size Deepika, the shot in the black negligee that got everyone going. Women cozy up to the King.

Anyway, this night I was cruising along the busy Mall Road in Civil Lines looking for a last fare when something about the peace of University Road pulled me in towards the Ridge. I left the rat race behind and sailed along past those sedate college gates in top gear, engine purring. All the walls have got higher since I was a student—maybe that's saying something, if only that I haven't got any shorter. I switched off the stereo. I was one of the first to install a system back in the twentieth century when the vehicle was new. There was always a Sufi fat-boy tape rolling to drown out the noise of the day. Nights nowadays you want to listen to the silence, when you can.

Directly opposite the main gate of Delhi University, where the road goes straight up to Flagstaff House, the hill stretch has been closed to traffic and an autorickshaw stand has sprung up at the barrier. A handful of peanut vendors and ice-cream carts congregate there during the day. At night, of course, it's deserted, and so it was this night, but sometimes you can pick up a late fare. I did a U-turn and drew up beside the gate. Ten or fifteen minutes under the entrance lights might be well spent, I thought. People don't like to walk along the Ridge.

The Delhi Ridge is a wilderness of rocks and thorn trees, nature's last stand in this grey city, the nearest thing we have to a forest. A hundred years ago they planted this barren upland with a Mexican tree that ran amok. Up along the crest are paved paths the municipality has laid in an attempt to tame the man-made jungle. Monkeys use the watershed as a safe base for raids down either side; peacocks honk at first light and then retire, leaving the field to a treepie with a harsh call—half heckle, half jeer. Morning walkers do their laughter therapy up there, and power joggers go by in pairs, tugging at the elastic bands of their tracksuit cuffs to consult expensive watches. But a careless jogger could ruin a pair of Nikes on the broken glass of last night's rumfest, if he's lucky. If he's unlucky, say he was working late, there's a higher price he could pay among the syringes and condoms and gutkha sachets that lie strewn in the red-brick dust. Even by day you'd jump if someone came up behind you on those paths. You don't go there after dark unless you have a minder. Or unless you are the minder.

Of course, lovers go there because there's nowhere else to go. Students mostly, from the DU campus. There are park benches where they can sit and make out by day. I used to go up the Ridge during my spell as a student, before the old man realized I was getting ideas and married me off. In the early days the wife and I took the boys there for a joyride once or twice, before they grew embarrassed about an outing on the workhorse. We'd sit and watch the monkeys by Flagstaff House. In winter they sun themselves and pick one another's fleas. A big male will turn up and simply roll over in front of a lesser creature, and the chosen one will leave whatever it was he was doing. I tend to believe in the chosen.

The newest tribe are the gardeners who arrived when Nehru Park was created. But joggers and gardeners and canoodlers and watchers tend to move on once the sun set. Everyone does except for the diehards, or those who blithely believe a special dispensation hangs over them like a royal parasol. And who knows, maybe they're mostly right.

They can be wrong. Every once in a while you read in the paper about a rape on the Ridge. I used to pay special attention to these snippets, partly because of my old association with the university.

That evening, I was parked outside the gates and starting to look at my watch when I heard the tube light smash. Few night sounds are more chilling, none more deliberate. After all, a tube light is something you carry with special care when you must, upright beside you. As if it were the body's ideal twin, smooth and colourless and fragile. It breaks in a shivering white cascade with the sound of heaven collapsing. If spirit had substance it would shatter like this, something between a gasp and a cry. And that's how it sounded, scary but somehow, how to put it, binding.

I sat up in my seat and looked at the Ridge, studying the darkness. Curiosity, of course, but partly the witching of that weird sound. In a minute I thought I heard a sort of cry, of earthly pain. I didn't stop to think. I started up the auto and zipped across the road, made an S around the double barrier, and headed uphill in the direction of the noise.

Next thing I heard was running feet, the *stamp-stamp-stamp* of cheap shoe leather. I pulled over and keyed off and waited. It was a young man and he ran straight into me. It wasn't that my dim headlight was switched off, just that he was running downhill and used the machine as a brake. He slammed into the windshield and stood there bent over and winded. Even by the faint light of my beam I could see he was bleeding. His face was cut up and he looked frightened. So frightened he'd lost his voice and could only point back up the road the way he'd come.

In the background I could see Flagstaff House like a stage set, a black cutout of a tower against the grey of the November sky.

'Get in,' I called, and shoved him into the passenger seat, 'and hang on!'

I turned the auto around on a five-rupee coin and was about to get the hell out of there, but he grabbed me by the shoulder and found his voice. 'She's still there!'

So he wasn't alone. I arm-locked the handlebar till we were facing uphill again and began the slow climb in the old machine. Come *on*, Bee! Even to me, the journey seemed to last forever.

At the top of the rise the boy jumped from the auto and ran a short way towards the tower calling the girl's name, but his wild turns of the head said she was not where he'd left her. I looped the tower in the machine, leaning on the horn and shouting words of support and threat into the dark. It must have been the purest gibberish, and a greater silence was the only reply.

'Get back in!' I called, and the boy obeyed but hung out of the auto scanning the side of the road. I've made a career of watching people's faces in the rearview mirror and his was intent, as if unaware of the volume of blood trickling down his forehead. The pain wouldn't have hit him yet. He seemed to be reading the night, willing it to disclose his harmed lover.

And then she appeared, or a figure appeared that the boy recognized, because he hopped out again and ran towards the brush. She was walking very slowly, smoothing down her kameez over and over again. The dupatta, if she wore one, was gone. The boy took her hand and led her tenderly towards the auto.

I took them straight downhill, jinked back around the traffic barrier, and turned left on to University Road. It was a clear run to the next corner where the road climbs back over the hill to the Hindu Rao Hospital. But a 212 bus coming the opposite way strayed across the white line and broke our momentum. 'Blue Line bastards!' I shouted, but we'd lost it and had to toil the rest of the way up to the crest. Then we raced down the other side of the Ridge and into the hospital gates.

Well, that's it, I thought as I headed home; you don't see people twice in the big city. But of course you do, maybe just the ones you think you won't. I told you I read people's faces in the mirror and it's true you can tell straight off the talkers, the tippers, the nasties, the frantic, the hunted, the doomed. I had watched the girl in the mirror whenever a car came the other way and I guessed she wouldn't report the crime. She still had her silver chain, a necklace of eyelets each with its little silver tear. The pain had finally struck the boy, and last I saw them she was leading.

The next day and the next day and the next I looked in the papers, but there was no report. Ah well, I thought, you lose some, you lose some.

Then there was a story: a savage rape on the Ridge. But the description didn't match and the date given was for a whole week after my little adventure. Over the next fortnight there were two more incidents; in both cases the girl's dupatta was taken and the boy's face messed up. The police issued descriptions of the assailants, two men in their late thirties. People were warned off the Ridge at night. An officer criticized trends in women's garments with words I remembered from twenty years before. And the general suggestion was that these things wouldn't happen but for the foolishness of the couples. Well, I thought, clipping out the stories, maybe all that hot young blood buzzing in the brain does make you a little careless.

Next morning I was tooling along University Road when I saw the boy. He didn't return my look—maybe after a month he really didn't recognize me—and ducked in at the main gate. On an impulse I parked the Bee and gave chase. It was him all right and he remembered me, but he didn't want to be reminded of that night.

'Well,' I said, 'you might want to forget it happened, but whoever did it hasn't stopped.'

He looked genuinely surprised, like he hadn't read the newspapers. I invited him for a coffee. In my day I used to wonder—a tea man myself—what drew people to the university coffee house, although the place had an undeniable glamour. I still don't know. As we sipped the filthy stuff I took out the news clippings from my pocket and laid them on the table between us. He glanced over them with a troubled expression. The scars were healing nicely on his face; with a bit of luck they would melt into worry lines.

'Look,' he said, when he had read enough, 'what's it to you?'

My heart contracted. I hadn't imagined he needed winning over. I'd steered clear of asking after the girl, figuring she'd slipped out of his life.

'Just tell me if you could recognize the guy again.'

'You mean the guys,' he said, stressing the plural in a defensive way.

'The guys. Have you ever seen them again?'

He searched my eyes as if looking for a handhold, then gave up. 'They hang around the metro station in the mornings. The main guy is built like a bouncer. He wears the same safari suit every day.'

'What colour?'

'Sort of a darkish colour.'

'Black?'

'Not quite black.'

'You mean like grey?'

'No, no, darker than that.'

'Sort of a blackish grey?'

'More like a greyish black. But darker.'

'Blackish black?'

'Ya!'

I saw I was getting nowhere fast. I made a last bid. 'Listen, I'll be at the metro station in the morning. Just point him out and I'll take it from there.'

'Take it where?' he asked despairingly, and I let the question dangle.

I didn't really expect him to be there the next day but he turned up.

'The safari wala's not here,' he said, 'but his sidey is. By the sweet potato vendor, pink shirt.'

He nodded towards the granite steps where the metro terrace descends to street level. A small sleek mongoose of a man in a red cotton overshirt was stabbing at a leaf plate with a toothpick while sweeping the concourse with his eyes. I dropped my gaze as the little pointed head swivelled towards us. When I looked up it had darted away. His hennaed hair was topped by a blue-and-white baseball cap worn the right way so the brim hooded his eyes. Fake Diesel pants with faded chaps were standard, I assumed, but the bulge in the back pocket could have been either a cell phone or a knife. I had never seen a man wear three shirts before, four if you counted the tee. True, it was winter and the outermost, a florist's dream of canna lilies, was zippered.

I was already moving towards him through the horde of students. I must confess I didn't stop to thank the boy, nor did he seem keen to stick around. I'm not a big sweet potato fan myself but the other item on the menu was a salad of yellow star fruit, sour to frizzling insanity, and I stood six inches from the sidey and ate it without turning a hair. Right then a Mallika bombshell went by and I groaned and staggered theatrically and caught the sidey's eye. He winked at me and grinned and I left it at that.

It wasn't till the next day that I saw the bouncer. I arrived there early and had to wait. I'd dropped off the schoolies and got rid of the little bench so I sat in the back of the Bee in the Deepika corner and smoked. Tobacco first, then a sweet cigarette. Back in the nineties I'd taken to sucking on them whenever I felt tempted and now I had two habits (three if you count stopping by the flat to trouble the wife when a fare brought me within stroking distance). Same red-and-white Phantom pack from childhood, little smooth white sugar sticks with a red dot on the end. Some days if you suck hard enough the red dot actually glows.

A cobra—that was my first thought when the villain appeared. I swear I expected all the cool university chicks to go into a frenzy of squawking the way forest birds do when a snake appears. But they stalked on by in their hiphug Pepe jeans with their smartphones gripped tighter than their textbooks, and the cobra watched them pass with a bland insouciance only the mongoose, stationed at the sweet-potato stand, could rightly read.

He wasn't in a safari suit (I felt a little vindicated in mine) but in white drill pants pressed to a knife's edge, elastic boots, and a black balloon jacket that gave him a slightly unmoored look. Or maybe that was his natural walk, weaving a little, like a wrestler, not a drunk. His mongoose sidekick fell into step beside him and the two of them walked up University Road without bothering to take in the scenery. I had expected to watch them ogling, but instead I found myself tailing them in the Bee, hanging a good way back. What surprised me more than their disinterest in the girls was the mongoose's face. You expect a planet to light up when the sun appears, but this sidey's face fell into a total eclipse. His eyes took an a haunted half-shadow and even the cap looked crestfallen.

I parked the Bee at the Flagstaff Road barrier, told the ice-cream wala the wife needed a branch of babul leaves, and trailed them uphill. At the top they took Magazine Road and I waited at Flagstaff House pretending to watch the monkeys. The road, once simply a path, follows the crest of the Ridge through the man-made jungle, low dusty thorn trees with twisted grey trunks and a canopy like mustard gas. On either side, beyond the park benches and half hidden by the thickets, are power substations and water tanks and gardeners' toolsheds like bunkers. Also ruins from Muslim times, tombs and such. Built with a prospect on what would have been a barren ridge, they now huddle blindly in the jungle, peculiarly functionless unless to conceal a walker caught short by nature. Turds and worse await the unwary foot.

Cobra and Mongoose sat down on a bench and looked about them. A gardener had set a hose with a ratchet spring to green the grass by the caged bougainvillea and was standing back to inspect the spray as it ticked around in a wide circle. Satisfied, he drifted off to join his fellow workers for a smoke by the garden gate at the far end of the park. Cobra and Mongoose watched him go, then idly observed the circling spray. Idly I imagined the eye of the hose coming around to fix them with a stare. Stray walkers came and went along the long park road; a couple sitting on a nearby bench rose and made for the far gate where the 212 goes by.

When the coast was clear Cobra got up and vanished into the forest, Mongoose trailing after. I stayed put. I knew there was no way out of the park except at the far end or back in my direction. They were gone ten, maybe fifteen minutes. I was getting restless when they reappeared and took the road to the far gate, strolling side by side as before.

I went to their bench and sat a moment to see what they saw. Rising out of the bushes opposite was one of those curly-wurly ochre buildings from Mughal days, a hunting lodge maybe. I sauntered into the forest and entered the ruin from behind to find two rooms; the roof of one had fallen in and the floor was grassed over. In the inner room hung a musky odour that told its own tale. A stair led up from the outer room but the way up was barred by an iron door; a heavy padlock hung from the hasp. I picked up the biggest stone I could find, lifted it with both hands, and brought it crashing down on the lock. The lock gave and hung there broken jawed; I unhooked it and threw it into the bushes downhill. (Any crook can remove a government padlock and replace it with his own, a nice rusty old job, and people will walk on by thinking: official.) Upstairs I found an empty chamber with a pillared balcony: nothing but fallen birds' nests on the floor. On the way back down I noticed a loose slab at the landing and lifted it. Underneath lay a yellow cement sack neatly folded in half. I took it back into the chamber and looked inside. There were four lengths of fabric in there, three coloured and one white. They were not new but clean, some printed, one embroidered; together they seemed a strange valueless hoard. It was only when I unfolded one that I realized what it was: a woman's dupatta.

I went straight back down to the Bee.

'Wife's going to be angry,' the ice-cream wala forecast.

'What?' As I drove off I realized he meant I'd forgotten the babul leaves.

I headed over the hill past the Hindu Rao Hospital along the 212 route, cursing the Blue Liners as I flew; sharing the same prey, buses and autos are natural enemies. I hung a left at the chai wala by the cell phone tower, then hard left again, and cut the engine, rolling to a halt right where the silk cotton tree that overhangs our whole neighbourhood is anchored. Just short of home.

I hadn't come to trouble the wife. I was there to call on a man dead five hundred years.

He stood nine feet tall, my ancestor, going by the grave. Called the Grave of the Nine-Foot Saint, always freshly painted green, it sticks out two feet into the main road, the remaining seven closing off the sidewalk. Huge blood-orange flowers flop down on it in summer, followed by a delicate rain of cotton that whitens the precinct like snow. It's an island of peace for a military man.

He was a military baba, my ancestor, the man whose name I bear. Baba Ganoush. Baba G., my wife calls me for short, or just Baba, though I don't really qualify. Babas were either plain holy or soldierly holy, and I'm neither. My military baba had no secret weapon: *He* was the weapon. He moved the army. He had a retinue of 786. Baskets of purple eggplants and potted marigolds moved before him in the field; caged songbirds and urns of rose water came behind. The night before a pitched battle, his linked light-boys dressed up as houris and oiled their bodies and did calisthenics for the host. Before the phalanx of warriors he drew a box in the dust with his ring finger and danced a victory dance that spun every watching soldier into heaven.

I stood by the grave and felt my shirt tail begin to lift and billow. His spirit clad me, sliding over my skin like a lover's hand. The air grew red and I was racked with pain and filled with heretical notions. Blood is our element, I remember thinking, not water. We swim in it from one life to the next, passing like a wet flame from wick to wick. So little to the body, I was thinking the other day while I bathed, the soaping is so quickly done, so little to do.

Go! I heard my Baba say. Fight, with love in your heart.

I went to a hardware store and bought a quarter-inch brush, a small tin of enamel blue, a cheap screwdriver, and a key ring with a red disc; on the sidewalk I bought a second-hand padlock. Then I went home and parked the Bee and kissed my wife. Not now, I said, detaching myself when she sent the boys out to play. I opened the paint can and outlined eggplants and marigolds on the nose of the Bee and rose water urns and caged bulbuls on her tail. The paint was still wet as I rode over the hill again and padlocked the iron door in the curly-wurly ruin.

Next day I tailed the pair again. They did a repeat of the hill walk and parted at the gate on the 212 route, sidey looking more depressed than ever. I tailed him home to a Maurice Nagar flat and made some inquiries with neighbours.

The morning after, I was at the DU metro station early. This time I braved the sweet potato but asked for an extra squeeze of lime.

Mongoose turned up in his floral jacket and ordered the same. We exchanged a wink when a Bips lookalike passed by on gel-pen refill heels. I chucked my leaf plate and ordered another.

'And one for my friend here!' I said.

'No, no,' he protested, but only formally. He was already chucking his leaf.

'Something else, these babes, no?' I said, strolling him gently away. He was walking before he knew it.

We drifted up University Road towards the gates. He seemed happy to get away from the metro, but kept looking back all the same.

'So, Mr Raju,' I began.

Mongoose stopped dead in his tracks. 'How do you know me?'

'Oh,' I brushed away an airy cobweb, 'we have our ways.'

At the *we* I drew myself up to my full height, laid a long finger on my shoulder, and tapped twice where some silver might adorn my epaulettes. He remained standing so I prodded him along with little shocks of home address and house history, even a little detail about a tiny nephew who might need a polio shot. (I picked that up from two door-to-door health workers.)

'And how is the, uh,' I gestured up the Ridge, 'shikar these days? Happy hunting?'

His eyes bulged. Sidies break down more or less right away so I was at pains to let him know I knew he was just the accessory.

'And your friend, the big gun?'

He was dumb and dry-mouthed. I walked him up the slope past the tower to the curly-wurly lodge in the forest.

'We've had to change the lock on your door, I'm afraid.' I produced the key ring that my older boy had painted in police blue-and-red; he had added off his own bat the sinister Delhi Police motto: *With You, For You, Always.* 'Go on, open it.'

He undid the padlock but lacked the strength to climb the stair.

'Don't you want to go and see?'

'I believe you.'

'All right. What can you tell us about your friend?'

Right then the cell phone rang in his cargo pants. He jumped where he stood. It was Cobra, I could tell. The timing shook me too; Mongoose simply came unhinged.

'If it's our friend,' I said, 'tell him you'll meet him tomorrow.'

He obeyed. But Cobra had other plans and after hearing him out, Mongoose hung up in an ecstasy of fear.

'What's up?'

He looked unseeingly at me, his finger and thumb worrying a burr on the cotton jacket.

'Hey!' I frowned and slapped him.

He began to whimper, edging away from me and then back as if pushed from the other side by an unnamed force. The phone slid into his pocket and he gripped the barred door like a prisoner who doesn't realize he's on the outside. 'He's crazy,' he wailed. 'He's mad!'

'What's this drama-shama?' I growled.

He slid down the door like a bad actor and squatted there with his forehead lodged between two bars.

'Hoy!' I booted him in the bum to no effect. I was aiming a harder kick when he began to speak.

'He's going to kill somebody. And he wants me to help.'

'Kill who?'

'Somebody. Anybody. He says no more fooling around. He says next time we use the knife. He says finish off the bastards. He says they need to be taught a lesson. They keep coming here and polluting the morals of the nation. But then he himself...'

'He himself what?'

He hung his head.

'He himself what?'

'Brings me here.'

'When are you seeing him?'

'He says we'll have a drink this evening. He says we'll want a bit of warming up. He wants me to bring a bottle of Walker. Where am I supposed to get the money?'

I thought for a bit. 'Okay, you get a bottle of Patiala whiskey and go to the rebottlers behind Kashmere Gate. They don't charge much. Your job is to get him drunk, okay? You don't drink in here?... Good. Get him drunk and then walk him to Flagstaff House. I'll be waiting there at ten o'clock. In an autorickshaw. We'll take him for a ride. Just get him drunk. And keep yourself sober. Do you think you can manage that?'

He seemed to come to life and we parted at the 212 bus stop.

'Ten o'clock!' I called as he climbed on his bus.

At half past nine, I was parked and waiting. I moved a couple of lovers on in a gruff policemanly voice and, as I watched them go, wondered where the knife would have gone in. Then there was nobody. I sat in the Bee and twiddled my thumbs and watched the night. The tower looked bleak and aloof, the Ridge close and unfriendly. Another feather of grey would have tipped the night sky into blackness. Brooding on my ancestor I realized that at this hour before the battle he would be drawing his mystic box in the dust and beginning his slow dance of death and transference. I simply sat and nibbled at a sugar stick. Before I knew it I had emptied the box.

At 10.10 I heard voices. They were singing but they were not houris. It was my quarry, drunk, both of them. Cobra was spitting threats at the world in between lines from an old song.

> *There's a boy* across *the river*
> *With a bottom like a peach*

'Get in, you idiot!' I whispered to Mongoose, who was busy playing the Lucknow game of After You. He obeyed and snuggled up to King Khan.

> *But alas I cannot swim!*

Cobra needed help and I brought myself to touch him. His balloon jacket felt dry and scaly, so I pinned him by the neck, bent him in two, and simply sprung him in. He turned to Deepika and began to slobber all over that sheer black negligee. I got in at once, started up, and took off, veering clockwise around the tower. My passengers were thrown left in a crazy centrifuge, Cobra leaning precariously out of the Bee.

'Hang on!' I yelled, and we plunged downhill, racing the way the boy had run the night of the tube light. Down below I jinked around the traffic barrier and left on to University Road. It was a repeat of the hospital ride, only this time I had the villains.

The Bee buzzed like the beauty she once was. I felt I was playing an instrument whose dark sweet drone underlaid the pair's drunken bawling. In days past I could tell the engine's semitones up and down the scale. I swooned to certain piston tremolos and awoke in time to pump a sweet glissando on the brake. Bee and I were partners in a dance whose music was in our blood. We moved in unison: I could trust her with any step, and she responded with an enabling precision; I could jiggle the schoolkids into giggling hysteria, sway a pouting beauty, or hit a bump at speed and bounce a snark straight up into the iron beam.

What to do, what to do? I was thinking as I sped through the dark. I had no plan. I watched the two men in my rearview mirror, but really I was looking a lot further back. Rape is a tricky business to judge. Unless you've felt the hot blunt thorn of it in your own flesh, your opinion isn't worth a lot. For a moment the mirror showed me just one face in the backseat, then as the bastard split in two I knew what I would do.

At the corner where the road goes over the hill, a northbound 212 was about to take the curve on to University Road. There's no median strip there and downhill buses always cut that corner. There's a moment when their headlights are shining clear up the Hindu College Road when in fact the bus is heading down towards the university gates. I switched off my light and spun the Bee around in a tight U-turn. Cobra flew out into the bus's path. Last in, first out.

I watched him go in the mirror and thought he flew a little further than I intended. Well, that's destiny, I thought: he's meant to lose a little less. I turned the U into an O and sped off into the night, but not before I saw the bus drive over his feet.

Enough to put him out of action, up on the Ridge anyway. Then I dropped the Mongoose home. As in *dropped.*

Well, that's that, I thought. You don't see faces twice in the big city. It's getting bigger all the time. That's progress: fluorescent lamps replacing tube lights, four-wheelers replacing three. But maybe a month later I did see the sidey again. I had a passenger so I couldn't stop, but he was looking fresh and expansive on the sidewalk and he gave me a long cool mongoose wink as I went by.

It wasn't till I arrived home that I got to thinking about it. The wife had made baba ganoush after scorching the eggplant skin on a naked flame in her painstaking way. It's the family favourite, picked up from an aunt in the Gulf, and it usually goes down in a great hurry, no chewing, but I was about to swallow when I saw that wink again and then all I remember is the wife and boys looking strangely at me because I just kept chewing on that mouthful.

I haven't seen Mongoose lately but I often see that shrewd little pair of eyes fixed on me. Then one of them closes in the deepest blackest wink, and I'm left wondering: O Holy Baba! *Which* of us was the sidey?

THE ELEPHANT AND THE TRAGOPAN

VIKRAM SETH

In Bingle Valley, broad and green,
Where neither hut nor field is seen,
Where bamboo, like a distant lawn,
Is gold at dusk and flushed at dawn,
Where rhododendron forests crown
The hills, and wander halfway down
In scarlet blossom, where each year
A dozen shy black bears appear,
Where a cold river, filmed with ice,
Sustains a minor paradise,
An elephant and tragopan
Discussed their fellow creature, man.

~

The tragopan last week had heard
The rumour from another bird
—Most probably a quail or sparrow:
Such birds have gossip in their marrow—
That man had hatched a crazy scheme
To mar their land and dam their stream,

To flood the earth on which they stood,
And cut the woods down for their wood.
The tragopan, good-natured pheasant,
A trifle shocked by this unpleasant
Even if quite unlikely news
Had scurried off to test the views
Of his urbane and patient friend,
The elephant, who in the end
Had swung his trunk from side to side
With gravitas, and thus replied:
'Who told you? Ah, the quail—oh well,
I rather doubt—but who can tell?
I would suggest we wait and see.
Now would you care to have some tea?'
'Gnau! gnau!' the tragopan agreed.
'That is exactly what I need.
And if you have a bamboo shoot
Or fresh oak-leaves or ginseng-root—
Something that's crunchy but not prickly...
I feel like biting something quickly.'
The elephant first brewed the tea
In silence, then said carefully:
'Now let me think what I can get you.
I fear this rumour has upset you.
Your breast looks redder than before.
Do ruffle down. Here, let me pour.'
He drew a lukewarm gallon up
His trunk, and poured his friend a cup.

~

A week passed, and the tragopan
One morning read the news and ran
In panic down the forest floor
To meet the elephant once more.
A cub-reporter bison calf
Who wrote for *Bingle Telegraph*
Had just confirmed the frightful fact
In language chilling and exact.

'Here, read it!' said the tragopan,
And so the elephant began:
'Bingle. 5th April. Saturday.
Reliable informants say
That the Great Bigshot Number One
Shri Padma Bhushan Gobardhun
And the Man-Council of this state,
Intending to alleviate
The water shortage in the town
Across our ridge and ten miles down,
Have spent three cartloads of rupees
So far upon consultants' fees—
Whose task is swiftly to appraise
Efficient, cheap, and speedy ways
To dam our stream, create a lake,
And blast a tunnel through to take
Sufficient water to supply
The houses that men occupy.'

~

'What do you think,' the tragopan
Burst out, 'about this wicked plan
To turn our valley blue and brown?
I will not take this lying down.
I'll cluck at them. I'll flap my wings.
I tell you, I will do such things—
What they are yet I do not know,
But, take my word, I mean to show
Those odious humans what I feel.
And the Great Partridge will reveal
—That Partridge, dwelling in the sky,
Who looks down on us from on high—
He will reveal to us the way—
So kneel with me and let us pray.'

~

The elephant said, 'Let me think.
Before we pray, let's have a drink.

Some bamboo wine—perhaps some tea?'
'No, no,' the bird said angrily,
'I will not give in to distraction.
This isn't time for tea but action.'
The wattled horns upon his head
Stood upright in an angry red.
The elephant said nothing; he
Surveyed the landscape thoughtfully
And flapped his ears like a great fan
To cool the angry tragopan.

~

'It's infamous, I know,' he said,
'But we have got to use our head.
Praying may help us—who can tell?—
But they, of course, have gods as well.
I would endeavour to maintain
Our plans on a terrestrial plane.
What I suggest is we convoke
The Beastly Board of Forest Folk
For a full meeting to discuss
The worst that can occur to us.'
And so, that evening, all the creatures
—With tusks or gills or other features—
Met at the river's edge to plan
How they might outmanoeuvre man.
Gibbons and squirrels, snakes, wild dogs,
Deer and macaques, three types of frogs,
Porcupines, eagles, trout, wagtails,
Civet cats, sparrows, bears and quails,
Bloodsucking leeches, mild-eyed newts,
And leopards in their spotted suits—
Stated their stances, asked their questions,
And made their manifold suggestions.
Some predators drooled at the sight,
But did not act on appetite.
The leopards did not kill the deer.
The smaller birds evinced no fear.

Each eagle claw sat in its glove.
The mood was truce, if not quite love.
At meetings of the Beastly Board
Eating each other was outlawed.

~

The arguments grew sharp and heated.
Some views advanced, and some retreated.
Some feared to starve, and some to drown.
Some said they should attack the town.
The trout said they were unconcerned
If the whole bamboo forest burned
So long as they had space to swim.
The mynahs joked, the boars looked grim.
They talked for hours, and at the close
At last the elephant arose,
And with a modest trumpet-call
Drew the attention of them all:

~

'O Beasts of Bingle gathered round,
Though in our search for common ground
I would not dream of unanimity
I hope our views may reach proximity.
I speak to you as one whose clan
Has served and therefore studied man.
He is a creature mild and vicious,
Practical-minded and capricious,
Loving and brutal, sane and mad,
The good as puzzling as the bad.
The sticky centre of this mess
Is an uneasy selfishness.
He rips our flesh and tears our skin
For cloth without, for food within.
The leopard's spots are his to wear.
Our ivory unknots his hair.
The tragopan falls to his gun.
He shoots the flying fox for fun.

The black bear dances to his whim.
My own tame cousins slave for him.
Yet we who give him work and food
Have never earned his gratitude.
He grasps our substance as of right
To quench and spur his appetite,
Nor will he grant us truce or grace
To rest secure in any place.
Sometimes he worships us as Gods
Or sings of us at Eisteddfods,
Or fashions fables, myths, and stories
To celebrate our deeds and glories.
And yet, despite this fertile fuss,
When has he truly cared for us?
He sees the planet as his fief
Where every hair or drop or leaf
Or seed or blade or grain of sand
Is destined for his mouth or hand.
If he is thirsty, we must thirst—
For of all creatures, man comes first.
If he needs room, then we must fly;
And if he hungers, we must die.
   Think what will happen, when his scheme
To tame our valley and our stream
Begins to thrust its way across
These gentle slopes of fern and moss
With axe, explosive, and machine.
Since rhododendron logs burn green
They'll all be chopped for firewood—
Or logged and smuggled out for good.
As every bird and mammal knows,
When the road comes, the forest goes.
And let me say this to the trout—
The bamboo will be slashed, no doubt,
And what the tragopan and I
Delight to eat, will burn and die.
But what will happen to your stream?
Before the reservoir, your dream

Of endless space, can come about,
The soot and filth will snuff you out.
What tolls for us is your own bell.
And similarly let me tell
The leopards who may fancy here
A forestful of fleeing deer—
After your happy, passing slaughter,
You too will have to flee from water.
You will be homeless, like us all.
It is this fate we must forestall.
So let me say to every single
Endangered denizen of Bingle:
We must unite in fur and feather—
For we will live or die together.'

~

All this made such enormous sense
That all except the rather dense
Grey peacock-pheasants burst out cheering.
The peacock-pheasants, after hearing
The riotous applause die down,
Asked, with an idiotic frown:
'But what is it we plan to do?'
A bison calf remarked: 'I knew
Those peacock-pheasants were half-witted.'
And everybody joshed and twitted
The silly birds till they were dumb.
'How typical! How troublesome!'
A monkey said: 'What awful taste!
How graceless and how brazen-faced,
When all of us are clapping paws,
To undermine our joint applause.'
Oddly, the elephant was the beast
Who of them all was put out least.
He flapped his ears and bowed his head.
'The pheasants have a point,' he said.

~

'Unfortunately,' he went on,
'The days of beastly strength are gone.
We don't have mankind on the run.
That's why he's done what he has done.
We can't, as someone here suggested,
Burn down the town. We'd be arrested.
Or maimed or shot or even eaten.
But I will not accept we're beaten.
Someone suggested that we flee
And set up our community
In some far valley where no man
Has ever trod—or ever can.
Sweet to the mind though it may seem,
This is, alas, an idle dream—
For nowhere lies beyond man's reach
To mar and burn and flood and leach.
A distant valley is indeed
No sanctuary from his greed.
Besides, the beasts already there
Will fight with us for food and air.
No, we must struggle for this land
Where we have stood and where we stand.
What I suggest is that we go
To the Great Bigshot down below
And show him how self-interest
And what his conscience says is best
Both tell him, "Let the valley be."
Who knows—perhaps he may agree,
If nothing else, to hear us out.
But we must take, without a doubt,
Firm data to support our prayer—
And in addition must prepare
Some other scheme by which he can
Ensure more water gets to man—
For, by the twitching of my trunk,
Without that we'll be truly sunk.'

~

And so it happened that a rally
Meandered forth from Bingle Valley
A few days later, up and down
The hills towards the human town.
With trumpet, cackle, grunt and hoot
They harmonized along their route,
And 'Long live Bingladesh' was heard
From snout of beast and beak of bird.
'Protect our spots,' the leopards growled;
While the wild dogs and gibbons howled:
'Redress our sad and sorry tale,
The tragedy of Bingle Vale.'
And there, red-breasted in the van,
Cluck-clucked the gallant tragopan—
Raised high upon the elephant's neck,
And guiding him by prod and peck.
The only absentees, the trout,
Were much relieved to slither out.
They asked: 'How can we wet our gills
Clambering up and down those hills?
The journey will be far too taxing;
We'd rather spend the time relaxing.
We'll guard the valley while you plead.'
'All right,' the other beasts agreed.

~

Meanwhile from fields and gates and doors
The villagers came out in scores
To see the cavalcade go by.
Some held their children shoulder-high
While others clutched a bow or gun
And dreamed of pork or venison—
But none had seen or even heard
Of such a horde of beast and bird,
And not a bullet or an arrow
Touched the least feather of a sparrow.
So stunned and stupefied were they,
They even cheered them on the way

Or joined them on the route to town—
Where the Great Bigshot with a frown
Said to his Ministers, 'Look here!
What is this thing that's drawing near?
What is this beastly ragtag army—
Have I gone blind? Or am I barmy?'

~

'Yes, yes, Sir—' said the Number Two.
'I mean, no, no, Sir—what to do?
They've not gone through the proper channels.
The Protocol Protection Panels
Have no idea who they are.
Nor does the Riffraff Registrar.
It's possible they don't exist.'
'Well,' said the Bigshot, getting pissed,
'Exist or not, they're getting near.
And you'll be Number Twelve, I fear,
Unless you find out what the fuss
Is all about, and tender us
Advice on what to say and do.
And think. And be. Now off with you.'
The Number Two was almost crying.
He rushed off with his shirt-tails flying,
Without a cummerbund or hat,
And flew back in a minute flat.
'Oh, Bigshot, Sir, thanks to your grace,
By which I'm here in second place,
Thanks to your wisdom and your power
Which grows in glory by the hour,
Thanks to the faith you've placed in me,
Which gives me strength to hear and see,
Thanks to—' 'Yes, yes,' the Bigshot said,
'Thanks to my power to cut you dead,
What is it you have come to learn?'
'Sir, Sir, they plan to overturn
Your orders, Sir, to dam up Bingle.
And, Sir, I saw some pressmen mingle

With the parade to interview
A clouded leopard and a shrew.
The beasts are all against your plan.
The worst of them's the tragopan.
His eyes are fierce, his breast is red.
He wears a wattle on his head.
He looks so angry I've a hunch
That he's the leader of the bunch.
And when I met them, they weren't far—
Oh Sir—oh no, Sir—here they are!'

~

For now a hoolock gibbon's paw
Was battering on the Bigshot's door
And animals from far and wide
Were crowding in on every side.
'Save Bingle Valley!' rose the cry;
'For Bingle let us do or die.'
'Wait!' screamed the Bigshot in a tizzy.
'Wait! Wait! You can't come in. I'm busy.
I'm the Great Bigshot Number One,
Shri Padma Bhushan Gobardhun.
I rule by popular anointment.
You have to meet me by appointment.'
'What nonsense!' cried the tragopan:
'You try to stop us if you can.'
The Bigshot sensed their resolution,
And turned from awe to elocution.
'Dear friends,' he said, 'regretfully,
The matter isn't up to me.
What the Man-Council has decreed
Is not for me to supersede.
It's true I, so to speak, presided.
But all—and none—of us decided.
That is the doctrine, don't you see,
Of joint responsibility.
But if next year in early fall
You fill, in seven copies, all

The forms that deal with such a case
And bring them over to my place
Together with the filing fees
And three translations in Chinese,
The Council, at my instigation,
May give them due consideration.
Meanwhile, my friends, since you are here
A little early in the year
—No fault of yours, of course, but still,
It's not the best of times—I will
Invite you to a mighty feast
Where every bird and every beast
Will sup on simply super food;
And later, if you're in the mood,
Please come to hear the speech I'm due
To give this evening at the zoo.'

~

At this pathetic tactless bribe
A sound rose from the beastly tribe
So threatening that the Bigshot trembled
And said to all who were assembled:
'My beastly comrades, bear with me.
You are upset, as I can see.
I meant the stadium, not the zoo.'
He gestured to his Number Two
Who scrawled a memo in his diary.
'Perhaps an innocent inquiry,'
The elephant said, 'may clear the air.
Please tell us all, were you aware,
Sir Bigshot, when you spoke just now,
That even if we did somehow
Fill out your forms and pay your fees,
Your cure would postdate our disease?
Before next fall our valley would
Have disappeared for ill or good.
The remedy that you suggest,
It might be thought, is not the best.'

~

A crafty look appeared upon
The Bigshot's face, and then was gone.
'Of course, my friends, it slipped my mind.
But then, these days, I often find
I have so many files to read,
So many seminars to lead,
So many meetings to attend,
So many talks, that in the end
A minor fact or two slips by.
But, elephant, both you and I
Appear to understand the world.'
And here the Bigshot's fingers curled
Around a little golden ring.
'This vast unwieldy gathering,
Dear Elephant, is not the place
Where we can reason, face to face,
About what can or should be done.
We should discuss this one on one.
To be quite frank, your deputation
Has not filled me with fond elation.
Tell them to leave; I'll close the door,
And we'll continue as before.'

~

Although the other beasts agreed,
The elephant declared: 'I'll need
My secretary and mahout
To help me sort this matter out.
Like all the rest, he's left the room,
But he can come back, I presume.
There's two of you and one of me—
So I expect that you'll agree.'
The Bigshot nodded: 'Call the man.'
Quick as a quack the tragopan
Opened the door and strutted in
To greet his buddy with a grin.
The Bigshot and his Number Two
Scowled as they murmured, 'How d'you do?'

~

Tea came; the Bigshot looked benign.
'Milk?' 'Thanks.' 'And sugar?' 'One is fine.'
'It's not too strong?' 'I like mine weak.'
At last the moment came to speak.
'You see, good beasts,' the Bigshot said,
'We need your water—or we're dead.
It's for the people that I act.
The town must drink, and that's a fact.
Believe me, all your agitation
Will only lead to worse frustration.
Go back, dear beasts, to Bingle now.
We'll relocate you all somehow
In quarters of a certain size.'
He yawned, and rolled his little eyes.

~

Immediately, the tragopan
Pulled out his papers, and began,
With fact and query and suggestion,
To give the Bigshot indigestion.
'You say the town is short of water,
Yet at the wedding of your daughter
The whole municipal supply
Was poured upon your lawns. Well, why?
And why is it that Minister's Hill
And Babu's Barrow drink their fill
Through every season, dry or wet,
When all the common people get
Is water on alternate days?
At least, that's what my data says,
And every figure has been checked.
So, Bigshot, wouldn't you expect
A radical redistribution
Would help provide a just solution?'

~

The Bigshot's placid face grew red.
He turned to Number Two and said

In a low voice: 'This agitator
Is dangerous. Deal with him later.'
Then, turning to the elephant,
He murmured sweetly, 'I'll be blunt.
Your friend's suggestion is quite charming,
But his naïveté's alarming.
Redistribute it night and day,
Redistribute it all away,
Ration each drop, and you'll still find
Demand will leave supply behind.'

~

The elephant first sipped his tea,
Then ate a biscuit leisuredly,
Then shook his head from side to side,
And, having cleared his trunk, replied:
'Well, even as regards supply,
I do not see the reason why
You do not use what lies to hand
Before you try to dam our land.
Even my short walk through this town
Shows me how everything's run down
During your long administration.
Your pipes cry out for renovation.
Your storage tanks corrode and leak;
The valves are loose, the washers weak.
I've seen the water gushing out
From every reservoir and spout.
Repair them: it will cost far less
Than driving us to homelessness
By blasting tunnels through our hills
And bloating your construction bills.
But that's just one of many things:
Plant trees; revive your wells and springs.
Guide from your roofs the monsoon rain
Into great tanks to use again.
Reduce your runoff and your waste
Rather than with unholy haste

Destroying beauty which, once gone,
The world will never look upon.'
The elephant, now overcome
With deep emotion, brushed a crumb
Of chocolate biscuit off his brow.

~

'Dear chap,' the Bigshot said, 'Somehow
I think you fail to comprehend
What really matters in the end.
The operative word is Votes,
And next to that comes Rupee-notes.
Your plans do not appeal to me
Because, dear chap, I fail to see
How they will help me gather either.'
He giggled, then continued: 'Neither
The charming cheques that generous firms
With whom the Council comes to terms
—Who wish to dam or log or clear
Or build—will come to me, I fear,
Nor votes from those who think my schemes
Will satisfy their thirsty dreams.
It's not just water that must funnel
Out of the hills through Bingle Tunnel.
Do animals have funds or votes—
Or anything but vocal throats?
Will you help me get re-elected?
You're speechless? Just as I suspected.
I've tried to talk things out with you.
Now I will tell you what to do:
Lift up your stupid trunk and sign
This waiver on the dotted line.
Give up all rights in Bingle Vale
For fur or feather, tusk or tail.
Sadly, since you're now in the know,
I can't afford to let you go.
Your friend will never leave this room.
The tragopan has found his tomb.

As for yourself, my Number Two
Will soon escort you to the zoo.
From this the other beasts will learn
Your lands are ours to slash and burn
And anyone defying man
Will be a second tragopan.'
He giggled with delight, and padded
His cheeks with air, and gently added:
'But if you go cahoots with me
I'll spare your friend and set you free.'
He stroked his ring. 'And I'll make sure
You'll be—let's say—provided for.'
Before you could say 'Pheasant stew'
The servile hands of Number Two
Grasped the bird's collar in a vice.
The elephant went cold as ice
To see his friend cry out in terror.
He would have signed the form in error
Had not the tragopan cried out:
'Don't sign. Gock, gock.' And at his shout
The Bigshot's son came running in
And struck the henchman on the chin.

~

While the foiled killer squealed and glared,
For a long time the Smallfry stared
With indignation at his father.
'Papa—' he said, 'I would much rather
Give up my place as Number Three
Than countenance such treachery.
Why can't we let the valley live?
Those who succeed us won't forgive
The Rape of Bingle. I recall,'
The Smallfry sighed, 'when I was small
You used to take me walking there
With Mama in the open air.
For me, a dusty city boy,
It was a dream of peace and joy.

Along safe paths we'd walk; a deer
Might unexpectedly appear
Among the bamboos and the moss
And raise its velvet ears and toss
Its startled head and bound away.
Once I saw leopard cubs at play
And heard the mother's warning cough
Before you quickly marched me off.
Until this day there's not a single
House or hut or field in Bingle.
How many worlds like this remain
To free our hearts from noise and pain?
And is this lovely fragile vision
To be destroyed by your decision?
And do you now propose to make
A tunnel, dam, and pleasure lake
With caravans and motorboats
And tourists at each other's throats,
Loudspeakers, shops, high-tension wires,
And ferris wheels and forest fires?
As the roads come, the trees will go.
Do villagers round Bingle know
What's going to happen to their lands?
Are they too eating from your hands?
I had gone snorkelling on the day
The Council met and signed away
The Bingle Bills. I know you signed—
But why can you not change your mind?
You talk of sacrifice and glory.
Your actions tell a different story.
Do you expect me to respect you—
Or decent folk not to detect you?
Where you have crept, must mankind crawl,
Feared, hated, and despised by all?
Don't sign, dear Elephant, don't sign.
Don't toe my wretched father's line.
Dear Tragopan, do not despair.
Don't yield the struggle in mid-air.

I'll help your cause. And as for you—'
(He turned towards the Number Two)
'This time your chin, next time your head—,'
Rubbing his fists, the Smallfry said.

~

The Number Two lay on the ground.
A snivelling, grovelling, snarling sound
Oozed from his throat. The Bigshot stood
As rigid as a block of wood.
He tried to speak; no words came out.
Then with an eerie strangled shout
He uttered: 'You malignant pup!
Is this the way I've brought you up?
Where did you learn your blubbery blabbering?
Your jelly-livered jungle-jabbering?
Your education's made you weak—
A no-good, nattering nature-freak
Who's snorkelled half his life away.
Who asked you to go off that day?
You've been brought up in privilege
With Coca-Cola in your fridge
And litchis in and out of season.
How dare you now descend to treason?
One day all this would have been yours—
These antlers and these heads of boars,
This office and these silver plates,
These luminous glass paperweights,
My voting bank, my Number Game,
My files, my fortune, and my fame.
I had a dream my only son
Would follow me as Number One.
I had been grooming you to be
A Bigger Bigshot after me.
You might have been a higher hero
And risen to be Number Zero—
But now, get out! You're in disgrace,'
He said, and struck the Smallfry's face.

~

The Smallfry, bleeding from the nose,
Fell, and the Number Two arose,
And slobbering over the Bigshot's hand
Called him the saviour of the land.
At this, the elephant got mad
And, putting down the pen he had
Clasped in his trunk to sign, instead
Poured the whole teapot on their head.
The water in a boiling arc
Splashed down upon its double-mark.
The Bigshot and his henchman howled.
The tragopan gock-gocked and scowled:
'You wanted water; here's your share.'
Then guards came in from everywhere—
And animals came in as well—
All was confusion and pell-mell
While news-reporters clicked and whirred
At limb of man and wing of bird.
The elephant stayed very still.
The tragopan rushed round—until,
Provoked by a pernicious peck,
The Bigshot wrung its little neck.

~

The tragopan collapsed and cried
'Gock, gock!' and rolled his eyes and died.
He died before he comprehended
His transient span on earth had ended—
Nor could he raise a plaintive cry
To the Great Partridge in the sky
Whose head is wrapped in golden gauze
To take his spirit in His claws.

~

What happened happened very fast.
The mêlée was put down at last.
The Smallfry cried out when he found
The pheasant stretched out on the ground.

The Bigshot too began repenting
When he saw everyone lamenting
The martyr's selfless sacrifice.
He had the body laid on ice,
Draped in the state flag, and arrayed
With chevron, scutcheon, and cockade—
And all the townsfolk came to scan
The features of the tragopan.
Four buglers played 'Abide with Me';
Four matrons wept on a settee;
Four brigadiers with visage grim
Threw cornflakes and puffed rice on him;
Four schoolgirls robbed the tragopan
Of feathers for a talisman;
And everyone stood round and kept
Long vigil while the hero slept.

~

A long, alas, a final sleep!
O, Elephant, long may you weep.
O, Elephant, long may you mourn.
This is a night that knows no dawn.
Ah! every Bingle eye is blurred
With sorrow for its hero-bird
And every Bingle heart in grief
Turns to its fellow for relief.
Alas for Bingle! Who will lead
The struggle in its hour of need?
Is it the grief-bowed elephant
Who now must bear the beastly brunt?
Or will the gallant martyr-bird
In death, if not in life, be heard?
Dare the egregious Bigshot mock
The cry, 'Save Bingle! Gock, gock, gock!'
And can a ghostly Tragopan
Help to attain a Bingle Ban?

~

For it undoubtedly was true
That suddenly the whole state knew
Of Bingle Valley and the trek
That ended in the fatal peck,
And panegyrics to the pheasant
In prose and verse were omnipresent.
Suggestions for a cenotaph
Appeared in *Bingle Telegraph*;
And several human papers too
Discussed the matter through and through.
The water problem in the state
Became a topic for debate.
The Bigshot, struggling with the flood,
Was splashed with editorial mud.
Then intellectuals began
To analyse the tragopan.
Was he a hothead or a martyr?
A compromiser or a tartar?
A balanced and strategic planner
Or an unthinking project-banner?
It seemed that nobody could tell.
And maybe that was just as well—
For mystery matched with eccentricity
Provides the grist for great publicity,
And myths of flexible dimension
Are apt to call forth less dissension.

~

This is a tale without a moral.
I hope the reader will not quarrel
About this minor missing link.
But if he likes them, he can think
Of five or seven that will do
As quasi-morals; here are two:
  The first is that you never know
Just when your luck may break, and so
You may as well work for your cause
Even without overt applause;

You might, in time, achieve your ends.
  The second is that you'll find friends
In the most unexpected places,
Hidden among unfriendly faces—
For Smallfry swim in every pond,
Even the Doldrums of Despond.

~

And so I'll end the story here.
What is to come is still unclear.
Whether the fates will smile or frown,
And Bingle Vale survive or drown,
I do not know and cannot say;
Indeed, perhaps, I never may.
I hope, of course, the beasts we've met
Will save their hidden valley, yet
The resolution of their plight
Is for the world, not me, to write.

thirty

~

FEAST

MANJULA PADMANABHAN

The vampire strolled into the Arrivals Lounge at New Delhi's Indira Gandhi International Airport, a faint smile on his face. He looked like a man of means, silver-haired, clean-shaven, handsome and tall. He had concealed the extreme pallor of his skin with touches of rouge, giving himself a complexion befitting a European traveller who spends his summers idling on the sunlit sands of expensive beaches. He was dressed in a light summer suit, pushing a luggage trolley with two cases on it, one large and one small, both in black hand-tooled leather.

It was 3 a.m., yet the brightly lit hall was seething with people awaiting the arrival of their friends and family. Metal barriers had been set up to create a channel through which the newly-arrived could pass through the hall. Standing right up against the barriers were a number of young men holding up placards with the names of arriving passengers variously scribbled by hand or printed out, in large black capital letters, in English.

A thin trickle of fellow passengers straggled along beside the vampire, their faces turned towards the waiting crowd, looking for anyone familiar. So plump! thought the vampire, as he too looked out into the throng, though not because he expected to recognize anyone. So eager. So innocent.

A moment later he had caught sight of a signboard with ANDREW
MORTON inscribed upon it. He had chosen to affect a plain-sounding,
forgettable name. The young man holding the board wore a dark maroon
uniform with gold embroidery on his pocket announcing the name of the
establishment to which he belonged. A hotel: the Maurya Sheraton.

'Ah—' said Morton, broadening his smile as he inclined his head towards
the young man. 'For me, I think?' He had an excellent repertoire of accents
but had chosen, for this trip, to affect a culture-neutral Anglophonic voice.

In another few minutes, the young man had introduced himself as 'Satish,
driver' and was trotting industriously ahead of the vampire, pushing the luggage
trolley. The mass of hot, sweating bodies parted briefly to let the two of them
through before surging back to fill the breach. To the monster's finely attuned
senses, the scent given off by the entities around him was as ripe and heady
as tropical fruit, pungent with a quality he could not quite identify.

Was it uniquely local? He wondered. *Or did it vary from one principality to
the next?*

There would be time enough to find out.

Then they were outside the terminal building and the mild humidity of
the interior of that over-bright Arrivals Lounge was revealed to be an air-
conditioned version of the sauna conditions outside.

'Oh... *my,*' said the vampire, temporarily nonplussed by the viscous quality of
the air. He dabbed his mouth with a fine white handkerchief. The atmosphere
was practically liquid! It was unlike anything he had ever encountered before.
Unpleasant, yes, but also intoxicating. Exotic. Rich with chemicals, human
airborne ejecta, germs, dust particles, scent molecules, pheromones.

Astounding!

A minute buzzing distracted him briefly. He snapped his head around to
discover its source and saw a tiny insect coasting like a surfer on the current
of warm air that flowed off Satish. Indeed, there was a cloud of them trailing
the young man. *Mosquitoes,* thought the vampire. *How quaint!*

Of course he had encountered such creatures before, but there was a subtle
difference. Here, in the Third World, they were still a force to be reckoned
with: carriers of disease, minuscule assassins. As he had no scent, they paid no
attention to him.

Then he was once more within the temperature-modulated environment of
the vehicle that Satish Driver was currently in command of. Once they were
on their way, the young man attempted to engage the vampire in conversation,

asking about his family, his home, his country. Morton listened and responded absent-mindedly. His keen intelligence was sucking in information about this new and unfamiliar environment with the same avid hunger that he turned upon his victims when he fed. *Why has it taken me so long to come here?* he wondered. *Why have I never thought of visiting before?*

The city streets were shrouded in darkness, streetlights spaced at long intervals. There were broad avenues lined with trees and bushes. Behind the trees reared the silhouettes of tall buildings, punctuated by the vertical strip of their lighted stairwells. Other vehicles careened by on either side. But whenever his own car paused to acknowledge a red light, permitting the vampire to sample his surroundings, he could sense the presence of countless warm bodies lying unguarded in sleep on the sidewalks. They revealed themselves as long and low bundles, giving off a faint glow.

The hotel was like a fist of brightness thrust up against the night sky. Satish saw Morton to his room, collected a juicy tip, and agreed to return the following evening in his capacity as driver and informal tour guide. By this time it was close to dawn so the vampire drew the heavy drapes across his windows, turned off all the lights then went into his bathroom and lay down, fully clothed, in the tub. It was the closest thing to a coffin he could find while travelling and he greatly preferred it to the suffocating embrace of a bed. He lay in the cold white container, listening to the murmur of water rattling in the plumbing and entered into that state of suspended animation which passed for sleep amongst entities of his kind.

For the next three evenings, he went out in the company of Satish Driver to the most densely populated areas of the city. Chandni Chowk was the name of one. The area around the shrine of Nizamuddin Auliya was another. The third was an underground shopping mall called Palika Bazaar. He immersed himself in the fragrant, living heat of his prey, stoking his hunger like a child teasing a caged animal, enjoying its keening protests.

In places where the teeming hordes were especially dense, squishy soft bodies pressed against him on all sides with a surprising lack of reserve. He had never before encountered such uninhibited yet anonymous physical contact. In Europe it was unheard of for strangers, even in the grip of football hysteria or Oktoberfest revelry, to tolerate intimacies at this level. When he managed to persuade Satish to introduce him to a commuter train, bursting at rush hour like a pod full of human peas, he practically swooned with the intoxication of being squeezed butt-to-butt, thigh-to-thigh, chest-to-chest against dozens of fellow passengers. With their slick-skinned brown faces inches from his own, the aroma they gave off filled him with quivering delight. Cardamom and clove, onion and ginger, mustard oil, pepper and...something else. In the beginning he decided that it was *innocence*.

Yes. It was innocence that produced a fragrance as precise and particular as that of a spice.

Returning from the third foray, sitting in the passenger seat beside Satish Driver, the smile on Morton's face as he acknowledged the pressure of the need building within his body had grown into a grimace. Every individual hair follicle was tense with longing. His hands were literally twitching in his lap. In his belly, the furnace of his craving burned so hot and bright that it was all he could do to avoid crying out. As it was, he had to keep his lips parted to allow him to pant softly. His canines thickened and grew long within his mouth, throbbing painfully as he reined them in, keeping them concealed.

At the hotel, he invited the driver to come up to his room. Within minutes of entering the chamber and locking the door, he had turned upon the lad and accomplished the deed. Satish didn't resist at all. He closed his eyes as the vampire embraced him, allowing himself to be bent backwards. Even as the gleaming white teeth punctured the soft, yielding skin of his throat, the driver merely stiffened and let out a tiny gurgling moan. Nothing more. For long minutes, Morton's senses reeled with the familiar blinding radiance that blotted out all thought, all awareness. He knew nothing of his surroundings until he had drained every last drop of the sweet, rich, life-sustaining elixir complete with all the other liquids that animated mortal frames. Only a dry, rattling husk remained in his arms when he was done.

Then he lay back on the carpet, sated.

Ah! he thought. *Ahhh. It has begun.*

His reign of passion in this new country. His personal, private bloodfest.

In his suitcase, he had a small electric saw with which he sliced up the late driver's body. The tub was an excellent location for the operation. Once he had rendered the body down to a series of two-inch-thick slabs, he packed the bits into sealable plastic bags, then scrubbed the tub clean of corpse-litter. He transferred the now unrecognizable remains of the driver to his cases, called for room service, went down, hired a taxi and checked out of the hotel.

By evening, he was in the Taj Mansingh, with a new alias and a passport to match.

He waited till after midnight to go on a long slow walk, carrying the sealable bags in his backpack and emptying a few of them into garbage dumps as he passed them. In order to avoid attracting attention, he assumed the shrunken frame of a scavenging beggar, complete with matted hair and dirty, swarthy skin. Changing shape always required an enormous expenditure of energy and by the time he had completed his task for the night, his stomach was growling again.

He was perhaps a half-mile from the hotel. Keeping to the deep shade of the shrubbery and creeping along the boundary wall of one of the gracious mansions lining the avenue, he came upon a sleeping human form, curled up like a dog on a patch of jute sacking. He did not attempt to find the tramp's neck, but bit down into the wrist at the pulse point.

The rank scent of the skin, unwashed perhaps for weeks and sticky with oozing wounds, did not deter him for even a second. He took his pleasures where he found them, which was frequently amongst the poor and the destitute, defenceless as they so often were.

When he had recovered from the storm of his own delight, he resumed his own shape, rearranged his clothes and rested for a few moments, with his back against the brick wall. Once again he had tasted that flavour, the one he had initially identified as innocence. It was sharp, sweet and unmistakeable, like a citrus fruit that concentrates all the passion of summer within the plump grains of its flesh. But what *was* it? He, who had drained countless containers of human blood in the course of his unnaturally long life, could not remember encountering this quality before. Was it connected to the climate? Or something specific to this culture? And if it was culture, then how curious that he could detect it even upon this castaway, this stinking derelict with no discernible society to call his own!

Leaving the desiccated remains where they lay, the vampire returned to his hotel feeling thoughtful.

In the course of the next couple of weeks, he continued his forays into the city's streets. He was astonished at the ease with which he found victims. The best locations were near garbage dumps, inside the many ruined tombs that dotted the city and under park benches. But public toilets and the stairwells of apartment buildings were equally well populated by unresisting targets. Never yet had he encountered such extreme submission, such a total absence of struggle. It was true that he'd always had a gentle touch, an almost surgical skill at locating the exact point on an artery that would produce the swiftest results and with the least discomfort to the owner. Even so, it amazed and slightly unnerved him that he should have such easy success.

Equally astonishing was the fact that no reports of mysterious deaths had appeared in the press. He had been expecting to cover his tracks with meticulous care, as he had done with Satish. But it wasn't necessary. If an encounter took place in a park, he buried the remains in a pit he dug in the same location. If in a building, he folded the dry husk into a sack and carried the small bundle away to the closest convenient water tank. He had even, on a couple of occasions, affected the appearance of an Indian coolie carrying a head-load in broad daylight. No one turned to look twice at him.

Though he fed only under cover of darkness, he no longer hid inside the hotel during daylight hours. Instead he travelled in buses and trains, walked through bustling market places and plunged into the crowds flocking outside shrines and other places of worship. He assumed a number of different disguises, finding that a young foreign male backpacker elicited the most positive responses from ordinary people. He talked to anyone who looked friendly, trying in various oblique ways to find answers to the questions that could never be asked directly of any mortals: *What is the meaning of this unique local taste? Why have I not encountered it before? Is it possible for culture to impart an actual physical fragrance? And if so, why have I never heard of such a thing before?*

In the hotel, he sought out fellow Europeans now and then, just to confirm that the scent or flavour or whatever it was did not stick to them. And it did not. Only long-time expatriates gave off a slight trace of it. Twice he'd been invited to Embassy receptions and had mingled with crowds of his own compatriots, feeling nostalgic for the familiar, uncomplicated scent he associated with European victims. It amused him to contemplate luring a fellow guest into the bushes for a quick nip, but he easily resisted the temptation. On the one hand, he was sated with daily feeding. On the other hand, he guessed that a disappearance from an Embassy party might result in alarm bells being rung. The tell-tale wounds would be recognized for what they were and the hunt would be on, with all the frenzy associated with it. How many years had it been since he'd last had to save himself from imminent exposure or fight off an attack from someone armed with The Weapons? He couldn't remember.

Back in Europe, he had got used to living so frugally, so mindful of the dangers associated with being a vampire that months would go by when he fed only on vermin, stray cats, crows plucked out of the air and even, on occasion, raw meat bought from a supermarket. He shuddered at the memory of the last time he had been driven to suck sustenance out of an inert lump of dead flesh bought at a Safeway counter. He now wondered how he was going to adjust back to his home environment. He'd grown flabby during his stay in this country, not just in terms of his physical dimensions, but in the loss of his vampire instincts. He simply didn't have to worry about exposure here. The majority of citizens were apparently unconscious of monsters of his kind and therefore had no defences against him.

But his six-month visa was running out and the stack of euros he'd brought with him had dwindled alarmingly. He could not own credit cards or any other financial instruments because of the peculiar problems associated with immortality. Not only did he need to suppress bank records which revealed the unnatural length of his life, but Morton, or Martin Payne as he now called himself, routinely struggled when handling currency notes that made direct reference to the religion of his culture. US dollars, for instance, were out of bounds for him because of that pernicious line: 'In God We Trust'. He had to wear gloves in order to avoid coming into direct contact with the notes before picking them up.

Sipping a cup of coffee in the hotel lobby restaurant one morning, fretting over the prospect of returning to Europe, he became suddenly aware that the hair on the back of his neck was prickling. He turned in his seat, to find a blonde woman standing near him, smiling in the way of someone requesting permission to join his table.

'May I?' she now said, as she came around to the empty chair opposite his.

Taken by surprise, Payne could not muster excuses quickly enough. 'Why... uh...yes, I suppose so—of course—' he said, beginning to stand up.

The woman stopped him. 'Oh, don't bother,' she said, sitting down. She was a little over middle age, with fashionably streaked hair and an amethyst necklace around her neck. Her crisp white linen jacket and skirt partnered with an ice-mauve silk blouse, spoke of expensive good taste.

'My name's Cindy,' she said, in an accent that owed more to America than England. 'Cindy Wright.' She did not offer her hand for shaking. 'You needn't bother looking around to see if there are empty seats elsewhere,' she said. She paused for him to absorb the implications of her remark. 'I'm here because I think I know what you are.'

The vampire raised his eyebrows. 'Oh?'

Already his mind was racing, estimating his chances of success if escape became necessary. Already he was berating himself for his lack of preparedness, his loss of caution. In a crowded room, in broad daylight, the only option was an extreme change of form—a bat, for instance—

'Relax,' said Cindy, as if she perfectly understood his train of thought. 'I'm not about to expose you.' She paused with that annoying, knowing smile playing upon her lips. 'I've been shadowing you for a couple of days so I'm pretty sure.' She shook her head at him as he continued to look blankly at her, 'Come on! Surely you've not lost *all* your instincts?' She leaned forward now and very quickly, in a gesture that shocked him because it was so unexpected, she flicked her upper lip with a finger, just enough that he could see one of her canines. Sure enough, there it was: the sharp unnatural point, unmistakeable even in its sheathed and quiescent state.

Relief flushed through Payne. 'My goodness!' he gasped. 'For a moment I thought...'

'I know, I know,' said Cindy. 'It happens to all of us. Exposure Panic Response! But that's the least of your worries. Don't just sit there gulping like a fish! Do you mean to say you really don't have the slightest idea what I'm talking about?'

'Well—I—' stammered Payne. 'This is so unexpected! I've not met anyone else for such a long time that I'd stopped hoping. One of us, I mean. Are... are there many? Of us?'

'Yes and no,' said Cindy. 'Very few with active teeth.'

'Huh?' said Payne, his eyes narrowing. 'I don't understand what you mean.'

Cindy glanced away. 'It's one of the final signs,' she said, 'that we've been here too long.' She looked back at him. 'Mine stopped descending after my first year. But there are other signs. Just the fact that we're both sitting here in the open, in daylight, for instance. Didn't that surprise you? No? You'll find that fresh-flowing water isn't much of a problem any more either. As for the reflections in mirrors, well—' She tapped one of the shiny metal spoons lying on the table. 'Pick one of these up and surprise yourself!'

Payne sucked in his breath, feeling deeply disturbed. 'What about...?'

Cindy pre-empted him. 'Each of us is different,' she said. She held up her right arm. 'Can you look at my charm bracelet?' Payne winced and averted his face, but not before he'd seen the small gold cross dangling from her wrist. 'There you go. You're still sensitive. But I can wear one now. You'll be able to as well, if you stay here long enough.'

'But...why?' said Payne in a hoarse voice. 'It makes no sense!'

On the contrary,' retorted Cindy, 'it makes all the sense in the world. We're products of a very specific belief system—I won't name it, since you're still too sensitive—and all our dark powers, even though they're forged in opposition to that system, require our absolute belief in it.' She stopped abruptly, staring at him with a searching expression. 'In fact, just by telling you this, I might be weakening your powers. Do you want me to stop?'

'No, no...' said Payne, in a low voice. He was finding it difficult to control his breathing. 'Please continue. I have indeed been puzzled and curious. Whatever it is, I need to know.'

'All right,' said Cindy, 'you asked for it.' She was no longer smiling. 'The belief system we belong to is an austere one. Think about it: one immortal soul, one life on earth, one chance for heaven or hell. Right? Within that system, by choosing to suck the life-essence out of our fellow creatures, we gave up our rights as mortals. In exchange we acquired supernatural privileges such as immortality. But in order to maintain our powers *we* must uphold our *own belief system*! Are you beginning to get my point? In order to be culture-specific monsters we ourselves have got to be True Believers!'

She gestured at the brightly lit, bustling hotel lobby in which they were both sitting. They were surrounded by guests and employees, a majority of whom were Indians. 'In this culture, the rules of faith are completely different. There's no precise heaven or hell. There's no immortal soul—not in the sense we understand it—and there's no single...uh...divine authority. Instead there are infinite births, infinite deaths, infinite divinities. It doesn't really matter whether they formally believe in reincarnation or what names they give their deities or whether they even believe in anyone or anything. Just by being here they get recycled. So when we come here, we're exposed to a system that's directly opposed to our own. Instead of a single life and a single fate, there's a raging torrent of lives and fates, truths and deaths!'

Payne listened intently, saying nothing.

'Within this system, though they have monsters, even one they call "vampire", there's nothing fixed and definite about them. Some say their feet are on backwards and that they lurk in lonely places. But there are no crucifixes or silver bullets with which to dispatch them, no well-defined appearance or behaviour. They're not feared in the same way we're feared back home.' She shrugged. 'After all, if one life slips away courtesy a ghoul or demon, there's always a chance of having better luck the next time around! In the *next* life. The next incarnation.' She sent him a searching look. 'It's a system based on infinite abundance, in which nothing and no one matters because there's always more where it came from. So when we, who belong to the one-life-one-chance system, come into contact with a multi-life-multi-chance system, we begin to drown. Not right away, but slowly, over time.'

Payne said, his voice plaintive, 'But they *succumb* to us! How can we be their victims?'

Cindy was shaking her head. 'Numbers matter. All they have to do is keep on succumbing until, finally, our resistance is overwhelmed. That sweet flavour you've noticed? It's not just a tropical spice. It's a lack of fear. They don't fear us because they *know,* in their deepest hearts, that their sheer numbers will prevail. Without active fear to define us, we cease to be monsters. Our powers wane. We begin to die.'

Payne could feel a churning sensation within himself. Already, he could sense the truth of what Cindy was saying. His teeth were aching strangely for instance. 'Surely it's reversible?' he whispered. 'Surely when we return to Europe we get our powers back?'

'No idea,' said Cindy. 'I've only met those of us who've chosen to stay.'

'But—but—mortality!' said Payne, grimacing. 'Don't you mind dying?'

'Ahh,' said Cindy, stretching luxuriously and leaning back in her chair. 'You get used to it after a while. And besides, once you've been here long enough, you find yourself starting to hope...' She was smiling once more.

'*Hope*! Hope for what?'

'Better luck next time!'

thirty-one

~

NURSING GOD'S COUNTRIES
GITHA HARIHARAN

In Quebec, the wind sings with a hoarse throat. It's because it sings the same song all the time that it sounds like a moan. It's my first winter.

In Munnar, rain sprays the fronds of the coconut trees, making them greener. The rain has settled down to a steady rhythm, the way it does after it has said its piece in one great burst of a downpour. Because the rain is calmer now, I can hear it better. Or I can make better sense of it. I can translate dripping whispers into words that wing from Munnar to Quebec. Monsoon lullaby, coconut lullaby. The wet leaves come so close to me they make a blur of green. The green leaks into my head whether I am asleep or awake.

My little girl must be watching the rain, listening to it. What does the rain tell Binny Mol that it doesn't tell me?

In the last photograph I got, she was not such a little girl. She is growing, she is now half-girl, half-woman. I prefer photographs to telephone calls or emails or Skype. I can look at the photograph alone at night and travel where I like. That's the time I learn my child again, make her completely mine.

There's a dream in her eyes; I recognize it because it lived in my head once, making me look like Binny Mol. I want to reach into the photo, tenderly spoon in a dose of bittersweet medicine: nurses should not dream too much.

Binny Mol pays no attention to me. I have to leave her with the rain, let her face dissolve into the leaves.

It's another mother and daughter I see now. I have become the daughter, and Ammai's words do not sound like rain at all. They are loud and clear as if she is still giving orders to the junior sisters though she is retired.

Ammai wanted, always, to go somewhere; at least as far as Delhi, Mumbai. But it never happened. Her cousins and sisters-in-law made their way to Muscat, Kuwait, Saudi, Manama, Dubai, names that slid down our tongues as easily as Munnar, as if these unseen places belonged to us or we belonged to them.

My going away was almost as good as Ammai's living as she wanted to live. We dreamt our separate dreams together.

We woke up every time we counted our notes. Hers for Good Angel Private Institute, mine for the agent. The bits of paper, the certificates and the rupee notes, rustled in my hands like promises. Almost a ticket. Ammai's ticket out of the country too, though she would stay at home. 'Go alone, you're a brave girl,' she said. 'I'll find you a good match later.'

But when she heard I was going to Bahrain, she said angrily, 'Why Bahrain? It's not a Christian country!'

Now, when she does not get angry, or ask sharp questions, I have an answer. Bahrain, Munnar, Delhi, Quebec, where are her Christian countries?

What I actually say to her these days is different, but also true. I sent her a picture of the church Annama, Sara and I go to most Sundays. We stood arm in arm for the photo. Sisters in arms, straight and serious, not laughing like the photos people send when they are on holiday, enjoying themselves. I described the inside of the church to Ammai because we didn't want to aim the camera at the altar and stained glass windows like tourists. I find photos are easier than words though; they say what is not so easily said.

All those years back, I didn't say anything when Ammai was worried about my going to Bahrain. Or when she put her arms around me before I left, saying, 'May Jesus travel with you.'

But by then I was *Flow-rens*, not she.

The first time she told me about Florence, I was a girl like Binny Mol. My mother told me stories of the lady who travelled with just a lamp and a mission as luggage. Every time she unpacked her bag, she healed someone.

My mother's stories always had something useful in them. I could pack them in plastic, seal them with a burning candle and take them with me. Like dried fish or pappadam or pickles, things that taste better when they keep. Or when you have made a vow to serve and send home the money they need.

'A nightingale is a bird,' my mother told me when I was a child. 'It's a little like a koel.' I imagined a koel, a nightingale in a white coat, singing a flowing lullaby. Tharaattu pattu. A lullaby that can put a baby to sleep because she is at home, safe and loved.

Some years later, when I had become Florence in Manama, it was I who sang to the koel. I watched my friend Sumithra next door and learnt to coo as I rocked Binny Mol. *Aaarero, raaro, raaro. Aaarero, raaro, raaro.* Her eyes would glaze as if I had given her the right drug. She would stare at me as if she would never let me go; then her eyes would finally close. My heart swelled till I could hardly breathe. It didn't matter whether I was nurse or mother or both.

The photos I send Binny Mol are different from what I send Ammai. I send her photos of myself, standing tall inside a snowy postcard of a wonderland, smiling and smiling. My smile may show her the future; for now, it is my unsung lullaby.

But sometimes I worry: is it easier to sing, or to be a mother, in a green place?

I have never seen so much white before, except on uniforms, or saris and mundus. On clothes.

Here it is skin; white skin. And it's not just people who wear white skin. It's the ground, the roofs, the trees. Some of the trees are naked under the white. Other trees still have leaves; hard leaves. Naked or clothed, they have to bear the weight of the snow. Their skin has to be thick enough for the layers of white, like the heavy make-up on a Kathakali dancer. Sometimes the doughy clumps make a pretend nose or false ears. The bush I see from my room has two flat piles of snow before it. A pair of floury feet. Any minute now, the bush may learn how to walk with those misshapen feet. It may walk slowly, as if recovering from an operation, take one small step at a time, swaying like a Kathakali dancer, and go in search of a warmer place.

I want that to happen by the time Binny Mol is here.

It's too soon to tell Binny Mol when she can come here. But I can almost see it: we will have a spick and span little Canadian home with heating. She will go to a free school. She will learn nursing at a place better than Good Angel, or study something else, even medicine. A good salary for not working herself to the bone. She will be like me, a career woman. But please Jesus she will do better than me, much better and more easily. When we come home in the evening, we will find Munnar in the big bottle of coconut oil, the appam pan, the beef in the fridge ready to be turned to stew.

If Paulose were not under Bahrain's brown earth, we would be a complete little family, the kind that laughs in Christmas card photos.

Paulose, I say to him sometimes in my head: I did it somehow. Our plan didn't fail, even though I went home without you. Ammai wrapped her arms round Binny Mol. She blessed me when I went back to Bahrain alone. 'You take God's country where you go,' she said.

The second time in Bahrain: saving all over again, sharing a tiny place with sisters without families. But Paulose, Binny Mol and I shared the photo frame sitting by my bed like a guardian angel. The angel reminded me of the day Paulose came home from the power plant to say he had got a job. The look on his face then—that's what I had to hold, like a photo, to prove our plan was beginning to work. It was no longer just my salary.

But too often that look of pleasure would slip in time, change into the look of a dying man. That actually came later, the pain, guilt, worry about me and the child, the irregular remittances home for months. Everything was getting gobbled up by his cancer, not just his body.

Sometimes I would pretend he was just another patient, another job to be done well. Bodies are all the same; so strange, because each body is different.

When Paulose and I met, two weeks before we were married, I could only see the difference. He was an engineer. He didn't have a degree, only a diploma. But his family lived in a house with a compound, on a nicer street than ours. It was time for me to marry, Ammai said. And all Paulose wanted to do was emigrate. That is how we made a match. If I were not in Bahrain, his family may not have approached my mother. I used to tease him about this later. Though he never denied it, he laughed and took me in his arms.

Once he joined me in Bahrain, found a job, got me pregnant, he became less of a stranger. We had something between us. We found it, or it found us, and it grew like a strong tree with spreading roots. If he was the man of the family, I was the one who had got him to Bahrain. I don't think he ever forgot that. But I did. I only remember it now because I have become the good match in place of Paulose. I still send money home to his parents and brother and sister, not just my mother and daughter. His family still calls me one of theirs. Once in a while, but only on a bad day, I think: remittance keeps even an in-law family together. The moment passes quickly. It's too late to quarrel with Paulose. I have to forgive him for dying.

Most of all, memory is what helps me forgive. It comforts me and lulls me to sleep in my single bed.

The memory I like, the one I have to ration and not take to bed every single night: it was one of those evenings when I got back from hospital later than usual. I let myself into our flat, the thought of cooking and bathing the baby making me feel even more tired. My back and legs were punishing me for standing or walking or bending all day. Then I noticed the fragrance from the kitchen that had reached the front room; dinner was waiting for me. And I could hear Paulose singing in the bedroom.

I tiptoed to the door and stood there, listening. He was singing one of the songs I had learnt from Sumithra. A royal lullaby, she called it.

The lullaby which once put a little prince to sleep in a palace in Kerala had made its way to Bahrain without a visa. It was now in the throat of the bear-hairy man who worked with machines all day and could still hold our baby with tenderness.

Omana-thingal kidavo... Both Paulose and I loved this song because it said nothing about sleep, but always, like magic, it put Binny Mol to sleep.

Binny Mol must have been fast asleep by now, dreaming good dreams, but Paulose was still singing. He was asking her, of all the lovely things in the world, which one she was. Was she his full moon? His lotus? His doe?

Our streak of luck. Our sweet singing bird. Ours, then only mine. My tender green leaf.

Green leaf, Binny Mol, Munnar.

When I went back to Bahrain alone, I would work all day with the Malayali nurses I stayed with. Every night, I dreamt in only one colour: green, all the shades of green. It was like a sickness, but it let me go home as easily as sleeping. Then the dreams stopped.

There was green in Bahrain too, but it all looked like pictures to me. The palm trees were real, but they were all in a line, obedient patients for the Indian and Pakistani gardeners to keep beautiful. It was clean like a good hospital. You know the brown is there, the dry skin of a sick old man. But you cream it, make it look fresh, because you have to keep up appearances even if you are dying. It makes everyone feel better, doctor, nurse, patient, family.

In any case, the manicured green faded where we Malayalis lived. Huddled together, we could make believe we were home and not home. We could be in Kerala and stay away from it.

When the papers came through for my going to Canada, the old green-dream sickness came back. Maybe it was all the expense and worry of applying again, emigrating again. Maybe it was because that was the time the nurses were striking back home in Kerala. I saw the pictures of nurses I knew in the midst of strangers, strangers who looked like me, dressed like me.

For a day or two I wanted to go back. Not to grow old, or die in my ancestral village, but to be a sister among sisters, hold up placards like them, or shout slogans, demanding my rights. I have never shouted a slogan in my life, but maybe I could do that in Kerala, make up for all the things I can be in Bahrain or Canada but not at home.

I didn't go back. I didn't learn to shout slogans. But I went all the way across a map glistening with blue-grey water, and found Quebec. One more of God's countries to nurse.

Some days, I feel I am in a strange dream alone. The dream is all white. It is quiet, or it speaks only French. But there are days when the work or the pay cheque or the time spent with other sisters shakes this feeling off me.

Green, brown, white. The changing colours pile up, they go from place to place. Bodies remain the same. They have to be taken care of.

I will work at it, my first Quebec winter, as I worked through all those green winters, the clinics in Munnar and Kochi and Bangalore and Bahrain.

By the time I get the papers for Binny Mol, I will learn how to sing even when the wind blows hard, turning green to brittle orange, brown, then white. It will be like learning a new language. Only the meanings of words will still be in Malayalam.

My certificates have piled up; the hours I have to be in the hospital have shrunk. I no longer change sheets or clean skin with a wet sponge. I have learnt to draw blood so the needle is light as a butterfly on the vein. 'Done already?' the pale young man asked me the other day. 'I didn't even feel it.'

I felt it though. I felt the glow of a woman even stronger than Ammai, and with a longer streak of luck.

PROPOSED FOR CONDEMNATION
CYRUS MISTRY

A jay had the body of a woman, soft and hairless. He was pale and handsome and he despised the company of men—at least of the men he had occasion to spend time with, the clerks and tellers at the bank where he worked, with their absurd hysterical cackling in Parsi Gujarati, their unabashed farting, their non-stop tea-drinking and discussion of the day's gambling figures. At twenty-eight, Ajay had no friends.

He had, once, before his father died. But when Gangaram Borkar, after twenty years' service in a bank, had retired and it had become common knowledge in the chawl that he had been forced to resign, two years before he was due to, because of a discrepancy in accounts, Ajay himself had cut off all contact with his friends. Two and a half years after this incident, his father died, a crumpled man. After he started working, there was even less reason for Ajay to meet anyone. He lived alone now, with his old mother, who was bedridden by arthritis and blinded by cataracts. In the morning before he left for work, he cooked her afternoon meal of bhat and dal and left it on a bedside table. When he came back home in the evening, he cooked their dinner.

Though Ajay bore the loneliness and the sickening dullness of his daily life with a dumb, almost noble placidity, and the neighbours praised his selfless devotion to his mother, at the heart of his being, a terrible, unrelieved hunger raged that set him on fire and drove him to distraction and tears and hatred, when he lay on his steel cot at night, alone with his mother's irregular snoring and the smells of old age and illness. It was a hunger for women. He wanted to possess them, love them as they had never been loved before, be engulfed by their love, smothered by their gorgeous overpowering smells, by their flesh which was soft as his own.

But Ajay did not know any women. He felt incapable of going through the motions of courting, of being witty, flattering, ingratiating. He saw an enormous falsehood in it all. He hated women who enjoyed flirtation, coquetry. The moment he saw any woman in the office interact with others, a bitterness tainted his desire and boredom blotted out any possibility of a relationship developing. He had no time for that. He felt that he could only completely love a total stranger, a woman whom he met by chance, who recognized in an instant the immediacy and intoxication of what he had to offer. Though he did not know any women, Ajay felt that he knew them better than anyone else. And what he had to offer was precious and unique, an enormous understanding of their beauty and gentleness, which existed in him because his own heart, he felt, was a woman's heart.

Every evening, returning home from work, Ajay lingered for a while on the platform at Churchgate before boarding a train, watching the women packed together like eels in the Ladies' compartment, all soft and slippery and pressed together. When he got off at Mahim, he stood near the end of the platform or sat, if a bench was vacant, and watched the trains pull in and pull out, the second-class women commuters tumble out in their saris and skirts, and others fight and push their way in and cling to each other, as the train started. As he watched, his heart would throb, his mouth would turn dry, and his penis would strain and cry out against this utter deprivation. He wanted to be among them, to be himself present among the ladies in the Ladies' compartment, crushed by their soft bodies, their perfumes, their silky materials, their proliferation of breasts. He knew he had to do it. He *had* to enact this greatest, most sublime fantasy that strained at every nerve in his body when he contemplated it, sent the blood rushing to his head, and at nights, when he was utterly alone, sat bestride him like a demon rider and squeezed his soul, till he gasped and hugged his genitals with both hands and rubbed his engorged monstrous cock till his desire was temporarily relieved.

Once he had decided, he went about collecting the articles he required, with an unhurried attention to detail and taste. In the dark evenings, every day for nearly a week, after he had finished cooking and feeding his mother, Ajay wandered the lamp-lit market streets in search of these objects which filled him with a cold excitement as he paid for each one.

The wig shops were mostly disappointing. Finally it was Maganlal Dresswalla, a supplier of stage props and costumes, who dug out just what he was looking for, a wig of just the right texture, not too dry and strawy, and of a brownish-black colour. Here he also found a theatrical bra with solid foam cups. He paid a deposit of sixty rupees and thirty rupees in rentals, but of course, he had already decided to forfeit his deposit.

Cosmetics were the easiest to acquire. He carefully selected his favourite shades of lipstick, eyeshadow and asked for them to be gift-wrapped. For a sari and blouse he had to return to Maganlal Dresswalla because his mother's, he found, were too faded and stained from lack of airing. He bought a pair of ladies' sandals without trying them on. At home he found they pinched him a little.

At last he was ready. He spent two nights sitting up in an armchair gazing at his image in the long cupboard-mirror in the dark, fully dressed and made-up as a woman... He was beautiful. He could not take his eyes off himself. Not that he needed to sit in the dark. His mother, who was sleeping, could hardly see beyond two feet even in broad daylight. No, the darkness calmed him. His mind was wide awake. No tremulous fears, no agonies of guilt, no tedious dialogues with self besieged him. He was charmed by his own beauty, amused and delighted. He felt larger than himself, more natural and confident than he remembered feeling ever before.

That Friday, his first day, he finished his household duties with great speed and dressed. He had no difficulty with his sari, for as a child he had often draped bedsheets around himself. His wig was held neatly in place with bob-pins, the locks of hair reaching down to his simulated breasts. His make-up was in impeccable taste. At the door, he told his mother:

'I'm going out, Tai...' Usually his mother received such information with only a grunt or no response but today she seemed to be in an agitated mood.

'Where? Where are you going out so late?... Come closer,' she said, reaching out with her arms. 'Come here... I want to hold you.'

Ajay remained where he was standing. 'I'm getting late,' he said, and peered out of the door. No one was in the passage. He stepped out quickly and shut the door.

No one saw him leave. No one stared at him in the street. He was calm.

But as he approached Mahim station, an insistent drumming took up in his chest and a tiny vein on his temple throbbed to its beat. When he boarded the first crowded Borivili local to come in, the blood left his feet and made them numb. He had difficulty breathing. A sudden spasm of fear gripped him that he might pass out in the Ladies' compartment and his disguise come undone. Then he would be at the mercy of the ladies in the Ladies' compartment...

But the train began to move and no one noticed him... All around him were women, bodies, smells, of stale perfume, sweat, hair, softness...oh heavenly! This was something... He stood close to the door on one side. Next to him, against the wall stood a plump young Gujarati girl in spectacles, with a pretty face, dressed in a salwar-kameez...

He had taken the precaution of wearing two pairs of tight underwear, but his crotch had swollen to a gigantic bulb. Just to lean against her, feel her soft skin rub against his in the swaying of the train... Oh God, this was heaven. His leg touched hers and he moved his knee a little, caressing her fleshy thighs under the thin salwar material. His lips brushed against her hair. He leaned into it and breathed deeply, his heart beating wildly at this unbelievable joy... He did not want anything more, God, just this, just this was enough...

But at the next station, Ajay could hardly believe his luck. An entire shoal of overweight Gujarati women climbed in and flattened him completely against his precious wonderful creature... O darling, don't mind me, my darling, you're wonderful!

O don't move, please don't move anybody.

Even in his crushed condition Ajay tried at first to hold his hips away from her, frightened by the extraordinary size and hardness of the bulge that might give him away. But in the rocking of the train it was not possible to avoid contact, and often there were violent jerks and he was flung with his full hardness against her. His cock screamed with hunger, but she showed no sign of having noticed anything amiss. He was surprised, but he grew bolder. He rubbed himself against her buttocks, trying to synchronize his movements to the jerking of the train. She was perhaps too young and innocent to recognize the significance of the lump that strained to bursting point in his sari... Oh God, how can I control myself any longer? With the back of his hand he moved his fingers very gently up the ridge of her back. They rested for a while against her slender waist and slid down again, describing the curve of her delicious bottom... I want you, I want you... A stream of joy rose in the pit of his stomach and made for his throat, but he cut it off with a deep intake of breath. At the next station, more women got in. Now he was completely crushed against his lover. He wanted to tell her how much he loved her, he wanted to cup her breasts in his hands and bury his lips in her neck and sway with her to the rhythm of the train. The young girl shifted her position. She now stood sideways, facing the wall. He moved himself, subtly. O! Push me more, squeeze me against her, go on, don't stop! The train started with a terrific jerk. Ajay's body was overcome by spasms of delight and a warm wet stickiness spread through his underwear.

Now Ajay began to roam the trains every evening. He bought himself a new railway pass valid from Churchgate to Virar, and he wrote his name in it as Kamini Kumar. At 7.15 every evening, after his domestic chores were done, he walked to Mahim station. He travelled all the way to Virar, then returned (though the southbound trains were relatively empty, he enjoyed the sensation of travelling in the Ladies', for its own sake) and caught another train from Churchgate, bound for Borivili or Virar. Sometimes he stayed out till 10 p.m. or later, even long after the compartment had quite emptied of women. His life at the office became a meaningless but necessary interlude to his life in the evenings, on the trains, as Kamini.

Sometimes he toyed with the idea of travelling fully dressed as a woman in the crowded general compartment, but this idea, though exciting, filled him with revulsion and fear. Sometimes his fantasy led him to want to be the sole passenger on a completely empty dark 'service' train flashing past the crowded stations without stopping, bound for the yard. But he was afraid he might be found and molested by the railway police.

One evening, when he climbed into a train at Mahim, he saw a dark attractive woman with long hair and shining eyes, staring at him. He went and stood beside her. She was probably over thirty, but she looked younger. She wore a short skirt and a 'sticker' which outlined the shape of her firm round breasts.

In the movement of the train, he allowed his hand to gently brush against her thigh. She did not move away, but seemed in fact to move closer to him. This made him brave, and he explored her naked leg where her skirt began. She was still. But now, after weeks of travelling as a lady, Ajay felt there were no limits to the pleasures he could take, there was nothing to fear here. With the thumb of his hand which held the steel bar on the side where they were standing, he fingered the curve of her breast and slowly let his thumb travel to her nipple. He felt her body quiver and her breath heave. Then as the train became more crowded, he lowered his hand again and let it slide up her skirt. Her thighs were smooth as silk, and she did not object. He discovered her panties, and the exquisite curve of her bottom. His fingers slid forward and crawled under the panties. His heart was pounding feverishly as he felt her hair which was damp like moss. His finger searched among the folds of flesh for the centre of her desire. Her body trembled under his hands and she pressed her legs together. He could control himself no longer. Fiercely, with his hand still in her, he pressed her with all his might against his hardness and spurted. In the same instant, she felt his throbbing creature flap against her and screamed. Her scream was drowned in the rattling of the train. She moved away from him abruptly, throwing him a look of horror and fright, and made for the opposite exit.

At the next station, she got off. He followed her. It was Vile Parle.

A stream of commuters was walking on the tracks, towards the level-crossing. She followed them down the sloping rear end of the platform and began to walk fast. Once she turned round quickly and her eyes met his and she quickened her steps on the wooden planks of the tracks. Once she stumbled and nearly hit the ground. Ajay was able to easily overtake the people in front of him, and then, he was walking beside her. He saw that her face was pale with terror, as if she had just seen a ghost. She was sobbing.

'Please...don't be scared,' he consoled her. He could not recognize his own voice which came out thin and quavering. 'I won't hurt you...'

'Go away,' she said hoarsely, throwing him another fearful look, her face streaming with tears.

'I love you. I swear I love you,' said Ajay in a whisper, his desperation giving him a courage he had never known himself to possess. His voice beseeched her to trust him, to trust that something great and wonderful could happen, *had* happened, between them.

Just then, a voice behind them shouted: 'Hoy! Do number—gaadi' and Ajay noticed that all the people walking in front of them had cleared the tracks. Maybe a hundred yards away, a train was approaching fast, its fierce yellow eye close enough to light up their faces. He pushed her off the track on to the side where there was enough place to stand, among the shrubbery, and held her comfortingly as a cold gust of wind and dust enveloped them, and the large wheels of the train rolled past with a mighty rumbling.

'Please come with me. We can just have a cup of tea here, at this restaurant,' he said to her when they reached the level-crossing. She stared into his eyes. Her own bloodshot and wild. She followed him wordlessly.

For a long time, they sat silently across a table in an Udipi restaurant. Tea was ordered, it arrived and cooled, but she would not look at him. She looked at the floor under the table, at her feet. At last she said, 'You should not have done this...'

'I did it for you, I did it because I love you,' said Ajay, not knowing what he was saying, but sure that he was speaking the truth.

'You have done a very bad thing...' she repeated, still looking at the floor. 'You have dressed as a woman and travelled in the Ladies'...and you are a man...'

'If I had come to you as a man would you have let me stand beside you and touch you? Oh, you've made me so happy today...'

'I hate men,' she said, looking at him for the first time in the restaurant.

'But I am a woman. Look at me. I love you as a woman. Look around. Does anyone know that I am not a woman?... I hate men too. I work in a bank. I hate all the men there. I hate travelling in the Gents', compared to that, the Ladies' is like a garden, like a paradise...'

They were silent again for a while. Then she spoke: 'I am a married woman... My husband is a salesman. He has to travel. He is never with me... Some days he comes home, late in the night with his small suitcase. He brings one or two friends along, and many bottles of whisky. They sit and drink while I prepare dinner for them. Then my husband says, "It's so hot, so hot..." And he takes off all his clothes and walks around naked in front of everyone. To shame me... And he won't let me go in. He insists I sit there with his friends, and he tells them, "Relax, relax...feel at home..." After dinner, they all sleep together in the hall. In the morning, he goes away again, with his suitcase. He leaves some money behind...'

Ajay's heart was moved by her story. He wanted her more than ever now. He wanted to soothe her, he wanted to assure her he would never hurt her. If he could have, he would have dropped to his knees under the table and buried his head in her skirt. 'Come with me,' he said. 'I live alone with my mother who is blind. We can spend the night quietly...'

But she shook her head firmly. Then she said: 'Don't ever try to follow me or find out where I live... I will leave you now. Don't move from here for another five minutes... We will meet only on the train...'

'But don't go yet,' Ajay panicked, as she stood up. 'Will you catch the same train every day? How will I find you? Tell me more. Tell me your name at least.'

She picked up her handbag and seemed not to have heard him. Then, without looking at him, she said, 'Lina.'

She had left her tea untouched.

Could it have been a dream? For the next three days, Ajay roamed the local trains, unable to amuse himself with other women, unable to find Lina. He began to be afraid he might never see her again.

On the fourth day, he spied her standing near the entrance of a Borivili-bound train, looking exactly the same, in a skirt and top, her hair loose, her eyes underlined with kaajal. She saw him, pretended not to know him. He climbed in and stood beside her in his green sari. Then his hands wandered, their bodies clung together and in the noise and roaring of the crowded train, they discovered a deep and frightening passion. They did not speak. He kissed her ears, caressed her body, he held her close to him. Then he backed her against the wall and let himself be crushed against her exquisite body. He rubbed her with his hand through her skirt, till she whispered to him, go on, go on, and her whole body shuddered and she squeezed him in his sari so hard that he stopped and cried out.

In this way, they met every day, or nearly every day. On days when he missed her, Ajay felt destroyed, hopeless. He would have liked then to return home immediately, but he had to stay out till ten at least, to ensure that his neighbours did not see him entering his room. Sometimes, after she got off at Vile Parle, he persuaded her to sit with him for half an hour or an hour, in the municipal park where there were tall hedges behind which couples crouched after dark. Sometimes they went back to the Udipi restaurant for a cup of tea.

But Ajay was most happy with her only on the train, in his garden of delights, where he was a woman himself, among the multiplicity of women. As the days passed, one obsession took hold of his mind and grew large and demanding. Nothing would satisfy him, he knew until he satisfied it: he wanted to make love to her on the train. Not muscle her and rub himself against her and come in his underwear. He wanted them to be naked, uninhibited, making passionate love, on a train.

But where? Ajay made a study of the railway yards at Borivili and Virar. There were rows of dark trains parked in the yard, among the signal posts and the overhead cables. But there were too many railway policemen about. Then he remembered a siding off Goregaon station where an old train had stood for many years. It was rusted and tattered. Its ochre-brown body looked like it had been scraped for a new coat of paint which it had never received. On one of the bogies, a bold hand had painted in white: PROPOSED FOR CONDEMNATION.

Before speaking to Lina about it, he went to Goregaon one evening and studied the locale. The train was completely unguarded, deserted. He found one bogie which stood beside a steel pillar which made it very simple to climb in. Inside, it smelt musty and the floor was covered with dust and bits of straw and the fans in the ceiling were shrouded with cobwebs. The bulbs had been removed, and also most of the metal straps.

That evening in the park, when he decided to speak to Lina about his plan, she was in a particularly depressed mood. Her husband had come home last night, alone this time, and left in the morning without saying a word to her. When Ajay asked Lina if she would come with him one evening to Goregaon, to make love to him in an abandoned train, she put her arm round him and kissed him. She squeezed his foam breasts and then her hand slid down to his cock which he had released under the sari from the constricting underwear, and she squeezed that too.

He pushed his hand up her blouse and under her bra and pinched and twisted her nipples. In a moment, they were nearly making love in a frenzy of passion behind the hedges of the municipal park. Then they were on a train to Goregaon.

The train was quite empty. An old blind woman was begging for alms as she sang a bhajan in Marathi. They stood near the door, taking the breeze, close to each other. A group of schoolgirls observed their intimacies and began giggling. But they laughed back, and continued their embrace.

When they reached Goregaon, they walked out of the station on the east side and alongside the tracks, on the road outside. It was dark, the rows of shops all had their shutters down, except for the paan shops. There were very few people about. At a certain gap in the fencing, they stepped in and continued walking on the tracks for a while, till they came to a section which branched off on to a dead, unused track, overgrown with shrubs, where the derelict train stood. Lina was shivering.

Ajay showed her where to place her foot, climbed in first and helped her in. It was like stepping into a cave. The long deserted compartment was cold and still, only very dimly-lit by the streetlight on the opposite side of the road, beyond a wall. They crouched on the floor, unmindful of the dust and cobwebs and crawled into a corner. They lay silently for a while in each other's arms, cold but not frightened, listening for sounds, listening to a train pass by... Then Ajay picked up Lina in his arms and stretched her out on the long wooden bench. She lay still, with her eyes shut. He slipped off her shoes and squeezed her legs. Then his hand lifted her skirt, he pulled down her panties. He buried his nose in her and rubbed her with it. She could no longer play dead. She clutched his hair and pressed his face against her. His hands reached up and played with her breasts...

They could not wait any more. An owl hooted loudly outside and the sound of an autorickshaw melted into the night. She slipped down on to the floor with him and spread out. Her juices had trickled down her thighs. They appetized him and he wanted to lick her more, but she wanted him inside her. He entered her. Flashes of light suddenly lit up the compartment and threw shadows. A fast train was hurtling by on one of the adjacent tracks. Lina was moaning with pleasure as he moved, throwing his whole weight into her. His breath came out in short gasps, his mouth was open and a little saliva drooled out of its corner, as he took her. He came, panting like a hungry dog. He collapsed on her and covered her face with kisses. But she lay still under him. Her eyes were open, and her face was staring somewhere, over his shoulder... Then suddenly the hair on Ajay's back stood on end and his spine froze, as Lina uttered a short, stifled scream and dug her nails into his neck.

He looked at what she saw: in the luminous, dust-particle-filled darkness of the compartment, three benches away from them, in full view, a starved-looked beggar with wild, unkempt hair was sitting very still, staring at them. He was naked except for a loincloth which had fallen aside. His face was expressionless, his mouth was open, eyes staring blankly like an idiot's. His cock was large and erect and shining in a beam of extraneous light.

thirty-three

~

TRYING TO DISCOVER INDIA
SHASHI THAROOR

It is strange, now, so many years later, to be telling my story, writing this journal. I think back to that other journal, the one Columbus kept, day by day, sitting at the desk in his cabin as the waves rocked the *Santa María*, his quill scratching away, recording it all—leagues sailed, the direction of the wind, his hopes and fears. Mine will lack that immediacy, because all I have are a few scraps of notes and my memories, the memories of an old man far away from his land and his people, writing in a language that is not mine, for readers in another world. But if I do not tell my story, who else will?

So let me begin as Columbus did, with the weighing of the anchors. It was, I remember, a Friday—I had already learned their days of the week, and this was the day of Venus, goddess of beauty, sprung from the foam of the sea—Friday the third of August, by the Christian calendar, the year 1492, when we set out from the bar of Saltes, at the confluence of two rivers, the Tinto and the Odiel, to sail into the unknown.

I remember it well, that last glimpse of Europe. It was a grey day, overhung by those European clouds that block the light without providing the blessed release of a real rain. Many families were at the quayside, the women in their long skirts and high collars trying to hold back their tears, their children waving, little hands receding with the shoreline. The men waved back, too, some of them, for most of Columbus's crew were from around these parts, and many did not know when, if ever, they would return to those who now bid them farewell. Then they went down to row, for the sails were limp, and we were counting on the ebb tide to take us downriver. I had no one waving me goodbye, but I stood on the deck, hearing the liturgy of the friars of La Rábida on our port side, chanting to the greater glory of their unforgiving God, as the men on deck knelt and crossed themselves in supplication. I was still there as we went over the bar, watching the coast of that young—and to me quite new—continent disappear, and I thought of my own departure from Kozhikode, silent like a pariah passing a sleeping Brahmin.

We had set sail at eight o'clock, an auspicious time by my calculations, just before the day entered the last four-fifths of the Panchami and the moon tenanted thirteen degrees of longitude in Sagittarius, which would have required, if you believed at all in my astrological accuracy, a series of rituals to propitiate the planets—rituals I am sure Columbus would not have had time for. So we cast off in time, and there was a strong breeze to sea, which took us southwards as the sun rose, a scarlet stain across a glowering sky.

The breeze took with it too another ship, not of our fleet, the last of many that were all supposed to have left by the previous day, carrying away from Spain the Jews who had failed to undergo the Catholic baptism decreed for them by Columbus's blessed Sovereigns. I had seen these people, kin to the Yehudis of Cranganore who had found safety in my homeland after an earlier European persecution (fleeing Romans then as they fled Christians now); I had seen them clogging the roads around Palos with their pathetic drooping donkeys and their overladen carts, choking the air with their chants of lamentation, stifling the sea-breeze with desperate sobs. I had seen them everywhere that Spanish summer, bundling their possessions on to carts, on to their backs, into their hearts, condemned to stop being what they were or stop being where they were, obliged to either betray their beliefs or abandon their homes. They were in our way at the port, screaming and wailing as they climbed pushed stumbled on board the ships that would take them away from the only homes they had ever known for generations. They suffered terribly, those Jews, suffered not just their immediate loss but even more the future loss of what they would never have again, their own country, the land to which they had given so much. I wanted to say to them, go in peace, my brothers and sisters, do not weep. I wanted to urge them to sail, sail away from this land where people think there is only one Truth (which they know), sail far without fear and without tears, for this Europe is a close-minded continent, full of intolerance and oppression, a place worth leaving behind. I wanted to say to them, there is a world beyond where men can be men whatever the colour of their skin or the shape of their nose, where you can grow your beards or put caps on your head or worship in your own way without anyone asking you to explain or defend yourselves. I know that world exists because I am from it, it is my world, it is India. Sail with me and I shall take you there.

But I could not, for it was not my place to say this, nor my hope to offer; and so on the same tide, sailing down the Rio Saltes, overladen with suffering humanity, the last Jew-ship left Spain, and I could hear the sad sounds of their mournful pipe and tabor fading as the Jews drifted away from us, towards the Levant, the Spanish sun rising behind them for the last time.

For ourselves, we headed southwest, towards our first stop, the islands lying on the latitude of the Indias, what these people called the Canaries. *Yellow birds*, I thought reflexively, but no, it was dogs these Latins had named the islands for, howling wild creatures they had usurped the land from. One day these Europeans will want to do the same to us, I knew, and they will treat us as dogs too, caging our treasures like yellow birds.

But the waters were calmer than my mind, and as our little fleet sailed placidly on, I went to look for our leader, Colon, the man they call Christophorus Columbus. I found him in good spirits, pacing the deck with a proprietary air, his ear cocked to an inaudible muse in the foam of the sea, his own personal Venus. 'Aha!' he exclaimed. 'El Indio! Is it not magnificent?'

I nodded, because agreement was clearly expected, and he smiled self-satisfiedly, looking out to sea, turning his face to the spray. 'I have been writing,' he declared, with that air he had of imparting information as if he were sharing a rare privilege. 'I am writing a journal.' He was enormously proud of this; it was as if the act of writing itself validated, in anticipation, the momentous importance of the voyage he was undertaking. He intended to present his journal to his generous Sovereigns, who in their infinite wisdom— he was careful not to mention cupidity—had made the voyage of Christopher Columbus possible.

I watched him strut about the deck, barking an order here, asking a question there, filled with the optimism of one who had finally been given the chance to be proved right, and wondered again, for the thousandth time, why he had embarked on this madness. There was, of course, this thing that men call glory, but that was surely not the most important thing to him: I did not believe Columbus would care very much whether his name lived on or not in the minds of men, whether it was whispered in the hush when men told great stories of adventure and heroism and the deaths of kings. No, for him the immediate was what mattered; personal gain, here and now, certainly meant more to him, for he lived in the present, tasted its disappointments like sour berries, and yearned for the sweetness of success on his palate today, not at the next meal. I would never understand his fondness for titles, something we never took seriously in my country: 'Don', now that he had been ennobled just in time for the voyage, and that ridiculous 'High Admiral of the Ocean Sea', which he had invented for himself and wanted everyone to use.

'Colon,' I said to him, 'I mean, Don Christophorus, what on earth do these words mean? It is like calling yourself Main Emperor of the Earthly Lands, or Crowned King of the Soil on the Ground: either you're an Admiral or you're not, and the rest is irrelevant.' But there was something about Columbus that needed the reassurance of those unnecessary words, and High Admiral of the Ocean Sea he remained.

He had convinced himself, had Columbus, that his precious Sovereigns would make him viceroy in perpetuity of all the lands he was going to find on his way to India. And that after him, oh he insisted on this, his eldest son would succeed to the same position, and so on and on; he was meticulously insistent, Columbus savouring the ascent of generations yet unborn. A dynasty of Columbuses to rule the waves, all by dispensation of a pair of monarchs who did not even know where he was going, and who kept their promises the way a sultan kept his wives, to be fulfilled or ignored as it pleased the monarch's fancy. Oh Columbus, I thought, but this time I did not say it, you are mad, you are naïve, you are addled deranged foolish demented crazy.

And he was, this great sailor and explorer, this High Admiral of the Ocean Sea, with his reddish cheeks and his far-seeing stare, he really was two lateens short of a topmast. He thought he was going to convert all of India to his holy faith; now that all the Jews had been driven out of Spain, Columbus wanted to build on this great Christian achievement by winning more magnificent victories for the Blessed Cause.

'My mission,' he told me, with that messianic glint in his cold blue eyes that must have frightened Isabella at first, 'my mission is to Christianize the Indians, bring enlightenment to the pagan.'

I tried to tell him we already had Christians in Kerala, converted by old Doubting Thomas the Apostle himself, Jesus's disciple and traveller for the faith, well before Columbus's own ancestors. 'Malayalis,' I said, 'were worshipping the cross when your so-called civilized Europeans were still feeding Christians to the lions.'

He looked at me, almost cross-eyed in disbelief. Then he found his refutation in my own idolatry: if Keralites were Christians, why wasn't I one?

I explained to him, patiently, that *some* Keralites were Christians, but most of us, perfectly satisfied with the ancient faith into which we had been born, were not.

'Aha!' he exclaimed, 'that proves it: they are not pagans, but heathens, which is worse, because pagans have never had the opportunity to see the light, whereas heathens have had the Truth offered to them and have spurned it to worship their false gods. The great and noble task remains to be accomplished, in the name of our glorious Sovereigns Ferdinand and Isabella, blessed be their thrones, princes devoted to the holy Christian faith and enemies of all idolatries and heresies...' I stopped listening to his benedictions. 'Accomplished, that is, by me.'

It was useless to argue with Columbus, because arguing with him was like pissing overboard in a high wind: it just blew back in your own face. Since the last time I raised matters of religion with him, explaining how in my country there are many ways of raising your hands to the stars, he began to threaten to instruct me in his one true faith. Me! Heir to four thousand years of belief and speculation, traveller across worlds, cynic and betrayer, I had found and lost more faiths than he had heard of.

'To speak of one true faith, Don Christophorus,' I had pointed out, 'is like speaking of one true way to get out of bed, or one true way to wipe your behind. The divine Creator has made it possible for men to reach out to him in a hundred, nay a thousand, different ways.'

But none of this would Columbus hear, and despite myself I envied him his fierce certitudes. This was a man who was often in error but never in doubt.

There were two more days of uneventful sailing, with the sun rising and the seas falling, and the seas rising and the sun setting, with the monotony that all sailors know, heading south by southwest in calm waters, the ships sighing across the waves. Columbus marked progress in his journal and gleamed: fifteen leagues on Friday, more than forty on Sunday, all as if it were going to make any difference. Only I knew, with a quiet despair I felt every time I looked across our mast to the sun sinking in the horizon, that we would never get to India this way.

For his calculations were all wrong. And he didn't like being told they were. It was soon after we met, when I had allowed the wine in my belly and the conviction in his voice to lead me to see him again. Out of Columbus's saddlebags had tumbled maps, books, sheets of paper. There were seven years of letters to and from some Italian called Toscanelli, who had sent him a map by Henricus Martellus showing the width of the ocean between the Spains and India. Or rather China, because it was the distance between Lisbon and Quinsay on the Cathay coast that Columbus had first worked out, figuring twenty-six spaces on the map. Since that was too far to landfall, he'd changed tack: it was Cipangu he was aiming for now, some 30 degrees east of China. Though I had never seen a map like this, this Cipangu struck me as oddly familiar: as I had first suspected, it had to be the island country that Indian Buddhist missionaries had taken their religion to, a cold and warlike place called Nihon, hardly the fabled land of gold and spices Columbus thought he was setting out for. I said nothing of this, of course: it was not in my interest to have him discover that, given the 'India' he was aiming for, I was not going to be the great asset he expected me to be on this journey. After all, I had to get out of Europe, and what did I care if it was China I reached at the end of the journey? There were ways to get to India from there, and with fewer obstacles than I would face heading east. No, if Columbus was going to sail, I wanted to be on his ship—provided, of course, he knew what he was doing.

So I said nothing about Cipangu, which he seemed to think of as an outlying Indian province, one of what he called 'the Indias'. What I did say, though, was altogether more practical, and quite unwelcome.

'Your figures don't make sense,' I ventured.

'By San Fernando, what do you mean?' he demanded redly. 'Look—it's quite plain. I have calculated the distance to India in degrees, based on the works of Martellus, of Marinus of Tyre, d'Ailly's *Imago Mundi*, and above all, infallibly, Pope Pius's *Historia Rerum Ubique Gestarum*. I have correlated all this to what is known from the accounts of Ser Marco Polo—look, here is his book. I have consulted the eminent Toscanelli, who has corroborated my conclusions. Now, when so many sources agree, there's really no room for dispute.' The pitch of his voice was high, but a note of authority insinuated itself into his tone: this was not the first time he was debating the point, nor, I knew and he knew, would it be the last.

He spread his hands in a gesture of patient accommodation. 'We agree that the earth, like any globe, is made up of 360 degrees. Marinus found that the known world worked out to 225 degrees; Marco Polo discovered another 28, and we know there are 30 degrees from the Cathay coast to Cipangu. That gives us 283 degrees, leaving 77 degrees unaccounted for between Cipangu and Europe. I propose to stop in the Canary Islands—9 degrees west of Europe—to reprovision myself. So when I restart my voyage, I will have just 68 degrees to traverse to reach Cipangu.'

'But...' I began.

'In fact, it's even better than that,' said Columbus. 'All the authorities agree that Marinus's degree is about 10 per cent too big. So instead of 68 degrees from the Canaries to Cipangu, it's really only about 60. Now the next question is, how long, in miles, is a degree? There your Moorish friend Alfragan has the answer: 56 and two-thirds miles, but this, of course, is on land, Roman miles, which work out on sea to 45 nautical miles.'

'Not really. You see...'

'I know what you're going to say. You're going to say that measurement applies only on the equator. Correct: if we sailed close to the equator, each degree would amount to 45 nautical miles. But we will sail at a latitude of about 28 degrees north, where the degrees are narrower—about 40 nautical miles. Which is even better. All we will have to do to arrive in Cipangu from the Canaries is to sail 60 degrees by 40 nautical miles, a small matter of 2,400 miles, not quite 750 leagues. Plain sailing.' He looked up, as if expecting applause.

'Not quite,' I interjected at last. 'First, I don't know how you can be so sure about how many degrees of land there are from here to China and this Cipangu of yours. The distance may be less, and across water it may be greater. But more important, Alfragan used Arabic miles, so when he spoke of 56 miles to a degree, it's really 66 nautical miles, not 45.'

This was clearly a new concept to Columbus, but he didn't even blink. 'I don't agree,' he said calmly. 'Why should Alfragan have used Arabic miles?'

'Because,' I pointed out not unreasonably, 'he's an Arab.'

'So what? Everyone uses Roman miles. Why wouldn't he have done the same? No, my dear Indian friend, your objection will not do. I will not listen to it. I have spent too many years already putting up with illiterate objections, the disbelief and the mockery of too many. Do you know what some foolish people have claimed as an excuse to turn down my request for support? They say the true distance is closer to 10,000 nautical miles, that my way will take us fourteen weeks on the high seas. Fourteen weeks!' He laughed sardonically, and I found I could not join in, because that figure was closer to my fears than Columbus's widely optimistic and inaccurate calculation. 'I'll be in India in three weeks. At most, four.'

In Cipangu, you mean, I thought. *If* you're right. Which you almost certainly are not, if you're proceeding from these figures. But if you are, and your madness might well be divinely blessed, I'll be closer to home. Somehow, some day, I'll make my way back to Kozhikode from there, more easily than I can from Spain.

He looked at me, watching me think, and waited for me to speak.

'Unless you run into some land on the way,' I suggested lamely. I had to say something.

'No chance of that,' Columbus laughed. 'Some little islands, perhaps. But even Aristotle opined that water runs from Pole to Pole between Spain and India.'

'Who is this Aristotle?' I asked innocently. 'A sailor?'

Columbus gave me an odd look. 'Aristotle,' he said, 'was the greatest of Greek philosophers.'

'Ah,' I nodded. 'In India, philosophers do not claim to know much about land or water. They confine themselves to the realm of the spirit.'

My barb had no effect. This man, this Christophorus Columbus, had a greater capacity for self-deception than anyone I'd ever met. I'd have to be wary of him. Let him fool himself if he wants to, I told myself, let him think he's fooled me; but let me not lose sight, as he seems all too ready to do, of terra firma.

I began, despite my better judgement, to plan for the voyage, to be friend philosopher guide translator interpreter slave to a madman with sails. As their Saint Augustine, a man I have read and have much respect for, put it: 'Melius est dubitare de occultis, quam litigare de incertis'—better to doubt the obscure than dispute the uncertain. So I tied my hopes, and my plans, to Columbus's. Until he walked into the tavern one day and threw his saddlebags on the floor at my feet.

'It's all over,' he said, 'before it has even begun.'

Columbus's enthusiasm, his optimistic figures, his messianic faith in himself, had failed to persuade the queen, the very queen who had sent him money for a mule and a fresh set of clothes to enable him to present himself to her. Her Committee of Experts—the third such committee to consider Columbus's ideas—had met at St Stephen's College in Salamanca, interviewed him, and advised her that the enterprise was unsound and unnecessary.

'These people of St Stephen's, may God take them,' Columbus swore, using the only oath in which he allowed himself to utter the name of the Lord in vain. 'They think they know everything.'

The fact was, they did; the committee's head, Friar Hernando de Talavera, was the queen's confessor, and he had gathered around him at St Stephen's College the foremost mathematicians, geographers and astronomers of Castile, who had unanimously decided that Columbus's numbers did not add up. So had I, though unlike Talavera's learned Stephanian professors, mine was the lesser college of experience. The ocean between what Columbus called the Spains and the Indias simply could not be as narrow as he thought it was. And yet....

Columbus was bitter. He had tried so hard and so long for Spain's backing. As he walked in the victory procession through Granada, with the royal banners of Ferdinand and Isabella flying silken-proud from the towers of the Alhambra, he had felt that his moment had come at last; that in this hour of triumph the King and Queen of all the Spains would not deny him the trifling sum of two million maravedis needed to equip him to find a new route to India. But they had, and his hopes lay crushed like groundnut shells at their feet.

'Never again,' he swore darkly, 'never again, by San Fernando, shall I set foot in Spain.' His disappointment smouldered in him like black coals, casting smokeless shadows on his soul. He was running out of wealthy monarchs to turn to. Portugal's Dom João II had lost interest when Bartholomew Dias sailed to the Cape of Good Hope; *there* was a way to India that seemed to make more sense than sailing westwards. England's Henry VII, to whom Columbus had sent his brother Bartholomew three years earlier, showed no interest at all.

'The British are so parsimonious they would rather let someone else find the route and then follow,' Columbus cursed, spitting out the words as he downed a draught of the tavern's most poisonous wine. 'For this reason they will never extend their rule beyond their own shores.'

That left only Charles VIII in France, and that is where Columbus wanted to go next to try his luck. Except that his brother Bartholomew had been there nearly two years already, with little to show for it. I looked at Columbus dubiously, seeing my hopes of escape sink into his worn boots.

'You are welcome to come with me to France,' he said to me in farewell, 'but you know I cannot afford to pay for you on the way.'

'That's all right,' I assured him, careful not to reveal how little I thought of his chances of success there. 'When your expedition is financed, send word and I shall come and join you there. I am hoping by then to have saved a little money. The queen's escribano de ración has taken some interest in my astrological skills.'

Columbus frowned his disapproval of my pagan practices. 'Luis de Santangel?' he asked speculatively. 'He is a good man. He seemed quite...sympathetic to my plans.'

'Yes,' I nodded, 'I believe he is.' And a plan began to form in my own mind.

Columbus set out on the mule his queen had paid for, his saddlebags bursting with the collective output of Martellus and Marinus, Seneca and Aeneas Sylvius, a fat friar by his side as his only companion. It was a long way to France. Suddenly I knew they would never get there.

I rushed to my new patron, Luis de Santangel, the man who controlled the royal purse. The day I cast his horoscope I had told him of his last three illnesses, described the state of his relations with his wife and portrayed his essential character in the most flattering terms.

'Fair master,' I announced breathlessly, 'I have seen the most remarkable sign in your astrological charts.'

'Indeed?' The keeper of the monarchs' money was no less immune to the appeal of the stars than the poorest Kozhikode fisherman.

'You have entered a most propitious period,' I told him, unfurling the parchment on which I had plotted the planets of his birth. 'One of great good fortune.' I could see the coins practically clanging together in Santangel's eyes. 'Look at the horoscope I have cast for you: your second, fourth, ninth and eleventh houses are all strong, and—this is what makes it special—the twelfth house is not stronger than the first or the eleventh, which would have given you great income but, alas, even greater expenditure.'

The escribano de ración looked relieved, and slightly amused. 'Tell me more, my learned friend. Where does this great good fortune I am about to get come from?'

'Here,' I said, stabbing emphatically at the rectangle I had drawn. 'In the eighth house. The eighth house,' I added, my voice dropping to a conspiratorial whisper, 'is the house of death. An inheritance, perhaps? A legacy? It is also the house of buried treasure, of good fortune rising from the bowels of the earth. Silver mines? Gold? Master, you have recently undertaken such ventures?'

Santangel shook his head, and the ghost of a smile haunted his avid face. 'I am too old to be expecting any inheritances,' he said. 'My parents, God rest their souls, passed away a long time ago, and I have not so much as a maiden aunt from whom to expect a legacy. What else does your eighth house portend?'

'A voyage, perhaps?' I offered tentatively. 'An expedition? Figuratively, shall we say, to the bowels of the earth? Our ancient astrological commentator, Rudra, wrote that fortune from the eighth house would often be fortune acquired through others, including through supporting the labour of others. Perhaps there is such a voyage you have financed, fair master?'

'No,' said Santangel, but I could see in his eyes that he had already thought of a voyage he might have financed but had failed to. He was looking at my face, at my fingers spreading the astrological charts in front of him, but he was seeing Columbus, saddling his mule in some way station on the dusty road to France. 'But go on.'

'Surely you must have, my exalted eminence,' I said. 'Look again: your fourth house is strong, indicating large properties, but your eleventh is also strong, suggesting recurring income. And then—look at your eighth house—this income will come not necessarily from those properties, but from others, perhaps islands far away, acquired for you, at your behest, by someone else. Someone financed, perhaps, by you? Answerable to you? Could there be such a man?'

'There could,' Luis de Santangel said, rising from his chair with a glitter in his eyes. 'My foreign friend, you do not know whereof you have spoken.' He clapped his hands for a servant. 'I only pray it is not too late.'

The queen's messengers caught up with Columbus at the village of Pinos Puente, just 10 miles from Granada; disappointment and overburdened saddlebags must have weighed him down, for he had not made much progress on the route. Her Catholic Majesty Isabella had changed her mind, for Luis de Santangel had offered to finance the expedition himself.

I did not tell Columbus about the part I had played in his destiny. This was not a man who liked the thought of being in anyone's debt. But as I watched him, smiling and content on the deck of the *Santa María*, his prematurely white hair blown about by the breeze, I felt a surge of almost paternal pride. In my own small way, I too had left a fingerprint on the pages of history.

'I have work for you, Indio,' Columbus said cheerfully. 'As I write my journal, so must you write.' He took me down to his cabin and carefully unwrapped an oilskin package. 'The royal letters of credence,' he said. 'In triplicate.' He opened a copy, and I read in the flowing script much favoured by the calligraphers of Aragon:

> *To the most serene prince [this space was left blank] our very dear friend, Ferdinand and Isabella, King and Queen of Castile, Aragon, Leon, etc., greetings and increase of good fortune. We have learned with joy of your esteem and high regard for us and our nation and of your great eagerness to receive information concerning our successes. Wherefore we have resolved to dispatch our noble captain Christophorus Colon to you, with letters, from which you may learn of our good health and prosperity....*

'Who on earth is this for?' I asked in disbelief. 'Great eagerness to receive information concerning our successes?' What makes your precious Sovereigns think there's any serene prince anywhere who's dying for news of their victories?'

'By San Fernando, you have an insolent tongue in your head, Indio,' responded Columbus, reddening. 'I did not ask you to question the Sovereigns. I merely want you to translate their Letter of Credence into the Indian language, that's all.'

'There are many Indian languages, Don Christophorus,' I replied.

'Well, translate it into all of them,' Columbus retorted shortly. 'These are the credentials I must present when we arrive in India.'

'And the blank space? Should it be addressed "To Whom it May Concern"?'

'May God take you, Indio. The space is to be filled in by Escobedo, secretary of our fleet, depending on whose domains we land in. The letters are intended to recommend me to all the lords and kings of India whose lands I will claim in the name of our blessed Sovereigns. The original is already addressed to Magno Cano, the Grand Khan.'

'There is no Grand Khan, Columbus,' I said wearily. 'I've tried to tell you since Granada. The Chinese haven't had a Grand Khan since before I was born.'

But Columbus would not listen. He had spent too much of his life trying to get others to listen to him; his own ears were closed, from habit, to anything he did not wish to hear.

And so I sat down, translating the letter into Malayalam and Arabic. I was writing a presumptuous document in an incomprehensible language, destined to be read by princes who did not exist, in a country we would never reach. But at least I was sailing away from Spain, though none of us on the *Niña*, the *Pinta* or the *Santa María* had the slightest idea where we were going.

~

DESOLATION, LUST

UPAMANYU CHATTERJEE

The slow train stopped yet again. Atri had read in a plaintive article somewhere that in the suburbs, many commuters, when near their houses, pulled the alarm chains to stop the trains. Then they scuttled across the tracks to their hovels in grey concrete blocks. Beyond the train window, two feet away, stood a sheet of corrugated iron. A hole had been torn in it. The back of a mirror at the base of the hole, around the mirror a man's sagging etiolated chest, with nipples dark pink, like suppurative insect bites. A bit of his right arm moved rhythmically beside the mirror. He was shaving. Here by the tracks, sharing an iron box with the rats of Bombay, surrounded by defecation, it was absurd, it seemed to Atri, that this man was shaving. With his smooth cheeks and chin, he must keep up appearances; Atri half-smiled at that thought. Once that man must've been new to the city, long before it made him its denizen. Work might've wrenched him away from home too, and dumped him here, one more inhabitant of one more city of encroachments and unauthorized constructions, where his glimpses of the sun were dependent on his shaving mirror and the unscheduled halts of trains. The train started again, but the desolation remained.

Atri felt completely empty and remote. He looked down at the dried mudprints of slippers on the compartment floor. This arrival in an alien unknown place was inevitable, like birth or death, this weaning away from home, inescapable, a second severed umbilical cord. He leaned his head against the window stains and watched the passing poles and wires and blocks of concrete. There is not a blade of grass in Bombay, Anand had said. Except the scrub rising out of the excrement by the tracks. God, the agonies one suffered just to stay sad and half-alive. Again Atri half-smiled at his self-pity.

Across from him, Sheela uncrossed her legs and continued to stare out. Atri looked at her gross, seductive face. Abruptly she said, 'Pultu, you should know something.' Atri hated his pet name but Sheela thought it cute. 'I might be pregnant.' The train was slowing down again. She looked at him a little tensely, as though prepared for almost any reaction.

How can anyone called Pultu make a girl pregnant, thought Atri fleetingly. 'You're sure?'

'I think so. I'm always regular, now I'm ten days overdue.'

Like a library book, thought Atri's disordered mind. They watched the train move into the pandemonium of Bombay Central. Hot coolies with large sweat patches and loud voices surged through the train. Atri and Sheela got off. He saw the queues at the ticket counter and the anxious unseeing railway-station faces and still wanted a ticket home. Outside they battled the rabble of hotel- and taxi-pimps. In the cab Atri said loudly, to beat the million horns, 'Have you done anything about this?'

'I thought you should know first.' Atri looked at her, her profile, at the dandruff in her hair. He patted Sheela's thigh and said, 'I'm sure everything'll be okay.' She nodded. The taxi turned into a gate. 'That wasn't long,' said Atri.

The Training Institute was shabby and one-storeyed. There was green slime on some of the many damp patches on the walls and an iron grille instead of a door into the lobby. A faint smell of dead rats and smegma which, in later years, Atri would always associate with Bombay. Just behind the grille sat a thin, genial man in an office chair, happily and loudly sucking in his tea from a saucer. 'I'll need help with my luggage,' said Atri.

'Oye Chhotu!' yelled the man into the corridor behind the water cooler. He turned back to Atri. 'You've come for the training sir?'

Atri and the taxi man took his luggage off. Sheela remained in the back seat. 'Don't know how long settling in will take. I'll come over as soon as I'm free. Should I telephone first?'

'It's just here.' Sheela gestured over her head. 'On Marine Drive itself. A lovely flat, you can almost walk it from here.' Atri looked at her wide purple lips and heavy jowls. They nodded and smiled awkwardly at each other. The taxi left. Atri remembered the June afternoon when he had suggested that they bathe together. Sheela had been shy and reluctant beneath her flippancy. Her banter had irritated him. Even when he had desired her he had felt alone.

Chhotu was sitting on Atri's blue suitcase. He was a small powerful man. 'Where's the office?' asked Atri. In reply Chhotu pushed his chin out towards the grille. He, the thin man and Atri shared the luggage. The thin man talked a lot. The hostel room was instantly depressing. A broken cement floor, with small craters. Atri stumbled over one. Old pink walls, two windows. One looked out on to a small field where some urchins played at cricket with a stick and an indistinguishable grey blob for a ball. Behind the field the rest of the universe was stopped by the nether floors of a skyscraper. Countless air-conditioners and some jagged holes in windowpanes. Beyond the other window Atri saw a bare narrow patch, a wire fence, then some sweating men in loincloths. 'Who are they?'

'The Youth Club, sir. People come for exercise,' said the thin man.

'This room's dirty. Can you arrange for a sweeper?'

Atri followed the thin man out and asked for the office. The open corridor looked on to a big cement courtyard. On the cement were the fading lines of a badminton court. The office lay beyond the courtyard. Its first room had an old desk and behind it a short, fat, bald, bespectacled and extremely unhelpful man. His breath seemed somehow to smell of farts. Atri said, 'What are the joining formalities? I have to sign in.'

Looking down on the bald head with its strands sprouting up above the left ear and stretching heroically across to the right, Atri wondered at himself. Here, in a strange building, in an unknown city, alone, signing up for a job I'm not interested in, having made a girl pregnant, what am I doing? He wanted to peel off the bald man's strands and let them hang beside his ear. They could reach, thought Atri, the pens in the shirt pocket, or they could gag his mouth and lessen its stench. In this daze Atri met the Director of the Institute and then returned to his room.

While settling in he thought disconnectedly of home and Sheela. In his dislocation he remembered the afternoon when, to lunch-hour music from the radio, he had taught Sheela how to inhale smoke. Earlier she used to puff away at cigarettes without inhaling. How dizzy and sick she had felt.

At lunch Atri met a Mrs Karve, the other trainee. 'Normally a batch of three or four, but this year you are small, just the two of you,' smiled the

Director. Mrs Karve was from some obscure Maharashtrian town. She had two children. Atri was immediately prepared to be disinterested in her. She was short and a little squat. Mr Karve was something in the State Bank.

In the late afternoon, before leaving to meet Sheela, Atri wrote a letter home. For a moment, he was tempted to write, 'Ma, Sheela is pregnant. Maybe she's lying to get me to marry her.' Sheela was staying with her sister and brother-in-law in a huge old flat on Marine Drive. According to Sheela her brother-in-law had often been offered unimaginable prices for it. She used to talk often of her holidays in Bombay.

Her sister Alka seemed friendly. 'I can't believe this is your first visit to Bombay. I mean, *everyone* has been to Bombay.'

'He's absolutely rural, yaar,' laughed Sheela. Atri was surprised at what he took to be her false gaiety. Compared to her sister's, Sheela's voice was even more loud and jarring.

Alka left for the kitchen to organize coffee. Atri moved out to the verandah. Sheela followed. The bright sun on a sea of rippled glass, the sea endless and oblivious, and below the distant toys bustling on Marine Drive. 'Your sister's very pretty,' said Atri.

'It runs in the family,' smiled Sheela.

On the breeze drifted in the nauseating smell of the sea. At last Atri said, 'You have to get an abortion, no?'

'Let's discuss that later.'

Atri was surprised. He moved closer, but she pulled away. 'We've never discussed marriage but this shouldn't force us into it.'

'Later.' Sheela was loud all evening. She played raucously with her nephew, an eight-month-old bore who resembled Winston Churchill. He urinated on her and she pressed her face into his stomach. Atri was by turns bored, irritated and worried. 'How dull you are not to like children,' said Sheela. Atri wondered whether she had told her sister about her pregnancy. Satish the brother-in-law arrived, short, soft and fat, something in some ad agency. His spectacles were tinted and he looked like a maker of hard porn. He fixed Atri a whisky. They pulled chairs out to the verandah and watched Sheela change the piddler's nappy. 'You married?' asked Satish.

'No.'

Satish nodded several times. Atri stayed to dinner. While Sheela distracted the child, he helped Alka lay the table. He noticed the thin wrinkled skin on the back of Alka's hands, it looked almost brittle. He wouldn't have associated that skin with the moist adhesive heat of Bombay.

'Sheela's good with the child.'

'Yeah, she's too stupid to be anything more in life than an ayah,' laughed Alka. The two sisters got on well and seemed to like each other.

Sheela drove him back in the Maruti. 'You drive very badly. You're really rash and blind,' said Atri.

'But I love driving. With my father I never get to touch the car. Here with my sister I can do what I like and Satish is very nice. He taught me driving when I was here four years ago. My parents are boring, yaar, compared to my sister.' Sheela talked on, unnecessarily. Atri had often told her that she spoke so loudly because she was a little deaf, and that was because she always listened to her cassettes on the headphones of her Walkman. At the pizza parlour near her hostel, she had as usual closed her eyes and begun nodding her head to the monotonous rock music. Then Micky had picked up the Walkman and whispered into it. Her beatific smiles had then turned genuine. Later Micky had said, 'I said some sexy stuff into that stupid machine. She wants it bad. I'll hump her, you joker, before you even get close enough to smell her skin.'

'When should we discuss the pregnancy?' Atri interrupted Sheela.

She said nothing, but took a right turn at great speed. Atri also kept quiet. His sense of desolation made him passive, quiescent in front of any outrage or complication that might involve him.

At the gate of the Training Institute Sheela asked, 'Coming home tomorrow?'

Atri poked his head in through the window. 'What about your pregnancy?' The word sounded alien.

'So?' Sheela's tone was aggressive, though she was looking away.

'We've to get a pregnancy test. Do you know anyone in Bombay?'

'No.'

'Should we tell your sister?'

'No.'

Atri got back into the car, reluctantly prepared for a long discussion. 'We'll have to return to Delhi as soon as we can.' Any reason, he thought, to get home.

Sheela said nothing. The silence was faintly hostile. Atri returned to his room and wrote a letter to a friend in Delhi. 'Anand, an emergency. Send me a telegram saying, Ma in hospital after a heart attack. Come. Sign it Papa. No jokes. Send it immediately, and Lightning. I'll explain why when we meet. Otherwise things are okay. This place promises to be really dull, maybe Madras and Jabalpur'll be better. I'm training with a female called Karve. But send the telegram.'

Atri went out right then, at midnight, to post the letter at the box by the gate. His letter of appointment to the job had said unequivocally that leave during training was impossible.

The cement courtyard was brightly moonlit. The surrounding corridors were black and faintly menacing. Atri could hear the distant eternal noise of the city. The corridors and courtyard reminded him of some ancient ancestral house, perhaps in Lucknow, where a huge, unmanageable, impoverished family would have lived and quarrelled together.

The fake telegram was an old and trusted ploy, but Atri wasn't particularly interested in its success. Nothing seemed to matter. He returned to his room, switched off the light and got into bed. With the darkness came desire. Once, just seven months ago, on an impulse, he had sent a telegram to Sheela at her hostel. The telegram had merely read, 'I lust you very much'. To Atri the act had been exhilarating. Sheela had been both touched and embarrassed. For in the hostel all telegrams were put up on the Mess Notice Board, open, their secrets unprotected. Everyone had read his message. It had then seemed silly.

He tried to explain to Sheela once how he felt, that love was a euphemism, that what everyone felt and eulogized was actually lust and the sense of possession. There were these phases in one's life, late teenage, or the middle twenties, when one expected of oneself to fall in love, when one was expected to be seen with a girl, attending seminars, catching buses, seeing movies in large outwardly happy groups. But lust was different, and not as transient. It was a biological necessity, like a part of one's metabolism. Then the personality and wishes of the object of lust didn't matter, and lust was somehow linked with the loneliness, which was infinitely more fundamental and long-lasting. Sheela had been offended, hadn't shown it, but had scoffed at his theorizing. She had said, 'You think it's very, what's the word, macho? to say you feel lust and no love and all that rubbish, because admitting to love you think will be sentimental.' Atri had had no wish to explain further because he had a core he found impossible to unveil. But he had disliked her for her assurance and the confidence with which she claimed him. I'm precious too, he had wanted to tell her, and I want to be alone and free; he had seen her as a threat to his dreams of dispossession.

He woke to the clink of metal. The room looked less ugly in the soft early-morning light. On the wall a patch of yellow showed up where the caked pink limewash had fallen off. The patch was shaped vaguely like Australia. Atri watched it unthinkingly.

The sounds were from the Youth Club beyond the window. After a while someone hammered at the door. Tea. Atri went over to the window. The tea was hot, sweet and sickening. A few feet away stood a sweating torso on pale, trembling pencil legs. With his mind elsewhere Atri watched the disproportionate figure tug at a barbell.

At Lucknow station, over a year ago, Atri had seen Sheela perfectly, it seemed to him, the oval head jigging with laughter at Micky's joke, surrounded by suitcases, leaning against a pillar proclaiming a new Hindi film. Dipping his stale rusk in his tea, with the roar of the chai wala's gas stove in his ears, Atri had remembered then that Micky had said, she's from Calcutta, yaar, all these Cal females are as loose as pyjamas, you can make her anytime. In the following months the confused Atri had often felt that if Sheela had been more affected, less open and friendly somehow, then he would have apologized for the grossness of his emotions.

At breakfast Atri sat beside Mrs Karve. Her sari was an eye-blinding green. 'It's sad that there are only two of us. If there had been more, then we could've written letters in class without their ever noticing us.'

Mrs Karve smiled politely. 'You're from Delhi?' Her Maharashtrian accent was strong.

'Yes.' Pause. 'How old are your children?'

'Dattatreya is five and Dinkar will be three this month. I am going to miss his birthday.'

'Tch tch.' They laughed at that.

The introductory lectures were excruciating. Before leaving for Bombay, Atri had heard his father yet again on marriage and loneliness. 'I didn't want to marry your mother, Pultu, you know that. On the day of the marriage, in the taxi in Egmore, but you have no idea of Madras, we were going to the Registry Office, I said to her in the taxi, we shouldn't marry because I'm not the right person to marry. That wasn't an original line, or even an original thought, it must've been said a million times before, but I meant it. She didn't like it, of course, how could she. Our marriage hasn't been very happy. I'm still lonely, I suppose so is she.'

After the afternoon session Atri automatically got into a taxi to go to Alka's flat. He didn't particularly want to meet Sheela, but he felt an obscure obligation. He again settled down in an armchair and watched her with her nephew who again wet himself. 'Does he have diabetes, you think? The amount he piddles.' The sisters laughed.

Later he asked Sheela for a walk down Marine Drive. 'Forget it, yaar,' she said, 'feeling too lazy.' Atri was irritated and said, 'I've written to Anand to send a telegram to me. As soon as it comes we should go back to Delhi and work out your pregnancy.' He immediately thought, I should've said the pregnancy, or perhaps our pregnancy.

Sheela said nothing and cooed to the child. Atri went out to the verandah to look at the sea, immense and impersonal. Once he had been excited about Bombay. He had thought of nights with Sheela away from her confining hostel and its Auschwitzean rules. But here with her sister and nephew she was another person. Strange. He had considered himself a successful strategist when she used to lie to her warden and come away with him. Now those once-exquisite assignations in borrowed rooms, the lying together with minds at rest, and gauging the passage of the hours by the glow of the sun on the pink bougainvillea by the window, all that seemed remote, sometimes a little sordid.

Sheela came out. Atri said, 'As soon as I get the telegram I'll ask the Director for leave. Then you say you forgot to submit your exam forms. If you like I'll telephone you from somewhere, and you pretend it's someone from Delhi, a classmate or something, say Raman, who telephoned about the exam forms. And the last date is imminent etcetera so you have to return.' Atri stopped listlessly. 'Why didn't you tell me this in Delhi, before we left?'

'I didn't want to spoil anything. I was scared.'

Atri didn't want to accuse or to hurt, not just then anyway. But he did want to escape the fetters of guilt. Yet away from her there seemed to be only desolation and lust; and with her there wasn't much else. He had always enjoyed being alone, but it seemed to him that Sheela had brought him a kind of emptiness. It couldn't be true, yet the moments of joy now seemed so remote. He picked up a cushion from the verandah floor and threw it on to the rocking chair. The chair began to move. To its creak, Sheela said, 'I'm still scared because I don't know what's going to happen.'

'You were very regular, no?'

'As regular as the bloody moon.'

Atri smiled at her reply and turned to her. But Sheela, with eyes closed, had lifted up her face to the warm sea breeze. He spoke to her profile. 'Maybe it's just delayed or something. But we'll feel more secure in Delhi, won't we? We know people there, we know...the city. Here,' he gestured helplessly at the skyline. Sheela was in her tight orange corduroys. He felt faintly the stirrings of desire. In school they had been made to study a poem called 'My Last Duchess' by Robert Browning. The poem had been about a Duke who kills his wife, or something like that. Atri had liked one bit:

Oh sir, she smiled, no doubt,
Whene'er I passed her; but who passed without
Much the same smile?

In school the lines had seemed to apply to his mother. Later Atri saw Sheela
also as absolutely undiscriminating in her friendships; 'at least holding hands
with anything in pants', Micky had once said. Atri had thought that his mother
and Sheela would hit it off together. After her first visit, Atri's mother had
summed up Sheela as cheap and vulgar. 'Lust,' she had said, clinking her bangles
while she poured out the tea, 'is making you see her as good company. She
has a good body, and a good skin, and a vulgar, attractive face. You two have
nothing in common. You're not serious about her, I hope?'

'I've told you so often, Ma, I only want to marry a divorced Muslim
transvestite whore.'

'She's probably that too.'

Atri's mother was intermittently inquisitive about Atri. She used to snoop
around in his room, and had once unearthed a book called *Your Contraceptions.*
As a joke Atri had bought it for Sheela and had written in it, 'In the years
to come, my lust, you are not going to say that I did not have foresight.' His
mother had said that the book confirmed her opinion of Sheela.

The shriek of slammed brakes drifted up from Marine Drive. From their
verandah, as from an eyrie, they watched the distant tableau. A white Fiat askew,
the disturbed traffic, the crowd forming, and in the background, as always,
the hush of the uncaring sea. 'We always used condoms. I mean I always did.
How did it happen?'

'One must have had a hole in it. Remember that joke about a bastard
perforating condoms in revenge.'

'Is it an accident or isn't it?' From that height they couldn't quite make out.

'Which do you mean?' They both laughed, a little startled by the unexpected
joy.

The cars began queuing. At the corner, the cleverer drivers, the old hands
who knew the traffic snarls of Bombay, were turning off Marine Drive into
the street with the ice-cream shop Sheela had raved about the night before.
She said, 'How's training?'

'As boring. I sat in class and dreamed about my PhD at MIT.'

'How are the others?' Sheela interrupted him. She had often suffered his
dreams of studying at MIT.

'There's only one other. She's okay. Prabha Karve. A mother of many children with unpronounceable names. She's short and has tufts of black hair around her midriff. Not just down, but thick wavy hair.'

'Again you judge by appearances. I've often told you, behind my sexy face lies a heart of gold.'

'Unlikely.' But Sheela did not laugh at that. She looked down at the traffic. Atri felt foolish and continued speaking to cloak his awkwardness. 'Mrs Karve's even been to Jabalpur. She said it's quite a nice place. I said I was petrified of going there for the training later. One more small town in this vast Indian hinterland. Just Karve and me, it's going to be really boring and lonely.'

'You can always make her pregnant.'

'Bitch.' They laughed.

'You'll get leave?'

'I'll lie, shouldn't be difficult.' Atri paused and moved closer to Sheela. He looked back into the room. Alka wasn't there. He bent down and gently kissed Sheela's shoulder. She shifted a little. 'I like your smell,' he said.

Alka appeared with coffee. They spoke of the weather and the traffic jam. 'How's the training?' she asked Atri.

'Teething troubles, that's this child's problem,' laughed Sheela. 'Can't grow into his new job. He once dreamed of a PhD and now can't wake up. But I think he will soon, poor boy, misses his mama's lap.'

Two days later the telegram arrived while Atri was in a lecture, absentmindedly eyeing Mrs Karve's midriff. It read, 'Ma in uppal with fart attack. Come.' The vulpine Director asked Atri to ring up Delhi from his office. 'We have an STD, go ahead.'

Silently cursing the grace with which the Director cross-checked the telegram, Atri dialled Alka's house, and then dialled three digits at random after the phone began to ring. 'Amazing,' Atri beamed at the Director, his fist holding the receiver, 'got through to Delhi first try.' A woman's voice answered. Atri hoped that it wasn't Alka and began in a high loud voice.

'Hallo, is that Delhi? Hallo? Is that Papa? Papa, Atri here.'

'What rubbish—'

'Hallo, Papa, I just got the telegram—'

'Oh hi Raman! Hi! What, from Delhi? How did you get my number?'

Atri turned his back to the Director to hide his suppressed giggles. But he couldn't contain the heaving of his stomach. A warmth at Sheela's quick uptake suffused him. 'How serious is it?'

'Oh yes, obviously. You always ring up for some boring reason or other.'

'Oh poor Ma, I'll get leave from here, I'll catch the first available train home.'

'Ohhhh! Of course, how horrible, ohhh now what do I do, huh, Raman, please?'

'No, don't bother about the station, I'll just take a taxi or something.'

'Yes, that's the only thing, but how disgusting of me to forget. Yes, I suppose that'll be the thing to do.'

'Don't worry, Papa, I'm coming.'

'Thanks a lot for calling Raman, you're a real friend, and just tell that bastard Maheshwari I'll get him the forms in time.'

'Okay, Papa, bye, and see you very soon.'

'Yes, I'll do just that, thanks once again, bye, be seeing you soon.'

Atri turned to the Director and said, 'A heart attack, sir. My mother's back home now, but she wants to meet me very much. Thank you, sir, for the phone.'

With one wave, the Director sent Atri's gratitude to the ceiling. 'You should leave as quickly as you can.'

'Thank you, sir. I'll go and check about the tickets.'

Again the taxi for Sheela's. Atri was galled even by this ritual, this meaningless daily seven rupees. The taxi stopped at a light near some massive telecommunications tower. He remembered, ages ago it seemed, the first film that he and Sheela had seen together, an insane Hindi extravaganza in a huge new hall in Lucknow. They had enjoyed the trash so much just because they had been together, knee touching knee, fingertips occasionally on forearm. She had laughed helplessly when he had breathed into her ear that the vamp in the film looked amazingly like her. Atri remembered fragments of their common past in abrupt flashes, and most of these memories confused him, because their magic seemed irretrievable.

Sheela opened the cream door. Her face glowed with conspiracy. 'I've to return to Delhi immediately. I forgot to fill up those bloody examination forms.' She stank of tobacco and her perfume. He had told her often how much he liked that smell. 'It's much simpler than bathing, yaar,' she had once said.

'You got leave? Talk,' she whispered, and gently patted his stomach.

'Be careful. There's a baby in there.'

'Bastard.' They laughed. Again between them a flicker.

'This stupid girl,' said Alka over tea, 'takes off on a holiday without bothering to find out whether she has to do anything important first.'

'Absolutely right,' said Atri.

'It's just an exam form,' laughed Sheela, 'I can always stay here and fill up my forms next year.'

'I might be going too,' said Atri. 'Three or four holidays coming up, some Marathi hero died or was born or something.' He had always concocted with ease. 'We could go together, but how do we get train tickets?'

'Oh Satish will get that done.'

The globular, lemon-green shade that hung from the ceiling danced in the salt breeze from the windows. 'I think we should leave tomorrow,' said Sheela, 'we'll never get tickets for tomorrow.' Again Atri registered her loud voice. His Second Standard class teacher had always said in class, 'Learn to speak softly. That is good manners.' Atri suddenly wanted to say so to Sheela, to jolt her out of her happiness. Then Churchill started shrieking from the bedroom. 'Time for the bastard's feed,' said Sheela.

Satish proved inquisitive. 'Which Marathi hero? I don't think there is any holiday this week.'

How do people ever plan things like murders with people like Satish around, thought Atri irrationally, and said, 'The Institute is closed for two or three days. Thought it must be a Marathi hero. Never bothered to find out why, just in case it was a mistake and they opened up again.'

A valedictory dinner that night. They all drank a little and laughed a lot. Atri saw Sheela's head jerk back. When she laughed, when her lips split and her belly shook, the sound followed later and sometimes never. She doesn't know the implications of pregnancy, thought Atri, the child.

Two mornings later, again the nondescript ugliness of Bombay Central. Atri watched the insane bustle on the local trains on another platform. The human mass at the mouths of coaches resembled fish eggs. Small dark agile men leaped into it with cane baskets and disappeared, like frogs into mud. Their train glided out on time. They waved to Alka and the child. Atri again noticed how much prettier than Sheela she was and felt vaguely guilty. From the window, passing glimpses of Bombay's suburban stations, with their odd attractive names; each name had character, with its strong regional tang. There seemed a clot in his stomach, like a lump of blood, at the thought of returning home. When he had arrived in Bombay he had felt that home had been lost forever. Now even though he was returning, he still thought it inaccessible; it had now withered into the destination of clandestine visits that had been arranged with endless petty lies, a place of abortions. Just for three or four days; he knew that the ecstasy of arrival would not compensate for the emptiness of departure.

Atri looked out at the ugliness, the absolute desolation of the city. Ten-storeyed grey box flats with slime climbing up the pipes that took the shit away, box verandahs of bored lonely women and men blankly surveying the

deformed world while waiting all day for TV to start. He looked across at
Sheela, at her oddly magnetic face, now a little unsure. 'Come and sit here.' She
did. They were alone in their compartment. 'How should we begin in Delhi?'

'I don't know.'

'First, by getting into bed.'

She tried to smile. 'I've no idea what's going to happen.'

Atri could say nothing. He put his arm around her. In Lucknow, at their
last parting, she had cried much, silently. He had been surprised and guilty.
'We'll hunt around, tomorrow itself.'

'Bombay was so nice.'

'You mustn't worry. This abortion thing will get over really fast.'

Sheela looked up into Atri's face. 'Yesterday I caught a local train for Bandra
to meet an old friend, Anjali, have you ever met her, she's pretty boring. In
the train, just above the last row of seats, there was a small board, an ad for
a clinic, it said abortions for only seventy.'

'But we couldn't've gone to some unknown dirty place.' There would've
been grime and guilt perhaps, in a small, harshly-lit waiting room, the lovelessness
in the faces of other almost-fathers by error, and a dusty telephone that didn't
work. Outside the world would have continued to move like the sea at Bombay,
an unending purposeless ebb and flow, totally heedless of the anguish of men.
'There are also some Marie Stopes Clinics.'

'Who was Marie Stopes?'

'I don't know, some do-gooder.'

Their conversation was staccato throughout the journey. The expressions
of tenderness and concern were awkward, almost false. The silences seemed
eternal, the depression insidious. When they caught themselves looking at each
other, their mouths would contort, as though bitter with remembrances. The
hours moved, the landscape changed, and the skies, Surat, Baroda, Ratlam, Kota,
fleeting fugitive visions of the hinterland, and for Atri the cloud heaved a little
only when he saw the polluted morning sky in the near distance over Delhi.

They went to the hostel first. Even Sheela looked gladder in the familiar
surroundings. 'Ugly as ever,' she said happily.

'I'll look in at about five.' Atri went home. His parents were surprised
and pleased. 'Some Marathi hero was born or died or something. Just a few
days off. I couldn't get through on the phone from Bombay, horrible lines, so
I just turned up.' Almost immediately he telephoned Anand at his office. His
mother grumbled, 'Not even half an hour at home and you want to meet
your friends. Especially that Anand.'

'I'll be back for lunch, Ma.'

'I'm honoured.'

Atri took the car keys from the table in his father's room. He felt happy to be driving again on Delhi's wide roads. Delhi was not particularly lovable, but it was familiar. And for him happiness had always been fragile and unexpected, even unsettling, like a time bomb. Birth, education, a job, marriage, the petty adjustments of marriage, a home of one's own, the breeding of children—he found the conventions of this circle meaningless, and the obtuseness of the world that arranged this insane succession galled him. He had grown up in one spot, had grown accustomed to it, like a finger to stroking a scar, and then had been wrenched away to an alien place. Rootlessness was agonizing enough within, yet the conventions of existence externalized it, and buttressed the agony with objective symbols for the eye.

Anand was an architect. His office was the size of the lavatory of a shabby restaurant. Five young architects sat in that small room. Their tables were six-inch ledges of some black synthetic thing. Everything was built in or recessed, 'maximum use of space'. Each architect earned vast sums.

'Let's go out for coffee somewhere.' The cafe they chose smelt of chhole and sour milk. Grease and forearms had dulled the scarlet decolam of the table.

'So you couldn't bear your training or what?'

'Your telegram read "Ma in uppal with fart attack". I hope that was unintentional.'

Anand laughed. 'It was a phonogram. Be happy it reached.'

'Sheela's pregnant, she thinks.' Anand's smile disappeared. 'We didn't know where to go in Bombay. You once said you knew some decent places.'

'That's bad.' Anand looked away for a while, out to the hive of offices across the dry fountain. 'You don't want to marry now?'

'No.'

'I know one good place. Costs about fifteen hundred, all told, air-conditioned and so on. You want to go tomorrow? I'll ring her up from office, for, what, eleven tomorrow morning?'

'What's her name?'

'Dr Singh. I call her Aarti. She's quite sexy. Has degrees from some funny-sounding place in Europe.'

'Anyone you know has been to her?'

'Ha ha.'

'Fifteen hundred is quite a lot.'

'You must pay for your sins.'

The flashy waiter brought coffee. Anand said, 'I feel sorry for Sheela, though I never really liked her. She must be having a bad time, tension etcetera.'

'Yet when you first met her,' Atri smiled, 'you'd hoped I wasn't serious about her. You'd said, remember that late-late-night coffee at the Emperor Oak, that "she's just not in our class, boy. It would be like marrying a servant, you know, someone one loses one's virginity to".'

'Yes, and you'd said, "male or female servant?"' They laughed. They were conscious of the baseness of their remarks, but also that there was no one else to hear them. Anand continued, 'She'd do very well for Micky.'

'Because you dislike him too.' They laughed again. Anand was an old friend.

Atri continued, 'Micky's very violent and…aggressive…in…expressions of love. In Lucknow sometimes he'd dash to some bush just to pluck a flower for her. An exhibitionist.' Which Sheela had seemed to like, he thought.

Anand lit a cigarette. 'Sheela's very young. And for most Indian girls, sex is not casual. Did you mislead her perhaps,' he looked away at the man behind the coffee percolator. 'Perhaps she assumed, or expected, marriage.'

'I don't know. We were always flippant.' Except that time in Lucknow, when she had wept. Most of the time Sheela had been so happy and somehow appealing, that Atri had sometimes wanted to cause pain, subtle and complex, to bewilder her so that she could reach, for a while, his emptiness.

Anand exhaled luxuriously. 'But good you aren't marrying. Take away that body, and she's quite thick.'

Anand is saying what Ma said, thought Atri, what an epitaph for Sheela, now she seemed to belong to the dead. He lit a cigarette, though he rarely smoked.

10.50 next morning. The opulence of the clinic was intimidating. It was in one of South Delhi's new mansions, vast and vulgar. Unsure of themselves, Sheela and Atri walked from the car close together, subtly touching each other in what would once have been erotic, back of hand against outside of thigh, breast against triceps. A large unweeded garden, a carved wooden door with a one-foot Khajuraho woman for a handle, the chill of effective air-conditioning, a cold hard woman at a white desk, Atri inhaled deeply to down the bobbing nervousness.

'We have an appointment with Dr Singh.'

The woman pushed her huge spectacles up and consulted two diaries. Sheela vanished through another door. Atri mumbled 'Good luck,' but she didn't acknowledge him in any way, perhaps she hadn't heard. She was in her bright yellow salwar-kameez. Leafing through glossies on the coffee table

Atri waited and wondered. Couldn't they tell from the urine these days that a girl was pregnant? He and Sheela were so ignorant. Then his mind wandered, pregnant women actually piddled out babies the size of cells, and pregnancy could be checked by examining piss under a microscope for these babies, who, wallowing happily in maternal urine but stifled under glass, waved at the monstrously distorted human eye above them. With each pregnancy actually a million babies were born but they all died when their heads exploded against the porcelain of the commode. The one who survived hung on to the moist walls of the womb with its claws.

Sheela emerged, expressionless. Atri stood up, but she didn't look at anybody. She said something to the cold hard woman, then they moved out, stopped outside at the Khajuraho handle. 'Well, what happened?' But he wasn't really interested. He wanted to say, maybe Dr Singh cooed to you and skilfully slipped in two fingers and pulled out a foetus with Micky's mind and Satish's face.

'We'll have to come back tomorrow to meet Dr Singh, she's only in in the mornings, but the report itself'll be ready by about four this evening.' That seemed to leave Atri in limbo. 'Perhaps we should go and have coffee somewhere,' he suggested, 'and get into bed or something, I don't know.' Sheela said nothing. While driving he looked at her often. Cocooned in himself, he had never considered her loneliness or her wishes. 'You like babies a lot, no? All that cooing and nephew's piddle. You'd've liked this child?' Sheela continued to look away. 'But marriage wouldn't've been possible, we'd've ended up like my parents.' He stopped, out of a kind of boredom.

After a while Sheela said, 'Do they, do your parents know about all this?'

'No, of course not.'

'I want to tell them. That you made me pregnant and that you want me to abort.'

'Okay, sure. And remember to tell yours too.' Atri parked badly under the derisive eye of the car park attendant. The cafe seemed to be full of schoolgirls. 'Your telling them will hurt them, that's all, perhaps confuse them a little.'

'So?' Sheela lit a cigarette.

'Remember how I taught you to inhale?' She exhaled into his face. He felt faintly aroused, and continued, 'Didn't someone say that abortion was the fruit of lust. If we had been in love, then we'd've had the baby. And if we hadn't lusted then we wouldn't have slept together.' She closed her eyes and leaned back, her hair disturbing the leaves of some potted shrub behind her. She stretched out her legs and continued to exhale in the direction of his face. Her cold face could've been tired, or disinterested; or a mask hiding her cold, controlled dislike. Atri again saw her perfectly, like he had once at Lucknow station, the brown throat, the firm, almost hard, upper arms. Ideally he wanted a sort of djinn or familiar, who would appear when required, in the guise that Atri desired, glut each of his sexual caprices, and then leave him absolutely alone; he would never impose himself on Atri; when not needed, he wouldn't even exist. Wanting to hurt and be hurt, Atri picked up Sheela's leg by her calf, pushed off her sandal, pinched her skin hard through her salwar. He rested her foot on his chair, pressed his crotch tense against her heel, his knees stubborn against her leg, and bent down and kissed, and bit, her unclean toes. Only such an act could silence the schoolgirls at the next table. Eyes still closed, but Sheela was smiling.

Of course they left instantly. They were laughing excitedly on the way to the car. Atri stopped for bananas. 'Here, have some phallic symbols.' They giggled and gobbled two each. This was an amazing way to end, thought Atri, of course it was ending, no more threats to those dreams of dispossession. In the car he licked the salt off Sheela's neck and felt her warm stomach beneath the kameez. 'Shall we name the foetus before they kill it? Let's call it Hawas, you know, Lust.'

Sheela said, 'If I'd kept my thighs crossed, I'd've had you begging for marriage from day one.'

'Why, I'll still beg in front of you, not for anything, just whimper and beg.'

'Then beg in front of me while I tell your parents about the mess we're in.' But this kind of conversation always embarrassed Sheela.

'Why don't we go to Anand's flat? Now we don't even need condoms.' Atri got out of the car to look for a phone. One chemist lied and said that his phone wasn't working, one huge general store simply refused. Atri succeeded at a bookshop run by a mother and her daughter. 'Anand, speaking from Connaught. Went there this morning. They'll give us the report at four or so. I say, is the key to your flat still under that brass thing?'

'You'll never learn.'

'Now we don't even need condoms.'

'My sister's been staying with me for weeks. She left her Hyderabad job and is now looking for another.'

'God, she can have mine anytime.'

'She might be out, in which case she'll have taken the key.'

'Any other place you can think of?'

'Your ugly Ambassador, you monkey.'

'Tch, bye now. I'll ring up again.'

'Good luck.'

When he returned to Sheela, she had changed. Again a wan dying face, eyes crinkling with cigarette smoke. 'You're not on heat any more?'

'I want to go back to the hostel.'

'Okay.' Atri moved the car out. Even the demands of lust are enervating, he thought; he would make no claims beyond them. 'When shall we go for the report? Tomorrow morning?'

'Okay.'

Atri returned home. 'See, I told you I'd be back for lunch,' he told his mother. At lunch his father asked, 'How's the job?'

'Dull and meaningless, but I expected nothing else.'

'It's important to like your job. If you really don't like this you should think again about your PhD.'

'Yes, let's see.' After lunch Atri and his father played two games of chess. Then Atri wandered from room to room, opening cupboards and unlocking memories; a bewildering afternoon in which he often thought that his past had been happy, and then wondered whether he thought so just because the past was irretrievable. In the evening he watched TV with his parents. He looked at them steadily gazing at the TV without registering any of its belches, and wondered about their youth. Perhaps they had been like him, perhaps he would finish up like them, in front of a TV, and at the point of his surrender to that uncaring sea, its hush would become a roar.

At the clinic the next morning the report said that Sheela was not pregnant. Dr Singh turned out to be fortyish, aggressive and revolting. 'Definitely not a pregnancy. Your period could be delayed for a variety of reasons. Mental tension or some illness. Get yourself thoroughly checked up. No, payment outside, please.'

Outside the clinic they paused uncertainly beneath a jacaranda in bloom. 'So are you ill, or what're you tense about?' Atri smiled.

Sheela bit her lip. 'It's really odd. I suppose this test was foolproof. Maybe I should really get that check-up.' In the car she said, 'I feel weird. All that tension for nothing.'

'Coffee somewhere I think. But not yesterday's cafe perhaps.'

Sheela laughed. 'No, let's go there.' They moved off. Atri was content to let things be. She repeated, half to herself, 'It's strange.' They chose the same table. He asked, 'Do you want to go to Bombay again?' She looked away. 'No, that would be very funny. I just want to sort of stay here and relax.' They had very little to say to each other, there seemed no cause for celebration.

After coffee they had nowhere to go so he dropped her back at the hostel. 'I'll probably go back tomorrow, I'll write from Bombay.'

'Stay a day or two more,' Sheela asked unconvincingly.

'No, we'd better leave each other alone for while, to recover from this one.' She laughed, waved and went lightly up the stairs.

On the train to Bombay, returning to the unknown city and an empty future, Atri was almost nauseous with depression. He curled in his berth and was assaulted by savage masochistic images of, among others, Mrs Karve. He wanted to be beaten senseless, to be deprived of his mind so that he didn't have to think.

Eight empty days in Bombay, the city of dead rats and smegma; on a few evenings rummy with Mrs Karve, on a few others watching the patch of Australia on the wall from his bed and absorbing the grunts and clinks from the Youth Club next door. Then, for their next phase of training, Atri and Mrs Karve moved to Madras. Madras was humid and alien-tongued. The new Training Institute there was on the edge of the city, close to the beach, girdled by fishermen. A small clean featureless room, 'maximum use of space', the same classes, in the evenings slow walks in the compound, watching disinterested gardeners trying to create a lawn out of rubble. Atri contracted dengue. The fever ravaged him for seven days. While Mrs Karve attended class, he lay alone watching the sanatorium-white ceiling, juggling with bittersweet images of his past. In the evenings Mrs Karve brought company and chatter, because she had nowhere to go. The fever procured an insidious intimacy between them. They listened to the delightful hyperbole of Urdu ghazals on her recorder and watched the evening pass the window. The songs spoke of loss and despair, the best of them creating a fleeting grace out of these ashes. *When I first saw you / the world moved on but I halted / and I haven't moved on since then, / waiting, still waiting / for you to move with me.* At moments, the music and Mrs Karve combined to stir him gently, like the memory, or the shadow, of another desire.

At the end of that week, they had some days off. Mr Karve emerged out of the hinterland, dull and good looking. Mrs Karve disappeared with him. One Sunday at twilight they took Atri to the beach in a noisy, smelling taxi. Madras, his parents had married here, he remembered. He sat timid and bent on a bench, like an invalid, knees together. The sea breeze revived his feverishness, the late-evening sky was a mess of black and gold. Behind him the Karves joined the babble of children around the kulfi wala. Here the sea was louder than at Bombay; it did not whisper, but roared its unconcern. In his tiredness, Atri half-smiled and prayed for the decline of all desire.

thirty-five

~

KAMA

VIKRAM CHANDRA

That summer I was heartbroken. I was weary of myself, of the endless details like prickly heat, and the smell of hopelessness in my armpits. It was after all very boring, nothing but something that someone else and I had thought would go on forever, and i had come apart savagely and with finality. It seemed so ordinary, so average in its particulars that I found it sordid to think about, and yet I could do nothing but think about it. I knew I was supposed to drink it away, but liquor just made me even more tired and sleep eluded me anyway. I went slouching sullenly about the city and waited for the monsoon to break, without faith, without belief in its powers, waiting only for something to change.

One evening they were talking of a murder. I say 'they' because lately I had been slumped over in my chair for weeks, silent but always nervous, shifting from one side to another incessantly. Subramanian had been watching me all this time. Now I was very interested in the details of the murder. I wanted to know how they had been killed It was a husband and wife and they had been found bloodied in their apartment in Colaba. The papers were full of it. The fact that there were no signs of a struggle gave me a particular satisfaction. nodded rapidly. The others watched me, uneasy.

'No great mystery there,' I said. 'It must have been love. Sex, you know.'

'Or gold,' Desai said. 'Property. It says that the police are questioning the servants.'

'Something like that,' I said. 'Simple and stupid.'

'Or the most complicated thing of all,' Subramaniam said suddenly.

'What's that?'

'Don't you know?'

He was smiling gently. I collapsed suddenly. I must have been insufferable, and they had been very patient and very kind.

'No, sir,' I said. 'Why don't you tell me?'

He laughed, his shoulders shaking. Then his face became serious, and he looked at me for a long time. He nodded with that peculiar motion of his, from side to side.

'All right,' he said. 'Listen.'

The body was almost submerged in the ditch, but what Sartaj noticed about it as he squatted beside it was its expression of pride. One arm was curved out of the water and rigid. The passer-by who had found the body was a driver on his way to the milk-stand for his memsahib's bottles, and he had seen through the rain a hand, reaching up out of the rushing stream as if for something. It had been raining for three days and three nights and through the morning now, and the water was actually roaring as it pushed below a culvert. The dead man was jammed in between the curving brick wall and the broken metal grill. The driver had stood next to the ditch and shouted until people in the nearby buildings had come out, and then he had stood guard until the police came. He seemed to think that somehow it was his responsibility since he had found it, but now he was trying not to look down at it.

The skin on the palm of the hand that emerged from the water was a strange bluish-grey.

Sartaj Singh, who was an inspector in Zone 13 and used to bodies, was squatting carefully next to the ditch, looking at the ground, but there was running water everywhere and it was likely that the body had drifted. He walked the few yards to the next culvert, feeling his boots sink into the mud. A gust of wind blew water into his face. The flashbulbs were freezing the drops in the air in their sudden glare. It was the first heavy rain of the monsoon, and he knew that the next few weeks would be miserable with mud, with clumsy raincoats and flooded streets, clothes that always seemed wet, and the impossibility of keeping a crease in one's pants. Anyway there was nothing to be found. The photographers had finished.

'All right,' Sartaj said. 'Get him out.'

They put a crowbar against the grill and pulled at the dead man's arm, his shoulders. Finally one of the constables, whose name was Katekar, shrugged, took off his boots, and got into the water. It beat against his waist and chest as he strained and finally the body came free. The driver gasped as they dragged the body out and over because under the chest, on the right side of the belly, the flesh had been eaten away. The rats had been at him before the water took him and covered him over. But the face was unconcerned and smug.

'Turn around,' Sartaj said to the driver. 'What's your name?'

'Raju.'

'Raju, take a look at him quickly. Do you know him?'

Finally Sartaj had to take Raju by the elbow and steady him. 'Look,' Sartaj said. Raju was in his early twenties and Sartaj knew that he had never imagined his own death. Now in the early morning he was looking at a corpse. He shook his head so violently that Sartaj felt the jerks in his arm. 'All right. Go over there and sit down. We'll need your statement.'

'Robbed, sir,' Katekar said as he pulled on his boots. He pointed with his chin at the white band of skin around the dead man's wrist, which showed up clearly even after his colour had drained away into the water. 'Must have been a good watch.'

'It was a big one anyway,' Sartaj said. 'Big enough to get him killed.'

He had been dead at least eight and not more than twelve hours when he was found. The cause of death was a single stab wound, under the sternum, only an inch and a quarter long but deep. The blade had pierced his heart.

'What was taken?' Parulkar said. Parulkar was Sartaj's boss. He had been promoted up to deputy commissioner from the Maharashtra police service, and still lived in Ghatkopar in what he called his ancestral abode.

'Wallet.' Sartaj said. 'And a watch.'

'Ah, I see,' Parulkar said. 'Bombay was never like this.'

Sartaj shrugged. 'It's a new world. He was fifty or thereabouts, no distinguishing marks.'

He looked up and Parulkar was smiling. Sartaj had been cleaning his boots with a moist rag, scrubbing away the mud caked around the tread and the ankles.

'It'll get on again as soon as you go out,' Parulkar said.

Sartaj nodded. 'Yes, sir. But the point is to keep trying?'

'Of course, of course,' Parulkar said, standing up and hitching his pants over his considerable belly. His uniform was always bunched up somehow, looking as if it had been made for someone else. 'Young fellows must be tip-top. Carry on, carry on.' But he was still smiling as he walked out of the room.

Sartaj stood up and walked across the room to a map of the state. In the glass, over the dark borders and the blue roads, he could see himself, and he checked his shoulders, the tuck of his shirt, the crispness of his crease. Now when the moisture hung in the air it was difficult to come close to the perfection he wanted to see in the glass, but he patted his turban and ran a finger over his sagging collar. He did not mind Parulkar's smile at all, because he was a dandy who came from a long line of dandies. His father had retired as a senior inspector in Zone 2, and every street urchin had recognized the swagger stick with the shining steel tips and the gleaming black boots. Sartaj's grandfather's upturned moustaches had been acknowledged as the most magnificent in all of Punjab, and he had died in service as a daroga, in a gun battle with Afghan smugglers near Peshawar. The legend went that when he was hit he was eating a dusseri mango. He sat down, not far from a babul bush, finished the mango, crossed his legs, held out his hand for a napkin that his seniormost havaldar was holding for him, wiped his fingers, dabbed at his mouth, twirled his moustaches, and died.

Sartaj had never been able to eat a mango without thinking of the old man, whom he had met only through the garlanded portrait that hung in his mother's house. Next to that picture was a portrait of Guru Nanak, another one of Guru Gobind Singh, and then one of Sartaj's father, who had made it to retirement and had passed one night in his sleep, resting on his back with his hands folded neatly on his chest. Sartaj was eating a mango now, holding a slice with his fingertips as he leafed through the reports from the Missing Persons Bureau with his other hand, stacking the probables to the left, face down, and the rejects to the right. He was looking mainly for the age, but also for the kind of man who would have wanted what the dead man wanted. He had it down to fifteen when the phone buzzed angrily. In the quietness after the rain it was very loud.

'Are you still in the office, you sad man?' a boy's voice said.

'Yes, I am,' Sartaj said.

'Doing what?'

'Eating the last alphonso mango of the season.'

'You should go home.'

'You're still up. You must be very happy or very sad.' It was Rahul, his wife's younger brother, who was now in his second year at Xavier's and therefore always falling in love with someone or out of it.

'Happy, actually,' Rahul said quietly. 'I bought a new shirt at Benetton today.' They had a mutual interest in clothes, although they mystified each other with their choices. They talked for a while about this shirt, and then suddenly Rahul said, 'When's the exam tomorrow?' Sartaj flipped over another report as Rahul talked nonsense about college. What that meant was that someone had come into the room, and Rahul was pretending that he was talking to a college friend. Finally Rahul said, 'See you at college tomorrow. Go to sleep soon. Night,' and hung up.

Ten minutes later the phone rang again. 'Hello, Sartaj,' his mother said. 'I just called home and of course you weren't there.' She lived alone in Poona, with a rose garden and one ageing Alsatian. When Sartaj's father had been alive, they called every Sunday, but now she allowed herself a daily call.

'Peri pauna, Ma,' Sartaj said.

'Jite raho, beta,' she said. 'Did you find a cook?'

'No, not yet, I've been busy.' Which, of course, was no excuse. Sartaj held the phone loosely against his ear, and turned pages, and his mother spoke at length about bad diets and nutrition. He could see clearly the sofa she was sitting on, the little table next to it, her small feet which he had just touched in devotion, her hands with which she had blessed him, and the sari wrapped around her plump shoulders, and the garlanded pictures on the wall.

'It's too late, Sartaj,' she said finally. 'Go home and rest.'

'Yes, Ma,' Sartaj said. But he stayed for another two hours before he walked home. Even then he walked slowly, stopping sometimes to watch the water as it roiled around the gutters and made whirlpools. He leaned against a wall and scanned the layered many-coloured mess of movie posters and political slogans, dominated by the latest broadside bearing the crossed spears of a right political party. He read all this with the concentration of an archaeologist smoothing away layers of ancient dust. What he was avoiding was the small bundle of foolscap paper that sat on his dining table, wrapped in a white ribbon. Rahul's sister had sent him these papers, and he couldn't bring himself yet to say the word for what she wanted. But he was supposed to be distinct now from her and her family, disengaged. He had been told that they considered him dead. Which was why Rahul made phone calls late at night: maybe that's when you talk to the dead.

Sartaj found the next of kin, whose name was Smt Asha Patel ('wife of missing person'), but not on the batch of missing persons reports that he had. It was in another stack that came three days later from the Missing Persons Bureau. The name of the dead man was Chetanbhai Ghanshyam Patel and the age was right, but what clinched it of course was the entry under distinguishing marks and features: 'Wearing a gold Rolex watch, value Rs 218,000.' Chetanbhai was then a man who liked people to know the make of watch that he was wearing and its exact and precisely calculated value. Sartaj did the missing property paperwork (in triplicate as required), and out of habit made some phone calls to his usual official and non-official sources, although he had no hope that he would ever see this grand keeper of time, but in this he was wrong because that same afternoon there was a phone call from the station at Bandra. Two of their constables, at seven that morning, had picked up one Shanker Ghorpade, a known bad character, beggar, suspected pilferer, and drunkard. The suspect had been observed in the very early morning hours to be staggering proudly through the bazaar at Linking Road, wearing an ornamental timepiece clearly beyond his means and needs. Since he was unable to provide a satisfactory explanation, they had brought him in, and had already been commended for their alertness.

The station house was hardly alert in the sudden, sultry heat of the afternoon when Sartaj held the watch up to the light. It was indeed a Rolex, large and heavy and very yellow, with a pleasing glistening feel under the thumb. Moitra, the Bandra inspector who had pulled it out of her desk, was leaning back and rubbing her eyes with the palms of her hands.

'Big fucker,' Moitra said. 'It should tell more than the time.'

Sartaj was turning it over and then over again. 'Like what?'

'I don't know,' Moitra said. 'Moon phases. The time in Tokyo. Whatever shit people think they need to know.'

'We all want to know the time in Tokyo. Come on. Let's see him.'

In the detection room Ghorpade groaned as he came through the door, hunched over and shuffling.

'Have you been third-degreeing him?' Sartaj said.

'For what?' Moitra said. 'Why to waste energy? He's a fucking bewda. By tonight, for a drink, he'll confess to killing Rajiv Gandhi. And haven't you received the memorandum from up high? No third degree, ask them questions with love and caring and tenderness.'

She laughed. It looked to Sartaj that the confession might actually come sooner than nightfall, judging from the trembling of Ghorpade's hands. 'Sit down,' he said flatly. It was his interrogation voice. He knew his head was leaning forward from his shoulders, and that his eyes had become opaque.

'Have fun,' Moitra said as she left. For a long moment afterwards Sartaj could hear her whistling down the corridor.

'We found the man, Ghorpade,' Sartaj said.

'What man?'

'The one you knifed.'

'I'm a bewda. I don't kill anyone. I just wore his watch.'

Sartaj had to lean closer to him to hear the words in the voice full of phlegm. Ghorpade had a small, lined face, dry lips, and days of grey stubble on his cheeks. He stank of sweat and monsoon damp.

'Why did he let you take it?' Sartaj said.

'He was lying down.'

'On his back?'

'No. Face down. So I took it.'

'Did you know he was dead?'

Ghorpade looked up with yellowed eyes. He shrugged.

'Was he in the gutter?'

'Yes.'

'Was it full of water?'

'No. It was just starting to rain. There was just a trickle.'

'Was there any blood?'

'No.'

'None?'

'No.'

'What time was it?'

Again Ghorpade shrugged.

'Did you look for a wallet?'

'It wasn't there.'

'Why didn't you sell the watch?'

'I was going to. A little later.'

'Later till what?'

'I just wanted to wear it for a while.' Ghorpade wrapped his arms around himself. 'It was a good gold watch.'

'Do I look like a fool to you?'

Ghorpade shook his head, slowly.

'But I must look like a fool to you. Otherwise why would you be telling me this fool's story?'

Ghorpade was quiet. His ruin seemed complete.

'All right,' Sartaj said. 'I'll be back to talk to you more. You think about what you've done. About this children's story you're telling me.'

Ghorpade was absolutely still, with his head lowered.

'I'll be back,' Sartaj said. He was almost at the door when Ghorpade spoke, and the words were indistinct, and his face was turned away. Sartaj took the three steps back and leaned forward, narrowing his eyes and blinking against the smell that hung over Ghorpade. 'What did you say?'

Ghorpade turned his face close to Sartaj, and Sartaj saw that he was really not very old, young perhaps, in his thirties. 'I'll be dead,' Ghorpade said.

'Nobody's going to kill you.'

'I'll die,' Ghorpade said. There was no fear in his voice. It was a statement of fact, and it required no sympathy in response, or any other kind of emotion. Sartaj turned and walked away.

The door to Chetanbhai Ghanshyam Patel's sixth-floor apartment was made of dark wood inlaid with criss-crossing copper bands and raised ivory studs, with Chetanbhai's name in gold at the centre. It was a door that belonged in some last-century haveli, with an elephant parked outside and durwans in safas. Now a window opened at the middle of it and a young face peered out through the bars.

'Yes?'

'Police,' Sartaj said. The boy's eyes took in Sartaj, and the bulk of Katekar's shoulders behind him. Sartaj was watching him carefully. This was something that Parulkar had taught him: go to their homes, watch their fear, and you will learn everything.

The door opened and Sartaj stepped in. 'Regarding the matter of one missing person Chetanbhai Patel... Your good name?'

'I'm his son. Kshitij Patel.' He was about nineteen, a little shaky.

'Who else is in the house?'

'My mother. She is sleeping, not well. She was very worried. The doctor has given her some medicine.'

Sartaj nodded and walked past him. The drawing room was large by Bombay standards, and cluttered with brass lamps and furniture and many-coloured hangings on the wall. The sofas were huge and an alarming red. On the wall to the left there was a long painting of a brilliant sunrise and another of a sad shepherd. Against the back wall there was an Apsara pouring water. Sartaj walked over to it and saw that she was almost life-size, with deep round breasts and huge eyes. She was all white, plaster, and quite startling to find in an apartment in the Narayan Housing Colony, far north and west of Andheri West.

Kshitij was watching him, and Sartaj felt the edge of his resentment without surprise. He used his ability to stalk into people's lives as another tool. What they felt about him was usually instructive.

'Please come to the morgue with us,' Sartaj said.

In the long moment then he saw recognition, regret, the usual struggle for control, and then Kshitij said, 'Yes.' But he did not move.

'Do you want to put on some shoes?' Sartaj said. He followed Kshitij into his room, which was shocking in its austerity after the gaudy brilliance of the rest of the house.

There was a shelf stacked neatly with books, a desk, a bed, and a calendar with a goddess on it. There was a window that opened out onto an expanse of swampy vegetation. It had begun to drizzle again. 'Is there somebody to take care of your mother?'

Kshitij looked up from his laces, startled. He blinked twice and then said, 'I will tell my neighbour.' Sartaj noted his thick black Bata shoes, his well-worn white shirt and brown pants. When they came out of the bedroom Kshitij shut the door firmly behind him. 'I'm ready,' he said.

'I need a photograph of your father,' Sartaj said.

Kshitij nodded, turned, walked away. The photograph he brought back was a picture of a happy family, Chetanbhai and his wife in front, stiff and square shouldered in a blue suit and green sari, and Kshitij behind, standing straight in a white shirt, his hand on his mother's shoulder.

He was remarkably steady at the morgue. Sartaj was impressed by his self-possession in the face of the damp walls, the yellow light, and the searing smell of formaldehyde that brought tears to the eyes. Sartaj forgave him a little then for his drab owlishness, his youth entirely lacking in dash or energy or charm. There was a sort of blunt and unprepossessing iron in him. Sartaj had brought some there who had been broken down by the dark corridors even before the room with the rattling metal trolleys and the atmosphere of congestion, but Kshitij identified his father without a tremor. He stood with his arms folded over his thin chest and said, 'Yes. Yes.' Outside, as they swayed in the police department Gypsy jeep, on the pitted roads, he asked, 'Has the police found out anything about the murder?' When Sartaj told him he couldn't talk about the investigation he nodded understandingly and lapsed into silence. But afterwards, at the station, he couldn't stop talking. He drank cup after cup of tea and told Sartaj that he was pre-med at Patekar. He was an only son. He wanted to specialize in neurology. He had been second in the state in the HSC exams, falling short by three marks mostly because of a bad mistake in Physics, which was his worst subject. It had been a sort of trick question in electricity. But other than that everything was moving according to plan.

When Sartaj asked about Chetanbhai Patel, Kshitij fell silent for a minute, a cup suspended halfway to his lips, his mouth open. Then, looking down into the cup of tea, he talked about his father. Chetanbhai was mostly a textile trader. He travelled often, to the interior sometimes, and they had thought he was late coming back from Nadiad this time, which is why they reported him missing two days after he was supposed to have returned. He did some export, mostly to the Middle East, but some to America and of course he wanted more. It was a long-established business, from before Kshitij's birth. Like many businessmen he had sometimes been the victim of petty crime. Once a briefcase with cash in it had been stolen from a local train.

'Did he seem afraid?' Sartaj said. 'Any enemies that you know of?'

'Enemies? No, of course not. Why would he have enemies?'

'Somebody in business that he had a quarrel with? Somebody in the locality?'

'No.'

'What about you?' Sartaj said. 'Do you have any enemies?'

'What do I have to do with it?'

'Sometimes people die because they get caught in their children's fights.'

Then there was again that flare of resentment in Kshitij, muted in the eyes but so strong in the shoulders and in the coil of his body that it was a kind of hatred.

'Do you have fights? Quarrels?' Sartaj said.

'No,' Kshitij said. 'Why would I?'

'Everyone has enemies.'

'I haven't done anything to make enemies.' He was now assured and confident and calm.

'Right,' Sartaj said. 'I think now we will look around your house. And I would like to meet Mrs Patel.'

In the jeep Sartaj considered his own vanity. He was sensitive to other people's feelings about him, and had still not learnt to be indifferent to the fear he caused, to the anger of those he investigated. He hid this uneasiness carefully because there was no place for it in an investigator's craft. To be hated was part of the job. But in college he had wanted to be loved by all, and Megha had teased him, you're everyone's hero. Then yours too, he had said. No, no, no, she said, and she shook her head, and kissed him. You have a terrible Panju accent, she said laughing, and your English is lousy, but you are just beautiful, and then she kissed him again. They had married out of vanity, their own and each other's. He had been the Casanova of the college, with a dada's reputation that her friends had warned her about. But she had been so very sure of herself, of her very good looks like a hawk and that shine she had of money, and they were so handsome together that people stopped in the streets to look at them. After they married they liked to make love sitting facing each other, his hair open about his shoulders so they were like mirror images, hardly moving, eyes locked together in an undulating competition towards and away from pleasurable collapse. The memory rose into his throat and Sartaj shook it away as the Gypsy rocked to a halt. A double line of young men in khaki shorts was plunging across the road.

'Bloody idiots,' Sartaj said. 'Won't even stay home in the rain.'

'They're Rakshaks, sir,' Katekar said, grinning. 'Tough boys. A little rain won't stop them. After all they want to clean up the country.'

'They'll all catch colds,' Sartaj said. The banner carried at the rear of the procession was soggy and limp, but Sartaj could see one of the crossed spears. 'And their mothers will have to wipe their noses.'

Katekar grinned. He rattled the gearshift to and fro and the Gypsy jerked forward. 'How is Mataji?' he said.

'She's very well,' Sartaj said. 'She remembers you often.'

Katekar was a great devotee of Sartaj's mother. Every time she stayed with Sartaj, Katekar made a special point of coming up to the flat, and touching her feet, not once but three times, bringing his hand up to his throat. Sartaj knew Katekar's mother had died just after Katekar had joined the force.

'Please tell her I said pranaam.'

Sartaj nodded, and looked over his shoulder. Kshitij was staring dully at the window and crying. His hands were locked together in his lap and the tears were sliding down his face. Now Katekar cursed softly as the jeep growled through a long patch of flooded road, leaving a wake behind. Sartaj turned away from Kshitij and shifted in his seat. Katekar was leaning forward, peering through the regularly spaced waves of water that the wipers were making on the windshield. He was cursing the water, the streets, and the city. His hands around the black plastic of the steering wheel were thick, with huge bulky wrists. He looked at Sartaj and smiled, and Sartaj had to grin back at him. In the rearview mirror, Sartaj could see Kshitij's shoulder, the line of his jaw, and he thought, it's always hard on the serious ones, they were always tragic with their earnestness and their belief in seriousness. He remembered two boys who were the grandsons of farmers in his grandfather's village near Patiala. He recalled them vaguely from a summer visit to the village, remembered them in blue pants and ties. There had been a celebration of their results in the seventh class exams, and he had tried to talk to them about the test match that everyone was listening to but had found them boring and uninformed. After that he had never seen them again and had not thought of them for years until his father had mentioned them during a Sunday phone call. They had been caught by a BSF patrol as they came over the border in the dunes near Jaisalmer, laden with grenades and ammunition. They had tried to fire back but had been neatly outflanked and machine-gunned. The papers had reported the death of two Grade-A terrorists and had reported their names and their affiliations. There had been a grainy black-and-white photograph of sprawled, bloodied figures with open mouths. Sartaj had never heard of their organization but had no doubt it was a very serious one.

The Apsara stood among a crowd of mourners, holding her pot tipped forward. The door to the apartment was open and Kshitij was surrounded by young men as soon as they stepped from the lift. In the front room neighbours sat and talked in whispers, and an older man embraced Kshitij for a long moment. Then Kshitij stood facing the door to the bedroom at the back of the house, and the seconds passed, and in his shoulders there was a huge reluctance, as if the next step were from one world into another. Finally the old man took Kshitij by the elbow and led him forward. Sartaj and Katekar followed behind closely, and over many shoulders Sartaj saw a woman sitting on the ground, surrounded by other women. They were holding her by the shoulders and arms, and she had one leg curled under her and the other straight out in front. She looked up with a blank face and Kshitij stopped. Sartaj wanted very much to see the boy's reaction, and he started to push gently past the old man but suddenly the woman started to keen, it was a long wailing sound that arched her back and the others strained to keep her still. It came again and Sartaj shivered, it was somehow quite expressionless, like a long blank wall stretching forever, and as stunning. Kshitij stood helplessly before it, and the room was very close, bodies pushed up to each other and the light broken up somehow into fragments of faces, and then Sartaj turned and walked out of the room. It was bad technique but he couldn't bring himself to look at them any more. The rest of the house was also filled and stifling, and Sartaj jostled shoulders and pushed until he was out.

Sartaj sat wrapped safely in the loneliness of his flat. It was very dark, moonless, and the small space between the gleams on the furniture held him comfortably in its absolute silence. He knew that if he disturbed nothing, not even the shadows on the floor, he could hold on to the madly delicate balance of peace that he had struggled himself into. He was trying not to think, and succeeding from moment to moment, and then the phone shrilled across the back of his neck. He held on for an instant, but on the third ring he turned his head and reached behind and picked it up. His hand was damp on the receiver and now he felt the sweat running down his sides.

'Ride?' Rahul said. Rahul practised a terseness which he had picked up from watching at least one American film a day on laser disc.

'I don't know, Rahul,' Sartaj said.

'I'm ten minutes away.'

Sartaj took a breath and tried to recall the thin quiet from a moment ago, but already there was laughter, and tinny music, spilling in over the windowsill. He was dizzy suddenly from the pounding in his head. 'All right,' he said. 'But I have to dress. Give me twenty.'

After a quick shower, and with the crisp collar of a freshly starched white kurta against his neck he felt cool again: it was a simple Lucknowi kurta, but the very thin gold chain threaded through the tiny gold studs made all the difference. The studs had belonged to his grandfather and he always stood a little straighter when he wore them. Rahul arrived with his customary honk downstairs, and he came up and they regarded each other acutely. This was a ritual of noticing each other's style, but this time Sartaj was aware only of the boy's long chin, of his nose which suggested his sister so strongly that Sartaj felt again that mixture of anger and longing. Finally he had to make a conscious effort to note the new haircut with sideburns, the loose red shirt, and the slightly flared black jeans. 'I have seen the look before,' Sartaj said. 'A long time ago.'

'Yeah?' Rahul said, without interest. 'I guess.'

'Yes,' Sartaj said. He thought, suddenly and apropos of nothing, that too young to know cycles is too young to know anything.

Rahul drove fast and well, with the assurance of the moneyed in a good car, or in this case a new red Mahindra jeep with a very good removable tape player. The music they listened to was completely foreign and remote to Sartaj, and as always it was played with a loudness that hovered on the edge of real pain.

'So how're your girlfriends?' Rahul shouted above the music.

'My what?'

'You know. Women.'

'I don't have any.'

'None? Such a big famous cop and all?'

Sartaj had been in the afternoon papers twice, both times for encounters with minor gangsters. The second confrontation had ended with gunfire, and a dead body on the floor in a dark corridor. Sartaj had fired six shots, and only one had hit. He had crouched, blinded and deafened and trembling, spilling shells onto the floor, but he had never told Rahul about that, or about the small spot of urine on the front of his pants. His picture, a formal studio portrait with retouched lips, had been in *Mid-day* the same afternoon. 'No, not even one woman. Slow down.' Rahul was speeding and then braking the jeep with a violence that was rattling the sleek little tape player in its housing.

'You're a real sad case, you know,' Rahul said.

Rahul had girlfriends and broke up with them and then had others with a speed and complexity that dazzled Sartaj, and he was worldly in a way that had been impossible all those years ago when Sartaj and Megha had been the talk of the campus. They had twisted against each other in cinema halls, desperate and hungry, but now Rahul and his friends were too bored with sex to talk about it. It had all changed and he had never seen it change. 'I'm just a poor old fogey, what to do, yaar?' Sartaj said with a laugh, and Rahul looked at him quickly but then had to swerve around a green Maruti 1000.

'Let's get a beer,' Rahul said.

'How old are you, sonny?' Sartaj said and Rahul laughed.

'Don't do the tulla thing on me now, Inspector sahib,' he said. 'I need a beer. You do too.'

'I do?'

Rahul ignored the question and sped past a timber and wood merchant's shop into a parking lot full of cars. A blue neon sign announced loudly that this was The Hideout, and inside the walls had been painted to look like the walls of a cave, and the floor was littered with barrels and crates. They were seated by a waiter in a black leather jacket, and above their table there was a large black-and-white print of Pran standing with legs wide apart, in black boots, flexing a whip. On the opposite wall a black-hatted foreign villain, one that Sartaj didn't know, glared over his left shoulder, caped and sinister.

'I arrested somebody on this street once,' Sartaj said.

'Yeah? A bad guy?' Rahul said, waving to somebody over the heads of the sleek and the young.

'Bad?' Sartaj said slowly. He was staring down at the price of the beer. 'Not really. He was greedy.' It was actually a quite dusty and unprepossessing commercial street, full of trucks and handcarts and the smell of rotting greens. The man Sartaj had arrested had been named Agha, and he had worked as a clerk for a company dealing in plastic goods. After they put the handcuffs on him he had looked at the owner of the company and said quietly, I have five children, and it was hard to tell whether that was an explanation for taking money or a cry for mercy, but it didn't matter anyway. 'I think he must still be in jail.'

'Everyone's looking at you,' Rahul said sullenly. 'Why do you dress like a Hindi movie?'

'This?' Sartaj said, running a finger over the collar of his kurta. 'You're the one who brought me here.'

The waiter brought their beer in what was obviously some designer's idea of roughneck tin mugs that belonged in a low den, and Rahul bent over his beer. Sartaj took a long gulp and was shocked by the pleasure of the thick cold curling against the back of his throat, and he wondered if things tasted better when you paid more for them. He took another long drink and sat up, revived, to look around and to listen to the pleasant buzz of music and the hum of voices that sounded sophisticated even when it was impossible to tell one word from another. He was trying to pin down what it was exactly, and after a while he decided it was that they sounded smooth, like there was a lubrication over it all, an oil that eased everything except that it was of course not greasy.

'She's getting married,' Rahul said.

'Who?' But even before he spoke it the frightening pitch and yaw of his stomach told him who it was.

'Megha.'

'To?'

'She told me not to tell.'

'I'll find out.'

Rahul looked up then. 'Yes, you will.' His Adam's apple ducked up and down and his face trembled, but then with a shake of his shoulders he said, 'Raj Sanghi. You know.'

Sartaj knew. This was the son of a friend of Megha's family, and Megha and Raj had known each other since childhood and the families had always thought that they were good together. He knew about all this. Now he sat with his hands on his thighs and found himself looking for a way to stop it, for a place where he could apply pressure until something snapped.

'Sorry,' Rahul said, and Sartaj saw that he looked frightened, terrified. He knew why: during one of their quarrels Megha had screamed at him that in his anger he had a face like a terrorist, looked as if the next thing he said or did would be complete and irrevocable, forever. He had looked at her then dumbly, made desolate and foreign by her choice of words. She had cried then and said she didn't mean *that* at all. They had broken parts of each other like that all through the time at the end, and he tasted these strange victories that left him empty and wishing for nothing more than endless sleep, like the last man on a battlefield where even the blades of grass were dead. Finally it had seemed better never to say anything at all.

'No, no,' Sartaj said. 'It's all right.' He reached across the table and awkwardly patted Rahul's wrist. He had to swallow before he could speak again. 'I don't think I should drink any more.'

It was morning and sparrows swirled madly through the arches of the police station. Ghorpade was sitting on the bench with his eyes closed when Sartaj and Katekar came into the detection room.

'Wake up,' Sartaj said, kicking one of the legs of the bench sharply. Ghorpade opened his eyes and Sartaj saw that he was not sleeping, or even sleepy, but that every moment was a struggle against some monstrous hunger. He had both his hands squeezed between his thighs, and he looked at Sartaj as if from some great distance. Katekar took his usual interrogation stance, legs apart and behind the suspect.

'Have you been thinking about what you did?' Sartaj said.

'I didn't do what you said,' Ghorpade said. His eyes were yellowish, rheumy.

'You better decide to tell me the truth,' Sartaj said. 'Or it'll be bad for you.'

Ghorpade shut his eyes again. Katekar widened his stance, and flexed his shoulders. But Sartaj shook his head, and said, 'Ghorpade, where do you live? Do you have a family?'

Ghorpade spoke without opening his eyes. 'I don't live anywhere.'

'Do you have a wife?'

'I had.'

'What happened?'

'She ran away.'

'Why?'

'I beat her.'

'Why?'

Ghorpade shrugged.

'How old are you, Ghorpade?'

Sartaj could hear the sparrows in the yard outside. Then finally Ghorpade spoke: 'I was born the year before the Chinese war.'

Outside, under the sky which was clouded again, Sartaj considered the slight possibility that he and Ghorpade shared birthdays. He had no idea why it seemed important. Now Moitra, whose first name was Suman, roared into the yard in her new Jeep. They had been batchmates in Nasik, and on the first day of the course she had let them know that she was twice as intelligent and thrice as tough as any of them. Sartaj had no problems with this, especially since it was probably true.

'Did he confess?' she said, bounding up the stairs. 'Closing the case?'

'I'm investigating,' Sartaj said. 'This is an investigation, Moitra. Remember?'

'Investigating what?' Moitra said over her shoulder as she sped down the corridor. 'Investigating whom?'

⌁

'Where is your mother?'

Kshitij was standing square in the middle of his doorway, his shoulders taut.

'What do you want with my mother?'

'Where is she?'

'She's here. Resting. She's not well. She's sleeping.'

'I want to talk to her.'

'Why?'

'This is a murder case. We talk to everyone concerned.'

'What does my mother have to do with a murder case?'

'She was married to your father,' Sartaj said, stepping forward. Kshitij stayed where he was and Sartaj put the palm of his hand against the boy's chest and pushed. Kshitij stumbled back, and Sartaj went past him into the drawing room.

'What do you want? Do you have a warrant? Why are you here? I heard you have a suspect in custody,' Kshitij said, following closely, but then Katekar had him by the arm and against the wall. Sartaj turned and Kshitij swallowed and subsided. Sartaj leaned forward and put his face close to Kshitij and watched him for a long moment, let him look and feel the pace of his anger as they listened to each other's breathing. Then he turned away abruptly and stalked through the room, towards the bedroom. Inside, the cupboards stood open, and the double bed and the floor were littered with paper. She was sitting on the balcony that opened out on to the swamp, and far away across muddy patches of green, the silver haze of the sea.

'Mrs Patel?'

When she turned to him her face was dense with grief. He cleared his throat and set forth briskly into his questions, when did you last see your husband, did he seem worried recently, were there any phone calls that upset him, were you aware of enemies and quarrels, were you aware of money difficulties, were you... She answered each time with a shake of the head, holding a hand at her throat. Her age was forty-nine, but her hair was a brilliant black, lustrous even in its disarray, and Sartaj looked at her, and thought that just a few days ago she must have been very pleasantly attractive, and that fact also settled into the confusion that surrounded the life and death of Chetanbhai Ghanshyam Patel.

Finally he asked: 'Can you tell me anything else? Is there anything else I should know?'

'No,' she said. 'No.' But the word was heavy with regret, and Sartaj followed her glance over his shoulder to the doorway to the bedroom, past Katekar, to the corridor where Kshitij's shadow lingered. When he turned back to Mrs Patel she was weeping, holding the end of her pallu to her eyes. And Sartaj, to his own surprise, felt a swell of emotion, rising like a knot in his chest.

In the bedroom, two of the cupboards were stacked full of shirts. Sartaj ran a finger up and down the row of the suits, rattling the wooden hangers against each other. The two other cupboards, against the opposite wall, were empty. He squatted and picked up a small booklet and flattened it out against his knee. It was a bank chequebook, with neat little tick marks in blue pencil next to the cheques and deposits. The closing balance was one lakh forty-six thousand rupees. He put the bank book in his pocket, and straightened up. There was a kind of grief in the wild litter across the room. Over this debris Sartaj began a quick but methodical survey, in a back-and-forth grid. In this practised routine there was a kind of relief.

'Aren't you supposed to have two witnesses for a panchnama if you do a search?' Kshitij said from the doorway.

'Am I doing a search?' Sartaj said.

'It looks like you are.'

'I'm just looking around. Why is all this on the floor?'

'I, I was just cleaning up. Sorting things out.'

'Yes, I see,' Sartaj said. After a murder some people tidied up. Others cooked, made huge quantities of food that nobody would or could eat. But every time there was an attempt to find one's way back to ordinary days. And all the paper on the floor was a record of the most innocuous kind of life: birth, insurance, deposits, loans, payments, the bills for hard-won purchases kept carefully for years. Now it was over. Sartaj looked across the room, towards the balcony, and she was staring out at the sea again.

Outside, in the hall, Sartaj ran his hand over the small row of books. The biography of Vivekananda, two Sidney Sheldon novels, *How to Be a Better Manager*.

'He was robbed, wasn't he?' Kshitij said, behind him. He looked tired and slight in his white shirt.

'We are investigating,' Sartaj said. The Apsara was gone, disappeared from her space by the wall.

'Is there any progress? What about this suspect? Who is he?'

Sartaj was thinking about the curve of the Apsara's shoulder. He turned his gaze with an effort and then said very directly, in his policeman's voice, 'We are investigating. We will let you know as soon as we find out anything.'

Outside, in the stairway that wound around the lift shaft, Sartaj leaned against the wall, breathless, hunted by something he had never known.

'Are you all right, sir?' Katekar said.

'I'm fine,' Sartaj said. But it was a loneliness so huge and so feral that he wanted to give up and collapse into the thick green swamp he could see far below, through a barred window. Even after the papers had come, after they had been sitting on their dining table for a week, he had never believed that the word 'divorce' meant something real. In his whole life he had never known anyone who had been divorced. He had never known anybody whose parents had been divorced. He couldn't remember a friend who had known anybody who had been divorced. Divorce was something that strange people did in the pages of *Society* magazine. His breath came dragging through the pain in his throat. He took a step forward and the green swirled dizzyingly under him, and he lurched forward and it pounded in his head, but then the white wall scraped across his shoulder. Katekar put a hand under his elbow. Sartaj followed his glance: on his own uniform, there was a streak of plaster from the wall, bright against the khaki.

Sartaj stood erect, feeling the muscles in his back, and he scraped the plaster off his shoulder with his right hand. He thumped it sharply and the white came off in little clouds. Sartaj shook his head. He tucked his shirt in, pulled up his belt, with both hands he felt the slant of his turban and corrected it. Then he smoothened his moustache, said, 'I'm all right. Come on.' He walked down the stairs, to his work.

~

'Mr Patel was a very helping kind of man,' Kaimal said. Kaimal was in his sixties, a retired merchant navy captain who lived two floors below the Patels in a flat of the same size and floor plan. 'We bought together,' Kaimal said. 'Seven years ago, almost exactly. Before that we rented in the same building in Santa Cruz.'

Mrs Kaimal brought out coffee in small steel tumblers, and Kaimal rubbed his forehead absently and ran a finger around the edge of the glass. A moment passed and Mrs Kaimal sat down beside her husband and put a hand on his wrist. Behind them Katekar, who was a tea drinker, sniffed suspiciously at his glass. 'He was very much younger than I am,' Kaimal said quietly.

'Was he well liked in the building here?' Sartaj said.

Kaimal looked up and nodded. 'He was president of the building society for three years in a row. He organized all our functions.' Leaning forward, Kaimal said decisively, 'He got jobs for the children of many people here.' He sat back, and said again, 'He was younger than I am.'

'You must have seen Kshitij grow up,' Sartaj said.

'Yes. He is a very intelligent boy.'

'Did his father think so?' Sartaj said.

Kaimal looked at him consideringly, and Sartaj could see the beginnings of distaste. This was familiar: the policeman's assumption of grief and deceit hidden in every happiness was frightening in its simplicity. It implicated everyone.

'Of course. He was very proud of him. Very proud.'

'And his mother is proud too?' Sartaj said.

'What else? Of course. Kshitij is a very good son to her. It is good to see a young man with such respect for his mother these days.'

'And were they happy together?'

'Who?' Kaimal said.

'Mr and Mrs Patel,' Sartaj said. 'Were they happy?'

'Happy?' Mrs Kaimal said, shaking her head, exasperated. 'They were husband and wife. What else would they be?'

Sartaj finished his coffee. It was very good coffee indeed. In people like this, decent and hospitable, loyalty to the departed was always the most unbreakable bond. They were telling a truth that had become sharp and clear in the sudden glare of death, and he knew he couldn't persuade them to turn them back towards the shadowed ambiguities that were so crucial to him. That would cause them to break with their obligations.

'I see,' he said slowly. 'I see.' He looked at Mrs Kaimal until she shifted uncomfortably, and then both husband and wife seemed to shrink against the faded lily pattern of their sofa. Then Sartaj said quietly, 'Were you aware if Mr Patel was in any kind of trouble? Did he seem afraid? Had he told you of any threats? Quarrels?'

'Threats?' Kaimal said. 'No.'

As Sartaj stood up they watched him apprehensively, turned their heads to watch Katekar's thumping walk. He thanked them for the coffee, told them to contact him if they remembered anything, and then he shut the door quietly behind himself and Katekar. They were a nice old pair, handsome and fine-drawn and cultured, but he had no regret for inflicting fear on them. It was what his job required of him, this distance from the rest of the world, their wariness of him, it was inevitable and necessary and he knew that very often it was this very thing that made it possible for him to grasp the truth, to see the secret and fix it, forever. Usually he thought nothing of it, never needed to, but today the click of the lock brought with it a bitter little bubble of loneliness in his mouth. He looked up and down the stairs, leaned towards the grilled door that covered the lift-shaft, and spat into the long pit. Then they went on to the next flat. As the day went on, as they walked down the stairs, from one home to another, Sartaj watched the sweat stain grow between Katekar's shoulder blades and spread across the large breadth of his back.

By late afternoon, from the fragments of many conversations, from hesitations and allusions and things left unsaid, he had teased out these unremarkable facts: the father was a genial man, full of humour, ready to backslap and also to come to one's help when needed; the son was known for his intelligence, for his first rank in every exam, for his quietness; the mother was a good cook, she doted on her son and laughed at her husband's jokes, the husband and wife went for drives every Saturday, long drives.

Sartaj learnt about Patel's passion for his red Contessa as he stood next to it in an incense-filled garage. Patel's driver, a tall, bulky man named Sharma, was polishing the car with a kind of melancholy patience, inch by inch and with many flourishes of a waxy rag. He had two agarbattis burning in front of a picture of Shiva—they let forth, now and then, undulating drifts of white smoke, full with the aroma of chameli, an aphrodisiacal essence of moonlight and river water and rain. Katekar strolled around the periphery of the room, looking at the cans on a shelf, the calendars on the wall.

'He liked to listen to ghazals in the car,' Sharma said. 'Every new cassette, we got. We just got new speakers.'

Between the driver's seat and the passenger's there was a box full of tapes. A tape with blue writing on it was in the deck.

'What's that one?' Sartaj said.

'His favourite. Mehdi Hassan. He listened to it again and again.' Sharma reached into the car and a moment later the song drifted out into the garage: *Voh jo ham me tum me karaar tha, tumhe yaad ho ke ya na yaad ho...*

'When was the last time he was in the car?'

'Last Saturday. I don't come for duty on Saturdays. But he must have gone for a ride.'

'Where?'

'I don't know. Around. But he must have gone.'

'Are you sure?'

'I've worked for him for thirteen years and every Saturday he went for a drive with memsahib. I used to pull the seat forward every Friday night for him. He liked cars. This one specially. Last weekend he even washed it.'

'Washed it?'

'Outside and in. It was shining clean when I came in on Monday morning.'

It was a very long car that filled up the garage so that Sartaj had to squeeze behind it. He opened a door and leaned in. The seats were spotless, and the interior smelt of soap and ammonia. The song was gentle and a little sad and very sweet.

'Was he worried in the last few days? Afraid of something? Upset?'

Sharma stopped rubbing at the metal and looked up at Sartaj. 'No. But I'm worried now. What will happen now to Kshitij baba?'

'Kshitij? Why?'

'He has to take care of his mother. Very much love between them. Often when he was tired and had a headache he would lie with his head in her lap. I saw this when I would go up to give the car keys back in the evenings. But he's very young. How will he manage? So young.'

Sartaj took the rag from him and sniffed at it. 'Are you afraid for your job?'

Sharma laughed aloud. He straightened up, away from the car, and he was at least three inches taller than Sartaj, and not at all afraid. 'Inspector, you know there are jobs for drivers in Bombay. No, I'm worried for him. He looked very tired today.'

'Did he take the car out?'

'No, no. When I came this morning I saw him taking rubbish out of the building. Then again an hour later.'

'Rubbish?'

'Out to the dump, over there behind the wall. He looked very tired.'

Now Mehdi Hassan sang the emperor's old complaint about *baat karni mujhe kabhi mushkil aisi to na thi*, and Sartaj searched the car. He scraped under the seats and rolled the grit off the floor mats between his fingers. The glove compartment held a receipt book from a petrol pump in Santa Cruz, vehicle booklets in a plastic wrapper, a pack of playing cards, and, held together by a black metal office clip, a stack of parking receipts.

'Where was his office?' Sartaj said.

'Andheri East. Near Natraj Studio.'

The receipt on top was rubber-stamped 'Colaba Parking'. Getting out of the car, Sartaj thumbed through the stack, and it seemed that Chetanbhai had gone to Colaba often. Sartaj opened the front passenger door again and knelt, trying to get as close to the floor mats as possible. A quick run of the forefinger over the underside of the front passenger seat, over the shiny metal at the very bottom, and then back, left an impression of some faint roughness, a texture across the metal. He reached into his pocket for the penknife on his keychain, and scraped with it and then held it up to the light, and there was a flaky rust on the blade, an innocuous brown the colour of farm earth after rain.

'Look at this,' he said, holding it up to Sharma.

'What is it?' Sharma said, peering. 'Rust?'

Sartaj's eyes moved, and Katekar caught Sharma by the collar, bent him suddenly forward at the waist, and with a full high turn of the shoulder hit him on the broad of his back with an open palm. The crack was shockingly loud and Sartaj hissed into Sharma's astounded face, 'Did you kill him? Why did you kill him?'

'I had nothing to do with it.'

Sartaj looked up at Katekar, and again Katekar's broad hand went up and came down like a piston.

'This is blood,' Sartaj said. This was not a fact, it was less than a theory, but Sharma believed it. His eyes were full of tears, and he was panting, holding his chest with both his hands. 'You didn't know it's very difficult to get blood cleaned, did you? No matter how much you clean there's always a bit left. Did you wash the car?'

'I tell you I don't know anything about this. I swear to you.'

'Why are you still here?'

'I get paid for the month on the thirtieth. Sahib, I'm just a poor man. I would have gone but I get my payment on the thirtieth. Nothing else.'

Sartaj was willing to wait and see who tried to flee. The guilty always ran. It was like starting an unknown animal out of a thicket. You tossed in a rock and waited to see what came out. 'All right,' he said, and Katekar let Sharma go and stepped away. But Sharma stayed bent over, huddled to his knees, his face red. 'But don't speak to anyone regarding this matter. Don't touch this car. Lock up the garage now and give me the key.' Sartaj said to Katekar, 'Watch the garage until we can get somebody from the forensics lab down here to take a look at the car.'

'Sir,' Katekar said, nodding, still looking at Sharma. He took a blue-and-white chequered handkerchief out of his pocket and wiped his mouth, his cheeks.

Sartaj bent over until his face was close to Sharma's. 'And you. Don't go anywhere, friend, don't leave the city. We'll be watching you.'

We'll be watching you, Sartaj said under his breath as he walked out behind the building. We'll be watching you. This was a lie he had learnt to tell easily. It was an illusion that suspects believed in easily, and it worked even when there was absolutely no truth in it. Parulkar had once forced a confession, and a nervous breakdown, from a domestic-murder suspect who had started to believe that there were policemen everywhere, on the roof of his house, in his bathroom, and behind his new Godrej fridge. The dump was on a road built out into the swamp, past the buildings under construction and the sodden mounds of earth. At the very end of the road, as it petered out into the thick green bushes, it was covered with a thick layer of paper, bones, and things liquefied and rotted. Flies buzzed around Sartaj's head as he carefully placed one foot after another. Sartaj walked by two children with bags over their

shoulders, picking out the plastic from the mixture. Further on, three yellow
dogs stopped eating to watch him over their shoulders, not moving an inch
as he edged past them with a flutter in his stomach. Ahead, there was a huge
blackened ring that smouldered across the road. He could see the edges of
it trembling with heat. He kicked at the wet surface of the waste and it fell
away, and underneath the fire was working relentlessly. He put his hands on
his knees, bent over, and walked slowly around the curve of the black circle.
Faces and old headlines blurred away as he watched. He straightened up, took
a large stride over the border, and went into the circle.

Now the ashes clung to his feet as he walked. Sartaj bent over a twisted
piece of plastic and turned it over. It was the casing from a video tape, half
melted away. The clouds shifted and suddenly the sun moved across the swamp.
The smell filled his head, rank but full and rich. Two egrets came gliding over
him and he turned his head full around to watch them. He was at the other
edge, near the water. Leaves scraped across his face. He bent down and peeled
a soggy curl of paper away from his shoe. It was a picture from a calendar, an
almost-nude model in designer tribal clothes, smiling over her shoulder. He
scraped at the muck with the side of his shoe and layers of paper parted. He
bent quickly, then, to peer at the handwriting on a blackened fragment. 'Patel,'
it said, in the neat script from the chequebook. He squatted, and tried to turn
it over, but it broke crisply in half. He took his ballpoint pen from his pocket
and used it to poke at the debris. Under a piece of plastic he found a torn
page with printing on it. He held it up between his forefinger and thumb. It
was thick paper, and he could make out the writing. '…having got up in the
morning and performed his necessary duties, should wash his teeth, apply a
limited quantity of ointments and perfumes to his body, put some ornaments
on his person and collyrium on his eyelids and below his eyes, colour his lips
with alacktaka, and look at himself in the glass. Having then…' As he read,
it tore across from its own weight and dropped. The sweat dripped down his
neck. He could hear the dogs barking.

The wall in front of Parulkar's house was crumbling even as Sartaj watched.
As he waited at the door, he watched the rain carry away tiny pieces of stone
and brick into the flowing gutter. But inside the floor was cool and polished
as he bent to take off his shoes. Parulkar's youngest daughter, Shaila, her hair
swinging in two enormous plaits behind her, watched him gravely. She was
fourteen, and she wore steel-rimmed spectacles at the very end of her nose.

When he had first met her she had loved to be tickled behind the ears. It made her weak with laughter. Now she was a very dignified young woman.

'You're starting to look like someone, Shaila,' Sartaj said.

'Who?'

'Oh, I don't know. A model? A movie star? Madhuri Dixit, maybe…'

She looked at him sideways, but with a little smile. 'I've decided I'm going to IIT,' she said. 'I'm going to do computers.'

'Really?' Sartaj said. 'No more Parulkars in physics?' Both her older sisters were physicists, one at ISRO and the other at Bombay University.

'Bo-ring,' Shaila said. 'That's old stuff, don't you know? Computers are hot, hot, hot.'

'Which I don't know anything about.'

'As if you know anything about physics.'

'You'll have to teach me.'

She took him, with both her hands, by the wrist and pulled him down the corridor to the drawing room, which was large and sparsely filled by its two divans and chatai. This forward-leaning walk, at least, was something she hadn't decided to outgrow yet.

'Sit,' she said, as she went back into the corridor. 'Papa will be out in a minute.'

Sartaj didn't mind the minute, or more, at all. He had always liked this room. It was long and opened out on one end into a small garden, which actually had a tree in it. The branches hung over the window and now the water dripped slowly, and the light was gentle and green. This light was the surprise the house hid behind its shabby walls, and today Sartaj was particularly glad for it.

'Anything new?' Parulkar said, as he came in, rolling up the sleeves of his kurta, which was a starched bluish-white. He looked very elegant when he was not in uniform. Sartaj told him about Ghorpade and Kshitij and the mother.

'The forensic report came from Vakola this morning,' Sartaj said. 'The interior of the Contessa seemed to have been washed with a three-ten mixture of hydrogen peroxide. There were no traces of blood, or any other suspicious substance. The car was scrubbed down, clean. It was wiped down very professionally. No fibres, nothing.'

'But what has the car to do with any of it? You could close the case really,' Parulkar said. 'You have the watch, and so physical evidence and a motive, and the suspect, what's his name, Ghorpade, places himself at the scene. What else could we need? What are you looking for?'

'Nothing, sir. I mean I don't know.'

'These are delicate times, Sartaj,' Parulkar said. 'Pushing too hard, without sufficient reason, on that family could lead to, let us say, sensitivities.' What he meant was that for an outsider, a Sikh, to push a little was to push a lot. It was true, even though the Patels were Gujaratis and so outsiders themselves. There were outsiders and outsiders. To say, I was born in Bombay, was very much beside the point. Sartaj nodded. 'Meanwhile,' Parulkar said, 'there's something else I have to speak to you about...'

He stopped as Shaila came in with a tray. Sartaj watched him as he bent forward to pick up his steaming mug of tea, because he had never known Parulkar to be delicate about business around his daughters. In this house filled with women they were clinically straightforward about death and mayhem.

'There is something else,' Parulkar said after Shaila had swung her plaits out of the room.

'Sir?'

'I got a call from Shantilal Nayak last night.'

'Sir.' Nayak was an MLA who lived in Goregaon. He was the sitting home minister, and he had come to Sartaj's wedding mainly as a guest of Megha's family.

'He mentioned some papers.'

'Papers?'

'That you were to sign?'

Sartaj felt, suddenly, a rush of hatred for the rich. He hated them for their confidence, their calm, how they thought everything could be *managed*. But he said, 'Yes, sir.'

'Sartaj,' Parulkar said, leaning forward a little. 'Sartaj. If I could, I would have given anything to change it all.'

'Yes, I know,' Sartaj said finally. It was true.

At the door Parulkar put a hand on his shoulder, and Shaila came running out to take his wrist again in both of her hands.

'Don't get wet,' Parulkar said. 'It's cold.'

Sartaj nodded and splashed down to the gate. Then he turned around and came back. 'What is alacktaka?'

'What is what?' Parulkar said.

'Alacktaka.'

'Never heard of it.'

'It's something you colour your lips with,' Sartaj said.

'My lips?'

'Not yours, sir. I meant generally. Men use it to colour their lips.'

'This is general somewhere?'

'That's what I want to know,' Sartaj said. He felt absurd standing there with his raincoat flapping around his knees and the water dripping from his eyebrows, and he turned abruptly and left them there, feeling their looks of enquiry on the back of his neck all the way down the lane. On the way home, under the bleak surges of anger, there was a nibbling doubt, an obsessive circling around something that was unknown and elusive. He now had facts about the deceased and his family, and these were quite ordinary and commonplace, repeated in any other family down the street or somewhere across the country. He knew something about the killer, if that was what Ghorpade was. They were two ordinary men who came together on a Bombay street corner one night. But Sartaj remembered the Rolex watch, and he was certain he knew nothing about the man who wore it, nothing that would explain the silky artifice of the thing, that would show how the commonplace and ordinary became Chetanbhai Ghanshyam Patel. He didn't know this and somehow it felt like a debt.

Sartaj had put aside notions of debt the next time he went to the Bandra station. There was no option other than winding up the investigation and charging Ghorpade with murder, this was clear. He said this to Moitra, who nodded and said, 'Yes, well, you'd better take a look at him.'

In the lockup the three other occupants were leaving Ghorpade well alone. He was lying on his side, his face against the wall, curled up. The air in the cell was cool, still between the old smooth stone of the walls.

'He can't keep any food in his stomach,' Moitra said. The whole holding area smelled of vomit.

'Ghorpade,' Sartaj said. 'Eh, Ghorpade.' Ghorpade lay unmoving.

'We'll send him to Cooper Hospital this evening,' Moitra said. 'But we have his statement anyway.' She meant that the hospital was no guarantee of his survival. 'Come on, I want to finish up early tonight.'

In her office, while they stamped and signed papers under a picture of Nehru, the phone rang. 'Hel-lo,' Moitra said softly. Sartaj looked up at her, and she changed to her usual clipped voice, 'Hello.'

'Is that Arun?' Sartaj said.

'Yes,' she said, holding a palm over the mouthpiece and regarding Sartaj with a steady glare that dared him to be funny. 'So?'

'I have a question for Arun.' Arun was her husband, a professor of history at Bombay University.

Moitra was still wary. 'You do?'

'Yes, ask him what alacktaka is.'

She mouthed the word, feeling it out to see if there was a joke in it. Then she took her hand off the mouthpiece and asked. 'He says he's never heard of it,' she said after a pause. 'He says what's the context?'

'I've forgotten,' Sartaj said. 'It's not important anyway.'

So after all, on paper, it was going to be an open and quickly shut case, not even worth a headline in the afternoon papers, but still Sartaj studied Chetanbhai Ghanshyam Patel's chequebook as the jeep wound its way north through afternoon traffic. The entries were routine, three hundred and eighty-three rupees for electricity to BSES on January 28, nine hundred and ninety-five rupees to the ShivSagar Co-op. Building Society on January 29, one hundred and twenty-five rupees to the Jankidas Publishing Company on February 1, two hundred and ninety-two rupees to the Milind Pharmacy the same day, and Sartaj bent the book and the pages flurried by, the same names and the same little amounts again, BSES, Hindustan Petroleum for cooking gas cylinders on April 15, one hundred and twenty-five again to the Jankidas Publishing Company on May 1, two hundred and forty-eight to Patekar College on May 14. There were no large amounts, no huge payments to cash, nothing that indicated danger, or even excitement. Finally Sartaj took a deep breath and closed the book and put it back in his pocket. It was time to give it up. There were other cases to follow, which meant, he knew, other puzzles that would distract him from himself, so there was no reason to cling to this one. It was time to give it up.

And yet when he faced Kshitij, Sartaj had to force himself to pull out the chequebook and hand it over. 'We will formally file charges tomorrow morning,' Sartaj said. 'I came to tell you. It is a man named Ghorpade.'

When Kshitij had opened the grandiose door to the apartment, he was holding a dumb-bell, and was dressed in a white banian and shorts which hung off his thin waist on to his angular hips. His chest heaved up and down as Sartaj told him about Ghorpade. His face was suffused and contorted, and when Sartaj was finished he nodded. He tried to speak and the breath came and went.

'I will get in touch shortly,' Sartaj said. They looked at each other and then Sartaj turned away. He paused, feeling as if he should say a word of comfort to this boy in his loneliness, and finally uttered with a smile that felt false, 'Building the body, haan?'

Kshitij nodded. 'Good,' Sartaj said. 'Good.'

As Sartaj shut the lift door he heard the boy's voice. 'Vande mataram.'

Sartaj paused with his finger on the button marked 'G'. Kshitij was standing with his hands by his sides, his back straight. Feeling slightly ridiculous, Sartaj came to attention, and Kshitij and he looked at each other through the metalwork of the door. It was an old-fashioned slogan Sartaj had heard all his life, mostly in movies, but he could never say it without a surge of belief. 'Vande mataram,' he said. Hail to the mother. And despite himself, unwillingly, he felt again, in his chest, a havoc of faith in the devious old mother he was saluting, and in the same moment, despair.

The sound of the rain was endless. It was still early afternoon but it was dark in the station house. Sartaj sat at his desk, loose-limbed, and watched water stream down the panes of his window. Under the torrent, there was a strange quiet in the station house. It was as if everyone and everything were waiting.

Parulkar came in and walked over to the window. The collar of his shirt was bunched up around his neck, and he looked damp and uncomfortable.

'I had another call from Nayak this morning,' he said. 'At home, this time.'

'I'm sorry, sir,' Sartaj said.

Parulkar walked around the room, nodding gently to himself. Finally, from behind Sartaj, he said, 'I don't understand. Why don't you just sign the papers and let it finish? What are you afraid of?'

And Sartaj, who was watching how the water on the window made the world outside a vague blur of brown and green, said without wanting or meaning to, 'I'm afraid of dying.'

Parulkar put a hand on his shoulder, for a moment, with an awkward sort of rumbling movement. Sartaj twisted in the chair but already Parulkar was walking towards the door. His shoulders were hunched up as he pulled violently at his belt, and all the way out of the door his face was turned away.

There was a weariness in Sartaj's arms and legs now, and his eyes, even after he closed them, felt hot and scratchy. Every breath was labour now, because he was afraid of the silence. He was too afraid even to feel contempt for himself.

Sartaj Singh lay flat on his back on the floor of his apartment, in a white banian and red pyjamas, arms wide to the sides, and contemplated death. He had these words in his head, 'to contemplate', and 'death'. Between them there was a kind of light, a huge clear fearful sky in which he was suspended. When the shrill of the doorbell called, it took him a full minute to descend from this thin and deadly atmosphere, to lift his weightless body off the floor. Then there was a stagger to the door as he rubbed his eyes. In his blankness he found the rubbing pleasurable, and he felt keenly his knuckle on the eyelid, so that when the door swung open against his shoulder and he saw her, he had trouble recognizing the scene he had imagined a thousand times. 'Hello, Megha,' he said finally, his hand still up to his face. She waited, until he understood the formalities now between them. 'Come in, please,' he said, and hated the words.

She walked stiffly, her shoulders high, and with a large black purse held hard against her hip. She stood next to the furniture they had chosen together, in a black skirt and high heels, stylish as always and with the closed face of a model on a runway. 'Sit, Megha,' Sartaj said. He pointed at the green sofa, and she arranged herself with her hands held in front of her, the purse standing straight up on the coffee table in front of her like a bulwark. Sartaj sat on a chair across from her and held his hands tightly across his stomach. He opened his mouth, and then shut it again.

'Rahul told me he told you,' Megha said.

'Told me what?' Sartaj said, even though he knew. His voice was loud. He wanted her to say it, the word. So that his pain would hurt her, as it always had. But she said it easily, as if she had been practising.

'I'm getting married.'

'Is that why you came? For that?' With the jerk of his head he meant the papers on the dining table behind him, but in the sudden snap of the motion he had also the policeman's brusqueness, the coiled promise of angry force. She shut her eyes.

'No, I didn't come for that,' she said. When she looked at him now her eyes were wet, and he felt inside the unhitching of pieces of himself, things drawing apart and falling away. 'I came because I thought I should tell you myself.' A tiny shrugging motion with her shoulder, and a hand drawing up and touching her mouth. 'I didn't want you to hear about it like that, from someone else.'

The sunlight in the room dappled the familiar sofa and made it unreal. Sartaj was aware now of the great distances to the surfaces of his body, the strangeness of the hand that lay like a knurled brown slab in his lap. He slumped in his chair, trembling.

'What are you smiling at?' Megha said hesitantly.

Sartaj thought about it. 'How did we get so old?'

He laughed then, and after a moment she with him, and the sound sped around the room, over the photographs, the few knick-knacks on the shelf, the stained dining table. They both stopped suddenly, at exactly the same moment.

'I'm sorry,' she said.

He struggled himself upright. 'Do you want some tea?' he said.

In the kitchen he had to wash the pot, and then the teacups as the water burbled. Then he stood ready with the sugar, alert and concentrated, and the smell of the heating milk and the leaves, and the wisps of steam, sent him reeling into the first morning of their marriage, the first time they had woken together, the profound heat of her skin against him, and her confession that she did not know how to make tea. I told you I can't cook, she giggled into his neck. But tea, Sartaj said, pretending to be angry, but after that he had always made tea in the morning. Now the heat from the stove spread across his knuckles, and he remembered the newspaper splayed across the table between them, and buttery kisses, and he felt his heart wrench, kick to the side like a living thing hurt, and he fell to his knees on the dirty floor, held his head between his hands, and wept. His sobs squeezed out against all the force of his arms, and the wooden doors on the cupboard under the washbasin rattled faintly as he bent and curled against them.

He felt Megha's hands on his shoulders, and her breath on his forehead as she whispered, 'Sartaj, Sartaj,' and he turned away from her, from his own embarrassment, but his strength was gone, and she pulled his head back, into the solid curve of her shoulder. He shook again and she held him tight, hard, and he felt with piercing awareness the pain of her forearm against the back of his neck. He was gone, then, vanished into the familiar fragrance of her perfume, unknown for so long, with its flowers and underlying tinge of salt. He was perfectly still. Her lips moved against his cheek, murmuring, something that he couldn't quite hear, and then he felt the brush on his mouth, a gift of softness and then the shifting suppleness, what he always experienced as a question. He kissed her desperately, afraid to stop or pause because then she would stop. But she wasn't stopping, she held his face in her hands, her long palms strongly on cheeks and chin, and sipped at him with little murmurs. Despite himself, Sartaj curved against her, an arm up and around, and he felt the weight of her breasts against his side, and she laughed into his mouth, not here, not here.

They almost made it to the sofa in the drawing room. He walked behind her, and he watched the sheer cloth of her skirt flap faintly against her legs, and her neck under the pinned hair, and the straight back under the expensive white cloth, and he reached with both hands to hold her by the upper arms. Again the vivid shock of the flesh. She fell back easily against him, offering him her neck. Under his nuzzling she squirmed and said, 'The curtains.' He stumbled back, dizzy, found the curtains and pulled. When he turned around into the sudden dimness she was sitting on the sofa, her hands together on her knees. 'I'm going to marry him,' she said, and her voice was small. Sartaj navigated towards her, one step and then another, and they peered over the sudden distance. He knew it was true and there was nothing he could say to it. He tamped down his entirely unreasonable anger and searched for words. Then she giggled. He followed her eyes and there was the unreasonable bulge in his pyjamas, his red shamiana she had called it once.

This time they found each other somewhere over the coffee table. He dragged her over it, one hand on her back and the other in her hair. Once they would have delighted in the lingering discarding of clothes, the slow fall of silk, the shifting of cotton and slow revelations, but now there wasn't the time. He laboured with the complications of her skirt as she shrugged off her blouse. Her pull at his nada dug into his side but his pyjamas came down efficiently with a single movement of her wrist. 'Sar-taj,' she said, and took his hands away from her skirt, and with two clear snaps it came away, and then she was against him. Now he dared to look at her face, and in the dark flush of her cheeks there was that concentration, that singular look of intent purpose he had not seen for a long time, and he was no longer afraid. Under his thumbs her nipples bloomed and she shivered helplessly and smiled.

But he buckled under the scrape of her fingernails on his thighs. She was arrogant now, full of secrets and sure, very sure, as he slid to his knees and on to his back. She pressed with her hands on his shoulders, pushing him down, straddling and hovering over him, her breasts a maddening lightness on his chest. Again her fingers moved over his stomach, scuffing, and his face contorted, saying take pity on it, my thing my muscle my cock, take pity on its loneliness, and she grasped him in her hand. She leaned low over him, breathing in the agonized relief in his exhalations. When she looked down along the length of their bodies, he looked with her, and saw her hand grasping hard, and springing from her fist, him, each pulse distinct. They had argued and talked and laughed about what to call their parts, she hated lund and chut, how vernac and crude and vulgar she said, cock and pussy and fuck felt foreign in his mouth, he said that to her and she laughed fondly and said all I want in your mouth is me and thrust her breast against his lips, *me by any other name.* But now she groaned, a curious groan mingling hunger and joy and defeat and yearning and she snaked down fast, moving like heat over his skin and she took him in her mouth. She reached with her neck, mouth wide open, and took him in. Me by any other name. His mind drunken reeled and she made greedy little noises, slobbering, and he heard his own voice calling, and her head bobbed and weaved, and in the confusion of pleasures he remembered *a long walk along a sandy beach the feeling of the sunrise ahead,* and raised his head and saw with wonder her lips on him. The stretch of the flesh beautiful and grotesque. His gasps in his mouth. A burning warmth against the side of his chest through the thin silky cloth on her hips. With his right hand he reached down and pulled pins from her hair. It uncoiled reluctantly and dropped slowly to his stomach.

When she looked up her face was blurred, her eyes hazy from wine. 'Condom?' she said. 'Condom?' He was still running back, retreating from the edge she had brought him to and relieved she had stopped, and as always the way she pronounced the word with the flat 'um' baffled him. 'Condom?'

'In the bedroom,' he said finally. He followed her, followed that movement of her haunches, that slight jiggle which still and now made his heart surge in tender ferocity. He found the unopened condom packet easily, in the table next to the bed. She lay back on the bed, twirled off her panties in a single arcing moment that bent her like a bow and back. Orange light spilled through the curtain on the west-facing window, across her belly and into the shadow below. His fingers fumbled at the plastic.

'Give,' she said. She took it from him as he tumbled on to the bed. She kissed his tip with a swirl of tongue, then rolled the rubber down. Then she was over him, squatting. She held him and he thought of the other man viciously. Look where she is now. Look. But who is the cuckold, which is the husband, and he felt despair in his throat, like black and bitter iron. But then he cried out in love, from the scalding oily embrace of her. She took him in, a fraction, just so much, so little. His hips bucked and she put a hand on his stomach. Don't move. He knew her pleasures. Her engulfing would last an eternity, little by little. She was absolutely still, not moving at all but yes slipping down inch by infinitesimal inch. On her face an expression of indescribable luxury. Even during the first time together, which had been her first time ever, she had been confident. Afterwards *she says ah that wasn't too good but I think it's going to get better and rolls herself in a yellow silk sari.* Sartaj saw now that this was the last time, and again flickering shadows of hopelessness chased the pleasure up his spine. He opened his eyes wide, to see her breasts light and golden in the slanting light, against the black brassiere, and he wanted to touch them but he knew not yet. Her mouth was open and he knew she was on the same razor edge between excruciating delight and impatience, holding by the will on to time. For in time there was joy. Children's shouts tumbling into the room but here the harsh breathing. In Benares *he is overwhelmed by time as an owl-faced shopkeeper displays for Megha his saris, he throws them in the air with a flourish and the silk billows, red and gold and deep blue and green, and Sartaj laughs as the colours float and fall but he is full of loss and afterwards on the street she asks. It's nothing, he says.* Nothing.

Now she exhaled, a wail, 'Sartaj it's so good.' He considered the nuances of 'it', distilled the traces of regret and exultation in her voice, but now with another grateful sigh she gave up, gave in, reached with a right hand to their joining, under her, above him, and with a finger strummed at herself, at the centre. Through her flesh he felt the vibration. Which he remembered. When *she finds out about him and someone else she cries and leaves for two weeks and three days. Later when he finds out about her, much later, he cannot believe it cannot see it in his head and then wants either to die or kill someone.* She flick flickered with her finger and he could hear it in her breathing and nothing else was moving and he raised his head to watch and she leaned down suddenly and kissed him and her tongue moved in and out of his mouth furiously. He felt fucked and was grateful. She spoke into his mouth now, a cry, something, and then shook and came down on him the rest of the way, and trembled defencelessly on him, and he held her. Until she was still.

But with his hands spread wide on her buttocks and his face in her neck, her shoulder, he found his rhythm. She stirred and moved with him. There was the fleeting awkwardness, a move this way and that and an unsatisfactory impact and a farting sound from between their bodies, but then she pushed herself up on his chest, palms spread, hair falling over his face, and together they had the movement, and he was moving in and out slicked from the sweet pocket of contentment, his thumbs on nipples pulled from the brassiere and rolled, and she made now small sounds on every stroke, halfway between protest and welcome, between all worlds, and Sartaj somewhere aware of the bed below, the roof, the building, and what they were doing high in the air above the earth, the eager grinding of the bodies, he in the body and out of it, mind moving and not moving, sweat on her forearms, *me by any other name,* the moving sun, and then she looked down at him with eyes shining with wonder, and he held her by the hip and strained up to her, rising off the bed and reaching in her, saying Megha, and she rolled down to meet him, and at the closest point of their meeting he felt the spill, ecstatic and alive, and in a last moment of thought he asked, is this me? Is this you?

The condom made a sad plop on the floor next to the bed. As he turned over Sartaj had the sensation of time starting to stir again. He lay on his side, put a hand on Megha's stomach and watched his fingers move with her breathing. She had an arm over her eyes, and he blinked hard, trying to read the set of her chin. He could feel the throbbing of his own heart. She turned to him suddenly. 'We must be mad,' she said, but there was no sadness in her. She smiled and touched his cheek with the tips of her fingers. 'I have to go.' He watched her walk across the bedroom, past the white wall with its filigree of shadow, and he knew he would remember this image forever, this person, this shimmering body moving away from his life. From the bathroom he heard the sound of rushing water.

Sartaj was no longer angry, or despairing, but as he lay on the sheets he was possessed of a certain clarity, and he could hear the world ending. In the huge distances of the red sky, in the far echoes of the evening he could feel the melancholy of its inevitable death. He ran a hand over his chest, and the slow prickling of the hair was distinct and delicious. He got up, walked into the bathroom. Megha was standing under the shower, his blue shower cap on her head, and was lathering her stomach absently. He took the soap from her, led her by the hand into the bedroom, to the bed. He had her lie down on her back, and smoothed away the soap from her body with his fingers. He bent his head to her breasts, and found cool beads of water and underneath an evasive smoothness. He tongued the nubbly brown nipples and she stirred under him restlessly. As he trailed down her body she tugged at his patka. It came off finally and with her fingers in his hair he put a hand under one knee and lifted the leg up, away. He heard her breath, sharp, and as always the close curl of her labia against each other, under the soft slope of black hair, was once more strangely unknown to him, familiar and yet astonishing. He kissed her thigh, in the crease, and there was the lush smell of her, round and full and loamy in his nostrils. The flesh to the centre flushed and trembled and thickened under his tongue. In and into the sudden salty heat he lapped, hungrily, following the twisting trail of her shakes, losing sight of the secret and finding it again. She held his head and moved him and herself to the place she wanted and then away from it. His fingers dabbed and stroked through the folds and in the plump fluttering confusion there was time and its thousand and one tales, *first flirtation, vanilla ice-cream eaten dripping from her fingers, and a*

Congress election poster outside the restaurant window while they quarrelled and he clung to none of them, they drifted and vanished and he sometimes himself and then vanished, his tongue moved and his lips and his fingers under her bottom, and then he heard her rising cry, and he knew she had her right index finger in her mouth, biting. Finally she drew him up and kissed him, licking his mouth and fingers. This time he put the condom on himself. They moved together and the bed creaked under them. His body bent over her and he looked back over his shoulder at their tangled shadows, rising and falling, and then down at her and at their hair mingling. He bent down to kiss her, and when he came away she was crying. With a groan he touched her cheek but she put a hand on his wrist. 'Don't stop,' she said. 'Don't stop. Don't stop.' And so he went on. In the twilight she raised a hand to his mouth and he could see her tears. And so then a moment when everything was lost, but her.

Afterwards it was dark and they said barely a word to each other. At the door she raised her cheek to his, and for a moment they stood like that. When she was gone he shut the door, and came back to the sofa, and sat on it, very still. He felt very empty, his mind like a hole, a black yawning in space, and he searched desperately for something to think about. He thought then of Kshitij, and his mother, and his father, and the boy's anger, the resentful line of his shoulders, and Sartaj began as if from a great distance to see a shape, a form. He sat on the sofa and thought about it. Outside the night came.

'Alacktaka.'

The word hurt Sartaj's ear. His heart was racing and he had no memory of picking up the phone.

'Alacktaka.'

'What?'

'We found out for you what it is.' It was Shaila, and she was whispering with great excitement.

'Who is we?'

'Me and my friend Gisela Middlecourt. We went to the library yesterday afternoon.' Then there was the sound of a struggle. 'Gisela, stop it.'

Sartaj waited for the giggling to subside, and then said, 'Why are you whispering?'

'Listen,' Shaila said. 'We're sort of interested in things to paint lips with.'

'You are?'

'Of course we are,' Shaila said. 'Be quiet and listen. So we wondered what alacktaka was. Gisela said we should look in the *Britannica*. It wasn't there. Then we looked in the *Oxford*. No. Then we thought, all right, the Urdu-English dictionary, which is down the reference shelf, you know.'

'No, I don't know,' Sartaj said, resting his head on his knees.

'Obviously you don't. It wasn't there. So then the Persian-English dictionary. Still no. Then we found the Sanskrit-English Dictionary, Ety, Ety-mologically and Phil-lolo-gically Arranged, by Sir Algernon Algernon-Williams, MA, KCIE, and Principal A. S. Bharve, published 1889.'

'Shaila, what's the point?'

'You're not grateful in the very least.'

'For what?'

'Alacktaka. Page 232 of the Sanskrit-English dictionary. Alakta, rarely Alacktaka. Red juice or lac, obtained from the red resin of certain trees and from the cochineal's red sap. Used by men and women to dye certain parts of their bodies, especially the soles of the feet and lips.' Now there was another bout of stifled laughter. 'R. 7.7. *Mk.* 4.15. *Km.* 5.34.'

'What's that?'

'*Rig Veda* 7.7. That's a reference. Sister Carmina told us. Do you know what *Km.* is? No, I'll tell you. Sister Carmina didn't want to tell us. It's the *Kama Sutra,* which she says isn't in the library. But Gisela's parents have a copy which they think is hidden away on top of their shelf. We looked it up. It's there, Chapter 5. Advice to the young gentleman, man-about-town. After your morning bath you put on balms and alakta, before you go out. I'll read it to you.'

'Shaila?'

'What?'

'Don't.' Sartaj was staring at the top of his own head, which he could see in the mirror on the wall. He was wondering what Chetanbhai Ghanshyam Patel put on in the morning. What was his aftershave? Sartaj rubbed the skin on his wrist, under his kara, remembering again the heavy silkiness of the Rolex. Where was Chetanbhai's copy of the *Kama Sutra*?

'Why are you quiet? Are you thinking? What are you thinking about?' Shaila chirped into Sartaj's ear, very interested.

'Never mind what I'm thinking about,' Sartaj said. 'You put that book back where you found it. And don't read any more.' He could hear them laughing as he hung up.

Taking a deep breath, Sartaj plunged into the swamp. Above, the morning sky was low and dark, heavy with black clouds. The water came up fast to his thighs and then to his waist, and he clutched dizzily at the reeds to keep his balance. Things moved under his feet and the water lapped against his shirt, but finally he was able to take a step, and then another. The surface of the water was covered with a foam-like scum, and there were rags and drabbles of paper stuck to the reeds. After another step the thick green plants closed behind him and he could no longer see the buildings across the road. He was trying to make a circle but he could no longer tell where he was.

He pushed ahead, moving aside the matted stalks with his hands. His face was covered with moisture and the breath burned in his chest. Something buzzed against his cheek and then he saw a flash of white to his left. He tried to turn and something gave way under him and he fell full length into the water. As he came closer he saw that the white was a smooth form, like stone, that came out of the liquid, and as he struggled he tried to place it in his memory, he was sure that he had seen it before, and then it turned over in his vision and took on a new form, like a cloud, and he saw that it was the Apsara's shoulder, and that she lay face down in the water, almost covered but held up by the criss-crossing reeds and the thrust of the swamp itself. He reached her and panted as he strained to turn her over. Her smile was eternal and untouched by what she had come to. He looked around for the buildings and the pathway, but could see nothing, and then he tried to imagine Chetanbhai's wife bringing out the Apsara and throwing her into the muck. It was impossible and improbable, but he could see without any effort Kshitij dragging out the white form, like a corpse, in the darkness of the early morning, not looking at the Apsara's eyes, her swollen lips, and tipping her into the water. What he couldn't form within himself was a logic for it, a first cause, a reason why. Not yet, he said, yet, and then he began to look for a way out.

~~~

'You smell,' Kshitij said when he opened the door to Sartaj's urgent knocking.

'Of shit, yes,' Sartaj said, walking into the apartment, which was clean now, neatened and stripped of its gaudiness. The paintings were gone, and the books on the shelf were a different lot, thicker with gold writing on the spines.

'Did you fall into a gutter or something?' Kshitij said, looking at the puddles of water on his floor. 'Hope you didn't swallow anything. You need medical attention if you did.'

'Not quite. Not quite.' The swamp was used as a gutter by the labourers who built the apartment buildings, and some of the servants who worked in them, and Sartaj understood Kshitij's pained look of distaste, but he was reading the titles on the shelf. '*A History of the Indian People*,' Sartaj said. 'Was your father a great reader?'

'Not really, no,' Kshitij said.

'Your mother?'

'No, she's not.'

'I see. And you?'

'Yes, I read. But is something wrong?'

'I'd like to see that chequebook I gave to you. If I may.'

'What is this? I thought you had the man.'

'I have the man. Where is it?'

There was a moment then in which Sartaj saw the possibilities clicking in Kshitij's eyes, and fear, and then Kshitij shrugged and laughed. 'All right. No problem.'

When he had it, a minute later, Sartaj held the chequebook by his fingertips, away from his body, and flicked through the pages. 'I'll keep this,' he said.

'Sure. But why?'

Sartaj looked at him, considering. Then he leaned forward and said deliberately, 'I'm very interested in your father's reading habits.'

The Jankidas Publishing Company turned out to be a man, a woman, and two computers in a garage. The garage was at the rear of an old four-storeyed building in a lane near Bandra station. Lines of fresh clothes hung from every balcony above the garage, and as Sartaj, now dry and no longer fetid, unlaced his shoes he was aware that at least three women were watching him from different homes above, their laundry forgotten. He had called Katekar from home, while he was changing, and had him waiting outside Chetanbhai's building in plain clothes, ready to shadow Kshitij. There was an excitement in his blood now, a hunter's prickling on his forearms. As he tugged at the lace on his left shoe there was again that darkened stirring in his mind, something falling into shape, barely recognizable yet. But moving. The 'Remove Shoes Please' sign had been done in a fancy curled typeface on red paper, and inside, Mr Jankidas was eating his lunch near his computer, under a purple sign that announced, 'We Believe in God and Cash. No Credit Please.' Mrs Jankidas who wore steel-rimmed spectacles like her husband and looked very much like him except for her full head of hair, held a tiffin from which she occasionally served out puris and bhaji. Mr Jankidas sat cross-legged in his chair, quite content, and it was as perfect a scene of domestic tranquillity as Sartaj had ever seen. He broke it with some satisfaction.

'I am aware of some transactions between you and a certain Mr Chetanbhai Ghanshyam Patel,' he said. 'Who is unfortunately in the position of being the deceased in a very serious case of murder.'

'It's more serious than the average murder case?' Mr Jankidas said. Sartaj liked him.

'It appears to be,' Sartaj said. 'It may turn out to be very complicated. Which is why I must appeal to you. This Mr Patel was in the habit of writing cheques to you. Monthly, that is, as far as I can tell.'

'He was our client,' Mr. Jankidas said.

'Satisfied, no doubt. What did he purchase from you?'

Mrs Jankidas tilted her head slightly and Sartaj saw the look of command that passed between them.

'We promise our clients confidentiality,' Mr Jankidas said. 'It is, you understand, part of the terms.'

Sartaj leaned back in his chair. He was quite comfortable in the air conditioning. 'Is this allowed, to operate a business in a garage in this building? According to the building society rules? What about municipal rules? I wonder.' He was addressing himself to Mrs Jankidas. 'I must remember to find out.' He turned his head to look at the other signs in the room. Mr Jankidas was a believer in signs. 'Dust Is the Enemy of Efficiency'. 'Customer Is our Joy'. When Sartaj looked back Mr Jankidas was ready to talk.

'We provide multiple services. Brochures. Business cards. Company papers. Legal typing. Invitations. Wedding cards. If you need, please.'

'And?'

'Also we publish a magazine.'

'Yes?'

'We function as a stage, you see. For the exchange of information. Mutual communications.'

'What kind of information?'

Mr Jankidas bent over in his chair and fished beneath the computer desk. He brought out a magazine with slick covers, all red and yellow, resplendent with many typefaces. On the front, a young woman looked straight into the camera, under the title in green, *The Metropolitan*. Sartaj had seen it before, at railway bookstalls, the pages stapled together securely. He took it from Mr

Jankidas and opened it and read, in the middle of a column, 'R-346. (M) *Bombay:* I am 32 year old, Engineer man working as a Class-I Gazetted Officer in a Central Govt. Establishment. 168 cms, 72 kgs, heartily welcome bc, cpn from frank, good-looking, bm ladies-couples l-nw, no bars. Cfndt assured-expected, pht appreciated. H: Games, nature, outing. LK: English, Hindi, Marathi.'

'What's a "bm"?' Sartaj said.

'Broad-minded,' Mrs Jankidas said suddenly. 'H for Hobbies, LK for Languages Known.'

'Languages. Of course,' Sartaj said. 'I see, I see.' Of course he didn't see at all, but he pushed on. 'I take it Chetanbhai was a customer for this.'

'Yes,' Mr. Jankidas said. 'They ran one ordinary ad every month.'

'They?'

Mr Jankidas picked up a ledger, ran a finger down a ruled page, and pronounced, 'M-434.'

Sartaj found M-434, towards the back of the magazine. '*Bombay:* In absence of true and loving friendship, life is nothing but one long process of getting tired. Let's join up! Come together to explore the fascinating future. An educated, charming, very friendly couple is inviting like-minded singles/couples to make the life a colourful span of sweet surprises, tender thrills. Melodious moments, fabulous felicities and warm welcome are all waiting for you. What are you waiting for?'

'It was the same ad every month,' Mr Jankidas said.

'It must have had results,' Sartaj said.

Mr Jankidas held up his hands. 'We don't know. We just forward the letters, for a small fee of course.'

'You see,' Mrs Jankidas said. 'Mutual communications.'

Sartaj drove his motorcycle to Colaba in a daze of wonder. He looked at the people passing by, at their faces, marvelling at their calm, their public banality. A woman in a red sari waited at a crossroad for a bus to pass, holding a netted bag full of potatoes. A taxi driver leaned on his cab, spilling tobacco into his palm. Noticing Sartaj's stare, a girl in a green and white school uniform looked away. But Sartaj was asking merely if she, they, how many of them had colourful spans of sweet surprises. He knew he had finally heard Chetanbhai Ghanshyam Patel's voice, and he felt revived and childlike, as if he knew nothing. 'Melodious moments,' he said to himself. 'Melodious moments.' After he parked he pulled the photograph of Chetanbhai and his family from the file, and sat looking at it, at Chetanbhai's round face, the sweetness of the smile on his face, the pride in his wife's eyes as she looked into the camera, secure in the thickness of her gold bangles and her family, confident of the future. And behind them Kshitij, unsmiling but earnest, and on the right of the picture, a flash of white just inside the frame, the curve of a soft hip.

Finding the parking lot that Chetanbhai had used was simple. Getting the two brothers who sat shifts in the wooden booth, handing out tickets and change, to remember the Patels was simple: they were excited by the thought of being part of an investigation, and the elder, who worked Saturday days, remembered, instantly, the red Contessa.

'That's them,' he said, when Sartaj showed them the photograph. 'Them two.'

'Which way did they go?' Sartaj said.

'Always there,' the man said, pointing with a rigid finger. 'There.'

Sartaj had a general direction, and so he started. The shopkeepers were busy with tourists, and there was commerce everywhere, from the stacked statues of Krishna and the men leaning in doorways muttering prices for illegal substances barely out of Sartaj's hearing. 'Who has time to look nowadays, baba?' one stall-keeper said, wrapping a pair of chappals for a Japanese couple. 'People come and go.' But then he looked at the photograph and remembered them. They had never bought anything but they had been walking by for years. They had become familiar to him because of their regularity.

Around the corner from the chappal seller, from his rows of footwear, there was an alley's length of beer bars. Sartaj went into the first, into the air-conditioned darkness, and found a patient line of women in shiny green and red churidars. They shook their heads at the photograph deliberately, back and forth and back again, and their lack of expression was complete and carefully maintained. Sartaj nodded, thanked them, and went on to the next bar. Finally, at the end of the road, near the seafront, there was a hotel used mainly by visiting Arabs. In the lobby, the men standing by the ceiling-high picture of an emir knew nothing, had seen nothing, weren't interested. 'We are guides, you see,' the oldest one said, as if that were a scientific explanation for not seeing anything. 'Guides for the Arabs.' He straightened his red tie and looked very serious. Sartaj smiled at the gentle mockery, nodded, and went on to the desk clerk, who knew nothing, but the durwan outside remembered the Patels. He had always thought they lived somewhere nearby.

By late afternoon Sartaj knew he was getting close to the destination. He overshot the trail only once, when they took a left, and it took him an hour to come back and find the direction again. Now he was striding briskly, along a back street lined with faded four-storey apartment buildings. Though the road was crowded with parked cars, it was quiet, shaded with trees, so that Sartaj could feel the lost elegance in the names of the buildings and in the circular balconies that must have been all the rage. To the left, in the shadow of a neem tree, behind a gate with a plaque—'Seaside Villa'—a man with white hair knelt among flowers. Sartaj walked by him, then turned and came back.

'You,' he said, crooking his finger.

The man pushed himself up and came to the gate, dusting his hands on old khaki shorts.

'Your good name, please,' Sartaj said.

'A. M. Khare, IFS retired.' Despite his torn banian there was the assurance of having travelled the world.

'Mr Khare, have you seen these persons?'

'Not once but many times. He talked to me about my flowers.'

'What kind of flowers are those?'

'Orchids. They are very hard to grow.'

'Did he know anything about flowers?'

'No, but he complimented me particularly.'

'Did you observe where they went after they talked to you?'

Khare shrugged. He seemed embarrassed. 'Yes.'

'Yes?'

'To the building there'—pointing with his chin—'called Daman.'

On the second floor of Daman, Sartaj found a boarding house, which was really a large flat, with thin partitions making tiny rooms let out mostly to trainees at the Taj. But, Mrs Khanna said, there was a deluxe suite, on the floor above, next to her own rooms, which she hired out very rarely and only to people known to her. Mrs Khanna wore a green caftan and smoked rapidly, and spoke in a no-nonsense style designed to intimidate tenants. She nodded at the photograph.

'Known them for years,' she said. 'Regulars. Nice people. Paid in cash, advance.'

'For what?' Sartaj said. 'Who came to see them?'

'I don't ask questions. Not my business.'

'But you notice things.'

She shook her head, deliberately. 'Not my business.'

'Let's see the room.'

There was a long passageway from Mrs Khanna's flat to the suite, with a locked door on either end. The inner door opened into a small room filled up by a coffee table and four old chairs. On the wall there was a painting, ruins on a cliff, over a river.

'See,' Mrs Khanna said. 'With attached bathroom. Very nice.'

Sartaj followed her into the bedroom. The green curtains were drawn and it was very dark, and Sartaj felt his head swim in the sudden quiet. Over the bed, a village belle flashed dark eyes at him over the edge of her stylized yellow dupatta. He reached down to the cassette player perched on the headboard and popped out the tape. There was no label. Sartaj put it back in the player and pressed a button. Mehdi Hassan sang: *Ranjish hi sahi...*

'Is this his tape?' Sartaj said.

'Yes. Mr Patel's tape.'

'And the paintings?'

'Also his. He said the room was very sparse.' She looked around the room, gesturing with a cigarette. 'He was, he was a very shaukeen type of person, you see.'

'Yes, a lover of the fabulous felicities.'

'What?'

But Sartaj was drawing back the curtains. Mrs Khanna watched keenly as he went through the bedroom and into the small bathroom, which was sparkling clean. She was clearly amused as he bent over to look behind the commode.

'It's cleaned every day. Or when it's used,' she said. 'Nothing left over. Nothing to find.'

'Very commendable,' Sartaj said. 'Are you sure you never saw any visitors?'

'No. Separate door there, opens out in front of the lift. They come and they go.'

'And this boy in the picture? Mr Patel's son? Have you ever seen him? Did he ever come here?'

'No.'

'He's dead, you know. Mr Patel is murdered. You know?'

'I read the paper.'

'What is your idea about it?'

Mrs Khanna was holding her cigarette carefully in two extended fingers. Nothing moved except the smoke. 'I'm not curious,' she said. 'Not my business. Don't want to know.'

Sartaj searched the room. The mattress was clean, the floor underneath swept, every surface was clean and polished, and the rubbish bin empty. Mrs

Khanna was a good housekeeper. In a drawer in the bureau next to the bed there was an opened packet of Trojan condoms, 'Ultra-Fine.'

'These? Mr Patel's?'

'Perhaps,' Mrs Khanna said.

'American. Very expensive.'

'He was shaukeen.'

Sartaj looked at the bureau, under it. He peered behind the headboard of the bed, then to the left, behind the bureau. It was close up to the wall. He put the tips of three fingers behind the wood and tugged. Then with a flicker of pain his fingers came away and the bureau sat, battered and unmoving. He squatted, gripped low by its legs and pulled. A grunt, another one, and it shifted. Now he was able to get behind it, to look. There was something between the green wood and the baseboard. He reached down into the crack, searched with his fingertips. He brought it up and saw that it was a photograph, a Polaroid. He wiped away the dust, and the colours in it formed an image, and he turned it this way and that, and saw then that it was a woman's body, naked but blurred, the brown of the skin hidden in parts by a moving smear of white, as if a sheet had been pulled off and she had turned away, all the frame filled with fast motion. Her face was hidden by a hand, an up-flung arm, the chin barely visible as a suggestion, but there was her hair, long and thick and luxuriant. And the curve of a naked hip.

'Do you know who this is?' Sartaj said.

Mrs Khanna considered the matter. 'No,' she said. She was bored. There was much in the world she didn't want to know about, and the naked body was no news to her.

'Did you ever see Mr Patel with a camera?'

'No.'

Sartaj was looking at the picture, trying to read what he could see of the forehead, the chin. Was it protest? Or laughter? Mrs Khanna watched him, and he saw that she was faintly amused by his attention to the picture. He held up the family photo of the Patels. 'Are you sure you never saw this boy?'

'Told you, I didn't.'

'This is a dangerous position you're in, you realize? Running a house of prostitution without license?'

'Nobody prostitutes anything here.'

'Still it would bear investigation.'

'Investigate if you want. I'm not involved.'

'You mean you don't want to be involved.'

She shrugged. Sartaj shook his head, and turned away. When he was at the door, she said, 'All right.'

He turned back to her. 'What?'

'I'll tell you,' she said. 'Because I didn't like the little bastard in that photograph. He threatened me. But if you try and get me involved I'll deny everything. I won't sign anything, and I won't appear anywhere.'

'Yes, fine,' Sartaj said. 'Tell me. What about the little bastard?'

'He came here,' Mrs Khanna said.

'When?'

'I don't know, nine, ten days ago.'

'Alone? What did he want?'

'Yes, alone. First he wouldn't tell me who he was, but said that he wanted to see the room that Mr and Mrs Patel hired. I said I didn't know any Mr and Mrs Patel, and why should I let him into any room. Then he began to argue, and I told him to take himself out of my house before I had him thrown out. Then he asked if they had left anything in there, anything at all, and I said I was tired of talking to him, and called Jaggan from downstairs to throw him out. Then he started shouting, abused me. Randi, he said. So Jaggan gave him a shake.'

'Then?'

'Then he said he would come back. With friends. Come back with your whole paltan, I said. We'll see you, he said. The Rakshaks will take care of your kind. They know you and your type. They know how to handle whores like you.' Mrs Khanna studied the end of her cigarette thoughtfully.

'He said the Rakshaks?' Sartaj said.

She tipped her head to the side. It was a strange gesture, full of resignation. 'Yes, that's what he said. They're absolutely mad, those bastards. Capable of anything. So I let him see the room. Why take trouble on unnecessarily?'

'True,' Sartaj said. 'What did he do in the room?'

'He went through it,' she said. 'Just like you, searched it, opened drawers, looked under the bed, in the bathroom. Looked in the rubbish bin. Like he was checking it for something. Evidence. Clues. Things left behind.'

'I see.'

'But he was shaking and his eyes were red. Speaking under his breath to himself. Mad.'

'Yes, mad. Did he find anything?'

'No, nothing. He was their son?'

'Yes.'

Mrs Khanna stubbed out her cigarette in an ashtray. 'Chetanbhai was a good man,' she said. 'Poor Chetanbhai.'

On his motorcycle, with metal trembling against his thighs, Sartaj thought of bodies. He tried to picture his own, and found it curiously blurred, his knowledge of it dulled and patchy. He had known it forever, but what were his shoulders like in pain? The back of his thighs in sleep? Megha he knew, the pulse on her wrist and the one that beat hard on her throat, but each time they settled against each other, had settled, he had felt compelled to take stock anew. Like a man afraid of memory dying. The loud blatting of the engine beat on his ears. He tried to imagine Chetanbhai Ghanshyam Patel naked, as he had been under the sheet at the morgue, a round face and a sloping chest, a paunch, thighs splayed apart. In his life had somebody watched him in the morning, awoken him with an invading caress? Sartaj tried to imagine his own parents, and his mind turned. It turned and stopped, as firmly as the bike stopped in grinding rush-hour traffic, in front of the temple at Mahalakshmi. His mother he saw in her gentle plumpness, her slow walk, her pleasure in afternoon talk and tea. He remembered his father's vigour, the energy the old man took such pride in at fifty, bounding up the stairs, roaring with unexpected laughter at his son's infrequent jokes. But had they lain against each other late at night, satiated but unable to sleep? Touching with a holding hand? These were only words, and Sartaj was unable to see it. It was impossible to imagine. There was no photograph that he could construct, rising the colours out of his childhood, marking here and there with lines until the shapes became bodies.

Katekar was waiting at the end of the lane when he reached the Narayan Housing Colony, smoking a bidi near a paan-seller's kiosk. Sartaj was wearing his dingiest civvies, but Katekar looked sceptical as Sartaj strolled up, relaxed and studiedly casual. Katekar was carefully nondescript in shiny rayon pants, and there was nothing in Sartaj's wardrobe that could equal the bland horror of his shirt. Sartaj accepted the criticism, because since training he had known he was not very good at shadowing, and his turban was only part of the problem. It was his walk and the cast of his shoulders—he had found it difficult always not to swagger, to fade away into the crowd.

'The mother's gone,' Katekar said, looking away and pulling on the bidi. They were two friends passing the time on a weekday evening. 'He drove her to Bombay Central this evening. She caught the Saurashtra Express to Okha. That's in Gujarat.'

'I know. They've started running,' Sartaj said. 'That's very good. Then?'

'He didn't wait for the train to leave,' Katekar said, his face stiff with disapproval. 'Just put her in the compartment and came back here and went up. Just got back. He's up there.'

'Right,' Sartaj said. 'Go back to the station house. Rest. Well done.'

Katekar nodded. 'With care, sahib. Something strange is going on here.'

'I know,' Sartaj said. He watched as Katekar pedalled away on an old Hero bicycle. In many ways, care had come hard to Sartaj, because in his pride he had believed that he would win the things he wanted by just asking. But now, older, he had found patience. He found a table behind a window at a restaurant called East Haven near the intersection and ordered dinner. He regretted the ragda-patis after the first bite. It had that searing hotness of all the cheap restaurant food that he and Megha had once lived on. It was incredible to him that they had actually enjoyed it, in Formica-panelled dives like this one. Now the place filled up quickly with t-shirted teenagers, and Sartaj listened intently to them, and pushed a piece of bread around his plate. The road outside was crowded with cars and scooters and through it all he could see the gate through which Kshitij must come, but an hour passed, and then two, and there was no sign of him. The groups of friends changed at the tables around Sartaj, and outside on the tiny patio, and it was a slightly older crowd now, less raucous, with none of the rapturous pubescent laughter of the early evening, but still there was no sight of Kshitij, or any mention of him in the lengthy gossip that wound and unwound in the smoky air. They all knew each other, as everyone had in Sartaj's colony, and he remembered sitting like this, shoulder to shoulder in an endless round of silly phattas and cups of tea. And the girls who came in and sat at their own tables but not too far away. Sartaj wondered if Kshitij would stay up there all night, behind the brassbound door, without even the Apsara for company, and then gave up the thought instantly: no one could stand that hideous loneliness of that apartment's four bare walls. He would come, sooner or later.

At eleven East Haven was nearly empty, and the waiters had stopped asking Sartaj if he wanted anything. The traffic was no longer the mad snarl of the evening, and yet Sartaj missed the car completely. It had passed and was already long out of sight behind a bus when Sartaj saw in his memory the red Contessa with the shadowed figure behind the wheel. He ran out to his motorcycle, leaving a hundred on the table, and cursed the machine as it coughed and hung. By the time he came out at the mouth of the bazaar, the Contessa was far ahead, a pair of red dots far ahead in the dark, and now his anger at himself for expecting a walking man made himself come up on it fast, too fast. Before the next intersection he found the right distance, two cars behind and to the left, and whenever there was a lit-up stretch of road he switched off his headlight, so that there would be no regular illumination in the Contessa's rearview mirrors to attract attention and suspicion.

But there was no sign that the driver ahead knew of the shadow: the car went steadily, not too fast or too slow, through the Andheri subway and north on the highway on the other side, past the Sahar turn-off, past the right turn for Film City, and still further north, and Sartaj exulted in a growing certainty: this was too far away and too late to be innocent. And in the next moment he remembered the joy of pointless and endless rides, not so long ago, and he rebuked himself for having become, completely and deep in the bone, what he had thought he would never become—a policeman like his father, and his father before him.

The car stopped in front of a new apartment building, in a development so new that there was still the rolling shape of a wooded hillside behind it. Sartaj switched off his engine and coasted gently to a halt, crunching across gravel as a figure got out of the Contessa and walked in through the double gates. There was a cool breath of air from the hill as Sartaj pushed down the kickstand and edged closer to the car, keeping his back against the low wall that ran around the building. There was that sound of night that was always lost in the city, insects and calling birds, and the size of the sky. Two men came out now, through the doors, each carrying something heavy, dark bundles that they loaded into the dickey and then the back seat. They went back and forth, carrying the square boxes to the car, and now Sartaj could see the glint of Kshitij's glasses. When the car was full, they got in, and now the sound of the engine was very loud, and the glaring circle of the headlights moved towards Sartaj. He took a deep breath and stepped out into the road, holding up his hand.

He leaned them against the hood of the car and patted under their arms and over their backs. He let them see the pistol under his bush shirt, and Kshitij frowned.

'What do you want?' he said. 'Why are you doing this?'

'Shut up,' Sartaj said. He walked around the car, and the back seat was full of the bulky packages. Sartaj's heart was thudding. The dickey opened smoothly, and Sartaj felt around in the box on top. There was only paper, pamphlets of some kind. He held the box by the edge and pulled it out, tipped it over on to the road, and the small stapled notebook spread across the asphalt. There was nothing else in the box.

'Why are you doing it?' the other boy said. He was larger than Kshitij, more confident in his body. 'That is only our literature.'

'Whose literature?'

'Us,' the boy said, gesturing at himself and Kshitij, and then at the building. 'Us, the Rakshaks.'

'You're a Rakshak?' Sartaj said to Kshitij. 'Yes,' Kshitij said, standing up straight. 'I am.'

All of Kshitij's resentment made sense now, his dense anger just under the innocuous surface, all of it barely concealed. Except, of course, to a man vain enough not to believe that he could be despised for what he was, for his beard, for his turban. But Kshitij no longer was hiding his contempt, his keen scalpel-edged hostility. His face was eager with it.

'You have no right to do this,' his friend said. 'We are a cultural organization.'

'This?' Sartaj said, bending down to pick up one of the pamphlets. 'This is your culture?' There was a line drawing of a goddess on the cover, superimposed over a map of India, and the words 'The Defender' underneath. He had seen the magazine before: it was a call to arms, a hearkening back to a perfect past of virtue and strength, and an explanation for every downfall. 'You defend nothing. You are attackers.'

'We attack only those who attack us. And those who attack our culture.'

'What is your name?'

'Pramod Wagle,' the boy said, puffing out his chest.

'Pramod Wagle. I want to see what's inside,' Sartaj said. He pushed them towards the building, and as they came up to it two men came out of a door and watched silently. On the ground floor, next to the unfinished lobby, there was a double door with the same goddess painted in full colour over the door, and she wore a sari of radiant white. Now there were three other men in the corridor, watching.

Inside, the apartment was divided into an office, which contained a large cyclostyle machine, and a dank gymnasium, with mirrors on the wall, and a rack of lathis.

'Everything is legal,' Pramod Wagle said. 'We are a registered organization.'

Sartaj was fingering the long length of a lathi, following the dark sheen of the wood. 'Yes,' Sartaj said. He had seen, four months before, a man killed with a lathi in a fight, his head had been split open, but Wagle was correct, it was legal. So was everything else. 'Come on,' Sartaj said to Kshitij, and turned to the door, but in the corridor outside it was they who watched him, a solid phalanx of dark faces, completely silent. They stepped aside to let him pass, but exactly that one moment too late that let him know that they could do anything they wanted, despite his pistol and any other thing.

'Drive to the police station,' Sartaj said. 'I will follow you on the motorcycle.'

'Why?' Kshitij said.

'There are certain questions we must ask you.'

'Am I under arrest?'

'No,' Sartaj said. Just behind Kshitij his friends made a crowd, pushing against each other's shoulders.

'Did your father own a Polaroid camera?' Sartaj said. In the boy's silence, his absolute stillness, there was absolute fear. Sartaj leaned over to him and spoke in his ear. He could smell, faintly, hair oil. 'I have the photographs, Kshitij. I could pull them out now in front of everyone and show them to you.'

'All right,' Kshitij said finally. His voice was loud. He turned to Pramod Wagle. 'It's all right. Just some questions.'

⌇⌇

But at the station he refused to answer questions. Somehow, during the drive to the station, with Sartaj's single light blinking and bobbing in his rearview mirror, he had decided that there were no questions he could answer. He sat with his arms folded across his chest, clenching his jaw. 'Why are you asking me anything? You have the man in jail already. I want to see a lawyer.'

'Come on, Kshitij,' Sartaj said, leaning back in his chair. 'Come on.' He had Kshitij sitting in his office, across the desk from him. Behind Kshitij, Katekar sat in a chair against the wall, slumped but alert. 'We know everything. We don't need you to say anything really. We know every last thing. We know what your mother and father did. Here, look at this advertisement. Strange, isn't it? Good, ordinary people. Doing this kind of thing. Unbelievable, if we didn't have it in black and white. Then, we have the woman who runs the house where they rented a room. A cheap whorehouse, really, it is. This woman, the manager, tells us all kinds of things. And also we have, finally, the photographs. Colour Polaroid photographs. Like life itself. Can you imagine? Disgusting photographs. Disgusting things they are doing with god-knows-who-all. To do things like that and then take photographs… I wouldn't have believed it myself if someone had told me. Only when one sees with one's own eyes can one believe. Is that what you were burning in the rubbish heap?'

Kshitij was staring at the brass plate on Sartaj's desk, which announced his name in blocky ornamental letters. He seemed to be reading it back and forth. Behind him, a slow and very faint smile spread across Katekar's face.

'So what happened, Kshitij?' Sartaj said. 'Did you find the photographs? Did you see them?'

'I don't know what you're talking about,' Kshitij said.

'Did you see your mother in these photographs?'

'I don't know what you're talking about.'

'With other men, Kshitij?'

'I don't know what you're talking about.'

'Doing things, Kshitij?'

'I don't know what you're talking about.'

'A mother is pure, Kshitij. After all she is a mother. But your mother, Kshitij. What is she? In a whorehouse? And your father? To take pictures? What did you see, Kshitij? We'll find out, you know, Kshitij. We're investigating, we have our people asking questions in Okha. We know she's gone to her brother's house near Dwarka. Samnagar, the place is called, isn't it? We'll bring her back. In handcuffs to jail. The world will know then. All your friends. Everyone will know everything about her. So you might as well tell us now. Maybe we can keep it all quiet. What happened? Did you see these photographs and get angry?'

'I don't know what you're talking about.'

'Did you see something horrible in the photographs, Kshitij? Of course it was horrible. Your parents. Your mother.' Sartaj stopped, swallowed. Then he leaned in close. 'Did you see your mother with some stranger? Sucking on him?' The light in the room was yellow, and outside the stillness of night, and small sounds from far away. Sartaj could see the outline of his own head, his turban, in Kshitij's eyes, and he knew he had on his interrogation face with the opaque eyes, and in his body, in his arms and legs, there was the uncurling of a virulent hunger, an angry need to know. There was also somewhere sympathy and disgust and horror, but all that was faint and far away and battened down, safely subterranean.

'I don't know,' Kshitij said. And then he stopped. There was no motion in his body, but a kind of rippling, like the surface of water in which no current can be seen. Through the night he grew more inert, like a stone sinking into the worn cloth of the chair, and yet somewhere in the sunken eyes, in the base of the throat, there was that agitation. At two in the morning it had begun to rain, and Sartaj left him to Katekar and walked outside, along the corridor that ran beside the offices. There was that usual late-night talk and movement at the front of the station, the drunks on their way into the lockup. Sartaj stretched, and reached with an open palm into the rain. His kara moved slowly on his wrist. There was the steady drip of water onto his skin. He and Katekar would keep the suspect up all night, taking turns, wearing at him with the repeated questions, beating at him with half-knowledge and insinuations, until in the early day he broke in exhaustion. It was likely. Many did.

He heard a shuffle behind him, the sound of feet. It was the head constable, bringing the night's phone messages. The first was a ten-thirty message from Rahul, asking for a call back. The second the usual late-evening call from Sartaj's mother, and the third was an intimation from Cooper Hospital. Ghorpade was dead. He had died at midnight after a day of discomfort and difficulty with breathing. Sartaj put the slips of paper in his pocket, scratched at his eyebrows, and then he walked back into his office. Katekar was leaning over the suspect, looming over him and letting him smell sweat and tobacco. But Kshitij had found from somewhere a small reserve of fortitude.

'If you are going to charge me with something, charge me,' he said. 'File an FIR, get a warrant. Otherwise what is all this? You're doing this because I'm a member of the Rakshaks.'

Sartaj sat behind his desk. He twisted his watch around his wrist, once, twice. His cheeks felt congested with rage. 'Katekar, this chutiya thinks we're idiots,' he said. 'And he thinks he's very smart. Take him down to the detection room and take some of his smartness out of him. Give him a good taste of what we do to smart chutiyas around here.'

Katekar had Kshitij by the scruff of his neck and out of his chair before the boy had time to react, even to open his mouth. As he turned Kshitij away, Sartaj forced himself to raise a hand: easy, no marks. Katekar slammed Kshitij through the swinging doors, and pushed him down the corridor. 'Chala,' he shouted, and there was a terrible anger in his voice.

Sartaj squared the papers on his desk and tried to work. It was raining heavily now, and the water softened all sounds. After a minute or two he gave up, sat back, and put his hands over his face. When the phone rang, he let it ring six times before he picked it up.

'Sartaj Singh.'

'Did you take money?' It was Rahul, and his voice was feathery and high.

'What's wrong, Rahul? Are you crying?'

'Ravinder Mama came for dinner today. They were talking about you. They asked if you had signed the papers yet.'

'Yes?'

'Megha said not yet. Then Ravinder Mama said you probably were waiting for money. He said you were all for sale. So I called him a name. Daddy told me to shut up. I threw a plate on the ground and left.'

Sartaj shut his eyes. 'No, I don't take money.' But Sartaj then remembered all the things he had been offered because of his uniform, that he had taken, suits at half-price from a tailor near Kala Ghoda, meals with Megha at a five-star restaurant, miraculous train reservations in the middle of summer. A smiling road contractor had once brought over a canister of ghee to Sartaj's grandfather's house as a Diwali gift, and the old man had tipped the canister over the contractor's head. But life moved in jerky half-moments, in the empty spaces between big decisions, and Sartaj had been unable to resist the specially discounted shoes at Lucky's. Italian style, the proprietor of the shop had said again and again, Italian style. That evening long ago Megha and Sartaj had taken Rahul out for a birthday dinner, the thirteenth, and Rahul had noticed the shoes, and Sartaj had promised him a pair, and then told him stories of detection. Sartaj sat in his chair, and his lips moved: And I arrested a man for a crime he didn't commit, and that man is dead, and life is very long, and investigation is one way to get through it, but to call it justice is only half the truth. 'No, I don't take money,' he said into the mouthpiece.

'Did you talk to Megha?'

'Yes, I talked to her. Rahul, it's, it's not going to work.'

'Why?'

'We're not suited to each other.'

'Like you couldn't communicate?'

'Yes. That's it,' Sartaj said, grateful for the phrase.

'You should've learned how to communicate.'

'Yes.'

'What was that?' Rahul said. There had been a sudden surprised yowl that echoed faintly down the corridor.

'A cat outside, I think.'

'A cat?'

'A cat.'

'Okay.' As always, Rahul believed him. Rahul had a whole and unadulterated faith that was beautiful in its clarity: he believed. Now Rahul was trying to help. 'These things happen in life, you know. Between men and women. I'll always be here, you know, to help or anything you need.'

'Thanks, Rahul,' Sartaj said, his voice thick. 'I know that.'

Megha hadn't believed. One morning at breakfast she had put down a newspaper crowded with angry headlines, and had asked, for the third time in a month, do you really hit people? Torture them? Her brow was heavy with doubt, and he knew that the easy answer would no longer service. Yes, he said, sometimes it's necessary. It's a tool, an instrument. That night and the next night she had slept on the very far edge of the bed. When he had touched the back of her neck at breakfast she said, without looking up, I hate the world you live in. He had wanted to say, it's your world also, but I am thirty-one years old and I live in the parts you don't want to see. I live there for you. But he had quietly picked up his briefcase from the table in the hall and had shut the door behind him without another word. That had been one of their many silences. Now Sartaj walked down the corridor, towards a certain room in his world. As he walked he could see the curving pools of light from the bulbs, fading into darkness in the yard, and from beyond the shuffling sound of leaves under the fall of rain. He stepped through a door, and then another one. Katekar had Kshitij strapped face down on a bench. The bare feet hung over one end. The room was bare but for that bench and a chair, and had curving ceilings and a single ventilator high on the wall and, high up, higher than a man's uplifted hands would stretch, a thick white metal pole that went from one wall to the other. In his hand Katekar had a lathi, with the wood shining a heavy brown in the yellow light.

Sartaj pulled the chair and sat in front of Kshitij, one leg crossed over another. Kshitij's face was red.

'I'm sorry that you're making me do this, Kshitij,' Sartaj said. 'I hate to do this. Why don't you just be sensible and tell me what you saw, what you did? Did you scrub down the car, Kshitij? Why? What was in it?'

Kshitij's eyes were amazed, as if he were seeing something that he had never imagined. He seemed to be thinking, contemplating some new but essential truth he had just discovered. Sartaj tapped him gently on the cheek.

'You know, Kshitij,' Sartaj said. 'I spoke to a lot of people about your father. Everybody loved him.' Now Kshitij looked up, straining his neck, his mouth working. 'Everyone liked him, you know. His business colleagues said he was dependable, hard-working, dedicated. They thought he had come far and was to go far. In your building, they said he took his neighbour's troubles on his shoulders like his own. Always he was willing to help, not only with advice but practically. At the weddings of other people's sons and daughters how much work he did, they said. In times of grief a good friend. Generous and happy. Fun to be around, always singing, always playing his ghazals, always ready for a movie or an outing. A good husband in a happy family they said.'

Kshitij's eyes were watery and a trickle seeped from his left nostril. 'He was *not* a good man.' His voice came out thick and anguished. In all his years Sartaj had never seen a face so full of pain as this one.

'What did he do, Kshitij?' Sartaj said, leaning over close. In his stomach there was a bubbling, nausea, but he had to go on. 'What did he do? Tell me. I know he wasn't good, he fooled them. What did he do?' It was the beginning of a confession and he felt it coming. But Kshitij teetered at the edge for a moment, found himself then, and with an appalling effort pulled himself back. Sartaj saw the struggle as the face settled, went from disarray to control.

'I have nothing to say,' Kshitij said. Sartaj sat back, shrugged. 'Then I can't help you,' he said. 'I'm very sorry for that.' He waited until Kshitij looked up, then nodded at Katekar. 'Go ahead,' he said, and got up.

He was halfway across the room when he heard Kshitij's voice, loud now. 'What's the matter, bastard? Can't hit me yourself?'

Sartaj turned, then looked around the room. Next to the door there was a row of black metal hooks, and from one of the hooks hung a worn strap, a piece of a heavy industrial belt meant for machinery, four inches wide, attached to a wooden handle. Sartaj felt in his arms a painful pulsing of blood. He took the patta, turned around, and with all the swing in his shoulder brought the strap up and around and on to Kshitij's buttocks. And then again. The sound it made was like two flat pieces of wood dashing together. He had his arm back again when he heard, through the rushing in his ears, Kshitij's voice. 'What?' he said. He stopped, took a deep breath, and stepped up to the bench. Finally he could make out the words.

'You can't hurt me,' Kshitij was saying.

'Oy, did you hear the noises you were making?' Katekar said.

'That was only the body,' Kshitij said, and Sartaj could see the drops of spittle darkening the dirty floor.

'I'll hurt you, bhenchod,' Katekar said.

'You can't hurt me,' Kshitij said. 'Or kill me. Only my body.'

Sartaj could see the eyes, shining and focused, looking straight ahead, straight through the grimy wall, at something a thousand miles and a thousand years away. He dropped the patta, stumbled to the door, which rattled under his shaking hand, and as he fled to the cooler outside air, he could hear Kshitij chanting, 'Jai Hind, Jai Hind.' But outside, in the corridor, the sound of the rain was loud, and the voice was lost in the water, and Sartaj leaned against a pillar, leaned lower and out above the bending hedges, and retched into the darkness.

When he was able to stand straight, he saw that Katekar was watching from the other side of the pillar.

'I'm all right,' Sartaj said.

Katekar nodded, then turned back to the doorway.

'Katekar,' Sartaj said. 'No more. Just talk to him.'

'No? You don't think he'll talk if we give him a little more?'

'Not this one.'

Katekar nodded. 'What could we do to him?' he said. 'He's already in hell.'

Sartaj sat on a bench in the corridor, one leg over the other, looking out at the greying sky. He watched Katekar walking up, stretching his right shoulder and then his left.

'This one's not talking, sir,' Katekar said.

'Yes, I know,' Sartaj said.

'He's one of those, sir,' Katekar said. 'Gets stronger.'

'It's all right,' Sartaj said. 'Sit.'

'Sir?'

'Sit down, Katekar.'

A moment, and then Katekar sat, his legs apart, his hands on his knees.

'Did you always want to be a policeman, Katekar?'

'My father was, sir.'

'Mine also.'

'I know, sir.'

The rain had stopped. There was a silence like Sartaj had never heard before.

'My back is going to hurt,' Katekar said.

'Is going to hurt?'

'It doesn't hurt now,' Katekar said. 'But it will. Not this month perhaps, or the next. But then someday soon it will. It'll start and get worse.'

'Is it the muscles? Or a disc?'

'The doctors say it's nothing. They give me pills and tablets and it still hurts. Then my wife sends me on pilgrimage to Pandharpur. I walk with the palkhi of Dnyaneshwar. There are hundreds of pilgrims. I tell nobody I am a policeman. Nobody can tell one from another. We walk during the day and it is very hot. For the first day and the next and after that my legs hurt and squeeze and become tight. My feet swell and blister and it is difficult to get up the next morning to walk. The sun is very hot and it is all a plain, no trees and just a straight road. The walking is hard. We walk all together. The days pass and it seems like it will never end. Everything is forgotten but the walking. At night the pilgrims sing songs. There are discourses. But usually I fall asleep early on the hard ground and dream of walking. Then I wake up in the morning and walk. Of course I don't believe in any of it. My wife sends me. But when it is over after fifteen days, I cannot remember when my back became better but always, it no longer hurts. I come back to the city, tired. But my back is all right. For a while. Then it starts to hurt again.'

Sartaj thought he should say something, but then the moment for speaking passed and they sat quietly next to each other. There were the distinct shapes of trees now, the walls, the top of a building across the way. Soon the colours would appear, the huge sweep of green, covering everything.

Samnagar was full of television aerials and pucca houses, and its share of modern amenities, Sartaj knew, was testimony towards the entrepreneurial and adventuresome spirit of its sons and daughters. Rupees and dollars and pounds sterling had updated everything except the old .303 Lee-Enfield his escort carried, which meant a certain reassuring traditionalism in the local crime. He couldn't quite decide whether the enigma he carried in the case file inside his briefcase was old or new. Go yourself, Parulkar had said. We don't have time to get her down here, we can hold the boy for only two days, three maybe, there is much political pressure. Already there are calls for your transfer.

So Sartaj had gone. 'Bad road,' the driver said. 'Water.' The house they were looking for was six miles away from the village, from the town it was becoming. The metalled road vanished after the first mile and a half, and between the fields of cotton a rutted path rose and dipped. Now, ahead, it vanished beneath a sheet of water.

'We walk,' Sartaj said. He stopped caring about his pants in the first three steps, and then splashed forward furiously. The clouds piled up ahead of him, black and dense. He had a sensation of feeling quite small, under this arching sky and long silence. They walked past a grove of trees and a ruin, a single wall with a door and a window in it. As he walked, Sartaj realized a bird had been calling, again and again.

The house, when they reached it in late afternoon, was set between hedges at the intersection of three fields. There was one young constable, dozing on a charpai in front of the house. He woke up scrambling for his cap, panicked by Sartaj's high official presence, but managed finally to say, 'She's inside.'

She sat alone in a room inside, on the ground in a corner, in a widow's white and also an attitude of despair so sharp that Sartaj stopped at the door, all his eager volition to get at the heart of it gone, vanquished completely. 'She hasn't said a word since she got here,' her brother whispered into Sartaj's ear. 'Not one.' It was obvious. Her hair hung around her face and to the floor. She was staring at a point on the ground a foot ahead of her, and she didn't look up as Sartaj took off his shoes, or as he came and squatted next to her. He leaned close to her and spoke into her ear. He told her he knew everything. He told her what he knew and then conjecture as truth, the old policeman's trick: yes, this is how it happened, Kshitij found out, he saw photographs, he quarrelled with his father, something happened, something was said, correct, do you remember, but now he saw the images as he told her and he was afraid she would say, yes, that story is how it happened, he was suddenly afraid. But she looked straight ahead at the floor and he wasn't sure she heard him at all. Then he had nothing else to tell her. He stopped, and he could hear the bird calling outside. He came close to her, so that her hair tickled his nose, and he said, 'I want to know how it happened. How you came to this. Kshitij said his father was not a good man.' She looked at him then, and her face was homely and common to any street. Sartaj put the Polaroid down on the ground in front of her. 'This is you. Did he force you to do this? Did your husband force you to go to that room in Colaba?'

She shook her head.

'Don't be afraid now,' Sartaj said. 'It's all over. He forced you?'

Her gaze was level, and through her grief Sartaj could feel her pride. 'No,' she said. 'No. He didn't force me. Nobody forced me.' She held him by the wrist and spoke to him, her breath hot against his cheek. She spoke fast in a Kutchi he didn't completely understand, but he understood that there was no compulsion, no solution so simple as a bad man, only a series of fragments, dinner at the Khyber with her husband and her son, their honeymoon long ago in Khandala, a train ride and an upper berth together in a crowded compartment, at breakfast he must have a glass of cold milk, a movie in Bangalore and a quarrel during interval, and Sartaj knew that what Chetanbhai and Ashaben had done together was as complete and as inexplicable as what had happened between him and Megha, real and true and impossible to tell afterwards. Somehow it had happened. Not somehow but anyhow. Things happened, Sartaj thought, one after the other, and what we want from it is a kind of shape, a case report. Now Ashaben looked up at him, holding him still, with that confessional need, and he had seen it before and he knew what it required. 'I understand,' Sartaj said, and understood nothing.

Before he left, he said, from the door, out of a sense of duty, 'Will you testify? Would you sign a statement?' She began to weep.

As they began the trek back to the jeep it began to drizzle. They splashed on, and then the gusts began to spray the water from the puddles up into their faces. Finally they ran for the grove of trees, past the ruined wall, and at the centre of the grove it was possible even to sit on the ground. It was damp but it was comfortable. Sartaj could see, still, through half-open eyes, the wall, the angle of the doorway. The bricks were small and oddly shaped. Sartaj had taken history in college, but he had no idea if the wall was fifty years old and English, or from a Mughal serai. Or from the other Dwarka, Krishna's ancient city, sunk now, the story said, below the waves somewhere to the south. As he sat on the ground, Sartaj could feel the earth against the back of his thighs, grainy against his calves.

'What kind of tree is this?' Sartaj said, leaning back against the trunk.

'Mango, janaab,' the seniormost of the constables said.

'Rest, sahib. There is plenty of time.'

Sartaj was thinking of the plaster Apsara, bobbing in the water near the Narayan Housing Colony, west and north of Andheri West, Bombay. He thought of her sinking and then rising a thousand years later to confound some historian's calculations, and he laughed. He thought of the curve of her shoulder and the drops fell through the leaves above him. His eyes closed. He thought of Megha, and he tried to answer the question, Rahul's question, his own, and he said what happened to us was that we loved each other, and we were unkind to each other, and impatient, and unfaithful, and disappointed, and yet we wanted it for forever, but these are only words, and then came a flowing stream of images, dense with colour and the perfume of her hair, and it carried him. He felt himself floating and it felt easy, but then a moment of wild fear, he was sinking, he clutched and held on to himself, tightly, tightly, but then he felt his pride quicken, the word reverberated suddenly, alacktaka, and he made himself, he let himself go, and he was plummeting, down, into darkness.

The Rolex slid easily between Sartaj's fingers. There was pleasure, distinct and unmistakable and undiluted, in the silky fall of it against his skin, in its weight, substantial and unexpected. Under his hands there was Chetanbhai's case file, closed. It was over, according to the file, and there was a murderer who had died in the hospital. Kshitij had left that morning, walking slowly to his friends at the front of the station, and they had watched Sartaj with careful regard as he watched them.

Sartaj picked up the phone and dialled. He swung around in the chair as he listened to the steady ringing, half a dozen rings, then ten. In the glass of the map case he could see the shape of his turban.

'Hello, Ma,' he said. 'Peri pauna.' She had walked painfully across the drawing room, holding her hip, he knew this.

'Jite raho, beta. Where have you been?'

'Casework,' he said. As she spoke he reached back far in memory, trying to find the earliest fleeting fragment of her. He remembered her in Dalhousie, a cold mountain day, her white sari in a white chair on a patio in the sunlight, the rising mountains behind, the cold white peaks far ahead. He running up to her. How old had she been then? Young, younger than Megha.

'Ma,' he said.

'Yes?'

He wanted to say something to her about his father. Something about the two of them together, what they had said to each other as they walked behind him down a twisting mountain road, under the unfamiliar hill trees, leaning towards each other.

'What, Sartaj?'

He swallowed. 'Nothing, Ma.'

'Is something wrong?'

'No, not at all, Ma.'

After he put the receiver down, Sartaj turned back to his desk, gathered up the file. As he stood up, the watch warming in his fist, he remembered suddenly his mother getting up from the kitchen table, walking behind his father, who bent over a newspaper and a cup of tea, and her hand as it brushed over the heavy shoulder, touching for a moment her husband's cheek. The small swing of the woman's hair as she walked away. And the small smile that flickered on the man's face.

There was a doorway in Sophia College Lane, across from Megha's building, where he used to wait when they had first met each other. He was hiding in it again, now, in a new uniform, and that other long-ago self felt foreign somehow, another Sartaj, faintly puzzling. There was music drifting down from a window above, a ghazal, '*ye dhuan sa kahaan se uthta hai*', and the swirling rush of cars below on Warden Road. He listened to the music, and when Megha first came out of the building he didn't recognize her. Her hair was cut short, above the shoulders, and she was wearing dark glasses and she looked very stylish and young. She paused with her hand on the door of the Mercedes, raised her head up, looked about as if she had heard something. Sartaj stepped back into the shadows. Then she got in, the door shut, and the car moved off quickly, past Sartaj. He had a glimpse of her profile, and then it was gone.

He straightened up. He walked across the road, to the gate where the same gatemen waited. The first time he had visited her in her home, in what he thought were his best and dazzling jeans, they had stopped him and made him wait while they checked upstairs.

'Sahib,' one said. 'You haven't come for a long time.' 'Yes,' Sartaj said. 'Will you give this upstairs? In memsahib's house?' This was the divorce papers, each page initialled, the last signed and dated and witnessed.

'Of course.' As Sartaj walked away the gateman called, 'Will you wait for a reply?'

'No need,' Sartaj said. He had Katekar and the jeep waiting below, at Breach Candy, but he wanted to walk for a while. A van passed with that ugly throbbing American music that Sartaj could feel in his chest. A school bus passed, and three girls in blue uniforms smiled toothily at him from the rear window. Sartaj laughed. He twirled his moustache. In the blaring evening rush he could feel the size of the city, its millions upon millions, its huge life and all its unsolved dead. A double-decker bus ground to a halt at the stop across the street, and people jostled in and out. On the side of the bus a poster for a new movie proclaimed: 'Love, Love, Love.' Somewhere, also in the city, there was Kshitij and his partymen, with their building full of weapons and their dreams of the past, and Sartaj knew that nothing was finished, that they remembered him as much as he thought of them. A light changed just as Sartaj was about to cross the road, and the stream of cars jerked ahead madly, causing him to jump back, and the sidewalk vendors and their customers smiled at him. He smiled also, waiting his moment. Then he plunged in.

thirty-six

~

WILD THINGS
ANJUM HASAN

Prasad swivels his head to watch a dragonfly that is moving too fast to allow him to fix its colour in his head. It is dazzling blue, but a moment later, transparent silver.

The September sun is beating down on the heads of the forty-two boys and nineteen girls of K.P. Kattimani High School. A map painted on the front wall of the school highlights this village in central Karnataka in large concentric rings, the circles radiating out towards the far borders of the state. The students sing one of Basava's vachanas for the morning assembly, standing on a caked mud yard and facing a peach cement building at whose crown, above the school emblem, sits the sacred bull, Nandi. He is always in profile, always in that restful stone pose regardless of the weather. The vachana, now and then, takes the form of a question: 'Isn't one mention of the name of Shiva enough? Isn't one mention of the name of God enough?'

Prasad has asked these questions many times in his last two years at the high school and for some reason—perhaps because of the enviable freedom of this dragonfly, perhaps because of the narrow, sweating face of the principal, Mr Hosaralli, whose glasses are of the kind that darken in the sun, and who opens his mouth wide to sing—he chooses not to repeat the questions this morning.

When the song is over, Mr Hosaralli says, 'One of our students was not singing.' The students know that the silence that follows this is not going to be filled with any subsequent remark. The principal is a man of few words which makes the little he says all the more ominous. His dark glasses add to his sinister aspect because outdoors no one knows who or what he's looking at. Indoors he looks at no one. He addresses remarks to walls and windows and ceilings when talking to people.

The dragonfly returns and Prasad sees that its wings are really blue. He chases it to the head of his row.

'Prasad Yelagodu!' shouts the principal in his weak voice. He turns his back and marches off the platform and the students can finally walk out of the sun, one row at a time. Prasad follows Mr Hosaralli to his office, which is a room with only a few things in it, but things that none of the other rooms have—a table fan, a maroon pincushion, a latch on the wooden cupboard on which hangs a shiny new lock.

Prasad bends and the principal whacks him on his behind with a thin cane. Once, twice, thrice. Prasad is not sure how many is the correct punishment, for this was a double crime—the crime of not singing compounded by the crime of breaking the line. Once, twice, thrice. Prasad remains bent because he is still not sure. Mr Hosaralli lays his cane down neatly on his desk and Prasad straightens and chants. 'I am sorry for my bad behaviour and I will reform myself according to the precepts of Basava.'

'Go and study,' says Mr Hosaralli to the pincushion. He has no interest in Prasad as such, even though he is the boy he canes most liberally. If he didn't call him out by name, Prasad wouldn't even be sure he recognized him. The principal can choose to cane any student he likes but he has conversations with none. Prasad has come to understand that this is the man's way of being neutral and powerful.

He goes to class. Usually canings put him in a good mood. In fact, they are effective. They curb his natural restlessness for a while till something—the way a classmate's teeth stick out, the longing for a second helping of buttermilk—makes him step out of line again. But today there is an itch in him that was not quelled by the cane. He kicks Srinivas on his bare, dusty ankle as soon as he sits next to him on the bench and Srinivas, displaying a belief in action before words, kicks him back and then asks him what his problem is.

Their teacher is building a wobbly column on the blackboard out of the names of all the Mughals. The column is straight till Akbar, then starts tilting to the right, and the history teacher extends his arm as though the letters have a magnetic force. Behind him the class talks as excitedly as a large family that has just been reunited after a long separation. When the teacher turns to face the class again, the noise subsides somewhat—all the same things are said but in milder tones.

Prasad and Srinivas kick each other discreetly and stab pencils into each other's ribs. By the end of class, Prasad is fighting to keep his seat, while his attack on Srinivas has repercussions further down the bench—Srinivas's jostling irritating the next boy, Kappanna, and Kappanna's violent response in turn affecting the boy sitting next to him. When the copper bell is rung in the yard by the peon, and the history teacher walks out leaving his leaning tower behind, all the rancour drains away from the faces of the fighting boys as if they were enacting a battle specifically for history while the subject of maths requires completely new tactics.

At lunchtime the students are fed rice and rasam on steel plates along with glasses of watery buttermilk. The nineteen girls sit in a regal huddle, separate from the boys. Their presence over the two years has diminished the boys' reactions from lofty disdain to mere contempt. In Prasad's class, one or two boys jeer when a girl puts up her hand to speak, but on the whole they ignore the girls. Anything more would be seen as disguised interest.

Prasad is sitting in a corner where the boys end and the girls begin. He doesn't know the name of the girl sitting near him, but next to her is Savitri from his class; she wears a bright silver nose ring and is always the first female to walk into the classroom. The other girls follow, all a little shyer than her.

'That rascal Bhagat,' says Savitri to Prasad. She masses the watery mix expertly in her cupped palm and eats at great speed. 'He gave me only one and a half cups of rasam. He pretends he's giving two but the second cup is always half empty.'

Prasad eats silently for a few moments, stunned by the way the girl has addressed him—exactly like his grumbling neighbour Gangamma talks to her old, handicapped husband when they sit in the evenings on the veranda of their home.

'What is this rasam and this rice?' he says then, imitating her tone. He stops eating and gestures at his food scornfully with an open palm. 'Nothing to bother about. I get enough of it at home.'

'Don't talk, Prasad Yelagodu,' says Savitri. 'I've seen you begging Bhagat for more buttermilk.'

The girl sitting between them makes an assenting sound. The two girls look at each other and snicker. Savitri licks her fingers clean with the thoroughness of a cat.

'I'm telling you, I can throw away this food. I'm eating it only because I don't want to eat the cane.'

'You're always boasting,' says Savitri. 'You're always saying the cane doesn't hurt you.'

'The cane is nothing. It's that donkey-face I don't want to see.'

'If you don't want the food, feed it to Pinkie,' the girl in the middle says suddenly, her tone authoritative. Pinkie is the school's vegetarian dog.

Savitri is one thing but now this nameless scrawny creature with frayed ribbons in her pigtails is heckling him too. Anger makes Prasad shake the rice from his fingers and, in the same instant, lift the plate with his left hand and fling it into the yard.

A moment of respectful silence settles on the students.

'He can be proud now,' says Savitri contemptuously to her friend, no longer interested in addressing Prasad directly. 'Now he can say, I am better than you because I threw away my rice.'

The younger boys say nothing, aware of how Prasad can come close to them and flick his fingernails at the cartilage of their ears in a peculiarly painful way, but Srinivas has scores to settle and calls out to the teachers eating lunch in one of the classrooms.

The maths teacher, Miss Pushpa, is confused about what happened despite the evidence of the rice grains scattered in the yard and the upturned plate. She finds it hard to believe that any of the greedy students of K. P. Kattimani High School would actually throw away food. Every noon, in fifteen minutes flat, there are sixty-one plates licked clean and sixty-one unsatisfied boys and girls.

'What happened, Prasad?' Miss Pushpa asks him with something approaching concern.

'I didn't feel like eating,' he says brazenly.

'Since when?' she asks, to which Prasad says nothing. 'Since when did you not feel like eating?'

'Miss, I asked him to feed it to the dog,' says the small but confident girl.

'Are you not the sons and daughters of farmers? Haven't you been taught that to waste a single grain of rice is a sin?'

The principal is called out. He shouts, 'Prasad Yelagodu!' and goes back to his lunch.

Prasad gets up and starts to follow Mr Hosaralli but instead of turning the corner of the verandah, jumps off it and runs away. The throwing of the rice, the hunger that remains in the shape of a circle in his stomach, the glint of Savitri's nose ring, all of these are making him think wild things. He is thinking of the dragonfly he couldn't catch. He is thinking of his cousin, Natesha, who works in a restaurant in Bangalore and sleeps in a backroom with his monthly earnings hidden in his underwear.

Bhagat, the bell-ringer, peon, lunch server and general dogsbody of the school, sits in the shade of the neem tree and indifferently watches Prasad run through the gate as if it's a sight he sees every day. It's only when he notices the principal shouting through the window of his office that he gets up and runs theatrically after Prasad.

Prasad sprints home. His parents are out in the fields; this is the time of the year when they harvest the bajra. They are lost among the tall, lush plants and will come home in the evening smelling of cut grass, with small gashes from the leaves all over their bare arms.

From a shelf in the kitchen he takes down the rusty tin of Glaxo milk powder in which his mother keeps a few hundred rupees rolled up among the beads and bangles. With the notes still in his fist, he heads to the bus stand.

Prasad tries to stand unobtrusively in one corner of the kitchen while Natesha wraps up two idlis and a vada in a square of plastic and then does up the parcel in newspaper and ties it with a piece of string from a spool hanging near his head. As soon as he has finished, he starts to wrap up some lemon rice. Then idlis and a vada again. He is a food-wrapping machine. There is a tall, sweaty dosai-making machine in a smoky corner of the kitchen and a noodle-frying machine next to him. Small barefooted boys smelling of the wet rags hanging from their shoulders come into the kitchen with orange tubs loaded with used dishes. They scrape leftover food into a plastic bin. Prasad is mesmerized by what they throw away—sometimes spoonfuls of the thick sludge of sambar, sometimes whole bowls of coconut chutney. Over and above the roar in the kitchen, he can hear the louder ruckus of the tiffin-eaters in the dining hall. That is why they don't finish their food, he thinks. Because they're talking so much. He waits for Natesha to offer him something but Natesha has only told him to stand quietly in a corner and wait.

'Natesha!' calls out Prasad now and then, but no one hears him. He goes to the toilet, recounts his notes and comes back. He asks a passing boy for a glass of water. Slowly the clamour of the tiffin-eaters dies out and Natesha says, 'You want to see Bangalore, macha? Then how can you keep standing here all night?'

That is the way Natesha is. He is always implying that Prasad has said things he hasn't. They go to the backroom where, among sacks of rice and plastic drums full of urad dal, Natesha changes into a pair of stone-grey jeans. A grimy bouquet is embroidered in red on the right leg of the jeans, enclosing the word 'Miracle'. He spits on to his comb and brushes his hair back, holding a speckled pocket mirror before his face.

'I'll show you such a place, you won't believe your eyes. Music, shops, food, girls....everything in one place.'

Prasad hasn't eaten anything since the small bag of peanuts he bought on the bus. He wouldn't mind even a cold idli but Natesha hates this food. He eats it morning, noon and night and dreams of bright orange chicken legs hanging on long skewers in the windows of the small eateries in Shivaji Nagar. Prasad watches Natesha silently. He cannot remember exactly how much rice he flung away. He wants it back and he wants something for having thrown it away—some kind of compensation for the tragedy.

'I ran away from school,' he tells Natesha. 'Donkey-face will cane me all morning when I go back.'

'When did I say you should go back? They need more boys to clean the tables here,' says Natesha casually.

Prasad considers this while Natesha puts on a pair of scuffed sneakers with 'Nike' written falsely in bold letters on their sides.

They go out into the cool night and reach pavements where there are men with ties mumbling into their phones and extremely thin girls in groups of three or four who talk so frantically to each other they have to keep stopping to catch their breath. Prasad gets even hungrier, seeing these people. But he doesn't look them in the eye. Natesha is all right but he is still wearing his scruffy blue uniform and rubber slippers in which he goes to school. He stays close to his cousin.

The last time he visited Bangalore, he and Natesha sat in Lalbagh, listening to music on his phone and smoking cigarettes. Natesha tried to call some girl whose number he had but the girl kept cutting him off. Then they went and ate a whole chicken. Now Natesha is a year older and has more swagger in his stride. He walks into the mall without pausing for the security guard, who rises a little from his seat and then sits down again, readjusting his oversized peaked cap. Prasad knows that he would have liked to stop them but doesn't have a reason to.

His heart leaps up with pride and the very lights in the enormous mall seem to him to belong in some secret way to Natesha, lights that are reflected a hundred times in shop windows bursting with things, many of which Prasad cannot even find names for. They stand stock still in the middle of this sky-high house of luxuries and laugh out loud. Then they take the escalator up.

After an hour, Prasad can no longer tell mannequins from humans. Everybody seems made of a glossier earth. He has marvelled at the shop just for neckties, the shop just for glittering ink pens, the shop just for biscuits. The cousins have taken taken pictures of each other against giant movie posters, or posed near make-up booths run by girls in men's suits who ignore them. They have seen people gathered at an ocean of tables, eating what the cousins could never afford, checked out the kinds of sports shoes boys their age are wearing, and the bare arms and jean-clad bottoms of girls who stand with their backs to them on the escalators without explanation or apology. It is the girls who bother Natesha the most, while Prasad looks at backlit pictures of burgers and chicken nuggets, and dwells on his hunger as if it were a dirty secret.

'What, guru,' whispers Prasad finally, as they go for up the third time, 'you can't touch anything here.'

Natesha clicks his tongue impatiently. 'You're talking like a villager, macha.'

When they reach the top floor, he turns abruptly and walks into a gigantic shop; they cannot see the ends of it. Everywhere are jeans they could fit into and jackets in whose pockets their hands could easily go.

'Let's buy something,' says Prasad before Natesha can suggest it. He actually wants to grab something and run. He's sure he can run many times faster than that fat man at the door, but what about Natesha? Will he follow him? Will he know what to do?

'This is not that kind of place, Prasada,' says Natesha kindly, suddenly the mature elder brother. 'You won't get anything for fifty rupees here. We're going now, okay. We're going to A1 Biryani Point.'

Prasad is about to suggest that they filch something but spots, right then, a sign announcing 'Rs 35' in large green letters.

They casually walk over to a shelf crowded with cans. Prasad glances at the other people in the store. He is careful to match his expression with theirs. Everyone looks bored or is talking very loudly.

'Scent,' he says. 'Only thirty-five.'

'Deo. You don't know what it is.'

Prasad smiles for the first time that day. 'Scent,' he repeats, examining the large spray-can with admiring fingers.

'You got lucky, macha,' says Natesha. 'You got an offer price.'

They head to the cash counter. A girl with a big mouth painted the colour of a bad bruise takes the can without looking at either boy and says in a cool voice, 'Five sixty-five.'

There is silence all around.

'One second,' says Natesha.

'No,' says Prasad, also in English. 'Thirty-five.'

'Five sixty-five,' says the girl in the same tone. 'Thirty-five rupees off. Actual price six hundred.'

She's looking at them now without actually challenging them. They could always return the thing. They could change their minds.

'I told you, macha. I told you about this place,' says Natesha and opens his wallet. 'Take three hundred from me.'

He wants to impress the girl. Prasad can't believe it. He wants to impress someone he's never going to see again. At least Savitri, for whom Prasad threw away his rice...

'Let's go,' says Prasad. 'We have to go urgently.'

'Give me your cash,' says Natesha coldly. Nothing in his tone indicates that for him three hundred is probably the savings of a full month.

Prasad slowly brings out the crumpled notes he stole from his mother's tin. Maybe Natesha is pulling some kind of stunt and in the end they'll get all their money back. But the money disappears beneath the counter and the girl smiles at them with her purple lips and puts the spray can in a plastic bag that's far too big for it.

'The money,' shouts Prasad.

'Chicken next time,' says Natesha.

'I don't want this,' Prasad tells the girl. 'Give me my money.'

'You idiot,' says Natesha and tries to push him away.

Prasad turns to face him and he sees that this is what the whole day's itch was leading to: he must fight his cousin in the bright lights of this mall with this girl watching—Natesha, who was his only hope when he ran out of the school gate.

He lunges at him head first.

Natesha thinks he's exaggerating. Only when he's slammed in the chest does he realize what Prasad means and he, the stronger and older of the two, catches the hair at the base of Prasad's neck and swings his face towards him. He delivers a hard slap on his cheek and then another. All at once, Prasad gives up; he becomes like the weeping child who looks up, for redress, at the face of the very same parent he is trying to resist.

By the time the salesgirl has gone across the huge store and come back with the security guard, by the time the curious people who'd gathered around to watch and comment in an English that Prasad didn't understand have drifted away, he is sitting on the floor, the heels of his palms pressed into his eyes. Natesha tucks the spray can into the waistband of his jeans.

The guard takes them all the way down and waits till they've crossed the asphalt and gone out of the gate.

'Fighting,' he says to his colleague at the main door. 'Fighting over what to buy.'

It's past nine o'clock. Prasad, still sniffing, and Natesha, grim, walk in silence. They pass the huge painted walls of the city, decorated with images of asymmetrical palaces and gap-toothed chimpanzees. There is no one but an autorickshaw driver peeing on the pictures, his empty vehicle parked nearby. Buses speed past, their rumble obscured by a sudden train on the railway bridge. The two boys can see the lighted windows in which other people are more content than them.

They walk for half an hour and, from nothing, go straight into the crowds. Everyone's steps are more hurried as they approach the intercity bus station, as if to reach faraway destinations faster. But Prasad and Natesha continue at the same slow pace.

Amid the pavement sellers of suitcases and bus timetables and wilting fruit, an old woman is frying bhajjis on a blue pushcart in the light of a hurricane lantern.

Natesha stops and says, 'When did I say we can't even eat bhajjis?'

They bring out all the coins and notes they have. After deducting Prasad's bus fare, there are about twenty-five rupees left.

They stand by the cart and wolf down egg bhajjis. Then they eat brinjal bhajjis and chilli bhajjis. After they have used up their money, the woman gives them one just-fried, too-hot-to-hold onion bhajji apiece.

'Eat,' she says and laughs, revealing red stubs of betel-nut-corroded teeth. They wipe their greasy hands on squares of old newspaper.

Afterwards, as Prasad is boarding the bus, Natesha pulls out the can from his waistband and chucks it over to his cousin who catches it one-handedly with the deftness of a wicketkeeper.

'What did I say?' Natesha asks indignantly, then disappears into the crowds in the station.

When the bus starts, Prasad is sleepily humming the Basava vachana of that morning. 'To erase thousands upon thousands of sins, isn't one mention of the name of Shiva enough? Isn't one mention of the name of God enough?...'

He thinks of donkey-face's cane and of kicking Srinivas when he isn't looking and the voice of his mother screaming when she discovers the missing money, and all the other things in his life that are both comforting and irritating for exactly the same reason.

He uncaps the can and wildly sprays his face with so much deodorant, the back of the bus smells of it all the way home.

STOLEN

AMRITA NARAYANAN

Parvathi, squat, generous-hipped, sweaty, is scraping seeds from the flesh of a papaya. She is working slowly, attentively, her brow furrowed in concentration, as she strokes and probes with her curled brown fingers, her hands tracing slow ellipses to pull the glittering seeds from their sticky embrace with the sometimes red, sometimes orange flesh. At its wet centre, the fruit is exactly the colour of her santra-red sari and blouse.

Sitting across from her, Meenakshi, equally full of figure, still dusted with talcum powder and carrying the Mysore-sandal scent of her morning bath, is speaking. She is talking about papayas: how this year's bumper crop has dropped the prices and rendered accessible to everyone the exotic fruit that is usually the preserve of the wealthy; how, if plucked early, the hard, sour fruit makes for good pickling. It is mostly a monologue.

Parvathi keeps working as she listens, but doesn't say much; all through her chatter Meenakshi's eyes are riveted on her companion.

Though she is silent, Parvathi speaks with her body. Her thighs flex and twitch under the tightly wound cotton sari; a roll of flesh slick with sweat trembles just below the edge of her blouse. And now her feet flatten and dig into the tiled kitchen floor as she begins to juice the pile of lemon-halves she has sliced earlier. She twists and grinds the hard yellow rinds on the mound of the ancient glass juicer, until the pulp yields its tart juice into the waiting saucer. As she works, her breasts move within the enclave of her blouse, the cotton alternately caressing and chafing her nipples.

The cool black floor the two women sit on is not their usual working place: today, their mistress, Mrs Subramanium, has dispatched them to help her daughter Sunita prepare for one of her high-society lunches, and though the kitchen is smaller, and it takes the women longer to find the vessels that they need, both are grateful for the respite from Mrs Subramanium's stern demeanour. It is almost a holiday.

'Where is Uma?' Meenakshi mutters to herself. She knows that if Uma, Sunita's own maid, does not show up this morning, it will fall to her to pick up the girl's cleaning tasks, with Parvathi at once chef and sous chef. The consternation Meenakshi feels distracts her from Parvathi's rocking buttocks and her hands that are deftly massaging the last of the lemons. She stands up and looks around the kitchen in dismay as the morning's sensuality seems to fade and she is reminded of the drudgery that kitchen work can be when performed under stress.

Although she is oblivious to the precise nature of Meenakshi's thoughts, Parvathi senses a soupçon of irritation has crept into the other woman's body, where before there was a languid stillness. She looks at Meenakshi standing, tapping her foot, and then instinctively, and without knowing why, she stands up and places the weight of her own foot on the tapping one, stilling it and instantaneously reclaiming the sensual atmosphere of the kitchen.

Meenakshi puts a kilo of potatoes to boil on the gas stove, and Uma enters the kitchen to the screams of the pressure cooker.

In her bedroom, Sunita, dressed in an emerald-green silk tunic—her final choice after an hour of looking-glass trial and debate—falls back on her bed and turns on the television, flipping channels till she finds something to hold her interest: a nature show about the mating habits of lions. Transfixed, she watches as the male lion brings himself and his partner to orgasm in what seems like a very frantic few seconds. If the television could gaze back at her it might see her thighs roll inward and her pubis lift, but as it is, there are no witnesses to Sunita's momentary arousal, the tiniest unfurling of her sex. She doesn't notice it herself. Then the thighs relax, the sacrum sinks, the moment passes.

Sunita switches the television off, briefly consults the clock and then heads down to the kitchen to check on lunch preparations, making a mental note to ask Uma to restore her room to its customary order after the meal.

In the kitchen, the sight of the counter strewn with coriander—Uma has left the partially de-leafed herb there to take her second bathroom break of the morning—annoys Sunita intensely. 'Ippodhaan chutney panna aarambichirukkiyaa?' she demands. 'Samosa filling readyaa?

'Uma yena time vanthaa?' Meenakshi snaps to attention and scurries around, propelled by the pitch of Sunita's voice. 'Juice aarambichchacha? Modhalai juice pannu.'

Parvathi continues to chop the boiled potatoes into tiny cubes, unaffected by the sudden commotion. Her unconcern, bordering on disdain, angers Sunita, but she finds herself unable to lash out at the woman.

Instead, she snaps, 'And where is Uma?'

Uma bustles into the kitchen just then with a triumphant announcement: 'Juice is ready, madam. Fridge-la irukku.'

Sunita opens the fridge and bends over to peer in. The juice that Parvathi had squeezed from the limes has been transformed into a frosty affair with ice, sugar and mint leaves. The cubes of diced papaya have gone into a salsa prepared from a recipe she had pulled off the internet. Sunita sighs in relief, and something in the way her shoulders relax under the green silk makes Uma walk up and stand behind her mistress. She leans forward to point out the chutneys that are also ready and poured into the green ceramic bowls reserved for guests, and as she does so, her soft belly fits perfectly into the curve of Sunita's lower back, and her breasts rest on the bump of Sunita's upper back.

The ringing of a phone breaks the moment and Sunita straightens up sharply, bumping into Uma. By the time she reaches her cell phone, charging on the bookshelf in the corridor, it has stopped ringing and she stares at it, unable to decide whether or not to return the call from a number she does not recognize.

Uma stands in the kitchen doorway watching the other woman, seized by how similar their shapes are. When Sunita finally looks up, another moment of wordless understanding passes between them.

The phone rings again, and habit trumps instinct: Sunita says roughly to the woman in whose presence she has just felt herself unfolding, 'What do you want?'

I want to keep looking at you, Uma says with her eyes, refusing to avert them or respond with words to Sunita's attempt at reasserting the hierarchy between them.

Sunita looks away and picks up the phone. It's her mother. Uma doesn't look away. As she watches Sunita transform into a schoolgirl talking to her Amma, one hand on her left breast, as if checking her heartbeat, she is reminded of an afternoon two summers ago. She had walked into the bedroom to clean it and found Sunita before the dressing table, trying on new bras. In the mirror, Uma had seen a pair of chai-coloured breasts topple out as Sunita peeled off a flesh-coloured bra to try on a lacy grey-and-black one. Uma had never seen a bra like that, one that cupped Sunita's breasts with the gentleness of a sensitive lover, that did not pull them up tightly like the stiff white bras Uma herself wore.

She had also noticed how similar their breasts were: firm yet supple, coloured like they were brewed from the same batch of tea leaves, and topped with identical dark nipples, the colour of areca nuts. Even the space between Sunita's breasts matched her own perfectly: a canal just wide enough for a man's hardness.

The object of Uma's reverie, meanwhile, is rapidly and busily checking off items on the lunch list, responding to her mother's directives like an anxious obstetric-medical intern who, having delivered their very first baby, must make sure the creature is intact: yes ten toes, yes ten fingers, yes two areolae, yes a fold of skin between the legs that curves out and then in…

Hours later, it seems like no time has passed, except that now Uma is exhausted— sitting on her haunches, her sari hoisted up to her knees, its folds bunched between her sweaty, tired thighs. She's watching Sunita on the phone with her mother again, in the aftermath of the party. This time it is Uma who is counting. She's calculating that it will take Sunita exactly five more minutes to render the post-mortem of the afternoon to her mother, another ten to fetch the hundred-rupee tips she will give Parvathi and Meenakshi as they leave, and then, perhaps, just two minutes more before she will ask Uma to make her a fresh cup of tea and come up to her room. About seventeen minutes in all. And then, maybe—who knows? Uma has a brief vision of her face buried between Sunita's damp thighs.

But something different is going on today. On the phone, Sunita is protesting. 'No, Amma, not necessary.'

But Mrs Subramanium always has her way, and Uma watches the daughter give in and say, 'Okay, okay, I will ask Parvathi for a massage, and yes, no oil on the head in the evening. I understand.' And then Sunita is calling for Parvathi and Parvathi is on the phone with Mrs Subramanium saying, 'Yes, madam, ah Dhanwantram thailam irukku, madam. Cheri, madam, okay, madam.'

When she understands what is about to happen, Uma's heart sinks. Suddenly, she is more exhausted than she had been after she cleared up the mess of the afternoon's lunch party. As the disappointment washes over her she hates herself, a top-worker, trained only to wash dishes and wipe floors, incapable of reading the labels on the bottles of Ayurvedic oil that Mrs Subramanium buys at Ayur Vaidya Shala and has chauffeur-delivered to her daughter. The hopelessness churns up from the pit of her stomach to the centre of her chest, and she can barely hear Sunita asking if she could stay until after the massage, because Parvathi will need someone to help her wipe down the oily rubber sheets and clear the massage table.

'Just rest and I'll call you when we are done,' Parvathi says kindly to Uma as Sunita tips Meenakshi, who is bone-tired and relieved to be leaving but cannot resist casting one last, longing look at Parvathi's wide buttocks.

Parvathi begins to prepare the double boiler in which she will warm the oil for the massage.

Uma is watching through the keyhole. At first she can only see Parvathi— drawing the heavy curtains so that the room is in darkness, lighting the fat orange-scented candle that sits beside the massage table, and then waiting, patiently, for Sunita to emerge from the bathroom. Sunita comes out wrapped in a Sarthy towel—one of those thin, mill-made affairs that absorb so much water they are soaked after one wipe—a superfluous covering because she must shed it almost immediately so that Parvathi can help her with the massage langoti, a thin piece of muslin that Sunita will pass between her legs and fasten around her waist just to cover the dark triangle between her thighs.

Mesmerized by Sunita's naked body, by the breasts that are testimony to their twinship, Uma bends further towards the keyhole. Were the women inside the room to look back at her, through the door, they would see the sensual curve of Uma's own breasts, optimally exposed at that angle, sweat pools gathering at either edge of her pink choli, and beads of perspiration sliding down to her navel.

Sunita sits on a stool and Parvathi begins with long, gentle strokes on her angular neck and shoulders. Her skin receives the oil gratefully, like a thirsty man receiving water. The buzz of the afternoon's conversation is ringing through Sunita's head. Her friends had been talking about men: how today's men were becoming more sensitive, less patriarchal, but still had a long way to go. She had been unable to focus on the little fragments of conversation that spiralled up from every corner of the room, torn as she was between wanting to be the perfect hostess that her mother would have wished her to be and wanting to participate in the happy chatter. But now, Parvathi's skilful massaging begins to calm her down.

Parvathi asks Sunita to lie on her belly on the massage table. Her strokes grow even longer, re-establishing for Sunita the reality of her body. She finds she can make more sense of the afternoon now, and as she does she wonders why so many of these ladies' lunches revolve around the decisions of their men: hers and hers and hers. Or their own decisions made without the knowledge or the approval of their men. Or their decisions to live without men. There seemed to be nothing outside the kingdom of men; even the space inside the head of every woman there appeared to be the kingdom of some man, or a few men.

'Saapaadu nalla irundhudha?' Parvathi asks, as she runs her hands up and down Sunita's back, and then her arms: from the shoulders down to her fingers, then up again, her thumbs grazing the sides of Sunita's breasts, not furtively but with confidence.

'Yes, the food was good,' says Sunita reassuringly, and then a sudden need to confide makes her say: 'Ellarum aambilai pathiye pesinaanga.' Parvathi is moved at being allowed in so close. She shuts her eyes for a second, picturing herself talking with the group of smartly dressed and perfumed women at lunch. Parvathi does not know what comes over her then, for she says: 'Eppovum appadithaan! It's always like that, every woman talking about men. Don't we have cunts after all? Kadaaseela aambilai paththeethaan paesuvom ellaam.'

Is that what it comes down to? Even as Sunita asks herself the question, the firm pressure of Parvathi's fingers on her inner thigh reminds her of something she has forgotten all day. She thinks better of pursuing the conversation, surrendering instead to the energy being pushed towards the place that Parvathi has just named. Then she gives herself over completely into feeling the one place that her mother has never given her any instructions about—except to cover it, close it, keep it tightly locked.

For the first time in a long while, Sunita allows herself to sink into the smell of herself. And then, slowly and deliberately, she distinguishes her own smell from the trifold scent of the Dhanwantram oil and two sets of armpits—her's and Parvathi's. Then, as Parvathi's fingers probe her flesh more urgently, she begins to recognize two Parvathi smells. The first comes from her unshaved armpits, sweaty after the day's labour.

And then there's the other Parvathi smell—unmistakeably the smell of cunt, slightly tangy, perhaps a little metallic, and carrying in it a whiff of damp earth.

Where do I remember that from, wonders Sunita, and even as she thinks about it Parvathi is rubbing more oil, her body is getting heavier and each stroke begins to separate her thoughts, allowing her to luxuriate in each of them individually. She begins to focus on one particular thought, which grows from the encouragement, like a fondled phallus. She remembers the smell. She remembers it rising from between her own legs the first time she climbed up into the loft of her parents' bedroom using the window grill as a foothold, looking in those black-and-white cabinets for Sidney Sheldon novels, which she flipped through until she found the words that had drawn her to that dark and secret place, words that released exactly this smell right there in her body, and released the moisture that accompanied it, this smell, and the sweetness that flooded her being when she rubbed the source of the smell—as she had done that day for the first time and brought her fingers to her mouth afterwards to taste.

Parvathi is massaging up from her ankles now, up past the back of her knees to the inside of her thighs and Sunita is wet but Parvathi is business-like. When Parvathi's thumbs come dangerously close to the langoti the second time Sunita absolves herself of the supervisory responsibility she took so seriously in the kitchen, thrusting the onus upon Parvathi and feeling quite sure that the maidservant can be counted upon to check herself. Just in time, each time.

Parvathi moves her thumbs in crescent-moon part circles that start at the back of Sunita's upper thigh and find their way slowly to her inner thigh. She barely touches the thin muslin that guards the lips between Sunita's legs before she lifts off and starts all over again at the outer thigh, each time advancing closer and very slightly deeper into the folds of the langoti. Sunita's eyes are firmly shut, and even as it occurs to her that she might be soaking the langoti with her juices, the thought is erased by the motion of Parvathi's kneading now on her buttocks, down and across, up and across and then in gentle circles around her sacrum.

Sunita can smell her own juices now, and as Parvathi straddles her to finish massaging her upper back, Sunita finds that if she loosens her belly, the weight of Parvathi's buttocks on her own, and the long, searching strokes that are now being applied from lower to upper back, create a delicious sensation in her cunt. As her pelvis jams into the weathered rubber mat that is on top of the massage table, subtle but nevertheless perceptible movements allow her to reap the harvest of oil that has collected in the mat from months of use—providing a complementary lubricant to the juices that are now matting the hair of her cunt. Just as she is beginning to pant a little, teeth grinding, mouth determinedly shut, stomach beginning to spasm, Parvathi suddenly stops.

Sunita is in an agony that could bring her to the verge of tears in minutes unless she finds meaning in this moment. 'Yenna aacchu?' Parvathi laughs softly. 'Please turn over, madam.' Motionless, Sunita waits till the hardness in her clitoris subsides, feeling the wail in her throat that she dare not release. In the stillness she accepts the loss of the place she had been visiting when Parvathi's movements stopped. Then, hoping that in the candlelight Parvathi will not notice anything, Sunita turns over.

The slight breeze of the November afternoon causes her to shiver slightly as she rolls her hips up and open to expose first her soaked, langoti-sheathed pubis, then the line of her waist and breasts bearing erect nipples as she lies down on her back and waits. Parvathi begins the dorsal massage at the legs, long lines addressing the points of pressure from thigh to toe, followed by a gentle pummelling and slapping that render first her right leg and then her left completely free of tension.

Her ankles give way and her feet fall to either side. Parvathi climbs on to the massage table and straddles Sunita's hips without putting her weight on them. The folds of her sari drop gently down, meeting the moisture of the langoti, and the folds of skin just beyond the paper-thin moist cotton skein. She massages Sunita's belly first, using her whole palm to rub across her waist in one direction then the other, eventually pressing her fingertips around Sunita's belly button to release the knots of tension there, the undigested pieces of lunch.

It's probably too soon after lunch to have a massage, Sunita thinks, the thought disappearing as soon as Parvathi pours a tiny quantity of oil into Sunita's belly button.

Another day she may have been squeamish, but today she feels no ticklishness, and her clitoris hardens in response to the delicate circular motion Parvathi is now using to pleasure the grooves of her navel that in turn sends electric ripples all the way down into her cunt. She moans very softly, and the thought that tells her to conceal the sound is wiped clean by the sensation of warm oil now being trickled on to her breasts. Sunita holds back a scream of relief; she feels Parvathi's hands grasp each of her breasts and hold them firmly for a few seconds before beginning to massage them one by one, in broad circular motions that encompass the whole breast. She sighs audibly as Parvathi works on her heavy breasts, pushing them from one side to another and watching how they fall as if admiring ripe fruit. If she were to look at Sunita's face instead of her breasts Parvathi would notice that her employer's daughter has ever so slightly opened her eyes.

The sight of Parvathi astride her, one breast in each hand, the line between her own neck and breasts exposed, glittering with sweat, pumps every nerve ending in Sunita's clitoris alive, bringing a fresh rush of dew to the petals of her cunt. In little crawling movements, like an insect making its way up a mud-hill, Parvathi is pinching the flesh of Sunita's breasts in smaller and smaller circles and ending in a very slight pinch to the nipple that causes Sunita to grasp either edge of the massage table tightly and whimper softly.

It is when Parvathi is repeating this movement for the third time, this edging up the breast, exploring and pinching, that she suddenly becomes aware of Sunita's arousal. She is struck by the pathos of the moment, the drama of power role reversal, but when she moves it is not from the munificence of one bestowing a favour.

There is not a second of planning, not an iota of forethought, but a boldness that can only come from an animal sense of knowing. Forbidden, in the rules of massage that she had studied that summer when Mrs Subramanium had decided that Parvathi would be a cheaper substitute for the Kerala-trained Ayurvedic masseuse, is pulling, and this is what Parvathi now does: she gently but firmly pulls both Sunita's nipples simultaneously, one in each hand, as she rocks back and forth, now sitting on her haunches, one foot on either side of Sunita's hips. The thick folds of her sari collect between her own thighs, and gently stimulate her via the movement while simultaneously, because the heavy cotton material is long enough—a six-yard Mangalgiri cotton inherited from Mrs Subramanium herself—it licks the folds of Sunita's langoti and pubis, creating an electrical impulse that along with the pulling of her nipples cause Sunita to break out in a sweat.

There is now no question of opening her eyes and also no turning back or denying what is happening.

Parvathi is rocking and Sunita gives in, allowing her buttocks to slightly lift off the table so her cunt can fully meet the folds of Parvathi's sari. The friction increases. Perceiving this shift Parvathi drops on to her knees and slides down the mat, then lowers her hips further so that her own cunt is grazing

Sunita's and her hands are on either side of Sunita's shoulders for support. She is moving like a man now and the folds of cloth between the two women's cunts are soaking. The effort of being silent is now far greater than the risk of making a sound, and an unmistakable panting and grunting breaks out. At the keyhole, Uma is sliding up and down the door almost exactly imitating Parvathi's movement. She is unable to hold back her screams but she has the thoughtfulness to scream noiselessly, mouth open, shoulders thrown back. She has formed the folds of her own sari into a thick cylindrical structure mimicking the thickness and firmness of a man's cock and she is rotating her hips round and round in circles, grinding them into the roll of cloth while she tightens her grip on the door handle and watches through the crack in the panelled wood.

Sunita is coming hard and intensely. My pussy is in ecstasy, she screams into her own heart because she knows she must not scream out loud. Her mouth is open and her face covered in sweat; her eyes are still closed, but Parvathi knows from the way her hips suddenly lift up off the massage table, the way her buttocks are tensing, toes pointing and her grip tightening on the table, to move just enough to bring her to one more shudder and to then seamlessly glide into a neck massage.

Uma too has released, panting, satiated. Eyes finally closed, she rests her forehead against the door oblivious to the rectangular stain of sweat that immediately appears where she touches it. Parvathi herself is not spent but she is less concerned with her own pleasure this afternoon than with fully and completely satisfying Sunita. Getting off the massage table she turns her back to Sunita and begins to deeply work the tissues of Sunita's belly, pressing downwards, releasing the air created by the orgasm and readying Sunita's body for what she will do next.

I don't want this to stop. I don't want this to stop.

The six-word ritornello that is echoing in Sunita's head is a direct communication from her cunt and, without words, Parvathi has heard it, and now her palms are together in a prayer position, ten fingers touching as she slides the edge of her folded hands from Sunita's navel down towards her pubis, then quickly lifting up and beginning again at the navel, pressing the edge of the hands into the belly and pushing down once more, first slowly and then rapidly. When she intuits Sunita's readiness by the curve of her now slightly relaxed buttocks she stops and begins to untie Sunita's langoti.

It's over, thinks Sunita and the disappointment falls like dead weight into every inch of her being. The fog of her depression descends so quickly and thickly that she does not at first feel what Parvathi is doing: she is releasing the back of the langoti so that she can manipulate with both her hands the cloth that is between Sunita's legs. In another minute she is pulling on either end of the muslin, with the motion of a milkmaid, and Sunita's whole cunt is dancing in glee, as the slippery muslin soaked with the fragrance of her juices wafts up to her such that she does not know what is more pleasurable, smell or touch, and she is coming again, this time thrashing and bouncing her hips up and down so hard that her upper body rises up off the table before falling back down.

'Geyser on panirku, ma,' says Parvathi announcing the readiness of the bath and brushing the hair from the younger woman's face. Sunita opens her eyes as she always does at the end of her massage, and thanks Parvathi out loud, gratitude shining in her eyes like stars in those places where city lights do not drown the magnificence of the night sky. Parvathi nods and then leaves the room, turning the doorknob quietly but nearly bumping into Uma as she opens the door.

She grabs her fiercely and for a second Uma thinks Parvathi is going to hit her although she can't understand why: after all, her insolence is no less than anyone else's. But it is not violence that Parvathi is craving and for a minute she contemplates Uma, holding her arm in a vice-like grip, breathing her in and smelling the sandwich of cilantro, onions, armpit and cunt.

Inside, Sunita is allowing herself one more minute of bliss, a marvelling at the miracle of her own body before she slides off the massage table, tiptoes into the bathroom and turns on the shower. The sound of cascading hot water enters the room and fills it. Outside, Uma is frozen in Parvathi's clutch. Stuck, she has a moment of enlightenment in the arms of her captor: it is she who longs to strike Parvathi, the capillaries of her soul are dilated with jealousy, she has been watching Sunita hungrily for months, and the rapture of seeing her beloved pleasured before her eyes is now overruled by the awareness of its cost: something precious was lost to her and her rage at her loss is immense. She grabs Parvathi in a wrestler's hold, locks her leg around the older woman's and pushes her to the ground. It is then that she smells Sunita on Parvathi and she half-yelps, half-shrieks: the sound of commingled violence and lust. Nostrils flaring, chest exploding, she follows the trail of the scent, ruthlessly pinning body parts as she moves, and this time it is Parvathi who is immobilized, as Uma's hands rip her sari and petticoat, and her head enters the opening she has created, grunting and panting, feverishly fumbling and, eventually, finding the destination her nose and tongue seek.

thirty-eight

~

THE GRAVESTONE

SHAHNAZ BASHIR

As soon as the mist lifts, which happens in the evenings these days, he'll dash straight to the graveyard. It promises to be a really chilly evening in the wake of the April rain. He is sure the drug-addicted gamblers, for whom the graveyard is a favourite place to play whist, won't be around today.

He's scared of the idea of doing it in the night. He can't sneak out in the presence of his vigilant family members. He usually goes to get his cheap Panama cigarettes from the village market in the evening. That's the time he can do it.

He lights a filterless cigarette and paces the short, narrow, flaking cement pathway outside his small mud-and-brick house. Cacti and dead geraniums, potted in discarded paint cans and small empty Fevicol buckets, are arrayed against the walls of the house. He finishes his cigarette, then slips back into the house, ascends the creaky wooden flight of stairs, and reaches for his old cobwebbed toolkit under the tin roof where the gable makes it difficult for a person to stand upright. His workman's fingers are calloused, the fingertips bristle with cracks, the fingernails are deformed. He rummages through the dusty tools, panting with the effort of climbing the stairs.

Shortly afterwards, he finds himself running towards the graveyard a kilometre away from the house. The pale light of dusk falls on the drenched asphalt road riddled with muddy, water-filled puddles. The roofs and leaves still drip.

The mist is all gone and the snow-shrouded mountains opposite him appear closer than they actually are. A gust breathes through the village and shakes the groves, orchards, mustard crops and grass. Scattered plastic bags balloon into the air.

Each wrinkle and crease on his pale and weathered face with its sunken cheeks tightens as he rushes on. Each strand of his grey hair stands on end. The pits in the road make him watch where he places his plastic-shod feet. His left hand, almost dysfunctional, trembles badly. He grabs the underside of his pheran to keep the hand steady.

His worn-out pheran smells of rain and stale smoke. He scurries on. The sound of dull clanking comes from under his pheran as the rusty hammer comes into contact with the blunt chisel. Plumes of smoke-stained breath burst from his nostrils. Occasionally, a bus or a scooter passes, tooting its horn.

Today he is welcomed again by the same stubborn clumps of nettle and thistle that are sprinkled throughout the graveyard. And beyond—there, near the graves—are assorted irises, their stamens powdering the white petals with yellow pollen dust on the insides. He traces the path to a grave whose epitaph in Urdu reads:

Shaheed Mushtaq Ahmad Najar

Muhammad Sultan was a talented carpenter before he fell from a roof while working and permanently injured his left arm. And just when the injured arm had begun to heal, against the doctor's advice, he went back to work. He had to take care of three daughters and an adolescent son, which is why he couldn't afford to rest the injured arm. The internal soft-tissue injury worsened to a haematoma. After a failed surgery, his arm was declared unfit for carpentry or any manual work.

He was a master khatamband, or lattice designer. Before cutting wood for use, he would smell it to gauge its quality. He specialized in mixing the classical with the modern, producing something that both old and new generations of Kashmiris liked. None in the entire village could rival his truss work. He was an expert on doors and windows—the thick part of work in carpentry. With each drag on the hookah, he came up with a new idea.

But Sultan had a shortcoming too: he was not a diligent worker.

He took on projects on a whim. He would hardly ever take on partners or apprentices. He liked to work alone. He normally worked for two days a week and took the rest of the days off. Sometimes he would disappear for days and later compensate for his absence by working overtime. His second romance, after carpentry, was accompanying the local militants around the village. He'd help them with anything, fetch them cigarettes, and even lavish money on them.

Carpentry was not only work or a source of livelihood for him, but an art form too. Art through which he expressed himself. Once at work, he would passionately sink into it. Sometimes he worked through the night. He would even work on Fridays, against custom, when all the carpenters and masons took the day off. He never cared for money until he really needed it. He was more of a dissolute artist than a mere time-bound carpenter. His hands are worth their weight in gold, that is what almost all his customers would say after marvelling at his work. It was because of his talent that people tolerated his truancies and wild habits.

The situation with his arm depressed him. It became impossible to work, or hold tools in his left hand. Then he started taking on projects on a contract basis, employing other carpenters and directing them. But their designs and work neither impressed nor satisfied his customers. Though his employees took all his directions, they never really followed them to the letter. They even cheated him of his share of the commission. And eventually his financial situation deteriorated.

His fate grew worse—his eldest daughter was returned by her in-laws, for the sixth time in four years of her marriage, for not fulfilling the demands of dowry. The last time she had returned with a swollen wrist and her snivelling, sick, one-year-old baby daughter. Her husband, the driver of a bus that somebody else owned, had wrung her wrist during an argument and thrown her out of his house. But at her father's house, the eldest daughter behaved as though she was just on an occasional visit. She even waffled on to her sisters about her in-laws, praising them as if they were good people and as if nothing had happened. She talked about the 'generosity' of her in-laws' neighbours. She described her brother-in-law, his tastes in food. These details bored and irritated her sisters.

Without work and with his broken arm, Sultan was struggling to deal with his two other daughters at home. His middle daughter was in her late thirties, a spinster, an almost illiterate woman, who, after her father's injury, had been supporting the household with her needlework skills. She worked well into the nights, tortured her eyes, strained and overburdened herself, and gained weight and premature wrinkles on her face. She was the most beautiful among her sisters, but there was an unattended fuzz of hair over her upper lip that made her look ugly at the same time. Lately, she was running short of work, and had taken to secretly begging at the shrine of Makhdoom Sahib on Fridays. She would slip on a long, dirty black burkha and leave home saying that she was visiting the shrine to pray to God to take away their hardships and grievances. But Muhammad Sultan had begun to suspect that her veiled expeditions were for another reason altogether—that she was out hustling to earn some money. He was furious about this but tried not to show it, especially as he never wanted to follow her to confirm his suspicions.

His youngest daughter had left school at the secondary level when Sultan's wife died of a colon haemorrhage. Since her mother's death, it was she who looked after the household, cooked and cleaned. Her presence in the house was the only presence Muhammad Sultan didn't feel. She almost didn't exist for him. She was just like the family cow she fed, washed, milked and cleaned. Whenever she cried, she didn't make a sound, not a single snivel or sob, just tears, which quietly streamed down her cheeks like melting pearls. She was born just before Mushtaq, her brother, Muhammad Sultan's only son, who died in adolescence.

Mushtaq distanced himself from school and books and instead followed a group of local militants like his father did. One day Mushtaq stayed behind with an armed group in a hideout—a posh house in Nishat. In the middle of the night, the hideout was raided, and he was the only one who was unable to escape. He was killed in the kitchen of the house. Next morning, Muhammad Sultan managed to go to the spot and see his dead son for the last time. After looking at the bullet-riddled body lying face down on the kitchen floor, he fainted. When he regained consciousness, he found that Mushtaq was being placed on a bier. The funeral was attended by the militants with whom the boy had fallen in. Wearing masks, they mixed with the funeral procession, and later also directed the graveyard management committee to get a quality marmoreal headstone chiselled for their friend. They wanted to have the headstone inscribed with an Urdu epitaph with the title 'Shaheed' before the boy's full name, which would signify that he was a full-fledged militant and had died for the cause of Kashmir's freedom. The outfit will be proud to bear all the costs, the area commander of the militants assured the graveyard management committee.

After the burial, some days later, a shiny black granite gravestone was erected at the grave, with a beautifully calligraphed Urdu epitaph in sparkling golden paint.

═⸗

It was some months after his son's death that Muhammad Sultan suffered the accident in which he damaged his left arm. Soon after the accident, his elder daughter was returned and the middle one lost her work. All possible sources of income disappeared. Vexed, he ate away at himself between sips from endless cups of salty nuun tea and puffs from filterless cigarettes.

He had often heard people mention how Sataar Wagay, one of his neighbours, whose son the army had tortured to death and dumped in the river, managed to get an ex gratia compensation of one lakh rupees sanctioned by the government. In the beginning, when he was not burdened by his own tragedies, Muhammad Sultan hated Sataar Wagay for accepting the compensation. He even called him a traitor for 'selling his son's sacrifice to the government'. But then Muhammad Sultan became confused when Rahman Parray, another of his neighbours, who had even been an active member of a separatist organization, distorted the facts surrounding his younger brother's killing by the army and accepted compensation of one lakh rupees.

Who was right, Wagay or Parray or himself? Baffled, Muhammad Sultan remained indecisive until his own condition forced him to think about applying for compensation too. Initially he hated himself for even thinking about asking the government for money. He wrestled with himself day and night.

He was already heavily in debt. Each morning the local grocer had begun to come to the house and threaten to take away Sultan's cow to settle his account. He even owed the neighbourhood baker and barber. He had stopped passing by their shops and now took longer routes, back and side paths, whenever he had to leave the village. But when his baby granddaughter was diagnosed with acute pneumonia, he gave up. Finally, he threw off the guise of commitment to the cause of freedom, ignored his guilt, and applied for compensation. He tried as much as possible to hide this from Gul Baghwaan, one of his close childhood friends, who had vehemently rejected an offer of compensation from the government after his son was killed in a crossfire incident.

Now, the only hurdle that came between Muhammad Sultan and his compensation was the word 'Shaheed', conspicuously engraved on his son's gravestone. If discovered anytime later, the word could ruin his chances.

═⸗

The graveyard is a plateau studded with gravestones and clumps of irises. It is away from the village houses and nestled on the edge of a vast expanse of paddy fields. As Muhammad Sultan sees the irises in the cloud-dimmed evening, the first stray thought that crosses his mind is how much better it would be to replace the cacti at his home with the irises.

He waits for the darkness to grow thicker. With the darkness comes a drizzle. Soon the voice of the muezzin from a nearby mosque floats into the air and mingles with the hissing rain. The wet mud of the graveyard sticks to the soles of his shoes, exposing patches of ochre under the upper layer of earth.

A few minutes later, when the rain stops, he holds the blunt chisel with his trembling left hand against the word Shaheed. He repeatedly strikes the head of the chisel. The sounds of metal clanking and the hammer's whumping travel through the earth and reach down to Mushtaq Ahmad Najar. At one strike the pointed end of the chisel slips, misses its mark, and scrapes off Mushtaq instead.

ELEPHANT AT SEA
KANISHK THAROOR

In the late summer of 1979, the Second Secretary of the Indian embassy to Morocco received a cable that undid his considerable years of training and left him floundering. The message read simply: 'Elephant en route.' Was it some sort of code? Further investigation only deepened his confusion. The cable had come from the customs office in Cochin, a port in the south of India. No, the customs officials reported back to him, it wasn't code. It was an elephant—an elephant, that along with its mahout, was now very much headed by ship to Casablanca. The Second Secretary probed: why send an elephant? Here at the customs office, the reply came, we handle only the movement of goods; for the movement of reasons, please refer your inquiry to the Ministry of External Affairs.

The Second Secretary telegrammed his colleagues in the ministry in Delhi. With telegrams to the ministry, it was important, first, to be terse so that you were considered economical and, second, to be sharp so that in the midst of reams of communication from outposts around the world, your message would be noticed. WHY SHIP ELEPHANT STOP EMBASSY ALREADY HAS CARS STOP. No one in the ministry seemed to know anything about the elephant. A flummoxed telegram returned to the embassy. WHAT ELEPHANT

STOP IS THIS CODE STOP. Embarrassed, the Second Secretary finally consulted the ambassador, who knew through long experience that it was pointless to question the whims of the capital. Marvellous, the ambassador said smoothing his moustache, an elephant, just what we need, and they couldn't even send it to us, no, they're sending it to Casablanca. You'll have to arrange for the thing to be met and picked up. He sprayed himself with cologne and mused: If an elephant can even be picked up.

That night the Second Secretary lay awake in bed, resenting the sheets, resenting the pillow, resenting the indifference of his work, resenting Morocco, resenting Arabic for its impossible, secret throatiness, and resenting, with what little bitterness was left to him, the unknown buffoon who would make diplomacy out of elephants.

The buffoon was not, as he imagined, some self-satisfied civil servant in South Block, but the princess of Morocco. Explanation arrived via telex from a friend in the ministry who owed him a few favours and so mustered the initiative to ask around.

The story of the elephant had begun six years earlier in the same Indian embassy in Rabat. At one of those habitual functions whose purpose seems so obvious in the preparation but disappears in the operation, the little Moroccan princess had come to the embassy and frozen before a picture of an elephant. It was among the many stock images—all approved by the ministry of tourism— that lined the lobby of the embassy: dawn over the Himalayan ranges; houseboats on the backwaters; the Taj Mahal rosy in its cushion of smog; a bright tractor devastating a field of wheat. The princess only had eyes for the elephant. Her wordless arm extended towards the picture, pointing. C'est un éléphant, said the embassy official tasked with escorting the princess. She remained transfixed. Vous aimez les éléphants? the unlucky man suggested. It seemed the princess did love elephants because she wouldn't move. The official, who had in previous posts offered counsel on trade policy with Indonesia and arms deals with the Soviet Union, looked around for help before lowering himself to her level. Mademoiselle, vous voulez un éléphant? he asked with the desperation stoked in him by all children—never mind the princess of Morocco. She turned, smiled, and gave him the smallest gift of a nod. It was enough. The official eventually spoke to the then ambassador who put in the request to Delhi, recommending the delivery of an elephant to satisfy the princess and to strengthen a bilateral friendship. The request passed through the appropriate channels at the usual speeds. Six years later, the creature was irrevocably on its way.

The Second Secretary received updates about the elephant's progress from various consular staff. In Yemen, it posed for photographs in front of the oldest coffee house in Aden. It bumped a football back and forth with boys on the beach in Alexandria. In Algiers, veterans of the war against the French held a reception in its honour; the Indian elephant was, in their words, a symbol of the ancient wisdom of a civilization that had inspired the global struggle against western imperialism. After it passed through Gibraltar, the Second Secretary got an excited telegram from the British naval high command. BRILLIANT STOP TOP PACHYDERM STOP.

The Moroccans were less enthused. What are we to do with it? they said. Casablanca's zoo has enough African elephants and there's no space for an Indian one. The Second Secretary protested. It's for the princess, he reminded them, she asked for it. Well, she may have, the Moroccans said, but she's away studying sociology in Paris now and has no interest in elephants. So you might have to take the thing back.

Only the Indian ambassador's persistence at a cocktail party won their grudging cooperation. It was agreed that the elephant would be specially housed in a portion of the royal gardens in Rabat. How it would get there was another matter altogether. The Moroccans insisted they had no trucks big enough to carry the creature between the two cities. This was a time of war and their heaviest military vehicles were all rumbling around the south. Worse, thanks to the mischief of Polisario terrorists, the single rail line between Casablanca and the capital was broken. But these inconveniences, the Moroccans claimed, shouldn't be a problem. After all, the elephant is its own means of transportation.

The Second Secretary was sent to Casablanca to escort the elephant back to Rabat. It took some time to find an appropriate launch in which to ferry the creature to port. When it finally disembarked, the small police band awaiting its arrival had grown sour in the heat. They rushed through the welcoming ditty and swiftly packed away their trombones. The reporters also sped through their work. They found the mahout altogether too clothed. As he perched on the elephant's back, they had him remove his shirt and roll up his trousers to look more convincing for the cameras, all knobby knees and gleaming skin. Instead of asking the mahout questions about the elephant, they surrounded the Second Secretary—he was wearing a suit. We are proud to share the joy of elephants with the people of Morocco, he said. The haathi belongs not to one nation, but to all.

What little the Second Secretary knew about elephants came from an urban childhood of zoos and encyclopaedias. As a measure of their robust memories, elephants hold grudges and harbour very finely developed notions of revenge. Elephants have sensitive feet capable of feeling through the earth, from long distances away, the approach of other elephants, or of rainstorms, or of bulldozers. Studies have shown that they can recognize their own reflections, suggesting that, however rudimentary, there may exist among elephants an amorphous theory of mind.

In sum, these scraps formed an altogether surreal idea of the elephant, one incommensurate with the full being in front of him, dappled with cooling splashes of mud, blinking restlessly and curling its trunk around the legs of its mahout. The mahout supplied more practical information. In its present ship-weary condition, the elephant could walk at most forty kilometres at a single stretch, probably no more than twenty-five. How far is it to the capital? he asked. He spoke no Hindi and the Second Secretary spoke no Malayalam, so they talked in a manner of English. A little more than ninety kilometres to Rabat, the Second Secretary said. Hoisting himself on to the elephant, the mahout surveyed the road leading out from the docks through the flat outskirts of Casablanca. For all the immensity of this unknown continent, the world always seemed more manageable from the back of an elephant. He smiled at the Second Secretary. Ninety of their Moroccan kilometres or ninety of ours? What are you talking about, the Second Secretary said, there's no difference. The mahout shook his head. You and I may not be able to tell the difference, but the elephant can.

In consultation with the two Moroccan gendarmes assigned to them, they agreed to break journey several times en route to Rabat. The convoy set out from Casablanca in the middle of the afternoon. The gendarmes led in their battered white car, its red lettering chipped and peeling. The Second Secretary brought up the rear in the embassy's sedan. In the middle, the mahout set a gentle tempo. The elephant wore an anklet that, wrapped around a man, would have had all the thickness of chains. Its every step tinkled with the jewellery of another land.

Perhaps because it was on its way to await the uncertain pleasure of a princess, or perhaps because it had already travelled so far, the elephant chose not to exert itself. But it quickened its pace whenever the coastal road veered west towards the Atlantic. The change would have been imperceptible to observers—and there were many on the busy highway—but the mahout felt it in his thighs. Each time the cobalt ocean wheeled into view, the creature's muscles seemed to quiver with new desire. It was an urge all the more palpable in its restraint; elephants are polite creatures of typically conservative

temperament. Yet it was enough of a rumble and churn for the mahout to
sense that his mount was already missing the sea.

At dusk, the elephant drank from a pond in the golf course of Mohammedia,
a pleasant enough beach town that brushed up against Morocco's only oil
refinery. The last of the day's players lofted their balls in long arcs overhead
before descending on the fairway and finding the creature asleep in a bunker.
It lay on its side, flanked by the apologetic guards, its heavy breathing raising
little tempests of sand. Nobody protested. Golf can only be improved by the
intrusion of an elephant, even a snoring one.

The Second Secretary was given accommodation in the clubhouse, in a
room glowing with trophies and the lidless glare of the nearby refinery. He
smoked and dipped at a plate of zaalouk. The mahout came in but declined the
invitation to share in the dish. He had already eaten with the policemen. The
Second Secretary gestured to a couch for the mahout to sleep on. The man
looked restless. No, I'll stay with the elephant, he said, this is our first night
back on land after weeks... It will rest poorly without me, and I without it.
The Second Secretary shrugged and returned to his puréed eggplant, only to
be surprised by the touch of the mahout's hand on his shoulder. In the parallel
universe of their own country, such contact would be almost unimaginable, a
movement far too intimate to cross the wide gulf of rank. Indians turn into
more equal beings when not at home.

Tell me truthfully, the mahout leaned forward, are there no other elephants
in Rabat? The Second Secretary sighed. I've told you already, it will have to
be kept by itself. The royal gardeners can only manage one elephant. It will
have all its creature comforts...don't worry. The mahout listened, see-sawing
his head from side to side. This was the strongest and happiest elephant he
had ever known, but he feared that it would struggle with its solitude. Like
humans, elephants yearn for other elephants. It will be lonely, he said, it will
need distractions. Annoyed, the Second Secretary promised they would make
sure that it was the most distracted elephant in Africa. He unlaced his shoes
and stretched out to sleep.

A few hours later, with the red and orange light of the refinery's towers
angling upon his face, he awoke to find the mahout sitting cross-legged in front
of him. I'm sorry to disturb you, the mahout said, but promise me something:
that you won't let them use it in the circus. In the circus? The Second
Secretary said. The mahout climbed to his feet and paced about the room. In
the circus, yes... I have heard about the way the firangs treat elephants, like
dolls, like puppets, like cartoons. He grew more animated still. They make them

dance, they make them ride cycles, they make them stand on their heads...
Sometimes, they think it is amusing to have the animals sit down to tea as
if elephants were old women...this can't be its fate. The Second Secretary sat
up. We won't let that happen, he said. Besides, you have nothing to worry
about; these people aren't firangs, they're Moroccans, they're so much like
us...just go to sleep.

But mahouts sleep as fitfully as elephants, and when the Second Secretary
rose at dawn to perform his ablutions, he found the man at the door of the
clubhouse, standing in a pose of total stillness, at war with the anxious writhing
of his eyebrows. The mahout burst into speech at the sight of the secretary.
These royal gardens, are they near the water? The water? The Second Secretary
blinked. Yes, the water, the ocean. I have no idea, the Second Secretary replied.
The mahout held the Second Secretary's hand in both of his. The elephant...
for it to be happy, it must be near the sea.

As he shaved, the Second Secretary muttered to himself about the mahout.
All the man has to do is deliver the elephant to Rabat and then the government
will give him a tidy cheque and send him home by plane. How many mahouts
ever see the inside of a plane? He'll never again get to be in a plane. He can
tell his parents, his wife, his children, if he has any, his grandchildren in the
future, that he was in a plane. And they'll tell all their friends and enemies and
the mahout will be famous forever throughout his village. Yet this madman
keeps me awake all night with his ridiculous demands for an elephant nobody
actually wants. If he fusses any more, we'll return him by boat. What glory
is there in a boat?

What the Second Secretary did not know—and what the mahout found
impossible to explain—was that, for the elephant at least, travel by boat was
utterly glorious. Before they left India, the mahout had worried about the
creature's well-being. How would it cope in steerage for all those cramped
weeks? Would it endure the din of the ship's innards, the engines and pipes
pumping at all times, soot-faced engineers swinging like monkeys from the
levers? Surely, the clamour of a mechanical universe would depress a creature
that loved nothing more at the end of the day than lowering itself into mud.
The only consolation the mahout could find was that he too was terrified of
the ship. There was a solidarity to be had between two beings who had never
travelled further than Kozhikode, two beings for whom the rusting expanse
of an ocean-going ship was only ever something to behold, not enter. In its
salty dark, the mahout imagined, they would comfort each other, leaning close,
pressing head to trunk.

It was not to be. While the mahout lurched from deck to deck vomiting, the elephant thrilled to life at sea. It trumpeted every time the captain sounded the great foghorn. The ship's sailors fawned over the creature, playing it music, showering it with nuts and chocolate. There were pleasures to be had even in its hold in the cargo bay, which rattled with the vigour of the ship's machinery. One morning, the mahout discovered the elephant rumbling. Serenely, it produced a low and eerie noise that seemed to come from its most interior parts. The mahout rushed to its side and held it as best as he could, trying to calm the animal. A moment's listening dispelled his fears; in perfect pitch, the elephant was merely mimicking the sound of the engines, as if through imitation it could bridge the divide between thought and matter and speak with the grey monstrosity of the ship.

On deck, the elephant stormed from side to side, relishing the heave of the ship, the rise and prostration of the bow as it carved its mass through the blue. The mahout studied the joy of the elephant with awe. He thought the elephant would grow bored—as he swiftly did—of the sea, but the wonder never wore off. As the wind sprayed it with foam, the creature seemed to admire the uninterrupted ocean in a kind of a rapture, a dervish-like ecstasy. It once occurred to the mahout that this might be the closest he would ever get to touching the divine: the elephant forgetting its elephantness in the vista of the sea, the veils of moksha parted, the creature poking its trunk into the beyond and feeling its way towards cosmic oneness. Then the motion of the vessel shook the mahout's insides loose. He staggered to the rim of the stern and emptied himself into the deep.

The sailors also felt the magic of the elephant's presence. Once, in the Indian Ocean, a solitary whale bobbed into view. This was hardly an unusual sight for the tanker's crew but they watched as the whale crested the surface and snorted through its blowhole. They hoped at that moment for a response from the elephant, a trumpet, a bellow, a spout of water from its trunk, some little signal of recognition. After all, what was the whale but the elephant of the sea? These two creatures were kin in bulk and grace, breathing the most air through the largest lungs in a world rightfully made for them. There could be no better omen than that shared understanding. For men who commit their bodies to the ocean, who surrender childhoods on paddy fields and factory floors for the education of currents and gales on the shipways, the communion of these beasts would be a vindication of their lives. But the sailors were disappointed. Standing on the starboard side, the elephant had not seen the whale at all, or if it had, it chose to ignore it, keeping its eyes fixed instead in contemplation of the water.

On the golf course, the mahout found the elephant by the pond, its trunk lingering at its feet. He massaged the hard knot of muscle on its lower back, the corbelled arch that lifted the creature's mass from the earth. As he readied the elephant for the onward march, he wondered whether a journey across the seas had the ability to change us. When the elephant regarded its reflection in the still water, did it see a being transformed? Could it? Maybe it was presumptuous of the mahout to think so grandly of the elephant's capacities, its self-awareness, its very sense of the possibility of a self. Perhaps this sad-eyed creature merely looked at the pond and thought: What a miserable excuse for a sea.

The convoy reached Skhirat in the early evening. On the way, the mahout had suggested to Adil, one of the Moroccan gendarmes (he had learned their names; the other was Marouane), that he come sit on the back of the elephant. Adil stripped off his uniform jacket, approached the elephant from the front, hesitated, crept around the side and kept creeping till he made a full circle, and looked imploringly up at the mahout. The mahout laughed. He scratched the back of the elephant's head and pressed one knee against its neck. It dipped to the ground. Adil tried to look the creature in the eye for reassurance, but it stared beyond him up the road, its ears flapping like fans. He grasped the mahout's forearm and heaved himself up, gripping the hairy hide with both hands as the elephant rose to its feet and lurched on.

Elephants respond to confidence, the mahout said, to certainty. You do not need to charm them so much as direct them… Like us, they are logical creatures, and like us, they understand that the order of the universe dictates to them a certain place, a certain rank, a certain dependence on the demands of others. The mahout spoke in Malayalam, but Adil listened to the tumbling words anyway, trying his best not to look at the road swaying beneath him. The truth is, the mahout continued, that driving an elephant does not require intuition or special intelligence, only a willingness to command…more than that, a belief in your command.

Command was in his blood. The mahout was raised to ride elephants, as was his father, and his father's father, and as far as he knew all the males of their line snaking back to some letterless past when man first wrestled the beast into obedience. No better life had presented itself to him than that of ordering elephants. He was aware that many in his village were jealous of his trade, the princely work that saved him from the drudgery of the fields. When the news arrived that he would be sent with the elephant to Morocco, his family lit many candles to fend off the evil eye. You'll come back a big man, they said, and nobody wishes well for big men. What nonsense, he laughed them off, I'll come back just the same…this isn't my journey, it's the journey of the elephant, I'm only an appendage of flesh. Adil squirmed behind him and cars passed, honking. How true, the mahout thought, I am commanded to command. I am an instrument of command. I am an instrument.

From the trailing sedan, the Second Secretary watched the spectacle of the Moroccan gendarme clinging to the elephant. He was surprised to feel a degree of envy. No invitation to mount the elephant had been extended to him. If anyone should first get a turn on the elephant, he thought, it should be me, not that fellow. He filed this grievance away as yet more proof of the strangeness of the mahout and as further evidence, if he needed any more, of the unending injustice that was the daily life of a Second Secretary.

At Skhirat, Adil slumped nauseous from the back of the elephant, attempted a few steps, and tumbled to his knees. The children of the village roared at his collapse and flocked about the elephant. Marouane, the other gendarme, dispersed them as best he could, but they remained bubbling in the corners of the village square, pantomiming the elephant and its minders. Skhirat's mayor, who was also its lead cleric, came to shake hands with the visitors and admire the elephant. In normal circumstances, the village was used to strangers passing through; it was a stop on the Casablanca-Rabat rail line. But since the interruption of rail service, the place had grown dustier and quieter. Its people were happy to produce a welcome fit for any occasion. They brought trellised tables into the square. Pitchers of fresh juices, cups of tea, and miraculous tagines came steaming from nearby houses. All the village's luminaries—its post office clerk, its librarian, its accountant, its letter-writer, its chief constable (its only constable), its doctor, its farm veterinarian (who kept his distance from the elephant, eyeing the creature with trepidation), and so on—assembled to have a meal with the Second Secretary, who was made to repeat, over and over again, in slow French, the basic facts of his life and of his world. The children cheered as the elephant munched carrots dipped in harissa. The gathering continued till late in the evening. When the day's last azaan interrupted proceedings, the elephant raised its trunk towards the minaret and bellowed in its own fashion the call to prayer. All the men of the village made their way to the mosque except for the librarian, who patted the Second Secretary on the back and smuggled him home to share a bottle of arak.

During the night, the Second Secretary snored drunk on the librarian's sofa. Adil slept on a bench. The mahout tucked his chin into his knees and dozed against the slumbering bulk of the elephant. Marouane stood awake, vigilant for any mischievous children lurking at the edges of the square. Nothing happened until an hour before the morning azaan. A shape formed in the provincial gloom and drifted towards the elephant. It was the cleric-mayor. Peace be upon you, Marouane said. And you, the cleric-mayor returned. He rolled up the sleeves of his robe and worked his way around the horizontal elephant. I just want to check, he said half to himself, I just want to check. Marouane watched him dubiously. Check what? The cleric-mayor had already knelt by the elephant's loins. He startled. Well, I'm just curious if the creature is Muslim. In three strides, Marouane had grabbed him by the collar, dragged him away, and dropped him to the ground. You fool, he looked down on the older man, you bumpkin... Be decent and keep your crazy ideas to yourself. Does a donkey have religion? Can a donkey be Muslim? How can this animal be Muslim? The cleric-mayor straightened up and jabbed a finger into Marouane's uniform. Boy, he snarled, have some respect...that peaceful creature is more of a man than any of your kind will ever be.

The commotion woke the elephant and the mahout. Alarmed by the animal's surging to its feet, the cleric-mayor made apologetic noises and ghosted away. The mahout saw Marouane's agitation. He pointed at Adil sleeping on the bench, urging the gendarme to follow his colleague's example. Marouane nodded and tried to get to sleep. The elephant snorted. It stamped. It wrapped its trunk around the mahout's waist, hugging the man close. Whatever beliefs it did possess, it certainly disliked being roused from its dreams.

The mahout stood for a little while, stroking the elephant's trunk until it subsided once more to its knees, then rolled on to its side. They were alone in the village square. At this time before dawn in the mahout's own village, the roosters would be outdoing one another, the potholed roads would already be clanging with traffic, his family in their multitude would be scratching and groaning and clucking in the shared sleeping space. Morocco had so much room, so much silence. The mahout watched Skhirat take shape in the leavening dark. There was an enviable modesty to the even spread of low buildings, the humble bakery warming its ovens at the edge of the square, the grace of the mosque's silhouetted minaret, the peace of all its obscurity. He knew that this was a tiny country pinned between desert and sea. He knew that his own country was large by any estimation. And yet the calm of this small place felt infinite.

The elephant nuzzled his hand and murmured in its sleep. He buried his face in its ear. He whispered: Sleep well, my beauty, sleep well, my prince. If you dream, don't dream of home and don't dream of me. Dream of the sea. You and I are now so alone in this world... Dream of the sea, my life, dream of the sea. The elephant slept, but its trunk remained wrapped around the body of the mahout. It refused to release him. Only as light began to escape down from the eastern mountains did the elephant loosen its grip and let the mahout go.

Dawn came with the first azaan. Adil shook awake, as did Marouane. The village crawled into its quiet, habitual motion, making the small adjustment for the sleeping elephant at its centre. The Second Secretary staggered to his sedan for his toiletries. By the time he finished brushing his teeth next to the well, both the gendarmes stood before him, delivering the news as best they could that the mahout had disappeared.

The Second Secretary was incredulous. Disappeared? Impossible. Skhirat was mobilized to find the mahout. Children swarmed over the rooftops. Scooters buzzed down the road in both directions. Farmers turned over their cauliflowers. The chief constable furiously blew his whistle. There was no trace of the man. At the post office, the Second Secretary sent a message to the embassy. MAHOUT ABSCONDED STOP PLEASE ADVISE STOP. The Indian ambassador rang the post office. He's vanished, has he? The Second Secretary said he had. That's a real pity, the ambassador lamented, but what to do. It's pure physics... You propel an object a certain distance and you just can't expect it to come to rest, it will keep going forward. You take a man this far from his benighted village and he'll lose all interest in going back... so it goes. But, sir, the Second Secretary interjected, what do we do about the elephant? I don't know, the ambassador said, Rabat isn't all that far. Yes, the Second Secretary agreed, but how do we move the elephant? Why are you asking me? The ambassador snapped. If I were a mahout, I wouldn't be having this bloody conversation with you, would I? Just do whatever it is you need to do.

The Second Secretary sat on the hood of his sedan, staring at the elephant. It looked back at him, long-lashed and indifferent. He imagined the various means at his disposal that would convince this mass of flesh to proceed down the last stretch of highway to the capital. Perhaps he could lay down a trail of carrots all the way to Rabat. Or maybe if all the children pushed hard enough together, they could inch the elephant up the road. Or better yet, why not just leave the elephant here for the people of Skhirat? Why not be generous and gift them the problem?

Pitying the glum resignation of the Second Secretary, Adil was stirred to provide the solution himself. He came forward to the elephant and placed his hand behind its head, speaking to the creature in Arabic. It bent down. Gingerly, he clambered on top. The elephant returned to its feet. Adil's prodding steered it on to the road. For the first time in two days, the Second Secretary smiled. Allons-y! he cried. Allons-y!

It was imperative to get going before Adil's luck ran out and the elephant decided to stop cooperating. The convoy reassembled and bid a hasty goodbye to Skhirat. All the villagers waved, except for the cleric-mayor. From the window of the mosque, he had seen the mahout slink away in the early hours, seen the alien gleam in his eyes, the rootless abandon of the wanderer. It was a sad spirit, one the cleric-mayor could not comprehend and dared not interfere with; who was he—who had never seriously left his village nor contemplated doing so—to judge the actions of a stranger come to a strange place? So he kept his peace. God be with you, the cleric-mayor said to the departing rump of the elephant, God be with you.

They had reached the outskirts of Rabat when the handlers from the royal gardens finally emerged and relieved them of the elephant. For the cameras, the ambassador posed in front of the creature with Adil, the hero who saved the day and strengthened the bonds between the people of India and Morocco. It was reported to the newspapers that the mahout had been incubating a mysterious tropical disease that had killed him en route to Rabat. The ambassador asked the Second Secretary to ensure that Adil's wife received a very tasteful flower arrangement. The Second Secretary did as he was told.

When the princess returned during her holidays, she was enchanted by the elephant. She would lie next to it and read aloud her books of philosophy and critical theory. She introduced it to champagne. The princess was so enamoured of the creature that she insisted it accompany her on a trip to the beach. The elephant hurried over the dunes at the sight of the ocean, the clarion call of its trunk warning the waves. Everybody laughed as it played in the surf. It seemed to enjoy being knocked over in the shallows, finding its feet, retreating to the beach, and then wheeling its bulk around for another charge against the sea. It all seemed a pleasure, but the elephant was sad that no matter how earnestly it plunged into the water, the tide always drove it back to shore.

ACKNOWLEDGEMENTS

This book wouldn't have come about without the passion, hard work and commitment to excellence of all those who helped me put it together—my wife Rachna Singh, who has worked with me on all my books, and my fellow-Alephs: Bena Sareen, Ritu Vajpeyi-Mohan, Aienla Ozukum, Simar Puneet, Pujitha Krishnan, Vasudha Iyer, Aruna Ghose and Rajkumari John. Other friends who willingly helped were Geeta Dharmarajan, Nilanjana Roy, Arunava Sinha, Githa Hariharan, Ina Puri Harish Trivedi, Ranga Rao, Amitava Kumar, O.V. Usha, Ramesh Menon and Kiran Desai—without their expert guidance, and assistance in a myriad ways, some of the finest stories in the book wouldn't be in it. However, I must make clear that while, in some ways, this was a collaborative effort, I am solely responsible for any omissions or eccentricities in the selection or any other shortcomings the book might have. My thanks to all those who helped make this book possible.

Grateful acknowledgement is made to the following copyright holders for permission to reprint copyrighted material in this volume.

'The Hunger of Stones' by Rabindranath Tagore translated by Amitav Ghosh. Translation copyright © Amitav Ghosh. Reprinted by permission of The Wylie Agency (UK) Limited and the translator.

'The Shroud' by Munshi Premchand translated by Arshia Sattar. Translation copyright © Arshia Sattar. Reprinted by permission of the translator and Pratilipi.

'A Horse and Two Goats' by R.K. Narayan. Copyright © The Estate of R.K. Narayan. Reprinted by permission of Indian Thought Publications.

'A Life' by Buddhadeva Bose translated by Arunava Sinha. Copyright © Buddhadeva Bose. Translation copyright © Arunava Sinha. Reprinted by permission of Damayanti Basu Singh and the translator.

'Toba Tek Singh' by Saadat Hasan Manto. Copyright © Saadat Hasan Manto. Translated by Khushwant Singh. Translation copyright © Mala Dayal. Reprinted by permission of Mala Dayal.

'The Flood' by Thakazhi Sivasankara Pillai. Copyright © Balakrishnan Nair. Translated by O.V. Usha. Translation copyright © O.V. Usha. Reprinted by permission of Balakrishnan Nair and the translator.

'The Blue Light' by Vaikom Muhammad Basheer. Copyright © Vaikom Muhammad Basheer. Translated by O.V. Usha. Translation copyright © O.V. Usha. Reprinted by permission of Fabi Basheer and the translator.

'The Somersault' by Gopinath Mohanty. Copyright © Gopinath Mohanty. Translated by Sitakant Mahapatra. Translation copyright © Sitakant Mahapatra. Reprinted by permission of Omkarnath Mohanty and the translator.

'Portrait of a Lady' by Khushwant Singh. Copyright © Mala Dayal. Reprinted by permission of Penguin Books India.

'Quilt' by Ismat Chughtai. Copyright © Ismat Chughtai. Translated by Rakhshanda Jalil. Translation copyright © Rakhshanda Jalil. Reprinted by permission of Ashish Sawhney and the translator.

'Stench of Kerosene' by Amrita Pritam. Copyright © Amrita Pritam. Translated by Khushwant Singh. Translation copyright © Mala Dayal. Reprinted by permission of Mala Dayal.

'Gold from the Grave' by Anna Bhau Sathe. Copyright © Anna Bhau Sathe. Translated by Vernon Gonsalves. Translation copyright © Vernon Gonsalves. Reprinted by permission of the translator.

'The Man who saw God' by D.B.G. Tilak. Copyright © D.B.G. Tilak. Translated by Ranga Rao. Translation copyright © Ranga Rao. Reprinted by permission of Satyanarayana Murthy and the translator.

'Inspector Matadeen on the Moon' by Harishankar Parsai. Copyright © Harishankar Parsai. Translated by C.M. Naim. Reprinted by permission of Katha Books.

'Draupadi' by Mahasweta Devi. Copyright © Mahasweta Devi. Translated by Gayatri Chakravorty Spivak. Reprinted by permission of Seagull Books.

'Countless Hitlers' by Vijaydan Detha. Copyright © Vijaydan Detha. Translated by Christi A. Merrill and Kailash Kabir. Translation copyright © Christi A. Merrill and Kailash Kabir. Reprinted by permission of the translators.

'Mirror of Illusion' by Nirmal Verma. Copyright © Gagan Gill. Translated by Geeta Kapur. Reprinted by permission of Gagan Gill.

'Reflowering' by Sundara Ramaswamy. Copyright © Kamala Ramaswamy. Translated by S. Krishnan. Reprinted by permission of Kamala Ramaswamy.

'Mouni' by U. R. Ananthamurthy. Copyright © U. R. Ananthamurthy. Translated by H.Y. Sharada Prasad. Reprinted by permission of Katha Books.

'Old Cypress' by Nisha da Cunha. Copyright © Nisha da Cunha. Reprinted by permission of the author.

'The Blue Umbrella' by Ruskin Bond. Copyright © Ruskin Bond. Reprinted by permission of Rupa Publications India and the author.

'Crossing the Ravi' by Gulzar. Copyright © Gulzar. Translated by Rakhshanda Jalil. Translation copyright © Rakhshanda Jalil. Reprinted by permission of the author and the translator.

'Games at Twilight' by Anita Desai. Copyright © Anita Desai. Reprinted by permission of the author, c/o Rogers, Coleridge & White Ltd., 20 Powis Mews, London W11 1JN.

'A Revolt of the Gods' by Vilas Sarang. Copyright © Vilas Sarang. Translated by the author. Translation copyright © Vilas Sarang. Reprinted by permission of Penguin Books India.

'In a Forest, a Deer' by Ambai. Copyright © Ambai. Translated by C. T. Indra. Reprinted by permission of Katha Books.

'Bhaskara Pattelar and My Life' by Paul Zacharia. Copyright © Paul Zacharia. Translated by Gita Krishnankutty. Reprinted by permission of the author.

'Tar Arrives' by Devanoora Mahadeva. Copyright © Devanoora Mahadeva. Translated by A. K. Ramanujan and Manu Shetty. Translation copyright © A. K. Ramanujan and Manu Shetty. Reprinted by permission of the author and Manu Shetty and the Ramanujan estate.

'The Elephant and the Tragopan' by Vikram Seth. Copyright © Vikram Seth. Reprinted by permission of the author.

'Feast' by Manjula Padmanabhan. Copyright © Manjula Padmanabhan. Reprinted by permission of the author.

'Nursing God's Countries' by Githa Hariharan. Copyright © Githa Hariharan. Reprinted by permission of The Abraaj Group Art Prize 2014: Garden and Spring and the author.

'Proposed for Condemnation' by Cyrus Mistry. Copyright © Cyrus Mistry. Reprinted by permission of the author.

'Trying to Discover India' by Shashi Tharoor. Copyright © Shashi Tharoor. Reprinted by permission of the author.

'Desolation, Lust' by Upamanyu Chatterjee. Copyright © Upamanyu Chatterjee. Reprinted by permission of the author.

'Kama' by Vikram Chandra. Copyright © Vikram Chandra. Reprinted by permission of Penguin Books India.

'Wild Things' by Anjum Hasan. Copyright © Anjum Hasan. Reprinted by permission of the author.

'Stolen' by Amrita Narayanan. Copyright © Amrita Narayanan. Reprinted by permission of the author.

'The Gravestone' by Shahnaz Bashir. Copyright © Shahnaz Bashir. Reprinted by permission of the author.

'Elephant at Sea' by Kanishk Tharoor. Copyright © Kanishk Tharoor. Reprinted by permission of the author.

We regret any errors or omissions in the above list of acknowledgements and would like to be notified of any corrections that should be incorporated in future editions of this book.

NOTES TO THE STORIES

David Davidar: Our Stories
All the Ramanujan quotes are taken from *The Collected Essays of A. K. Ramanujan* (Oxford: Delhi 1999).

p xvii **All writing may be**: The R.K. Narayan quote is taken from his foreword to *Under the Banyan Tree* by R.K. Narayan (Penguin: New York 1995).

p xviii **Why is Anton Chekhov**: This William Boyd quote is taken from his article in the issue of *Prospect* magazine dated 10 July 2006.

p xviii **We all come out**: The quote about Gogol that is attributed to Dostoevsky is attributed by some to Turgenev.

p xix **I see that it is invariably**: George Orwell's quote is taken from his 1946 essay, 'Why I Write'.

p xxiii **Indians have lived**: This quote is from Vikram Chandra's essay 'Cult of Authenticity' published in the *Boston Review* on 1 January 2000.

p xxiv **There's a musicality**: This quote is from the Gregory Rabassa interview with Susan Bernofsky that appeared in *Rumpus* (19 September 2013).

Rabindranath Tagore: The Hunger of Stones
This story was first published in 1895 as 'Kshudhita Pashan'. Its translator, Amitav Ghosh, says of it: 'As an allegory of the colonial condition, it is a work of extraordinary suggestive power and atmospheric richness'.

Munshi Premchand: The Shroud
Premchand often wrote his stories in both the Urdu and Devanagari scripts. This story is believed to have been first published in 1935 as 'Kafan' in *Chaand,* and has been repeatedly translated and anthologized. As with many of his stories, it shows Premchand's engagement with the social issues of his time.

The translator would like to thank Rahul Soni and Giriraj Kiradoo.

R.K. Narayan: A Horse and Two Goats
First published in the *Hindu* in 1960, this story became popular after it became the title story of a collection published in 1970. It was inspired by an American friend turning up at Narayan's house one evening with a clay horse 'crammed into his station wagon'.

Buddhadeva Bose: A Life
First published in 1957 as 'Ekti Jibon', this story is believed to be one of the turning points in the evolution of modern Bengali literature. Bose's short stories often present the lives of individuals—or moments in the life of one individual—capturing their thoughts, emotions and contradictions in relation to events and people around them. The story's translator, Arunava Sinha, says 'A Life' is 'the finest example of Bose's craft, a moving but unsentimental celebration of a heroic life devoted to knowledge'.

Saadat Hasan Manto: Toba Tek Singh
First published in 1955, this is Manto's best-known story. It has been widely translated and anthologized. It is a satire on the relationship between the newly formed India and Pakistan and on bureaucratic processes that often seem absurd.

Thakazhi S. Pillai: The Flood
Titled 'Vellappokkathil' (In the Flood) in Malayalam, the setting for the story is Thakazhi's native Kuttanad, which is periodically ravaged by floods.

Vaikom Muhammad Basheer: The Blue Light
It was originally published as 'Neela Velichcham' in a short story collection (*Paavappettavarude Veshya*) in 1952. In 1964 it was made into a movie called *Bhargavi Nilayam* for which the author wrote the script. The movie became so successful that 'Bhargavi Nilayam' is often used in Malayalam as shorthand for a haunted house.

Gopinath Mohanty: The Somersault
The story was first published in Oriya in *Naa Mane Nahin* in 1968. The English translation was first published in the collection *The Ant and Other Stories* in 1979 by United Writers, Calcutta. The author's storytelling technique was renowned for its clarity and deceptive simplicity.

Khushwant Singh: Portrait of a Lady
This semi-autobiographical story was first published in *The Collected Short Stories of Khushwant Singh* in 1989. It has since been anthologized extensively.

Ismat Chughtai: Quilt
Titled 'Lihaaf' in Urdu, this is Chughtai's most celebrated story and perhaps the most

controversial. Two years after this story was published in *Adab-i-Latif* in 1942, she was sued for obscenity by a Lahore court. She chose to contest the case rather than apologize, and won.

p 95 **bhangan**: A sweeperess or one who cleaned people's homes and latrines, and belonged to the untouchable class; used as a derogatory term.

p 95 **Ayatul Kursi**: 'The Throne Verse' is a verse from the Holy Quran that is recited in times of danger or to ward off misfortune. Children are encouraged to recite it when they are afraid. The fragment referred to here occurs, literally, in the middle of the verse.

Amrita Pritam: Stench of Kerosene
Khushwant Singh picked this as one of his favourite stories from India and it was included in the anthology (*Our Favourite Indian Stories)* he edited with Neelam Kumar (Jaico: Mumbai 2002). This story depicts the plight of women in orthodox Indian households.

Anna Bhau Sathe: Gold from the Grave
Originally published in Marathi as 'Smashaanathil Sona', this story deals with the difficulties that Dalits face in finding work and the hardships they have to overcome just to survive.

D. B. G. Tilak: The Man Who Saw God
This story was first published as 'Devunni Choosina Vaadu' in *Jyothi* in 1961. Tilak's short stories were influenced by Maxim Gorky and Tagore.

Harishankar Parsai: Inspector Matadeen on the Moon
'Inspector Matadeen on the Moon' was first published in Hindi as 'Inspector Matadeen Chand Par' in 1968.

p 129 **When it's time for**: The literal translation of the original would read: 'When it's time for delivery, send your bed to lend a hand.' In certain segments of the society Parsai has written about the words 'house' and 'bed' were loosely used to refer to a wife.

p 129 **Pravisi nagara**: This line is from a verse from Tulsidas's *Ramcharitmanas,* recited by Lankini, the residing deity of Lanka, inviting Hanuman to enter Lanka when he arrived there in search of Sita.

Mahasweta Devi: Draupadi
The protagonist in this story is named for the most celebrated heroine of the Mahabharata, but is called Dopdi, her Santhali name, rather than the Sanskrit Draupadi. While Draupadi in the Mahabharata was protected by God, who prevented her from being unclothed, the Dopdi in this story is gang-raped by the police.

The translator is grateful to Soumya Chakravarti for his help in solving occasional problems of English synonyms and in archival research.

p 146 **Champabhumi** and **Radhabhumi** are archaic names for certain areas of Bengal. 'Bhumi' is simply 'land'. All of Bengal is thus 'Bangabhumi'.

p 146 **You fucking jackal:** The jackal following the tiger is a common image.

p 147 **No comrade will**: Modern Bengali does not distinguish between 'her' and 'his'. The 'her' in this sentence can therefore be considered an interpretation.

p 148 **Her own cloth:** A sari is what 'proper' Indian women wear. Dopdi wears a much-abbreviated version, without a blouse or underclothes. It is referred to simply as 'the cloth'.

Vijaydan Detha: Countless Hitlers
Detha was well-known for drawing his fiction from stories he heard from friends, family and neighbours. Titled 'Alekhun Hitler' in Rajasthani, this story is believed to have been overheard by the author on a bus. He said in a 2011 interview with the *Hindu*: 'There is a Hitler in every one of us. It draws its strength from condescension for another being and the realization of the power to overpower and destroy it.'

Nirmal Verma: Mirror of Illusion
First published as 'Maya Darpan', this story was the basis for an acclaimed feature film made in 1972 by Kumar Shahani. In his stories, Verma explores the themes of love, separation, alienation and nostalgia.

Sundara Ramaswamy: Reflowering
Originally published as 'Vikasham' in the Tamil edition of *India Today* (January 31–February 5 1990), Ramaswamy won the Katha Award for Creative Fiction in 1991 for this story.

U. R. Ananthamurthy: Mouni
This is one of Ananthamurthy's best-known short stories and one of his favourites. It was made into a critically acclaimed movie in 2003. Speaking about language, he said: 'Our analytical and conceptual articulation comes from the English language. But only our mother-tongue can provide us with

metaphors that describe emotional states. ...the living language is the language that is spoken in the streets. ...Kannada has been preserved by non-literates...but my articulation, my analytical self, my ideational self, all these come from my familiarity with the English language.'

Nisha da Cunha: Old Cypress
'Old Cypress' came out of a trip the author made to Ooty and Coonoor, which included visits to tea estates and wildlife sanctuaries in the Nilgiris.

Ruskin Bond: The Blue Umbrella
This story was first published in 1972. The author wove a story about a girl called Binya and her brother Bijju who lived in a village close to his home in Mussoorie. He has called this story one of his favourites 'mainly because the children in it were real people, and my love for them and for the mountains comes through'.

Gulzar: Crossing the Ravi
This story was born out of the author's traumatic experience of Partition. He said in a 1987 interview to *India Today*: 'I haven't yet distanced myself from that'. In the same interview he said that the story was a metaphor for his own childhood trauma: 'It is I who have been left behind and drowned, it is I who have come this side.'

p 255 **sharanarthis**: literally, those who seek refuge; the term was used in the early days of the Partition for Hindus and Sikhs from Pakistan who crossed the border to seek refuge in India. In much the same way, the Muslims who went to Pakistan from India were called muhajirs in their new homeland, meaning immigrants.

p 256 **Sankrant or Sankranti**: Refers to the transmigration of the Sun from one constellation to the next. There are 12 Sankrantis in a year. Makar Sankranti, marking the movement of the Sun into Capricorn, is celebrated with especial fervour in Punjab.

Anita Desai: Games at Twilight
'Games at Twilight' was first published in Anita Desai's 1978 collection of the same name and has since become one of her most widely anthologized stories. Reviewing the collection, the writer Victoria Glendinning called the story 'a jewel. It recounts something that has happened in one way or another to nearly everyone in childhood.'

Vilas Sarang: A Revolt of the Gods
Vilas Sarang is an accomplished translator of his own works from Marathi into English. This story is taken from the anthology *The Women in Cages* about which the author wrote: 'Most of the stories in this volume were written in Marathi. Subsequently, they were re-done in English. I say "re-done" because what I did cannot exactly be called translation. The final draft of each story was done without consulting the original for at that stage, my main concern was to see that the story worked in English. This was rendered easier because when I write in Marathi, I often mentally translate sentences that were formed in my mind originally in English.'

Ambai: In a Forest, a Deer
This story was published in Tamil as 'Kaatil Oru Maan'. As a child, Ambai spent a lot of time in her grandparents' house in Coimbatore and draws from these memories for this and many of her other stories. The author is highly regarded for her innovative story-telling, wit and lyrical writing.

Paul Zacharia: Bhaskara Pattelar and My Life
This story was originally titled 'Bhaaskara Pattelaarum Ente Jivithamum' in Malayalam. The man who provided the inspiration for Pattelar's character used to live in a village in south Karnataka when the author lived there.

Devanoora Mahadeva: Tar Arrives
'The process of writing this story changed my outlook', says the author. A story by Purnachandra Tejasvi, 'Abachurina Postafisu', which deviated from the Navya (modern) way of writing fiction that was then in vogue in Kannada literature, is an influence on 'Tar Arrives'.

Irwin Allan Sealy: Last In, First Out
Allan Sealy wrote this story for a book called *Delhi Noir,* which was part of a series of books published by Akashic Books in the US—each of the books comprised noir stories set in various cities or locations. This story explores the unexpected consequences of vigilantism.

Vikram Seth: The Elephant and the Tragopan
'The Elephant and the Tragopan' is taken from Vikram Seth's *Beastly Tales from Here and There.* The author tells us how this tale and the others in the book took shape: 'Because it was very hot

in my house one day and I could not concentrate on my work, I decided to write a summer story involving mangoes and a river. By the time I had finished writing [it]...another story and other animals had begun stirring in my mind. And so it went on until all ten of these beastly tales had been born—or reborn.'

Manjula Padmanabhan: Feast
'Feast' builds on the question: what happens when a non-Christian is attacked by a vampire? The author says: 'That idea led to the view that, in a certain sense, a classical Dracula-style vampire must also be a believer.' She explores what happens when such a being encounters an opposing belief-system.

Githa Hariharan: Nursing God's Countries
In this story the author was responding to the artist Anup Mathew's photographs of Kerala nurses in different parts of the world. She says: 'The first thing that came to mind was the variety of physical landscapes in a travelling nurse's life: the lush green of Kerala, the desert or pruned and artificial green of the gulf countries, the snowy landscape of a place like Canada. The physical landscapes merging one into the other were incomplete without the layered mental landscapes, of memory and lived experience, and hopes for the future. All these landscapes, real and imagined, merged with actual photos in the story.'

Cyrus Mistry: Proposed for Condemnation
This story took as its inspiration 'a dark, deserted, crumbling, rust-eaten skeleton of a compartment of a train at a railway siding, on which were painted in white the words "Proposed For Condemnation"'—something that the author saw when he was a frequent traveller on the local trains in Bombay. This combined with 'lascivious feelings towards compartment-loads of women crushed together' resulted in this rather erotic story.

Shashi Tharoor: Trying to Discover India
This story was to have been the beginning of a novel ('The India of Discovery') that the author started writing in 1992—the five hundredth anniversary of Columbus's maiden voyage. It was to contain two juxtaposed narratives—one of an Indian seaman who, 500 years ago, accidentally discovers America, and one of an Indian in America, 500 years later, who accidentally discovers India (and himself).

Unfortunately, Tharoor's responsibilities in handling the Yugoslav Civil War did not leave him time to complete the project.

Upamanyu Chatterjee: Desolation, Lust
This story was first published in the October 1986 edition of *London Magazine*.

Vikram Chandra: Kama
This story is taken from Vikram Chandra's collection of stories, *Love and Longing in Bombay* (Penguin: Delhi 1997). The narrative explores Bombay's underbelly. It is here that we first encounter Sartaj Singh, the indomitable Bombay cop, who is the protagonist of Chandra's great novel, *Sacred Games*.

Anjum Hasan: Wild Things
Anjum Hasan's fiction is influenced by Raymond Carver—she loves his ability to draw a story out of ordinary, even humdrum lives. She also admires R. K. Narayan for his entirely natural sounding and yet completely invented Indian English idiom and Guillermo Arriaga for his sense of drama.

Amrita Narayanan: Stolen
This story was first published in *A Pleasant Kind of Heavy* (Aleph: New Delhi 2013). In an interview with *Mid-day*, the author said: 'So often erotica is restricted to the briefest of moments in the bedroom when in fact, it is everywhere where there is life: in trees, flowers, fruit, and human moments of various kinds. Writing about the erotic is writing about life.' She attributes the opening frames of this story to a scene she witnessed in a friend's kitchen in Mysore.

Shahnaz Bashir: The Gravestone
This story was first published in the July 2014 issue of the *Caravan*. It deals with the existential compulsions of people who give up their passions for their political ideals. It was influenced by novels like *Life & Times of Michael K* and *Disgrace* by J. M. Coetzee and Chekhov's stories.

Kanishk Tharoor: Elephant at Sea
This story draws from the quasi-folkloric tone of writers like Italo Calvino, Ryszard Kapuściński, Jorge Luis Borges and José Saramago. It is based on a story told to the author when he was a child by a diplomat friend of his parents. He has always believed the story was true although he hasn't been able to confirm its veracity.

NOTES ON THE AUTHORS

Ambai (born 1944) is the nom de plume of Dr C. S. Lakshmi who is a Tamil writer, feminist, historian, and independent researcher in women's studies. Her stories have been translated into English in the volumes *A Purple Sea* and *In a Forest, a Deer*. She is presently the Director of Sound & Picture Archives for Research on Women (SPARROW) in Mumbai.

U. R. Ananthamurthy (1932–2014) was a Gandhian socialist and one of the pioneers of the Navya (new) movement in Kannada literature. He published five novels, one play, eight short-story collections, as well as anthologies of poetry and essays—his work has been translated into several languages. His best known work is his 1966 novel, *Samskara*, which was a finalist for the Man Booker Prize. Another of his novels, *Bharatipura*, was shortlisted for the 2011 Hindu Literary Prize and the DSC Prize for South Asian Literature.

Vaikom Muhammad Basheer (1908–1994) was a humanist, freedom fighter, novelist and short-story writer who revolutionized Malayalam literature through sarcasm, satire and black humour. His notable works include *Balyakalasakhi, Shabdangal, Pathummayude Aadu, Mathilukal, Ntuppuppakkoranendarnnu, Janmadinam* and *Anargha Nimisham*. He was awarded the Padma Shri in 1982. He is fondly remembered as the Beypore Sultan.

Shahnaz Bashir (born 1980) teaches creative journalism and literary reportage at the Central University of Kashmir, where he is the coordinator of the media studies programme. His debut novel, *The Half Mother*, was published in 2014.

Ruskin Bond (born 1934) has been writing for over sixty years, and now has over 120 titles in print—novels, collections of stories, poetry, essays, anthologies and books for children. His first novel, *The Room on the Roof*, received the prestigious John Llewellyn Rhys award in 1957. He has also received the Padma Shri, and two awards from the Sahitya Akademi—one for his short stories and another for his writing for children. In 2012, the Delhi government awarded him its Lifetime Achievement Award.

Buddhadeva Bose (1908–1974) one of the most celebrated Bengali writers of the twentieth century, was a central figure in the Bengali modernist movement and was widely considered to be the successor to Rabindranath Tagore. Bose wrote numerous novels, short story collections, plays, essays, criticisms, memoirs and volumes of poetry. He started and edited the renowned poetry magazine *Kavita*. He was also the acclaimed translator into Bengali of Baudelaire, Hölderlin and Rilke. Bose was awarded the Padma Bhushan in 1970.

Vikram Chandra (born 1961) divides his time between Bombay and Washington D.C. His debut novel, *Red Earth and Pouring Rain*, was awarded the David Higham Prize for Fiction and the Commonwealth Writers Prize for Best First Published Book. His collection of stories, *Love and Longing in Bombay*, was published in 1997 and won the Commonwealth Writers Prize for the Eurasia region. It was also shortlisted for the Guardian Fiction Prize. His most recent novel, *Sacred Games*, was published in 2007 and was shortlisted for the Encore Awards. His latest work, *Geek Sublime: Writing Fiction, Coding Software,* deals with his twin interests of writing and coding.

Upamanyu Chatterjee (born 1959) joined the Indian Administrative Service in 1983. His published works include short stories and the novels *English, August: An Indian Story* (1988), *The Last Burden* (1993), *The Mammaries of the Welfare State* (2000), which won the Sahitya Akademi Award for writing in English, *Weight Loss* (2006), *Way to Go* (2010) and *Fairy Tales at Fifty* (2014). In 2008, he was awarded the Officier of Ordre des Arts et des Lettres by the French Government for his contribution to literature. He is married and has two daughters.

Ismat Chughtai (1915–1991), the grande dame of Urdu literature, was a novelist, short-story writer and essayist who represented the birth of a revolutionary feminist politics and aesthetics in Urdu literature in the twentieth century. Although a spirited member of the Progressive Writers' Movement in India, she spoke vehemently against its orthodoxy and inflexibility.

Nisha da Cunha (born 1934) studied English Literature at Miranda House and Cambridge. She has directed several plays, and published four short story collections: *Old Cypress: Stories, The Permanence of Grief, Set My Heart in Aspic,* and *No Black, No White: Short Stories.*

Anita Desai (born 1937) is one of the country's foremost novelists and the author of sixteen works of fiction, including *Clear Light of Day, In Custody,* and *Fasting, Feasting*—all shortlisted for the Booker Prize, as well as *Baumgartner's Bombay* (1988). She received the Sahitya Akademi Award in 1978 for her novel *Fire on the Mountain;* and won the Guardian Prize for *The Village by the Sea* in 1983. She is the Emerita John E. Burchard Professor of Humanities at the Massachusetts Institute of Technology and divides her time between the US and Mexico.

Vijaydan Detha (1926–2013), also known as Bijji and 'the Shakespeare of Rajasthan' has more than 800 short stories to his credit, including *Bataan ri Phulwadi* (A Garden of Tales), a fourteen-volume collection of stories that draws on folklore and the spoken dialects of Rajasthan. His stories and novels were adapted for many plays and movies including Habib Tanvir's *Charandas Chor,* Amol Palekar's *Paheli* and *Duvidha* by Mani Kaul. He was co-founder of Rupayan Sansthan, an institute that documents Rajasthani folklore, arts and music; he was a recipient of the Padma Shri and Sahitya Akademi awards.

Mahasweta Devi (born 1926) is a noted social activist and Bengali writer. Her first book, *The Queen of Jhansi,* was published in 1956. She has published twenty collections of short stories and close to a hundred novels, primarily in Bengali. She has also been a regular contributor to several literary magazines such as *Bortika,* a journal dedicated to the cause of oppressed communities within India. She has won several literary prizes including the 1979 Sahitya Akademi Award for her novel *Aranyer Adhikar;* the Padma Shri in 1986; the Jnanpith Award in 1996; the Ramon Magsaysay Award for Journalism, Literature and the Creative Communication Arts in 1997; and the Padma Vibhushan in 2006.

Gulzar (born 1936) is a poet, storyteller, director, scriptwriter and lyricist—a towering figure in contemporary Indian literature and cinema. He has been awarded the Padma Bhushan, the Sahitya Akademi Award and the Dadasaheb Phalke Award. He has also won several Indian National Film Awards, Filmfare Awards, an Academy award and a Grammy.

Githa Hariharan (born 1954) has written novels, short fiction and essays over the last three decades. Her acclaimed work includes *The Thousand Faces of Night* which won the Commonwealth Writers Prize for Best First Book in 1993; the short story collection *The Art of Dying;* the essay collection *Almost Home: Cities and Other Places;* and the novels *The Ghosts of Vasu Master, When Dreams Travel, In Times of Siege* and *Fugitive Histories.*

Anjum Hasan (born 1972) is the author of the novels *Neti, Neti* and *Lunatic in my Head;* an anthology of short stories, *Difficult Pleasures;* and a book of poems, *Street on the Hill.* She is Books Editor at the *Caravan* and lives in Bangalore.

Devanoora Mahadeva (born 1948) is a renowned Dalit writer and activist who writes in Kannada. He has published a collection of short stories, *Devanooru,* and a long short story *Odalaala,* which has been made into a film. Several of his works have been translated into various languages.

Saadat Hasan Manto (1912–1955) is the most widely read and, arguably, the most controversial short story writer in Urdu; he was born in Samrala in Punjab's Ludhiana district. In a literary, journalistic, radio scripting and film-writing career spread over more than two decades, he produced twenty-two collections of short stories, one novel, five collections of radio plays, three collections of essays, two collections of personal sketches and many scripts for films. He was tried for obscenity half a dozen times, thrice before and thrice after Independence.

Cyrus Mistry (born 1956) began his writing career as a playwright, freelance journalist and short-story writer. His play *Doongaji House,* written in 1977 when he was twenty-one, has acquired classic status in contemporary Indian theatre in English. One of his short stories was made into a Gujarati feature film. His plays and screenplays have won several awards. His novel, *Chronicle of a Corpse Bearer* (2012), won the DSC Prize for South Asian Literature 2014. His collection of short stories, *Passion Flower,* was published in 2014.

Gopinath Mohanty (1914–1991) was one of the greatest Oriya writers of the mid-twentieth century. His first novel, *Mana Gahirara Chasa*, was published in 1940, which was followed by *Dadi Budha* (1944), *Paraja* (1945) and *Amrutara Santan* (1947). He published twenty-four novels and ten collections of short stories in addition to three plays, two biographies, two volumes of critical essays, and five books on the languages of the Kandh, Gadaba and Saora tribes. He also translated Tolstoy's *War and Peace* and Tagore's *Jogajog* into Oriya.

Amrita Narayanan (born 1975) is a native of Tamil Nadu who now lives in Goa. Her first book, *A Pleasant Kind of Heavy and Other Erotic Stories*, was published in 2013 under the pen-name Aranyani, and was a finalist for the Shakti Bhatt First Book Award.

R. K. Narayan (1906–2001) was one of the country's greatest writers, illuminating the human condition through small-town life. He created the fictional town of Malgudi, which he introduced in his first work of fiction, *Swami and Friends*; it was the setting for many of his works. In 1958 Narayan's work *The Guide* won the Sahitya Akademi Award and was adapted for film and Broadway. He won numerous awards, including the Padma Vibhushan, and was nominated to the Rajya Sabha.

Manjula Padmanabhan (born 1953) is an author/artist living in the US and India. *Harvest*, her fifth play, won the 1997 Onassis Award for Theatre. Her books include *Getting There, Escape* and *Three Virgins*.

Harishankar Parsai (1924–1995) was a noted satirist and humorist of modern Hindi literature, a writer who was almost completely defined by his chosen genre: satire. He wrote a number of short story collections, among them *Premchand ke Phattey Joote*, and *Viklang Sharaddha ka Daur* for which he won the Sahitya Akademi Award in 1982.

Thakazhi Sivasankara Pillai (1912–1999) was a Malayalam novelist and short-story writer, popularly known as Thakazhi, after his place of birth. He wrote several novels and over 600 short stories. His most famous works are the epic novel *Kayar* (Coir, 1978), for which he

won India's highest literary award, the Jnanpith; and *Chemmeen* (Prawns, 1956).

Munshi Premchand (1880–1936) was a pioneer of modern Hindi and Urdu fiction. He wrote nearly 300 stories and novels. Among his best known novels are *Sevasadan*, *Rangmanch*, *Gaban*, *Nirmala* and *Godan*. Much of Premchand's best work is to be found in his 250 or so short stories, collected in Hindi under the title *Manasarovar*.

Amrita Pritam (1919–2005) is acclaimed as the doyenne of Punjabi literature. Her best-known works are the poem 'Aaj Aakhan Waris Shan Nu' and the novel *Pinjar* (The Skeleton), which was made into an award-winning film in 2003. In 1956, she became the first woman to win the Sahitya Akademi Award for her magnum opus, a long poem, *Sunehe* (Messages); she received the Jnanpith in 1982 for *Kagaz Te Canvas* (The Paper and the Canvas). In 2004, she was awarded the Padma Vibhushan, as well as the Sahitya Akademi's Lifetime Achievement Award.

Sundara Ramaswamy (1931–2005), a Tamil writer, was a great modernist and a dazzling stylist who wrote poetry, novels, short stories, plays and essays. He was known as Su. Ra. in literary circles. Each of his three novels is recognized as path-breaking, and he also edited and published a notable literary magazine, *Kalachuvadu*, which became a forum for new writing and literary debate in Tamil.

Vilas Sarang (born 1942) was educated in Mumbai, and at Indiana University. He has taught English in various countries and was the head of the English department in Mumbai University for several years. His short stories have been collected in *Fair Tree of the Void* and *The Women in Cages*. His novel *Tandoor Cinders* was published in 2008.

Anna Bhau Sathe (1920–1969) was a writer and social reformer. Despite a lack of formal education, he wrote thirty-five novels in Marathi, of which the most famous—*Fakira*—is in its nineteenth edition. He also wrote numerous collections of short stories, screenplays, ballads, and a travelogue on Russia. Much like his life, his stories embody the human struggle against immense odds.

Irwin Allan Sealy (born 1951) was born in Allahabad and educated in Lucknow and Delhi.

He is the author of *The Trotter-Nama, The Everest Hotel, The Brainfever Bird* and other novels; *The Small Wild Goose Pagoda*, an almanac, shortlisted for the 2014 Tata Literature Live! Book of the Year Award (Non-Fiction); and a travelogue, *From Yukon to Yucatan*. He lives in Dehradun, where he is apprenticed to a bricklayer.

Vikram Seth (born 1952) is the acclaimed author of three novels: *The Golden Gate, An Equal Music* and *A Suitable Boy*, one of the most beloved and widely read books of recent times. He has also published several books of poetry, an opera libretto, a book of other libretti, and two highly regarded works of non-fiction, *From Heaven Lake* and *Two Lives*. He is presently at work on *A Suitable Girl*.

Khushwant Singh (1915–2014) was, arguably, India's best-known and most widely read author, columnist and journalist. He was the founder-editor of *Yojana*, and editor of the *Illustrated Weekly of India*, *National Herald* and the *Hindustan Times*. He wrote several books, including the novels *Train to Pakistan*, *I Shall Not Hear the Nightingale* and *Delhi*; his autobiography, *Truth, Love & a Little Malice*; and the two-volume *A History of the Sikhs*. He also translated from Hindi, Urdu and Punjabi. Khushwant Singh was a member of the Rajya Sabha from 1980 to 1986. In 2007, he was awarded India's second highest civilian honour, the Padma Vibhushan.

Rabindranath Tagore (1861–1941) was a Bengali poet, philosopher, artist, playwright, composer and novelist. India's first Nobel laureate, Tagore won the 1913 Nobel Prize in Literature. He composed the text of both India's and Bangladesh's respective national anthems.

Kanishk Tharoor (born 1984) is a writer based in New York City. His work has appeared in several international and Indian publications; his short fiction was nominated for a National Magazine Award. He studied at Yale, where he graduated magna cum laude and Phi Beta Kappa with BAs in History and Literature, and at Columbia, where he was a FLAS fellow in Persian and South Asian studies.

Shashi Tharoor (born 1956) is the bestselling author of fourteen books, both fiction and non-fiction, besides being a noted critic and columnist, a former Under Secretary-General of the United Nations and a former Indian Government Minister. His columns have been published widely, including in the *New York Times, Washington Post, TIME, Newsweek*, and the *Times of India*. He is a well-known speaker on India's economics and politics, on the freedom of the press, human rights, Indian culture and international affairs. He has previously served as Minister of State for Human Resource Development and Minister of State for External Affairs in the Government of India. He served as Under-Secretary General of the United Nations during Kofi Annan's leadership. He is a member of the Indian National Congress, and has been twice elected to Parliament from Thiruvananthapuram. He chairs Parliament's External Affairs Committee.

D. B. G. Tilak (1921–1966) was an influential Telugu poet, novelist and short-story writer. He is best known for his collection of poems *Amrutham Kurisina Ratri* (The Night When Nectar Rained) published in 1969.

Nirmal Verma (1929–2005) was a pioneer of the Nayi Kahaani Movement in Hindi literature. He wrote five novels, eight collections of short stories and nine volumes of essays and travelogues. Following his work at the Oriental Institute at Prague in the 1960s, he undertook translations of contemporary Czech writers such as Milan Kundera, Bohumil Hrabal and Vaclav Havel in Hindi, much before their work became popular internationally. He won the Sahitya Akademi Lifetime Achievement Award (1985), the Jnanpith Award (1999) and was awarded the Padma Bhushan in 2002.

Paul Zacharia (born 1945) is an eminent Malayalam fiction writer and essayist. He was awarded the Kendra Sahitya Akademi Award in 2005 for his short-story collection, *Zachariyayute Kathakal*. His works that have been translated into English include *Bhaskara Pattelar and Other Stories, Reflections of a Hen in Her Last Hour and Other Stories* and *Paul Zacharia: Two Novellas*.

NOTES ON THE TRANSLATORS

Amitav Ghosh (born 1956) was born in Calcutta and grew up in Bangladesh, Sri Lanka and India. He studied at the universities of Delhi and Oxford and has taught at a number of institutions, most recently Harvard, and written for many publications. He is the author of several novels including the bestselling *Sea of Poppies* which was shortlisted for the Man Booker Prize in 2008. He currently divides his time between Calcutta, Goa and Brooklyn.

Vernon Gonsalves (born 1957) is a social and political activist. He was on the editorial board of *Thingi Kamgar Masik*, a Marathi language monthly magazine for workers, from 1981 to 1986. He has been associated with the Marathi magazine *Jahirnama* since the eighties; until his arrest in August 2007 (on charges of being a Naxalite), he was also involved with a bilingual Hindi-Marathi magazine *Kamgar*. He was released from prison in 2013.

C. T. Indra is the former Head of the Department of English, Madras University. She has been the recipient of an Australia-India Council Special Award for outstanding partnership, a Fulbright Postdoctoral Fellowship to Harvard University, and a Faculty Enrichment Award to Canada from the Shastri Indo-Canadian Institute. She received the Katha Award for Translation for this translation of Ambai's story.

Rakhshanda Jalil (born 1963) is a well-known writer, critic and literary historian. She is best known for her book on Delhi's lesser-known monuments, *Invisible City: The Hidden Monuments of India*, as well as her biographies of Urdu writers Qurratulain Hyder and Rashid Jahan.

Kailash Kabir (born 1954) is an award-winning translator and poet of Hindi and Rajasthani who makes his home in Jodhpur, India.

Geeta Kapur (born 1943) is a noted art critic, art historian and curator. One of the pioneers of art criticism in India, she taught in the Humanities and Social Sciences department of IIT Delhi from 1967 to 1973. She has held fellowships at the Indian Institute of Advanced Study, Shimla and Nehru Memorial Museum and Library, New Delhi.

S. Krishnan (1925–2005) was a prominent journalist and translator. He served as Cultural Advisor to the United States Information Service in Chennai for many years.

Gita Krishnankutty is a scholar, critic and a well-known translator of Malayalam and French literature.

Sitakant Mahapatra (born 1937) is a literary critic in Oriya as well as English. Previously of the IAS, he has also been Chairman of National Book Trust, New Delhi. He has published poetry collections, essay collections, a travelogue, and many contemplative works, and numerous translations. He was awarded the 1974 Sahitya Akademi Award in Oriya for his poetry collection, *Sabdar Akash* (The Sky of Words), the Jnanpith Award in 1993, the Padma Bhushan in 2002 and the Padma Vibhushan in 2011 and has also been the recipient of several other prestigious awards.

Christi A. Merrill is an assistant professor of South Asian Literature and postcolonial studies at the University of Michigan. Her translations from Hindi, French, and Rajasthani and essays on translation have appeared in journals such as *Genre*, *Studies in Twentieth Century Literature*, *The Iowa Review*, *Modern Poetry in Translation*, and *Indian Literature*, the Sahitya Akademi's bi-monthly journal.

C. M. Naim (born 1936) is an American scholar of Urdu language and literature. He is currently Professor Emeritus at the University of Chicago. Naim is the founding editor of both *Annual of Urdu Studies* and *Mahfil* (now *Journal of South Asian Literature*), as well as the author of the definitive textbook for Urdu pedagogy in English.

H. Y. Sharada Prasad (1924–2008) began his professional life as a journalist. He went on to edit the Planning Commission's journal *Yojana* and then became Indira Gandhi's media advisor and speech writer. He was the Joint Editor of the *Selected Works of Jawaharlal Nehru* and has authored several books, the last of which was *The Book I Won't Be Writing*.

A. K. Ramanujan (1929–1993) was a poet,

scholar, author, philologist, folklorist, translator, poet and playwright. His academic research ranged across five languages: Tamil, Kannada, Telugu, Sanskrit, and English. He published works on both classical and modern variants of the literature of these languages and also argued strongly for giving local, non-standard dialects their due.

Ranga Rao (born 1936) is the author of three novels *Fowl-Filcher*, *The Drunk Tantra*, and *The River is Three Quarters Full;* and of the short story collection, *An Indian Idyll and Other Stories.* He has also edited and translated into English two anthologies of Telugu stories, *Classic Telugu Short Stories* and *That Man on the Road.* He is currently Visiting Faculty at Sri Sathya Sai Institute of Higher Learning (Deemed University), Prasanthi Nilayam (Andhra Pradesh).

Arshia Sattar's (born 1960) English translations include Valmiki's *Ramayana* and *Tales from the Kathasaritsagara.* She has a PhD from the Department of South Asian Languages and Civilizations at the University of Chicago and her interests are Indian epics, mythology and the story traditions of the subcontinent.

Manu Shetty has worked with the Committee on Social Thought, University of Chicago, on 'Tulu Oral Narratives'. He was a student of

A. K. Ramanujan and collaborated with him in translating the works of contemporary Kannada writers.

Arunava Sinha (born 1962) translates classic, modern and contemporary Bengali fiction and non-fiction into English. Thirty of his translations have been published so far. Twice the winner of the Crossword translation award, for Sankar's *Chowringhee* (2007) and Anita Agnihotri's *Seventeen* (2011), respectively, and the winner of the Muse India translation award (2013) for Buddhadeva Bose's *When The Time Is Right*, he has also been shortlisted for The Independent Foreign Fiction prize (2009) for his translation of *Chowringhee*. Besides India, his translations have been published in the UK and the US in English, and in several European and Asian countries through further translation. He was born and grew up in Kolkata, and lives and writes in New Delhi.

Gayatri Chakravorty Spivak (born 1942) is a literary theorist, philosopher and professor at Columbia University, where she is a founding member of the school's Institute for Comparative Literature and Society. She is best known for the essay 'Can the Subaltern Speak?' considered a founding text of postcolonialism. She received the Padma Bhushan in 2013.

O.V. Usha (born 1948) is a Malayalam poet and novelist. She has published four volumes of poems and a novel. Her articles have appeared in various journals. She served the Mahatma Gandhi University, Kottayam, as its director of publications. She won the Kerala State Film Award for Best Lyrics for *Mazha*, a Malayalam film released in 2000.